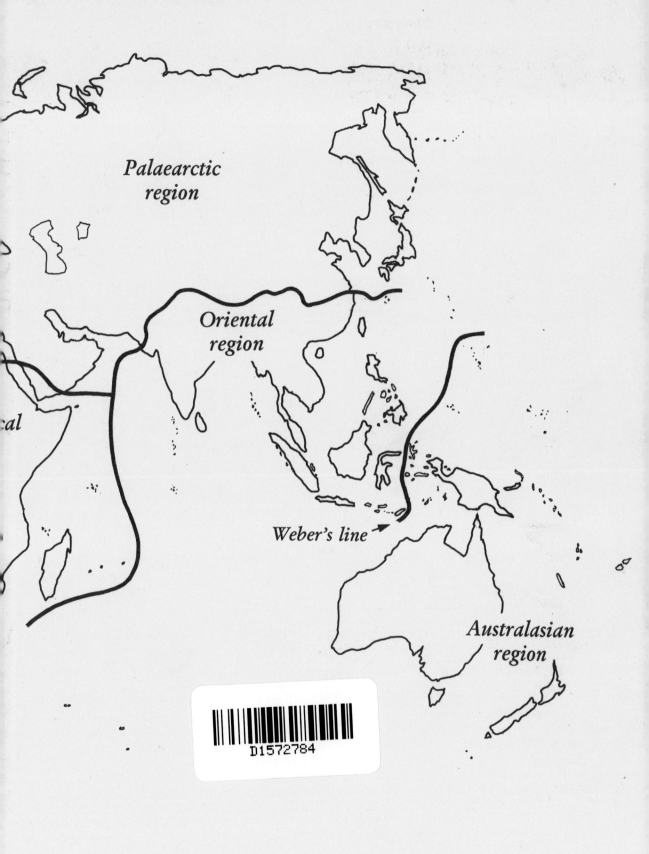

MEDICAL INSECTS AND ARACHNIDS

THE NATURAL HISTORY MUSEUM

MEDICAL INSECTS AND ARACHNIDS

Edited by
RICHARD P. LANE *and* ROGER W. CROSSKEY

*Department of Entomology,
The Natural History Museum
(British Museum (Natural History)) London, UK*

CHAPMAN & HALL
London · Glasgow · New York · Tokyo · Melbourne · Madras

Published by Chapman & Hall, 2–6 Boundary Row, London SE1 8HN

Chapman & Hall, 2–6 Boundary Row, London SE1 8HN, UK

Blackie Academic & Professional, Wester Cleddens Road, Bishopbriggs, Glasgow G64 2NZ, UK

Chapman & Hall Inc., 29 West 35th Street, New York NY10001, USA

Chapman & Hall Japan, Thomson Publishing Japan, Hirakawacho Nemoto Building, 6F, 1-7-11 Hirakawa-cho, Chiyoda-ku, Tokyo 102, Japan

Chapman & Hall Australia, Thomas Nelson Australia, 102 Dodds Street, South Melbourne, Victoria 3205, Australia

Chapman & Hall India, R. Seshadri, 32 Second Main Road, CIT East, Madras 600 035, India

First edition 1993

© 1993 British Museum (Natural History)

Typeset in 10/12pt Palatino by Expo Holding, Malaysia
Printed in Great Britain at the University Press, Cambridge

ISBN 0 412 40000 6

Apart from any fair dealing for the purposes of research or private study, or criticism or review, as permitted under the UK Copyright Designs and Patents Act, 1988, this publication may not be reproduced, stored, or transmitted, in any form or by any means, without the prior permission in writing of the publishers, or in the case of reprographic reproduction only in accordance with the terms of the licences issued by the Copyright Licensing Agency in the UK, or in accordance with the terms of licences issued by the appropriate Reproduction Rights Organization outside the UK. Enquiries concerning reproduction outside the terms stated here should be sent to the publishers at the London address printed on this page.

The publisher makes no representation, express or implied, with regard to the accuracy of the information contained in this book and cannot accept any legal responsibility or liability for any errors or omissions that may be made.

A catalogue record for this book is available from the British Library

Library of Congress Cataloging-in-Publication data available

The use of the term 'man' is not indicative of a specific gender and should be taken to refer to both men and women throughout.

Contents

Contributors	*xi*
Preface	*xiii*

1 General introduction
R. P. Lane and R. W. Crosskey — 1

Classification and nomenclature	1
Species and species complexes	12
Distribution and zoogeographical regions	16
Vectors and vector status	19
Collection, storage and handling of specimens	23
References	27

2 Introduction to the arthropods
R. P. Lane — 30

Structure	30
Classification	32
The principal arthropod groups	33
Further reading	46

Part One Diptera — 49

3 Introduction to the Diptera
R. W. Crosskey — 51

Morphology	52
Taxonomy	63
Biology	72
References	75

4 Sandflies (Phlebotominae)
R. P. Lane — 78

Recognition and elements of structure	79
Classification and identification	89
Biology	100
Medical importance	103
Control	107

	Collecting, preserving and rearing material	111
	References	113

5 Mosquitoes (Culicidae) — 120
M. W. Service

Recognition and elements of structure	121
Classification and identification	139
Biology	173
Medical importance	196
Control	208
Collecting, preserving and rearing material	216
References	221

6 Blackflies (Simuliidae) — 241
R. W. Crosskey

Recognition and elements of structure	242
Classification and identification	249
Biology	259
Medical importance	267
Control	277
Collecting, preserving and rearing material	279
References	282

7 Biting midges (Ceratopogonidae) — 288
John Boorman

Recognition and elements of structure	290
Classification and identification	294
Biology	298
Medical importance	299
Control	302
Collecting, preserving and rearing material	303
References	304

8 Horse-flies, deer-flies and clegs (Tabanidae) — 310
John E. Chainey

Recognition and elements of structure	310
Classification and identification	316
Biology	323
Medical importance	325
Control	327

	Collecting, preserving and rearing material	327
	References	328
9	*Tsetse-flies (Glossinidae)* A. M. Jordan	333
	Recognition and elements of structure	334
	Classification and identification	340
	Biology	365
	Medical importance	372
	Control	379
	Collecting, preserving and rearing material	381
	References	382
10	*Stable-flies and horn-flies (bloodsucking Muscidae)* R. W. Crosskey	389
	Recognition and elements of structure	389
	Classification and identification	391
	Biology	396
	Medical importance	398
	Control	399
	Collecting, preserving and rearing material	400
	References	400
11	*House-flies, blow-flies and their allies (calyptrate Diptera)* R. W. Crosskey and R. P. Lane	403
	Recognition of calyptrates and elements of structure	403
	Identification of medically important calyptrate flies	407
	Biology of synanthropic flies	419
	Medical importance of adult calyptrate flies	423
	Control	425
	References	426
12	*Diptera causing myiasis in man* Martin J. R. Hall and Kenneth G. V. Smith	429
	Recognition and elements of structure	430
	Review of dipterous families associated with human myiasis	431
	Maggot therapy	459
	Control	459
	Collecting, preserving and rearing material	462
	References	464

viii Contents

Part Two Other insects — 471

13 Cockroaches (Blattaria) — 473
N. R. H. Burgess

- Recognition and elements of structure — 473
- Classification and identification — 474
- Biology — 479
- Medical importance — 480
- Control — 481
- References — 482

14 Bedbugs and kissing-bugs (bloodsucking Hemiptera) — 483
C. J. Schofield and W. R. Dolling

- Recognition and elements of structure — 484
- Classification and identification — 485
- Biology of the Triatominae — 502
- Medical importance — 506
- Control — 508
- Collecting, preserving and rearing material — 509
- References — 510

15 Lice (Anoplura) — 517
Joanna Ibarra

- Recognition and elements of structure — 518
- Biology — 521
- Medical importance — 523
- Control — 525
- References — 527

16 Fleas (Siphonaptera) — 529
Robert E. Lewis

- Recognition and elements of structure — 529
- Classification and identification — 534
- Biology — 555
- Medical importance — 557
- Control — 568
- Collecting, preserving and rearing material — 569
- References — 570

17 Insects of minor medical importance — 576
Kenneth G. V. Smith

- Diptera — 576
- Hymenoptera — 581

Coleoptera	583
Lepidoptera	585
Thysanoptera (thrips)	588
Incidental insects	589
Insects and hygiene	589
References	589

Part Three Arachnids   595

18 Ticks and mites (Acari)   597
M. R. G. Varma

Recognition, structure and classification of the Acari	597
Review of the suborder Ixodida (ticks)	600
Biology of ticks	616
Medical importance of ticks	623
Introduction to mites: recognition and structure	631
Classification and identification of mites	632
Biology and medical importance of mites	634
Control of ticks and mites	648
Collecting, preserving and rearing material of ticks and mites	649
References	654

19 Spiders and scorpions (Araneae and Scorpiones)   659
J. L. Cloudsley-Thompson

Recognition and elements of structure	660
Classification and identification	666
Biology	670
Medical importance	672
Control	678
Collecting and preserving material	679
References	680

Scientific names index   683

Subject index   706

Contributors

Mr John Boorman
6 Beckingham Road, Guildford, Surrey GU2 6BN

Dr N. R. H. Burgess
Department of Military Entomology, Royal Army Medical College, Millbank, London SW1P 4RJ

Mr John E. Chainey
Department of Entomology, The Natural History Museum, Cromwell Road, London SW7 5BD

Professor J. L. Cloudsley-Thompson
Department of Biology, Medawar Building, University College London, Gower Street, London WC1E 6BT

Dr R. W. Crosskey
Department of Entomology, The Natural History Museum, Cromwell Road, London SW7 5BD

Mr W. R. Dolling
Brook Farmhouse, Elstronwick, Hull HU12 9BP

Dr Martin J. R. Hall
Department of Entomology, The Natural History Museum, Cromwell Road, London SW7 5BD

Mrs Joanna Ibarra
Community Hygiene Concern, 32 Crane Avenue, Isleworth, Middlesex TW7 7JL

Dr A. M. Jordan
Tsetse Research Laboratory, University of Bristol, Langford House, Langford, Bristol BS18 7DU

Dr R. P. Lane
Department of Entomology, The Natural History Museum, Cromwell Road, London SW7 5BD

Professor Robert E. Lewis
Department of Entomology, Iowa State University, Ames, Iowa 50011, USA

Dr C. J. Schofield
7 rue Maclonay, Pregnin, St. Genis - Pouilly, 01630 France

Professor M. W. Service
Vector Biology and Control Group, Liverpool School of Tropical Medicine, Pembroke Place, Liverpool L3 5QA

Mr Kenneth G. V. Smith
70 Hollick Wood Avenue, London N12 0LT

Professor M. G. R. Varma
c/o Department of Medical Parasitology, London School of Hygiene and Tropical Medicine, Keppel Street, London WC1E 7HT

Preface

Surprising though it seems, the world faces almost as great a threat today from arthropod-borne diseases as it did in the heady days of the 1950s when global eradication of such diseases by eliminating their vectors with synthetic insecticides, particularly DDT, seemed a real possibility. Malaria, for example, still causes tremendous morbidity and mortality throughout the world, especially in Africa. Knowledge of the biology of insect and arachnid disease vectors is arguably more important now than it has ever been. Biological research directed at the development of better methods of control becomes even more important in the light of the partial failure of many control schemes that are based on insecticides – although not all is gloom, since basic biological studies have contributed enormously to the outstanding success of international control programmes such as the vast Onchocerciasis Control Programme in West Africa.

It is a *sine qua non* for proper understanding of the epidemiology and successful vector control of any human disease transmitted by an arthropod that all concerned with the problem – medical entomologist, parasitologist, field technician – have a good basic understanding of the arthropod's biology. Knowledge will be needed not only of its direct relationship to any parasite or pathogen that it transmits but also of its structure, its life history and its behaviour – in short, its natural history. Above all, it will be necessary to be sure that it is correctly identified.

Close attention to biological details often reveals the presence of species complexes of vectors, which if they are to be analysed for correct identification of component sibling species, require the application of sophisticated techniques extending beyond the ubiquitous morphological approach to identification. Study of such complexes leads to understanding of speciation mechanisms, and has begun to take medical entomology into evolutionary biology. In a similar manner, there is no clear division between medical entomology and veterinary entomology – very often the same groups of insects, mites or ticks transmit parasites to domestic animals. Clearly, medical entomology does not stand in isolation.

Unfortunately, while medical entomology is comparatively rich in works which deal with particular groups of arthropods at the level needed by specialists, it is rather less well supplied with more general literature which broadly summarizes information and meets the needs of students in tropical medicine and allied fields.

The book we present here, under the short title *Medical Insects and Arachnids*, brings to fruition a project we initiated several years ago when it became clear to us that there was no up-to-date and comprehensive introduction to arthropods important in human health. The march of events in medical entomology over the past 20 years has outdated the well-known book *Insects and Other Arthropods of Medical Importance* that was edited by our colleague Mr K. G. V. Smith and published by the British Museum (Natural History) in 1973 – one need think only of how the importance of sibling species complexes in arthropod-borne vectors of human diseases has come to be fully realized and of the varied approaches to vector taxonomy and identification this has brought in its train.

This book is rooted in group-by-group systematic coverage, and on this account we have (as editors) prepared the chapters in generally similar format so that the structure and biology of the different groups can be easily compared. So far as is practicable, we have balanced the chapters to reflect the biological complexity and proved medical importance of the groups concerned. Control is covered only briefly, with the focus mainly on the biological rationale for the interventions adopted.

We have paid particular attention to the identification keys that are included (revising manuscripts where necessary) because we are conscious of how difficult students find keys when these are written in the jargon of taxonomists. The keys are presented in a completely uniform style and in what we believe is more than usually friendly language, avoiding unnecessary technical terms.

As part of the essential aim of the book, stress is laid on 'systematic' aspects of medical entomology, that is to say on the structure and biology of each medically related group of arthropods and upon how the important families, genera and species can most easily be identified. Special mention is made of non-morphological taxonomic techniques, and the scientific background to methods and concepts applicable in medical entomology is summarized in a general introduction. (We have omitted the names of those who described the genera, species and subspecies mentioned in the text because these do not form part of the scientific names concerned and serve no purpose in a book of this kind.)

Major effort has been applied to the illustrations, many of which have been specially drawn or modified to meet the needs of the text. Labelling of illustrations is to a uniform style and in full, thereby obviating the need (which we have always found irksome) to refer to lengthy legends or to memorize abbreviations. Roman numerals have been everywhere avoided as we see no point in imposing this unnecessary system on those who might not be familiar with it; larval instars and abdominal segments, for example, are labelled in arabic numbers.

Many medical entomologists working in the field without access to libraries depend on literature obtained in photocopy or through library loan. Correct bibliographic citation is therefore essential. To meet this need, we have verified all the references cited in the book. The references provide leads for readers wanting more information on morphology, taxonomy, biology, vector/parasite relations, vector control and faunistics. (In connection with the last, we observe that we did not anticipate the rapid demise of the Soviet Union while this book was being completed and the now- historical abbreviation USSR is used.)

This book could not have been prepared without the keen cooperation of our 13 fellow contributors, and we take this opportunity to thank them not only for their contributions but for allowing us to modify their texts when this seemed desirable for consistency and balance between chapters. Others have helped us behind the scenes, and it is our pleasure to acknowledge assistance received from Dr Anne Baker (on mites) and Mr Paul Hillyard (on spiders), from librarians Julie Harvey and Kathy Martin who obtained literature not immediately to hand in the British Museum (Natural History), and from Jan Townsend and Angel Hathaway who typed parts of the text. Professor M. W. Service provided us with some helpful bibliographic information.

Some illustrations for which the British Museum (Natural History) does not own copyright have appeared in previous literature, and we are grateful to the following publishers for permission to reproduce the figures indicated: American Museum of Natural History (Fig. 14.2*a*); Blackwell Scientific Publications (Figs 5.2*d*, 5.7*b*, 5.9, 5.12, 5.16*b*,*d*, 5.20); CAB International (Fig. 5.10*f*); Canada Agriculture (Figs 11.5*a*, 12.1.2*b*,*c*, 17.2*b*); Commonwealth Scientific and Industrial Research Organization, Canberra (Figs 3.9*a*, 3.9*d*, 11.5*b*,*c* originally drawn by T. Binder); Edward Arnold (Fig. 5.14); Entomological Society of America (Fig. 5.25); Faune de Madagascar (Fig. 5.22); Oxford University Press (Fig. 5.29); Unwin Hyman's Ltd (Fig. 5.10*c*); Walter Reed Biosystematics Unit (Figs 5.8, 5.10*a*,*b*,*d*,*e*, 5.11, 5.13*a*,*b*, 5.16*a*,*c*, 5.17, 5.19, 5.24, 5.26); World Health Organization (Fig. 5.30). Figs 4.2, 4.6, 6.3, 8.4 and 10.2*b*,*d*,*e* are reproduced courtesy of the Wellcome Trust from Boris Jobling's *Anatomical Drawings of Biting Flies* (1987); Figs 11.5*a*, 12.12*b*,*c* and 17.2*c* are from Agriculture Canada's *Manual of Nearctic Diptera*, volume 1 (1981), and are reproduced with the permission of the Minister of Supply and Services, Canada, 1992. Mr M. Smith kindly supplied the photograph used for Fig. 15.3.

Acknowledgements made by individual chapter contributors are as follows. J. L. Cloudsley-Thompson (Araneae and Scorpiones) thanks Maureen Hakney and Helen Wilson for deciphering his handwriting and typing his manuscript; R. E. Lewis (Siphonaptera) thanks Mr F. G. A. Smit for the splendid job done by him in the flea chapter published previously in the book edited by K. G. V. Smith (see above); C. J. Schofield and W. R. Dolling (Hemiptera) thank Professor Randall T. Schuh and Professor Carl W. Schaeffer for advice on higher classification of bugs, and Dr H. Brailovsky for helpful discussion of recently described triatomine species; M. W. Service (Culicidae) thanks Professor A. N. Clements, Professor P. S. Corbet, Dr. E. L. Peyton and Dr H. Townson for comments on draft text, Professors L. de Vos, G. Josens and B. Vray for the photograph used in Fig. 5.4, Dr M. Coetzee and Dr R. Hunt for the photograph used in Fig. 5.31, Dr L. P. Lounibos for the photograph used in Fig. 5.28, and Dr. R. A. Ward for information on the number of world mosquito species.

We conclude this preface with appreciative thanks to our wives for their forbearance in tolerating long periods when we appeared to be 'incommunicado' while in the throes of editing and preparing this book. Special thanks go to Peggy Crosskey, and to her sister Jean Johns, who shouldered the considerable task of compiling the draft index to scientific names.

Richard P. Lane and Roger W. Crosskey

British Museum (Natural History), London, January 1992

CHAPTER ONE

General introduction

R. P. Lane and R. W. Crosskey

Anyone already experienced in the biology and classification of medically important insects or arachnids will probably use this book by going directly to a chapter which particularly interests or concerns them – without the need for a general introduction. However, students and others less familiar with arthropods concerned in human health might find it helpful to be provided with background reading of a general nature that is relevant to all the specialist groups of noxious arthropods covered in later chapters. This introductory chapter briefly considers the following general topics:

Classification and nomenclature
Species and species complexes
Distribution and zoogeographical regions
Vectors and vector status
Collection, storage and handling of specimens

CLASSIFICATION AND NOMENCLATURE

Systematic zoology is a multifaceted activity which endeavours to order the rich diversity of the animal world and to develop methods and principles to make this task easier (Crowson, 1970; Mayr and Ashlock, 1991). At present, well over a million animal species have been formally described and, given the increasing quantity of information on these and all the species awaiting description, it is clear that a system has to be developed in which the species can be arranged in a useful manner. In any system it is important that it is easy to retrieve information, and this is best done if similar animals are grouped together. Under these conditions the system becomes predictive, enabling us to infer that a species about which we know little is likely to possess certain features (e.g. details of structure or aspects of biology) because of what we already know about species close to it in the classification.

If a few moments' thought are given to the contributions of systematic zoology to biology in general, it becomes clear that systematics has had considerable impact in several ways – as a unifying principle paralleling that of molecular biology, as underpinning for applied biology through analysis of variation and correct identification of species, as a major activity in theoretical biology (including the

Medical Insects and Arachnids Edited by Richard P. Lane and Roger W. Crosskey.
Published in 1993 by Chapman & Hall ISBN 0 412 40000 6

introduction of the population approach), and as a valuable counterbalance to attempts to simplify the complexity of biology to a series of chemical processes (the reductionist approach).

The task of systematics is not just to provide a means of identifying or pigeon-holing specimens, such as through the production of identification keys, but also to act as a synthesizing activity taking into account comparative studies from the molecular level to the population level. There are two basic stages in systematics: first, recognition of biologically operational units (species, subspecies, etc.) which are the basic units of any classification, and secondly the ordering of these units into higher categories or levels (genera, families, etc.)

Within any one group of organisms at least three stages of taxonomic development can be recognized. The alpha stage (= alphataxonomy) is that in which new species are being described and provisionally arranged into genera; this is followed by the beta stage (= betataxonomy), in which relationships between the species and between the higher categories are explored; and finally, the gamma stage (= gammataxonomy), in which attention is focused on intraspecific variation, evolutionary processes and functional or causal explanations for the observations made. Most groups of arthropods of medical importance are either in the beta or the gamma stages, although parts of the groups (usually those containing the species that are not medically important) can still be at a quite crude and basic alpha stage.

Interest in systematics over the past few decades has focused on the need to obtain better tools to assist with the classification and ordering of our observations, and to some extent of the associated processes which have given rise to the diversity of species. For applied biology, such as medical entomology, this interest has resulted in more effective means of recognizing species. To fulfil the aims of systematic zoology it has been necessary to develop techniques with which we can search for other attributes of organisms besides the merely anatomical (morphological) on which to characterize them (for use in both classification and identification). This has led to the increased use of isoenzyme electrophoresis and chromatography of biologically active compounds (chemotaxonomy), chromosome analysis (cytotaxonomy) and DNA/RNA studies (molecular systematics). There has also been considerable development in the theory of classification and identification to make the processes more objective (i.e. more scientific and testable) as a move away from the subjective art which taxonomy used to be. Some major movements have been numerical taxonomy, phylogenetics and cladistics.

At this point it is worth defining and contrasting some of the main terms used in systematics. Taxonomy is defined as the theory and practice of describing, naming and classifying organisms; classification is the process of delimiting, ordering and ranking of taxa within a hierarchical series of groups (species, genera, families, etc.). Both these aspects require a large database and are essentially comparative in nature. The terms systematics and biosystematics are used to indicate a wider interest than classification and taxonomy, viz. in the diversity of organisms and the processes which generate it. Identification is a precise term describing the allocation of an unknown specimen to a predefined group, e.g. the recognition of the species to which an individual mosquito specimen belongs (also called 'determination'). The term 'speciation' is one of the most commonly misused technical words in biology: it means the evolutionary process by which species multiply, but it is often grossly misused as if it means identification (to speciate is to evolve into more species, *not* to determine a specimen's identity). Nomenclature is the non-scientific but essential component of taxonomy concerned with the *names* of organisms and their correct use. It is essential for maintaining the effective communication between all biologists of knowledge and ideas about organisms. The word taxon (plural, taxa) is used for any taxonomic group (category) that is sufficiently distinct to be worthy of a collective term, e.g. subspecies, species, genus, family. The expression 'higher taxa' is often used as a collective expression for the taxa above the generic level.

Methods and ranks in classification

The basic building block of systematics is the species, and the various approaches to how species are defined and used in practice are discussed below. However, having decided on a species concept there are many different ways of addressing the problem of how to classify species. The principal methods are:

- **Pheneticism**, which classifies those organisms together that are most similar overall, usually giving equal weight to each character.
- **Cladism**, in which organisms are grouped according to recency of common ancestry; groups are joined on the grounds that they possess the same derived (apomorphic or highly evolved) characters, and not on primitive (plesiomorphic) characters.
- **Evolutionary classification**, which draws on the theory of evolution to justify the quest for finding groups produced by evolution. In many ways this approach is a synthesis of several other methods.

Whatever method is used to classify organisms, it is essential to recognize that there is no single 'right' classification. Therefore, there is no such idea as *the* classification of animals, only *a* classification of animals. All classifications are theories, suggestions of how organisms are related, which if clearly presented can be tested using different types of characters (e.g. molecular data to test a morphologically based classification). By and large, the various classifications of the animal kingdom are relatively similar, usually differing in details or in the emphasis put on different characters. Most classifications attempt to represent evolutionary 'relatedness' and to show that the most closely related groups are more recently evolved than distantly related groups.

For more than 200 years taxonomists have used six basic categories into which, in ascending order, to classify the animal kingdom, viz. species, genus (plural, genera), family, order, class and phylum (plural, phyla). With time, as more organisms were discovered, the need arose for various intermediate ranks in the classification, such as the subgenus (between species and genus), subfamily (between genus and family) and suborder (between family and order). There is no fixed limit to the number of levels (categories at different hierarchical rank) which can be introduced into a classification scheme and taxonomists working on many complicated groups of animals have used many other terms when these seemed necessary for showing better the supposed evolutionary relationships – for instance, categories called species complex, species group, tribe and superfamily are now widely used. The basis, whatever categories are employed, is that each is subordinate to another.

Textbooks dealing with outline classification almost always show different levels in the hierarchy in a manner that reads downwards from the highest to the lowest, e.g. from phyla down to genera. A category is often said to be 'divided' or 'subdivided' into lower categories, e.g. 'the family Mycetophilidae is divided into five subfamilies', but this is a most misleading way of looking at classification. It is important to understand that, in principle, animal (and plant) classifications are built up from the bottom by the assembly of species (the 'real' units in nature) and that taxa at all levels above the species are relatively arbitrary creations of taxonomists which attempt to show how assemblies of species are related to each other – assemblies of species incorporated into a classification are more distantly related to one another at each higher rung of the ladder (rank in the hierarchy). In this way classification is very different from identification; classification is from the bottom up while identification is from the top downwards.

A list of all the successively higher ranks needed to show in full the classification of a species can be extensive. The following is an example, showing the classification of *Simulium soubrense* in the phylum Arthropoda.

Sibling species: *Simulium soubrense* (species in nature)
  – which belongs, with some other sibling species, to the
Subcomplex: *Simulium sanctipauli*
  – which belongs, with some other subcomplexes, to the
Complex: *Simulium damnosum*
  – which belongs, with some other morphospecies, to the
Species group: *Simulium damnosum*
  – which belongs, with a few other species groups, to the
Subgenus: *Edwardsellum*
  – which belongs, with many other subgenera, to the
Genus: *Simulium*
  – which belongs, with another genus, to the
Tribe: Simuliini
  – which belongs, with another tribe, to the
Subfamily: Simuliinae
  – which belongs, with another subfamily, to the
Family: Simuliidae
  – which belongs, with some other families, to the
Superfamily: Chironomoidea
  – which belongs, with another superfamily, to the
Infraorder: Culicomorpha
  – which belongs, with some other infraorders, to the
Suborder: Nematocera
  – which belongs, with another suborder, to the
Order: Diptera
  – which belongs, with several other orders, to the
Division: Endopterygota (Holometabola)
  – which belongs, with another division, to the
Class: Insecta
  – which belongs, with several other classes, to the
Phylum: Arthropoda
  – which belongs, with many other phyla, to the
Kingdom: Animalia

Such classifications can be confusing and daunting to students and even experienced medical entomologists, but need not be since for everyday purposes it is usually only necessary to have a basic idea of how any medically important arthropod species is classified at the species-group, subgeneric/generic, and subfamily/family levels. In general, levels below that of species are the specialist area of cytotaxonomists and chemotaxonomists, and levels above the genus are mainly the concern of conventional (museum) taxonomists.

It must be emphasized that none of the categories above species is defined in the critical and testable way that species (at least in theory) are defined. The consequence of this is that specialists often disagree about the number, rank and limits of categories above the species level (the supraspecific taxa) to be recognized. Such disagreement shows up to a considerable extent in medically important insects, especially at the level of genera and subgenera. One specialist often ranks a cluster of species as a subgenus when another (while not disagreeing on the affinity of these species) will prefer to rank the cluster as a genus; yet another specialist might consider that the cluster merits only species-group rank within some quite different subgenus or genus. There are no objective criteria for deciding between these points of view. In some ways the argument is over the 'grain' of the classification, some preferring a 'coarse grain' and others a 'fine grain' approach. The most useful criterion could very well be ease of communication, genera being treated in a fairly conservative way (e.g. as is usual in *Simulium*); then ample use can be made of the subgeneric and species-group categories to express the presumed relationships between species. This approach is particularly useful in applied biology. Obviously, if new characters reveal quite important differences between groups of species that were formerly grouped into subgenera, then a case might be made to elevate the subgenera to genera. Nevertheless, this does not always deter specialists from sterile discussions of rank even though they are essentially reassessing only the same (usually morphological) data.

To illustrate the problem, here are two alternative ways in which the chief African vector (*damnosum*) and the chief Central American vector (*ochraceum*) of human onchocerciasis are classified in the family Simuliidae by different specialists.

Species: *Simulium damnosum*
Subgenus: *Edwardsellum*
Genus: *Simulium*
Tribe: Simuliini
Subfamily: Simuliinae

Species: *Edwardsellum damnosum*
Genus: *Edwardsellum*
Tribe: Wilhelmiini
Subfamily: Simuliinae

Species: *Simulium ochraceum*
Subgenus: *Psilopelmia*
Genus: *Simulium*
Tribe: Simuliini
Subfamily: Simuliinae

Species: *Psilopelmia ochracea*
Genus: *Psilopelmia*
Tribe: Eusimuliini
Subfamily: Simuliinae

Nomenclature

Nomenclature is the aspect of biology that deals with the names applied to organisms, not with how organisms are classified. Oddly to taxonomists – those who classify and name organisms – the topic often causes apprehension among other biologists, either because they appreciate its importance but do not understand the guiding principles and governing rules or because they dismiss it as an arcane or unimportant adjunct to the 'real' business of biology. Certainly, a considerable jargon has developed in the subject (just as in many other subjects) and it has abstruse aspects of little concern to most biologists: but the essential aim of nomenclature is simple – to ensure that names are applied consistently to the same biological entity. Communication of biological information, without misunderstanding, relies totally on stable and generally agreed names, and it is not an exaggeration to say that nomenclature is the one aspect of biology that all biologists use in their day-to-day working lives – whether they are conscious of the fact or not. Without scientific nomenclature, biology would effectively cease to be international and would quickly become mired in confusion.

An excellent introduction to the basics of nomenclature for biologists at all levels of experience is provided by Jeffrey (1989) and is the most useful learning source for beginners. More detailed coverage is given in taxonomic textbooks by Blackwelder (1967) and Mayr and Ashlock (1991). The *International Code of Zoological Nomenclature* (see below) is a work essential for practising taxonomists but is too highly technical for general introduction into nomenclature. However, Crosskey (1988) provides a short explanation of some of the Code requirements of most interest for medical entomologists.

Scientific names

The familiar binomial system for naming animals and plants was devised by the Swedish naturalist Carl Linnaeus, and the tenth edition of his great work titled the *Systema Naturae* (published in 1758) marks the beginning of scientific biological nomenclature. All names used for organisms before that date have only historical interest; they have no formal status in 'modern' classifications. The name of a species consists of two words, the first word being the generic name (used for all member species of a genus) and the second word being the specific name. The two words form a combination or binomen which is unique for each species and constitutes the scientific name. The latter is often called the Latin name, but this term is best avoided as many organism names are rooted in other languages.

Scientific names, however, must be written in the Latin (Roman) alphabet, the one used in this book, and (by convention) are usually printed in italics or underlined when they apply to genera, subgenera, species or subspecies. Under the *International Code of Zoological Nomenclature*, every generic name in the animal kingdom must have a capital (upper case) initial letter and every specific name a small (lower case) initial letter; hence it will be seen that even specific names that are based on the names of persons (e.g. *Culicoides grahamii* or *Chrysomya bezziana*) begin with a lower case letter. Other rules are that names must be spelt without a break or hyphen (e.g. *Anopheles nuneztovari*, not *nunez tovari* or *nunez-tovari*, *Culex demeilloni* not *de meilloni*) and without any umlaut, accent or apostrophe (e.g. *Aedes* not *Aëdes*, *Wohlfahrtia seguyi* not *séguyi*, *Tipula oneili* not *o'neili*). Specific names should never be abbreviated. Generic names can be shortened to their initial letter (e.g. *A. funestus* for *Anopheles funestus*) but this should never be done in the title of a work and only elsewhere if the generic name has previously been given in full. Generic names should only be abbreviated to two letters when there is real risk of confusion between different genera with the same initial letter – for instance, if *Lutzomyia* vectors and *Leishmania* parasites are under discussion in the same piece of text.

A scientific name (especially of a species) is often seen in the literature to be suffixed with the name(s) of its original author(s), and sometimes also with the year date of first publication. However, this is a much overdone convention, followed even by many experienced biologists under the false impression that it is 'correct form'. Authors' names are *not* part of a scientific name and their citation (as also that of the year date) is *always* optional. Books and research articles, unless catalogues or checklists, or works concerned strictly with taxonomy, are best not cluttered with such items – which are redundant for general purposes. No describers' names or year dates are given for any scientific name in this book.

A point concerning describers' names that can cause confusion is that when cited after species names they are sometimes in parentheses (round brackets) and sometimes not. This is due to a mandatory rule of nomenclature – thought essentially valueless by many taxonomists (including this book's editors) – that parentheses must enclose the name(s) of the author(s) whenever a species is no longer put in the genus in which it was placed when first described (and so now has a different binomen). The Code recommends that, if the year date is also cited, this is placed in the same parentheses, e.g. *Aedes aegypti* (Linnaeus, 1762). (Transfer of a species into a genus to which it has never been assigned previously produces a 'new combination', the binomen not having been used before.)

Rules for specific names apply equally for subspecies, rules for generic names equally for subgenera (Principle of Coordination). The citation of subgeneric names is always optional, but whenever one is given it must follow the generic name and be placed in parentheses, e.g. *Simulium (Edwardsellum) damnosum*.

International Code of Zoological Nomenclature
In its details, nomenclature is a complex subject and different sets of rules exist for viruses, bacteria, plants and animals. Zoology (which includes entomology) is served by a work known familiarly to taxonomic zoologists simply as 'the Code' – that is to say, the *International Code of Zoological Nomenclature*, now in the third edition (International Commission on Zoological Nomenclature, 1985). The Code provides the framework for naming taxa of the animal kingdom and covers such matters as how names are to be published, formed (spelt and treated), interpreted from 'types' (specimens specially designated for nomenclatural purposes), and dealt with when matters of synonymy (different names for the same taxa) or homonymy (same names for different taxa) arise. The Code is published by the International Commission on Zoological Nomenclature (ICZN) in definitive English and French versions, but there are now Russian, Czech and Romanian translations and German, Italian, Japanese and Spanish versions are in preparation.

The Code consists of topic chapters divided into sections called articles. These contain provisions that are mandatory (rules that must be followed) or advisory (guidelines to be followed as good practice and called recommendations). Articles are written in a very precise manner (some would say in pseudo-legal jargon), but this is to cover all eventualities and avoid ambiguity. Experienced taxonomists meet many tangled situations over names and know how much such careful wording is needed if these are to be unravelled, but it must be admitted that the Code inevitably appears daunting to those who are uninitiated in its technicalities. Medical entomologists, however, can usually manage with a quite modest knowledge of the Code, since their taxonomic work (if any) is mostly focused on species, and relatively few of the 88 articles (other than those of a general nature) bear only on species names. (The Code, it should be noted, deals only with names of taxa ranked from subspecies to superfamily level. Names for infrasubspecific taxa, such as 'forms', and for suborders and above are outside its scope.)

Nomenclatural difficulties arise from time to time which either cannot be resolved by using the Code or which would entail undesirable upsets in names if its provisions were adhered to strictly. The commonsense solution is often not the one that is correct under the Code, and then it is important to have an adjudication by the International Commission on Zoological Nomenclature (ICZN) that will be the least disruptive and most generally welcome. An application (case) is put to ICZN, which then considers the problem and publishes a decision upon it (called an *Opinion*) in its *Bulletin of Zoological Nomenclature*. (An example from parasitology is that in Opinion 508, published in 1958, the ICZN ruled that the specific name of the 'river blindness' filarial nematode is *volvulus*, not *volvulas* as it was originally spelt, and that Leuckart, not Manson, is to be accepted as its describer.)

It is necessary to be clear that, although moves are afoot to extend the function of ICZN to registration of all names, the main purpose of ICZN is not to vet and pass each scientific name, or to examine the scientific merit of each new species, but to provide a code of practice – and ensure so far as possible that this is appropriately applied in the interests of stable and universally acceptable names. The Commission in no way encroaches on the freedom of zoologists to classify animals as they wish, and its decisions are usually accepted and adopted. It has, however, no powers by which it can ultimately enforce its decisions if renegade zoologists refuse to abide by them.

Describing new species
Some of the most important provisions of the Code for the working medical entomologist are best explained by outlining the steps that need to be taken when describing a new species (still a common undertaking in lesser known groups such as the Phlebotominae).

The objective of describing a new species (or any other taxon) is to establish it as a distinct biological entity and give it a formal name which scientists can use to refer to that organism in future – and one that they can apply without ambiguity. It is therefore wise to remember that a name once established in the literature cannot be removed and will be recorded for posterity in the life science databases. It is easy to burden the literature with names for supposedly new taxa which might prove unsound in the light of later research. Taxonomists are in a sense prophets without honour, since describing new species (or any other taxa) is a responsibility to other biologists coming afterwards; it bestows no honour on the describer.

Describing a new species is basically a two-part process. The first is strictly scientific in nature and involves deciding whether the species is distinct and truly new (undescribed and unnamed) – i.e. can it be distinguished by whatever characteristics are appropriate from other (known) species? – and whether it is worthy of species rank. Despite the plethora of species concepts, biological or other, these decisions are nearly always subjective even when rooted in a theoretical understanding of what a species is in nature. The second stage in describing a species is 'non-scientific' and concerns the appropriate use of a scientific name – the nomenclatural stage. Both stages always benefit from the input of advice from a taxonomist experienced in the group concerned.

The name chosen for a new species must be a combination of Latin letters usable as a word, even if arbitrary in combination or derived from a language not using the Latin alphabet. In medical entomology a common practice (much overdone if excusable in some circumstances) is to name new species after the geographical places from which the original specimens came or after persons associated with their collection (patronymic names); a preferable approach is to base the name on some characteristic feature of the species (e.g. as *Phlebotomus argentipes* means the *Phlebotomus* with silvery legs).

When choosing a name for a new species it is essential to be sure that it has not already been used for another species in the same genus, otherwise there will be homonymy. If the genus is large, it might be necessary to consult a specialist familiar with all names and catalogue sources before deciding. It is also essential to make sure that the chosen name does not stand alone when first published; a previously unused name unaccompanied by a description is a *nomen nudum* and as such is illegitimate (in Code terminology, 'unavailable' for use); such bare names are potential sources of future trouble in nomenclature and should be rigidly avoided. (If a manuscript name has been chosen it should be 'kept under wraps' until formally published with the description in case others inadvertently use it earlier in print.)

The description that accompanies the name when a new species is described should give a sufficiently comprehensive written and illustrated account of the appropriate characters as to distinguish it from other named species, with emphasis on the features that are considered unique (diagnostic). The name and description must be published in a way that ensures that the name will meet criteria of publication in the Code; this usually means publication in a printed journal intended for permanent scientific record and issued in multiple identical copies. Only then, and providing various other criteria are satisfied such as correct spelling, will the name be formally usable in scientific nomenclature ('available'). The name will not be available if (for example) the description is issued on computer printout or on microfilm or is contained in a thesis.

The author of a name is the person who first publishes it in a way that satisfies the criteria of publication (or those who do so if two or more authors are involved). However, in works with many authors it is usually only one or two who have been responsible for naming the new species and drawing up its description. Unfortunately, the tendency to publication of papers with multiple authorship by ten or even more named persons is leading to an increasing problem when such papers include the description of new species and also cover a range of topics on the biology and medical importance. In these cases the best course is to cite after the name of the new species only the person(s) responsible for the taxonomy; then the entire list of the paper's authors need not be given in formal taxonomic revisions and catalogues where author citation is standard; all authors receive credit for their contribution to the work without cluttering the literature with long strings of authors' names attached to a scientific name.

The date of publication (as for any scientific name) will be that on which the work containing the new species name and description is first distributed. This is not always the year date given on a journal, as it is now common for many journal parts not to appear (or even be printed) until well into the year after that stated on their covers or title pages. (Special care is needed about dates when working only from reprints.)

Type specimens

Detailed study of a supposedly single 'species' sometimes shows it to comprise two or more species, recognized perhaps by use of more refined (often non-morphological) techniques. This situation has arisen many times in medical entomology, in particular with the discovery of species complexes. It poses the problem as to which newly recognized species should bear the original name. Often this question can be resolved by comparison with the 'type', i.e. the specimen (usually designated at the time of description) which provides the reference point for how the name should correctly be used. This name-bearing specimen, or sometimes series of specimens, has usually been deposited in a major reference collection, such as a museum (as recommended in the Code). There are various kinds of type, and an account of them, with special emphasis on the type concept in relation to sibling species diagnosed on chromosomal or other non-morphological criteria, can be found in Crosskey (1988).

Type specimens are almost always designated for new species of arthropods, but (contrary to general belief) it is not mandatory under the Code for a type to be designated by the describer(s) of a new species. However, the specimens on which a new species is based form its 'type series' from which a single specimen, the 'holotype', is usually selected to bear the name of the species – the others (if any) being called 'paratypes' and having no name-bearing function. The Code recommends that a holotype is designated, but if this is not done then all the specimens on which the author(s) based the species concerned at the time of description have equal status and are called 'syntypes'. From such a syntype series, one specimen (the most suitable) can then be designated as the 'lectotype' and become the single name-bearing specimen equivalent to the holotype; once this is done, any other syntypes lose any potential importance as name-bearing specimens and become 'paralectotypes' equivalent to paratypes. Sometimes all the original specimens of a species are lost or destroyed, and in this circumstance any specimen of the species can be designated as 'neotype' to fulfil the name-bearing role; however, the Code provides strict rules about this, including mandatory deposit of any neotype in a museum or similar institution. Types in the categories holotype, syntype, lectotype and neotype are known as 'primary types' because they are all actually (or in the case of syntypes, potentially) name-bearing specimens. (Loosely, any individual primary type is called simply 'the type'.)

Some other categories of type than those already mentioned are sometimes used and named in the literature, such as allotype (a designated specimen of opposite sex to the holotype) and topotype (an original or subsequently collected specimen from the place of origin of the primary type), but none is recognized by ICZN or regulated by Code provisions. (Designation of allotypes remains popular in medical entomology but serves no nomenclatural purpose.)

To 'identify' a specimen as belonging to a certain named species means notionally that it is considered to be conspecific with the primary type specimen of that species. In practice, however, direct comparisons of specimens with types are made only by taxonomists with access to type specimens in reference collections, and routinely identifications are assumed to be correct when specimens conform to characters given in descriptions and illustrations, or agree with other specimens presumed to be correctly identified. In other words, use of type specimens has little to do with the biological species concept which defines species in nature in terms of genetically distinct populations. Types are merely a device for stabilizing the use of names. Failure to understand this simple basic function has been a rather persistent cause of confusion.

Changing of names

Nomenclature impacts on the biologist most when it involves change in the familiar names of species. Many biologists do not understand why such changes are necessary and, paradoxically, see them simply as irksome byproducts of taxonomic activity even when they recognize the fluid nature of taxonomy. Taxonomists are often said to 'love changing names' but it is far from the truth because (often coping with hundreds of names) they have a stronger need for stability than anybody else. Better knowledge and new ideas continually modify taxonomic concepts – so names can never be absolutely fixed.

Four main reasons for name changes will be briefly mentioned here:

1. Synonymy
2. Homonymy
3. Combination
4. Gender

1. Synonymy   The specialist who studies a large or neglected group, or the fauna of a little-known area, or critically evaluates old types, or has new biological or genetic data, often concludes (on the evidence available at that time) that two or more putatively different named species are not in fact distinct. Two or more names (synonyms of each other) then apply to the newly redefined species. Under the Code's Principle of Priority, the first-published (oldest) name takes precedence as the 'senior synonym' and the others are suppressed as invalid 'junior synonyms'. (The same principle applies at all levels from subspecies to superfamily, the oldest name being valid for the taxon and any others invalid junior synonyms.) Those who knew the taxon by a name now junior and invalid, perhaps not knowing the reasons for change, will probably assume the taxonomists have been up to their name-changing tricks again.

2. Homonymy   The rules of nomenclature are designed to prevent any two species having the same binomen (combination of generic and specific name). Similarly, different genera cannot have the same name even if they are in quite different parts of the animal kingdom. If homonymy is discovered, the later-published name involved (junior homonym) must be replaced by another name. If the taxon concerned has a synonym, this comes into use instead of the 'preoccupied' junior homonym, but if not a special new name (*nomen novum*) will be needed. Either way, the valid name will change. Example: the name of the Congo floor-maggot was *Auchmeromyia luteola*, but when the Afrotropical Diptera were catalogued it was found that *luteola* is a junior homonym (it was originally published in the binomen *Musca luteola* but this already existed for a quite different species); hence *luteola* was replaced by *senegalensis*, a junior synonym of *luteola*, and the Congo floor-maggot became *Auchmeromyia senegalensis* instead of *A. luteola*. In this case the change was dictated strictly by rules of nomenclature, not by any new scientific insight.

3. Combination A name change occurs in the generic element of a binomen when it is decided (for whatever reason) to transfer a species from one genus to another. Often there is resistance to change in a familiar name, particularly if this comes about because a 'splitting' taxonomist uses a multitude of restricted genera and not because the species had hitherto been misclassified. Examples: the binomen *Sarcophaga crassipalpis* becomes *Parasarcophaga crassipalpis* or *Jantia crassipalpis* if an author follows one of the restricted-genera classifications of Sarcophagidae (Chapter 11); the binomen *Simulium erythrocephalum* becomes *Boophthora erythrocephala* if an author follows the restricted-genera system for Simuliidae.

4. Gender The changed binomen just mentioned is an example of altered spelling caused by the rules of nomenclature. These require that if a specific name is an adjective its spelling must accord with the gender of the generic name with which it is validly combined. Hence the change of specific name ending in that case from *-um* (*Simulium* neuter) to *-a* (*Boophthora* feminine). Few have much sympathy with this rule, which is specially troublesome when it causes major spelling changes such as *major* to *magnum* or *ater* to *atra*.

SPECIES AND SPECIES COMPLEXES

The species concept

The cornerstone of systematics for all concerned with applied biology is the species concept. This is the fundamental point of reference for all comparisons between taxa, since the species is the basic biologically operational unit. It is the unit with which the medical entomologist is primarily concerned because individuals of a species are generally similar in their ability to bite (if bloodsucking species) or transmit parasites.

The biological species concept has attracted considerable attention since it was developed by Dobzhansky in 1937. A tremendous amount has been written about it since then. The early definitions of Dobzhansky and Mayr were that 'species are groups of interbreeding natural populations that are reproductively isolated from other such groups' (Mayr, 1966, 1971). This definition stresses the population aspect and as such it is the definition that is biological rather than the species. This definition distinguishes biological classification from classification of inanimate objects and emphasizes their relational properties and not their intrinsic qualities. In practice very few species have ever been tested for fitting the biological definition but it is nevertheless an expression of a remarkably useful concept – even though not easy to apply for a variety of operational reasons; for example, it is notoriously difficult and often impossible to cross-mate the sexes of a supposedly single species under observable (e.g. laboratory) conditions in which the fertility of the progeny could be determined. Several other definitions of species have been proposed as variations on the theme of reproductive separation, such as Paterson's definition that applies to bisexual animals and states that 'members of a species share a common mate-recognition system'.

In practice, there are various kinds of 'species' whose definition essentially depends on the nature of the evidence or criteria used for their recognition. Four principal kinds of criteria are derived from morphology, ecology, physiology and genetics, and can be briefly described as follows.

1. Museum criteria   These are usually based on morphological data used to distinguish species (morphospecies) and rely on the taxonomist's ability to differentiate between variation induced by the environment and inherent genetically determined variation. It is important, however, to appreciate that although morphological data are used, these are estimates of genetic differences. There is considerable reliance on the taxonomist's opinion (or intuition), but after some experience the approach is easy and quick to apply.

2. Ecological criteria   These are used when species are distinguished on the basis of differences in habitat and observed variations in behaviour. Taking such criteria into account sometimes kindles suspicions that what has been deemed one species might be a mixture of several closely similar species (see 'species complexes' described below). In the case of human disease vectors, this is often later proved by chromosomal studies or other means.

3. Physiological criteria   These are based on the fact that within a species there is continuous variation in physiological characteristics. Any discontinuities potentially reveal the existence of biologically distinct species (e.g. *Anopheles gambiae* and *A. merus* larval tolerance of salinity in the habitat).

4. Genetic criteria   These are the most reliable of all specific criteria as they directly assess the biological species concept. Cross-sterility tests between putative species reveal infertility in the offspring, thus signifying the existence of post-mating barriers; hybrid fertility, however, is an equivocal condition. Measures of genetic difference can also be obtained from chromosome studies, isoenzyme electrophoresis, and other techniques in which comparisons are made between observed differences in frequency and those expected from the Hardy Weinberg equilibrium.

To these several kinds of criteria can be added one which is perhaps best termed the cynic's criterion. It is lightheartedly summed up in the taxonomist's phrase, in which there is more than a kernel of truth, that 'a species is what a good taxonomist says is a species'.

Within species, not all populations are exactly alike. There is often considerable variation associated with geographical, environmental, seasonal and genetic factors. This is reflected taxonomically in the recognition of various 'infraspecific' categories for which a quite complicated terminology exists (see e.g. Mayr, 1966). The principal infraspecific category (and the only one that needs to be referred to here because mentioned in some chapters of this book) is the subspecies. Taxa at subspecies level are widely recognized in the animal kingdom but not always understood in the same way. The term 'subspecies' tends to be much misused, often being applied to infraspecific variants which – as in the case of the so-called subspecies of *Musca domestica* (Chapter 11) – are not defined on the basis of differing geographical distribution. Strictly speaking, in the generally accepted definition, subspecies are geographically based taxa within a species (usually a little different from each other morphologically) which have largely separate distributions but whose populations interbreed where they meet and overlap. In insect groups covered by this book, subspecies are only recognized to any considerable extent among the fleas (Siphonaptera), and even in these their recognition seems to be largely unwarranted (Chapter 16).

Nomenclaturally, each subspecies is named by using a trinomen, i.e. a third scientific name is added to the species binomen. This is identical with the specific name if the subspecies is that to which the type specimen of the species belongs (e.g. *Aedes aegypti aegypti*) and different from it if not (e.g. *Aedes aegypti formosus*).

Species complexes

The existence of 'species complexes' in some insect vectors of human disease, especially among the Culicidae (Chapter 5) and the Simuliidae (Chapter 6), has become known through intensive research on these families over the past 25 years or so, and their existence in other medically important insects is extremely probable (Service, 1988; World Health Organization, 1977; Wright and Pal, 1967). In Africa, a full understanding of the relationship between man, parasite and vector is essential for the correct interpretation of the epidemiology of malaria and onchocerciasis. In other words, species complexes are not simply phenomena of interest to taxonomists: they have fundamental importance in the biology of disease and must be taken into account by medical entomologists and parasitologists. The following is a short summary of what complexes are and how they are usually recognized. It refers essentially to the Diptera, as comparable complexes have not yet been unmasked in other groups of medically important arthropods.

Animals of many different groups have a propensity to produce sibling species, defined as pairs or groups of closely related species that are morphologically indistinguishable ('isomorphic') but reproductively isolated, and which frequently live in the same area (i.e. are sympatric). Among insects, sibling species are particularly common in the true flies (Diptera), and the tropical African blackfly *Simulium damnosum* complex – in which about 40 have been recognized – apparently holds the record for the highest number of siblings known among any organism in the animal kingdom. A salient point about sibling species, however, is that there is negligible proof of their existence from what might be thought to be the obvious route – experimental genetics. Information about sibling species and complexes is inferential rather than experimentally proved. Blackflies are virtually impossible to colonize and cross-mate in the manner that would be required, and even mosquitoes (considerably more amenable laboratory insects than blackflies) have proved difficult and laborious material for crossing experiments able to detect species complexes (Chapter 5).

The taxonomist's conventional species are almost always defined on the basis of morphological differences that provide 'diagnostic characters' for distinguishing between entities presumed to be single species; they are morphospecies for which (usually) there are few or no genetic data to prove their purity as single species or their reproductive isolation (inability to cross-mate *and* produce fertile offspring) from other such morphospecies. However, it is often observed in the field that a morphologically defined species is not biologically uniform but shows, in different parts of its range or at different seasons, puzzling inconsistencies in its behaviour or ecology (including susceptibility to parasites). This is a clue that apparently single species might actually be a mixed collection of genetically isolated species in nature – in short, a species complex.

Sibling species are by definition isomorphic or virtually so, and other means than morphology must therefore be used to unmask their existence in a complex, from the basis of their description, and provide the means by which they can be identified and distinguished. In practice, this mainly means using cytological attributes of specialized chromosomes (cytotaxonomy) or patterns of isoenzymes (chemotaxonomy). The Diptera are cytologically unusual as they have fewer pairs of chromosomes than normal in insects and often have polytene chromosomes, massive chromosomes of a kind unknown in other orders and formed by repeated lengthwise multiplication of DNA. Polytenized chromosomes are best developed in certain cells (e.g. adult ovarian nurse cells in mosquitoes and larval silk glands in blackflies) and, when stained, show conspicuous banding patterns and other micromorphological landmarks characteristic of each species (Figure 6.6). Stepwise rearrangements of the chromosomal structure provide specialists with insights into how species are evolutionarily related ('cytophylogenies' can be developed) but in practice the fundamental importance of chromosomal characteristics for medical entomology is to show that speciation has occurred in a particular vector morphospecies – i.e. in revealing a complex. The term cytospecies is often used for convenience to indicate that a species (usually a sibling species of a complex) is diagnosed and recognized primarily by chromosomal criteria. A cytotype or cytoform is a chromosomally recognized constituent of a morphospecies whose status is still uncertain: it might be a sibling species or only a polymorphic variant.

Mosquitoes (Culicidae) and blackflies (Simuliidae) nearly always have three pairs of chromosomes ($2n = 6$). In these families, the commonest chromosomal features which provide usable cytotaxonomic characters are rearrangements called inversions – sections of chromosome turned around between break points so that the linear order of the genes is reversed on that part of the chromosome – but there can also be interchanges of whole arms (i.e. exchanges of the terminal parts of non-homologous chromosomes). The nature of the inversions, whether both elements of the pair are intimately bound band for band or whether there is a failed pairing at some point on one or other chromosome pair, often makes it possible to decide if specimens belong to the same or different species and to detect natural hybrids between species. Sex-related chromosomal features can often be recognized but the sex chromosomes are not necessarily always overtly differentiated. Many inversions, however, are linked to the sex chromosomes and changes in their position are correlated with speciation.

Sibling species have in every way the same natural status as any other biological species and differ chromosomally by the same criteria. The principal characteristics by which they are (usually) differentiated therefore include fixed sequential differences in inversions and interchanges, sex-chromosome differences and differences in polymorphic (so-called 'floating') inversions – the latter being alternative banding sequences confined to one constituent of a chromosome pair. However, finding that such differences exist in the chromosomes of a morphospecies does not prove that each chromosomally different entity is a species and therefore part of a complex. The differences seen might manifest only a chromosomal polymorphism within one species (intraspecific variation), which is unrelated to speciation. Much depends on whether the populations possessing the cytological differences are geographically separated (allopatric) or mixed (sympatric). If the latter, and there is no evidence of hybrids, the sibling species status is warranted on the assumption that there is no cross-mating and gene flow between the populations. The allopatric situation is more difficult, since there is no gene flow; any chromosomal differences observed in the geographically separate populations could represent no more than isolated points on a gradient of chromosomal variation that possibly exists. On islands it is still more difficult to determine sibling species status since reproductive isolation is enforced by geography.

Within any population there will be variation between individuals in the structure of certain enzymes which is genetically determined and can be revealed by the technique of electrophoresis. In this technique, there is differential migration of proteins in an electric field associated with differences in molecular structure and charge. The various proteins form bands on the supporting medium (e.g. starch gel, polyacrylamide) which can be made visible by stain-substrates. Using this method, the genetic make-up of an individual can be seen – whether it is homozygous or heterozygous for a particular isoenzyme. By looking for the absence of heterozygotes, or even a dearth of heterozygotes relative to the proportion predicted by the Hardy Weinberg equilibrium, evidence for the absence of random mating (panmixis) can be obtained. The analysis of isoenzymes, with its very direct genetic interpretation, is particularly useful in those vector groups – such as *Culex* and *Aedes* mosquitoes, phlebotomine sandflies, tsetse-flies and ticks – which do not have well-developed polytene chromosomes.

Other techniques, such as gas chromatography of hydrocarbons extracted from the waxy layer of the body surface cuticle, have also been used in studies of species complexes but the results have been confirmatory rather than analytical. No doubt the application of the techniques of molecular genetics will have a great impact on future analysis of species complexes.

DISTRIBUTION AND ZOOGEOGRAPHICAL REGIONS

The human diseases transmitted by arthropods are mainly tropical or subtropical and the contents of this book are inevitably heavily biased towards insects and arachnids found in the warmer parts of the globe. Still, even though arthropod vectors of human diseases are mostly equatorial and limited in their geographical distribution – the tsetse (*Glossina*) vectors of human trypanosomiasis (sleeping sickness), for instance, are confined to tropical Africa – many of them have close relatives in temperate or even arctic regions whose biology is essentially similar even though they do not act as vectors of disease in humans. In summer, the northern tundra wastes are alive with bloodsucking mosquitoes and blackflies which often belong to the same genera as are found in the tropics and bite humans viciously if they have the chance – but harmlessly except for the extreme discomfort. Warm temperatures are usually not sustained consistently enough in these high latitudes for the successful development to the infective stage of parasites that affect man.

Much can be learned that is important in relation to the biology of tropical and subtropical disease vectors by studying the biology of their relatives in other areas, and the reader will notice on entering the specialist chapters of this book the frequency with which contributors mention the important research done outside the tropics and refer by name to the various zoogeographical regions, particularly in respect of the faunal and taxonomic literature. A brief description is therefore needed here of the background to, and extent of, these regions.

Since the mid-nineteenth century, zoologists have divided the world into six zoogeographical regions (sometimes called realms), Antarctica excluded. Each region is centred on the whole or the greater part of a continent and contains many animal species that are peculiar to it, together with others that are shared with one or more neighbouring regions. Each regional fauna contains many endemic elements (species, genera, families) and also elements which are indigenous (occur there naturally) but not endemic. The regions are usually delimited by physical boundaries such as deserts, mountain ranges, seas and oceans, of which by far the most effective in zoogeographical terms for isolating the regions are the oceans and seas. However, the faunal boundaries are never quite as distinct as the physical barriers between the regions, and among some animal groups there has been considerable spread of faunas between regions. Briefly summarized, the zoogeographical regions cover the following areas (Fig. 1.1):

- **Afrotropical region** Africa south of mid-Sahara, southwestern Arabia, Madagascar and other islands of the southwestern Indian Ocean, Cape Verde Islands. (The same areas as covered by the obsolete term Ethiopian region.)
- **Australasian region** Australia and Tasmania, New Zealand, Melanesia, Micronesia, Polynesia.
- **Nearctic region** North America and its arctic islands, Greenland, northern Mexico.
- **Neotropical region** South America, Central America including southern Mexico, Caribbean islands.
- **Oriental region** Asia east of Pakistan and south of the Himalayas and central China, Taiwan, Sri Lanka, and South East Asian archipelago eastwards to include Sulawesi.
- **Palaearctic region** Europe (including the British Isles), North Africa, Asia north of the Himalayas and central China, Iceland, mid-Atlantic islands, Japan.

Broadly speaking, the regional limits are well accepted, although some zoologists treat Madagascar and associated islands as a distinct region (Malagasy) and some recognize an Oceanian region separate from Australia. More ambiguous is the boundary between the Oriental and Australasian regions, a question which endlessly vexes zoologists. No one line is satisfactory for all animals. Wallace's Line between Borneo and Sulawesi is widely used because it is best for vertebrates (especially mammals), but Weber's Line between Sulawesi and Maluku (Moluccas) gives a better fit for most insects (including Diptera) and is the one recognized in this book.

In the most widely used zoogeographical system each of the six main regions is divided, on various faunal grounds, into four subregions; however, subregional terminology does not feature much in this book and therefore no details are given here. The subregions are named and their limits shown on a map in Lincoln et al. (1982, p. 271). In the Palaearctic region the 'eremic zone', approximately equating with the Mediterranean subregion and consisting of the warm deserts of North Africa, the Middle East and southern USSR, is a particularly important feature in the distribution of *Phlebotomus* vectors of leishmaniasis.

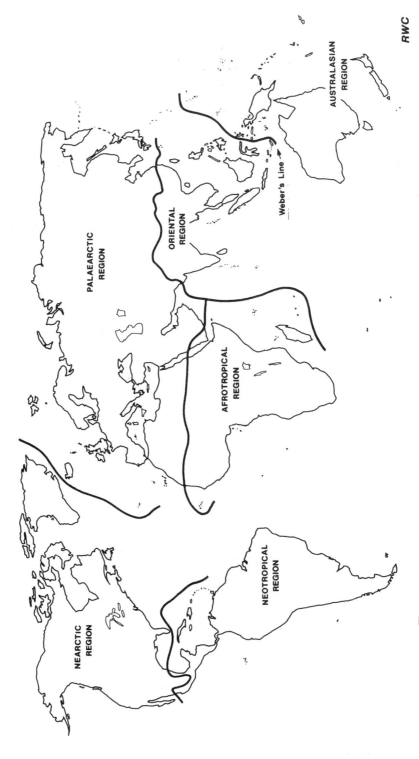

Figure 1.1 World map showing the names and approximate limits of the zoogeographical regions.

The term Holarctic should be explained here because it is used in several chapters. Zoogeographers often combine the Nearctic and Palaearctic regions to form one large realm characterized by shared faunal elements distributed around the more northerly (boreal) parts of the northern hemisphere. Some far northern species of biting flies actually have circumboreal Holarctic distributions which have resulted from faunal interchange across areas which appear widely separated on conventional flat maps but are not far apart in global geography. In general, the regions in the southern hemisphere are more distinct than those of the northern hemisphere.

The distribution of arthropods, like that of other organisms, is dynamic and continually changing. Present distributions reflect the effects of past geological events. The now widely accepted theory of continental drift asserts that the present positions of the major landmasses are the product of ancient break-up of a large supercontinent (Pangaea) into smaller continents which, sliding on a viscous base, have drifted gradually apart into their present positions. Sometimes the drifting proto-continents have fragmented, parts either becoming major islands such as New Zealand and Madagascar, or becoming connected to another proto-continent – as in the case of peninsular India now joined to Asia. Interpretation of animal (and plant) distributions in the light of such continental drift – rather than by invoking older theories of dispersal from centres of radiation by wind, sea or locomotion – is called vicariance biogeography (Nelson and Rosen, 1981).

The continents are terrestrial parts of larger tectonic plates, which, as the result of collision and buckling at their meeting edges, can give rise to high mountain ranges, e.g. the Andes and Himalayas are still forming where one tectonic plate is forcing itself beneath another. When all landmasses were united as Pangaea, which palaeogeographers conclude was from 245 to 180 million years ago, terrestrial animals were free to move over the large supercontinental surface; afterwards, the sundered faunas evolved independently and so developed much of the diversity of genera and families we see today. Large continents have sometimes met as the result of drift and become united into a continuous landmass, so allowing animals to spread and become intermingled in areas they did not previously occupy. North and South America, for instance, became joined in relatively recent geological times at the isthmus of Central America.

Useful texts for further reading on zoogeography or general biogeography are Darlington (1957), George (1972) and Müller (1986).

VECTORS AND VECTOR STATUS

Most insects and arachnids of medical importance are studied because they transmit parasites or pathogens to humans, and so are referred to as vectors of infections. Most vectors are haematophagous (feed on blood) and acquire or transmit the infection when they feed. There are varied mechanisms by which parasites are transmitted and these can be arranged in a series of categories of increasing evolutionary or ecological complexity. Such an arrangement is not an evolutionary pathway such that one category gives rise to the next over time, but is a convenient way of describing the many different vector–parasite or vector–pathogen combinations.

Modes of transmission

There are two basic methods by which parasites and pathogens are transmitted by vectors (mechanical and biological transmission), differing in how essential the vector is in disseminating the infective organisms.

Mechanical transmission

The simplest form of transmission is mechanical transmission in which the vector spreads the parasite by contact with successive hosts without any obligatory development of the parasite in the vector. Transmission is mainly fortuitous, by contamination, and is usually only an alternative to other methods of transmission such as by drinking contaminated water or contagion (contact with an infected person). Omnivorous insects frequenting human dwellings, such as house-flies (*Musca*) or cockroaches which feed on faeces and human food, have the potential to transmit enteric pathogens mechanically. Transmission of pathogens from sores is also possible, e.g. *Musca sorbens* can transmit the agent of trachoma although this is usually transmitted by contagion. There are numerous reports, mostly from laboratory investigations, of bloodsucking insects mechanically transmitting parasites in a small drop of blood on their mouthparts. The main limitation to mechanical transmission is the viability of the pathogen in the hostile environment of the mouthparts or elsewhere on the external surface of the vector. However, the transmission of myxomatosis between rabbits is solely by mechanical means via the mouthparts of mosquitoes, blackflies or fleas. In this case the aetiological agent is viable on insect mouthparts for up to three months, which is sufficient advantage for the pathogen to cause great epizootics of myxomatosis. There is no evidence that HIV is transmitted by insects, mechanically or otherwise.

Biological transmission

Most vectors are biological vectors, that is to say they are an essential stage in the life cycle of a parasite which undergoes either multiplication or development in the vector before reaching an infective stage capable of invading a new host. There is a cycle of transmission between a vector and a vertebrate. The time between the uptake and deposition of a parasite is the latent period (also called the extrinsic incubation period) and is an important factor in the epidemiology of a vector-borne disease.

Transmission by eating the vector The simplest form of biological transmission is where the parasite is transmitted when the host eats the vector. Only one host can be infected for the life of each vector. Two examples fall into this category. Eggs of the dog tapeworm *Dipylidium caninum* are taken up by flea larvae and transmission takes place when adult fleas are eaten by dogs or accidentally by children; and *Borrelia* spirochaetes which cause louse-borne relapsing fever are transmitted when the infected lice are 'cracked' between the fingers or teeth of their human hosts.

Transmission during or after bloodsucking Most cyclically transmitted parasites are acquired by their vectors when these feed on the blood of a mammalian host. However, there are various mechanisms by which the parasites get out of the vector and into the host.

Proliferation in the gut and transmission in faeces. The parasites enter the vector with the blood-meal and develop in the gut, sometimes undergoing several developmental stages. *Rickettsia prowazeki* develops in the gut cells of clothing lice and the cells rupture some eight to ten days after the infection has caused the pathogens to be excreted with the louse faeces. The faeces can remain infective for up to three months and invade the human by being scratched into a wound (louse bite) or through mucous membranes. Perhaps the most important transmission cycle in this category is that of the protozoan *Trypanosoma cruzi* in triatomine bugs. The metacyclic stage parasites are excreted during the bug's diuresis (elimination of excess fluid) following a blood-meal or in the faeces; either way, the infected bug excretions can be rubbed into the eyes or ingested by a human.

Proliferation in the vector's gut and transmission by bite After ingestion with the blood-meal the parasites remain in the gut before moving forwards into the fore-gut for transmission when the vector next feeds on blood. The simplest example is the multiplication of *Yersinia pestis* (the plague bacillus) to form a plug in the fore-gut of a flea. This plug has to be broken up and expelled by regurgitation when the flea pierces a new host, otherwise it is unable to ingest fresh blood. A more elaborate version of this type of transmission cycle is when the parasites travel further down the gut and undergo division and multiplication before actively migrating forward to the fore-gut or mouthparts. This is exactly what happens with *Leishmania* parasites in their sandfly hosts. The parasites can attach to the cells of the vector's gut but they never penetrate to the haemocoel.

Penetration of vector's gut and transmission by bite Many parasites do not stay in the vector's gut to become involved in the aggressive effects of digestive enzymes and entanglement in the peritrophic membrane (an envelope formed around a blood-meal) but rapidly penetrate the cells of the gut wall. The parasites then become intimately involved with the vector and are exposed to the arthropod's immune system. Little is known of these vector defence reactions and the evasive tactics of the parasites but current research is revealing a surprising 'arms race' between the two groups of organisms (Ewald, 1983). Having entered the haemocoel, either after development or multiplication (or both), the parasite has to get out of the vector and invade a new host. Filarial parasites such as *Onchocerca* in *Simulium*, and *Wuchereria* in mosquitoes, enter their vectors as microfilariae and moult two or three times to reach the infective stage before migrating to the thick, fleshy, labium of the mouthparts; then, when the vector feeds, the mobile worms burst out and on to the new host's skin, which they rapidly penetrate via the hole made by the vector's mouthparts. There is no multiplication of the filariae in the vector.

Many other parasites, on invading the haemocoel, replicate and then circulate throughout the vector before finally concentrating in the salivary glands. This is an effective strategy because saliva is produced each time the vector feeds on blood, and therefore the parasite can be inoculated directly into the host. Some of the most widely known parasites have evolved this strategy of transmission, including the malaria parasites *(Plasmodium)*, the trypanosomes causing sleeping sickness in man *(Trypanosoma brucei)*, and all the arboviruses (e.g. yellow fever and dengue viruses). In some cases there is interference by parasites with the production of saliva to maximize the chances of their transmission.

Vector incrimination

The criteria for incriminating species as vectors depend very much on the biological background to the development and transmission of the parasite or pathogen. In the case of mechanical vectors it is often very difficult to obtain conclusive evidence on the role of putative vectors in pathogen transmission, especially when there are several other alternative routes of infection. There is considerable controversy in the scientific literature on the exact role such vectors play in disease epidemiology. Certainly the isolation of viruses from bloodsucking insects and acarines during an epidemic of an arboviral disease is not proof of their vector status, for it is the ability to 'allow' multiplication and transmission of the virus that is the key to incriminating an arbovirus vector (World Health Organization, 1967). For protozoan parasites there have been criteria discussed and guidelines published which essentially concern the nature of the man–vector contact and the ability of the vector to support the (often complicated) development of the parasites after the complete digestion of the blood-meal (e.g. Killick-Kendrick, 1990). The physiological ability of a vector to support the development of particular parasites is termed its vector competence or vector potential. The situation in the field can be different from that determined in the laboratory; for instance a species might be infected with a parasite and even transmit it experimentally even though it might bite man only rarely. A refractory vector is physiologically unable to support parasite development (cf. a susceptible vector) or will support it to only a limited extent. The extent of refractoriness can vary not only within a species but also within a conspecific population and seems usually to be genetically determined.

Numerous terms have been coined to describe the relative importance of a species as a vector; a primary vector (= principal, main, or major vector) is the species responsible for most transmission of a disease-causing organism in any one place. In some situations there can be several primary vectors (i.e. they are sympatric) but their seasonality, ecology and behaviour might be very different. Secondary vectors (= subsidiary vectors) are considered less important and although they regularly transmit parasites it is usually at such a low level that they are unable to maintain the disease cycle in the absence of the primary vectors. By far the most effective method of comparing the relative importance of different vector species is to use a quantitative measure of their transmission. The simplest estimates are made by combining the man-biting rate (number of vector individuals biting per unit time) with the infection rate (proportion of vectors infected with the parasite) to give the infective biting rate, usually expressed per year or month. The Annual Transmission Potential (ATP) used to monitor the huge Onchocerciasis Control Programme in West Africa is essentially this estimate. A more precise measure of the importance of a species in the transmission of infection is its vectorial capacity. This is a precise quantity and mathematically is the rate of increase in the number of human cases of an infection transmitted by vectors. It can be represented by the equation:

$$C = \frac{m\, a^2\, p^n}{-\ln p}$$

where m = the number of vector individuals per human, a = the number of blood-meals per day per vector individual on humans, p = the daily survival rate, and

n = the extrinsic incubation period. There has been some debate recently over exactly how reliable estimates of the vectorial capacity really are given the difficulty in measuring some of the parameters (Dye, 1990). However, despite these concerns the concept remains extremely useful in focusing attention on the need to approach the problem quantitatively. If used comparatively, however, such as at the beginning and end of a control campaign, the vectorial capacity is a powerful tool to measure change. The term vectorial efficiency is often used as a subjective expression of the vectorial capacity and cited as high, medium or low capacity.

Many biological factors affect the ability of a species to be an efficient vector, including the man-biting rate, seasonal distribution, survival rate, diurnal biting cycle, dispersal and host-biting preferences. Such factors affect horizontal transmission (from one individual host to another at one 'period' of time) but vertical transmission (from one vector generation or stage of metamorphosis) also occurs. Many prokaryotic disease-causing organisms (viruses, bacteria, rickettsiae) can be transmitted from the adult of a given vector through its egg and immature stages to the next generation of adults. Where such vertical transmission takes place a disease can be maintained in an area without passing through a human, except perhaps during an epidemic in which the human population acts as an amplifying host. Good examples of this are sandfly fever virus in phlebotomines and Congo-Crimean haemorrhagic virus in ticks. When the immature stages of a vector also feed on blood (e.g. lice, mites) they can become infected and this infection can be passed from one instar (= developmental stage) to another, the pathogen surviving the moulting process. This is known as trans-stadial transmission.

In practical field studies, the incrimination and subsequent control of a disease vector involves many different scientific disciplines, but central to the whole process is the correct and precise identification of vector species.

COLLECTION, STORAGE AND HANDLING OF SPECIMENS

Every medical entomologist has to obtain and handle arthropod specimens at some time during field research. Usually this involves collecting and preserving specimens by special methods which apply mainly, sometimes only, to the particular group being studied; trying to collect, preserve and study sandflies by the methods used for blackflies would be futile. Nevertheless, routine procedures used for insects in general by entomologists are to a large extent relevant for medically important insects in particular. What follows is therefore a short account of some basic points about the collection, storage and handling of specimens aimed at providing some background to the special accounts given in the specialist chapters. Aspects covered bear closely on the very important matter of voucher material mentioned later. Specimens, it should be noted, are often called 'material', and the two words are virtually synonymous to entomologists.

Detailed information on collecting and preservation can be found in Martin (1977, arachnids in addition to insects), Oldroyd (1970, insects, general), Stehr (1987, immature insects), Steyskal et al. (1986, insects, general) and Walker and Crosby (1988, insects, general). An account in French is provided by Colas (1988). The work of Martin (1977) is recommended for its comprehensiveness, clear illustration and introduction to the many kinds of insect trap. Lane (1974) deals with preserving and mounting medically important insects.

Basics of collecting and handling material

Collecting equipment

Equipment for obtaining insects ranges from the simplest, such as entomological nets and collecting tubes, to very sophisticated traps used for special purposes. Sampling is the standardized collection of specimens for quantitative purposes and can employ a wide range of equipment; however, the term is used loosely as more or less equal to collecting in qualitative surveys. Here it is possible to mention only the simple and basic items routinely used and needed for collecting. The many kinds of insect trap used in specialized research, some catching individuals mechanically (e.g. suction traps), others luring them to their fate (e.g. traps which light up or emit carbon dioxide or mimic host animals), cannot be considered here but the reader is referred to Muirhead-Thomson (1991), Service (1993) and Martin (1977) as starting points about traps as sampling devices.

Modern entomological hand-held nets usually consist of a bag of fine-mesh fabric fitted to a circular frame, the latter being of aluminium and hinged so that it can be collapsed when not in use. Nets, however, have limited value in medical entomology, though examples of their use are the collection of mosquitoes by sweeping vegetation and the capture of individual tsetse-flies from their hosts (a swift arm sweep accompanied by a deft flick of the wrist twists the bag over the frame edge and traps flies as they take wing). Purpose determines the appropriate type of net. Insects of interest to medical entomologists show up best in white bags (the black bags favoured by lepidopterists are useless).

Aspirators, unlike nets, are important equipment for medical entomologists and are used (for instance) for the collection of mosquitoes and sandflies from their resting places, and even for picking up ectoparasites encouraged to leave their hosts. They are simple but effective pieces of apparatus into which small insects are sucked, the suction being applied either by mouth of the human collector, by various rubber bulb devices or more commonly by small battery-powered fans. The simplest consists of a length of glass or clear rigid-plastic tubing (about 25 cm long) with a gauze over one end to which is also attached a length of flexible tubing. Mosquitoes and the like are drawn into the aspirator by sucking on the flexible tube and are easily expelled by blowing. Another type consists of a wide glass or plastic tube (about 10–15 cm long) with each end closed by a bung through which passes a narrow-bore glass tube open at both ends; suction is applied to a length of rubber tubing (often provided with a glass mouthpiece) attached to one of the narrow-bore tubes and insects enter via the end of the other narrow tube, the intake tube; a piece of gauze over the inner end of the suction tube prevents inhalation of specimens from the wide tube in which they become trapped.

Many medically important insects can be captured even without a net or aspirator by simply placing a collecting tube over them while they rest or are engaged in bloodsucking – a time when they are not easily distracted. Many sizes and types of tube or vial are obviously suitable for this, but flat-bottomed glass tubes provided with corks are best (the deeper the cork the better). Small insects (e.g. sandflies, *Culicoides* midges) can become stuck in plastic tubes by static electricity, especially tubes made of clear polystyrene.

Ectoparasitic insects and acarines require specialized methods which are discussed in the relevant chapters.

Killing and preserving specimens
The most important point to bear in mind when killing and preserving specimens is that the methods used are dictated by the types of characters to be examined for identification. Failure to consider this point at an early stage can render large amounts of carefully collected specimens quite useless, e.g. mosquitoes should not be collected or preserved in liquids because the scales covering the body and wings will be lost, making identification almost impossible.

Many insects and other arthropods if left to die make poor-quality specimens; for identification or long-storage as voucher specimens they should be killed by exposure to one of the standard killing agents. These include ethyl acetate, ether, tetrachloroethane and chloroform, which can be absorbed on to cotton wool in a glass tube or bottle. Insects can be killed by freezing if such facilities are available in the field, or as a last resort by exposing specimens to tobacco smoke (but beware of contaminating equipment for subsequent use with live material) or the hot sun. Larvae can be killed by immersion in alcohol, as can adults of small arthropods destined for storage in alcohol or slide mounting. Large dipterous maggots (e.g. muscids or calliphorids) are best killed in boiling water to retain their shape. Vapour from fluid killing agents tends to condense on the walls of killing bottles and tubes, especially in warm tropical conditions, and it is important to remove specimens for pinning as soon as they are dead. Specimens should be dealt with rapidly once dead, pinned or otherwise preserved, to avoid their becoming too shrivelled or brittle – particularly any specimens that have been exposed to chloroform, as this killing agent quickly makes insects stiff.

Three basic methods of general preservation are used: dry, immersion in fluid and slide-mounting. Dry preservation is satisfactory, in some cases essential, for several groups of vectors (e.g. mosquitoes, blackflies, horse-flies and blow-flies) their hard exoskeleton preserving most external morphology, and these are commonly pinned. Large specimens (e.g. *Tabanus* horse-flies) can be impaled directly by long pins, but small specimens need first to be impaled on fine micro-pins and then 'staged', i.e. the micro-pinned specimens mounted on small pieces of material such as polyurethane ('plastazote') which are then impaled on a long carrier pin (see Fig. 6.15c). For safety, the specimen should be fixed near the middle of the micro-pin, otherwise it will be vulnerable to loss or damage. The method sometimes seen in which the micro-pin is first passed through the stage and the specimen then skewered on the point of the micro-pin should *never* be used. Stainless steel micro-pins are specially made in several fine calibres for specimens of different sizes; such pins require handling with forceps. The method for small specimens of glueing them to pieces of card, celluloid or other material should be avoided (specimens are often later lost when the adhesive becomes brittle).

Wet preservation is necessary for many medically important arthropods and is usual for larval insects, certain adult insects (e.g. ectoparasites such as fleas and lice) and for arachnids. The specimens are submerged in a fluid which prevents decomposition, usually ethyl alcohol (ethanol). Other preservatives include formalin and methyl alcohol (methanol), but the former decolorizes specimens and the latter makes specimens brittle. Ethanol (at 80% strength, without glycerine) is the best general-purpose fluid preservative. With time, all preservatives bleach specimens, but keeping them in darkness helps to slow the process. Many kinds of vial or tube are suitable for preserving specimens in ethanol but it eventually evaporates from any container however tight the cap or bung; vials and tubes should therefore be put for safety into larger storage jars which are also filled with alcohol and have tightly sealing caps (see Fig. 6.15*a*). (Glass jam jars with plastic-lined twist-grip metal caps are excellent: cheap, easily available and much superior to expensive equivalents from laboratory suppliers.)

Some medically important arthropods, such as midges, sandflies, fleas and mites, need to be slide-mounted for identification and subsequent permanent storage, although they can be stored temporarily in alcohol. There are two main types of mountants: water based, such as Berlese's and Puri's medium, or organic solvent-based media such as Canada balsam and Euparal. Mountants differ in their refractive indices, so some are more suitable for one group of arthropods than another. Appropriate techniques and mounting methods are described in the relevant chapters. Specimens are most conveniently studied and compared, with minimal loss of time for location, if they are mounted in the same position on each slide. Placing a microscope slide on a card-frame with the required location of the specimen marked as the first stage in preparation is a simple method of ensuring standardization.

Collections of pinned insects can be troublesome to keep free of pests and mould, especially in the tropics. It is useless to expect specimens to remain unharmed if steps are not taken for their safety. The first precaution is to ensure that they are kept in store-boxes that are as air-tight as possible. Tightly closing purpose-made store-boxes from entomological suppliers, with wooden carcasses and a plastozote lining are best (cork linings become very hard and so make it difficult to insert or extract pins). Tight closure stops any pests entering and eating specimens. Treatment of the box with fungicide (e.g. merthiolate) is usually enough to prevent mould, but extra protection can be obtained when specimens are at risk in extremely damp places by wrapping each store-box tightly in a plastic bag containing some loose silica gel. An insecticide in the box, such as dichlorvos impregnated into an inert plastic block (as commonly sold for controlling domestic pests) can be very effective but must be renewed regularly.

Specimen data

Specimens without labels are worthless. Every specimen or sample should therefore have data labels providing the following information – where and when the specimens were collected, and the host for ectoparasites – as a bare minimum. Geographical information should comprise the country, country subdivision (province, district or county), orientation relative to a feature such as a village or river, the altitude, and grid reference. The data should be given to day-date accuracy. The labels should show the data in full (i.e. not cryptically by some code referring to data stored elsewhere) and accompany the specimens, i.e. be fixed to the pins of dry-preserved specimens, placed *inside* the vials in the case of fluid-preserved specimens, and affixed to slides of slide-mounted specimens.

Mailing specimens

Specimens must often be mailed if identification by a specialist is needed or for other purposes, such as deposit in a reference collection. Unfortunately many specimens reach their destination in broken or irretrievably damaged condition, but this is almost always due to poor packing before despatch rather than carelessness on the part of the postal services. Damage can be avoided by taking simple precautions.

Specimens in fluid should be mailed in plastic or glass vials with a screw cap (e.g. 'bijou' bottles used in parasite culture and microbiology), not in tubes with cork or plastic stoppers. Glass vials are as safe as plastic vials if properly packed, but whatever type is used the caps should be taped to eliminate the risk of leakage and desiccation of contents. Hollowed wooden containers are commercially available for mailing single vials of various sizes. Each vial must be plugged inside (e.g. with lightly scrolled tissue paper) to prevent specimens slopping back and forth during transit (a common cause of fragmentation) and air bubbles must be rigorously excluded.

For safe mailing of dry-pinned insects it is essential to pack them in a rigid cardboard, plastic or wooden box which is then packed into a much larger carton. Nearly all damage to pinned specimens in the mail is caused by the lack, or inadequate thickness, of outer packing around the specimen-box. A minimum thickness of 8 cm of soft packing material (e.g. bubble-wrap or wood-wool) should be used around each surface of even a small specimen-box, more for large store-boxes. If specimens are staged, a pair of long pins should be fixed cross-wise over each mount to prevent it moving during transit; such specimens should also be spaced so they will not touch one another even if mounts accidentally swing round. A piece of cotton-wool should be pinned in a corner of the specimen-box to trap any specimens or broken parts that break loose.

The contents of a mailed package are commonly marked on the outer wrapper or any required declaration form as 'specimens for scientific study: no commercial value'. This makes it unlikely the package will be opened, and the contents possibly damaged, when scrutinized through customs.

Voucher material

It cannot be stressed too strongly that in any study of medically important insects, whether field or laboratory based, a well-preserved and annotated sample of each recognized taxon should be retained. These voucher specimens have paramount importance both during a study, for comparison with other reference collections, and afterwards – when later taxonomic findings often make it necessary to re-examine old material and discover its precise identity. Because of such retrospective importance, voucher material should be deposited in major regional or international collections where it will be safely preserved, made available for study, and be a permanent resource both for the field medical entomologist and the scientific community.

REFERENCES

References are grouped in the following list under the headings used in the foregoing text.

Classification and nomenclature

Blackwelder, R. E. 1967. *Taxonomy: a text and reference book*. xiv + 698 pp. John Wiley, New York.
Crosskey, R. W. 1988. Old tools and new taxonomic problems in bloodsucking insects. Pp. 1–18 in Service, M. W. (ed.), *Biosystematics of haematophagous insects*. xi + 363 pp. Clarendon Press, Oxford.

Crowson, R. A. 1970. *Classification and biology.* 350 pp. Heinemann Educational Books, London.
International Commission on Zoological Nomenclature 1985. *International Code of Zoological Nomenclature.* Third edition: adopted by the XX General Assembly of the International Union of Biological Sciences. xx + 338 pp. International Trust for Zoological Nomenclature, London.
Jeffrey, C. 1989. *Biological nomenclature.* Third edition. ix + 86 pp. Edward Arnold, London.
Mayr, E. and Ashlock, P. D. 1991. *Principles of systematic zoology.* Second edition. xx + 475 pp. McGraw-Hill, New York.

Species and species complexes

Mayr, E. 1966. *Animal species and evolution.* xiv + 797 pp. Harvard University Press, Cambridge, Massachusetts.
Mayr, E. 1971. *Populations, species, and evolution: an abridgement of animal species and evolution.* xv + 453 pp. Belknap Press of Harvard University Press, Cambridge, Massachusetts.
Service, M. W. 1988. New tools for old taxonomic problems in bloodsucking insects. Pp. 325–345 in Service, M. W. (ed.), *Biosystematics of haematophagous insects.* xi + 363 pp. Clarendon Press, Oxford.
World Health Organization 1977. Species complexes in insect vectors of disease (blackflies, mosquitos [sic], tsetse flies). Mimeographed document WHO/VBC/77.656 and WHO/ONCHO/77.131, 56 pp. World Health Organization, Geneva.
Wright, J. W. and Pal, R. (eds) 1967. *Genetics of insect vectors of disease.* xix + 794 pp. Elsevier, Amsterdam.

Distribution and zoogeographical regions

Darlington, P. J. 1957. *Zoogeography: the geographical distribution of animals.* xi + 675 pp. John Wiley, New York, and Chapman & Hall, London.
George, W. 1972. *Animal geography.* x + 142 pp. Heinemann, London.
Lincoln, R. J., Boxshall, G. A. and Clark, P. F. 1982. *A dictionary of ecology, evolution and systematics.* [viii] + 298 pp. Cambridge University Press, Cambridge. [Contains maps of biogeographical regions and subregions and a table of the geological time scale.]
Müller, P. 1986. *Biogeography.* x + 377 pp. Harper & Row, New York.
Nelson, G. and Rosen, D. E. (eds) 1981. *Vicariance biogeography: a critique.* xvi + 593 pp. Columbia University Press, New York.

Vectors and vector status

Dye, C. 1990. Epidemiological significance of vector–parasite interactions. *Parasitology* **101**: 409–15.
Ewald, P. W. 1983. Host–parasite relations, vectors, and the evolution of disease severity. *Annual Review of Ecology and Systematics* **14**: 465–85.
Killick-Kendrick, R. 1990. Phlebotomine vectors of the leishmaniases: a review. *Medical and Veterinary Entomology* **4**: 1–24. [Provides criteria for incrimination of vectors.]
World Health Organization 1967. Arboviruses and human disease. Report of a WHO scientific group. *World Health Organization Technical Report Series* **369**: 1-84.

Collection, storage and handling of specimens

Colas, G. 1988. *Guide de l'entomologiste: l'entomologiste sur le terrain – préparation, conservation des insectes et des collections.* 329 pp. Éditions N. Bourbée, Paris.

Lane, J. 1974. The preservation and mounting of insects of medical importance. Mimeographed document WHO/VBC/74.502, 20 pp. World Health Organization, Geneva.

Martin, J. E. H. 1977. *Collecting, preparing, and preserving insects, mites, and spiders.* 182 pp. Agriculture Canada, Ottawa. [= Part 1 of The insects and arachnids of Canada.]

Muirhead-Thomson, R. C. 1991. *Trap responses of flying insects: the influence of trap design on capture efficiency.* xi + 287 pp. Academic Press, London.

Oldroyd, H. 1970. *Collecting, preserving and studying insects.* Second edition. 336 pp. Hutchinson, London.

Service M. W. 1993. *Mosquito ecology: field sampling methods.* Second edition. xiii + 988 pp. Elsevier Applied Science, London and New York.

Stehr, F. W. 1987. Techniques for collecting, rearing, preserving, and studying immature insects. Pp. 7–18 in Stehr, F. W. (ed.), *Immature insects*: vol. 1, xiv + 754 pp. Kendall/Hunt, Dubuque, Iowa.

Steyskal, G. C., Murphy, W. L. and **Hoover, E. M**. 1986. Insects and mites: techniques for collection and preservation. *United States Department of Agriculture Miscellaneous Publications* **1443**: 1–103.

Walker, A. K. and **Crosby, T. K.** 1988. The preparation and curation of insects. Revised edition. *DSIR Information Series* No. 163, 90 pp. New Zealand Department of Scientific and Industrial Research and Entomological Society of New Zealand.

CHAPTER TWO

Introduction to the arthropods

R. P. Lane

The phylum Arthropoda, which includes crabs, shrimps, spiders, millipedes, centipedes and insects, is invariably described with superlatives. More than 85% of animal species are arthropods, over one million species of insects have already been described and estimates of the total number of species range up to 30 million (based on extrapolation from tropical forest studies). The phylum has radiated into every habitat, from the ocean floor to high altitudes, and its members are extraordinarily diverse in their structure and life histories. The biomass of arthropods is colossal; some calculations suggest it is greater than all other animals combined.

Arthropods are characterized by a segmented body covered in a resistant covering of cuticle which is often hardened to form an exoskeleton. Flexible cuticle between sections of the limbs and body segments forms joints and allows movement by muscles attached to the cuticle. The principal component of the cuticle is the carbohydrate chitin, the second most abundant polymer on earth, after cellulose.

Other distinguishing features of arthropods include a segmented body showing bilateral symmetry and a variable number of body segments with paired appendages having different functions depending on the part of the body from which they originate. The circulatory system comprises a dorsal heart and vascular spaces making up a haemocoel, and the central nervous system consists of an anterior supra-oesophageal centre or brain and a ganglionated ventral nerve cord. The muscles are principally of the striated fibre type and there is a general absence of ciliated epithelial cells.

The arthropods are an ancient group, for example the trilobites had already widely radiated to become the dominant arthropods in the Early Palaeozoic seas (about 550 million years ago) and then became extinct by 280 million years ago. It is presumed that the arthropods had a long Precambrian period of development during which they differentiated from their annelid ancestors. The earliest arachnids appeared about 400 million years ago and the insects, millipedes and sea-spiders appeared in the Devonian period, about 350 million years ago.

STRUCTURE

The exoskeleton, so typical of arthropods, is the key to their successes and limitations. It has several functions – to support the body, protect the animal from the

Medical Insects and Arachnids Edited by Richard P. Lane and Roger W. Crosskey.
Published in 1993 by Chapman & Hall ISBN 0 412 40000 6

external environment, reduce water loss, store energy and (through the development of limbs and wings) assist in locomotion. The external integument consists of a thin, impermeable, non-chitinous outer layer (the epicuticle) and a thick, elastic, permeable and laminated inner layer (endocuticle) composed mainly of protein and chitin. The outer layers of the endocuticle can be hardened either by the incorporation of calcium carbonate such as occurs in many marine crustaceans or by sclerotization or 'tanning' of proteins in the cuticle of insects and arachnids. The hardening given to the mouthparts has generated a considerable diversity of structures, thus enabling arthropods to exploit a tremendous range of food sources, e.g. the incorporation of metals such as zinc into the mandibles of some plant-eating insects reduces abrasion and wear. The exoskeleton also extends into the gut (fore-gut and hind-gut) and lines the tracheae, the tubes used in respiration by the myriapods and insects.

The combination of hardened plates with soft membranes between them gives both strength and rigidity to the body as well as flexibility. Thus the construction of an arthropod as a series of jointed tubes gives it considerable mechanical advantage; the skeleton has greater power of resistance to bending than the endoskeleton of vertebrates. For the same cross-sectional area of muscle and skeleton, a solid endoskeleton would be nearly three times weaker than a hollow exoskeleton. Similarly, to have the same strength as an exoskeleton, an endoskeleton would need to be considerably thicker, leaving little additional space for musculature. Because of its mechanical efficiency, and being composed of remarkably flexible material, the exoskeleton of arthropods has been expressed in an astonishing range of body forms and structures unrivalled anywhere in the animal kingdom.

However, the presence of an exoskeleton has a number of limitations: it puts a maximum limit on the physical size of an organism, it limits growth, thereby necessitating the shedding of the outer covering for any increase in size, and the skeleton needs to be perforated with sensilla to monitor the outside world. Despite the size limitations, extinct aquatic arthropods grew to 1.8 metres in length but most modern species are considerably smaller. A few extant marine crustaceans (giant spider-crabs) can grow to almost twice that size and weigh 6.4 kg but most arthropods are relatively small. Those in water are able to attain a larger size than those on land. The largest terrestrial insects and spiders do not weigh more than about 100 g and the largest insects such as the goliath beetles measure 15 cm long and 10 cm wide. The smallest arthropods include some parasitic wasps, beetles and mites that are less than 0.25 mm in length; these, despite their complex structures and behaviour, can weigh less than the nucleus of a large cell.

Growth can only be achieved in organisms with a hardened outer covering by the process of moulting. The inner layers of the cuticle are digested by a series of enzymes to separate the old 'skin' from the newly formed cuticle. After moulting, the arthropod swallows water or air to inflate its flexible cuticle until this hardens. The process of moulting is controlled by a complicated interplay between special hormones.

The process of respiration has been solved in different ways by the various groups of arthropods. Insects and myriapods have small tubes, tracheae, through which oxygen diffuses to all parts of the body. The physical constraints governing the diffusion process are such that an increase in size of an insect or myriapod is not accompanied by a proportional increase in the rate at which oxygen is delivered to the tissues; this is one of the principal factors governing the maximum size of insects. However, respiration via tracheae can be very efficient and is able to operate with a very small difference in the partial pressure of gases between the tissue end of the tracheae and the outside atmosphere. Tracheal respiration delivers oxygen to insect muscle, which is the most active tissue in the animal kingdom. Aquatic arthropods either use gills (especially the Crustacea) or diffusion across the cuticle. Arachnids have lung-books which are enclosed gills.

Despite the many limitations put on them by having an exoskeleton, the arthropods have been extraordinarily successful.

CLASSIFICATION

The higher classification of the arthropods is a matter of considerable controversy and several different schemes have been proposed since the concept of the Arthropoda was first introduced some 140 years ago. There is no single accepted scheme. The principal difference between the alternative schemes lies in the significance given to particular features, such as the articulation of the mandibles and position of the gonopore (genital opening), and how these features are used to produce a hierarchical classification. The various classifications fall into two types. One view suggests that the arthropods are monophyletic in origin, i.e. they all arose from a single common ancestor, and this is the approach most commonly described in the literature (especially in textbooks). The alternative view disputes this simple interpretation and holds that the arthropod form arose several times and that the different groups came to resemble one another by convergent evolution; that is, the arthropods are polyphyletic (see excellent works by Manton, 1973, 1977). Each alternative scheme uses its own names for the subdivisions of the arthropods. An introduction to the topic can be found in Davies (1988) and Clarke (1988). Unfortunately there is little convincing fossil evidence to support the proposed evolutionary steps in any of the classifications and therefore it is unlikely that a consensus will emerge in the near future.

The classifications which assume a monophyletic origin of the arthropods differ in the ranking they give to the principal groups – arachnids, crustaceans, myriapods, pentastomids, diplopods, insects, trilobites etc., but all assume that arthropods arose from an annelid-like ancestor which was an arthropod with clear body-segmentation. One scheme (Fig. 2.1a) divides the arthropods into three subphyla: (1) the Trilobitomorpha, containing the extinct trilobites; (2) the Mandibulata, containing seven classes with articulated mandibles, with antennae, and the thorax with appendages of various types; and (3) the Chelicerata, containing the remaining five classes, with a prosoma (head) without antennae, but with pincer-like chelicerae and the thoracic appendages only used as walking legs. This scheme leaves out the Onychophora, Pentastomida and Tardigrada on the grounds that their relationships to the rest of the arthropods are unclear. A similar classification (Fig. 2.1b) divides the arthropods into six groups, although it is not specified which of these groups are of class rank. The main difference from the previous scheme lies in the equal ranking given to the Trilobita, Crustacea, Chelicerata, Onychophora, Myriapoda and Insecta.

In the classification based on a polyphyletic origin of the arthropods (Fig. 2.1c), there are three groups: the Crustacea, Chelicerata and the Uniramia (consisting of the Onychophora, Myriapoda and Hexapoda). Each group is a separate phylum,

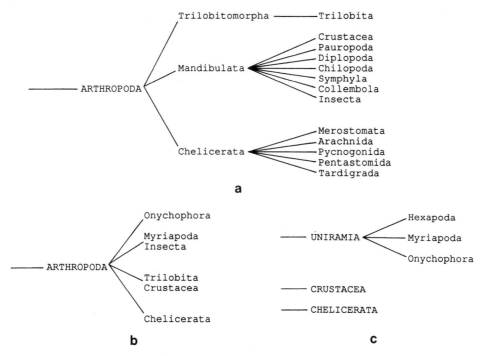

Figure 2.1 Three alternative classifications of the arthropods: (*a*) and (*b*) assuming they have a monophyletic origin; (*c*) assuming they are polyphyletic.

and therefore the phylum Arthropoda as such is not recognized. The position of several groups is left uncertain; the trilobites are thought to be sufficiently distinct to represent a phylum of their own, the Tardigrada are provisionally placed in the Uniramia and the Pentastomida are left unplaced. However, in addition to the reordering of the arthropods, the polyphyletic interpretation of Manton makes substantial changes to the classification of the insects, which she suggests also have several evolutionary origins. The primitive insect groups Collembola, Diplura and Protura have long been recognized as distinct from the rest of the insects (the Thysanura and Pterygota) and Manton suggests they should be ranked equally to the Thysanura and Pterygota and merit separate class status. More recently, Manton (1977) has suggested that the Thysanura and Pterygota should themselves be separate classes. Under the latter scheme, the class Insecta disappears and is replaced by an assemblage called the Hexapoda.

THE PRINCIPAL ARTHROPOD GROUPS

The Trilobita were marine animals whose fossil remains are common in rocks of the Cambrian and Silurian geological periods. There are no modern representatives.

Onychophora are extraordinary animals, generally considered intermediate between the arthropods and annelids because of the regular body segmentation and are much studied as 'evolutionary missing links'. Their arthropod-like features include paired limbs ending in claws, respiration by tracheae, a haemocoelic body cavity, and the general nature of the reproductive system. The similarities to annelids are found in the segmentally repeated coelomoducts (excretory organs), structure of the eyes, rudimentary structure of the head and the presence of unstriated muscles. The Onychophora, of which the most well known is the genus *Peripatus*, live in permanently damp habitats and feed on decaying vegetation. They have no medical importance.

Crustacea

The majority of the 25 000 species of Crustacea are marine (lobsters, crabs, barnacles, shrimps), with some living in freshwater, and one group (Isopoda, woodlice) completely terrestrial. The crustaceans are characterized by two pairs of antennae, a pair of mandibles, and at least five pairs of legs. In the more advanced forms (e.g. Decapoda – lobsters etc.) the body is divided into a cephalothorax and a segmented abdomen. The appendages are usually specialized to perform a number of particular functions. Breathing is by means of gills or through the cuticle.

Many crustaceans (crabs, lobsters) are a source of food for humans but can be toxic if they have recently eaten fruits or leaves of particular plants. Tropical crabs (xanthid and coconut crabs) can acquire neurotoxins from feeding on dinoflagellates of the genus *Gonyaulax*, especially when the flagellates are sufficiently abundant to cause 'red tides'.

More importantly, some crustacean species are involved in the transmission of parasites to man. Parasites such as the lung-fluke *Paragonimus westermani* are acquired when freshwater crabs (*Potamon* and *Cambarus*) and crayfish are eaten raw, an activity common in parts of the Far East, such as Japan. Other *Paragonimus* species infect humans in Africa and the neotropics. When the encysted parasite metacercariae (an immature stage) enter the stomach of man their cyst wall is digested and the undeveloped cercariae emerge and penetrate the gut lining before migrating to the lung bronchioles; here they can remain for up to 20 years.

Small freshwater copepods of the genus *Cyclops* (Fig. 2.2) and its relatives are important as intermediate hosts of several helminths, the most important of which is *Dracunculus medinensis* (Guinea worm). Dracunculiasis is found in much of West Africa and parts of East Africa, western India, Pakistan and the Middle East, where transmission centres on pools of freshwater and open step-wells. The *Dracunculus* larvae are released by the female from an infected person into freshwater, whereupon they are pursued and eaten by *Cyclops* attracted to the jerky larval

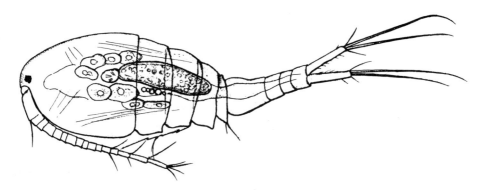

Figure 2.2 A copepod crustacean of the genus *Cyclops* containing immature stages of *Dracunculus*, the medically important Guinea worm.

motion. Several species of *Cyclops* (e.g. *C. quadricornis*) are effective intermediate hosts, as are two species of *Mesocyclops*, and, in the true tropics, *Tropocyclops* species. The parasites undergo two or three moults in the haemocoel of the crustacean before they become infective to humans when *Cyclops*-contaminated water is drunk. Cyclopoid crustaceans are also intermediate hosts of several other helminths which can infect man, such as *Diphyllobothrium* (a tapeworm from parts of Europe and North America, and cooler parts of South America) and *Gnathostoma* (a nematode from South East Asia).

Freshwater crabs of the genus *Potamonautes* are the phoretic carriers of larvae and pupae of the *Simulium neavei* group, the adults of which transmit onchocerciasis in East Africa.

Myriapoda

The Myriapoda have a five-segmented or six-segmented head with a single pair of antennae. The remainder of the body is composed of many leg-bearing segments of similar appearance. Myriapods have a tracheal system for respiration and Malpighian tubules for excretion, similar to those present in insects. The number of body segments increases with each moult, the additional segments arising from division of the penultimate segment.

The taxonomic ranking of the Myriapoda (i.e. as a class or subclass) depends on whether the arthropods are considered to be monophyletic, as discussed above. In most classifications the myriapods are divided into four groups – the Pauropoda, Symphyla, Diplopoda and the Chilopoda. The first two groups are small arthropods living in the soil and other hidden places and have no medical importance.

The Diplopoda or millipedes are elongate myriapods with many apparent segments each of which has two pairs of legs (Fig. 2.3a). The head has a single pair of antennae, a pair of mandibles, and a broad plate formed from the maxillae (gnathochilarium) which is used to define the subgroups of Diplopoda. There are more than 8000 known species. Millipedes are predominantly herbivorous or saprophagous (feeding on decaying vegetation) and therefore do not possess poisonous biting or piercing structures. However, many species produce defence secretions from glands on the sides of each body segment. Some species are able to project these secretions several centimetres when attacked or handled. The secretions are unpalatable and toxic to predators as they contain hydrogen cyanide and a variety of aldehydes, esters, phenols and quinonoids. Several tropical genera, including *Rhinocricus* (Caribbean), *Spirobolus* (Tanzania), *Spirostreptus* and *Iulus* (Indonesia) and *Polyceroconas* (New Guinea) contain species which are injurious to humans. The presence of the secretions on the skin can cause an immediate smarting which is then followed by skin swellings or blistering. More commonly, there is no immediate pain and the skin is simply stained brown or purple, a condition which either resolves without further clinical problems or worse can subsequently blister and peel. If the secretions are introduced into the eyes by squirting, or passively via the hands, intense conjunctivitis results, sometimes even leading to corneal ulceration and blindness.

Centipedes (Chilopoda) are of more immediate medical importance than millipedes because of the poisons they can inject from 'pincers' or 'poison claws' modified from the first pair of legs. There is a single pair of legs on each segment (Fig. 2.3b) and the head bears many-segmented antennae, mandibles and two pairs of maxillae. Some of the 2800 species can grow to as much as 26 cm long. The centipedes are divided into four groups: Geophilimorpha, Lithobiomorpha, Scolopendromorpha and the Scutigeromorpha. In general, centipedes are nocturnal predators feeding on other arthropods, killing them with the poison from the modified fore legs. The larger species apparently also feed on small vertebrates. It is principally the tropical scolopendromorphs that cause problems to humans, mainly by virtue of their large size – their poison claws being able to pierce the skin. Most large centipedes are venomous and their bite very painful to humans at the time of puncture; the bite can subsequently lead to marked swelling (oedema), blistering and even necrosis. Systemic effects such as vomiting, headache and cardiac arrhythmias are extremely uncommon and there are very few authentic reports of fatalities.

Pentastomida (tongue-worms)

The affinities of these curious endoparasites are not at all clear; they have been variously considered as highly modified helminths, nematodes, crustaceans, annelids or arachnids. Although pentastomids have some arthropod-like features, they lack sufficient structures (respiratory and circulatory systems, and appendages) to determine clearly their relationships. The body of the adult (Fig. 2.3c,d) is elongate and worm-like with distinct annulations (secondary segmentation). Either side of the mouth is a pair of retractile sclerotized nooks (Fig. 2.3c, arrowed), considered to be claws by some specialists. The males are smaller than the females. Both sexes are found in the nose, nasal sinuses and respiratory tract of mammals, birds or reptiles. The larvae usually develop in an intermediate vertebrate host.

The Pentastomida are divided into two groups, but only the Porocephalida (which contain two superfamilies, the Porocephaloidea and Linguatuloidea) affect mammals. The Linguatuloidea include only the family Linguatulidae. Both adults and larvae are confined to mammals. Human infestations with *Linguatula serrata* (Fig. 2.3d) are known from throughout the world, all from the upper respiratory tract. The adult female is 80–120 mm in length and produces eggs which exit the definitive host in nasal secretions. The eggs are subsequently ingested by the intermediate hosts, which for *L. serrata* are small herbivorous mammals such as rabbits and rodents. The larvae in the viscera of the intermediate hosts are finally ingested by the canid definitive host. In man the larvae appear unable to develop to adults.

Most human infestations by *Armillifer* (Porocephaloidea, Armilliferidae) are asymptomatic and recognized only at autopsy. The two commonest species affecting humans, *A. moniliformis* (China, South East Asia) and *A. armillatus* (tropical Africa) (Fig. 2.3c) live in the respiratory tract of snakes and infection takes place by eating inadequately cooked snake meat or drinking water contaminated by snake secretions or faeces. Prevalence can be as high as 45% in some populations, such as Malay aborigines, but again the parasites do not appear to be unduly pathogenic.

Arachnids and relatives (Chelicerata)

The Chelicerata contain some familiar animals such as spiders, ticks, mites and scorpions as well as a number of rather aberrant groups. They are characterized by having the body divided into two parts, the cephalothorax and the abdomen. There are no antennae but a pair of prehensile chelicerae are present in their place which

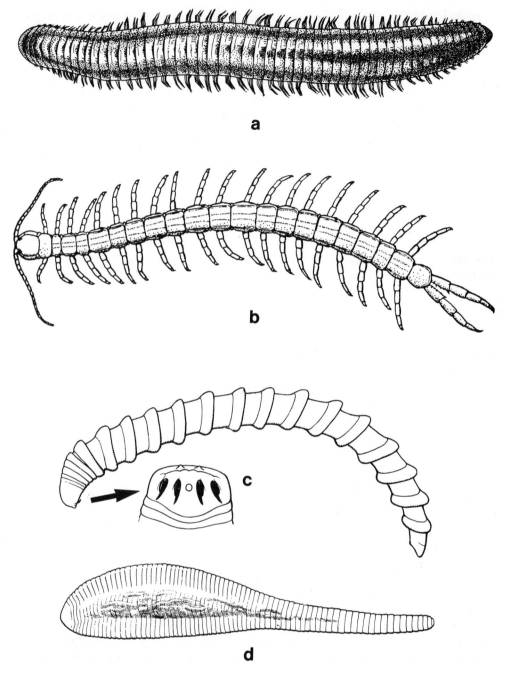

Figure 2.3 Some medically important arthropods other than insects: (*a*) a typical millipede (Diplopoda); (*b*) a scolopendromorph centipede (Chilopoda); (*c*) adult of the tongue-worm *Armillifer armillatus* (Pentastomida), with enlargement of the paired hooks present each side of the mouth that are typical of Pentastomida; (*d*) larva of the tongue-worm *Linguatula serrata* (Pentastomida).

serve a variety of functions. The mouthparts do not have true jaws (such as mandibles in insects and crustaceans), although some of the head appendages have been developed for handling and breaking up the food. Four pairs of legs are present.

There is not universal agreement on which groups are placed in the Chelicerata. Some schemes include the Tardigrada (water-bears) and Pentastomida (tongue-worms). However, three classes, Merostomata (containing the king-crabs), Pycnogonida (sea-spiders) and Arachnida (spiders, mites), are closely related and are the principal components of the Chelicerata. Only the Arachnida and pentastomids have any medical importance, and the latter are considered separately for convenience.

The class Arachnida is divided into several orders, namely Araneae (spiders), Scorpiones (scorpions), Acari (ticks and mites), Thelyphonida (whip-scorpions), Amblypygi (tailless whip-scorpions), Solifugae (wind-scorpions or camel-spiders) and Pseudoscorpiones (false-scorpions). Only the Acari, Araneae and Scorpiones have any medical significance.

The Acari contain tick and mite vectors of pathogens to man and sometimes actively colonize the skin, as in the case of the follicle mites (*Demodex*). Most spiders give poisonous bites, but few have jaws sufficiently powerful to penetrate human skin: and, even among these, the venoms, with a few important exceptions, are harmless. In the case of the larger tarantulas (Mygalomorphae), the urticating hairs that clothe the body can cause more pain than the bite. The stings of scorpions, on the other hand, are often very painful and, in several species, can even endanger human life.

Insecta (= Hexapoda)

The insects can readily be distinguished from other arthropods by the division of the body into head, thorax and abdomen, the presence of only three pairs of legs (in adults) and usually two pairs of wings. Like the myriapods they respire by means of tracheae. Insects constitute more than 90% of all arthropod species described to date, and are currently being described at the rate of about 8500 new species every year.

There are several aspects of the insect biology which have made them so very successful. Some of these, such as the presence of an exoskeleton, are common to all the arthropods but other features, particularly the development of flight, are peculiar to insects.

Metamorphosis is a dominant feature of the insect life-style and is of two types: incomplete (hemimetabolous) and complete (holometabolous). In the former, the egg hatches into an immature stage usually referred to as a nymph. The nymph superficially resembles the adult, although it is not sexually mature, and the wings are represented by small external wing buds. Most importantly, the nymphs are usually found in the same place as the adults feeding on the same type of food. In the case of blood-feeding insects (e.g. lice, triatomine bugs) this means that all stages are potentially involved in the transmission of parasites. The wings, when these are present, develop fully at the final moult and the insect becomes sexually mature; there is no intermediate resting stage. In insects with a holometabolous metamorphosis the egg hatches into an immature stage which is substantially different from the adult in structure and biology. The immature stages are usually referred to as larvae, although more specific terms relating to their appearance are used, such as caterpillar, grub or maggot. Larvae usually live in places quite different from the adults and have a completely different diet, thus there is little competition for resources between the immature and adult insect. The larvae of medically important insects are usually unattended by the adults but in some insects, such as tsetse-flies, the larvae (developing one at a time) are retained inside the female and nourished in the uterus until fully grown. The larval stage of holometabolous insects is a period of considerable growth which culminates in a moult into a pupa. The function of the pupal stage is to act as a bridge between the growth phase of the larva and the dispersive and reproductive phase of the adult. The pupa does not feed but undergoes a remarkable transformation in which many organ systems are broken down and reorganized into a completely different form from that of the larva, in preparation for adult life. The wing buds, which have been hidden within the larva, now become external in those insects which have winged adults. The adult emerging from the pupa is radically different in form and physiology from the larva.

Classification
As indicated above, there is much debate on the origins and classification of the insects. Most of the discussion revolves around the most primitive insect orders. These ideas need not be considered further here as they have little bearing on the biology and classification of insects of medical importance. The simplest arrangement, which is pragmatic rather than fully phylogenetic, assembles the insects into three groups (Table 2.1, see Davies, 1988).

1. Apterygote insects are primitively wingless, have pregenital appendages on the adult abdomen and show a very slight metamorphosis although they do moult after sexual maturity. Five orders, including the Collembola (springtails), Zygentoma (=Thysanura s.str., silverfish), and Archaeognatha (bristletails).
2. Exopterygote insects or Hemimetabola are winged or secondarily wingless, and have metamorphosis of the incomplete type. Compound eyes are present along with externally developing wing-pads and genital appendages. This division contains numerous orders of familiar insects including Odonata (dragonflies), Orthoptera (locusts, grasshoppers, crickets), Isoptera (termites), and orders containing medically important species in the Dictyoptera (cockroaches), Hemiptera (true bugs) and Phthiraptera (lice).
3. Endopterygote insects or Holometabola are winged or secondarily wingless; metamorphosis is of the complete type with a pupal stage. The wings and external genitalia develop internally in the larval stages. The dominant insect groups belong to this division, e.g. Coleoptera (beetles), Lepidoptera (butterflies and moths), Hymenoptera (bees, ants and wasps) as well as the principal orders containing medically important insects, namely the Diptera and Siphonaptera.

External structure and physiology
The nature of insect integument has an enormous impact on the insects as a group – the physiological functions of the cuticle define the environments in which insects can live and the life-styles they can adopt. Similarly, the integument has had a greater influence on the development of entomology than any other aspect of insect structure or biology. The myriad of forms and structures into which the integument has been moulded throughout the long history of the group is the basis of insect classification and identification.

Table 2.1 The orders of living insects

The table provides a synopsis of living insect orders in systematic sequence. The Apterygotes are included although they are considered by some specialists to be separate classes from true insects (Insecta). The medical significance of an order (if any) is indicated as follows:* order of minor medical importance; ** order includes some vector species; *** order includes many vector species. Numbers in the right-hand column show how many valid species (approximately) are at present described and named in each order; they are based on up-to-date information supplied by authorities in the British Museum (Natural History) and are often considerably higher than outdated figures given in entomological textbooks.

Division and order	Included insects (common names)	No. of species
Apterygote insects		
Diplura	Diplurans (no common name)	600
Protura	Proturans (no common name)	200
Collembola	Springtails	1500
Archaeognatha	Bristletails	250
Zygentoma	Silverfish	330
Exopterygote insects (= Hemimetabola)		
Ephemeroptera	Mayflies	2000
Odonata	Dragonflies, damselflies	5000
Plecoptera	Stoneflies	2000
Grylloblattodea	Grylloblattids (no common name)	20
Orthoptera	Grasshoppers, bush-crickets, crickets	20 000
Phasmida	Stick-insects, leaf-insects	2500
Dermaptera	Earwigs	1800
Embioptera	Web-spinners ('foot-spinners')	400
Dictyoptera*	Cockroaches, praying mantises	7000
Isoptera	Termites ('white ants')	2300
Zoraptera	Zorapterans (no common name)	25
Psocoptera	Booklice, psocids	2500
Phthiraptera**	Lice (biting lice and sucking lice)	3300
Hemiptera**	Bugs, cicadas, aphids, scale insects	81 000
Thysanoptera	Thunderflies (thrips)	5000
Endopterygote insects (= Holometabola)		
Megaloptera	Alderflies, hellgrammites	400
Neuroptera	Lacewings, antlions	5000
Coleoptera*	Beetles	370 000
Strepsiptera	Stylopids (no common name)	370
Mecoptera	Scorpionflies	400
Siphonaptera**	Fleas	2000
Diptera***	Flies (true flies)	120 000
Lepidoptera*	Butterflies, moths	150 000
Trichoptera	Caddisflies	7000
Hymenoptera*	Ichneumons, ants, wasps, bees	150 000

The exoskeleton of insects consists of three main non-cellular layers secreted by an epidermis (Fig. 2.4). The outermost layer, epicuticle, is less than 4 μm thick and is predominantly protein but also contains waxes (hydrocarbons and fatty acids). The next layer, the exocuticle, is a thick layer composed of chitin and protein. Chitin is a nitrogenous polysaccharide consisting mainly of acetylglucosamine in

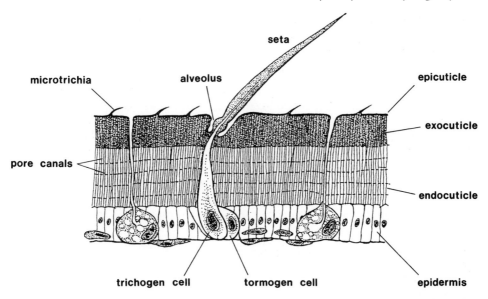

Figure 2.4 The integument of an insect in semi-schematic form to show the different cuticular layers and the different formation of microtrichia and macrotrichia (setae), the latter originating from a socket (alveolus). (Reproduced from Davies (1988) with relabelling.)

long chains and is resistant to a wide range of alkalis and mineral acids. Minute fibrils of chitin are laid down at different angles within the protein matrix, rather like fibre-glass in resin, and usually in a regular manner (like plywood). Consequently the cuticle has considerable structural strength as well as the material strength of its constituents. The protein of the exocuticle can become tanned to form a hard substance called sclerotin. Tanning takes place by the cross-linking of protein molecules by quinone or acetyl-dopamine. The inner endocuticle is the thickest layer and also consists of chitin and proteins but the latter are not tanned. Both the exocuticle and the endocuticle are perforated by pore canals which carry waxes secreted by specialized cells on to the external surface of the cuticle.

The integument of insects has many outgrowths in the form of scales, spines, hairs, bristles and spurs. These outgrowths fall into two categories, those which are non-articulated and are simply fine folding of the outermost layer of the cuticle (microtrichia), and those which are articulated such as setae (macrotrichia) and scales. The articulated setae and scales are attached to the cuticle by a thin membrane and sit in a pit, the alveolus. They are hollow outgrowths of the exocuticle and epicuticle and are secreted products of a single epidermal cell (trichogen cell) whereas the socket is secreted by another cell (tormogen cell). Scales are simply modified setae and are often longitudinally ribbed for added strength. The non-cellular microtrichia can be very fine, even less than the wavelength of light, and therefore give distinctive patterns or shading known as pollinosity or tomentum (e.g. in Simuliidae and Muscidae). These patterned areas can appear to change shape and colour with changes in the direction of reflected light. Structural colours caused by layered, ultrastructurally patterned cuticle (diffraction gratings) are found in many insects and give a metallic blue or green appearance (e.g. in greenbottles of the genus *Lucilia* and in *Haemagogus* mosquitoes). Other colours are caused by the presence of pigment or by a combination of pigment and colour-producing structures.

As mentioned earlier, the body of insects is divided into head, thorax and abdomen. Fundamentally, the whole insect consists of a series of segments each with an associated pair of appendages, a condition most clearly seen in the embryo. In the adult it is only the head and thoracic appendages which are most obvious. Internal divisions of the body segmentation can be seen in the nervous, respiratory and circulatory systems, and body musculature. The body segments can be either clearly demarcated, such as those of the abdomen, or fused such as those forming the head. However, where the body segments are relatively distinct four main areas can be seen in the exoskeleton – a dorsal tergum, lateral pleural plates on either side and a ventral sternum. In most adult insects these four areas consist of hardened cuticle; plates on the dorsal surface are then referred to as tergites, on the pleural surfaces as pleurites and on the ventral surface as sternites. In immature stages, especially of Diptera, the cuticle is not usually differentiated into hardened plates.

By examination of the head appendages during embryonic development it has been concluded that the head is composed of several segments fused into one capsule. The head-capsule has several important landmarks in the form of sutures, which delimit slabs of sclerotized cuticle, and the head appendages (antennae and mouthpart structures). On the front of the head is usually a suture shaped like an inverted Y which separates the vertex on the top of the head from the genae on the sides and the frons in the centre. Below the frons is the clypeus, which is often greatly expanded in sucking insects because it acts as the attachment for large muscles of the pumping organs (e.g. cibarium) used in drawing up liquid food. Inside the head is a set of strengthening struts collectively termed the tentorium.

The antennae are paired structures which vary greatly between groups of insects and often are different in males and females (sexually dimorphic). Muscles are attached from the head to the first antennal segment (scape) and from there to the second segment (pedicel); the remaining antennal segments constitute the flagellum and have no internal musculature except in very primitive insects. The antennae are principally sensory organs bearing an enormous array of sensilla which detect temperature, humidity, infra-red radiation and mechanical movement (including sound) and chemicals. The chemicals can be cues for host location in the case of bloodsucking insects or pheromones to detect mates or even rival males.

The groundplan of insect mouthparts consists of an anterior pair of mandibles with paired maxillae and a fused labium behind (the labium represents the second pair of maxillae). Each maxilla consists of five parts, the cardo at the base, the stipes in the middle (to which is attached a segmented maxillary palp) and two apical structures, the lacinia and galea. The lacinia is usually a cutting structure. A single tongue-like process, the hypopharynx, arises from the floor of the mouth and lies immediately in front of the labium. In most insects a plate from the front of the head, called the labrum, stretches down over the front of the mandibles. The mouthparts of insects show a wide range of adaptations to different foods and feeding mechanisms. One of the simplest arrangements of the mouthparts is found in cockroaches, insects in which the mandibles are used in opposition for chewing and the maxillae and hypopharynx used to manipulate the food against the labrum and labium. In bloodsucking insects, such as mosquitoes, the mandibles, maxillae (more precisely, the laciniae), labrum and hypopharynx have developed into strong, slender blades which interlock to form a tube. This tube, the syntrophium (fascicle), is wrapped in the fleshy labium when not piercing a host. Blood-feeding is often associated with the production of saliva containing many active compounds which cause vasodilatation and inhibit clotting by blocking second-messengers.

Insects have two types of eyes, simple single-lens structures called ocelli found on the top of the head and the more well-known compound eyes consisting of multiple, similarly shaped lenses. Despite their wide distribution throughout the insects, little is known about how ocelli work, but in contrast there is much interest and research on the compound eye. The basic element of a compound eye is the ommatidium, a simple eye composed of a lens (of cuticle), a crystalline cone for focusing, retinal cells to receive light impulses and two sets of iris cells. One set of iris cells controls the amount of light falling on the retinal cells and the other cells (secondary iris cells) control the amount of light passing sideways, i.e. into neighbouring ommatidia. Perception of the outside world is made by integrating the information received by each ommatidium and is usually likened to the spots of print used to make up a newspaper picture (or for that matter the pointilliste impressionist pictures of Seurat). The clue to the functioning of the eye comes as much from the integration of nerve impulses through a hierarchical processing network as from the physical structure of an ommatidium. Insect eyes can perceive colour, and many insects see in the ultra-violet end of the spectrum.

The thorax consists of three segments which are much more apparent externally than are the head segments. The segments are referred to as the prothorax, mesothorax and metathorax. The thoracic segments are modified in the adults of different insect groups, the most obvious being the expansion of the mesothorax and metathorax in association with the development of wings. Each thoracic segment has a pair of legs used for locomotion, walking, swimming etc., and in some cases to catch food (e.g. mantids, predaceous bugs). Most legs have the same basic structure: a coxa at the base with a small triangular trochanter separating it from the femur, a tibia, and finally a segmented tarsus. At the tip of the tarsus are claws (usually paired) and various small structures such as the paired pulvilli and a single empodium. Insects are able to climb up smooth surfaces by using adhesive organs on their tarsi which exploit molecular adhesive forces set up in secretions from fine hairs.

Wings are a notable feature of insects and have had remarkable impact on their success as a group. They give insects the ability to migrate, locate food sources with ease, escape predators and find mates. Most insects have two pairs of wings, but the Diptera have a single pair originating from the mesothorax and the hind wings are modified to form a pair of remarkable balancing organs, the halteres (singular, halter), which operate as gyroscopes. Receptors in the halteres apparently measure shearing forces set up in the cuticle during the figure-of-eight movement of the knobbed structures during flight; the forces can be compared between the two halteres to determine relative movement. The venation of the wings is of great importance in understanding the relationships between different groups of insects, mainly because it is not usually modified for particular ways of life as are other structures (e.g. mouthparts, legs). Thus wing venation represents a series of conserved and readily homologized characters. A general groundplan has been devised from which all other expressions of the venation have developed. This hypothetical system is most clearly seen in the primitive winged insects such as the Odonata and Ephemeroptera (Table 2.1) and is based on the tracheal development in the pupal or nymphal wing-pads and the venation of fossil insects. This hypothetical system divides the venation into six basic regions or branching systems. At the anterior part of the wing is the costa (denoted C) and behind this the subcosta (Sc), then the radial sector (denoted Rs) which has up to five main branches. In the centre of the wing is the medial sector (denoted M) with up to six main veins often separated into the anterior and posterior medial sector because they originate from two main branches. In the hind part of the wing are the last two branching systems – the cubital sector (Cu) and the anal sector (denoted A) which are the most often modified in different groups. The abbreviation for each wing vein consists of letters representing its sector (R, Cu etc.), followed by the number within that sector; thus R5 is the fifth vein in the radial sector (Rs). Cross-veins, small veins joining longitudinal veins, are named according to the veins they join and wing cells are named from the vein immediately in front of them (as the fly flies!). The advantage of this universal system is that it allows the venation of different groups to be compared and the modifications to be made explicit. (More on the venation of dipterous wings is given in the next chapter.)

The dynamics of insect flight has captured the imagination of many physiologists and morphologists. The movement of the wing surface during flight is very complex but for simplicity can be described as a twisting surface in which the wing tip circumscribes a figure-of-eight movement inclined forward relative to the longitudinal axis of the insect. This gives both lift and forward propulsion. Manipulation of the angle of this figure-of-eight movement will allow the insect to hover, or even fly backwards. Power for flight comes from two sets of thoracic muscles (in most higher insects) which change the shape of the thorax – a structure which can be thought of as a box in which the lid (the mesonotum and metanotum) is pulled down into the box by vertical muscles, thereby in turn pulling down the bases of the attached wings. Having distorted the box by this action, longitudinal muscles now flip the thorax back into shape and so flip the wing bases upwards again. Thus the thorax begins to resonate, mostly by muscular action but partly by resistance to bending of the cuticle. Certain small muscles are involved in twisting the wing to change its angle of attack to the air. The frequency of wing-beat varies greatly, depending on the type of wings an insect has and the way in which it flies. For example, butterflies have a low wing-beat frequency because they use the wing-clap technique of flying in which large vortices are set up giving lift when the wings are wrenched apart. Mosquitoes, using a more conventional system, beat their wings at 280–310 beats per second and *Musca* (house-flies) at 180–200 beats per second.

Wings have been secondarily lost in many ectoparasitic insects living on mammals, e.g. lice, fleas, parasitic earwigs (*Hemimerus*). This condition is associated with several other modifications such as the possession of strong claws, rows of specialized spines (ctenidia of fleas), lateral or dorsoventral compression in body form etc. Some ectoparasites have wings to help them locate their hosts but shed them soon after they have fed (e.g. *Hippobosca*).

The abdomen of insects shows the fundamental body segmentation most clearly, but even here there are many modifications, especially at the tip of the abdomen where the external genitalia are found. The genitalia are composed of structures which originated from simple abdominal appendages. Understanding these origins allows structures to be homologized (i.e. considered equivalent because of their origin) and thus different groups can be compared in a similar manner to the comparisons made by using the wing venation. The male and female genitalia of many species appear to operate as a lock-and-key device and thus act as a barrier to between-species matings. The behavioural repertoire of the sexes prior to and during copulation is clearly as important as the morphological structure of the genital apparatus. However, despite these behavioural dimensions, minute differences in the genitalia still remain extremely important in distinguishing and subsequently identifying species. The basic structure of the male external genitalia consists of lateral claspers which grasp the female during copulation and a medial intromittent aedeagus (penis) accompanied by a pair of parameres. Each of these structures can be modified in extraordinarily diverse ways. The female external genitalia are usually rather simpler in structure. The end of the abdomen is often elongated to form an ovipositor. Internally, the female has one or more spermathecae for storing sperm after mating; these are ectodermal in origin (i.e. from cells of the external surface of the insect) and are usually sclerotized. The ovipositor of many bees and wasps is modified for stinging.

Internal anatomy
The main organ systems of insects are the respiratory system, digestive system, nervous system, excretory system and reproductive system.

The respiratory system of most insects consists of an elaborate network of tubes known as tracheae. The tracheae connect directly between the atmosphere through openings (spiracles) and individual cells, being able to deliver oxygen to within 5 μm of the mitochondria in highly respiring muscle cells. The tips of the tracheae can be as little as 1 μm in diameter. In some immature aquatic insects there is a closed tracheal system and oxygen is acquired by diffusion through the thin cuticle of the body surface or specialized external gills.

The digestive system of most insects, including the blood-feeding forms, is divided into three regions: the fore-gut, mid-gut and hind-gut. The fore- and hind-gut are ectodermal in origin and this gives rise to a great difference in their structure from the mid-gut. Only the mid-gut has cells with microvilli, the other parts of the gut being lined with cuticle – an important aspect when considering the attachment or penetration of parasites. The fore-gut of sucking insects (including bloodsucking species) is very different in form from that of chewing insects which have an enlarged gizzard; the latter is no more than a valve in sucking insects. In most insects the food is separated from the lining of the mid-gut by a thin layer of 'cuticle' – the peritrophic membrane. In some insects this membrane is secreted continuously (e.g. tsetse-flies) and in others it is secreted each time a blood-meal is taken (e.g. mosquitoes). The peritrophic membrane is essentially a sieve which allows enzymes and breakdown products to move freely through it; it is thought to protect the mid-gut cells from abrasion because insects do not secrete mucus for this purpose. The hind-gut usually contains rectal pads which are important in regulating salt and water balance. At the junction of the mid-gut and hind-gut are long, slender outgrowths of the gut called the Malpighian tubules which function as the principal excretory organs. Within the gut, insects have three main classes of enzymes, carbohydrases for the breakdown of carbohydrates to simple sugars, proteases for the digestion of proteins and lipases for catalysing fat digestion. Glycosidases in the mid-gut break down complex sugars, on which many blood-sucking insects feed and which fuel their flight (however, tsetse-flies fly on proline metabolism). Circulation within the insect body is relatively simple. The body cavity is a blood-filled 'haemocoel' which is often divided into sections by thin sheets of tissue. Circulation of the blood (haemolymph) occurs through the action of a longitudinal contractile tube (heart) lying against the roof of the body cavity. The haemolymph is usually colourless and contains various classes of cells, some of which are involved in immune responses to pathogens and parasites. There is evidence of a wide spectrum of peptides involved in the humoral response to infection.

The female reproductive organs consist of ovaries, accessory glands and the spermathecae. The ovaries have considerable interest in medically important Diptera because they sometimes indicate the age of the insect expressed in terms of the number of egg batches laid and, therefore by inference, the number of blood-meals taken. The ovary consists of a series of egg tubes, the ovarioles, in which an apical germarium contains the germ cells producing oocytes. The vitellarium consists of a long series of developing eggs, the youngest being at the apex, nearest to the germarium. As the eggs develop they acquire yolk (vitellin) which is manufactured in the fat body as vitellogenin and is transferred through the haemolymph to the ovaries where it is taken up by pinocytosis (particle uptake by cells). Each developing egg is covered with a layer of follicular epithelium which secretes the chorion (egg shell). Many insect eggs have a characteristic external pattern.

The male reproductive system is relatively simple, with testes generating the sperm and the accessory glands producing the seminal fluid (and in some groups the spermatophore or 'sperm packet' which is passed during copulation to the female).

FURTHER READING

The following is a selected list of textbooks and other works suggested for readers needing further information about arthropods, particularly insects. Text-cited references are included here.

Atkins, M. D. 1978. *Insects in perspective.* vii + 513 pp. Macmillan, New York, and Collier Macmillan, London.

Borror, D. J., Triplehorn, C. A. and **Johnson, N. F.** 1989. *An introduction to the study of insects.* Sixth edition. xiv + 875 pp. Saunders College Publishing, Philadelphia.

Brusca, R. C. and **Brusca, G. J.** 1990. *Invertebrates.* xviii + 922 pp. Sinauer Associates, Sunderland, Massachusetts.

Bücherl, W. and **Buckley, E. E.** (eds) 1971. *Venomous animals and their venoms*: vol. 3, *Venomous invertebrates.* xxii + 537 pp. Academic Press, New York and London.

Chapman, R. F. 1982. *The insects: structure and function.* Third edition. xiv + 919 pp. Hodder & Stoughton, London.

Clarke, K. U. 1988. Impact arthropodisation. *Antenna* **12**: 49–54.

Daly, H. V., Doyen, J. T. and **Ehrlich, P.** 1978. *Introduction to insect biology and diversity.* x + 564 pp. McGraw-Hill, New York.

Davies, R. G. 1988. *Outlines of entomology.* Seventh edition. vii + 408 pp. Chapman & Hall, London and New York.

Elzinga, R. J. 1981. *Fundamentals of entomology.* Second edition. x + 422 pp. Prentice-Hall, Englewood Cliffs, New Jersey.

Gillott, C. 1980. *Entomology.* xviii + 729 pp. Plenum Press, New York and London.

Goldsworthy, G. J. and **Wheeler, C. H.** (eds) 1989. *Insect flight.* 371 pp. CRC Press, Boca Raton, Florida.

Gupta, A. P. (ed.) 1979. *Arthropod phylogeny.* xx + 762 pp. Van Nostrand Reinhold, New York.
Hennig, W. 1981. *Insect phylogeny.* xxii + 514 pp. John Wiley, Chichester.
Herreid, C. F. and **Fourtner, C. R.** (eds) 1981. *Locomotion and energetics of arthropods.* viii + 546 pp. Plenum Press, New York and London.
Hinton, H. E. 1981. *Biology of insect eggs*: vol. 1, ix + pp. 1–473; vol. 2, xviii + pp. 475–778; vol. 3, xvii + pp. 779–1125. Pergamon Press, Oxford.
Kerkut, G. A. and **Gilbert, L. I.** (eds) 1985. *Comprehensive insect physiology, biochemistry and pharmacology*: vol. 1, *Embryogenesis and reproduction,* xvi + 487 pp.; vol. 2, *Postembryonic development,* xvi + 505 pp.; vol. 3, *Integument, respiration and circulation,* xvi + 625 pp.; vol. 4, *Regulation: digestion, nutrition, excretion,* xvi + 639 pp.; vol. 5, *Nervous system: structure and motor function,* xviii + 646 pp.; vol. 6, *Nervous system: sensory,* xvi + 710 pp.; vol. 7, *Endocrinology I,* xvi + 564 pp.; vol. 8, *Endocrinology II,* xvi + 595 pp.; vol. 9, *Behaviour,* xvi + 735 pp.; vol. 10, *Biochemistry,* xviii + 715 pp.; vol. 11, *Pharmacology,* xiv + 740 pp.; vol. 12, *Insect control,* xiv + 849 pp.; vol. 13, *Cumulative indexes,* xxiv + 314 pp. Pergamon Press, Oxford.
King, R. C. and **Akai, H.** (eds) 1982. *Insect ultrastructure*: vol. 1, xxii + 485 pp.; vol. 2, xxv + 624 pp. Plenum Press, New York and London.
Manton, S. M. 1973. Arthropod phylogeny – a modern synthesis. *Journal of Zoology* **171**: 111–130.
Manton, S. M. 1977. *The Arthropoda: habits, functional morphology, and evolution.* xx + 527 pp. Clarendon Press, Oxford.
Matsuda, R. 1965. Morphology and evolution of the insect head. *Memoirs of the American Entomological Institute* **4**: 1–334.
Matsuda, R. 1970. Morphology and evolution of the insect thorax. *Memoirs of the Entomological Society of Canada* **76**: 1–431.
Matsuda, R. 1976. *Morphology and evolution of the insect abdomen with special reference to developmental patterns and their bearings upon systematics.* viii + 534 pp. Pergamon Press, Oxford. [Volume 56 of *International Series in Pure and Applied Biology (Zoology Division)*, ed. G. A. Kerkut.]
Meglitsch, P. A. and **Schram, F. R.** 1991. *Invertebrate zoology.* Third edition. vi + 623 pp. Oxford University Press, New York and Oxford.
Richards, O. W. and **Davies, R. G.** 1977. *Imms' general textbook of entomology.* Tenth edition. vol. 1, *Structure, physiology and development,* viii + pp. 1–418; vol. 2, *Classification and biology,* viii + pp. 421–1354. Chapman & Hall, London, and John Wiley, New York.
Sims, R. W. (ed.) 1980. *Animal identification: a reference guide*: vol. 2, Land and freshwater animals (*not* insects). x + 120 pp. British Museum (Natural History), London, and John Wiley, Chichester.
Stehr, F. W. (ed.) 1987. *Immature insects*: vol. 1, xiv + 754 pp. Kendall/Hunt, Dubuque, Iowa. [Coverage not stated on title page but work covers 27 orders, including orthopteroid insects, lice, Trichoptera, Lepidoptera and Hymenoptera.]
Stehr, F. W. (ed.) 1991. *Immature insects*: vol. 2, xvi + 975 pp. Kendall/Hunt, Dubuque, Iowa. [Coverage not stated on title page but work covers ten orders, including hemipteroid and neuropteroid insects, Coleoptera, Siphonaptera and Diptera.]
Taylor, F. and **Karban, R.** (eds) 1986. *The evolution of insect life cycles.* x + 287 pp. Springer-Verlag, New York.
Wigglesworth, V. B. 1964. *The life of insects.* xii + 360 pp. Weidenfeld & Nicolson, London. [Easy and instructive reading at a semipopular level.]
Willmer, P. 1990. *Invertebrate relationships: patterns in animal evolution.* ix + 400 pp. Cambridge University Press, Cambridge.

Part One:
DIPTERA

CHAPTER THREE

Introduction to the Diptera

R. W. Crosskey

The classical scholar Aristotle appreciated how distinct flies are among living organisms and gave them the collective Greek name 'Diptera' in the fourth century BC. It means 'two-winged' (*di* + *pteron*), and is very apt because flies differ from almost all other insects in having only one pair of functional wings. Linnaeus did not need to invent a new name when, in the mid-eighteenth century, he introduced the system of biological classification used today: he merely adopted Diptera as the name for his order of insects containing all the true flies. It has remained the universal name.

The wings of Diptera are the counterpart of the front pair of wings in four-winged insects, and flies have no hind wings as such; these are present, however, in a highly modified form as a pair of club-like balancing organs called halteres (singular 'halter', from Latin for a weight). The basic recognition features of Diptera are the presence of an anterior pair of wings and posterior halteres (Fig. 3.1). From a practical point of view, the halteres are really the diagnostic test of what is a fly because they are always present, whereas some flies have actually become wingless in the course of evolution from winged ancestors. A point to note, though confusion between Diptera and stylopids (Strepsiptera) is unlikely, is that male stylopids are constructed the other way about from flies: they have large hind wings and fore wings reduced to halter-like organs. Some male coccid bugs which have only one pair of wings look at first glance like small Diptera of the suborder Nematocera but can readily be distinguished on close examination by having no halteres.

Diptera are holometabolous insects (Chapter 2) with complete metamorphosis (egg, larva, pupa, adult) in which the larva almost always develops in an environment totally different from that of the adult: as Oldroyd and Smith (1973) remark, 'it is as if the fly lived two completely different lives, with different structure, physiology, senses and different powers of movement'. As a result of this double life, species that cause problems in medicine, veterinary science, agriculture or horticulture hardly ever do so as larvae *and* adults.

Flies are by far the most important insects from the medical point of view, even though most of the 120 000 described species are entirely harmless to man. The Diptera include the only insects responsible for transmitting several major human diseases, including malaria and filariasis (Chapter 5), onchocerciasis (Chapter 6), leishmaniasis (Chapter 4) and African trypanosomiasis (Chapter 9). These diseases are caused by pathogens or parasites which occur in human blood or skin and are

Medical Insects and Arachnids Edited by Richard P. Lane and Roger W. Crosskey.
Published in 1993 by Chapman & Hall ISBN 0 412 40000 6

52 *Introduction to the Diptera*

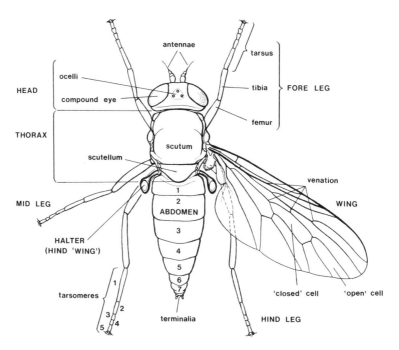

Figure 3.1 Basic morphology of a typical adult fly (Diptera), left wing and right mid leg omitted.

transmitted by flies which bite man. About 11 000 species (9% of Diptera) belong to families in which the adult female flies (in some cases also the males) are adapted structurally and physiologically to a bloodsucking mode of life, although relatively few among this number are actually vectors of disease or important as biting pests of man. Some Diptera are medically important, however, not because of the bloodsucking habit but because the adults can directly contaminate human food with pathogenic organisms (e.g. the house-fly, *Musca domestica*) or because their larvae sometimes invade the human body by accident or as the result of deliberate parasitism. Myiasis, infestation of the body with dipterous larvae, is treated in detail in Chapter 12.

The origins of the Diptera are obscure. The earliest fossils of two-winged insects which are unequivocally flies are from Triassic rocks 220–210 million years old (Wootton, 1981). Fossils are mainly of adult wings. Immatures are very rare, but include remarkably well-preserved Cretaceous blackfly larvae and a Jurassic blackfly pupa. Many extinct genera and families have been described, especially from the Jurassic period (210–140 million years BP), a time when most modern families of Nematocera appear also to have existed. Cretaceous and Tertiary Diptera are scant (Rohdendorf, 1964), but the latter include Baltic amber fossils and tsetse-flies from North America.

MORPHOLOGY

The following general account of dipteran morphology is intended mainly for orientation into features important for classification above family level. Readers

Morphology 53

may find it useful to use it in conjunction with the 'elements of structure' sections in Chapters 4–10, which contain information and illustrations for the families containing bloodsucking species able to transmit human disease.

Adult flies

The body has the three main divisions usual in insects – head, thorax and abdomen (Fig. 3.1). It is often very bristly, especially in the Muscomorpha (see later), flies in which the strongest bristles (setae) stand in a definite groundplan (chaetotaxy). Antennae, legs and halteres are always present, but not always the wings; on some species these are reduced or have been completely lost.

The head is occupied largely by the compound eyes (Fig. 3.1) (usually just called the eyes), which are often so big (especially in males) that they virtually meet in front and obliterate the strip of head (frons) which would otherwise separate them; the head is then called holoptic (Fig. 3.2b), in contrast to the eyes-separated dichoptic condition (Fig. 3.2a). The top of the head (vertex) commonly bears three ocelli, i.e. simple round 'eyes' like shiny beads arranged in a triangle and raised above surface level (Fig. 3.2a). The mouthparts are suspended below the head and form a more or less tubular proboscis (Fig. 3.2a,d) adapted for sucking (not chewing) and terminating in spongy labella through which fluids and fine-particle solids (such as pollen) can be imbibed. In most bloodsucking flies certain mouthparts are modified into toothed 'stylets' for biting the animal hosts (see Chapters 4–10), and in some the whole proboscis is a slender piercing organ directed forwards instead of downwards. The antennae vary enormously in number and shape of the segments and have great importance in taxonomy, as do their many sensilla. The first segment (scape) is sometimes much reduced (e.g. mosquitoes), the second (pedicel) much swollen (many Nematocera), and the remainder (flagellum) often highly modified. In many Brachycera the segments beyond the third are coalesced into a stylus or a bristle-like arista borne on the end of the outer edge of the third antennal (first flagellar) segment (Fig. 3.2c,d). Diversity of the antenna in bloodsucking flies can be seen in Fig. 3.3.

The thorax (Fig. 3.4) is characterized by massive development of the mesothorax, which houses enormous muscles associated with flying on only one pair of wings, and by reduction of the prothorax and metathorax. Almost the entire upper surface is formed by the scutum of the mesothorax (sometimes wrongly called the 'mesonotum'); the scutum is to a varying extent divided into anterior and posterior parts by the transverse suture (a feature used in taxonomy). The scutellum, directly behind the scutum, almost always forms a well-developed convex lobe. The katepisternum is very large and forms most of the undersurface between the bases of the fore and mid legs. An important landmark in many higher flies is the greater ampulla (= infra-alar bulla), a knob below the wing attachment perhaps associated with the ability of these flies to make a buzzing sound. Constant landmarks in all flies are two pairs of thoracic respiratory openings, the anterior (mesothoracic) and posterior (metathoracic) spiracles. The halteres are rooted to the metathorax behind and above the posterior spiracles.

The wings of each species have a remarkably constant venation, i.e. arrangement of veins supporting the wing blade and dividing it into cells (membrane areas enclosed by veins). The cells are called 'open' if they reach the wing margin and

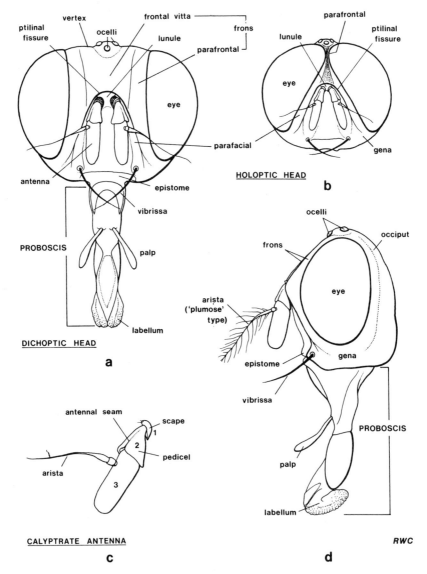

Figure 3.2 Principal features of an adult dipteran head, as seen in a typical muscomorph fly of the subsection Calyptratae (see text): (*a*) facial view of dichoptic (eyes-separated) head, including proboscis; (*b*) facial view of holoptic (eyes-meeting) head, proboscis omitted; (*c*) typical antenna of calyptrate fly, showing characteristic dorsal antennal seam of second segment (pedicel); (*d*) left side view of head and proboscis.

'closed' if they do not (Fig. 3.1). The venation is extremely important for classification and identification, and some familiarity with its terminology is necessary for studying Diptera even at very basic level. The modern terminology of McAlpine (1981a) supplants earlier systems and is used here. It is rooted in a very old system in which six primary veins are recognized in insect wings: costa (C), subcosta (Sc), radius (R), media (M), cubitus (Cu) and anal vein (A). Veins R, M, Cu and A are

Morphology 55

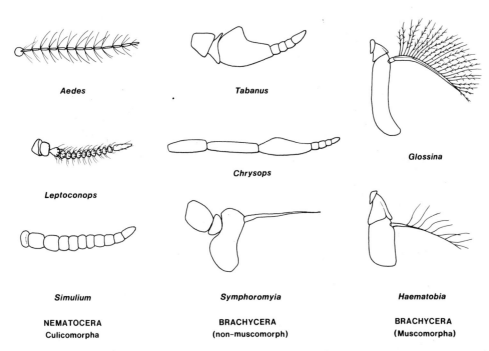

Figure 3.3 Diverse types of adult antennae of Nematocera and Brachycera seen in important genera of bloodsucking flies of the families Culicidae (*Aedes*, female), Ceratopogonidae (*Leptoconops*, female), Simuliidae (*Simulium*), Tabanidae (*Tabanus* and *Chrysops*), Rhagionidae (*Symphoromyia*), Glossinidae (*Glossina*) and Muscidae (*Haematobia*). All shown in right side view.

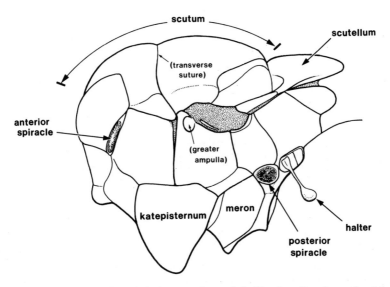

Figure 3.4 Principal features of an adult dipteran thorax, left side view. Based on a fly of the calyptrate Muscomorpha in which greater ampullae are present and the transverse suture complete. The shaded central area indicates diagrammatically the position of the wing attachment.

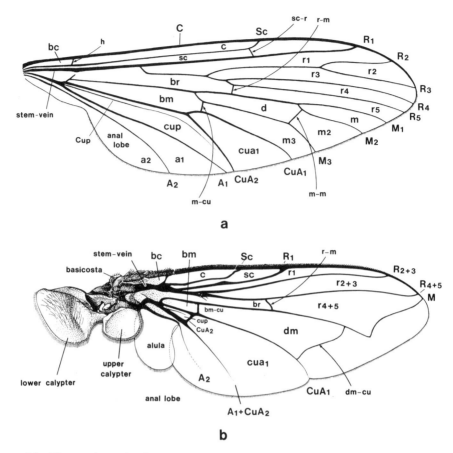

Figure 3.5 Wings and associated structures in Diptera: (*a*) groundplan wing venation; (*b*) wing, alula and calypteres of a muscomorph calyptrate fly, drawn from the blow-fly *Calliphora vicina*. (Terminology of veins and cells, and groundplan illustration, after McAlpine, 1981a.)

branched in the groundplan of the dipteran wing (Fig. 3.5*a*) and in places connected by short cross-veins (e.g. r–m). In some flies the groundplan is almost totally lost and in most Muscomorpha the veins are greatly reduced (Fig. 3.5*b*). Literature on the latter often uses terms such as 'fourth vein' and 'sixth vein', but these are obsolete and should only be used if accompanied by the modern standardized terminology (cf. Figs 9.2*a* and 9.3). The diversity of wing venation found in bloodsucking flies can be seen in Fig. 3.6 (see also Chapters 4–10). The membrane of the wing is usually covered with microscopic unsocketed hairs (microtrichia), and some veins bear socketed hairs or spinules (macrotrichia); in mosquitoes, the latter are modified to form broad scales. The wings are usually clear (hyaline) but sometimes marked (e.g. with a dark cross-band in *Chrysops* and a rosette pattern in *Haematopota*). The alula and calypteres (Fig. 3.5*b*) are three rounded membranous lobes associated with the wing base and important in higher classification. They are absent in Nematocera but present in many Brachycera – the calypteres (or squamae) occurring particularly in Tabanidae and Muscomorpha Calyptratae. The alula arises from the hind margin of the wing stalk and is demarcated from the wing proper by a deep notch. The lower calypter (thoracic squama) is attached along its inner edge

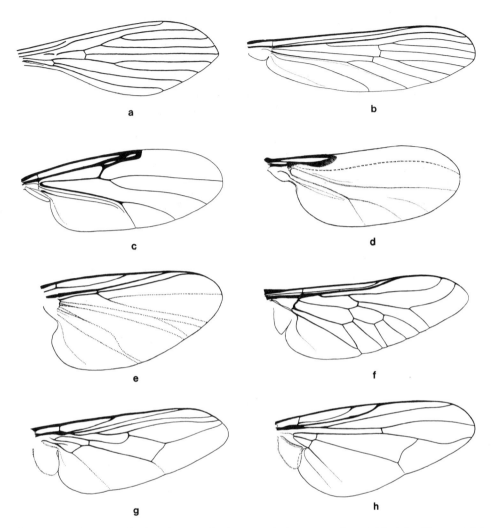

Figure 3.6 Diversity of wing shape and wing venation amongst the principal families containing bloodsucking Diptera: (*a*) Psychodidae (*Phlebotomus*); (*b*) Culicidae; (*c*) Ceratopogonidae (*Culicoides*); (*d*) Ceratopogonidae (*Leptoconops*); (*e*) Simuliidae (*Simulium*); (*f*) Tabanidae (*Tabanus*); (*g*) Glossinidae (*Glossina*); (*h*) Muscidae (*Stomoxys*). The characteristic hairs of the sandfly wing (*a*) and scales of the mosquito wing (*b*) have been omitted so as not to obscure the vein configuration (see figures in Chapters 4 and 5).

to the wall of the thorax and the upper calypter (alar squama) mainly to the extreme base of the wing – the effect of these attachments being that the upper calypteres concertina over the lower calypteres when the wings are closed and stretch away flat from them when the wings are spread (as in Fig. 3.5*b*).

The legs are seldom much modified, but all five parts (coxae, trochanters, femora, tibiae, tarsi) sometimes show minor adaptations. Almost without exception the tarsi have five segments (tarsomeres), the last of which carries on each leg a pair of claws. The acropod (so-called 'pretarsus') between the claws usually comprises a pair of lateral flap-like pulvilli and a median empodium. Presence, absence and shape of these structures can be taxonomically useful: lobate and bristle-like types of empodium are shown in Figs 7.3*a* and 8.5*f*.

The abdomen ranges in shape from very long and slender (as in Tipulidae and Asilidae) to almost globular (as in many calyptrates). The basic number of segments is 11 but the terminal segments are much modified in association with the genitalia and the basic number is always more or less obscured. In most higher flies the externally visible abdomen consists of only the first few segments; the upper segmental plates (tergites) are enlarged and wrapped round the sides and venter of the abdomen, sometimes partly covering the lower plates (sternites) and incorporating the spiracles; the apparent first segment is formed by the fusion of first and second tergites (T1 + 2). There are typically seven pairs of spiracles, but reductions occur in many families and the maximum number of eight pairs is found only in females of some Nematocera and lower Brachycera. The female genital opening lies between the eighth and ninth sternites, the orifice of the male copulatory organ (aedeagus, phallus) immediately behind the ninth sternite. The aedeagus and associated copulatory structures form compact terminalia (hypopygium), sometimes exposed – especially in Nematocera – but often largely concealed within the end of the abdomen; tergite 9 is often a large arched plate (epandrium) from which the terminalia are partly suspended. The male genitalia of many flies have become partly or entirely rotated on the long axis of the abdomen, a fact which has substantial bearing on posture during mating (McAlpine, 1981a, pp. 56–59). Terminal segments of females are often modified to form a telescopic ovipositor through which the eggs are laid. In females the basic number of spermathecae (internal capsules of the reproductive system for storing sperm) is three, but in some families (e.g. Simuliidae) there is only one spermatheca or the number varies (e.g. one, two or three in Ceratopogonidae).

Immature stages

Egg These show much variation in structure. The commonest shape is elongate-ovoid, but eggs are sometimes virtually spherical or boomerang-shaped. Long eggs often have a dorsal groove (e.g. Fig. 10.1a). Sometimes the egg bears paired or multiple respiratory horns or a long single or double whip-like filament. In certain Oestridae there is a ventral flange by which the egg is stuck to a hair of the vertebrate host. The shell (chorion) of terrestrial eggs, unlike most aquatic eggs, is often heavily sculptured and frequently has respiratory air-spaces (aeropyles). Eggs of aquatic species tend to be very simple (e.g. Simuliidae, Fig. 6.1b), but in anopheline mosquitoes there are large lateral floats (Fig. 5.12a).

Larva Typically the larval body is soft and fairly clearly segmented (Fig. 3.7a-e) but there is little or no obvious distinction between the thorax and abdomen except in Culicidae and their relatives. There are three thoracic segments and seven to nine abdominal segments, but in Culicidae and Simuliidae the thoracic segments are fused together and in therevid and scenopinid larvae (Fig. 3.8d) there appear to be 20 segments because of secondary subdivision. The head is a well-defined cranium in most nematocerous larval (Fig. 3.8a–c) but a much reduced structure (sometimes retractable) in other larvae and is more or less vestigial in the Muscomorpha – an infraorder in which the larvae are known as maggots when the front end is pointed and the hind end truncate (Fig. 3.8e), and as grubs when the swollen body has a more or less oval outline.

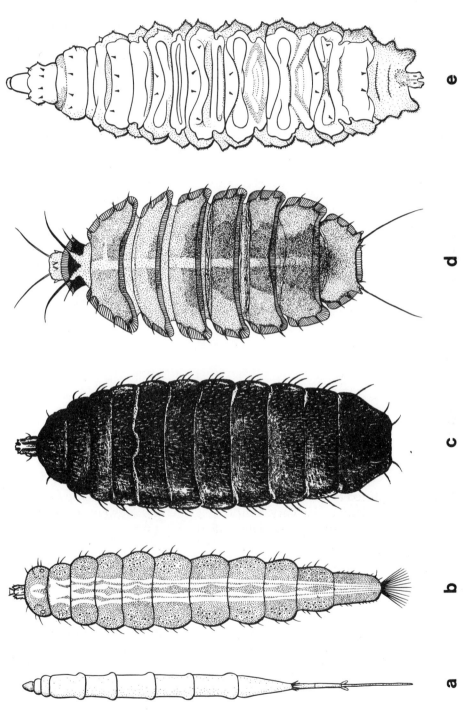

Figure 3.7 Some varied types of dipterous larvae, dorsal view: (a) *Ptychoptera* (Nematocera, Ptychopteridae); (b and c) *Odontomyia* and *Hermetia* (Brachycera, Stratiomyidae); (d) *Lonchoptera* (Muscomorpha-Aschiza, Lonchopteridae); (e) *Pipiza* (Muscomorpha-Aschiza, Syrphidae). All illustrations are of fully grown larvae.

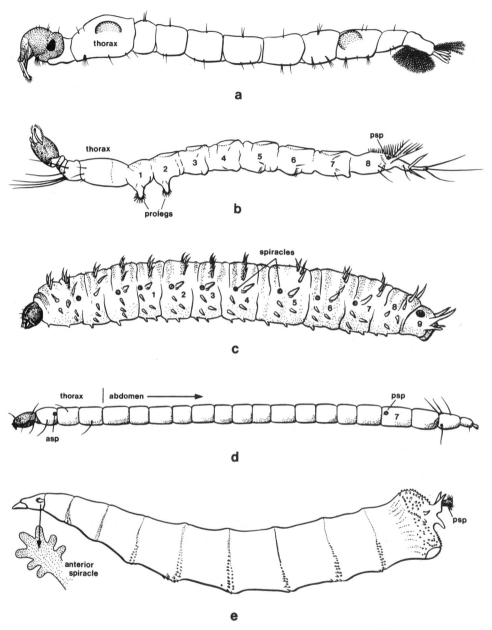

Figure 3.8 Larvae of some Diptera to illustrate various features mentioned in the text, left side view: (a–c) respectively, *Chaoborus* (Chaoboridae), *Dixa* (Dixidae) and *Bibio* (Bibionidae), showing the well-formed head capsule typical of Nematocera; (d) larva of *Scenopinus* (Scenopinidae) showing the high number of apparent abdominal segments; (e) typical maggot of an acalyptrate fly, drawn from *Sepsis* (Sepsidae) with enlargement of the digitate anterior spiracle. asp = anterior spiracles, psp = posterior spiracles. Three types of larval spiracular system are shown (see text), namely metapneustic (*Dixa*, b); holopneustic (*Bibio*, c); amphipneustic (*Scenopinus* and *Sepsis*, d and e). All illustrations are of fully grown (last instar) larvae.

The head above the mouth is formed by the labrum, a wedge-shaped or sometimes snout-like lobe best developed in the Nematocera; in mosquitoes and blackflies the outer parts of the labrum have been subdivided to form highly elaborate mouth-brushes and fans for larval filter-feeding. The antennae are slender and segmented in many Nematocera, and the mandibles horizontal grasping organs, but in other larvae the antennae are usually represented by small papillae and the mandibles orientated vertically. In muscomorph larvae the normal mouthparts are atrophied in conjunction with the 'headless' condition and replaced by an internal framework of articulated sclerites (cephalopharyngeal skeleton) in which the foremost part is formed by a pair of downwardly directed mouth-hooks.

Dipterous larvae never have true legs but some have fleshy pseudopods or prolegs that act as locomotory organs (e.g. *Dixa*, Fig. 3.8b). Chironomid and simuliid larvae, for instance, have a foot-like ventral thoracic proleg crowned with hooks. In tabanid larvae there are circlets of blunt pseudopods on each abdominal segment (Fig. 8.6a), and larvae of *Fannia* have long tapering segmental processes (Fig. 12.10). Most maggots and grubs have creeping welts, raised areas covered with tiny spicules or fine parallel ridges, and some have belts of large spines.

Larval spiracles have great importance in classification and identification since they vary in number, distribution and structure in different families. There are no open spiracles (apneustic condition) in aquatic larvae which remain permanently beneath the water surface and respire through the cuticle (e.g. Chironomidae and Simuliidae). The maximum number of ten pairs (holopneustic condition) occurs in Bibionidae (Fig. 3.8c). Between these states are other arrangements, the commonest of which consists of anterior and posterior pairs of spiracles (amphipneustic condition), the former on the prothoracic segment just behind the head and the latter at the end of the abdomen. This condition occurs in most Brachycera, including the maggots and grubs of Muscomorpha. Larvae of Culicidae and Tabanidae have only posterior spiracles (metapneustic condition), these usually protruding on a structure called the siphon (e.g. Fig. 5.9). A rare state, found mainly in some Mycetophilidae, is the presence only of anterior spiracles (propneustic condition). Anterior and posterior spiracles usually have very different structure. In muscomorph maggots the anterior pair protrude as small digitate processes (Fig. 3.8e) but the posterior pair (usually) take the form of large semicircular discs perforated with three (often winding) slits or numerous holes. The posterior spiracles of sarcophagids are recessed in a deep atrium and those of Glossinidae protrude as two big bosses (polypneustic lobes, Fig. 9.7).

Pupa and puparium The pupal stage is manifested in Diptera in two ways, as a true pupa and as a puparium. In almost all Nematocera and non-muscomorph Brachycera the last larval instar sheds its skin and transforms into a pupa on which the appendages are externally visible (Fig. 3.9a,d) and either free (exarate condition) or sealed to the body (obtect condition); the adult escapes from the pupa through a longitudinal slit in the thoracic cuticle. The pupation of the Muscomorpha is quite different and is technically called puporiation because it involves the formation of a puparium within which the pupal stage and developing adult fly lie concealed and

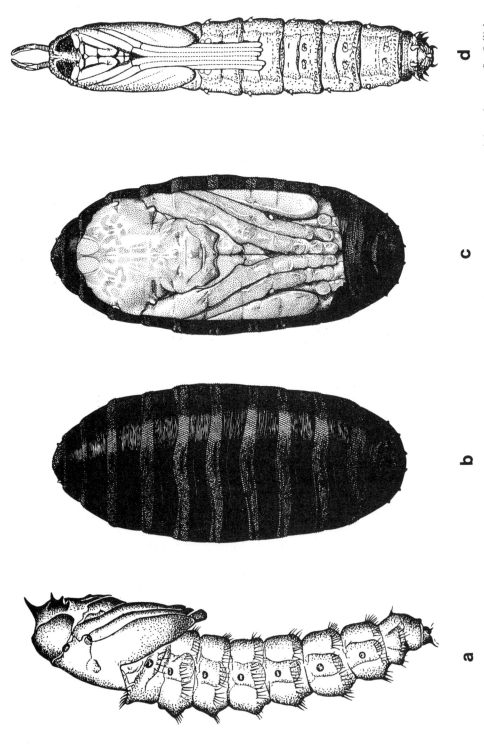

Figure 3.9 Pupae and puparia of Diptera: (*a*) typical pupa of Asilidae (lower Brachycera), right side view; (*b*) puparium of *Calliphora vicina* (Muscomorpha); (*c*) puparium of *Calliphora vicina* with a section removed to show the pupa within; (*d*) typical pupa of Tipulidae (Nematocera), ventral view. Other pupae can be seen in Figs 5.8*a*, 6.1*c* and 8.6*b*.

completely immobile (Fig. 3.9b,c). The puparium, typically shaped like a barrel with convex ends, is a shell formed by the hardened last larval skin on which larval structures such as the posterior spiracles are still visible. The emerging adult fly ruptures the puparium with the ptilinum (Fig. 9.7d), a bladder-like organ temporarily extruded from the face and never used again.

Pupae sometimes have conspicuous spiracular gills, cuticular outgrowths of the body wall near a spiracle designed for respiration of the developing adult. These structures occur, for instance, in some Tipulidae and are universal in Simuliidae (Chapter 6).

Pupae are rarely able to move, but those of mosquitoes have terminal abdominal paddles which enable them to swim (Fig. 5.8a).

Morphological literature

Morphology and associated terminology are covered in superbly illustrated detail for adults by McAlpine (1981a) and for larvae by Teskey (1981a), but students requiring information at less specialist level should consult Richards and Davies (1977, pp. 951–965). Hennig (1973), in German, is very valuable for all stages. Hinton (1981, pp. 724–762) covers egg structure, and Ferrar (1987) provides extensive illustration for eggs and larvae of Muscomorpha.

TAXONOMY

Higher classification

There is now considerable agreement about how Diptera should be classified, but it is nevertheless common to find that even major catalogues differ in whether two or three suborders are recognized – Nematocera, Brachycera and Cyclorrhapha, or only the first two – and in the limits and positions given to many of the families. These differences reflect the enormous size and complexity of the order, and problems in presuming from all life stages the evolutionary relationships (phylogeny). For this book, a system is used which will be called the 'McAlpine classification' because it has been made available in the *Manual of Nearctic Diptera*, a massive three-volume work for which J. F. McAlpine is lead author or editor (McAlpine et al., 1981; McAlpine, 1987, 1989a). This classification, largely developed from a previous phylogenetic groundplan of the Diptera provided by Hennig (1973), seems likely now to supersede other systems. It is shown down to family level in Table 3.1.

At the highest level the McAlpine classification differs from the system usually used in modern times by the recognition of only two suborders, the Nematocera and Brachycera. This results from ranking the erstwhile third suborder Cyclorrhapha as an infraorder (now called Muscomorpha) within an enlarged Brachycera. In effect, the highly evolved Muscomorpha are now the 'higher' Brachycera, in contrast to less specialized flies of the 'lower' (orthorrhaphous) Brachycera equating with the second suborder in the three-suborder system. However, an important point to note is that in McAlpine (1989a), the concluding volume of the *Manual*, higher classification within the Brachycera is not agreed

Table 3.1 Classification of the families of Diptera

Families are listed according to the classification in McAlpine et al. (1981) for the lower Brachycera and in McAlpine (1989a) for the Nematocera and Muscomorpha (= Cyclorrhapha): see text explanation. The suffixes '-morpha', '-oidea' and '-idae' denote names of infraorders, superfamilies and families respectively.

Suborder NEMATOCERA
  Tipulomorpha
    Tipulidae
  Blephariceromorpha
    Blephariceroidea
      Blephariceridae
      Deuterophlebiidae
    Nymphomyioidea
      Nymphomyiidae
  Axymyiomorpha
    Axymyiidae
  Bibionomorpha
    Pachyneuroidea
      Pachyneuridae
    Bibionoidea
      Bibionidae
    Sciaroidea
      Mycetophilidae
      Sciaridae
      Cecidomyiidae
  Psychodomorpha
    Psychodoidea
      Psychodidae
    Trichoceroidea
      Perissommatidae
      Trichoceridae
      Anisopodidae
      Scatopsidae
      Synneuridae
  Ptychopteromorpha
    Tanyderidae
    Ptychopteridae
  Culicomorpha
    Culicoidea
      Dixidae
      Corethrellidae
      Chaoboridae
      Culicidae
    Chironomoidea
      Thaumaleidae
      Simuliidae
      Ceratopogonidae
      Chironomidae

Suborder BRACHYCERA
  Tabanomorpha
    Tabanoidea
      Pelecorhynchidae
      Tabanidae
      Athericidae
      Rhagionidae

Tabanomorpha (*cont.*)
    Stratiomyoidea
        Pantophthalmidae[a]
        Xylophagidae
        Xylomyidae
        Stratiomyidae
Asilomorpha
    Asiloidea
        Therevidae
        Scenopinidae
        Vermileonidae
        Mydidae
        Apioceridae
        Asilidae
    Bombylioidea
        Acroceridae
        Nemestrinidae
        Bombyliidae
        Hilarimorphidae
    Empidoidea
        Empididae
        Dolichopodidae
Muscomorpha (= Cyclorrhapha)
    Section Aschiza
        Platypezoidea
            Platypezidae
            Lonchopteridae
            Ironomyiidae
            Sciadoceridae
            Phoridae
        Syrphoidea
            Syrphidae
            Pipunculidae
    Section Schizophora
    Subsection ACALYPTRATAE[b]
        Nerioidea
            Micropezidae
            Neriidae
            Cypselosomatidae
        Diopsoidea
            Tanypezidae
            Strongylophthalmyiidae
            Psilidae
            Somatiidae
            Nothybidae
            Megamerinidae
            Syringogastridae
            Diopsidae
        Conopoidea
            Conopidae
        Tephritoidea
            Lonchaeidae
            Otitidae
            Platystomatidae
            Tephritidae
            Pyrgotidae
            Tachiniscidae
            Richardiidae

Muscomorpha (= Cyclorrhapha) (*cont.*)
    Pallopteridae
    Piophilidae
  Lauxanioidea
    Lauxaniidae
    Eurychoromyiidae
    Celyphidae
    Dryomyzidae
    Helosciomyzidae
    Sciomyzidae
    Ropalomeridae
    Sepsidae
  Opomyzoidea
    Clusiidae
    Acartophthalmidae
    Odiniidae
    Agromyzidae
    Fergusoninidae
    Opomyzidae
    Anthomyzidae
    Aulacigastridae
    Periscelididae
    Neurochaetidae
    Teratomyzidae
    Xenasteiidae
    Asteiidae
  Carnoidea
    Australimyzidae
    Braulidae
    Carnidae
    Milichiidae
    Risidae
    Cryptochetidae
    Chloropidae
  Sphaeroceroidea
    Heleomyzidae
    Mormotomyiidae
    Chyromyidae
    Sphaeroceridae
  Ephydroidea
    Curtonotidae
    Camillidae
    Drosophilidae
    Diastatidae
    Ephydridae
Subsection CALYPTRATAE
  Hippoboscoidea
    Glossinidae
    Hippoboscidae
    Streblidae
    Nycteribiidae
  Muscoidea
    Scathophagidae
    Anthomyiidae
    Fanniidae
    Muscidae
  Oestroidea
    Calliphoridae

Muscomorpha (= Cyclorrhapha) (*cont.*)
    Mystacinobiidae
    Sarcophagidae
    Rhinophoridae
    Tachinidae
    Oestridae

---

[a] Neotropical family additional to McAlpine et al. (1981).
[b] Extra levels of 'suprafamily' and numbered 'subgroup' used in McAlpine's (1989b) list of Acalyptratae omitted for simplicity.

between Woodley (1989) and McAlpine (1989b): the former promulgates a controversial classification in which the infraorder Muscomorpha is expanded to include three superfamilies of the old Brachycera (e.g. Asiloidea), whereas McAlpine rejects this and accords Muscomorpha exactly the same scope as the old Cyclorrhapha. Woodley's ideas being untested, the classification of lower Brachycera families shown in Table 3.1 is that of McAlpine et al. (1981), but the classification of Nematocera and Muscomorpha is from McAlpine (1989a).

The following outline of the higher classification serves to show the hierarchical arrangement of categories above superfamily level mentioned for their medical relationship in the following paragraphs:

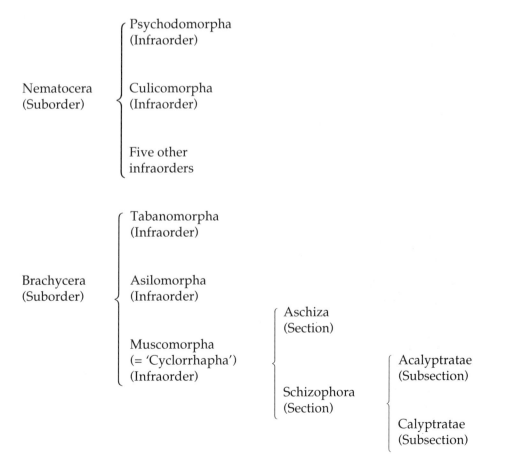

Suborder Nematocera

This suborder includes mainly soft-bodied flies such as gnats and the crane-flies (daddy-longlegs) of the family Tipulidae. Body and legs are usually slender, the wings often long and narrow; the venation of the wings is typically much branched (in the Tipulidae quite similar to the dipteran groundplan) but much reduced in many families. The antennae have numerous generally similar segments (Fig. 3.3), except commonly for enlargement of the pedicel, and the terminal segments are not consolidated (as in many Brachycera) to form a stylus or arista. The palps are usually pendulous (not in mosquitoes) and typically have four or five segments. The larva has a completely formed and non-retractile head capsule with opposable biting mandibles. The pupa is either obtect or exarate, the adult escaping through a dorsal longitudinal slit in the thoracic cuticle.

The suborder is classified into seven infraorders. Two of these, Psychodomorpha and Culicomorpha, contain many bloodsucking species. Two superfamilies form the Psychodomorpha, of which the Psychodoidea contains only the family Psychodidae – tiny Nematocera of moth-like appearance, hairy body and wings, second antennal segment (pedicel) not longer than the first segment (scape), and three-branched media. The subfamily Phlebotominae is medically important (Chapter 4).

The infraorder Culicomorpha includes mosquitoes, blackflies and various biting and non-biting midges. The males often have bushy antennae specialized for hearing the females, and the pedicel in both sexes is often bulbously enlarged (though neither characteristic occurs in the Simuliidae). The media of the wing has fewer than three branches, and the hind legs of the pupa form an S-shaped curve. The infraorder is divided into two superfamilies, Culicoidea and Chironomoidea, but no simple character instantly distinguishes all members of one from the other. Larvae of Culicoidea have open spiracles but lack thoracic prolegs, whereas larvae of Chironomoidea lack open spiracles (except in Thaumaleidae) and often have thoracic prolegs (always in Simuliidae and Chironomidae). Ancestral larvae probably had labral feeding brushes or fans (Wood and Borkent, 1989), though these are absent in Ceratopogonidae and Chironomidae and occur only in Culicidae and Dixidae of the Culicoidea and in Simuliidae of the Chironomoidea.

Suborder Brachycera

This enormous suborder is extremely varied. It includes familiar hard-bodied and compact flies such as horse-flies, hover-flies and house-flies. Many slender-bodied forms occur, but the flies tend to be short and robust, with sturdy legs and fewer exposed abdominal segments than in Nematocera. The antennae usually have no more than three obvious segments, the others being reduced and consolidated into a stubby stylus or a bristle-like arista borne on the end or top of the third segment (Fig. 3.3). The palps have only one or two segments and are often directed forwards. The larva has a reduced and retractile head or is a maggot with vestigial head represented by an internal cephalopharyngeal skeleton (Muscomorpha) (Fig. 12.1); the larval mandibles are sickle-shaped or hook-like and operate vertically instead of horizontally. The pupa is either obtect or formed within a puparium (see Morphology).

The medically most important Brachycera are the Tabanidae, Glossinidae, Muscidae and certain Oestroidea. The family Tabanidae (Chapter 8) belongs to the

Tabanomorpha, an infraorder of lower Brachycera with bulbous adult face and the larval head capable of retraction into the thorax; it differs from all other families of the infraorder by possessing enlarged lower calypteres like those of the calyptrate Muscomorpha – the infraorder to which the other families just mentioned all belong. Lower Brachycera other than Tabanidae have scarcely any medical importance, but some Rhagionidae (notably *Symphoromyia*) and Athericidae – families belonging to the superfamily Tabanoidea in the Tabanomorpha – occasionally bite man.

A brief account must be given here of muscomorph classification, as medical entomologists should know the difference between flies of the Calyptratae and the Acalyptratae. The Muscomorpha is a huge infraorder of highly evolved flies in which the adult antennae usually have an arista borne on the third antennal segment (Figs 3.2c, 3.3, 10.2a), the palps have only one segment, the larva is a typical maggot with mouth-hooks, and the pupa is formed in a puparium from which the adult escapes by pushing off a circular cap. The infraorder is classified into two easily definable sections of very unequal size, the Aschiza and Schizophora. Adult schizophoran flies have two facial landmarks, the ptilinum and lunule, which are absent in the Aschiza. The ptilinum is a sac which the emerging adult of a schizophoran fly everts from the front of the head and uses to break out of the puparium and force a passage through the soil or other surrounding medium before escaping into the air. The organ is withdrawn soon after emergence, but its position remains visible on the fly's head as the ptilinal fissure (Fig. 3.2a,b), a fine arched groove between frons and antennae which extends downwards each side of the face towards the mouth-edge (epistome). The lunule is a small sclerite, shaped like an inverted gardener's edging tool, situated between the arch of the ptilinal fissure and the insertions of the antennae (Fig. 3.2a).

The section Aschiza, including Syrphidae (hover-flies), Phoridae (scuttle-flies) and certain smaller families, has almost no medical importance – except inasmuch as larvae are an occasional cause of accidental myiasis – and further remarks on Brachycera classification can be confined to the Schizophora. This section is classified into two subsections, Acalyptratae and Calyptratae, which (disregarding a few unimportant exceptions) can be distinguished as follows:

Acalyptratae: second antennal segment (pedicel) without groove (antennal seam) along outer edge. Sides of thorax without greater ampullae. Calypteres absent or feebly developed.

Calyptratae: second antennal segment (pedicel) with groove (antennal seam) along outer edge (Fig. 3.2c). Sides of thorax with greater ampullae (Figs 3.4, 9.5). Calypteres present, usually very large (Fig. 3.5b).

The subsection Acalyptratae is a vast assemblage of mainly small muscomorph Diptera classified into nine superfamilies and nearly 60 families. They have little medical importance. The larvae (e.g. of Drosophilidae) sometimes produce accidental myiasis, and adult Ephydridae, Chloropidae and Sphaeroceridae have minor importance as sweat-flies, eye-flies and lesser dung-flies occasionally involved in the mechanical transmission of harmful organisms (Chapter 17).

The subsection Calyptratae is classified into the superfamilies Hippoboscoidea, Muscoidea and Oestroidea. Each contains one or more families of moderate to very high medical importance. The four families of Hippoboscoidea contain ectoparasitic Diptera whose adults feed on the blood of birds and mammals and whose larvae develop one at a time in the mother's uterus until voided when ready to pupariate. The old name 'Pupipara' was applied to these viviparous flies, other than the tsetse-flies (Glossinidae), but it is now accepted that the Glossinidae should be included in the group (now Hippoboscoidea): all evidence points to Hippoboscidae being the nearest living relatives of the tsetse-flies (see McAlpine, 1989b). The Nycteribiidae and Streblidae are highly modified flies which live as ectoparasites on bats and are provided (like the Hippoboscidae) with heavily toothed claws; the nycteribiids are wingless and spider-like. Hippoboscidae (louse-flies, Fig. 17.3d) are much-flattened and leathery flies which sometimes bite humans but are not medically important.

The calyptrate superfamilies Muscoidea and Oestroidea include the familiar house-flies, blow-flies and flesh-flies, and also the bot-flies whose larvae occur as parasites in the bodies of vertebrate animals (myiasis). A full understanding of why the four families of Muscoidea and the six families of Oestroidea (Table 3.1) have been assembled on phylogenetic principles into separate superfamilies requires intimate knowledge of muscomorph morphology, so here it must suffice to note the single most obvious outward character by which these two superfamilies can (with rare exceptions) be distinguished: meron of the adult fly thorax (Fig. 3.4) bare or virtually so in Muscoidea, with a vertical row of meral bristles in Oestroidea (Fig.11.3). Another character with some practical usefulness is that in Oestroidea the wing vein M (the so-called 'fourth vein') is bent strongly forwards not far from the wing margin (Fig. 3.5b); in most Muscoidea (not Muscinae) vein M runs almost straight to the margin.

The larvae of various muscoid and oestroid Calyptratae are a common cause of human myiasis, either accidental myiasis (e.g. as sometimes caused by members of the muscoid family Fanniidae) or truly parasitic myiasis as caused by certain members of three families of the Oestroidea – Calliphoridae, Sarcophagidae and Oestridae – in which development takes place in vertebrate animals (Chapter 12). The oestroid families Tachinidae and Rhinophoridae, on the other hand, are completely harmless to man: the species are parasites of insects and woodlice, respectively. Adult feeding habits in some oestroid and muscoid families render the flies liable to contaminate human food with pathogenic organisms (Chapter 11). A few Muscoidea (unlike any Oestroidea) are bloodsucking and for this reason have a minor medical role (Chapter 10).

Family recognition

The Diptera require a very large number of categories for their satisfactory classification and 125 families are listed in Table 3.1. Most families are found worldwide, and are therefore listed in the regional catalogues; for example, catalogues for the Afrotropical (Crosskey, 1980) and Australasian faunas (Evenhuis, 1989) list 95 and 116 families respectively. The discrepancy, however, does not reflect significant difference in the faunas: it results from authorities differing in the taxonomic limits they give to families. Some specialists, for example, make five families of the fungus-gnats (Mycetophilidae) instead of the usual one. Such splitting has had the serious consequence of greatly complicating family recognition with little or no compensating advantage.

Many small families, especially among the Acalyptratae, are so obscure that even experienced specialists find them very difficult to identify. On the other hand, the adults of large and worldwide families are often recognizable at a glance from their appearance. Sometimes, however, a small and geographically restricted family is easier to recognize than a large cosmopolitan one: the tsetse-flies (Glossinidae) of Africa, for example, can be easily recognized by the unaided eye of the beginner whereas it is a more expert matter to distinguish Muscidae from Calliphoridae reliably without the aid of a microscope. Sometimes the living fly offers clues to identity which (depending on how the wings are positioned after death) are not always evident on preserved specimens. In some families, or parts thereof, the wings are closed in a characteristic way: in Haematopotoni, for example, the wings at rest do not overlap but meet in roof-like posture. In Glossinidae, Hippoboscidae and Simuliidae the wings at rest close fully in scissors-like fashion and conceal the abdomen, whereas in most Muscidae, Calliphoridae and many other Muscomorpha they diverge behind when at rest and expose part of the abdomen.

Although some modern identification keys to families exist (see references below), they are difficult to use successfully without considerable experience. Inevitably they are lengthy, complex and extensively illustrated, because of the difficulty of providing reliably for many misfit 'exceptions' – genera or species not typical of the families to which they certainly belong. Whereas Simuliidae are instantly recognizable by their remarkably uniform wing venation, this is not the case with Ceratopogonidae (see wings in Fig. 7.3). So-called 'simplified' keys, or keys to 'principal' families, have a misleading simplicity and high potential for misidentification. In practice, what medical entomologists usually want is to be able to recognize families of potential medical importance – is this fly a simuliid? – rather than a complicated key for identifying, say, all flies present in a light-trap catch. The reader is therefore referred to the 'recognition' sections in the illustrated family or subfamily treatments in Chapters 4–10. A key is provided, however, in Chapter 11 to help with identification of the calyptrate families.

Faunal and taxonomic literature

The following is a categorized list of the most important modern works. Bibliographical details are given in the References. All are useful beyond the geographical limits stated in their titles.

Bibliographies: Worldwide (Hollis, 1980, pp. 90–119); Britain and western Europe (Sims et al., 1988, pp. 130–160). Foote (1991) includes a massive bibliography for immature stages (pp. 880–915).

Catalogues: Afrotropical (Crosskey, 1980); Australasian (Evenhuis, 1989); North America (Stone et al., 1965, under revision for Nearctic including Mexico); Neotropical (Papavero, 1966–1984); Oriental (Delfinado and Hardy, 1973–1977); Palaearctic (Soós and Papp, 1984–1991, incomplete). Computerized world catalogue planned.

Classification: McAlpine et al. (1981); McAlpine (1989a).

Family identification keys: Borror et al. (1989, adults); Cole (1969, western North America); Foote (1991, larvae); Hennig (1973); McAlpine (1981b, adults); Oldroyd (1970, adults, Britain); Smith (1989, immatures, Britain); Teskey (1981b, 1984, larvae).

Faunal monographs (coverage to identification of species level): Britain (see Handbooks series issued by Royal Entomological Society); Nearctic (Griffiths, 1980–1991, incomplete); Palaearctic (Lindner, 1924–1985, incomplete). There are no faunal monographs for other zoogeographical regions. Local faunal series are available for some European and Asiatic countries but not comprehensively for all families (see *Faune de France, Fauna d'Italia, Fauna Hungarica, Fauna of the USSR, Fauna of India, Insecta Japonica*, etc.: consult entomological librarian for further guidance).

Phylogeny: Hennig (1973); McAlpine (1989a).

Semi-popular faunas (excellent illustrated overviews): Britain (Colyer and Hammond, 1968); North America (Cole, 1969, western states, includes keys to genera); Africa (Skaife, 1979).

Textbooks: Borror et al. (1989); Colless and McAlpine (1991, Australia); Richards and Davies (1977, good coverage for Diptera).

BIOLOGY

The dipterous egg, larva, pupa/puparium (see above) and adult nearly always have an independent existence, eggs being freely laid and hatching away from the parent flies. Sometimes, however, as in some calliphorids and tachinids, the eggs hatch in the uterus of the reproductive system and the female deposits newly hatched larvae. The Hippoboscoidea (which include the tsetse-flies) have evolved a peculiar reproductive method in which the free larval stage has been abolished and a single larva develops in the uterus until ready to leave the parent female and pupariate. In the Cecidomyiidae (both phytophagous or predatory) some extraordinary species even reproduce directly from larvae without intervention of the adult (paedogenesis).

Dipterous eggs are usually not very conspicuous, and for many flies they have been little studied. They are laid in varied situations in anticipation of the natural development medium for the ensuing larva. The eggs of each developed batch are usually laid together, often in strings or neat layers and sometimes sealed in a gelatinous secretion. In many flies they remain dormant (diapause) for weeks or months before hatching if a resting phase is necessary to ensure that the larvae hatch when circumstances are most favourable for their development. Larval growth proceeds through several stages (instars) between each of which the old cuticle is shed. In most flies the number of larval instars is constant, e.g. three in all muscomorph flies and four in most Nematocera, but the Simuliidae and Tabanidae are unusual in having a high and variable number (as many as nine and possibly more).

Dipterous larvae have a general need for water or dampness. Those that are not aquatic very seldom expose themselves to the air in the manner of caterpillars and tend to live concealed in a humid medium or a damp place. The larvae of phlebotomine sandflies, for example, though these delicate little flies often occur in desert regions, must have traces of moisture in the burrows they inhabit if they are to survive. Larvae of many flies are wholly or partly adapted for aquatic life. Those of Chironomidae and Simuliidae live permanently submerged and get oxygen from the water (spiracles are abolished) whereas the more mobile aquatic mosquito larvae spend time at the surface breathing air through a special spiracular siphon. The larvae of certain Syrphidae and Ephydridae (also a few mosquitoes) breathe by inserting special devices into the stems of submerged plants: with some justice they have been amusingly described as 'no more aquatic than a skin-diver'. Larvae in some aquatic families are provided with specialized mouthpart structures (e.g. mouthbrushes of mosquitoes) for gathering food. In many larval Nematocera the hypostomium is a toothed 'chin' apparently used to loosen edible material from a solid surface.

Many larvae develop in semi-aquatic habitats, the Tabanidae in soft mud at the edges of ponds and streams. Muscomorph maggots also generally require a development medium that is soft and moist. Those of many Muscidae, for example, inhabit rotting vegetable matter and those of Sarcophagidae mainly develop in decomposing animal matter (from which has apparently evolved the habit in *Wohlfahrtia* of feeding in living flesh). Development in excrement is the habit of Sphaeroceridae, Scathophagidae and various Muscidae: in the muscid horn-flies (*Haematobia*) the habit is so strong that the females even try to oviposit on wet faeces while these are being voided.

Larval food and feeding methods are very varied. Some aquatic larvae depend on grazing algae from immersed surfaces, but the specialized larvae of Simuliidae living in rivers and streams are filter-feeders which passively strain suspended particles out of the water current with sophisticated head fans and use the digestible component of the filtrate as food. Larvae in soil usually feed on decomposing vegetable matter, but some (e.g. 'leatherjacket' larvae of Tipulidae) attack the roots of living plants. The larvae of most Mycetophilidae feed on fungi. Predation by larvae is very common throughout the order. Asilid larvae feed mainly on immature insects, and tabanid larvae are sometimes so carnivorous as to attack living vertebrate flesh. It seems surprising, but 'headless' maggots with mouth-hooks can be just as much predators as nematocerous larvae provided with pincers-like opposable mandibles swung from a hard head-capsule. Sometimes predation is very highly specialized; larvae of some bombyliids subsist on a diet of grasshopper eggs.

Feeding dominates larval life because flies mainly depend on nutrients accumulated at this time to nourish the developing eggs when they become adult. True legs are absent, so movement in the development medium is achieved with pseudopods or prolegs (see Morphology section) or, in maggots with 'creeping welts', swollen areas with spinules or ridges. Aquatic larvae in still (lentic) water habitats are often active swimmers (e.g. mosquitoes) but those of running (lotic) water habitats must anchor themselves to a submerged substrate to prevent being swept away; blepharicerid larvae do this with large ventral suckers, and simuliid larvae with crowns of proleg hooks latched into silk pads secreted on the substrate.

Many Diptera have a parasitic nutritional life-style, either as larvae or adults but not both. Larval parasitism occurs mainly among Muscomorpha and, depending on the family, can affect invertebrate or vertebrate animals. Larvae of the vast family Tachinidae (8000 known species) develop as endoparasites in other insects (caterpillars and beetle grubs being the main hosts) and those of Rhinophoridae similarly in woodlice (terrestrial isopod Crustacea). Some of the Oestroidea, including all Oestridae, are obligate parasites whose larvae develop in warm-blooded vertebrates, and are sometimes the cause of human myiasis (Chapter 12). Adult parasitism is manifested by female Phlebotominae, Culicidae, Simuliidae, Ceratopogonidae and Tabanidae, and by male and female Glossinidae, Hippoboscidae and Stomoxyinae, which consume blood from vertebrate animals (see Chapters 4–10) – and in the case of Ceratopogonidae sometimes from other insects. The flies are winged and visit the host whenever they need blood, feeding externally as 'intermittent ectoparasites'. The Nycteribiidae and Streblidae are in this category but are less dispersive (wingless in all nycteribiids and some streblids) and spend more time on their bat hosts; the peculiar *Ascodipteron* (Streblidae) has even become an endoparasite, the female embedding in the host's skin and transforming into a wingless and legless pouch for her developing larva. Adults of the Asilidae, Empididae and certain other families are accomplished predators which pounce upon smaller insects and suck them dry.

Mating and egg-laying are the dominant concerns of adult life – except for a very few species in which there is no mating because the male sex has been abolished and females reproduce parthenogenetically from unfertilized eggs. Adult flies often live several weeks, taking food and sometimes maturing several egg batches (in most bloodsucking species with the aid of proteins from ingested blood), but it is also very common for adult life – especially of males – to be very short. Males only need to live briefly for their role of inseminating the females, as even when females mate only once (as is often the case) they can store the sperm in their spermatheca(e) to fertilize all the eggs they produce. The life span of females, too, can be much reduced if they emerge from pupae with sufficient stored nutrients inherited from the larva to nourish the eggs, and in Chironomidae, for example, adult life of both sexes is extremely brief – the midges mating, ovipositing and dying without having fed (except perhaps on a little honeydew) within a day or two of emergence. Mating habits are varied and some flies have evolved quite complex courtship rituals (usually involving the wings or antennae). Males of many Nematocera form mating swarms to attract females, but copulation is not necessarily aerial. Various mating postures are used, largely depending on how the male's genitalia are rotated on the abdomen. In many flies the male secretes a container (spermatophore) around the ejaculate before this is transferred to the female.

Wings provide most adult flies with great aerial mobility, allowing them to disperse in search of mates and egg-laying sites (also hosts in the case of bloodsucking species), and to occupy different habitats from the immature stages. However, species that associate closely with vertebrates, such as streblids on cave-bats and horn-flies on cattle, can be dispersed more by hosts than their own volition. Powers of flight, however, and the ability to be dispersed by winds are strong, and some blackflies can migrate for hundreds of kilometres. The general mobility is well evidenced in that Diptera are much more abundant over the sea than any other insects. There are even a few species (Chironomidae) that have colonized the marine environment. The Diptera truly justify their reputation as the most biologically varied – and to many the most interesting – of all insects.

Biological literature

The enormous literature on dipteran biology is in need of modern synthesis. An informative and eminently readable book by Oldroyd (1964) on general natural history is an excellent beginning but it is unfortunately now out of print. An older French work by Séguy (1950), developed from a card index and not a general read, is still valuable as a well-indexed mine of information.

These works aside, information is mainly available on a family-by-family basis in entomological textbooks, specialist manuals, regional faunas and catalogues, and (for some families) comprehensive books. The following are helpful for starting-point information: Colless and McAlpine (1991), Griffiths (1980 onwards, incomplete), Lindner (1924–1985), McAlpine et al. (1981), McAlpine (1987), Richards and Davies (1977). An excellent semi-popular treatment for African Diptera is Skaife (1979). Family introductions in regional catalogues (see 'Faunal and taxonomic literature' above) contain pithy summaries of key aspects of biology. Biology of eggs is covered in Hinton (1981), muscomorph immatures by Ferrar (1987), and Diptera of forensic importance by Smith (1986).

Books of which readers should be aware because they cover the biology of families or subfamilies concerned in human disease are: Bates (1949, Culicidae); Buxton (1955, Glossinidae); Clements (1992, Culicidae); Crosskey (1990, Simuliidae); Gillett (1971, Culicidae); Skidmore (1985, Muscidae); Zumpt (1965, Oestridae) and Zumpt (1973, Stomoxyinae). There are no equivalent comprehensive books on Ceratopogonidae, Phlebotominae or Tabanidae.

REFERENCES

Bates, M. 1949. *The natural history of mosquitoes*. xv + 378 pp. Macmillan, New York. [Facsimile by Harper Torchbooks, 1965.]

Borror, D. J., Triplehorn, C. A. and **Johnson, N. F**. 1989. *An introduction to the study of insects*. Sixth edition. xiv + 875 pp. Saunders College Publishing, Philadelphia.

Buxton, P. A. 1955. *The natural history of tsetse flies: an account of the biology of the genus Glossina (Diptera)*. xviii + 816 pp. H. K. Lewis, London.

Clements, A. N. 1992. *The biology of mosquitoes*: vol. 1, *Development, nutrition and reproduction*. xii + 509 pp. Chapman & Hall, London.

Cole, F. R. 1969. *The flies of western North America*. xi + 693 pp. University of California, Berkeley and Los Angeles.

Colless, D. H. and **McAlpine, D. K**. 1991. Diptera (flies). Pp. 717–786 in Naumann, I. D. (ed.), *The insects of Australia: a textbook for students and research workers*. Second edition. Vol. 1, xvi + pp. 1–542; vol. 2, vi + pp. 543–1137. Melbourne University Press, Carlton, Victoria.

Colyer, C. N. and **Hammond, C. O**. 1968. *Flies of the British Isles*. Second edition. 384 pp. Frederick Warne, London and New York.

Crosskey, R. W. (ed.) 1980. *Catalogue of the Diptera of the Afrotropical region*. 1437 pp. British Museum (Natural History), London.

Crosskey, R. W. 1990. *The natural history of blackflies*. ix + 711 pp. John Wiley, Chichester.

Delfinado, M. D. and **Hardy, D. E**. (eds) 1973–1977. *A catalog of the Diptera of the Oriental region*: vol. 1, *Suborder Nematocera*, 618 pp. (1973); vol. 2, *Suborder Brachycera through Division Aschiza, suborder Cyclorrhapha*, 459 pp. (1975); vol. 3, *Suborder Cyclorrhapha (excluding Division Aschiza)*, x + 854 pp. (1977)..University Press of Hawaii, Honolulu.

Evenhuis, N. L. (ed.) 1989. *Catalog of the Diptera of the Australasian and Oceanian regions*. 1155 pp. Bishop Museum Press, Honolulu, and E. J. Brill, Leiden.

Ferrar, P. 1987. A guide to the breeding habits and immature stages of Diptera Cyclorrhapha. *Entomonograph* **8**: 1–907.

Foote, R. 1991. Diptera. Pp. 690–915 in Stehr, F. W. (ed.), *Immature insects*: vol. 2, xvi + 975 pp. Kendall/Hunt, Dubuque, Iowa.

Gillett, J. D. 1971. *Mosquitos* [sic]. xiii + 274 pp. Weidenfeld & Nicolson, London.
Griffiths, G. C. D. (ed.) 1980–1991, onwards. *Flies of the Nearctic region* [incomplete]. E. Schweizerbart'sche, Stuttgart. [Faunal work equivalent to, and issued by the same publisher as, Lindner (1924–1985) for Palaearctic region: appears irregularly at wide intervals as parts of family texts are completed. To date Anthomyiidae and Bombyliidae available, little else.]
Hennig, W. 1973. Diptera (Zweiflügler). In Kükenthal, W. (ed.), *Handbuch der Zoologie*: vol. 4 (2) (2) (31), 337 pp. Walter de Gruyter, Berlin and New York.
Hinton, H. E. 1981. *Biology of insect eggs*: vol. 2, xviii + pp. 475–778. Pergamon Press, Oxford.
Hollis, D. (ed.) 1980. *Animal identification: a reference guide*: vol. 3, *Insects*, viii + 160 pp. British Museum (Natural History), London, and John Wiley, Chichester.
Lindner, E. (ed.) 1924–1985. *Die Fliegen der palaearktischen Region* [incomplete]. E. Schweizerbart'sche, Stuttgart. [Faunal work in German issued irregularly since 1924 with individually numbered families, some still unfinished.]
McAlpine, J. F. 1981a. Morphology and terminology – adults. Pp. 9–63 in McAlpine et al. (1981).
McAlpine, J. F. 1981b. Key to families – adults. Pp. 89–124 in McAlpine et al. (1981).
McAlpine, J. F. (ed.) 1987. *Manual of Nearctic Diptera*: vol. 2, vi + pp. 675–1332. Research Branch, Agriculture Canada (Monograph No. 28).
McAlpine, J. F. (ed.) 1989a. *Manual of Nearctic Diptera*: vol. 3, vi + pp. 1333–1581. Research Branch, Agriculture Canada (Monograph No. 32).
McAlpine, J. F. 1989b. Phylogeny and classification of the Muscomorpha. Pp. 1397–1518 in McAlpine (1989a).
McAlpine, J. F., Peterson, B. V., Shewell, G. E., Teskey, H. J., Vockeroth, J. R. and **Wood, D. M.** 1981. *Manual of Nearctic Diptera*: vol. 1, vi + 674 pp. Research Branch, Agriculture Canada (Monograph No. 27).
Oldroyd, H. 1964. *The natural history of flies*. xiv + 324 pp. Weidenfeld & Nicolson, London.
Oldroyd, H. 1970. Diptera I. Introduction and key to families. Third edition. *Handbooks for the Identification of British Insects* **9** (1): 1–104.
Oldroyd, H. and **Smith, K. G. V.** 1973. Eggs and larvae of flies. Pp. 289–323 in Smith, K. G. V. (ed.), *Insects and other arthropods of medical importance*. xiv + 561 pp. British Museum (Natural History), London.
[Papavero, N. (ed.)] 1966–1984. *A catalogue of the Diptera of the Americas south of the United States* [incomplete]. Secretaria de Agricultura, São Paulo (1966–1969) and Universidade de São Paulo, São Paulo (1969–1984). [Issued as individual family fascicles under authorship of contributor without addition of editor's name.]
Richards, O. W. and **Davies, R. G.** 1977. *Imms's general textbook of entomology*. Tenth edition: vol. 2, *Classification and biology*. viii + pp. 421–1354. Chapman & Hall, London, and John Wiley, New York.
Rohdendorf, B. B. 1964. Historical development of the Diptera. *Trudy Paleontologicheskogo Instituta* **100**: 1–311. [In Russian: English translation in Hocking, B., Oldroyd, H. and Ball, G. E. (eds) (1974) *The historical development of Diptera*. xv + 360 pp., University of Alberta Press, Edmonton.]
Séguy, E. 1950. La biologie des Diptères. *Encyclopédie entomologique* (A) **26**: 1–609.
Sims, R. W., Freeman, P. and **Hawksworth, D. L.** (eds) 1988. *Key works to the fauna and flora of the British Isles and north-western Europe*. xii + 312. Clarendon Press, Oxford.
Skaife, S. H. 1979. *African insect life*. New edition (revised J. Ledger). 279 pp. Country Life Books, London.
Skidmore, P. 1985. *The biology of the Muscidae of the world*. [xii] + 550 pp. W. Junk, Dordrecht.
Smith, K. G. V. 1986. *A manual of forensic entomology*. 205 pp. British Museum (Natural History), London, and Cornell University Press, Ithaca, New York.
Smith, K. G. V. 1989. An introduction to the immature stages of British flies; Diptera larvae, with notes on eggs, puparia and pupae. *Handbooks for the Identification of British Insects* **10** (14): 1–280.

Soós, A. and **Papp, L.** (eds) 1984–1991. *Catalogue of Palaearctic Diptera*: [vol. 1, Trichoceridae-Nymphomyiidae, in preparation]; vol. 2, Psychodidae-Chironomidae, 499 pp. (1991); vol. 3, Ceratopogonidae-Mycetophilidae, 448 pp. (1988); vol. 4, Sciaridae-Anisopodidae, 441 pp. (1986); vol. 5, Athericidae-Asilidae, 446 pp. (1988); vol. 6, Therevidae-Empididae, 435 pp. (1989); vol. 7, Dolichopodidae-Platypezidae, 291 pp. (1991); vol. 8, Syrphidae-Conopidae, 363 pp. (1988); vol. 9, Micropezidae-Agromyzidae, 460 pp. (1984); vol. 10, Clusiidae-Chloropidae, 402 pp. (1984); vol. 11, Scathophagidae-Hypodermatidae, 346 pp. (1986); vol. 12, Sarcophagidae-Calliphoridae, 265 pp. (1986); [vol. 13, Anthomyiidae-Tachinidae, in preparation]; [vol. 14, Index, in preparation]. Elsevier, Amsterdam.

Stone, A., Sabrosky, C. W., Wirth, W. W., Foote, R. H. and **Coulson, J. R.** (eds) 1965. *A catalog of the Diptera of America north of Mexico*. Agriculture Handbook 276, iv + 1696 pp. US Department of Agriculture, Washington. D.C.

Teskey, H. J. 1981a. Morphology and terminology – larvae. Pp. 65–88 in McAlpine et al. (1981).

Teskey, H. J. 1981b. Key to families – larvae. Pp. 125–147 in McAlpine et al. (1981).

Teskey, H. J. 1984. Aquatic Diptera: Part One. Larvae of aquatic Diptera. Pp. 448–466, in Merritt, R. W. and Cummins, K. W. (eds), *An introduction to the aquatic insects of North America*. Second edition. xiii + 722 pp. Kendall/Hunt, Dubuque, Iowa.

Wood, D. M. and **Borkent, A.** 1989. Phylogeny and classification of the Nematocera. Pp. 1333–1370 in McAlpine (1989a).

Woodley, N. E. 1989. Phylogeny and classification of the 'Orthorrhaphous' Brachycera. Pp. 1371–1395 in McAlpine (1989a).

Wootton, R. J. 1981. Palaeozoic insects. *Annual Review of Entomology* **26**: 319–344.

Zumpt, F. 1965. *Myiasis in man and animals in the Old World: a textbook for physicians, veterinarians and zoologists*. xv + 267 pp. Butterworths, London.

Zumpt, F. 1973. *The stomoxyine biting flies of the world. Diptera: Muscidae. Taxonomy, biology, economic importance and control measures*. viii + 175 pp. Gustav Fischer, Stuttgart.

CHAPTER FOUR

Sandflies (Phlebotominae)

R. P. Lane

Phlebotomine sandflies are delicate, hairy flies with long slender legs. Of the 700 or so species, only about 70 species are thought to be involved in the transmission of disease to man. The flies are easily distinguished from other small Diptera when alive by the characteristic manner in which they hold their pointed wings above their body (like a vertical V), especially from other members of the family Psychodidae to which they belong.

It is important to distinguish phlebotomine sandflies from other small biting flies known colloquially as 'sandflies' in certain parts of the world, especially midges of the genus *Culicoides* which abound in coastal areas of the southeastern United States, Central America and the Caribbean, and Simuliidae in Australasia. These other flies have very different biologies and medical importance from phlebotomines.

Only female sandflies suck blood, and two genera contain anthropophagous species: *Phlebotomus* in the Old World and *Lutzomyia* in the New World. Rarely, some species of the genus *Sergentomyia*, which feed principally on reptiles, will bite man but there is no evidence to suggest they are ever capable of transmitting human parasites.

Phlebotomines are mostly known as vectors of *Leishmania*, protozoan parasites which cause visceral leishmaniasis (kala-azar) and various forms of cutaneous leishmaniasis (oriental sore, espundia etc.) in man. However, they also transmit bartonellosis (Oroya fever, Carrion's disease) in northwestern South America caused by the bacterium *Bartonella bacilliformis*, and sandfly fever virus throughout North Africa and the Middle East.

Sandflies are found mainly in the tropics and subtropics, with a few species penetrating into temperate regions in both the northern (to 50°N) and southern hemispheres (to about 40°S). There are no sandflies in New Zealand or on Pacific islands. In the Old World, man-biting sandflies (and therefore leishmaniasis) are confined to the subtropics, there being very few anthropophilic species in Africa south of the Sahara and none in South East Asia (although species of *Phlebotomus* are present). In contrast, the transmission of leishmaniasis in the New World is principally in the tropics.

Sandflies occur in a very wide range of habitats from sea level (a major focus of leishmaniasis around the Dead Sea is below sea level!) to altitudes of 2800 m or more in the Andes and Ethiopia, and from hot dry deserts, through savannas and

Medical Insects and Arachnids Edited by Richard P. Lane and Roger W. Crosskey.
Published in 1993 by Chapman & Hall ISBN 0 412 40000 6

open woodland to dense tropical rain forest. In general, each species has fairly specific ecological requirements and in a few cases these encompass the conditions in and around the dwellings of man or his domestic animals. The majority of peridomestic species are vectors of infections to man. The highly focal nature of many leishmaniases is undoubtedly a result of ecological constraints on the vectors and probably, to a lesser extent, on the mammalian reservoirs. For example, in central Asia sandflies are restricted to the natural foci provided by colonies of gerbils (murid rodents of the subfamily Gerbillinae) and by soil texture and moisture. Even in apparently homogeneous tropical rain forests, sandflies are not uniformly distributed.

In the Old World most foci of leishmaniasis, particularly of cutaneous leishmaniasis, are in dry, semi-arid areas – in contrast to the New World where the disease is mainly transmitted in forests (although visceral leishmaniasis in South America is primarily in savanna areas).

Sandflies are a difficult group to study in the field, particularly in relation to disease because the adults are small and can be hard to find; finding larvae is almost impossible. This has made the incrimination of vectors during epidemiological studies particularly difficult and the vector status of many species therefore remains uncertain. Compared to other dipterous families containing vectors (e.g. Culicidae and Simuliidae) the biology of sandflies is poorly known. This is a major constraint to understanding the epidemiology of the diseases they transmit.

RECOGNITION AND ELEMENTS OF STRUCTURE

Phlebotomine sandflies belong to the family Psychodidae which contains some of the most primitive Diptera. The family is recognized by the presence of a dense clothing of hairs on the wings, and a characteristic wing venation consisting of numerous parallel veins running to the wing margin (Fig. 4.1c). The wings are usually pointed. A full family diagnosis is in Quate and Vockeroth (1981).

Phlebotomine sandflies are distinguished from other subfamilies of the Psychodidae (Trichomyiinae, Sycoracinae, Psychodinae and Bruchomyiinae) by the absence of an eye-bridge and the presence of five-segmented palps, biting mouthparts which are at least as long as the head, antennal segments almost cylindrical, and a five-branched radial vein (i.e. R1 and four-branched radial sector, Fig. 4.5a). When alive, phlebotomines hold their wings above the body in a characteristic 'V' shape (Fig. 4.1a) in contrast to most other members of the family which hold their wings tent-like (an inverted 'V') over the abdomen (Fig. 17.2c). Species of *Nemopalpus* in the Bruchomyiinae look superficially like large phlebotomines but lack biting mouthparts. The Sycoracinae contain the only other bloodsucking species in the Psychodidae.

General accounts of sandfly morphology and anatomy are given by Abonnenc (1972), Davis (1967, some terminology not generally accepted), Forattini (1973) and Jobling (1987).

The head is elongated with the mouthparts pointing down to the substrate, and contains many taxonomically important characters, especially in the cibarium, pharynx and antennae (Fig. 4.2). The mouthparts consist of the labrum, mandibles, laciniae (part of the maxillae) and hypopharynx lying in the anterior labial gutter (Fig. 4.2c) (Jobling, 1976). The laterally placed laciniae hold the mandibles, hypo-

80 Sandflies (Phlebotominae)

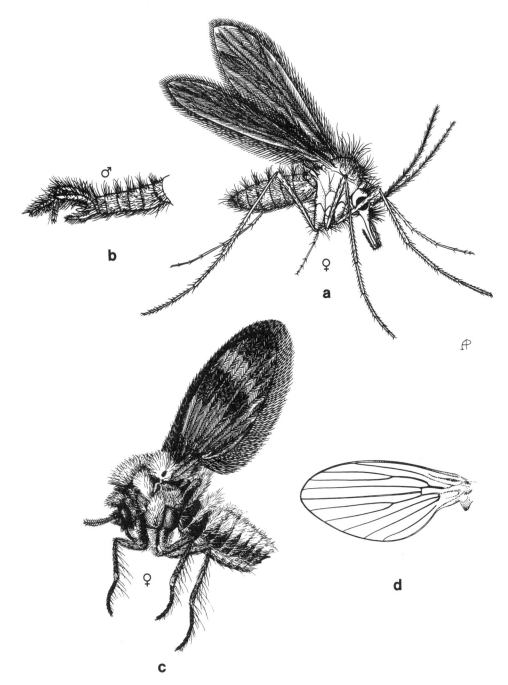

Figure 4.1 Typical appearance of adult Psychodidae: (*a*) female of the sandfly *Phlebotomus papatasi*, a vector of leishmaniasis; (*b*) terminal abdominal segments of the male of *Phlebotomus papatasi*, right lateral view; (*c*) female of a *Pericoma* species (Psychodinae) with a wing artificially raised to show the hairy appearance of the wings in Psychodidae; (*d*) wing of *Pericoma* with vestiture removed to show the venation.

pharynx and labrum together by means of a secretion produced by a small gland at the base of the laciniae (Jobling, 1976). The laciniae exert an outward force during feeding, in unison, not independently, allowing the mandibles to cut into the host's

Recognition and elements of structure   81

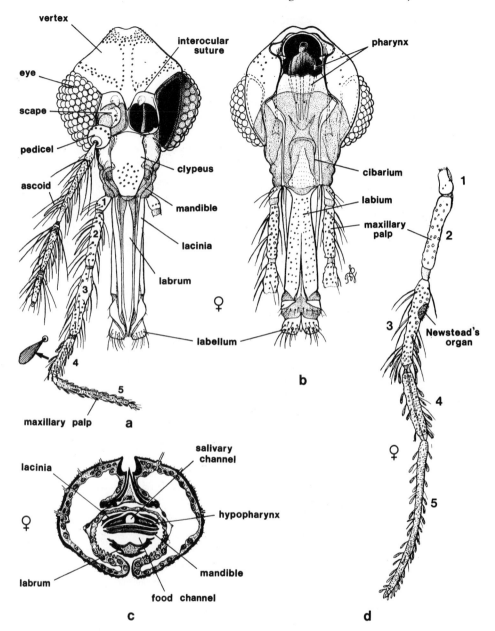

Figure 4.2 Basic morphology of the head of a female of *Phlebotomus papatasi*: (*a*) head in facial view with part of left eye removed and antenna not complete; (*b*) head in posterior view with pharynx partly exposed and both maxillary palps incomplete; (*c*) cross-section of the proboscis near its lower end; (*d*) maxillary palp. (Reproduced with relabelling from Jobling (1987) courtesy of the Trustees of the Wellcome Trust.)

skin. The mandibles are thicker than the laciniae and have saw-toothed inner margins at their apex which differ in shape between mammal-feeding and reptile-feeding species (Lewis, 1975). At their base the mandibles are curved and articulate, with a conspicuous condyle on the head capsule. Unlike the mandibles and laciniae, neither the hypopharynx nor the labrum are paired but both have tooth-like projections at their tip. The hypopharynx has salivary ducts running through it and opening at its tip. Numerous sensilla are present along the length of the labrum, the distribution of which differs between groups of species (Lewis, 1975). The male mouthparts are similar to those of the female except that mandibles are absent and the other structures are thinner and weaker.

The maxillary palps are five-segmented and taxonomically important. The relative lengths of the segments are expressed as the palp formula in which the segments are listed in ascending order of length. On the ventral surface of the third segment are bulb-shaped chemoreceptors termed Newstead's organ (Fig. 4.2d) which are applied to the surface of the host during feeding. The antennae are similar in both sexes and consist of 16 segments, namely scape, pedicel, and a flagellum of 14 segments. The relative lengths of segments are used to distinguish closely related species. Thin-walled, finger-like receptors, termed ascoids, are present on most segments and are of considerable taxonomic interest since they vary in shape, and in size relative to the segment (Fig. 4.3c-e), especially in the genus *Lutzomyia*.

The food canal or channel (Fig. 4.2c) continues from the mouthparts into the cibarium. This chamber has large muscles attached to the clypeus which contract to raise the dorsal surface of the cibarium and hence draw up blood and pass it to the next section – the pharynx. The cibarium has two series of teeth. The 'anterior' or 'vertical' teeth project down into the cavity and appear as a series of dots when viewed ventro-dorsally on slide-mounted specimens. The 'posterior' or 'horizontal' teeth project backwards towards the pharynx (Fig. 4.3a,b). The number, size and arrangement of cibarial teeth are of considerable importance in distinguishing genera, species-groups and species. The pigment patch is a patch of thickened cuticle on the dorsal wall of the cibarium. The pharynx consists of three plates enclosing a cavity, triangular in cross-section. The shape of the pharynx varies considerably between species as do the size, shape and distribution of the teeth on the posterior portion of the pharynx, i.e. the pharyngeal armature (Fig. 4.3f,g). The function of the often elaborate teeth of the cibarium and pharynx is unknown; possibly they break up the cells in the blood-meal.

The dorsal surface of the thorax is usually covered in long slender scales, giving the characteristically hairy appearance of sandflies. Species of *Lutzomyia* vary to some extent in the pigmentation of the pleural sclerites and this is sometimes an aid to the rapid sorting of sandflies in areas where detailed taxonomic studies have already been made. The distribution of scales on the post-spiracular region and the anepimeron is used to distinguish species of *Sergentomyia (Grassomyia)* and *Lutzomyia* (Fig. 4.4a). Wing venation is used extensively in sandfly taxonomy, both to distinguish genera and (by means of various ratios) to differentiate species (Fig. 4.5). In general, *Sergentomyia* has narrower and more pointed wings than *Phlebotomus* and many species of *Lutzomyia*.

The posterior margins of the abdominal tergites bear different types of setae which differentiate the Old World genera. Species of *Phlebotomus* have erect hairs articulated in large round sockets on all segments whereas most *Sergentomyia* species (except in the subgenus *Sintonius*) have recumbent hairs on tergites 2–6 arising from narrow oval sockets (Fig. 4.4b,c). In the abdomen of some species are specialized organs with connections to the exterior. In male *Lutzomyia longipalpis* elaborate secretory cells produce sex pheromones which are released via small

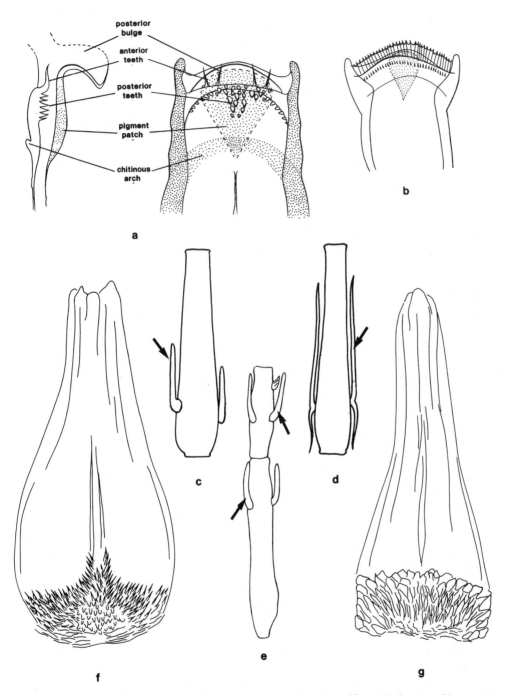

Figure 4.3 Features of the cibarium and antennae in some female sandflies: (*a*) cibarium of *Lutzomyia panamensis* in longitudinal section (left) and ventral aspect (right); (*b*) cibarium of *Sergentomyia* (*Grassomyia*) species; (*c*) third antennal segment of a typical sandfly with simple antennal ascoids (arrowed); (*d*) third antennal segment of a typical sandfly with branched antennal ascoids (arrowed); (*e*) ascoids (arrowed) of third and fourth antennal segments of *Phlebotomus (Paraphlebotomus) alexandri*; (*f*) pharynx of the female of *Phlebotomus (Adlerius) arabicus*; (*g*) pharynx of the female of *Phlebotomus (Paraphlebotomus) alexandri*.

84 *Sandflies (Phlebotominae)*

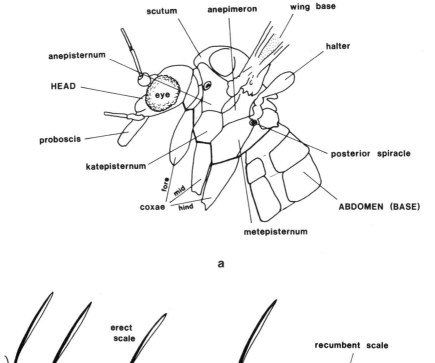

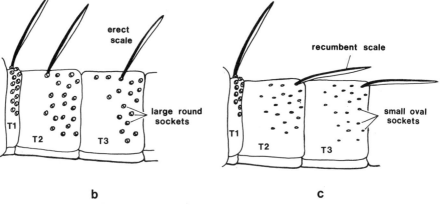

Figure 4.4 Some thoracic and abdominal characteristics of adult Phlebotominae: (*a*) thorax of a typical sandfly in left lateral view showing the principal features; (*b,c*) second and third abdominal segments of female *Phlebotomus* (*b*) and *Sergentomyia* (*c*) showing the difference between the genera in tergite scale posture and socket size and shape (one scale only shown for each tergite).

sclerotized ducts (Lane and Bernardes, 1990). Other species have structures which are presumed to be sensory, such as the trumpet organs in *Chinius julianensis* and the bell-shaped organs of species of the African *Sergentomyia* (*Spelaeomyia*); curiously, both of these taxa contain cavernicolous species. In some species of *Sergentomyia* (e.g. *S. schwetzi*), there is intraspecific variation in the size of the abdominal tergites which might also contain secretory glands.

In females, the internal genital structure consists of a sclerotized furca (= genital fork) arising from the genital orifice and supporting paired spermathecae (Fig. 4.6*a,b*). The spermathecae and their ducts show enormous diversity of shape (Fig. 4.6*b–h*). The base of the common spermathecal ducts has recently been used to separate species of *Phlebotomus* (*Larroussius*), which were hitherto very difficult to distinguish (Léger et al., 1983). At the top of the spermatheca are small hair-like tubules connected to nutritive cells.

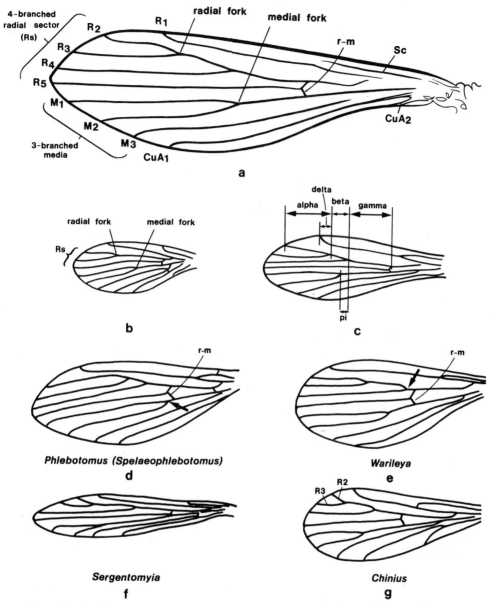

Figure 4.5 Wings of Psychodidae to show the branching patterns of the veins (vestiture omitted): (*a*) typical *Phlebotomus*, showing the terminology of the venation; (*b*) in subfamily Trichomyiinae (Sycoracinae similar); (*c*) vein configuration ratios used in phlebotomine taxonomy; (*d*) venation of the *Phlebotomus* subgenus *Spelaeophlebotomus*, showing the position of the M2 and M3 fork (arrowed) relative to r–m; (*e*) venation of *Warileya*, showing the position (arrowed) of the first fork of the radial sector (Rs) relative to r–m; (*f*) narrow and pointed wing characteristic of *Sergentomyia*; (*g*) venation of *Chinius*, showing short vein R2 and its strong angulation relative to R3.

The external genitalia of males are conspicuous and consist of the claspers (style and coxite), parameres, sclerotized aedeagus and surstyles (Fig. 4.7). There is some debate over the homology of these structures (Davis, 1967), e.g. the structure referred to as the 'aedeagus' in sandflies is not an intromittent organ, as it is actually the long genital filaments lying between the two halves of the 'aedeagus' which

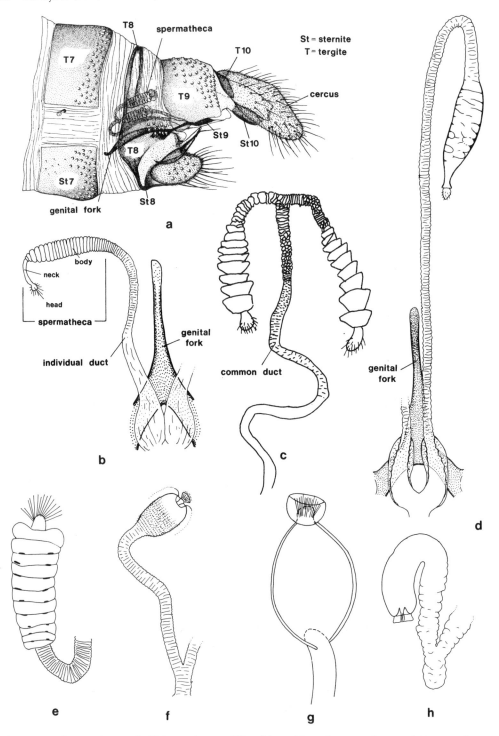

Figure 4.6 Spermathecae of phlebotomine sandflies: (*a*) position of spermathecae relative to other features of the terminal segments of the female abdomen, left lateral view; (*b*) *Phlebotomus (Larroussius) kandelakii*; (*c*) *Lutzomyia (Psychodopygus)* sp.; (*d*) *Phlebotomus (Adlerius) chinensis*; (*e*) *Phlebotomus (Phlebotomus) papatasi*; (*f*) *Sergentomyia (Grassomyia) squamipleuris*; (*g*) *Sergentomyia (Parrotomyia) palestinensis*; (*h*) *Sergentomyia (Sergentomyia) punjabensis*. (Part (*a*) reproduced with relabelling from Jobling (1987) courtesy of the Trustees of the Wellcome Trust.)

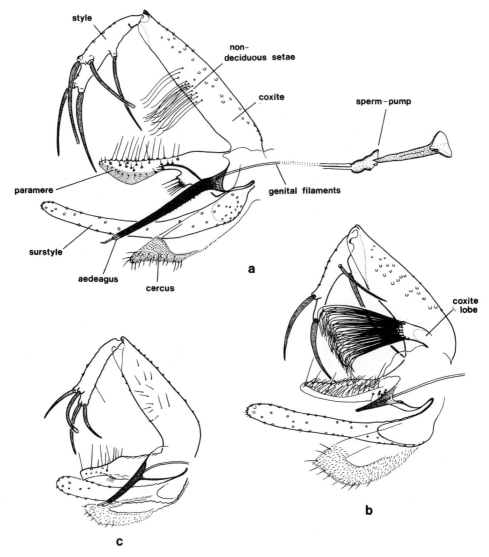

Figure 4.7 Male genitalia of phlebotomine sandflies, right lateral view: (a) *Phlebotomus (Larroussius) kandelakii*; (b) *Phlebotomus (Paraphlebotomus) nuri*, showing the large coxite lobe; (c) *Sergentomyia babu*.

penetrate the female (see Rioux and Golvan, 1969, for a diagram of flies in copula). The style (= dististyle) bears strong spines, varying in their number, length and position. The coxite (= basistyle) has numerous sensory hairs; these are sometimes grouped in patches ('non-deciduous setae') (Fig. 4.7a), or on small ventral button-like discs or on fleshy lobes (Fig. 4.7b). Internally, the genital filaments are connected to the sperm-pump, a remarkable structure consisting of a plunger and barrel (similar in structure and function to a hypodermic syringe) which ejects the sperm in a spermatophore.

The eggs are elongated ovoids with an elaborate chorionic sculpturing. There is considerable variation in the pattern of this sculpture but little congruence between the type of pattern and generic or subgeneric groupings (Fausto et al., 1991; Ward and Ready, 1975). The larvae of phlebotomines are very distinctive (Fig. 4.8a). They have a well-sclerotized head capsule bearing robust mandibles

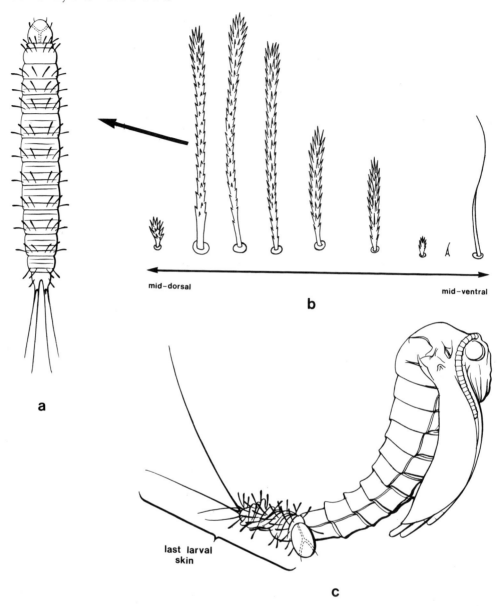

Figure 4.8 Immature stages of *Phlebotomus* sandflies: (*a*) mature (fourth-instar) larva in dorsal view, showing 'matchstick' hairs and paired caudal setae; (*b*) setae of larval metathorax as they round the body from the mid-dorsal to the mid-ventral surface; (*c*) pupa and its adhering last larval skin.

(Fig. 4.9), and the whole body bears rows of multibranched club-shaped ('matchstick') setae; in addition, two pairs of long caudal setae are present in the second to fourth instars. The first-instar larvae have only one pair of caudal setae, as curiously do later instars of *Phlebotomus tobbi*. The caudal bristles are almost as long as the body in mature larvae of some species; their function is unknown. Possibly, they are used in defence, since they are hollow and have minute pores in their surface which appear to be connected to basal secretory cells. A system for naming the larval setae of Neotropical species proposed by Ward (1976) is now being adopted with some

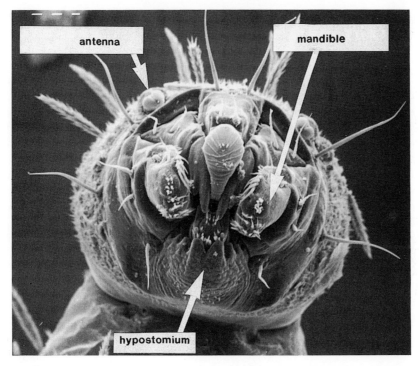

Figure 4.9 Head of mature larva of *Phlebotomus* in facial view as seen by the scanning electron microscope.

modification for the Old World fauna. The pupa is attached to the substrate by the skin of the last larval instar and this has a characteristic appearance (Fig. 4.8c).

CLASSIFICATION AND IDENTIFICATION

The higher classification of sandflies is still a matter of some controversy and there is no universally accepted system. The principal differences lie in ranking of taxa rather than their composition or affinities (see reviews in Abonnenc, 1972; Lane, 1986a; Lewis et al., 1977). Some specialists have considered sandflies to be a separate family (e.g. Abonnenc and Léger, 1976; Lewis, 1973a; Perfil'ev, 1966) mainly following Rohdendorf (1964), but most retain subfamily status within the Psychodidae (Fairchild, 1955; Lewis, 1978, 1982; Theodor, 1958). Forattini (1973) retained subfamily status for the sandflies but increased the number of genera to ten, eight of them in the New World alone and transferred the rather aberrant taxon *Hertigia* to the subfamily Bruchomyiinae. The most substantial departure from the single subfamily status of sandflies was that of Abonnenc and Léger (1976) who proposed a new 'rational' classification in which the family Phlebotomidae is divided into three subfamilies, one Old World, one Old and New World and one New World only. This has not received much support in the literature.

In an attempt to stabilize the higher classification of sandflies, Lewis et al. (1977) presented a 'unified' system based on practical criteria such that the subfamily Phlebotominae comprises five genera: *Warileya* (two subgenera); *Phlebotomus* (ten subgenera); *Sergentomyia* (seven subgenera with 54 species unplaced); *Brumptomyia* and *Lutzomyia* (26 subgeneric taxa with 19 species unplaced). This somewhat conservative approach has been generally accepted (e.g. Vianna Martins et al., 1978).

The most recently published revision of phlebotomine sandfly higher classification is that of Artemiev and Neronov (1984), who divided the subfamily into 14 genera. Within *Phlebotomus*, subgenera *Austrophlebotomus*, *Spelaeophlebotomus* and *Idiophlebotomus* (all non-man-biting), and within *Lutzomyia* the subgenus *Psychodopygus*, are elevated to genera.

Few arguments on higher classification have been explicit in the relative emphasis placed on particular characters, notable exceptions being Abonnenc and Léger (1976), Ready et al. (1980), and, to a rather limited extent, Lewis et al. (1977). Immature stages have not yet been used in higher classification. Some attempts to bring 'logic' into the argument have been ineffective because absolute ranking of taxa above the species level has little scientific basis (i.e. hypotheses are not refutable).

There is little doubt that the curious and recently described *Chinius* (Leng, 1987) from China warrants generic status.

It is likely that within the next few years there will be a substantial change in the higher classification of the phlebotomines with proposals for the introduction of tribes and the elevation of many subgenera to genera.

Genera

The relatively conservative approach to genera adopted here means that the subfamily consists of only six genera: *Phlebotomus*, *Sergentomyia* and *Chinius* in the Old World and *Warileya*, *Brumptomyia* and *Lutzomyia* in the New World. However, the large genera (i.e. *Phlebotomus*, *Sergentomyia* and *Lutzomyia*) are classified into subgenera. Often these divisions are in the form of subgenera but in the case of *Lutzomyia* there is a jumble of subgenera, species-groups and below them 'series' (of species) (Table 4.1). Unfortunately, the different systems employed by various specialists, especially in the New World, often do not differ simply in the ranking of supraspecific taxa but also in their definition, making it difficult to compare alternative systems.

Species

Many species groups are recognized within *Lutzomyia* and *Sergentomyia*, but as yet there are no species complexes in phlebotomines comparable to those known in the Culicidae and Simuliidae. *Lutzomyia longipalpis*, the principal vector of visceral leishmaniasis in South America, has been shown by cross-mating experiments to have at least two genetically distinct groups of populations which can only be differentiated by their male pheromones. Several other species (e.g. *Phlebotomus argentipes*) have populations of sufficiently different behaviour or ecology, and morphometrics, to be composed of sibling species. Further research is necessary to clarify the status of these species.

The subspecies category has been used in the past as a means of expressing in formal taxonomy the minute differences in morphology seen between samples from different localities (e.g. Theodor, 1958). Now that much more collecting has been done, particularly for the vector species such as Mediterranean *Phlebotomus*, most of these erstwhile subspecies are treated as species.

Table 4.1 Synopsis of a conservative classification of phlebotomine sandflies

The classification of the genus *Lutzomyia* is being revised: currently there is no clear distinction in ranking between its included subgenera and species-groups. Species-groups are recognized within the subgenera of *Phlebotomus* and *Sergentomyia* but are omitted from the table.

New World sandflies	Old World sandflies
Genus *Brumptomyia*	Genus *Phlebotomus*
Genus *Lutzomyia*	[*Abonnencius* (subgenus)][a]
aragaoi species-group	*Adlerius* (subgenus)
baityi species-group	*Anaphlebotomus* (subgenus)
Coromyia (subgenus)	*Austrophlebotomus* (subgenus)
Dampfomyia (subgenus)	*Euphlebotomus* (subgenus)
delpozoi species-group	*Idiophlebotomus* (subgenus)
dreisbachi species-group	*Kasaulius* (subgenus)
Evandromyia (subgenus)	*Larroussius* (subgenus)
Helcocertomyia (subgenus)	*Paraphlebotomus* (subgenus)
lanei species-group	*Phlebotomus* s.str. (subgenus)
Lutzomyia s.str. (subgenus)	*Spelaeophlebotomus* (subgenus)
migonei species-group	*Synphlebotomus* (subgenus)
Micropygomyia (subgenus)	*Transphlebotomus* (subgenus)
Nyssomyia (subgenus)	
oswaldoi species-group	Genus *Sergentomyia*
pilosa species-group	*Capensomyia* (subgenus)
Pintomyia (subgenus)	*Demeillonius* (subgenus)
Pressatia (subgenus)	*Grassomyia* (subgenus)
Psathyromyia (subgenus)	*Neophlebotomus* (subgenus)
Psychodopygus (subgenus)	*Parrotomyia* (subgenus)
rupicola species-group	*Parvidens* (subgenus)
saulensis species-group	*Sergentomyia* s.str. (subgenus)
Sciopemyia (subgenus)	*Sintonius* (subgenus)
Trichopygomyia (subgenus)	*Spelaeomyia* (subgenus)
verrucarum species-group	Unplaced species
Viannamyia (subgenus)	
Ungrouped species	Genus *Chinius*
Genus *Warileya*	
Hertigia (subgenus)	
Warileya s.str. (subgenus)	

[a] Not recognized here as a valid subgenus.

Currently, all species of phlebotomines are defined morphologically but newer analytical techniques have been applied throughout the group to distinguish problematical species (Lane, 1986a: review). Multivariate morphometric techniques (multiple discriminant analysis) have met with limited success (Lane and Ready, 1985), but biochemical techniques, such as enzyme electrophoresis and gas chromatography of cuticular hydrocarbons, have been more successful at distinguishing between morphologically similar species and are becoming increasingly used in phlebotomine studies. One notable example is *Lutzomyia carrerai*: two colour morphs have been shown to be genetically distinct species on the basis of isoenzymes, and one of them (*Lu. yucumensis*) subsequently incriminated as a vector of *Leishmania braziliensis* (Caillard et al., 1986). The genetic relationships of species within the *Lu. verrucarum* group (determined by genetic distance using isoenzyme polymorphisms) are generally concordant with the results of morphological studies

(Kreutzer et al., 1990). Analysis of cuticular hydrocarbons has been used both to differentiate species with isomorphic females (e.g. Phillips et al., 1990a) and to explore geographic variation within a species (e.g. Phillips et al., 1990b).

Unfortunately phlebotomines do not have well-developed polytene chromosomes as found in other nematocerous Diptera and therefore cytology has been of little use in resolving taxonomic problems. DNA probes have been developed to identify established species as well as the parasites they harbour (e.g. Ready et al., 1988) and to distinguish females which are morphologically similar (Ready et al., 1991). There has been little use of molecular genetics to analyse 'infraspecific' variation but no doubt this will prove to be a valuable research tool in due course.

Identification

Identification of sandflies is based mainly on internal structures and therefore requires specimens to be mounted on microscope slides. The following key is to genera and subgenera, except *Lutzomyia*; a monograph on the Nearctic and Neotropical species is in press (Young and Duncan) which incorporates revised subgeneric groupings, an extensive identification key, species descriptions and a plethora of illustration.

Identification key to the subfamilies of Psychodidae

As the main aim of this key is to facilitate separation of adult Phlebotominae from other Psychodidae no features are given for distinguishing Trichomyiinae and Sycoracinae, subfamilies which differ mainly in the larval stage. Females of at least some Sycoracinae are bloodsucking (feeding on frogs) but man-biting is unknown for this rare subfamily and specimens are unlikely to come before the medical entomologist.

1   Wing with two longitudinal veins (R4 and R5) between radial and medial forks which run to wing margin, i.e. radial sector (Rs) four-branched (Fig. 4.5a).... 2
— Wing with one longitudinal vein between radial and medial forks which runs to wing margin, i.e. radial sector (Rs) three-branched (Fig. 4.5b) .... Trichomyiinae and Sycoracinae
2   Antennal segments cylindrical (Figs 4.1a, 4.2a) or barrel-shaped. Eyes widely separated from one another, though sometimes almost meeting dorsally 3
— Antennal segments either strongly swollen basally and thence tapering to cylindrical shape (nodiform) or cylindrical but swollen centrally (fusiform). Eyes separated or meeting in midline of head above antennae .... Psychodinae
3   Mouthparts as long as or longer than height of head (except in *Warileya*) (Figs 4.1a, 4.2a,b) .... Phlebotominae
— Mouthparts shorter than height of head .... Bruchomyiinae

Identification key to genera and subgenera of Phlebotominae (excluding *Lutzomyia* subgenera)

1   New World sandflies 2
— Old World sandflies 5
2   Wings with distinct point; venation with radial sector (Rs) forking well beyond position of r–m cross-vein (Fig. 4.5a). Male style shorter than coxite (Fig. 4.7a). Mouthparts more than half head height 4

— Wings broad and rounded; venation with radial sector (Rs) forking near r–m cross-vein, either before, level with or slightly beyond cross-vein position (Fig. 4.5e, arrowed). Male style longer than coxite. Mouthparts short.................. .... **Warileya** 3
3 Palp segment 5 shorter than segment 3. Antennal segment 3 longer than total palp length... **W. (Hertigia)**
— Palp segment 5 much longer than segment 3. Antennal segment 3 much shorter than total palp length .... **W. (Warileya)** s.str.
4 Interocular suture complete (Fig. 4.2a). Male style with five large spines of which two are on a basal tubercle. Female cibarium with four longitudinal rows of posterior teeth, anterior teeth absent... **Brumptomyia**
— Interocular suture incomplete. Male style with one to six major spines (if five spines present then all separate and not arising from a common tubercle). Female cibarium with one row of posterior teeth (Fig. 4.3a), anterior teeth often present.. **Lutzomyia**
5 Wing distinctly pointed; vein R2 as long as or longer than R2+3, vein R2 almost parallel to R3 (Fig. 4.5c). Abdominal tergites without posterior processes..... 6
— Wing broadly rounded; vein R2 shorter than R2+3 and at a sharp angle to R3 (Fig. 4.5g). Abdominal tergites with small posterior processes. [China, one species, occurring in caves and probably not man-biting]..................... **Chinius**
6 Abdominal tergites 2–6 with numerous erect setae which arise from large round sockets of the same size as those on tergite 1 (Fig. 4.4b). Female cibarium usually without teeth or pigment patch. Male style with three to five major spines .... **Phlebotomus** 7
— Abdominal tergites 2–6 with horizontal (recumbent) setae which arise from small oval sockets (Fig. 4.4c) (in subgenus *Sintonius* a few erect setae present on posterior margins of tergites). Female cibarium with teeth and pigment patch (except in subgenus *Parvidens*). Male style with four (Fig. 4.7c) to six major spines .... **Sergentomyia** 27
7 Wing with distance between bases of R4 and R5 not more than a quarter of wing width. Third antennal segment very long, longer than total palp length. Palp sensilla not spatulate. Sperm-pump accompanied by one pair of parallel sclerotized rods. Male styles very long. Spermathecal ducts usually short.... 8
— Wing with distance between bases R4 and R5 relatively long, more than a third of wing width. Palp sensilla spatulate. Sperm-pump without associated sclerotized rods 9
8 Wing with vein M1+2 forking at level of r–m cross-vein (Fig. 4.5d, arrowed). Female cibarium without teeth. Third antennal segment 2.3–2.5 times as long as labrum. Palp segment 3 not enlarged at base, with scattered sensilla. Male style with four long spines. [Afrotropical region, occurring in caves].................... .... **P. (Spelaeophlebotomus)**
— Wing with vein R2+3 forking beyond r–m cross-vein. Female cibarium with teeth covering large area. Third antennal segment more than three times as long as labrum. Palp segment 3 enlarged at base, with sunken patch of sensilla. Male style with three to five thick spines .... **P. (Idiophlebotomus)**
9 Male style with four or more spines. Female cibarium without teeth. Hypopharynx with teeth. Spermatheca usually segmented 10
— Male style with three spines. Female cibarium with row of five to ten teeth. Hypopharynx with few or no teeth. Spermatheca thin-walled and unsegmented. [Australasia] .... **P. (Austrophlebotomus)**

10 Males 11
— Females 18
11 Coxite with a basal process, sometimes very small and button-like, bearing several strong setae. Genital filaments 1.3–2.3 times as long as sperm-pump.... 12
— Coxite without a basal process bearing strong setae. Genital filaments more than three times as long as sperm-pump 14
12 Basal process of coxite flat and button-like and bearing only a few strong setae... 13
— Basal process of coxite in form of a fleshy lobe bearing a tuft of setae (Fig. 4.7b)... .... **P. (Paraphlebotomus)**
13 Style long, cylindrical, with five short spines (of which three in apical position); parameres with two dorsal processes .... **P. (Phlebotomus) s.str.**
— Style short, with long slender spines (of which two in apical position); parameres without dorsal processes (simple) .... **P. (Synphlebotomus)**
14 Style with five long spines 15
— Style with four spines (except five in one species with cone-shaped aedeagus bearing semicircular appendages). [Paramere either simple or with one or two lobes; aedeagus with one or two basal appendages] .... **P. (Anaphlebotomus)**
15 Paramere simple, without lobes 16
— Paramere with one or two lobes .... **P. (Euphlebotomus)**
16 Paramere not conical, slender or finger-shaped, without lateral spine 17
— Paramere conical and with lateral spine. [India, one species with unusually narrow wings and long legs] .... **P. (Kasaulius)**
17 Genital filaments more than six times as long as sperm-pump; aedeagus with small triangular subterminal tubercle .... **P. (Adlerius)**
— Genital filaments three to five times as long as sperm-pump; aedeagus without subapical tubercle .... **P. (Larroussius)** and **P. (Transphlebotomus)**
18 Individual spermathecal ducts separate to base, no common duct 19
— Individual spermathecal ducts joined to form common duct 23
19 Spermathecae clearly segmented (Fig. 4.6b,e) 20
— Spermathecae indistinctly segmented, outline smooth (Fig. 4.6d)... **P. (Adlerius)**
20 Spermatheca with head embedded in body, without neck 21
— Spermatheca head on long neck. [Pharyngeal armature consisting of series of semicircular punctuated lines] .... **P. (Larroussius)** (part)
21 Spermatheca with apical segment not much larger than preceding segments. Pharyngeal armature with scales or series of ridges 22
— Spermatheca with apical segment much larger than preceding segments. Pharyngeal armature consisting of a series of large and distinct scales (Fig. 4.3g) .... **P. (Paraphlebotomus)**
22 Spermathecal head broad, separated from last segment by high ridge. [Spermathecal body consisting of segments of equal size (Fig. 4.6e)] .... **P. (Phlebotomus) s.str.**
— Spermathecal head narrow, not separated from last segment by high ridge .... **P. (Synphlebotomus)**
23 Individual spermathecal ducts narrow; spermathecae distinctly or indistinctly segmented but always with well-defined body. Pharyngeal teeth in series of concentric lines or scales with minute backwardly-facing teeth 24

— Individual spermathecal ducts broad, joining at genital opening; spermathecae in form of simple striated tubes, not clearly demarcated into segments, head small. Pharyngeal teeth pointing forwards.................. **P. (Transphlebotomus)**

24 Spermatheca with neck shorter than spermathecal head, either clearly or indistinctly segmented 25

— Spermatheca with neck several times as long as spermathecal head, clearly segmented (Fig. 4.6b) .... **P. (Larroussius)** (part)

25 Spermatheca with apical segment considerably longer than other segments but of more or less equal width; segmentation either regular or indistinct 26

— Spermatheca with apical segment rarely longer but considerably narrower than other segments; spermathecal head large, often with a long neck (spermatheca forming a long slender tube in one species)..
....**P. (Anaphlebotomus)**

26 Spermathecae long, slender and worm-like. Pharyngeal armature of long teeth. Antennal flagellar segments and pedicel very long. [India, one species only (*P. newsteadi*)].. **P. (Kasaulius)**

— Spermathecae clearly segmented. Pharyngeal armature consisting of semicircular concentric lines.. **P. (Euphlebotomus)**

27 Female cerci not unusually long (Fig. 4.6a). Male style with at most four spines
.... 28

— Female cerci unusually long. Male style with six spines. [Afrotropical region]......
.... **S. (Demeillonius)**

28 Anepisternum (Fig. 4.4a) with two groups of scale-like setae (one behind anterior spiracle, other on lower front corner).. 29

— Anepisternum usually bare, at most with one small group of scale-like setae......
.... 30

29 Spermatheca in form of round sclerotized capsule with small spicules (Fig. 4.6f). Antennal segment 3 without ascoids. Female cibarium with teeth in distinct convex row (Fig. 4.3b). Male style with four spines; paramere not bifid (simple) .... **S. (Grassomyia)**

— Spermatheca in form of a simple thin-walled tube. Antennal segment 3 with two ascoids. Female cibarium without teeth. Male style with five spines; paramere bifid. [Afrotropical region, three species] .... **S. (Parvidens)**

30 Spermathecal duct without a large bulge at junction with spermatheca. Cibarial teeth pointed or in a series of ridges. Legs normal. Antennal segment 5 without papilla; antennal segments 3–15 of male each with one ascoid. Male style with four spines 31

— Spermathecal duct with large bulge at junction with spermatheca, latter a large sac. Cibarial teeth long and pointed. Legs unusually long. Antennal segment 5 with papilla; antennal segments 3–15 of male each with two ascoids. Male style with two spines. [Afrotropical region, occurring in caves].........................
.... **S. (Spelaeomyia)**

31 Abdomen with segments 2–6 each bearing a few erect hairs, these arising from large round sockets on posterior margin (better developed in male than female). Spermatheca distinctly segmented. Male aedeagus slender, sharply pointed... 32

— Abdomen with segments 2–6 each bearing horizontal (recumbent) hairs, these arising from small oval sockets on posterior margin. Spermatheca not clearly segmented, sometimes with vague transverse striations 33

32　Spermathecal segments symmetrical, almost equal in size. Aedeagus short, narrowing uniformly towards tip .... **S. (Sintonius)**
— Spermathecal segments asymmetrical, convoluted, apical segment larger. Aedeagus elongate and cone-shaped, ventral surface of apical quarter curved upwards to its rounded and translucent extremity. [Southern Africa]............... .... **S. (Capensomyia)**
33　Spermatheca usually in form of a capsule (Fig. 4.6g), never tubular. Wing with vein R1 longer than R2+3. Antennal segment 3 longer than segments 4 and 5 together, usually longer than labrum.. 34
— Spermatheca tubular, with smooth walls of uniform width along their length (Fig. 4.6h). Wing with vein R1 usually 0.3–0.8 as long as R2+3. Antennal segment 3 shorter than segments 4 and 5 together, usually shorter than labrum. [Aedeagus stout; style with four spines (two terminal and two subterminal)] (Fig. 4.7c) .... **S. (Sergentomyia)** s.str.
34　Female cibarium with teeth in concave row, or if in a comb-like row then teeth not narrow. Pharynx not narrowing abruptly. Spermatheca not capsular. Aedeagus not sharply pointed; paramere not hooked...................................... 35
— Female cibarium with comb-like row of strong parallel teeth. Pharynx with distinct narrowing posteriorly (lamp-glass shape). Spermatheca in form of a smooth spherical or ellipsoid capsule (Fig. 4.6g). Aedeagus slender, triangular and narrowing gradually to a sharp point; paramere hooked. [Style either with all spines terminal or two spines terminal and two subterminal] .... **S. (Parrotomyia)**
35　Female cibarium with teeth usually parallel, often equal in size and not very narrow. Pharynx slender along its length, with teeth or scales (teeth absent in some species). Antennal segment 3 long (longer than segments 4 and 5 together) and often one to two times as long as labrum. Wing with vein R2 usually longer than R2+3. Spermatheca in form of a thin-walled capsule, usually with transverse striations. Aedeagus slender and with blunt tip; paramere hooked; style with two terminal spines and two subterminal or medial spines..
.... **S. (Neophlebotomus)** [*Rondanomyia* in Lewis (1978), synonym]
— Without such features present simultaneously .... unplaced species of **Sergentomyia**

Principal anthropophagic genera and subgenera

Phlebotomus

This genus contains almost all man-biting sandflies, and the only vectors of pathogens to man, in the Old World. It is relatively homogeneous morphologically. Although males can be identified to subgenus, this is more difficult for females, which unfortunately is the stage most encountered in epidemiological studies. In many subgenera it is almost impossible to identify females unambiguously to species, although they can often be identified to species pairs.

Phlebotomus (Adlerius)　Males of this subgenus have very long genital ducts, several times the length of the sperm-pump, simple parameres and a patch of non-deciduous setae on the inner surface of each coxite. Females have a characteristic, incompletely segmented, spermatheca with a tapering neck (Fig. 4.6d) and minute head and pharyngeal armature of spiculate teeth (Fig. 4.3f). Females of the different species are almost impossible to differentiate. Most of the 17 species are found in rocky habitats from the eastern Mediterranean to the Himalayas, and several are suspected vectors of Leishmania infantum.

Phlebotomus (Euphlebotomus) The male is recognized by its characteristic three-lobed parameres; the female is recognized by the long-necked shape of the many-segmented spermatheca, the pharyngeal armature, small size and dark appearance. Only one species of the subgenus, *P. argentipes*, is a major vector of visceral leishmaniasis in the villages and towns of the alluvial plains of northeastern India, Bangladesh and the Nepalese Terai.

Phlebotomus (Larroussius) In this subgenus the pharyngeal teeth of the females are punctiform and the spermatheca possesses rounded segments and a long finger-like neck and head (Fig. 4.6b). Males have simple parameres and relatively short spermathecal ducts (three to five times as long as the sperm-pump) (Fig. 4.7a). Species are difficult to separate from one another as females but the base of the common spermathecal duct provides useful characters. Several species are proven or suspected vectors of parasites causing visceral leishmaniasis in the Mediterranean basin and mountainous areas from the Middle East to northern Pakistan. In the lowland *Acacia* forests of southern Sudan, *P. orientalis* transmits a form of *Le. donovani* responsible for epidemic visceral leishmaniasis, and in the highlands of East Africa, two species transmit *Le. aethiopica* causing cutaneous leishmaniasis.

Phlebotomus (Paraphlebotomus) The pharynx of the females in this subgenus has large scale-like teeth with smooth margins which appear like a network (Fig. 4.3g). The spermathecae usually have the terminal segment much larger than the preceding segments. The males have characteristic fleshy lobes on the inner surface of the coxite bearing tufts of long setae and the styles have only four spines (Fig. 4.7b). Many species are associated with rodent burrows, different species seemingly adapted to different soils, and transmit *Le. major* to rodents. One species, *P. sergenti*, is frequently peridomestic and transmits anthroponotic cutaneous leishmaniasis caused by *Le. tropica*, and another, *P. alexandri*, is a vector of *Leishmania* causing visceral disease in western China (Xinjiang = Sinkiang).

Phlebotomus (Phlebotomus) Females of this subgenus have a pharynx armed with large teeth fringed with minute denticles (microtrichia) and spermathecae with similarly shaped segments (Fig. 4.6e). The males have short spines on the long cylindrical style, parameres with long appendages and a small button at the base of the coxite bearing a few long setae. The most widespread species, *P. papatasi*, is peridomestic and bites man avidly. All species are either proven or suspected vectors of *Le. major* and are found in arid areas along the margins of the Sahara desert (Sahel and Mediterranean littoral) through the Middle East to northern India.

Phlebotomus (Synphlebotomus) In this Afrotropical subgenus the males have a lobe on the coxite, and five spines on the short bulbous style. Females are not easily distinguished from other *Phlebotomus*. Three species are found in termite hills in the scrubland of East Africa (Kenya, southern Ethiopia and Somalia) where at least two of them transmit visceral leishmaniasis to man. Until recently, females of these species were considered indistinguishable. In southern Africa, the geographically variable *P. rossi* transmits an undescribed species of *Leishmania* in Namibia, causing cutaneous disease in man.

Sergentomyia
This is a huge genus distributed throughout the Old World; the fauna is particularly rich in tropical areas where *Phlebotomus* is scarce – Africa south of the Sahara, South East Asia and the islands of the western Pacific rim (Philippines etc.). It is not an easy genus to define and some subgenera such as *Sintonius* show intermediate characteristics with *Phlebotomus*; broadly, it shows greatest affinity to the New World *Lutzomyia*. A few species bite man; in East Africa *Sergentomyia garnhami* can have a biting rate as high as local species of *Phlebotomus* but there is no evidence that it transmits pathogens to man. Some species of the subgenus *Sintonius* also bite man. However, most species of the genus feed on reptiles. These include vectors of *Sauroleishmania* – protozoan parasites formerly placed in *Leishmania* but now considered to be trypanosomes.

Lutzomyia
This is a genus of some 350 species distributed throughout North, Central and South America. There is greater morphological variation between species than in either of the Old World genera. The subgeneric groupings are still not satisfactory. Most human-biting species are confined to a few subgenera, the remaining species being scattered throughout several species-groups and subgenera.

Lutzomyia (Lutzomyia) In females of this subgenus the spermatheca is cylindrical and clearly segmented, and in males the coxite bears a basal tuft and the paramere has strong curved setae. *Lutzomyia longipalpis* is the most widely distributed of the New World species and is the vector of visceral leishmaniasis. It is found in dry, rocky or scrubby habitats.

Lutzomyia (Helcocertomyia) Females of this subgenus have a long fifth palp segment, the spermatheca (of vector species) has a spherical terminal segment, the remaining segments are small and form a cylindrical body, with long individual ducts. Males have five spines on the style, strong setae on the coxite and a simple paramere. *Lutzomyia peruensis* and other species of the *peruensis* series are vectors of *Le. peruviana* in the Andes, and *Lu. hartmanni* is a suspected vector in lower altitude forests of Colombia and Panama.

Lutzomyia (Nyssomyia) The females of this subgenus have a large terminal knob on the spermatheca, and palpal segment 5 as long as segment 3. The male genitalia are simple. Most species are found in tropical forests; some, such as *Lu. olmeca* and *Lu. flaviscutellata* in wet forests, others (e.g. *Lu. umbratilis*) in dry, so-called 'terra firme' forest. Only one anthropophilic species, *Lu. intermedia*, is truly peridomestic. Several species are vectors of *Leishmania* to man, causing cutaneous disease.

Lutzomyia (Psychodopygus) In females of this subgenus the spermatheca is markedly imbricated (overlapping segments laid like tiles on a roof) and has heavily ridged ducts (Fig. 4.6c), and palp segment 5 is very short. In most species some of the anterior teeth in the cibarium are enlarged. Most species are found in wet forests and several are important vectors of *Leishmania* to man in both northern and central South America.

Faunal and taxonomic literature

General: Seccombe et al. (1993) give a checklist of the Old World sandflies and Vianna Martins et al. (1978) of the New World fauna, although the latter is now becoming out of date. There are catalogues or checklists to all the faunal regions, but some are old and in need of updating. The literature on sandflies (distribution, biology etc.) and their relation to leishmaniasis transmission in the Old World has been reviewed in a series of very useful publications by Zahar (1979, 1980a,b, 1981a,b).

There is a network of taxonomists preparing a taxonomic database for all the 386 currently recognized species in the New World (Computer Identification of Phlebotomines of the Americas group, CIPA). The CIPA database includes information on 90 morphological characters for each species, distribution and all bibliographic references, and should be completed by 1993. Much of the CIPA nomenclature and basic taxonomy follows the new monograph on Latin American sandflies being produced by Young and Duncan.

The genus *Phlebotomus* has been revised by Lewis (1982) and Artemiev and Neronov (1984).

Palaearctic: Catalogue (Wagner, 1991); regional work (Theodor, 1958, keys); *Phlebotomus* (*Larroussius*) species (Léger et al., 1983, Mediterranean basin); *Phlebotomus* (*Adlerius*) species (Artemiev, 1980); literature review (Zahar, 1979). Afghanistan (Artemiev, 1978, *Phlebotomus* key, illustrations; 1983, distribution); Algeria (Dedet et al., 1985, illustrations); Arabia (Killick-Kendrick and Peters, 1991, annotated bibliography); China (Leng, 1987; 1988, checklist, references); Egypt (Lane, 1986b, key, illustrations); France (Rioux and Golvan, 1969, illustrations, distribution); Greece (Léger et al., 1986a,b); Iran (Theodor and Mesghali, 1964, key, descriptions; Javadian and Mesghali, 1975, checklist); Iraq (Abul-Hab and Ahmed, 1984); Jordan (Lane et al., 1988, key); Morocco (Bailly-Choumara et al., 1972, illustrations); Pakistan (Lewis, 1967); Saudi Arabia (Lewis and Büttiker, 1982, key, distribution; 1983, habitats); Spain (Gil Collado, 1977); Tunisia (Croset et al., 1978, illustrations); Turkey (Houin et al., 1971); USSR (Perfil'ev, 1966 [1968], keys, illustrations).

Afrotropical: Catalogue (Duckhouse and Lewis, 1980); regional work (Abonnenc, 1972, keys, descriptions, biology; Abonnenc and Minter, 1965, bilingual keys). Central African Republic (Grepin, 1983); Ethiopia (Ashford, 1974); southern Africa (Davidson, 1981, *Phlebotomus* (*Anaphlebotomus*); 1983, *Sergentomyia* (*Capensomyia*); 1990, *Sergentomyia* (*Sergentomyia*); Sudan (Quate, 1964, key, well illustrated); Yemen (Lewis, 1974).

Oriental: Catalogue (Lewis, 1973b); regional fauna (Lewis, 1978, keys, brief descriptions, references).

Australasian: Catalogue (Duckhouse and Lewis, 1989); *Phlebotomus* (*Austrophlebotomus*) (Lewis and Dyce, 1982); Australia (Lewis and Dyce, 1988, *Sergentomyia*); Papua New Guinea (Quate and Quate, 1967, monograph).

Nearctic: Monograph (Young and Perkins, 1984, key, well-illustrated descriptions).

Neotropical: A new monograph is being prepared by Young and Duncan (in preparation) and is expected to be published in 1993/4. Regional fauna (Forattini, 1973, keys, well illustrated, classification outdated; Theodor, 1965, higher classification; Vianna Martins et al., 1978, checklist, distribution). Brazil (Ryan, 1986, Pará only); Colombia (Young, 1979, key, well illustrated); Costa Rica (Murillo and Zeledón, 1985); Ecuador (Young and Rogers, 1984); French Guiana (Lebbe et al., 1987, illustrations and database); Peru (Young et al., 1985, references); Venezuela (Feliciangeli et al., 1988, key).

BIOLOGY

Life history

Relatively little is known about the life history of sandflies and their terrestrial breeding sites, in striking contrast to other nematocerous flies such as Culicidae, Simuliidae, Ceratopogonidae and Chironomidae. Breeding and adult resting often takes place in the same microhabitat as far as is known, such as the soil accumulated in cracks in walls or rock, in animal burrows and shelters, caves, or in damp leaf litter in forests. The main requirements for breeding sites are moisture and the presence of organic detritus etc. on which the larvae can feed. Almost all available information on immature stages has been gained from laboratory-rearing of eggs obtained from wild caught females.

Between 30 and 70 eggs are scattered about the potential breeding site by the ovipositing female and hatching occurs one to two weeks later. In the laboratory the eggs are often forced into small holes or scratches in the rearing substrate and a putative oviposition pheromone has been demonstrated in *P. papatasi* (Elnaiem and Ward, 1991). Breeding sites are always terrestrial, and relatively cool and damp, and include the forest floor in tropical forests (Hanson, 1961), rodent burrows in deserts, soil in animal pens and dens, caves, and (presumably) in termite hills and cracks under rocks. Although it is necessary for the breeding site to be damp the larvae will migrate if it becomes too wet; *P. argentipes* larvae, for instance, move up the mudbrick walls of houses in northern India with the seasonal floods and down again as the waters recede. There are four larval stages. The food consists of organic detritus and perhaps also micro-organisms. Larvae of some species burrow through the substrate while others graze on the surface. Diapause occurs during the fourth instar of several species inhabiting areas with cool winters (e.g. with *Phlebotomus ariasi* in the Mediterranean basin and *P. longipes* at high altitude in East Africa). The duration of the larval instars varies greatly, both between and within species (in the laboratory at least), and is regulated mainly by temperature. The pupa is inactive and usually hatches within five to ten days. The period from oviposition to adult eclosion is 20–40 days, but up to several months in diapausing species. Adults emerge from the pupa during the hours of darkness, often just before dawn. In males, the terminalia rotate through 180° during the 24 hours immediately after emergence.

When they are not active, adults seek out cool and relatively humid, but not wet, dark niches. Resting sites such as caves, tree holes, tree trunks, cracks in rocks and cavities between boulders, fissures in the ground, buildings, termite hills and animal burrows are commonly used. In tropical forests the space between the buttresses which support huge, shallow-rooted trees are a particularly important resting site of many vectors (Christensen and de Vasquez, 1982). By withdrawing to daytime sites, sandflies are able to survive in very hot and dry climates, conditions which would otherwise rapidly kill them. Adults become active at night when the ambient temperature drops and humidity rises. Resting sites determine the ecological distribution of different species and the study of these is therefore very important in the study of sandflies. Various traps are used in fieldwork to intercept sandflies moving to and from resting sites.

Several abiotic factors affect sandfly numbers, but temperature and rainfall are the most important. Most species in temperate regions have only one generation per year, and consequently a single peak of activity and transmission, but the same species can have two or three generations per year in climatically more favourable areas. In some tropical environments, sandflies react more to rainfall than temperature cycles and consequently are more prevalent in either the wet or dry season, depending on species. Several anthropophilic species can be present in any one area, each with its own annual cycle of activity and potential for parasite transmission.

Mating

In the very few species so far studied in the field, mating is associated with a host rather than the aerial swarms so typical of other nematocerous Diptera. Groups of males accumulate either on the host or nearby, awaiting the arrival of females seeking a blood-meal. These 'swarms' form a mating lek, e.g. in *Phlebotomus argentipes* (Lane et al., 1990), a species in which individuals are spaced regularly and there is considerable competition for position manifested by continuous jostling whenever a new individual arrives in the array. Species-specific pheromones are produced by males of *Lutzomyia longipalpis* and one genetically distinct form of this species produces a diterpenoid whereas the other produces a homofarnasene compound (Ward et al., 1988). The pheromones are biologically active in conjunction with host odours (Morton and Ward, 1989), attracting females to hosts on which males are assembled. The males of *Lu. longipalpis* also produce 'courtship songs' consisting of pulses of wing vibration to attract and encourage females to mate (Ward et al., 1988).

Feeding behaviour and hosts

Only females feed on blood, using the nutrients to develop eggs. Feeding takes place on exposed parts of the body by thrusting the tiny mouthparts (length 0.15–0.57 mm) into the skin and (with the minutely toothed mandibles used in a scissors-like manner) creating a small pool from which the blood is sucked. An extremely potent vasodilating peptide is injected into the wound to induce formation of an extravascular pool of blood (Ribeiro et al., 1989). Blood taken in this manner is directed into the mid-gut. Liquids taken by other means (e.g. sugar-feeding) are directed to the crop for sterilization and then to the mid-gut. Both males and females feed on sugars. These can be obtained from aphid honeydew or from plants. The sugar turanose is produced in the sandfly gut by hydrolysis of another sugar, melizitose, secreted by aphids (MacVicker et al., 1990). Plant sugars are either acquired passively by feeding from extrafloral nectaries, fallen fruits etc.

or after active piercing of leaves and stems (Schlein and Yuval, 1987). The presence of sugars is essential for the full development of *Leishmania*.

Sandflies are typically crepuscular or nocturnal, biting at different times of the night according to the species but, they will bite during the day when disturbed (e.g. in dense forest, caves or buildings). Only a few species are endophilic; these are mostly peridomestic species, such as *P. papatasi*, *P. chinensis*, *P. sergenti* and *Lu. longipalpis*. It is a relatively small ecological step from animal burrows, caves or rock piles to human dwellings and the distinction between peridomestic and 'wild' species therefore has little meaning in sparsely populated areas where buildings are made of local materials and often without entire walls. However, most species prefer to bite and rest outside (exophilic) near their probable breeding and resting sites.

The biting rates of sandflies vary greatly, up to a maximum of about 1000/hour, but the attack rate of some vectors in major foci can be surprisingly low, e.g. in northeastern India the mean rate for *P. argentipes* is 0.65 bites/hour. Climatic factors affect activity markedly; biting for instance does not usually take place below 20°C in tropical species. *Phlebotomus papatasi* is active between 45% and 60% relative humidity but other species not below 75–85% RH.

Most species probably have a narrow range of preferred hosts. Knowledge is limited, however, by two main factors – the small numbers of flies caught which contain fresh blood-meals and the few antisera to species of wild mammals yet available for use in blood-meal analysis. Species of *Sergentomyia* feed predominantly on reptiles, as do some *Lutzomyia* and *Phlebotomus* species, but members of the latter genera feed mainly on warm-blooded vertebrates, especially mammals. Several peridomestic, and mammal-feeding sandflies (e.g. *P. papatasi*, *P. sergenti* and *Lu. longipalpis*) will readily feed on poultry. From what little is known of host-seeking behaviour, sandflies move up odour plumes in a similar manner to mosquitoes. Once on the host they move in a series of small hops to locate preferred feeding areas; on rodents these are the ears and feet, on dogs the nose, and on cattle the belly. There is laboratory evidence that questing sandflies are attracted to *Leishmania*-infected skin of rodents. A feeding pheromone issuing from the palps of engorging females attracts other females, and might function as a mark of a groom-free zone on the host (Schlein et al., 1984).

Most species are gonotrophically concordant, taking one blood-meal for each batch of eggs matured. However, autogeny, the ability to lay eggs without a blood-meal, is found in populations of some man-biting species (e.g. *P. papatasi*). Oviposition usually takes place 3–8 days after a blood-meal. The highest parous rates occur in populations towards the end of the 'sandfly season' (i.e. peak seasonal abundance) when sandfly infection rates are at their highest and transmission is most likely. Unfortunately, it is not easy to determine accurately how many times a female sandfly has laid eggs (in contrast to mosquitoes). The residual secretions in the accessory glands (at the base of the oviduct) have been used for age-grading but are not very reliable. At present, the established method of searching for follicular relics in the ovarioles still needs to be developed beyond simply distinguishing parous and nulliparous sandflies.

Dispersal and movement

Sandflies usually have a short hopping flight, especially when close to prospective hosts. Little is known of their long-range movements, although they can fly up to

2.2 km over a period of a few days in open habitats (Killick-Kendrick et al., 1984). They only fly at night, but in a single night can move several hundred metres in their search for a host and for subsequent resting and breeding sites (e.g. Yuval and Schlein, 1986). The flight range of some species (e.g. *P. sergenti*) is greater in humid than dry climates. Although slight air movement aids the detection of hosts along odour plumes, wind speeds of greater than 1.5 m/s inhibit flight, which ceases altogether in light winds of 4–5 m/s. In forests, sandflies such as the Amazonian *Lu. umbratilis* exhibit regular vertical movements in addition to horizontal movement patterns. The flies rest in the tree buttresses during the day, migrating to the canopy in search of a host at night (Ready et al., 1986). Furthermore, within a forest, there is often a marked vertical zonation in which different species of sandflies are active, so that some species remain close to the ground, feeding on ground-dwelling rodents (e.g. the South American vectors *Lu. flaviscutellata* and *Lu. olmeca* feeding on *Proechimys* and *Oryzymus* rodents). Thus the transmission of some parasites between the normal hosts is predominantly at ground level (e.g. *Le. amazonensis*) while others are transmitted in the canopy (e.g. *Le. panamensis*). Man, of course, as a non-arboreal animal, usually acquires infection at ground level.

MEDICAL IMPORTANCE

Biting pests

Phlebotomine sandflies can occasionally be significant biting pests, for instance in the Middle East where the peridomestic *P. papatasi* bites avidly. Persons newly exposed to the bites of this and other species often experience a severe urticarial reaction known as 'harara'. This period of sensitization can later be followed by desensitization.

Disease vectors

Sandflies transmit viruses and bacteria, but are most widely known as vectors of several *Leishmania* species to man.

Viruses
Phlebotomines transmit several viruses to man of which those causing sandfly fevers are the most important. Sandfly fever (papataci fever, three-day fever) is caused by two distinct virus serotypes (Naples and Sicilian) and results in acute febrile illness in man, lasting two to four days – and sometimes for much longer periods (Tesh, 1988). It is common during the summer months throughout the Mediterranean basin, the Middle East, Pakistan and parts of India and Central America. In Mediterranean areas where the disease is endemic most of the population is thought to be infected during childhood, suffering at that time only a mild illness. The disease has been considered of military importance because up to 75% of non-immune adults arriving in an endemic area can be affected.

Phlebotomus papatasi is the vector in Egypt and is generally thought to transmit sandfly fever throughout the Old World range of the disease. Natural infection rates of sandflies are between 0.015% and 0.5%. No natural vertebrate reservoir is known but infected humans can infect flies and thus have some amplifying effect during epidemics. It is most likely that the principal transmission mechanism is by transovarial transmission along genetically susceptible lines of the vector.

Sandflies transmit several other phleboviruses. *Phlebotomus perniciosus* transmits Toscana virus in the northern and western Mediterranean, and in the New World *Lutzomyia trapidoi* and *Lu. ylephiletor* transmit Chagres and Punta Toro viruses. Some 11 different vesiculoviruses have been isolated from sandflies and the transmission cycles established for some of these; trans-ovarial transmission maintains infection in sandflies. Six different viruses causing vesicular stomatitis have been found in phlebotomines. The disease they cause is clinically indistinguishable from foot-and-mouth disease in animals and in man has caused encephalitis. The remaining vesiculoviruses (e.g. Isfahan, Chandipura) cause mild, self-limiting fevers of three to five days' duration.

Bacteria

Infections due to *Bartonella bacilliformis* (Oroya fever, Verruga peruana) are found only in the central Andean Cordilleras of Peru, Colombia and Ecuador. The disease is endemic in valleys between 750 and 2700 m above sea level, apparently being altitudinally restricted by the ecological requirements of the vectors. Little is known of the transmission cycle but there is apparently no known animal reservoir and the vector presumably acquires the pathogen only from infected humans. The lack of a described development cycle of *B. bacilliformis* in the vector (it occurs in the gut and on the mouthparts) suggests that transmission is mechanical. In Peru, *Lu. verrucarum* is considered the vector, although the disease exists in the absence of this species. The closely related *Lu. colombiana* is thought to transmit the disease in Colombia.

Leishmaniases

The leishmaniases are diseases caused by parasites of the genus *Leishmania* (Protozoa, Trypanosomatidae) and are transmitted solely by sandflies. They have received considerable attention in the past decade and knowledge has increased dramatically. Modern summaries can be found in Hart (1989), Peters and Killick-Kendrick (1987), Walton et al. (1988) and World Health Organization (1990).

Leishmaniasis manifests itself in man in four main forms: cutaneous, mucocutaneous, diffuse cutaneous and visceral leishmaniasis. At one time the clinical manifestation in man was the principal criterion on which the disease and its epidemiology was studied. With the advent of sophisticated biochemical methods for identifying the parasites it was found that clinical criteria were not always reliable predictors of the infecting parasite and now the key to studying the disease is the accurate identification of parasites. For example, dry cutaneous lesions in the Old World were considered to be caused by *Le. tropica* but now it is clear that similar lesions are caused by *Le. infantum* (also a cause of visceral leishmaniasis) and by *Le. major*, a parasite earlier thought to cause only 'wet lesions'. Each parasite has a different transmission cycle.

In cutaneous disease an ulcer forms at the site of the infecting bite, often with a clearly raised margin in which the parasites abound. Lesions can be wet with a shiny central area of necrotic tissue and serous liquid or dry with a crusty scab. Usually parasites are confined to the lesion, but some *Leishmania* parasites such as *Le. guyanensis* will invade the lymphatic system and produce a series of skin lesions along the lymph duct. In the Old World, the disease is found mostly in arid regions:

North Africa, the Mediterranean littoral, the Middle East to northwestern India and central Asia, East Africa, and in several small foci in the Sahel and southern Africa. In the New World it occurs mainly in forests from Mexico to northern Argentina.

Several different parasites are responsible for cutaneous infections in man, principally *Le. major*, *Le. tropica*, *Le. aethiopica* in the Old World and *Le. braziliensis*, *Le. guyanensis*, *Le. panamensis* and *Le. mexicana* in the New World. Often, cases heal spontaneously after several months. A related condition, post kala-azar dermal leishmaniasis (PKDL), follows incomplete cure of visceral leishmaniasis caused by *Le. donovani* and is characterized by an abundance of parasites in nodules and skin thickenings.

Mucocutaneous leishmaniasis, also known as espundia, is a severely disfiguring disease caused by *Le. braziliensis* and *Le. panamensis* invading and eroding the cartilaginous tissues of the nose and palate. The infection begins with a small cutaneous lesion anywhere on the body which eventually heals but in about 5% of patients is followed by metastatic spread and eruption of lesions in the naso-pharynx – sometimes several years after the initial infection. Mucosal lesions never heal spontaneously and death from secondary infections is common in untreated patients.

Some species of *Leishmania*, in particular *Le. amazonensis* and *Le. aethiopica*, cause widespread cutaneous papules or nodules over the body, a condition known as diffuse cutaneous leishmaniasis. The condition does not heal spontaneously and is often difficult to treat.

In visceral leishmaniasis the parasites invade the cells of the spleen, bone-marrow and liver, and the infection is often fatal if not treated. Two parasites, *Le. donovani* and *Le. infantum*, are responsible for causing the disease. However, in South America the causative agent is usually called *Le. chagasi* but the name is considered by some specialists to be synonymous with *Le. infantum*. In the Old World, visceral disease caused by *Le. infantum* occurs sporadically from the Mediterranean area through the Middle East and Central Asia to northern China, and is usually associated with rocky areas; it affects children, as the specific name implies. In contrast, the major focus in northeast India, Bangladesh and Nepal involves young adults living on the plains. In East Africa two areas are affected, Kenya and Ethiopia, and the southern Sudan. In the New World cases are thinly dispersed throughout Central and South America, but the main focus is in northeastern Brazil.

Leishmania exists in mammals as an amastigote form (without a flagellum) which develops and reproduces intracellularly in macrophages in the skin or in the reticuloendothelial system. The parasites are able to avoid the respiratory burst by which macrophages usually kill other unicellular pathogens. There are almost 30 species of *Leishmania*, at least 21 of which infect man (Lainson and Shaw, 1987).

Leishmaniasis is usually a zoonosis, involving a mammal other than man, and in many cases man is simply an accidental 'dead end' host. There are a few examples, however, where there is no reservoir host, the parasite being found only in man (anthroponotic leishmaniasis) and sandflies. The transmission of *Le. tropica* (cutaneous disease) and *Le. donovani* in India (visceral disease) are of this type. Numerous kinds of mammals act as reservoirs, including Canidae, edentates, marsupials and rodents.

Occupational differences exist between those acquiring infections. Cutaneous disease in the Neotropical region is primarily associated with working in the forest environment, whereas in the Old World the disease affects mainly people in rural savanna and urban areas. The distribution and abundance of sandflies can alter with changing land use. Deforestation in the New World has led to a marked reduction in cutaneous leishmaniasis (after the initial increase in disease during land clearance), but whether such cleared areas will provide opportunities for the extension in the range of the visceral leishmaniasis vector *Lu. longipalpis* and subsequent transmission of visceral disease remains to be seen. Visceral leishmaniasis has become more prevalent in urban areas of Brazil in recent years. In the Old World there has been a marked increase in cutaneous leishmaniasis (caused by *Le. major*) associated with development programmes, particularly in the Middle East. However, in Egypt a small focus of visceral disease was associated with town expansion. By raising the water table, irrigation projects considerably increase the breeding of some vector species, e.g. *P. papatasi*.

It is important to understand natural transmission cycles for two reasons: first, enzootic cycles can pose an unknown threat in remote areas when these are being developed (e.g. for mining or agriculture), and, secondly, because it is possible that transmission is maintained by several species of which only one transmits the infection to man (for example, *P. mongolensis*, *P. caucasicus* and *P. andrejevi* transmit *Le. major* from rodent to rodent in the southern USSR, but only *P. papatasi* transmits the parasite to man). If vector control is considered appropriate it need only be directed at one species.

With 21 species of human-infective parasites, numerous reservoir and vector species, in a wide range of topographically different foci, the ecology and epidemiology of the leishmaniases are extremely diverse – without doubt the most diverse of all the vector-borne diseases. Ashford and Bettini (1987) review the ecology and epidemiology of leishmaniasis in the Old World, and Shaw and Lainson (1987) for the New World.

After ingestion by the sandfly vector, the amastigotes in an infected blood-meal undergo metamorphosis to the promastigote form and multiply within the sandfly gut, first within the digesting blood-meal and then either attached to the mid-gut or hind-gut wall. A further metamorphosis takes place to an infective metacyclic promastigote which is highly motile and immunologically prepared for invasion of the vertebrate host. The site in which the parasite development takes place varies between different groups of *Leishmania* and is related to the micromorphology and biochemistry of the sandfly gut. The attachment mechanisms of *Leishmania* are adapted to the surface structure of the section of the gut to which they adhere: nectomonad promastigotes with long flagella attach to microvilli of the mid-gut; paramastigotes have modified flagella to attach to cuticle lining the fore-gut (Molyneux and Killick-Kendrick, 1987). The parasites are thought to modify the levels of proteolytic enzymes (trypsin and chymotrypsin) in the sandfly gut. The relationship between parasites and the vector is further complicated; enzyme activity in the mid-gut following blood-meals from different hosts differentially affects the survival of *Leishmania*.

The location of the parasites in the gut during their development has been used in parasite classification. Basically, parasites of the subgenus *Viannia* (*Le. braziliensis*, *Le. panamensis* and *Le. guyanensis*) develop in the hind-gut before they migrate forward through the mid-gut for transmission by bite, and those of the subgenus *Leishmania* (all other mammal-infective *Leishmania*) develop entirely within the mid-gut prior to anterior migration.

Although transmission occurs through the bite of a sandfly the exact mechanism by which parasites are taken up or deposited in the skin of a new host is unclear.

Infection with parasites changes the behaviour of a sandfly; a heavily infected fly probes much more frequently than an uninfected fly in a manner analogous to a flea infected with *Yersinia*. *Leishmania* can be readily transmitted during each probe, which may only last a few seconds, and this is probably the origin of the multiple lesions seen in some patients (particularly those with *Le. major* infections). Infection does not appreciably alter the dispersal of sandflies but it reduces their longevity and fecundity.

Incrimination of a sandfly species as a vector is difficult as many criteria have to be satisfied before a species can be unambiguously incriminated. The literature is full of vagaries which have been summarized in an excellent review by Killick-Kendrick (1990). Remarkably few species fulfil all the vectorial ability criteria. These are based primarily on the discovery of natural infections in wild-caught flies and experimental transmission studies, but include evidence of contact between the sandfly and man, contact between the sandfly and the reservoir host (where known), and the life cycle of the parasite in the fly. As there is often no sharp distinction between important and minor or occasional vectors, it is impossible to draw up a definitive list of vectors, but Table 4.2 gives a synopsis of the proven and suspected vectors of *Leishmania*. Currently, the incrimination of sandflies as vectors in the field is a most laborious procedure involving the dissection of thousands of sandflies because infection rates are very low (typically less than 1%). New techniques employing DNA probes and monoclonal antibodies for the *in situ* detection of parasites in vectors look promising and should increase the pace of field research towards quantitative epidemiological studies.

Most subgenera of *Phlebotomus* contain vectors or suspected vectors. Species of *Phlebotomus* s.str. are associated with *Le. major* transmission in arid environments of East Africa, the Middle East and the USSR. The subgenus *Paraphlebotomus* contains many species which live in rodent burrows in Central Asia and transmit *Le. major* between rodents and occasionally to man. One species, the peridomestic *P. sergenti*, is a vector of *Le. tropica* in western Asia and the Middle East. Vectors of visceral leishmaniasis in the Old World belong to several subgenera: *Larroussius* (Mediterranean basin and Sahel), *Synphlebotomus* (East Africa), *Euphlebotomus* (India), *Adlerius* (Near East and northern China) and *Paraphlebotomus* (western China). The genus *Lutzomyia* is much more diverse than its Old World counterpart but vector species are confined to relatively few important subgenera: *Nyssomyia*, *Psychodopygus* (both contain cutaneous leishmaniasis vectors), and *Lutzomyia* s.str. (*Lu. longipalpis* the vector of visceral leishmaniasis in South America). The remaining vector species are distributed through many other subgenera and species-groups.

CONTROL

Disease control

The drugs available for treatment are injected arsenical compounds which are both expensive and difficult to administer routinely in a rural health clinic. Antifungal compounds such as paromomycin are currently being tested in clinical trials against both cutaneous and visceral infections. There is evidence of cell-mediated immunity to infection, and vaccines are therefore being sought using whole killed parasites and recombinant antigens aimed at producing a sub-unit vaccine. Disease control,

Table 4.2 Proven or strongly suspected vectors of *Leishmania* species infecting man (based on Killick-Kendrick, 1990)

The distribution given for each sandfly species does not imply that any species is a vector over the whole of its range.

Vectors are graded according to the quality of evidence incriminating them in transmission of the parasites to man: 1 = proven vector, i.e. parasites isolated and typed several times, man-vector and reservoir-vector contact established, parasites develop in sandfly gut, experimental transmission by bite in some cases; 2 = anthropophilic species in which only a few parasite isolations have been made and typed or in which flagellates have been seen in wild-caught flies but not typed (most such species will probably be confirmed as vectors); 3 = suspected vectors in which females are morphologically indistinguishable from closely related species or in which parasites have only been observed in the presence of a blood-meal (laboratory evidence, including experimental transmission, exists in some cases). Several species of *Leishmania* have been isolated from human infections but have not been sufficiently well characterized to be included in the table.

Sandfly	Parasite (*Leishmania* sp.)	Distribution of sandfly	Grade of vector
OLD WORLD VECTORS			
Phlebotomus (Adlerius)			
P. chinensis	*Le. infantum*	Northern and central China	2
P. longiductus	*Le. infantum*	North Africa to Central Asia	3
Phlebotomus (Euphlebotomus)			
P. argentipes	*Le. donovani*	Bangladesh, Nepal, India	1
Phlebotomus (Larroussius)			
P. ariasi	*Le. infantum*	Western Mediterranean	1
P. kandelakii	*Le. infantum*	Iran, Afghanistan, USSR	3
P. langeroni	*Le. infantum*	Egypt–Tunisia littoral	2
P. longicuspis	*Le. infantum*	North Africa	3
P. longipes	*Le. aethiopica*	Kenya, Ethiopia	1
P. neglectus	*Le. infantum*	Eastern Mediterranean	2
P. orientalis	*Le. donovani*	Sudan	2
P. pedifer	*Le. aethiopica*	Kenya	1
P. perfiliewi	*Le. infantum*	Mediterranean basin	1
P. perniciosus	*Le. infantum*	Western Mediterranean	1
P. smirnovi	*Le. infantum*	Central Asia	3
P. tobbi	*Le. infantum*	Eastern Mediterranean	3
P. transcaucasicus	*Le. infantum*	Caucasus	3
Phlebotomus (Paraphlebotomus)			
P. alexandri	*Le. donovani*	North Africa to western China	1
	Le. major		2
P. caucasicus	*Le. donovani*	Iran, southern USSR	3
P. sergenti	*Le. tropica*	Middle East	1
Phlebotomus (Phlebotomus)			
P. duboscqi	*Le. major*	Sahel of Africa	1
P. papatasi	*Le. major*	North Africa, Middle East	1
P. salehi	*Le. major*	Iran, Pakistan	2
Phlebotomus (Synphlebotomus)			
P. ansarii	*Le. major*	Iran	2
P. celiae	*Le. donovani*	Kenya, southern Ethiopia	3
P. martini	*Le. donovani*	East Africa	1
P. vansomeranae	*Le. donovani*	Kenya, Somalia	3
P. near *rossi*	*Le.* sp. Namibia	Namibia	2

Table 4.2 (contd)

NEW WORLD VECTORS			
Lutzomyia (Helcocertomyia)			
Lu. hartmanni	Le. panamensis	Panama, Colombia	3
Lu. peruensis	Le. peruviana	Northern Andes	3
Lutzomyia (Lutzomyia)			
Lu. longipalpis	Le. infantum (= Le. chagasi)	Central and South America	1
Lu. evansi	Le. infantum	Colombia	2
Lu. diabolica	Le. 'Texas'	Southern USA	3
Lutzomyia (Nyssomyia)			
Lu. anduzei	Le. guyanensis	Northern South America	1
Lu. flaviscutellata	Le. amazonensis	Northern South America	1
	Le. pifanoi		3
Lu. intermedia	Le. braziliensis	Southern Brazil	2
Lu. olmeca	Le. venezuelensis	Northern South America	3
Lu. olmeca nociva	Le. amazonensis	Amazon basin	2
Lu. olmeca olmeca	Le. mexicana	Central America	1
Lu. trapidoi	Le. panamensis	Central America	1
Lu. umbratilis	Le. guyanensis	Amazon basin	1
Lu. whitmani	Le. braziliensis	Eastern Brazil	2
	Le. guyanensis		3
Lu. ylephiletor	Le. amazonensis	Central America	3
	Le. panamensis		2
Lutzomyia (Pintomyia)			
Lu. pessoai	Le. braziliensis	Southern Brazil	3
Lutzomyia (Psychodopygus)			
Lu. amazonensis	Le. braziliensis	Northern Amazon basin	3
Lu. ayrozai	Le. braziliensis	Southeastern Brazil	3
Lu. carerrai	Le. braziliensis	Western Amazon basin	1
Lu. complexa	Le. braziliensis	Pará, Brazil	3
Lu. llanosmartinsi	Le. braziliensis		3
Lu. panamensis	Le. braziliensis	Central and northern	3
	Le. panamensis	South America	2
Lu. paraensis	Le. braziliensis	Northern South America	3
Lu. wellcomei	Le. braziliensis	Pará, Brazil	1
Lu. yucumensis	Le. braziliensis	Bolivia	3
Other species:			
Lu. aracuchensis	Le. peruviana	Peru	3
Lu. christophei	Le. 'Dominica'	Dominican Republic	3
Lu. gomezi	Le. panamensis	Central and northern South America	2
Lu. migonei	Le. braziliensis	South America	3
Lu. spinicrassa	Le. braziliensis	Colombia	3
Lu. verrucarum	Le. peruviana	Northern Andes	3
Lu. youngi	Le. garnhami	Venezuela	3

where it is justified by the public health significance of the infection, is by reservoir control, treatment of human cases in anthroponotic disease, and by vector control. Where the reservoirs are colonial murid rodents such as *Psammomys* (sand rats), *Rhombomys* (great gerbils) and *Arvicanthis* (African grass rats) the animals can be killed by rodenticides or by deep ploughing their burrows. Domestic dogs acting as reservoirs of visceral disease can be examined and the infected dogs destroyed. There is little prospect of controlling reservoirs among wild animals in forests or mountainous areas.

A publication from the World Health Organization in 1990 gives a detailed description of control strategies recommended for each 'nosogeographical' situation, i.e. each ecological habitat differing in physical structure, vector species, parasite species and reservoir. This publication also describes the steps to be taken for establishing a national control plan.

Sandfly control

In many foci of leishmaniasis, with notable exceptions, sandfly control is *ad hoc*, even in foci where sandflies are peridomestic and therefore most amenable to control. There are relatively few evaluations of intervention and much of the information available is anecdotal (Lane, 1991).

The principal methods used in sandfly control have been the application of insecticides, sometimes in conjunction with environmental management. Control of larvae is often impossible, as the breeding sites of many vectors are either unknown or inaccessible. Sandflies are able to penetrate the standard insect-screening used on houses and normal mesh bed-nets; specially fine bed-nets therefore have to be used and these are uncomfortable to sleep under in humid tropical climates. Chemical repellents (DEET, DMP) applied to clothing and bed-nets are effective but when applied to the skin they are frequently lost through perspiration. This is especially so during manual labour such as agricultural work, road building and military operations.

The main emphasis of insecticide control has been spraying with residual insecticides in the peridomestic environment, especially in houses, against the vectors of anthroponotic leishmaniasis. DDT is still one of the most common compounds in use, especially as there is only one report of insecticide resistance (*P. papatasi* in northern India). While there have been control programmes directed specifically at sandflies in several countries (Yugoslavia, Peru, USSR, China and, more recently, Saudi Arabia and Egypt), effective control has occurred in some countries as a by-product of antimalarial spraying (e.g. India, several Mediterranean foci). Leishmaniasis has increased with the cessation of such spraying. Insecticide control of adults is only feasible where peridomestic transmission occurs in discrete and well-populated communities. Where sandflies are exophilic, or bite away from human habitations, insecticide control is often not viable; there has been little success with attempts to control forest sandflies by barrier spraying in the Neotropical region, for example.

Some of the most successful control programmes against sandflies have involved integrated control of both vector and reservoir. The reservoirs have been either colonial rodents in arid regions of the Old World, or dogs in various Old and New World foci. The great gerbil (*Rhombomys opimus*) is a reservoir of *Le. major* in Central Asia, where it lives in networks of burrows in clay soils of the loess type. Attacking the *Rhombomys* by rodenticides in conjunction with deep ploughing of the burrows in soft soils has proved most effective, reducing the source of parasites and bloodmeals and the vector breeding sites.

COLLECTING, PRESERVING AND REARING MATERIAL

Because of their small size and retiring behaviour sandflies require a range of specialized collecting methods (Maroli and Fausto, 1986; Killick-Kendrick, 1987). Numerous methods are available for qualitative studies on sandflies, but as for many other biting flies, many lack accuracy and repeatability when used in quantitative studies.

The simplest methods are direct searching of daytime resting sites with a torch and aspirator, e.g. inside houses, animal burrows, tree trunks, tree-holes and buttresses, caves, wells and under rocks etc. Several types of aspirator are used, normally with filters attached to mouth aspirators or mechanical aspirators to avoid the potential risk of histoplasmosis. These methods are particularly useful for collecting blood-fed or gravid flies required for establishing colonies and can be standardized by searching for a fixed time. The tent-like damasceno trap can be placed over animal burrows or tree buttresses in dense forests and the resting sandflies forced up into the trap in which the collector is sitting by the use of smoke or agitation with a flexible stick. Spray catches in houses, where white sheets are placed on the ground and over furniture and knock-down insecticide sprayed into the room to collect resting sandflies, have been little used but offer the opportunity of standardized sampling.

Man-biting, or where appropriate man-landing, catches remain an important method of determining which species are anthropophagic, their diel rhythm, and relative and seasonal abundance. Such methods must be standardized for comparative studies, usually with two collectors exposing their arms, legs and torso, and collecting from each other.

Traps for sandflies are either interceptive or attractive. The commonest interceptive trap, indeed the commonest trap in sandfly studies, is the sticky-trap or castor-oil trap. This is simply a sheet of paper impregnated with castor-oil supported vertically on a stick like a sail, or attached to a rigid surface (e.g. glass, plastic sheeting or cardboard). The skill in the use of these traps lies in positioning them in the presumed path of sandflies moving to or from resting sites. Traps are best placed at an angle against walls or rocks, underneath large rocks or stones, and at the entrance to animal burrows, caves or crevices. In the Russian steppes, sticky papers are attached horizontally to a frame and allowed to swivel in response to the wind direction. A simple yet effective cylindrical trap for collecting flies entering and leaving small animal burrows has a central screen and replaceable sticky papers at each end (Yuval and Schlein, 1986). In humid climates the paper loses its rigidity and hence the trap's effectiveness. Furthermore, in wet forests sandflies are often too dispersed to be intercepted in sufficient numbers to make these traps worthwhile. Traps are usually placed out in the late afternoon and collected the following day, when adhering sandflies are removed with a needle and the oil removed either in a dilute detergent solution, lactophenol or alcohol.

A variety of light-traps has been used for phlebotomines, their effectiveness varying according to the species being studied and the habitat in which they are being used; light-traps are most inefficient in open desert habitats (e.g. in North Africa or the Middle East) but are the mainstay of work on vectors in the Mediterranean area and parts of South America. The most widely used trap is the miniature

CDC light trap with a tungsten bulb, not ultra-violet tubes. Chemical light sticks have been effective in some circumstances, especially when combined with sticky-traps.

Because of the importance of reservoir hosts in leishmaniasis epidemiology, considerable attention has been paid to the species of sandflies attracted to wild or domestic mammals. The Disney trap (Disney, 1966), a large tray covered in castor-oil with a caged bait animal at it centre, is widely used in Central and South America. Several other traps based on access to the bait animal via funnels or boxes lined with sticky paper or polythene sheeting are also successful. Direct collecting from tethered animals (e.g. dogs or cattle) is useful for obtaining specimens but does not yield reliable biting rates because any attractant effect of the collector cannot be easily eliminated.

Several traps combine attraction cues. The Shannon trap consists of a series of vertical white sheets with a shallow box roof and uses light and human or animal bait. Entrance/exit traps over animal burrows measure attractiveness to the owner of the burrow, as well as the burrow itself, as a resting site.

Collecting or sampling of larvae is extremely labour-intensive and often proves remarkably unsuccessful because specific breeding sites are unknown. Larvae can be extracted by flotation techniques from soil samples and the lining of rodent burrows. Emergence traps are useful for locating breeding sites.

Preservation and preparation for identification

Sandflies can be stored in 80% alcohol or Val Andre solution (see below) but long-term storage in alcohol discolours specimens and makes them difficult to clear, while storage in Val Andre solution over-clears them and makes them fragile. Alternatively, specimens can be stored dry in a vial under gentle pressure from a twist of tissue paper to avoid movement. Cotton wool should not be used as it can cause the antennae, legs etc. to become entangled and liable to be broken when specimens are removed. Sandflies usually need to be slide-mounted for accurate identification and there is considerable debate over the most appropriate mounting medium, the choice being between either water-soluble mounts based on gum-chloral or Canada balsam mountants (especially the phenol-balsam form). Berlese's medium was strongly recommended by Lewis (1973a) because it is very simple and quick to use and has a refractive index which allows easy examination of delicate structures (such as spermathecae). However, it has serious shortcomings if not perfectly prepared and stored at a precise humidity (crystallization, discoloration, reduction in viscosity). These are not alleviated by ringing the coverslip because the mountant can become heavily discoloured (sometimes almost black) from the interaction of mountant and ringing medium, especially if Euparal is used.

For water-based media such as Berlese's medium, flies can be mounted directly from 80% alcohol or water following detergent processing of sticky-trap samples. The speed at which specimens can be mounted makes this a preferred method for ecological studies with short-term interests but not for taxonomic studies where reference material is required. During the preparation of the ingredients (gum arabic 12 g: chloral hydrate crystals 20 g: glacial acetic acid 5 ml: 50% w/w glucose syrup 5 ml: distilled water 30–40 ml) it is essential that they are not heated above 40–50°C, since heating predisposes the mountant to crystallization.

For mounting in phenol-balsam, the specimens are treated with cool potassium hydroxide solution (5%) until soft, washed in water, dehydrated in an alcohol series to absolute alcohol, transferred to phenol/alcohol (saturated solution of phenol in absolute alcohol) and mounted in phenol-balsam (pure Canada balsam, without xylene, dissolved in phenol/alcohol). The phenol-balsam makes permanent preparations in which all the structures used in identification can be clearly seen.

Whichever mountant is chosen, the same procedure for dissection is used: the head is removed from the body and mounted ventral side uppermost to display the cibarium and pharynx, and the thorax and abdomen mounted laterally. To identify females of *Phlebotomus* (*Larroussius*) species it is necessary to dissect out the base of the common spermathecal duct in Berlese medium without the gum arabic (=Val Andre's solution).

Larvae and pupae can be stored and examined in Val Andre's solution. Where iso-female broods have been reared (i.e. the progeny of individual females), a few larvae should be slide-mounted for close examination of the head and measurement of the setae. Eggs are best stored dry, or slide-mounted without the use of organic solvents which might destroy the waxy exochorion.

Rearing and colonization

Wild-caught and blood-fed sandflies can usually be induced to oviposit on damp filter paper in small glass tubes. The flies are fed sugar solution on cotton wool and kept in a humid environment. For taxonomic studies, the progeny of individual females should be kept separately to ensure the correct association of larvae and adults (both males and females).

Sandflies have been reared in the laboratory since the early part of the century but establishment of productive colonies only began in the 1970s with the use of semi-natural larval media and rigorous attention to detail during both the establishment and maintenance of the colony. See Killick-Kendrick (1987) for details of colonization procedures.

REFERENCES

Abonnenc, E. 1972. Les Phlébotomes de la région éthiopienne (Diptera, Psychodidae). *Mémoires de l'ORSTOM* **55**: 1–289.

Abonnenc, E. and **Léger, N.** 1976. Sur une classification rationnelle des Diptères Phlebotomidae. *Cahiers ORSTOM* (Entomologie médicale et Parasitologie) **14**: 69–78.

Abonnenc, E. and **Minter, D. M.** 1965. Bilingual keys for the identification of the sandflies of the Ethiopian region. *Cahiers ORSTOM* (Entomologie médicale) **5**: 1–63.

Abul-Hab, J. and **Ahmed, S. A.** 1984. Revision of the family Phlebotomidae (Diptera) in Iraq. *Journal of Biological Sciences Research* (Baghdad) **7**: 1–64.

Artemiev, M. M. 1978. *Sandflies (Diptera, Psychodidae, Phlebotominae) of Afghanistan.* iv + 87 pp. Ministry of Public Health, Malaria and Leishmania Institute, Kabul.

Artemiev, M. M. 1980. A revision of sandflies of the subgenus *Adlerius* (Diptera, Phlebotominae, *Phlebotomus*). *Zoologichesky Zhurnal* **59**: 1177–1193. [In Russian with English summary.]

Artemiev, M. M. 1983. Sandflies (Diptera, Phlebotominae) of Afghanistan. Part 2. Distribution according to landscapes. *Meditsinskaya Parazitologiya i Parazitarnye* **52**: 25–33. [In Russian with English summary.]

Artemiev, M. M. and Neronov, V. M. 1984. *Distribution and ecology of sandflies of the Old World (genus Phlebotomus)*. 207 pp. Institute of Evolutionary Morphology and Animal Ecology, USSR Academy of Sciences, Moscow. [In Russian.]

Ashford, R. W. 1974. Sandflies (Diptera: Phlebotomidae) from Ethiopia: taxonomic and biological notes. *Journal of Medical Entomology* **11**: 605–616.

Ashford, R. W. and Bettini, S. 1987. Ecology and epidemiology: Old World. Pp. 365–424 in Peters, W. and Killick-Kendrick, R. (eds), *The leishmaniases in biology and medicine*: vol. 1, *Biology and epidemiology*. xxv + 550 + XXVIII (Index). Academic Press, London.

Bailly-Choumara, H., Abonnenc, E. and Pastre, J. 1972. Contribution à l'étude des Phlébotomes du Maroc (Diptera, Psychodidae). Données faunistiques et écologiques. *Cahiers ORSTOM* (Entomologie médicale et Parasitologie) **9** (1971): 431–460.

Caillard, T., Tibayrenc, M., Le Pont, F., Dujardin, J. P., Desjeux, P. and Ayala, F. J. 1986. Diagnosis by isozyme methods of two cryptic species, *Psychodopygus carrerai* and *P. yucumensis* (Diptera: Psychodidae). *Journal of Medical Entomology* **23**: 489–492.

Christensen, H. A. and de Vasquez, A. M. 1982. The tree-buttress biotope: a pathobiocenose of *Leishmania braziliensis*. *American Journal of Tropical Medicine and Hygiene* **31**: 243–251.

Croset, H., Rioux, J. A., Maistre, M. and Bayar, N. 1978. Les Phlébotomes de Tunisie (Diptera, Phlebotomidae): mise au point systématique, chorologique et éthologique. *Annales de Parasitologie humaine et comparée* **53**: 711–749.

Davidson, I. H. 1981. The subgenus *Anaphlebotomus* of *Phlebotomus* (Diptera: Psychodidae: Phlebotominae) in southern Africa. *Journal of the Entomological Society of Southern Africa* **44**: 259–264.

Davidson, I. H. 1983. The subgenus *Capensomyia* of *Sergentomyia* (Diptera: Phlebotominae): two new species from South Africa and Namibia, with a key to all known species. *Zeitschrift für angewandte Zoologie* **70**: 217–224.

Davidson, I. H. 1990. *Sandflies of Africa south of the Sahara: taxonomy and systematics of the genus Sergentomyia*. 75 [+3] pp. South African Institute for Medical Research, Johannesburg.

Davis, N. T. 1967. Leishmaniasis in the Sudan Republic. 28. Anatomical studies on *Phlebotomus orientalis* Parrot and *P. papatasi* Scopoli (Diptera: Psychodidae). *Journal of Medical Entomology* **4**: 50–65.

Dedet, J. P., Addadi, K. and Belazzoug, S. 1985. Les phlébotomes (Diptera, Psychodidae) d'Algérie. *Cahiers ORSTOM* (Entomologie médicale et Parasitologie) **22** (1984): 99–127.

Disney, R. H. L. 1966. A trap for phlebotomine sandflies attracted to rats. *Bulletin of Entomological Research* **56**: 445–451.

Duckhouse, D. A. and Lewis, D. J. 1980. Family Psychodidae. Pp. 93–105 in Crosskey, R. W. (ed.), *Catalogue of the Diptera of the Afrotropical region*. 1437 pp. British Museum (Natural History), London.

Duckhouse, D. A. and Lewis, D. J. 1989. Family Psychodidae. Pp. 166–179 in Evenhuis, N. L. (ed.), *Catalog of the Diptera of the Australasian and Oceanian regions*. 1155 pp. Bishop Museum Press, Honolulu, and E. J. Brill, Leiden.

Elnaiem, D. A. and Ward, R. D. 1991. Response of the sandfly *Lutzomyia longipalpis* to an oviposition pheromone associated with conspecific eggs. *Medical and Veterinary Entomology* **5**: 87–91.

Fairchild, G. B. 1955. The relationships and classification of the Phlebotominae (Diptera, Psychodidae). *Annals of the Entomological Society of America* **48**: 182–196.

Fausto, A. M., Maroli, M. and Mazzini, M. 1991 Ootaxonomy investigation of three sandfly species (Diptera: Psychodidae) from Italy. *Parassitologia* **33** (Supplemento 1): 225–228.

Feliciangeli, M. D., Ramirez-Perez, J. and Ramirez, A. 1988. The phlebotomine sandflies of Venezuelan Amazonia. *Medical and Veterinary Entomology* **2**: 47–65.

Forattini, O. P. 1973. *Entomologia Médica: 4, Psychodidae. Phlebotominae. Leishmanioses. Bartonelose*. 658 pp. Universidade de São Paulo.

Gil Collado, J. 1977. Phlébotomes et leishmanioses en Espagne. *Colloques Internationaux du Centre National de la Recherche Scientifique* (Paris) **239**: 177–190.

Grepin, G. 1983. Phlébotomes (Diptera – Phlebotominae) de la République Centrafricaine. *Annales de Parasitologie humaine et comparée* **8**: 85–90.

Hanson, W. J. 1961. The breeding places of *Phlebotomus* in Panama (Diptera, Psychodidae). *Annals of the Entomological Society of America* **54**: 317–322.

Hart, D. T. (ed.) 1989. *Leishmaniasis: the current status and new strategies for control.* xiv + 1041 pp. Plenum Press, New York and London.

Houin, R., Abonnenc, E. and **Deniau, M.** 1971. Résultats d'un sondage Phlébotomes du sud de la Turquie. *Annales de Parasitologie humaine et comparée* **46**: 633–652.

Javadian, E. and **Mesghali, A.** 1975. Check-list of phlebotomine sandflies ('Diptera: Psychodidae') of Iran. *Bulletin de la Société de Pathologie exotique* **68**: 207–209.

Jobling, B. 1976. On the fascicle of blood-sucking Diptera. In addition a description of the maxillary glands in *Phlebotomus papatasi,* together with the musculature of the labium and pulsatory organ of both the latter species and also of some other Diptera. *Journal of Natural History* **10**: 457–461.

Jobling, B. 1987. *Anatomical drawings of biting flies.* 119 pp. British Museum (Natural History), London.

Killick-Kendrick, R. 1987. Methods for the study of phlebotomine sandflies. Pp. 473–497 in Peters, W. and Killick-Kendrick, R. (eds), *The leishmaniases in biology and medicine*: vol. 1, *Biology and epidemiology.* xxv + 550 + XXVIII (Index). Academic Press, London.

Killick-Kendrick, R. 1990. Phlebotomine vectors of the leishmaniases: a review. *Medical and Veterinary Entomology* **4**: 1–24.

Killick-Kendrick, R. and **Peters, W.** 1991. Leishmaniasis in Arabia: an annotated bibliography. *American Journal of Tropical Medicine and Hygiene* **44** (Supplement): 1–64.

Killick-Kendrick, R., Rioux, J.-A., Bailly, M., Guy, M. W., Wilkes, T. J., Guy, F. M., Davidson, I., Knechtli, R., Ward, R. D., Guilvard, E., Perieres, J. and **Dubois, H.** 1984. Ecology of leishmaniasis in the south of France 20. Dispersal of *Phlebotomus ariasi* Tonnoir, 1921 as a factor in the spread of visceral leishmaniasis in the Cévennes. *Annales de Parasitologie humaine et compareé* **59**: 555–572.

Kreutzer, R. D., Palau, M. T., Morales, A., Ferro, C., Feliciangeli, D. and **Young, D. G.** 1990. Genetic relationships among phlebotomine sand flies (Diptera: Psychodidae) in the *verrucarum* species group. *Journal of Medical Entomology* **27**: 1–8.

Lainson, R. and **Shaw, J. J.** 1987. Evolution, classification and geographical distribution. Pp. 1–120 in Peters, W. and Killick-Kendrick, R. (eds), *The leishmaniases in biology and medicine*: vol. 1, *Biology and epidemiology.* xxv + 550 + XXVIII (Index). Academic Press, London.

Lane, R. P. 1986a. Recent advances in the systematics of phlebotomine sandflies. *Insect Science and its Applications* **7**: 225–230.

Lane, R. P. 1986b. The sandflies of Egypt (Diptera: Phlebotominae). *Bulletin of the British Museum (Natural History)* (Entomology) **52**: 1–35.

Lane, R. P. 1991. The contribution of sandfly control to leishmaniasis control. *Annales de la Société belge de Médecine Tropicale* **71** (Supplement): 65–74.

Lane, R. P. and **Bernardes, D. de S.** 1990. Histology and structure of pheromone secreting glands in males of the phlebotomine sandfly *Lutzomyia longipalpis. Annals of Tropical Medicine and Parasitology* **84**: 53–61.

Lane, R. P., Abdel-Hafez, S. and **Kamhawi, S.** 1988. The distribution of phlebotomine sandflies in the principal ecological zones of Jordan. *Medical and Veterinary Entomology* **2**: 237–246.

Lane, R. P., Pile, M. M. and **Amerasinge, F. P.** 1990. Anthropophagy and aggregation behaviour in the sandfly *Phlebotomus argentipes* in Sri Lanka. *Medical and Veterinary Entomology* **4**: 79–88.

Lane, R. P. and **Ready, P. D.** 1985. Multivariate discrimination between *Lutzomyia wellcomei,* a vector of mucocutaneous leishmaniasis, and *Lu. complexus* (Diptera: Phlebotominae). *Annals of Tropical Medicine and Parasitology* **79**: 469–472.

Lebbe, J., Vignes, R. and **Dedet, J. P.** 1987. *Computer aided identification of phlebotomine sandflies of French Guiana (Diptera: Psychodidae)* [French title also]. 165 pp. Institut Pasteur de la Guyane Française, Cayenne. [Issued with computer floppy disk.]

Léger, N., Pesson, B. and **Madulo-Leblond, G.** 1986a. Les phlébotomes de Grèce. 1re partie. *Bulletin de la Société de Pathologie exotique* **79**: 386–397.

Léger, N., Pesson, B. and **Madulo-Leblond, G.** 1986b. Les phlébotomes de Grèce. 2ᵉ partie. *Bulletin de la Société de Pathologie exotique* **79**: 514–524.

Léger, N., Pesson, B., Madulo-Leblond, G. and **Abonnenc, E.** 1983. Sur la différenciation des femelles du sous-genre *Larroussius* Nitzulescu, 1931 (Diptera – Phlebotomidae) de la région méditerranéenne. *Annales de Parasitologie humaine et compareé* **58**: 611–623.

Leng, Y. J. 1987. A preliminary survey of phlebotomine sandflies in limestone caves of Sichuan and Guizhou Provinces, south-west China, and description and discussion of a primitive new genus *Chinius*. *Annals of Tropical Medicine and Parasitology* **81**: 311–317.

Leng, Y. J. 1988. A review of phlebotomine sandflies and their transmission of leishmaniasis in China. *Japanese Journal of Sanitary Zoology* **39**: 323–337. [In English.]

Lewis, D. J. 1967. The phlebotomine sand-flies of West Pakistan (Diptera: Psychodidae). *Bulletin of the British Museum (Natural History) (Entomology)* **19**: 1–57.

Lewis, D. J. 1973a. Phlebotomidae and Psychodidae (sand-flies and moth-flies). Pp. 155–179 in Smith, K. G. V. (ed.), *Insects and other arthropods of medical importance*. xiv + 561 pp. British Museum (Natural History), London.

Lewis, D. J. 1973b. Family Phlebotomidae. Pp. 245–254 in Delfinado, M. D. and Hardy, D. E. (eds), *A catalog of the Diptera of the Oriental region*: vol. 1, *Suborder Nematocera*. 618 pp. University Press of Hawaii, Honolulu.

Lewis, D. J. 1974. The phlebotomid sandflies of Yemen Arab Republic. *Tropenmedizin und Parasitologie* **25**: 187–197.

Lewis, D. J. 1975. Functional morphology of the mouth parts in New World phlebotomine sandflies (Diptera: Psychodidae). *Transactions of the Royal Entomological Society of London* **126**: 497–532.

Lewis, D. J. 1978. The phlebotomine sandflies (Diptera: Psychodidae) of the Oriental region. *Bulletin of the British Museum (Natural History) (Entomology)* **37**: 217–343.

Lewis, D. J. 1982. A taxonomic review of the genus *Phlebotomus* (Diptera: Psychodidae). *Bulletin of the British Museum (Natural History) (Entomology)* **45**: 121–209.

Lewis, D. J. and **Büttiker, W.** 1982. Insects of Saudi Arabia: the taxonomy and distribution of Saudi Arabian phlebotomine sandflies (Diptera: Psychodidae). *Fauna of Saudi Arabia* **4**: 353–397.

Lewis, D. J. and **Büttiker, W.** 1983. Some ecological aspects of Saudi Arabian phlebotomine sandflies (Diptera: Psychodidae). *Fauna of Saudi Arabia* **5**: 479–530.

Lewis, D. J. and **Dyce, A. L.** 1982. The subgenus *Austrophlebotomus* Theodor of *Phlebotomus* Rondani and Berté (Diptera: Psychodidae). *Journal of the Australian Entomological Society* **21**: 37–54.

Lewis, D. J. and **Dyce, A. L.** 1988. Taxonomy of the Australasian Phlebotominae (Diptera: Psychodidae) with revision of the genus *Sergentomyia* from the region. *Invertebrate Taxonomy* **2**: 755–804.

Lewis, D. J., Young, D. G., Fairchild, G. B. and **Minter, D. M.** 1977. Proposals for a stable classification of phlebotomine sandflies (Diptera: Psychodidae). *Systematic Entomology* **2**: 319–332.

MacVicker, J. A. K., Moore, J. S., Molyneux, D. H. and **Maroli, M.** 1990. Honeydew sugars in wild-caught Italian phlebotomine sandflies (Diptera: Psychodidae) as detected by high performance liquid chromatography. *Bulletin of Entomological Research* **80**: 339–344.

Maroli, M. and **Fausto, A. M.** 1986. Metodi di campionamento e montaggio dei flebotomi (Diptera: Psychodidae). *Raporti Istisan* 86/11: 1–74.

Molyneux, D. H. and **Killick-Kendrick, R.** 1987. Morphology, ultrastructure and life cycles. Pp. 121–176 in Peters, W. and Killick-Kendrick, R. (eds), *The leishmaniases in biology and medicine*: vol. 1, *Biology and epidemiology*. xxv + 550 + XXVIII (Index). Academic Press, London.

Morton, I. E. and **Ward, R. D.** 1989. Laboratory response of female *Lutzomyia longipalpis* sandflies to a host and male pheromone source over distance. *Medical and Veterinary Entomology* **3**: 219–223.

Murillo, J. and **Zeledón, R.** 1985. Flebótomos de Costa Rica (Diptera: Psychodidae). *Brenesia* **23** (Suplemento): 1–137.

Perfil'ev, P. P. 1966. Sandflies (family Phlebotomidae). *Fauna of the USSR*. New Series. No. 93, *Insects, Diptera* 3(2). 382 pp. [In Russian: English translation titled *Phlebotomidae (sandflies) (Moskity)*, x + 363 pp., Israel Program for Scientific Translations, Jerusalem, 1968.]

Peters, W. and **Killick-Kendrick, R.** (eds) 1987. *The leishmaniases in biology and medicine*: vol. 2, *Clinical aspects and control*. xxv + pp. 551–941 + XXVIII (Index). Academic Press, London.

Phillips, A., Le Pont, F., Desjeux, P., Broomfield, G. and **Molyneux, D. H.** 1990a. Separation of *Psychodopygus carrerai carrerai* and *P. yucumensis* (Diptera: Psychodidae) by gas chromatography of cuticular hydrocarbons. *Acta Tropica* **47**: 145–149.

Phillips, A., Milligan, P. J. M., Maroli, M., Lane, R. P., Kamhawi, S., Broomfield, G. and **Molyneux, D. H.** 1990b. Intraspecific variation in the cuticular hydrocarbons of the sandfly *Phlebotomus perfiliewi* from Italy. *Medical and Veterinary Entomology* **4**: 451–457.

Quate, L. W. 1964. *Phlebotomus* sandflies from the Paloich area in the Sudan (Diptera, Psychodidae). *Journal of Medical Entomology* **1**: 213–268.

Quate, L. W. and **Quate, S. H.** 1967. A monograph of Papuan Psychodidae, including *Phlebotomus* (Diptera). *Pacific Insects Monograph* **15**: 1–216.

Quate, L. W. and **Vockeroth, J. R.** 1981. Psychodidae. Pp. 293–300 in McAlpine, J. F., Peterson, B. V., Shewell, G. E., Teskey, H. J., Vockeroth, J. R. and Wood, D. M. (eds), *Manual of Nearctic Diptera*: vol. 1, vi + 674 pp. Research Branch, Agriculture Canada (Monograph No. 27).

Ready, P. D., Fraiha, H., Lainson, R. and **Shaw, J. J.** 1980. *Psychodopygus* as a genus: reasons for a flexible classification of the phlebotomine sand flies (Diptera: Psychodidae). *Journal of Medical Entomology* **17**: 75–88.

Ready, P. D., Lainson, R., Shaw, J. J. and **Souza, A. A.** 1991. DNA probes for distinguishing *Psychodopygus wellcomei* from *Psychodopygus complexus* (Diptera: Psychodidae). *Memorias do Instituto Oswaldo Cruz* **86**: 41–49.

Ready, P. D., Lainson, R., Shaw, J. J. and **Ward, R. D.** 1986. The ecology of *Lutzomyia umbratilis* Ward & Fraiha (Diptera: Psychodidae), the major vector to man of *Leishmania braziliensis guyanensis* in north-eastern Amazonian Brazil. *Bulletin of Entomological Research* **76**: 21–40.

Ready, P. D., Smith, D. F. and **Killick-Kendrick, R.** 1988. DNA hybridizations on squash-blotted sandflies to identify both *Phlebotomus papatasi* and infecting *Leishmania major*. *Medical and Veterinary Entomology* **2**: 109–116.

Ribeiro, J. M. C., Vachereau, A., Modi, G. B. and **Tesh, R. B.** 1989. A novel vasodilatory peptide from the salivary glands of the sand fly *Lutzomyia longipalpis*. *Science* **243**: 212–214.

Rioux, J. A. and **Golvan, Y. J.** 1969. Epidémiologie des léishmanioses dans le sud de la France. *Monographie de l'Institut National de la Santé et de la Recherche Médicale* **37**: 1–220.

Rohdendorf, B. B. 1964. Historical development of the Diptera. *Trudȳ Paleontologicheskogo Instituta* **100**: 1–311. [In Russian: English translation in Hocking, B., Oldroyd, H. and Ball, G. E. (eds) (1974) *The historical development of Diptera*, xv + 360 pp., University of Alberta Press, Edmonton.]

Ryan, L. 1986. Flebótomos do estado do Pará, Brasil. (Diptera: Psychodidae: Phlebotominae). *Documento Técnico Instituto Evandro Chagas, Belém*: vol. 1, vii + 154 pp. Ministerio de Saúde, Belém.

Schlein, Y. and **Yuval, B.** 1987. Leishmaniasis in the Jordan Valley IV. Attraction of *Phlebotomus papatasi* (Diptera: Psychodidae) to plants in the field. *Journal of Medical Entomology* **24**: 87–90.

Schlein, Y., Yuval, B. and **Warburg, A.** 1984. Aggregation pheromone released from the palps of feeding female *Phlebotomus papatasi* (Psychodidae). *Journal of Insect Physiology* **30**: 153–156.

Seccombe, A. K., Ready, P. D. and **Huddleston, L. M.** 1993. A catalogue of Old World phlebotomine sandflies (Diptera: Psychodidae, Phlebotominae). *Occasional Papers in Systematic Entomology* **8**: 1–57.

Shaw, J. J. and **Lainson, R.** 1987. Ecology and epidemiology: New World. Pp. 291–363 in Peters, W. and Killick-Kendrick, R. (eds), *The leishmaniases in biology and medicine*: vol. 1, *Biology and epidemiology*. xxv + 550 + XXVIII (Index). Academic Press, London.

Tesh, R. B. 1988. The genus *Phlebovirus* and its vectors. *Annual Review of Entomology* **33**: 169–181.

Theodor, O. 1958. Psychodidae – Phlebotominae. *Die Fliegen der palaearktischen Region* **9c**: 1–55.

Theodor, O. 1965. On the classification of American Phlebotominae. *Journal of Medical Entomology* **2**: 171–197.

Theodor, O. and **Mesghali, A.** 1964. On the Phlebotominae of Iran. *Journal of Medical Entomology* **1**: 285–300.

Vianna Martins, A., Williams, P. and **Falcão, A. L.** 1978. *American sand flies (Diptera: Psychodidae, Phlebotominae).* 195 pp. Academia Brasileira de Ciéncias, Rio de Janeiro.

Wagner, R. 1991. Family Psychodidae. Pp. 11–65 in Soós, A. and Papp, L. (eds), *Catalogue of Palaearctic Diptera*: vol. 2, Psychodidae – Chironomidae. 499 pp. Elsevier, Amsterdam.

Walton, B. C., Wijeyaratne, P. and **Moddaber, F.** (eds) 1988. *Research on control strategies for the leishmaniases. Proceedings of an international workshop held in Ottawa, Canada, 1–4 June 1987.* viii + 374 pp., IDRC, Ottawa.

Ward, R. D. 1976. A revised numerical chaetotaxy for Neotropical phlebotomine sandfly larvae (Diptera: Psychodidae). *Systematic Entomology* **1**: 89–94.

Ward, R. D. and **Ready, P. A.** 1975. Chorionic sculpturing in some sandfly eggs (Diptera, Psychodidae). *Journal of Entomology* (A) **50**: 127–134.

Ward, R. D., Phillips, A., Burnet, B. and **Marcondes, C. M.** 1988. The *Lutzomyia longipalpis* complex: reproduction and distribution. Pp. 257–269 in Service, M. W. (ed.), *Biosystematics of haematophagous insects*. xi + 363 pp. Clarendon Press, Oxford.

World Health Organization 1990. Control of the leishmaniases. *World Health Organization Technical Report Series* **793**: 1–158.

Young, D. G. 1979. A review of the bloodsucking psychodid flies of Colombia (Diptera: Phlebotominae and Sycoracinae). *University of Florida Agricultural Experiment Stations Bulletin (Technical)* **806**: 1–266.

Young, D. G. and **Duncan, M. A.** (In press) Guide to the identification and geographic distribution of *Lutzomyia* sandflies in Mexico, the West Indies, Central and South America (Diptera: Psychodidae).

Young, D. G. and **Perkins, P. V.** 1984. Phlebotomine sand flies of North America (Diptera: Psychodidae). *Mosquito News* **44**: 263–304.

Young, D. G. and **Rogers, T. E.** 1984. The phlebotomine sand fly fauna (Diptera: Psychodidae) of Ecuador. *Journal of Medical Entomology* **21**: 597–611.

Young, D. G., Pérez, J. E. and **Romero, G.** 1985. New records of phlebotomine sand flies from Peru with a description of *Lutzomyia oligodonta*, n. sp., from the Rimac valley (Diptera: Psychodidae). *International Journal of Entomology* **27**: 136–146.

Yuval, B. and **Schlein, Y.** 1986. Leishmaniasis in the Jordan Valley III. Nocturnal activity of *Phlebotomus papatasi* (Diptera: Psychodidae) in relation to nutrition and ovarian development. *Journal of Medical Entomology* **23**: 411–415.

Zahar, A. R. 1979. Studies on leishmaniasis vectors/reservoirs and their control in the Old World. Mimeographed document WHO/VBC/79.749, 88 pp. World Health Organization, Geneva.

Zahar, A. R. 1980a. Studies on leishmaniasis vectors/reservoirs and their control in the Old World. Part III Middle East. Mimeographed document WHO/VBC/80.776, 78 pp. World Health Organization, Geneva.

Zahar, **A. R**. 1980b. Studies on leishmaniasis vectors/reservoirs and their control in the Old World. Part IV. Asia and Pacific. Mimeographed document WHO/VBC/80.786, 85 pp. World Health Organization, Geneva.

Zahar, **A. R**. 1981a. Studies on leishmaniasis vectors/reservoirs and their control in the Old World. Part V. Tropical Africa. Mimeographed document WHO/VBC/81.825, 198 pp. World Health Organization, Geneva.

Zahar, **A. R**. 1981b. Studies on leishmaniasis vectors/reservoirs and their control in the Old World. Supplement to parts I–V. Mimeographed document WHO/VBC/81.827, 14 pp. World Health Organization, Geneva.

CHAPTER FIVE

Mosquitoes (Culicidae)

M. W. Service

The mosquitoes have perhaps attained greater public notoriety than any other arthropods. They include the only organisms able to transmit human malaria, and apart from carrying this and other diseases are almost unrivalled as irritating biting pests. All mosquitoes belong to the family Culicidae, which at present contains about 3450 species and subspecies. These are documented in world catalogues (see Ward, 1992) which are updated every few years. Superficially the adults resemble certain other Nematocera, especially the Chironomidae, Dixidae and Chaoboridae, which like mosquitoes have aquatic immature stages. Mosquitoes, however, are readily distinguished from such similar looking flies by their conspicuous forwardly projecting proboscis and scales on the wing veins.

Mosquitoes have an almost worldwide distribution, being found throughout the tropics and temperate regions and even well beyond the Arctic Circle; they are absent only from Antarctica and a few islands. Their great diversity of habitats and life-history strategies has allowed them to colonize many contrasting environments. For example, mosquito larvae are found in ponds, swamps, salt-water marshes, polluted waters of septic tanks, rice fields, discarded domestic containers, rock-pools, tree-holes, plant axils and pitcher plants, and in a variety of other aquatic habitats. Adults are encountered in almost all types of ecological zones, from equatorial rain forests, urbanized conglomerates, cultivated lands to semi-arid areas.

Mosquitoes have a much longer lineage than man. There are indications that some of today's major taxonomic divisions were established by the Cretaceous period (65–140 million years BP), and fossil mosquitoes are known from the Eocene (38–54 million years BP) – whereas the earliest hominids, of which man is the sole descendant, split from other primate lineages only about four million years ago. Most fossil finds of mosquitoes have been from the Oligocene period (26–38 million years BP) (Lutz, 1985), by which time two of today's common genera, *Aedes* and *Culex*, can be recognized, and probably also *Mansonia*. Some of the richest fossil finds are in the so-called 'insect bed' (Oligocene period) on the Isle of Wight. Although the genera *Anopheles* and *Toxorhynchites* are regarded as primitive, they have not yet been found as fossils. It is believed that mosquitoes originated in the Jurassic geological period and that, in the absence of mammals, they fed on reptiles, amphibians or birds, as do some species today; there is little evidence for this, however, and ideas about their origins must remain speculative. From comparative chromosomal study of nematoceran genera, it appears that mosquitoes evolved

Medical Insects and Arachnids Edited by Richard P. Lane and Roger W. Crosskey.
Published in 1993 by Chapman & Hall ISBN 0 412 40000 6

from the Chaoboridae and that the Toxorhynchitinae is the most primitive of the three culicid subfamilies derived from a *Mochlonyx*-like ancestor. The Anophelinae and Culicinae evolved along separate lines.

There has probably been more research on mosquitoes than on any other family of insects. This is because, apart from being a biting nuisance in many parts of the world – so severe sometimes as to make outdoor activities almost impossible – they are vectors of disease. Patrick Manson, a Scottish physician regarded as the 'father' of tropical medicine, was working in China when in 1877 he made the historic discovery that mosquitoes harboured the causative agent of Bancroftian filariasis. This was the first time that an insect had been implicated as a possible vector of disease to man and can be considered the 'birth of medical entomology' (Service, 1978). This event was followed 20 years later by findings of Ross and Grassi that mosquitoes transmit malaria, and a little later to their incrimination in the transmission of yellow fever, dengue and many other arboviruses, as well as other species of filaria.

Despite such intense professional interest, mosquitoes have not generally received vernacular names, although it is not uncommon for *Aedes aegypti* to be called the yellow fever mosquito, *Culex pipiens* the northern common house mosquito, and *Culex quinquefasciatus* the southern house mosquito. Some North American species have been given more entertaining names such as the brown-striped wood mosquito (*Aedes excrucians*), the irritating mosquito (*Coquillettidia perturbans*), and the daylight mosquito (*Anopheles crucians*), while *Aedes sierrensis* is often known as the western tree-hole mosquito. More recently, studies on *Wyeomyia smithii* in North America have earned it the name of the pitcher-plant mosquito, although there are several other species that breed in pitcher plants. The rather nondescript *Culex quinquefasciatus* has been called the shitty brown mosquito – aptly, considering its dingy appearance and habit of breeding in cesspits!

In Sierra Leone the mosquito has distinguished itself in the Medal of the Mosquito (Civil and Military Division) awarded for special gallantry.

Among general books on mosquitoes are Bates (1949) on the natural history, an entertaining classic; a non-specialist introduction by Gillett (1971); and works by Mattingly (1969) and Horsfall (1972) which, in quite different styles, analyse and discuss mosquito biology in relation to disease.

RECOGNITION AND ELEMENTS OF STRUCTURE

Adults

Mosquitoes are generally 3–6 mm long, some of the largest belonging to the mainly tropical *Toxorhynchites* (19 mm long, 12–24 mm wing spread). A general description of an adult is given below (Fig. 5.1). The terminology used is that proposed by Harbach and Knight (1980, 1981) with only a few changes; for example the term **claw** and not unguis is used for the tarsal claws.

Head
The head is globular and covered dorsally to a varying extent with decumbent and erect or semi-erect dark or pale scales. The two large and conspicuous compound eyes are composed of some 350–500 ommatidia, but in *Armigeres* 963 ommatidia have been recorded. There are no ocelli. Dorsally the head comprises the occiput,

122 *Mosquitoes (Culicidae)*

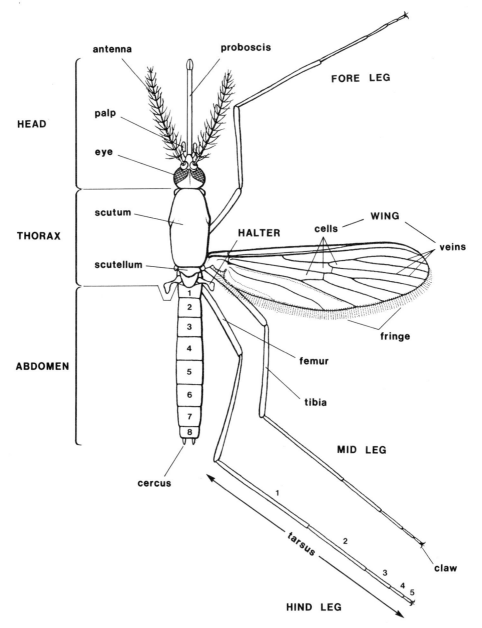

Figure 5.1 Basic morphology of a generalized adult female mosquito, left wing and legs omitted. Note that abdominal segment 8 is not visible in all species and that all tarsi have five segments (tarsomeres) as numbered.

which merges into the vertex anteriorly (Fig. 5.2). Below and between the eyes is the frons, a small area from which arise the antennae. Each antenna consists of a narrow basal ring, the scape, a bulbous segment termed the pedicel (or torus) (Fig. 5.2a), which is almost entirely filled by Johnston's organ, and the 13–14-segmented filamentous flagellum. The flagellar segments (flagellomeres) in females are usually about of equal length (Fig. 5.2a), but in males of most species the two apical segments are longer than the others. In females whorls of five to seven setae

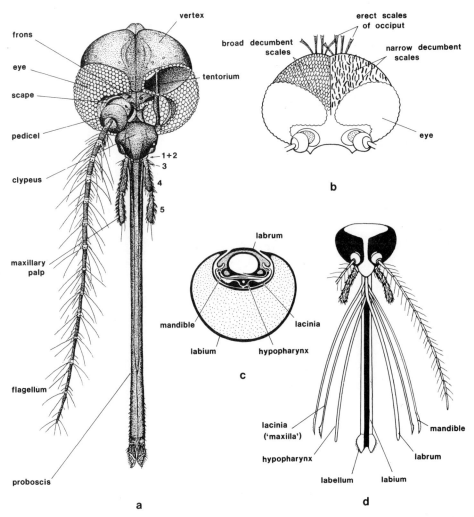

Figure 5.2 Head structure of the female mosquito: (*a*) facial view of head and head appendages of *Aedes aegypti* (Culicinae), part of one eye removed and showing internal supporting strut (tentorium); (*b*) diagram of head in facial view showing upright (erect) and prone (decumbent) types of scale sometimes occurring on vertex and occiput; (*c*) cross-section of proboscis near middle of its length (semi-diagrammatic); (*d*) stylets of the syntrophium (fascicle) spread out from the channel in the labium of the proboscis in which they are enclosed when at rest (semi-diagrammatic). Note that on the maxillary palp only four palpomeres are externally evident. (Part (*a*) reproduced with relabelling from Jobling (1987) courtesy of the Trustees of the Wellcome Trust.)

arise from the bases of most segments; in males of most species these setae (fibrillae) are both more numerous (30–40) and conspicuously longer than in the females (Fig. 5.3). At emergence the fibrillae are recumbent but after two to three days they become erect and make the male antennae plumose or brush-like (e.g. Fig. 5.23). In some species the fibrillae remain erect, while in others they are recumbent during the day and erect only at night, or at dusk or dawn.

Adults can usually be sexed by examining their antennae, but there are some mosquitoes in which males do not have such long hairs (e.g. *Malaya, Hodgesia, Heizmannia, Deinocerites*). In male *Culex* (*Lophoceraomyia*) the antennae have peculiar tufts of blade-like scales on many segments.

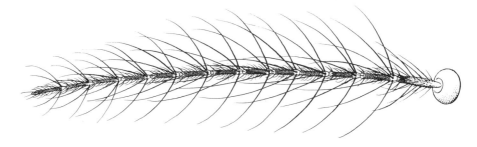

FEMALE

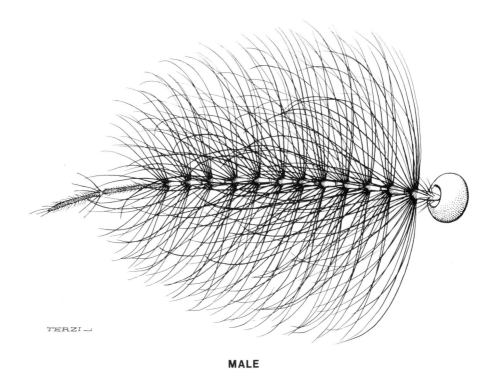

MALE

Figure 5.3 Antennae of typical female and male mosquitoes showing the conspicuous sexual difference in length and density of the sensory hairs.

Very important sensory organs are located on the antennae. Johnston's organ (Belton, 1989) housed in the pedicel contains large numbers of scolopidia which detect minute movements in the basal plate of the flagellum occurring when the male flagellum resonates in response to sound waves emitted by the female's wing beat. (Johnston's organ is less developed in the female and the importance of hearing in this sex remains obscure.) In addition there are several types of sensilla mostly concerned with chemoreception. In *Anopheles* these include short and long thin-walled sensilla (sensilla trichodea), short and long setae (sensilla chaetica), peg-like organs (sensilla basiconica), campaniform sensilla, and olfactory pits with peg organs (sensilla coeloconica) (McIver, 1982, review). In culicines the coeloconic sensilla are very small or absent. (Some of these sensory structures detect stimuli from hosts and oviposition sites including host odours, CO_2, water vapour and convection currents.)

The maxillary palps, usually covered with scales, arise below a small triangular clypeus and lateral to the base of the proboscis and run alongside it. They are composed of five segments (palpomeres), though usually only four are visible in females because palpomeres 1 and 2 are fused (Fig. 5.2a); in a very few genera only two or just one palpomere is visible. The length, shape and hairiness of the palps differ according to species and sex. In anophelines they are long in both sexes but in male culicines they are usually longer than in females (Fig. 5.14). In some genera, however, the palps are short in both sexes, e.g. *Aedeomyia, Ficalbia, Deinocerites, Heizmannia, Malaya, Phoniomyia, Trichoprosopon* and *Uranotaenia*; in *Limatus* they are minute in both sexes and easily overlooked. Four structures are found on the female palps, namely scales usually restricted to the basal palpomeres, short microtrichia densely covering the palpomeres, long sensilla chaetica and thin-walled peg-like organs (sensilla basiconica). Most of these structures are apparently olfactory, for example the sensilla basiconica detecting carbon dioxide and other sensilla detecting humidity gradients.

Male mosquitoes have the same types of sensilla on the palps and antennae as females, but fewer of them. They are used to locate sugar-meals and resting sites.

The mosquito's mouthparts are highly modified for piercing and form a prominent proboscis which extends forwards from the bottom of the head (Figs 5.1 and 5.4). Usually the only structure that is visible is the large fleshy labium which ends in the paired labella and more or less ensheathes the other structures. The other components of the proboscis (Fig. 5.2d), the stylets, collectively known as the syntrophium or fascicle (Fig. 5.4), consist of a dorsal labrum that is grooved underneath, a pair of hair-like mandibles which often have fine teeth (30–50 in some species), a more ventral and slightly stouter pair of laciniae which curve apically and have a series of about 11–15 rather coarse teeth, and a single hollow stylet termed the hypopharynx. Saliva is pumped down the middle of the hypopharynx, while the maxillae are the principal piercing structures. The arrangement of the components of the proboscis is shown in Fig. 5.2c.

When a female takes a blood-meal, the labrum, mandibles, maxillae and hypopharynx are thrust into the skin, while the labella rest on the surface of the skin with the labium bending backwards (Fig. 5.5). The labrum is the only stylet to have sense organs, there being two or more peg-like structures at its tip. The mechanics of feeding and function of the various mouthparts were described by Gordon and Lumsden (1939); Ribeiro (1987) discusses the role of saliva.

Male mosquitoes have a prominent proboscis (Fig. 5.23) with a well-developed labium and labrum which superficially resemble those of the female, but the maxillae and mandibles can be reduced in size or the mandibles absent. Because male mosquitoes do not take blood-meals they have no need for these structures.

Thorax

The thorax (Fig. 5.6) is composed of the prothorax, mesothorax and metathorax, though this is not very obvious. Each of these segments bears a pair of legs. A pair of wings arises from the mesothorax, and the metathorax has a pair of knob-like halteres. The prothorax is reduced to a pair of small antepronotal lobes lying either side of the neck (Fig. 5.6b); in *Wyeomyia, Sabethes* and *Haemagogus* they are unusually large (Figs 5.16b and 5.18b). The mesothorax is much larger and the dorsal part, the scutum (or dorsum as it is sometimes called) is greatly enlarged to accommodate flight muscles. Behind the scutum is the scutellum, which has the posterior edge rounded or trilobed (Fig. 5.15); posteriorly is the shiny dome-shaped

126 *Mosquitoes (Culicidae)*

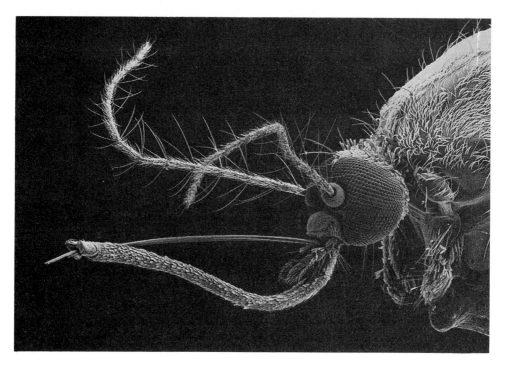

Figure 5.4 Scanning electron microscope photograph of the head of an adult female culicine mosquito showing the separated main elements of the proboscis: slender stiletto-like syntrophium (fascicle) and curved and scale-covered labium.

mesopostnotum. Both the scutellum and mesopostnotum are part of the mesothorax. The metathorax is greatly reduced and comprises a narrow sclerite (metanotum) (Fig. 5.6c).

Laterally the thorax is divided into several distinct sclerites whose size and structure varies somewhat in different genera. A pair of mesothoracic and metathoracic spiracles are present laterally. Figure 5.6c shows a generalized configuration of the side of a mosquito thorax and Fig. 5.6a a representative thorax showing setae and scales. The dorsal and lateral surfaces of the thorax have varying numbers of setae and scales, their location, shape, colour and form being of great taxonomic importance in identification of genera and species (see later key).

The legs are covered with (usually) appressed, dark and/or pale scales, the combination of which can give the legs a banded or speckled appearance. Occasionally, tufts of semi-erect scales are present. The tibiae and femora are about equal in length. The tarsi are long and five-segmented; the last segment (fifth tarsomere) has one or two claws which normally have at least one tooth. Between the claws is a seta called the empodium, and in *Culex* species there is in addition a pair of pad-like pulvilli (Fig. 5.16d).

The wings are long, narrow, transparent and membranous, and have a very consistent arrangement of veins (Fig. 5.7a). The veins are covered dorsally and ventrally with closely appressed scales. A fringe of narrow outstanding scales extends from the apex along the posterior border of the wing (Figs 5.1 and 5.7b).

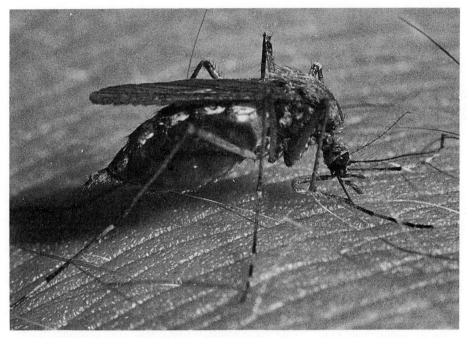

Figure 5.5 A female culicine mosquito (*Aedes*) in the act of bloodsucking.

The shape and arrangement of dark and pale scales on the wings are of great taxonomic importance, particularly in the genus *Anopheles* (Fig. 5.7*b*). Wilkerson and Peyton (1990) provide a standardized nomenclature for the costal spots on *Anopheles* and other spotted-wing mosquitoes.

Each halter has a white, yellowish or brownish-scaled knob-like head, the capitellum, joined by the narrow scape to the slightly swollen basal portion called the scabellum. The halteres are gyroscopic organs and house mechanoreceptors called campaniform sensilla.

Abdomen
The abdomen is composed of ten segments, of which only the first seven or eight are usually visible (Fig. 5.1), the last two to three being modified for reproductive purposes. Each visible segment consists of a tergum joined by the pleural membrane to the sternum. The elasticity of these pleural membranes allows the female abdomen to become distended when the stomach is full of blood or the ovaries well developed. In unfed females and males the terga curve round the sides and beneath the abdomen and partially obscure the sterna. Both terga and sterna are often covered with appressed scales, and in a few species there are lateral projecting, and often brightly coloured, tufts of scales on some terga. In females, segment 8 is the last complete segment; segments 9 and 10 are usually much reduced, and a pair of apical flap-like or finger-like cerci are attached to segment 10. In some genera such as *Culex* and *Mansonia* the terminal segments are withdrawn and the cerci short and sometimes invisible (Fig. 5.18*d*), so the abdomen appears blunt-tipped, whereas in others such as *Aedes* the terminal segments protrude, the cerci are longer and clearly visible, and the abdomen appears pointed (Fig. 5.18*c*).

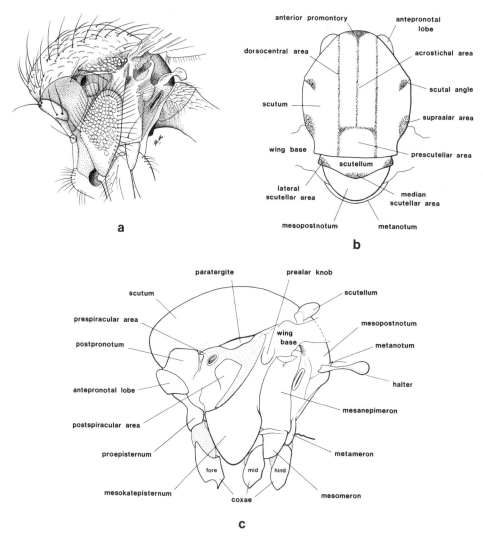

Figure 5.6 Thorax of the adult mosquito: (*a*) left side view in a representative species (*Uranotaenia caeruleocephala*) complete with its vestiture of setae and scales; (*b*) landmarks of dorsal surface with vestiture omitted; (*c*) landmarks of lateral (pleural) surface with vestiture omitted. The two oval respiratory spiracles are shown on (*c*), mesothoracic spiracle behind the postpronotum and metathoracic spiracle behind the mesanepimeron.

In male mosquitoes segment 8 is not unduly small, but segments 9 and 10 are modified to form the male external genitalia (hypopygium, terminalia). After a male has emerged from the pupa, segments 8–10 gradually rotate and after about one day have turned through 180°, so that the sternites 8–10 become dorsal. Segment 9 bears a pair of jointed claspers, each with a swollen basal gonocoxite and a slender distal gonostyle, with which the male grasps the female during copulation. The heavily sclerotized phallosome, consisting of basal pieces, parameres and aedeagus and sometimes (as in *Culex*) the lateral plates, is between the proctiger, gonocoxites and sternum 9. There is great diversity in the shape of the claspers, phallosome and various associated structures of the male genitalia, making them of particular value for species identification.

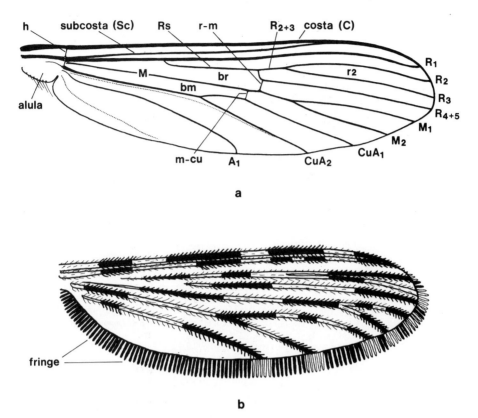

Figure 5.7 Wing of the adult mosquito: (*a*) basic structure with veins and some cells labelled on the notation of McAlpine et al. (1981); (*b*) wing of *Anopheles* showing its characteristic pattern produced by blocks of dark and pale scales on the veins.

Pupae

Mosquito pupae are aquatic and comma-shaped (Fig. 5.8*a*), and unlike pupae of most insects are active when disturbed. The integument (cuticula) is well sclerotized and forms a semi-rigid case enclosing and protecting the developing adult structures.

Cephalothorax
The head and thorax are fused to form a relatively large bulbous cephalothorax. Dorsally the thorax has a pair of double-walled moveable thoracic respiratory trumpets, through which the pupa breathes. The outer wall is hydrophilic, whereas the inner wall is lined with minute setae and spicules and is hydrophobic and thus traps air and is not wetted when the pupa submerges. Each trumpet is divisible into a basal tubular meatus and a distal more open pinna. The meatus can be divided into a proximal tracheoid portion having numerous concentric rings, and a distal reticulate part covered with a fine network. The shape and length of the trumpets vary considerably between species; for example in some *Mimomyia* species the trumpets can be very long (Fig. 5.8*c, e*) and in some *Wyeomyia* species the pinnae are extraordinarily long and filamentous (Fig. 5.8*d*). In *Anopheles* the trumpets are short, and in some species such as *A. barbirostris* the opening of the pinna is wider than the length of the trumpet. In *Mansonia* (Fig. 5.20*a*), *Coquillettidia* and *Mimomyia pallida* of

130 Mosquitoes (Culicidae)

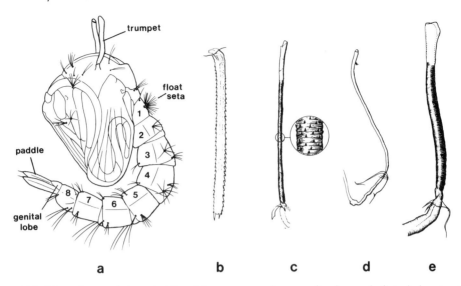

Figure 5.8 Mosquito pupal characteristics: (*a*) appearance of a generalized pupa (culicine) showing the comma-like shape, thoracic respiratory trumpets, much-branched 'float' seta on the first abdominal segment, abdominal paddles and genital lobe; (*b*) paddle of *Mimomyia deguzmanae*; (*c*) trumpet of *Mimomyia deguzmanae*; (*d*) trumpet of *Wyeomyia circumcincta*; (*e*) trumpet of *Mimomyia chamberlaini*.

Africa, and *Mimomyia hybrida* of South East Asia, the trumpets are modified for piercing submerged aquatic vegetation. In *Hodgesia* and some *Anopheles* an additional flap-like lobe, called the tragus, seems to assist in holding the trumpets at the water surface.

Conspicuous dark large compound eyes of the developing adult are clearly visible through the pupal integument, posterior to which are the smaller larval ocelli (stemmata). The proboscis, folded legs and wings of the developing adult can also be seen through the pupal integument. There are about 12 pairs of setae on the cephalothorax.

Abdomen

The abdomen is composed of ten segments, but only the first eight are visible and moveable on each other. Each segment is a completely sclerotized ring. Each tergum and sternum has simple and branched setae, about eight pairs on most terga and five pairs on the sterna. A dendritic seta, often called the palmate or float seta because it helps maintain the pupa at the water surface, is conspicuous on the first tergum (Fig. 5.8*a*). The last visible segment (8) has a pair of flattened and usually overlapping oval-shaped paddles, the margin of which is often fringed with, usually short, setae or fine denticles. There is also usually a short apical seta and sometimes a short dorsal or ventral accessory seta. In some mosquitoes, e.g. species of *Aedeomyia*, *Deinocerites* and *Eretmapodites*, the apical seta is about as long as the paddle. The paddles usually have a thickened midrib, but in some species of *Udaya* and in *Opifex* this is virtually absent. In *Mimomyia* (*Ingramia*) the paddles are very narrow and ribbon-like (Fig. 5.8*b*), presumably as an adaption to pupae life in plant axils and pitcher plants. Between the paddles is a small pouch-like process (genital lobe) which contains the developing external genital processes of the adult mosquito and which, being larger in the male than the female pupa, enables pupae to be sexed. (Male pupae are also usually rather smaller than female pupae of the same species.)

A few Neotropical sabethines and culicines which inhabit bromeliad plant axils have luminescent patches on the cuticula. These persist for some time after adult emergence and have no known function. Some *Aedes* (*Finlaya*) from Mindoro Island in the Philippines exhibit iridescent blue flashes behind the trumpets, and at least one species of *Wyeomyia* (*W. luna* from Jamaica) has similar iridescent spots.

The chaetotaxy of the cephalothorax and abdomen, and shape of the paddles and respiratory trumpets are used in generic and species identification.

Larvae

All larvae are clearly divided into a head, thorax and abdomen (Fig. 5.9). Apart from the head, and if present the siphon, the larval integument is mostly soft and membranous to allow the characteristic sinuous movements made when larvae are swimming.

Head

The head is well sclerotized and often rather dark, and in a few species almost black. It may be roundish, broader than long, or longer than broad. In larvae of *Toxorhynchites*, *Topomyia* and *Psorophora* s.str. the head is more or less square, whereas in some species of *Aedes* and *Culex* the posterior ocular bulges make the head appear almost triangular. The head of *Deinocerites* larvae is unique in having broad subantennal or anterolateral pouches which, together with the absence of ocular bulges, make the head broadest at about the level of the antennae. At the posterior part of the head is a blackish cervical collar, which in *Anopheles* decreases in thickness in succeeding instars. Anterolaterally the head bears the mouthparts, among the most elaborate of any insects. The most conspicuous components are the paired mouthbrushes (lateral palatal brushes) with numerous setae arranged in rows – sometimes as many as 100 or more simple or serrate setae in each mouthbrush.

In some predatory mosquitoes such as *Toxorhynchites* the mouthbrushes are reduced to only about ten flattened non-serrated blade-like setae. Other predators such as *Culex* (*Lutzia*) *tigripes* have about 50–60 bristle-like setae with coarse serrations, whereas *Psorophora* (*Psorophora*) larvae have some 20 thickish pectinate filaments. Mouthbrushes help scrape vegetation from various surfaces, or sweep fine food particles towards the mouth. The morphology and functions of the mouthbrushes of mosquitoes are described by Rashed and Mulla (1990), and the comparative morphology by Harbach (1977).

Lateral to the mouth are paired mandibles and, ventrally, the maxillae, both of which are flattened and heavily sclerotized. The mandibles have serrated molar and incisor-shaped teeth, and various stout and thin setae, some of which are arranged to form the mandibular brush. The maxillae are less developed and generally lack teeth but have numerous setae some of which form the maxillary brush. A pair of short, almost cylindrical, palps arises from the base of the maxillae. The mouthparts can be extraordinarily developed: for example, in *Johnbelkinia* the maxillae have a well-developed horn, and in *Wyeomyia confusa* they have a peculiar long tubular horn-like projection considerably longer than the head (Fig. 5.10d). The function of these modifications is unknown. Larvae of *Tripteroides* (*Rachisoura*) have extra-

132 Mosquitoes (Culicidae)

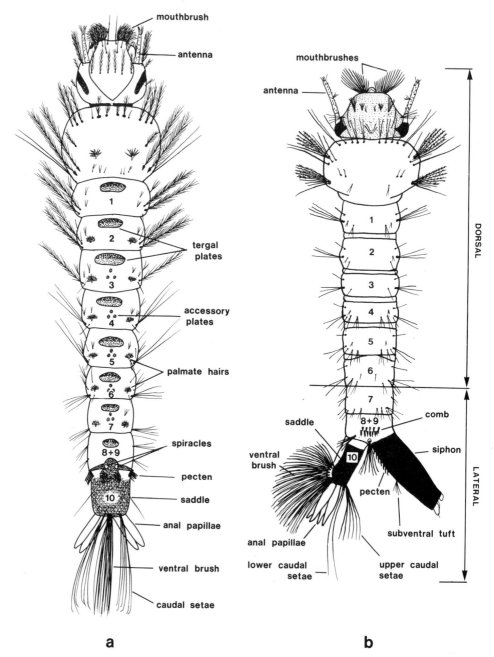

Figure 5.9 Basic morphology of mosquito larvae: (a) *Anopheles*, subfamily Anophelinae; (b) typical Culicinae.

ordinary large maxillary spines associated with their predatory habit (Fig. 5.10c). The mouthbrushes, mandibles and maxillae beat at about 180–240 beats/min, and food particles can be caught by all three structures. Between the maxillae and ventral to them is a heavily sclerotized plate-like dorsomentum (mentum), which is part of the much reduced labium. It is usually triangular and has several teeth, the shape, number and arrangement of which is important taxonomically.

Recognition and elements of structure 133

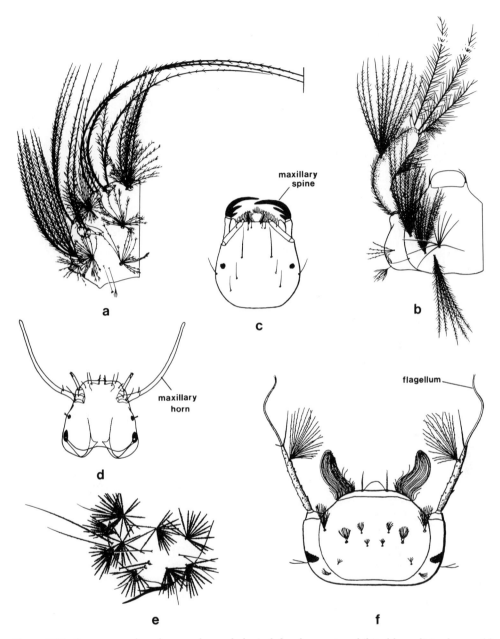

Figure 5.10 Some examples of unusual morphological developments exhibited by culicine larvae: (*a*) part of thorax of *Aedeomyia catasticta* to show stellate setae and enormously long setae (tips of latter not fully complete as indicated by vertical bar); (*b*) part of head of *Aedeomyia catasticta* to show plumose setae and greatly swollen antenna; (*c*) head of *Tripteroides filipes* showing large maxillary spines; (*d*) head of *Wyeomyia confusa* showing long maxillary horns; (*e*) part of thorax of *Tripteroides powelli* to show stellate setae; (*f*) head of *Coquillettidia* showing long flagella of antennae. Note that (*e*) has been partly rotated for figure arrangement and that in natural posture the long paired hairs (left side) point forwards.

The antennae, which are attached anterolaterally to the head, vary greatly in length and form. They can be short and cylindrical, long and curved, or greatly swollen and flattened (e.g. *Aedeomyia*) (Fig. 5.10*b*), denuded of spicules or heavily spiculate. In *Mimomyia* they are two-segmented with a freely articulated distal portion, whereas in *Coquillettidia* they are modified by having an extra segment (the flagellum, Fig. 5.10*f*).

The head capsule has 16 pairs of setae. Their form, undivided, branched or spine-like, and position in relation to one another, are very important for species identification.

First-instar larvae have a pair of small ovoid blackish lateral eyes (stemmata), and much bigger compound eyes which become increasingly larger and more pigmented in succeeding instars. First-instar larvae are easily recognized by a small, dark, cone-like projection surrounded by a paler area near the middle of the dorsal surface of the head. This is the egg-breaker or egg-burster which, as its name suggests, aids the larva in breaking through the egg shell when it hatches.

Thorax
The prothoracic, mesothoracic and metathoracic segments are completely fused to form an ovate structure considerably wider than the head. Sutures are not apparent but the three segments can be recognized by their three distinct groups of setae. The thoracic integument can be smooth, aculeate or spiculate. There are 42 pairs of setae which vary greatly in length and development in different species; some setae in larvae such as those of certain *Tripteroides* and *Aedeomyia catasticta* are stellate (Fig. 5.10*a, e*) and give the larva a hairy appearance. In *Mimomyia* (*Ingramia*) the lateral setae of the mesothorax are at the ends of peculiar finger-like protuberances which are believed to assist larvae in crawling over wet surfaces, e.g. from one plant axil to another.

The prothorax of *Anopheles* larvae has a dorsal pair of transparent bilobed ('notched') organs which assist in supporting the larval body parallel to the water surface. When larvae are not at the surface the organs are retracted into the thorax.

Abdomen
The abdomen consists of ten segments but only nine are visible externally, because segment 9 is fused either to segment 8 or to the last segment (referred to as segment 10) (Fig. 5.9). The first seven abdominal segments are about equal in size and shape and have 13 or 15 pairs of setae of various length and degree of branching. Some species (e.g. of *Anopheles*) have palmate hairs (seta 1) on some, or all, of segments 1–7 (Fig. 5.9*a*). Sometimes, as in *Aedeomyia, Tripteroides*, some *Mimomyia* and *Aedes*, numerous stellate setae cover the abdomen (and thorax) and give the larvae a very hairy appearance. In other species, such as some *Eretmapodites*, the more lateral setae arise from dark sclerotized conical bosses. In *Anopheles* oval sclerotized tergal plates, and in some species from one to three tiny accessory plates (Fig. 5.9*a*), are present dorsally on segments 1–8+9. Small sclerotized plates sometimes occur in other genera. For example, in *Toxorhynchites* they are present laterally on most segments (and also the thorax) and bear setae, and in some *Orthopodomyia* tergal plates are present on segments 7 and 8+9; a lateral sclerite is present on segment 8+9 of most *Uranotaenia* and some *Psorophora*.

Mosquito larvae are metapneustic, their one pair of spiracles located on segment 8+9 and surrounded by the spiracular apparatus. This consists of five movable flap-like, more or less triangular, perispiracular valves which help to prevent water entering the spiracles when the larva submerges. In the Anophelinae the spiracles are almost flush with the surface of the segment, but in Toxorhynchitinae and Culicinae the tracheae have a single orifice at the end of a sclerotized siphon

(Fig. 5.9b). Larvae of *Mansonia* and *Coquillettidia*, and some species of *Mimomyia* such as *M. pallida* and *M. hybrida*, do not come to the water surface to breathe and the spiracular apparatus is modified for piercing submerged vegetation and obtaining oxygen from plant tissues (Fig. 5.20).

The culicine siphon frequently has a series of fringed spines of various shape, the pecten (Fig. 5.9b), which extends for a variable distance along each side of the siphon. The structure is absent in several genera, including *Armigeres*, *Coquillettidia*, *Mansonia*, *Orthopodomyia*, *Phoniomyia*, *Sabethes*, *Toxorhynchites*, *Trichoprosopon*, *Wyeomyia* and *Mimomyia* (*Ingramia*), and sometimes in species of other genera. In the Anophelinae the pecten is different, consisting of a row of spine-like teeth arising from a triangular sclerotized plate situated on segment 8+9 (Fig. 5.9a). The culicine siphon has one or more pairs of subventral setal tufts (1–S) and sometimes lateral and dorsal groups of setae as well. In some mosquitoes, such as *Phoniomyia fuscipes*, *Topomyia gracilis* and *Tripteroides* the siphon is almost completely covered with conspicuous setae (Fig. 5.11d). The shape and number of pecten spines, the chaetotaxy of the siphon and the ratio of its length to width (siphonal index) are important characters used in identification.

On each side of segment 8+9 in the Culicinae there is a comb consisting of a variable number of pointed or rounded sclerotized scales or spines arranged in one

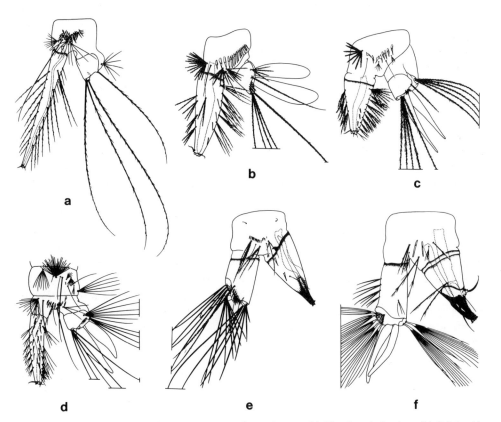

Figure 5.11 End of larval abdomen in some culicine larvae: (*a*) *Phoniomyia fuscipes*; (*b*) *Tripteroides powelli*; (*c*) *Tripteroides stonei*; (*d*) *Topomyia gracilis*; (*e*) *Mimoyia hybrida*; (*f*) *Mansonia uniformis*. Tips of very long hairs are not all fully complete (as indicated by vertical or horizontal bars).

or more rows, or as a patch. Occasionally comb scales or spines arise from a sclerotized plate, for example in most *Uranotaenia* and *Aedeomyia*, and in some species in other genera including *Psorophora* and *Aedes*. The comb is absent in *Toxorhynchites*, some species of *Trichoprosopon*, and in Anophelinae (except that a small number of close-set finger-like scales in front of the pecten represent this structure in the first-instar larva).

Abdominal segment 10 (anal segment) has a sclerotized saddle (Fig. 5.9*a*) which is either dorsolateral and truly saddle-like or completely encircles the segment. The genus *Deinocerites* is unusual because normally, in addition to the dorsal saddle, there is a smaller ventral saddle. The saddle can be smooth, reticulate or spiculate and its posterior margin spiculate or armed with rather large spines. It also has a lateral seta (1–X). Ventral to the saddle is a conspicuous series of many-branched tufts of long setae, mostly paired, and comprising the ventral brush or fan (4–X) which acts as a rudder during swimming. This is particularly well developed in *Anopheles, Chagasia, Psorophora, Toxorhynchites,* many *Aedes* and *Culex* and in *Opifex fuscus*. In contrast, in species that swim little, and those living in tree-holes and plant axils, the ventral brush is usually greatly reduced in length; in sabethine genera the brush is represented only by a few long branched setae. In the Anophelinae the ventral brush (4–X) arises from a grid of transverse bars known as the barred area (or grid bar). In the Culicinae tufts of setae arising from the barred area are called the cratal setae, while any unpaired tufts anterior to this area comprise the precratal setae. Dorsally and distal to the saddle are two groups of long setae, the upper (2–X) and lower caudal setae (3–X). In the Anophelinae the lower caudal setae (3–X) are usually hooked apically enabling larvae to grasp surface vegetation. Segment 10 terminates in two pairs of transparent and flexible anal papillae ('gills'). The dorsal and ventral pairs can be unequal in size, and in a few mosquitoes such as *Wyeomyia smithii* there is only one pair. The anal papillae are commonly finger-shaped; they can be exceptionally long and are also sometimes (as in *Aedes dupreei*) very narrow. In some *Aedes* (*Neomacleaya*) they superficially appear segmented. In others, especially species occurring in salt-water, they are very small and rounded. Their length also varies considerably within a species, being short in larvae reared in water of high ionic concentration and longest when larvae are kept in distilled or rain water. The anal papillae have an osmoregulatory function.

Larvae, unlike pupae, cannot readily be sexed, but the external genitalia of adult males can be detected in third-instar larvae as dark imaginal buds in abdominal segment 10. These develop in the fourth instar into conical lobes which occupy most of the segment. No such development is seen externally in female larvae. Dissection reveals developing testes surrounded by a layer of fat-body in segment 6 of male larvae, and ovaries (without fat-body) in segment 6 of female larvae.

Eggs

Mosquito eggs are usually 1 mm or less long. They are generally ovoid, but some are subtriangular (*Sabethes, Wyeomyia* and *Aedes* such as *A. rusticus*); others are almost spherical (e.g. *Toxorhynchites*) or have terminal filaments (*Mansonia* (*Mansonioides*)). Anteriorly, the egg is pierced by the micropyle through which spermatozoa enter as the eggs leave the female during oviposition. The egg shell is composed of a thin, colourless, transparent and delicately sculptured outer chorion (easily damaged and rubbed off). In many mosquitoes, in particular *Aedes* and *Psorophora* species, the outer chorion is covered with bosses (tubercles) and an interconnected series of ridges (outer chorionic reticulum). This gives the egg a polygonal pattern of whitish lines enclosing black areas, in the centre of which are chorionic tubercles. In aedine eggs the lower surface can lack chorionic tubercles and be covered with a transparent gelatinous material which swells up when wetted to surround the entire egg. The gelatinous covering enables eggs to adhere to surfaces on which they are laid, such as the damp walls of tree-holes or the inside walls of earthern water-storage pots. Not all *Aedes* eggs have the gelatinous coating.

Below the outer chorion is the much thicker inner chorion which immediately after oviposition is whitish but within a few hours tans to become brown or black; it is then relatively impermeable to water and provides the main protective covering or shell of the egg. The inner chorion is smooth in anophelines, toxorhynchitines and many culicines, but in most aedines it is covered with a fine polygonal sculpture made by a series of raised ridges which form the so-called 'inner chorionic reticulum' giving the eggs a mosaic pattern (see *Aedes* eggs, Fig. 5.12b). Finally, after oviposition, the transparent serosal cuticle is secreted by the embryo to form a very thin layer beneath the inner chorion.

Most *Anopheles* eggs are shaped like the hull of a boat (Fig. 5.12a) and the dorsal surface is referred to as the deck. The outer chorion is in many, but not all, *Anopheles* covered with a fine network of polygonal markings which gives the egg a characteristic pattern. The outer chorion is modified in most *Anopheles* eggs to form both the frill around the deck and a pair of conspicuous lateral floats. These consist of distinct air cells (Fig. 5.12a) and help to keep eggs afloat on the water. Floats are present only in anophelines, and are multiple in *Chagasia*. A frill is also present on the eggs of *Hodgesia* and *Mimomyia*, and in *Orthopodomyia* the egg is surrounded by a different and more complex, but probably homologous, structure termed the flange.

Aedine eggs such as those of *Aedes*, *Haemagogus* and *Psorophora* are blackish and often cigar-shaped, but their shape varies greatly according to species; some are almost triangular whereas others are very thin (e.g. *Aedes woodi*). Eggs of other species, such as most *Culex*, *Coquillettidia*, *Trichoprosopon*, *Uranotaenia* and *Culiseta* (*Culiseta*), are more brown than black and lack the conspicuous sculpturing so characteristic of aedine species. Eggs of these mosquitoes are nearly always laid vertically in several rows that are held together by surface tension into a more or less compact egg raft (Fig. 5.12b). The anterior (bottom) end is hydrophilic and this helps to keep the eggs upright. The shape of the raft is sometimes diagnostic of the species; in general *Coquillettidia* egg rafts are much narrower and longer than those of *Culex*. Female *Mansonia* are exceptional in that they lay their eggs in a sticky cluster that is glued to the undersurface of floating vegetation (Fig. 5.12b). Most mosquitoes (e.g. anophelines and aedines), however, do not deposit their eggs in a sticky mass or in an egg raft but lay them individually.

Internal anatomy of adults

We need to know something about the internal anatomy of mosquitoes to understand their biology and how they have become so efficient as disease vectors. Jobling (1987) gives an excellent guide to internal anatomy.

The food channel is formed by the inverted gutter-shaped labrum and leads into the fore-gut containing the cibarial and pharyngeal pumps. The rapid expansion

138 *Mosquitoes (Culicidae)*

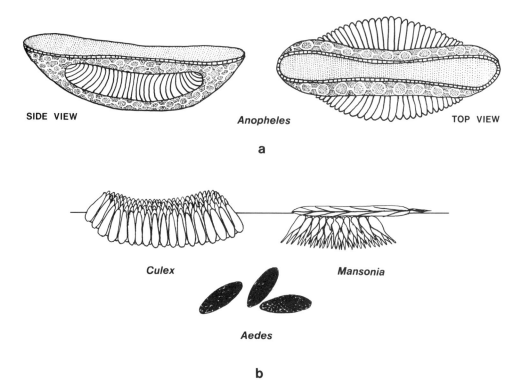

Figure 5.12 Mosquito eggs: (*a*) *Anopheles* egg showing (left) hull-like shape and (right) lateral floats; (*b*) egg raft of *Culex*, egg mass of *Mansonia*, and typical *Aedes* eggs. The eggs of *Mansonia* are glued to the underside of a floating leaf.

and contraction of these pumps enables blood to be sucked into the oesophagus and passed directly to the stomach. There are two oesophageal dorsal diverticula and a much larger, elongated, ventral diverticulum (crop) into which sugary secretions are shunted. Later, sugary meals are passed to the stomach. In most mosquitoes a peritrophic membrane is secreted in the stomach 20 minutes to eight hours after blood-feeding and completely encloses the blood-meal. If a second blood-meal is taken this is surrounded by another peritrophic membrane. To some extent the peritrophic membrane impedes passage of ingested parasites across the gut wall. There are five excretory Malpighian tubules at the end of the mid-gut (stomach) in contrast to the four that are usual in Diptera.

Paired salivary glands lie ventrally in the thorax just above the fore legs. Each gland consists of three finger-like lobes (acini) of which the central one can be smaller than the others. The salivary ducts, one from each gland, unite into a common salivary duct in the head which passes down the hypopharynx. The main effect of saliva being pumped into a bite is to increase the victim's flow of blood. The saliva, at least in some species, contains anticoagulants which prevent blood clotting and blocking of the food channel. Saliva of some species contains agglutinins, haemagglutins and haemolysins (Ribeiro, 1987; Titus and Ribeiro, 1990). Saliva from *Anopheles quadrimaculatus* prevents clotting when diluted 1 to 10 000 volumes of human blood. It has even been suggested that persistent attacks from mosquitoes and other bloodsucking insects might help prevent coronary thrombosis. Sometimes the saliva apparently lacks both agglutinins and anticoagulins but the species with such 'inert' saliva nevertheless feed successfully. One of the more important constituents of saliva is the enzyme apyrase, which is antihaemostatic and produces haematomas in the skin which facilitate blood location during probing and reduce the time required for feeding; in fact the main role of saliva is to accelerate feeding (Pappas et al., 1986). Other substances in saliva appear to be anaesthetic and help reduce pain inflicted by the mosquito's bite, so reducing defensive reactions by the host. But these substances are also partially responsible for the typical allergic skin reactions following a mosquito bite. Salivary glands of male mosquitoes lack both agglutinins and anticoagulins. Gooding (1972) presents a useful literature review of blood digestion in insects.

Each of the female mosquito's two ovaries has some 50–200 ovarioles, the number depending on the species and body size. Development of the eggs (described later) is usually dependent on previous uptake of blood. An important feature of Culicidae is that they have giant polytene chromosomes which are specially well developed in the ovarian nurse cells. In *Anopheles* the banding patterns are very clear and are invaluable for distinguishing closely related species within species complexes.

The female has a single spermatheca in *Anopheles, Uranotaenia* and some *Aedes*, two in *Mansonia* and *Coquillettidia* and three in *Culex* and most *Aedes*.

In males, paired accessory glands secrete a fluid which passes out with the spermatozoa during copulation. This secretion forms the mating plug encountered in female *Anopheles* and a related structure in at least some culicines. In *Aedes aegypti*, and certain other species, the accessory glands produce a substance called 'matrone,' which when introduced into females during copulation induces sexual refractoriness. This and other constituents from the accessory glands facilitate oviposition, modify the biting pattern, stimulate fertility of the sperm, and increase the amount of blood ingested by females (see Young and Downe, 1987, for references). Matrone has been found to induce autogeny in the salt-marsh mosquitoes *Aedes taeniorhynchus* and *Aedes sollicitans*, and a method for extracting the substance from mosquitoes has been patented because of its possible implications for mosquito control.

The fat-body, something of a misnomer because the cells contain glycogen in addition to fat, is very poorly developed in male mosquitoes and usually also in females. However, in hibernating adults the abdomen is often greatly distended with fat reserves. Females of some non-hibernating species, e.g. *Aedes aegypti*, can accumulate fat by feeding solely on sugar solutions. Apart from being an energy source, the fat-body also has an excretory function, storing uric acid as a waste product.

CLASSIFICATION AND IDENTIFICATION

The dixid and chaoborid midges used to be placed together with mosquitoes in the Culicidae, but each group now has family status (Dixidae and Chaoboridae) and the family Culicidae includes only mosquitoes.

The Culicidae are classified into three subfamilies, the Toxorhynchitinae, Anophelinae and Culicinae, of which the last is by far the largest and is conveniently subdivided into two tribes, the Culicini and Sabethini (Table 5.1). However, many specialists classify the Culicinae into ten tribes, some of which contain only a

Table 5.1 Classification of the genera of mosquitoes

Subfamily TOXORHYN-CHITINAE	Subfamily ANOPHELINAE	Tribe SABETHINI	Subfamily CULICINAE Tribe CULICINI	Alternative placement
Toxorhynchites (3) (mainly Cosmotropical)	Anopheles (6) (Cosmopolitan) Bironella (3) (Australasian) Chagasia (Neotropical)	Sabethes (4) (Neotropical) Wyeomyia (10) (New World, mainly Neotropical) Phoniomyia (Neotropical) Limatus (Neotropical) Trichoprosopon (Neotropical) Tripteroides (5) (Oriental, mainly Australasian) Topomyia (2) (mainly Oriental, Australasian) Maorigoeldia (New Zealand) Malaya (Afrotropical, Oriental, Australasian) Johnbelkinia (Neotropical) Runchomyia (3) (Neotropical) Shannoniana (Neotropical)	Aedeomyia (2) (mainly Cosmotropical, Australasian) Aedes (43) (Cosmopolitan) Armigeres (2) (Oriental, Australasian) Eretmapodites (Afrotropical) Haemagogus (2) (Neotropical) Heizmannia (2) (Oriental) Opifex (New Zealand) Psorophora (3) (New World, mainly Neotropical) Udaya (Oriental) Zeugnomyia (Oriental) Culex (22) (Cosmopolitan) Deinocerites (New World, mainly Neotropical) Galindomyia (Colombia) Culiseta (7) (Cosmopolitan, mainly temperate Old World) Ficalbia (Afrotropical, Oriental, Australasian) Mimomyia (3) (Afrotropical, Australasian, Oriental) Hodgesia (Afrotropical, Oriental, Australasian) Coquillettidia (3) (Cosmopolitan) Mansonia (2) (mainly Cosmotropical) Orthopodomyia (Cosmopolitan) Uranotaenia (2) (Cosmopolitan)	AEDEOMYIINI AEDINI CULICINI CULISETINI FICALBIINI HODGESIINI MANSONIINI ORTHOPODOMYIINI URANOTAENIINI

Note: For each genus classified into subgenera the number of these currently recognized is shown in parentheses after the name.

single genus. Within the tribes there are differences of opinion on the ranking of some genera. For example, a few specialists recognize *Coquillettidia* as a subgenus of *Mansonia* whereas most consider it a genus. Here, 37 genera are recognized (Knight and Stone, 1977; Knight, 1978; Ward, 1984, 1992). Many genera have recognized subgenera, but only a few out of the total of 129 subgenera need be mentioned in this chapter; the number in each genus is given in Table 5.1.

Some mosquitoes occur as species complexes (Chapter 1), of which the most famous, and first to be recognized, was the *Anopheles maculipennis* complex. Frequently species in a complex cannot be distinguished morphologically, or only with great difficulty. Usually, however, such species can be distinguished by their chromosomes and/or their isoenzymes. Quantitative differences in the cuticular hydrocarbons (Phillips et al., 1988) can sometimes be used to distinguish between sibling species, such as those in the *Anopheles gambiae* and *Anopheles culicifacies* complexes. The latest technique involves identification by DNA probes, such as has been done for the *Anopheles gambiae* complex (Collins et al., 1988; Gale and Crampton, 1987), *Anopheles quadrimaculatus* (Cockburn et al., 1988) and *Anopheles dirus* (Panyim et al., 1988a, b). Finally, there are a few mosquitoes recognized as distinct species only by their behaviour or existence of mating barriers (e.g. members of the *Culex pipiens* complex).

Subfamily Toxorhynchitinae

Toxorhynchites is the only genus within this subfamily and contains the world's largest mosquitoes. There are some 76 mainly tropical species, but a few occur in Japan, the maritime areas of the Far Eastern USSR and in eastern USA northwards into Canada. The genus is considered the most primitive in the Culicidae and in some respects resembles the sabethine *Trichoprosopon*. (Two other very primitive mosquito genera are *Bironella* and *Chagasia*.)

Adults are readily recognized by their large size (up to 19 mm long, and 24 mm wing spread) and prominent recurved proboscis (Fig. 5.13a,b). The body is covered with green, purplish or reddish iridescent scales, and the posterior abdominal segments have lateral tufts of black, white, red or yellow hair-like scales (Fig. 5.13c). Females are incapable of blood-feeding and both sexes suck nectar and other naturally occurring sugary substances. *Toxorhynchites* is consequently not involved in disease transmission.

The species are basically forest mosquitoes and their principal larval habitats are tree-holes and bamboo, but a few breed in leaf axils, pitcher plants and rock-pools. Some species utilize man-made container-habitats such as water-storage pots and water-filled discarded tyres as breeding places. Eggs are dropped onto water while the female is in flight, and can be 'thrown' into rather inaccessible places. The eggs, which float on the water surface, are subspherical, yellowish or white and have a granular appearance. They hatch within two to three days and cannot withstand desiccation. Larvae are obligatory predators, mainly on mosquito larvae of other species, but in the absence of suitable prey they often become cannibalistic. Larvae are very large (12–18 mm long) and often reddish in colour. They are recognized by their highly modified mouthbrushes, consisting of six to ten non-serrated, thick, curved blade-like setae, and by the absence of both a comb and pecten. Like the Culicinae, the larvae have a siphon, but in *Toxorhynchites* it is always short and conical.

142 *Mosquitoes (Culicidae)*

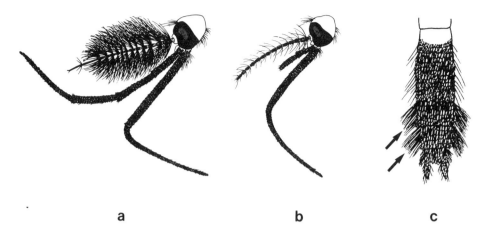

Figure 5.13 Important adult features of *Toxorhynchites* (subfamily Toxorhynchitinae): (*a*) head of male; (*b*) head of female of *T. splendens* showing the strongly recurved proboscis distinctive for the genus; (*c*) characteristic lateral scale tufts (arrowed) of the abdomen.

The biology of *Toxorhynchites* is summarized by Steffan and Evenhuis (1981), and an annotated bibliography is presented by Steffan et al. (1980) and in its supplement by Manning et al. (1982). Taxonomy of subgenus *Toxorhynchites* in the Australasian, eastern Palaearctic and Oriental regions is covered by Steffan and Evenhuis (1985).

Subfamily Anophelinae

Figures 5.14 and 5.15 illustrate the criteria which distinguish the Anophelinae and Culicinae.

Anopheline eggs (Fig. 5.12*a*) are laid singly on the water surface and in most species are characteristically boat-shaped; they have air-filled floats which prevent them from sinking. Eggs of a few species, such as the desert-inhabiting *Anopheles cinereus* of East Africa and the Palaearctic *A. plumbeus*, lack floats and hang perpendicularly in the water. In contrast, eggs of *Chagasia* have multiple floats. Anopheline eggs are unable to withstand desiccation.

The siphonless larvae (Fig. 5.9*a*) have abdominal tergal plates and well-developed abdominal palmate setae (setae 1) which, together with the thoracic notched organs (Nuttall and Shipley's organs), help to maintain larvae parallel to the water surface. The Anophelinae are surface filter-feeders in which feeding is facilitated by turning the head through 180° to allow the mouthbrushes to sweep the undersurface of the water. Pupae are not readily differentiated by eye from those of culicines, but the respiratory trumpets are always short and the pinnae broad and split almost to the base (a few culicines have short and rather broad trumpets but these are not split along their length). Pupae of most species have a spine-like seta (seta 9) laterally at the corners of abdominal segments 2–7, 3–7 or 4–7, and a small accessory seta (2–P) on the paddles anterior to, and in line with, the apical seta (1–P).

The Anophelinae comprise the genera *Chagasia*, *Bironella* and *Anopheles*. Adults of most *Anopheles* rest with the head, thorax and abdomen in a straight line and held at an angle of 30°–45° to the surface (Fig. 5.14). Some species, such as *A. crucians* and *A. sinensis*, stand almost upright and a few (e.g. *A. culicifacies*) adopt a posture more or less parallel to the surface – a position more characteristic of the Culicinae. Most *Anopheles* can be recognized by the clustering of dark and pale scales into dark and

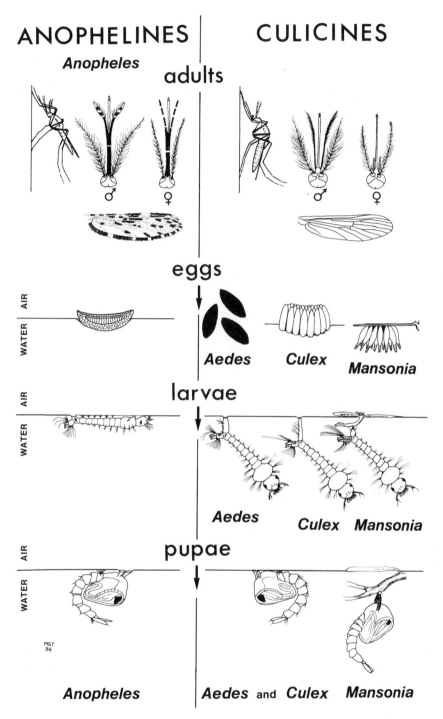

Figure 5.14 Chart of the main differences between mosquitoes of the subfamilies Anophelinae (left) and Culicinae (right).

pale blocks on the wing veins, especially along the costa (Fig. 5.7b). Some species, however, such as the North American *A. barberi* and the Palaearctic *A. claviger*, have veins covered with more or less uniform brownish scales. A more reliable method of identifying anophelines is by their palps, which are long in both sexes (except in *Bironella*), and usually blackish with narrow pale transverse bands. In males the palps are thickened apically giving them a clubbed appearance (Fig. 5.14). Palps of females sometimes have suberect scales which make them rather shaggy. The scutellum has the posterior edge smoothly rounded (except in *Chagasia*) and its setae are evenly distributed (Fig. 5.15). The most reliable method of recognizing the subfamily is by the abdominal sternites, which are devoid of scales or almost so. The tergites also usually lack scales. Females have only one spermatheca. The hum from the wing beat of Anophelinae is low-pitched and almost inaudible to humans unless the mosquito is hovering close to the ears.

There are four *Chagasia* species, three restricted to South America and one (*C. bathana*) extending northwards into southern Mexico. Adults are distinguished from other Anophelinae by their slightly trilobed scutellum and its arrangement of setae in three distinct groups. Larvae have peculiarly shaped abdominal palmate setae (seta 1) and tend to bend their bodies in the form of a U, rather like larvae of Dixidae; they occur in forest streams. The genus *Bironella* (nine species) is recognized by the wavy course of veins M_{1+2} and M_{3+4} and by often having the palps of the male considerably shorter than the proboscis. Larvae have a pair of palmate setae (1–M) on the mesothorax, and the inner clypeal hairs (2–C) are always very close together; they are found in shaded swamps. *Bironella* occurs in Australia, Papua New Guinea, the Solomon Islands, the Moluccas and Melanesia. Tenorio (1977) has revised the genus. Both *Chagasia* and *Bironella* are considered primitive culicid genera and neither has any species of medical importance.

The genus *Anopheles* has an almost worldwide distribution, but it is absent from Micronesia (except for Guam) and Polynesia. There are 422 species, some of which are found in cold northern temperate regions, others in the hot humid tropics. Adults generally bite at night inside or outside houses, and the larvae are usually found in ground collections of water, often with vegetation. *Anopheles* first gained notoriety when it was found that these mosquitoes are vectors of human malaria; later they were incriminated in the transmission of filariasis and certain arboviruses. There are six subgenera differentiated mainly on the number and position of certain spines on the gonocoxites of the male genitalia.

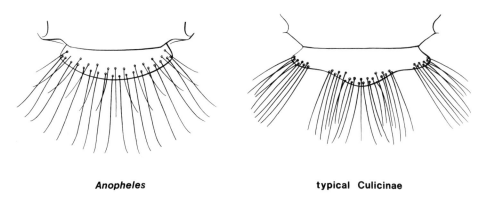

Anopheles **typical Culicinae**

Figure 5.15 Difference in posterior outline shape of the scutellum of anopheline and typical culicine mosquitoes.

Subgenus *Kerteszia* contains 11 species of Neotropical mosquitoes, all essentially forest dwellers breeding in axils of epiphytic bromeliad plants (except for *A. bambusicolus* which breeds in bamboo). Several species bite man and six have been shown to transmit malaria; however, only *A. bellator* in Trinidad and *A. cruzii* in Brazil are important vectors. *Anopheles bellator* also transmits Bancroftian filariasis.

The subgenus *Nyssorhynchus* contains 28 species and includes the dominant anophelines in the Neotropical region. The *A. albimanus* section is covered by Faran (1980). The subgenus occurs from southern USA through Central America to Argentina, and also in the West Indies. Larvae are found in ground collections of water. Several species, e.g. *A. albimanus*, *A. argyritarsis*, *A. darlingi*, *A. nuneztovari* and *A. aquasalis*, are important vectors of malaria and some also transmit Bancroftian filariasis. *Anopheles albitarsis* and *A. aquasalis* are also arbovirus vectors. The counterpart of this subgenus in the Old World is *Cellia*, which is the largest subgenus and has some 189 species. White (1979) opines that sibling species are common, so there are probably many cryptic species yet to be recognized. *Cellia* species are thought to form a relatively recent offshoot of the genus which originated in the Oriental region and slowly radiated outwards. Among the important malaria vectors are *A. aconitus*, *A. arabiensis*, *A. culicifacies*, *A. dirus*, *A. farauti*, *A. funestus*, *A. fluviatilis*, *A. gambiae*, *A. leucosphyrus*, *A. maculatus*, *A. minimus*, *A. nili*, *A. punctulatus*, *A. sergentii*, *A. stephensi* and *A. superpictus*. This large subgenus has been divided, by the arrangement of the cones and rods in the pharyngeal (cibarial) armature of the females, into the series 'Cellia', 'Neocellia', 'Myzomyia', 'Neomyzomyia', 'Paramyzomyia' and 'Pyretophorus'. Although these names have no formal nomenclatural status they are nevertheless useful for grouping species with similar biology. Each series contains vectors of malaria and usually of filariasis.

The subgenus *Anopheles* contains 175 typically rather large species and is divided into six series, only two of which, 'Myzorhynchus' and 'Anopheles', contain disease vectors, e.g. *A. freeborni*, *A. hyrcanus*, *A. maculipennis*, *A. pseudopunctipennis* and *A. sinensis*. They tend to form species groups of several similar-looking mosquitoes which can nevertheless be epidemiologically quite distinct. For example, one species of a group can be a vector of human malaria, while another almost indistinguishable species is a forest entity transmitting simian or rodent malaria.

The remaining subgenera, *Stethomyia* and *Lophopodomyia*, are Neotropical and contain in total only about 11 species of which none has medical importance.

Subfamily Culicinae

This is much the largest subfamily, containing some 2925 species in 33 genera (Table 5.1) for which it is difficult to generalize the morphological distinguishing features and biology.

Culicine eggs vary greatly in shape and method of oviposition but none possesses air-filled floats. In some genera the eggs are oval and black, laid singly and able to withstand desiccation (e.g. *Aedes* and *Psorophora*). In contrast, eggs of *Culex*, *Coquillettidia* and *Culiseta* (*Culiseta*) are longer and thinner, brownish, formed into egg rafts and cannot withstand desiccation. In *Mansonia* the eggs are stuck together in a jelly-like mass on the undersurfaces of floating aquatic vegetation (Fig. 5.12*b*).

All culicine larvae have a siphon, more than ten (usually many more) setae in the mouthbrushes and almost always a comb on abdominal segment 8+9 (Fig. 5.9*b*).

Larvae characteristically suspend themselves from the water surface by their siphons and thus hang down at an angle, although in some mosquitoes with very short siphons (e.g. *Uranotaenia*) they adopt a resting position similar to that of anophelines. Larvae of *Mansonia*, *Coquillettidia* and a few *Mimomyia* insert their siphons into plants. Larval heads do not rotate through 180° for feeding as do those of *Anopheles*. A few larvae are predaceous, but most are either filter-feeders or browsers.

Pupal trumpets can be short and conical but they are generally rather long and cylindrical, and in some species extremely long (e.g. many *Mimomyia* and some *Wyeomyia*). The openings of the trumpets do not extend as a split to the base as in *Anopheles*. The paddles either lack an accessory seta (2–P), or if one is present then it arises on a level with and lateral to the apical seta (1–P); nearly all species lack a spine-like seta (seta 9) on the corners of the abdominal segments.

When at rest adults nearly always assume a position in which the body lies parallel to the surface. The wing scales are usually uniformly brown or blackish, though areas of white, yellow or silvery scales can sometimes be present, especially at the bases of the veins. Sometimes, as in *Mansonia*, there is a mixture of pale and dark scales giving the wing a 'salt and pepper' appearance, whereas in *Mimomyia* (*Ingramia*), a few *Culex*, and some *Orthopodomyia*, the scales are arranged in pale and dark 'blocks' reminiscent of those of many *Anopheles*. The trilobed scutellum has the setae arranged in three distinct groups (Fig. 5.15). In females the palps are usually considerably shorter, in many species much shorter, than the proboscis (Fig. 5.2a). In males the palps are usually as long as or a little longer than the proboscis, pointed and with numerous fine long hairs. They are not expanded apically into clubs, but are often directed upwards (Fig. 5.23). In some genera, however, such as *Limatus*, *Phoniomyia* and *Sabethes*, the palps of males are very short.

The most reliable characters for separating all genera from the Anophelinae are the sternites and tergites which are densely covered with scales, and the palps of females which are not more than one-third as long as the proboscis.

The many culicine genera are here conveniently grouped into two main tribes, Sabethini and Culicini.

Tribe Sabethini

This tribe (see Table 5.1) comprises 12 genera found mainly in the Neotropical, Oriental and Australasian regions. The genus *Malaya*, however, occurs in Africa as well as in the last two regions. There are some 220 New World and 173 Old World species, most of which are tropical, but a few are found in New Zealand, Australia and China; one species, *Wyeomyia smithii*, occurs from northeastern USA into Canada as far as Newfoundland. It is not easy to state morphological characters that distinguish the Sabethini from the Culicini, because most of the distinctive features also occur in individual culicine genera. However, the reduction of the larval ventral brush (4–X) to one, or more rarely two, pairs of setae is more or less diagnostic for Sabethini. Many larvae have the thorax and abdomen covered with large stellate setae. (A few aberrant culicine species also lack the ventral brush or have it greatly reduced and some species have numerous stellate setae.) The curious ability of larvae to crawl from one leaf axil to another is a common trait in the Sabethini and the peculiar adult habit of arching the hind legs over the body both when in flight and at rest is a useful field character.

Some Sabethini have undergone unusual morphological modifications. The Neotropical *Sabethes*, for example, are beautifully coloured mosquitoes, with bodies covered in metallic green, blue and purplish scales, and legs clothed in black, white and purplish scales. The antepronotal lobes are exceptionally well developed (Fig. 5.16*b*). In certain species some or all tarsi have rows of long outstanding scales giving them an unusual, paddle-like, appearance (Fig. 5.16*a*). Extraordinary morphological developments have occurred in the immature stages of some sabethines; for example, the pupal trumpets of the South American *Wyeomyia circumcincta* are exceptionally long and tube-like (Fig. 5.8*d*) and in *W. confusa* the larval maxillae have a long tubular projection (Fig. 5.10*d*). Many *Tripteroides* larvae are covered with thick stellate setae and have the maxillae curiously modified into large clasping organs (Fig. 5.10*c*) thought to help them catch their prey. Somewhat similar modifications are found in larvae of some *Topomyia* and *Malaya*. Brightly coloured adults of *Tripteroides* have silvery scales on the thorax, abdomen and legs, and bear a striking resemblance to *Eretmapodites*, a genus with predaceous larvae occurring solely in Africa, but the dull-coloured *Tripteroides* resemble in several respects the culicine genus *Armigeres*. The New World genus *Trichoprosopon* (15 species), counterpart of the Old World *Tripteroides*, is difficult to separate from the latter except by the different distribution. The Neotropical genera *Johnbelkinia* (three species), *Runchomyia* (11 species) and *Shannoniana* (three species) look very similar to *Trichoprosopon*.

Adult *Malaya* are unusual in having the proboscis swollen apically in association with their extraordinary habit of feeding on regurgitated stomach contents of ants.

Many sabethines breed in leaf axils, including those of bromeliad plants, others in flower bracts and spathes, and some (e.g. *Wyeomyia smithii*) in pitcher plants. The larvae of most species are predatory. Many Old World species of *Tripteroides* and some *Topomyia* are found in tree-holes or bamboo, others utilize plant axils, *Nepenthes* pitcher plants and coconut husks, while a few breed in domestic and other artificial containers. Larvae of the Neotropical *Limatus* are predominantly

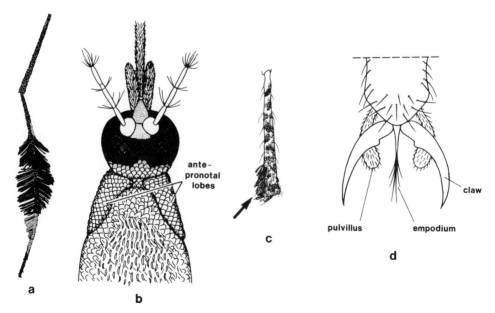

Figure 5.16 Some taxonomic characters of adult Culicinae: (*a*) a typical tarsal 'paddle' of long scales occurring on one or more pairs of tarsi of female *Sabethes*, illustrated from the mid leg of *S. belisarioi*; (*b*) showing the large antepronotal lobes of *Sabethes* which almost meet in the midline of the thorax (and can appear to do so in perfect specimens because of the dense scale cover); (*c*) hind femur of *Aedeomyia catasticta* showing large scale tufts; (*d*) tip of last tarsal segment in *Culex* showing the large pulvilli (semi-diagrammatic).

container-breeders (coconut husks, snail shells, bamboo, domestic containers), while larvae of *Trichoprosopon* occur mainly in water-filled fallen fruit husks.

The genus *Maorigoeldia* contains a single, large, beautiful blue species with silvery markings. It is found only in New Zealand. Adults are unique among Old World sabethines in having well-developed scutal setae. The large larvae, found in tree-holes and assorted man-made containers, have the thorax and abdominal base whitish and the remainder of the abdomen yellowish. Adults do not usually bite man.

Phoniomyia contains about 23 species that are all Neotropical. Adults have metallic golden and violet scales, and the proboscis is long, slender and curved apically. They bite man as well as other animals. Larvae are unusual in having a long tapering siphon with numerous long unbranched, but slightly plumose, setae, both ventrally and dorsally (Fig. 5.11*a*). Larvae occur in bromeliad axils, and more rarely in cut bamboo.

Although adult females of most Sabethini take blood-meals and sometimes bite man they are not important as vectors of disease. *Sabethes chloropterus*, however, is believed to have a role in yellow fever transmission in Central America, and in the Neotropical region various species of *Limatus*, *Trichoprosopon*, *Johnbelkinia* and *Wyeomyia* are implicated in the transmission of arboviruses to man.

Tribe Culicini
There are 21 non-sabethine genera in the subfamily Culicinae which are often placed in an all-embracing tribe, the Culicini. Many specialists, though, split the Culicinae into ten tribes and restrict the Culicini to include only *Culex*, *Deinocerites* and *Galindomyia*. This system is too complicated for the present account. Still, the problem arises as how best to group the genera. No method is entirely satisfactory because many genera have similarities to several apparently unrelated genera, and considerable morphological plasticity occurs even within a genus. Here the scheme used by Mattingly (1969), based mainly on biological attributes, is adopted and the Culicini classified into four main groups.

1. Aedine genera This group comprises *Aedes*, *Psorophora* and *Opifex*, species of which have eggs capable of withstanding many months of desiccation. Larval habitats are very diverse, and include ground collections of water such as flooded meadows, small ponds, rock-pools, leaf axils, tree-holes, bamboo and a variety of man-made container-habitats, such as abandoned tin cans, tyres and water-storage pots. Most habitats are temporary, and are liable to dry out for parts of the year. In arctic regions the habitats include pools which are covered with snow and ice for much of the year.

Aedes: This is a very large genus (962 species) containing several medically important species such as *A. aegypti*, an important vector of yellow fever, dengue and other arboviruses, and a few other species that transmit Brugian and Bancroftian filariasis. Many *Aedes* species are nocturnal biters but the disease vectors include many that bite during the day or in the early evening. *Aedes* mosquitoes are usually the principal cause of biting nuisances in northern temperate regions. Adults are commonly clothed with dark brown or black scales and contrasting white or silvery scales which give the legs and abdomen a conspicuously banded appearance and the scutum a pattern diagnostic for the species. These markings on *A. aegypti* are particularly characteristic and taxonomically diagnostic for this species (Fig. 5.17). Some species, however, have yellowish scales on the body and legs and others are almost entirely dark. The tip of the abdomen is characteristically pointed (Fig. 5.18*c*).

Eggs are black with fine reticulate sculpturing (Fig. 5.12*b*). They can withstand desiccation for months and in some species for several years. Larvae usually have short barrel-shaped siphons and there is only one pair, usually branched, of siphonal subventral tufts (1–S) (never situated near the base). Larvae are found in most types of water, including marshes, forest pools, snow-melt pools, rock-pools, natural and artificial container-type habitats such as tree-holes, bamboo, leaf axils, tin cans, water-storage pots; a few species occur in salt-water.

The genus is divided into 43 subgenera, of which only a few contain disease vectors, notably *Stegomyia* (117 species). This subgenus contains several important vectors, including the 'yellow fever mosquito', *A. aegypti*, with an almost cosmotropical distribution and one of the world's most widespread mosquitoes. *Aedes aegypti* exists in several infraspecific forms, characterized both morphologically and behaviourally. There is a dark (totally black abdominal tergites) African feral subspecies, *A. aegypti formosus*, confined to Africa south of the Sahara and mainly zoophagic; it rests and feeds out of doors and its larvae are mainly found in treeholes. *Aedes aegypti aegypti* has white scales on the tergites, is anthropophagic and peridomestic and oviposits in water-storage pots and tree-holes. Finally, the palest form, *queenslandensis*, is even more domestic, its larvae occurring in water-storage pots inside houses and its adults having the greatest propensity for biting man. There is complete gradation of colour forms from pale *queenslandensis* through typical *aegypti* to the dark *formosus*.

Other important virus vectors include *A. africanus* and *A. bromeliae* transmitting yellow fever in Africa, and *A. albopictus*, a dengue vector in the Oriental region. *Aedes albopictus* also occurs in Australasia and on some Afrotropical islands (including Mauritius, Madagascar and the Seychelles) and has recently been reported from mainland Africa (Cornel and Hunt, 1991; Savage and Miller, 1991). It recently

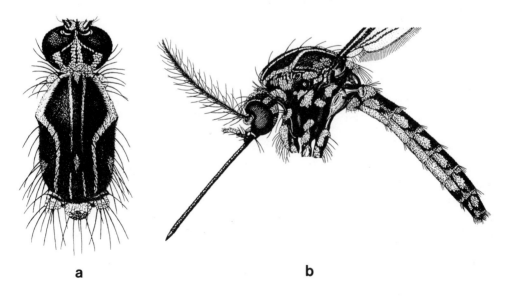

Figure 5.17 Female of *Aedes aegypti*, the yellow fever mosquito: (*a*) upper surface of head and thorax showing the distinctive pattern formed by white scales; (*b*) left side view of the body, wing and legs omitted, showing the conspicuous patches of dense white scales and pattern of the scutum when seen in profile. (Reproduced from Tanaka et al., 1979.)

150 Mosquitoes (Culicidae)

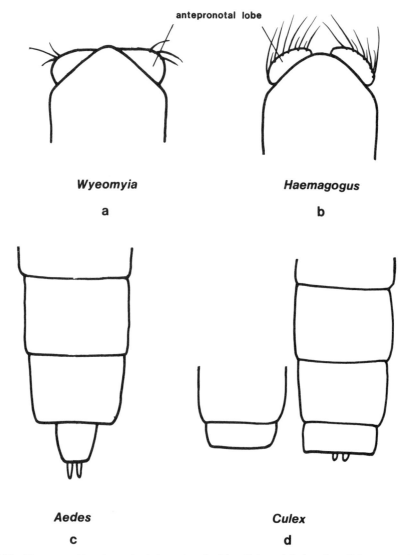

Figure 5.18 Diagrams of two important characters for identifying adult female culicine mosquitoes: (*a* and *b*) antepronotal lobes of large size in two representative genera with this feature (see key); (*c*) tip of abdomen of retractile type with conspicuous cerci, pointed and typical of *Aedes*; (*b*) tip of abdomen of non-retractile type with inconspicuous or invisible cerci, blunt-ended and typical of *Culex*.

gained notoriety by invading the USA, apparently through importation (probably from Japan) of vehicle tyres containing dry but viable eggs. By the end of 1991 it had established itself in 23 states and extended its range to Mexico. It has also spread independently to Brazil and (in 1990) to Albania and Italy (Sabatini et al., 1990). Other important and related vector species are *A. scutellaris*, *A. pseudoscutellaris* and *A. polynesiensis* of the Pacific Islands, all vectors of dengue and the latter two also of Bancroftian filariasis. Many *Stegomyia* species oviposit in tree-holes.

Aedimorphus is a large subgenus (100 species) which is mainly Afrotropical but includes a few species in all other zoogeographical regions. *Aedes vexans* has an extensive distribution, occurring not only in tropical regions (Africa, Central America, South East Asia) but also in northern temperate regions (Europe, USSR,

USA and Canada northwards to latitude 62°N). Its larvae commonly occur in flooded meadows and its adults are annoying and persistent biters which in Europe can be involved in arbovirus transmission. *Aedes vittatus* is interesting because until recently it was considered an aberrant *Stegomyia* mosquito. It oviposits in rock-pools, bites man and a variety of animals, and is widely distributed in the Oriental and Afrotropical regions, as well as occurring in the Mediterranean area.

The subgenus *Finlaya* contains 199 species almost all of which are Oriental and Australasian but some Afrotropical and a very few (e.g. *A. geniculatus*) Palaearctic. With the exception of *A. togoi*, reported from northwestern USA and southwestern Canada (where probably introduced from Japan), the subgenus is absent from the New World. Adults have dark scales on the wing veins and a few species (*A. kochi* group) have spotted wings. A conspicuous tuft of scales occurs ventrally on the abdomen in several species. Larvae of most species occur in tree-holes and plant axils, but those of the mainly Oriental *A. togoi* live in rock-pools, including many that are saline. A few species, e.g. *A. poicilius*, *A. togoi*, *A. kochi* and *A. niveus*, are important vectors of filariasis or arboviruses in the Oriental region.

Diceromyia is a small subgenus (27 species) restricted to the Oriental and Afrotropical regions. Adults are rather small mosquitoes rather similar to *Stegomyia* but with dark tarsi and broad wings with pale and dark scales reminiscent of *Mansonia*. Larvae occur mainly in tree-holes. Two species, *A. taylori* and *A. furcifer* (indistinguishable as adults), are implicated in yellow fever transmission in West Africa and in the spread of Chikungunya virus.

Most Nearctic and Palaearctic *Aedes* belong to the subgenus *Ochlerotatus* (199 species) of mainly northern temperate distribution. A few species, such as *A. diantaeus*, *A. hexodontus*, *A. pionips* and *A. punctodes*, are circumpolar and found in cold northern parts of the Nearctic and Palaearctic regions. *Ochlerotatus* is the most widely distributed *Aedes* subgenus, several species occurring in Australasia and the Neotropical region and a few in the Oriental and Afrotropical regions. Larvae of several species, including *A. taeniorhynchus* and *A. sollicitans* in North America, *A. vigilax* in New Caledonia and Australia, and *A. detritus* and *A. caspius* in Europe and the Middle East, live in salt-water; the adults can be troublesome biters. Larvae of a very few species, e.g. *A. sierrensis*, live in tree-holes and those of *A. atropalpus* in rock-pools. There are few medically important *Ochlerotatus* species, but *A. vigilax* transmits Bancroftian filariasis in New Caledonia and several viruses (including Ross River virus) in Australia. *Aedes sierrensis* and *A. atlanticus* are important vectors of dog heartworm disease (*Dirofilaria immitis*) in North America. A few other species are involved in the transmission of various arboviruses, especially those causing the encephalitides.

Thirty-three of the remaining 36 *Aedes* subgenera have fewer than 25 species each. The subgenus *Mucidus* is found in the Oriental, Australasian and Afrotropical regions. The predaceous larvae live in ground collections of water and the adults are rather mouldy-looking insects because their legs are covered with erect or semi-erect scales.

The subgenus *Protomacleaya* is almost entirely Neotropical, but a few species range into southern USA, and the tree-hole species *A. hendersoni* and *A. triseriatus* are exclusively Nearctic; the latter is found as far north as Greenland. *Aedes triseriatus*, but not apparently *A. hendersoni*, is an important vector of the California group of encephalitis viruses.

Finally, mention should be made of *Aedes (Huaedes) wauensis* from Papua New Guinea because it is unique among culicines in having the larval ventral brush (4–X) composed of plumose setae similar to those of *Toxorhynchites*; the larvae are reddish, another feature of Toxorhynchitinae.

Psorophora: The genus *Psorophora* has 47 species assembled into three subgenera. It is basically Neotropical but each subgenus contains a few species, such as *Psorophora (Grabhamia) columbiae, P. (Grabhamia) discolor, P. (Janthinosoma) ferox* and *P. (Psorophora) ciliata*, whose distributions extend into North America. *Psorophora* mosquitoes can be most vicious pests, and several species are vectors of arboviruses, such as Ilheus virus transmitted by *P. ferox* and Venezuelan equine encephalomyelitis virus spread by *P. ferox* and *P. confinnis*. Most species are floodwater mosquitoes, so-called because their eggs, laid in mud and debris of meadows, ricefields and woodlands, can withstand months of desiccation and hatch when flooded – often giving rise to enormous biting populations. Adults are recognized by having both spiracular and postspiracular setae and larvae are characterized by a row of precratal tufts in the ventral brush (4–X). Larvae of the subgenus *Psorophora* are predaceous.

Opifex: This genus contains one species, *O. fuscus* found only in New Zealand. Larvae occur in coastal rock-pools containing brackish water. Adults are large and strange-looking blue mosquitoes that were thought to be Tipulidae when first described in 1902. Males are unusual in having non-plumose antennae (Fig. 5.19a). The claim to fame of the species is the precocity exhibited by the adult males, which copulate with the females before these have fully emerged from the pupae. Adults are autogenous. When ovipositing, the female thrusts the tip of the abdomen under the water surface to place her eggs on rocks. Eggs can withstand desiccation for up to about six months. *Opifex* is generally regarded as the result of primitive divergence from the main aedine stock.

2. Quasi-sabethine genera The genera *Haemagogus, Heizmannia, Zeugnomyia, Armigeres, Udaya* and *Eretmapodites* comprising this group have a mixture of aedine and sabethine characters. Essentially they are container-breeders, larvae being found in tree-holes, bamboo, leaf axils (including those of bromeliads), rock-pools, empty snail shells, coconut husks, fallen forest leaves and domestic utensils. Eggs are not usually drought-resistant but those of many *Haemagogus* species can withstand partial desiccation, remaining viable in a humid atmosphere for several months; eggs of some tree-hole-breeding *Haemagogus* inhabiting dry regions of Mexico are fully drought-resistant.

Haemagogus: This genus is closely related to the genus *Aedes* but the small adults are readily distinguished by being more brightly coloured. The scutum and abdomen, for example, are covered with flat metallic green or blue scales, and the antepronotal lobes, which are also covered with metallic scales, are exceptionally large (Fig. 5.18b). Adults live in the jungles of Central and South America, feeding by day on monkeys in the tree canopy but sometimes descending to bite man in the jungle clearings. At least *H. janthinomys, H. equinus, H. leucocelaenus, H. spegazzinii* and *H. capricornii* are vectors of sylvatic yellow fever. Larvae are found mainly in tree-holes and bamboo; in urban areas, however, they can sometimes be found in artificial containers. Larvae of yellow fever vectors are exceptionally hairy, and this feature distinguishes them from most of the medically unimportant species. Eggs of some species can withstand desiccation for at least seven to ten months.

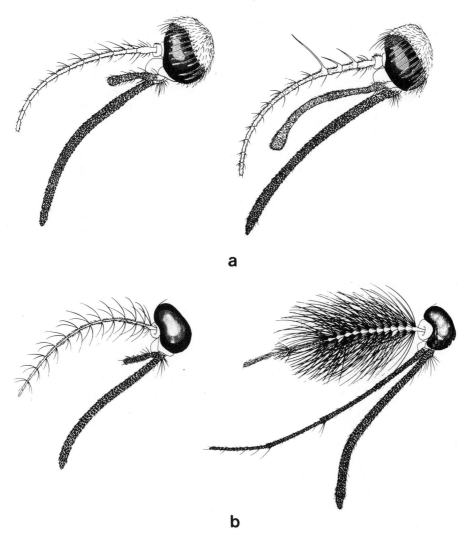

Figure 5.19 Adult heads in profile: (*a*) *Opifex fuscus*, female (left) and male (right); (*b*) *Armigeres subalbatus*, female (left) and male (right).

Heizmannia: These mosquitoes are morphologically similar to *Haemagogus* and are the Oriental counterpart of this genus. Eggs can tolerate desiccation, but apparently for shorter periods than those of *Haemagogus*. Larvae occur mainly in tree-holes and bamboo, and resemble those of *Aedes*. Adults are brightly coloured insects with the scutum covered with matt golden and bright iridescent scales, while the abdominal tergites are often silvery. Although adults commonly bite man during the day no species has yet been incriminated as a virus vector.

Armigeres: There are some 49 species in the genus *Armigeres*, which is confined to the Oriental and Australasian regions where it is both common and widespread. Adults are morphologically similar to *Aedes* but differ in having the proboscis slightly curved downwards (Fig. 5.19*b*) and laterally compressed. In the subgenus *Armigeres*, eggs are laid singly and can withstand limited desiccation. Those of *Armigeres (Leicesteria)*, however, are laid as a raft and cannot tolerate drying. In a few species, such as *A. (Leicesteria) dolichocephalus*, the thorax is abnormally narrow, allowing females to enter very small bore holes (1.5 mm diameter) in bamboo to lay their eggs on the water. Larvae are unusual in lacking a pecten; they occur chiefly in bamboo, but also in tree-holes, coconut husks, domestic utensils, *Colocasia* (cocoyam) leaf axils and *Nepenthes* pitcher plants. *Armigeres subalbatus* is a semi-domestic mosquito found in polluted waters, including latrines. Adults bite man during the day and night, and although this is not a proven vector it is suspected that they transmit *Wuchereria bancrofti*. Sepik virus has been isolated from an unidentified *Armigeres* species.

Eretmapodites: This is the only mosquito genus that is confined to the Afrotropical region. All 44 species are basically forest mosquitoes, although many have successfully adapted to life in banana plantations. Adults are thin and yellowish-brown mosquitoes with patches of broad silvery scales on the head, thorax and abdomen; they somewhat resemble *Aedes* species. The male genitalia are very complex, perhaps in relation to the unusual mating habit in which the male hangs head down from the female for over an hour – an exceptionally long time for a mosquito – and keeps twitching violently throughout its sexual encounter. The pale to dark brown eggs, rather like those of *Aedes*, are deposited above the water-line and can withstand limited desiccation. Larvae are found in water-filled fallen leaves, fruit husks, plant axils, snail shells (mainly those of *Achatina*), domestic containers, bamboo and more rarely tree-holes. *Eretmapodites subsimplicipes* and *E. grahami* larvae occur in water accumulating in the concave tops of polyporaceous fungi. *Eretmapodites* larvae are identified by having four or fewer pecten spines and the ventral brush (4–X) composed of a few single or double setae. Larvae of the *E. chrysogaster* group are immediately recognizable by the lateral setae of abdominal segments 1–6 arising from conspicuous sclerotized bosses. Larvae of most species are facultative predators. Some arboviruses, including that of Rift Valley fever, have been isolated from a few species and early laboratory experiments (in 1928) showed that *E. chrysogaster* can transmit yellow fever.

Udaya and *Zeugnomyia:* There are only three species in the genus *Udaya*, all of which are Oriental; adults resemble *Aedes* species. The genus *Zeugnomyia* is also Oriental and contains four species. Adults are rather small and easily recognized by a vertical broad stripe of silvery scales on the thoracic pleurae. Neither genus contains vector species.

3. Genera with vegetation-seeking species This group contains the genera *Aedeomyia*, *Coquillettidia*, *Ficalbia*, *Hodgesia*, *Mansonia*, *Mimomyia* and *Uranotaenia*. These are dissimilar morphologically, but all breed in larval habitats with dense vegetation. For species of *Mansonia* and *Coquillettidia*, and a few *Mimomyia*, aquatic vegetation is obligatory because the larvae and pupae pierce plants to obtain oxygen (Fig. 5.20).

The genera *Coquillettidia* and *Mansonia* have an almost worldwide distribution. Ranges of both extend into northern temperate regions.

Coquillettidia: Some taxonomists still regard *Coquillettidia* as a subgenus of *Mansonia*. Females are mostly rather large and yellowish mosquitoes, some almost

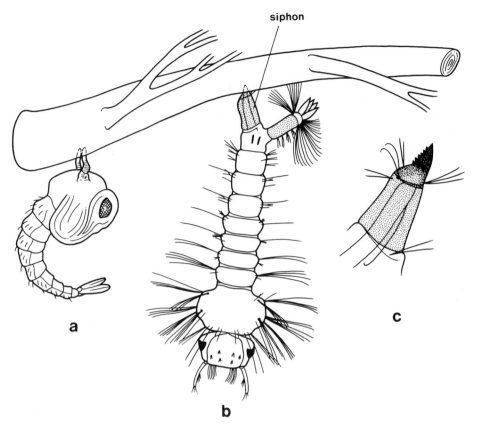

Figure 5.20 Early stages of *Mansonia* attached to a water-plant stem: (*a*) pupa; (*b*) larva; (*c*) enlargement of larval siphon.

entirely bright yellow, which bite during the day and night. Eggs are laid on the water surface and formed into rafts. Larvae are unusual in having a long apical flagellar extension hinged to the basal part of the antenna (Fig. 5.10*f*). The genus has three subgenera, *Austromansonia* confined to New Zealand, the Neotropical *Rhynchotaenia* and the mainly Afrotropical *Coquillettidia* s.str.; there are, however, some Oriental and Australasian species. *Coquillettidia perturbans* is unique in being found in the New World (Mexico to Canada), where it is a serious pest of man and livestock and a vector of Eastern equine encephalomyelitis. *Coquillettidia richiardii* is common in Europe and extends eastwards into Siberia and Syria; in many areas it causes a serious biting nuisance. In South America *C. venezuelensis* is a vector of several arboviruses.

Mansonia: *Mansonia* adults are readily recognized by their speckled wings, the veins covered with broad and asymmetrical pale and dark scales, and by their pale-banded legs (Fig. 5.21). Eggs are drawn out apically into a short filament (Fig. 5.12*b*). Females thrust the tips of their abdomens under the water and lay their eggs in a rosette-shaped mass glued to the underside of floating vegetation. Adults are mainly nocturnal biters. There are two subgenera, *Mansonia* (Neotropical and Nearctic) and *Mansonioides* (Old World). The most important man-biting species of the subgenus *Mansonia* is *M. titillans*, which is found throughout South and Central

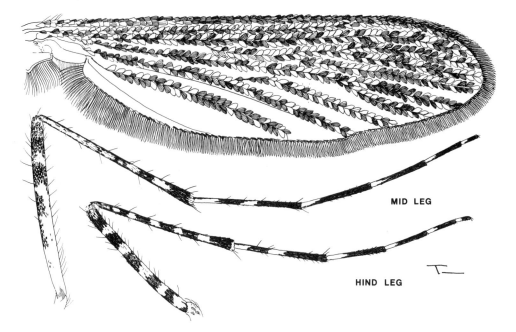

Figure 5.21 Wing and legs of a typical species of *Mansonia* mosquito showing (a) speckled appearance of the wing caused by admixture of broad dark and pale scales along the veins, and (b) the characteristic dark and pale banded legs. The species illustrated is *M. uniformis*.

America and in some southern states of the USA; it is a vector of various arboviruses, including Venezuelan equine encephalomyelitis.

The most widespread species of the subgenus *Mansonioides* is *M. uniformis*, which occurs from West Africa, through the Indian subcontinent eastwards to Japan and the Australasian region. Several arboviruses are transmitted by *Mansonioides* species, but their greatest importance is as vectors of Brugian filariasis in India and South East Asia; the most important vectors are *M. annulata*, *M. annulifera*, *M. bonneae*, *M. dives*, *M. indiana* and *M. uniformis*. They are generally refractory to the *Wuchereria bancrofti* parasite, except that *M. uniformis* is a vector of Bancoftian filariasis in western New Guinea (Irian Jaya).

Mimomyia and *Ficalbia*: *Mimomyia* has been considered a subgenus of *Ficalbia*, but is now treated as a distinct genus with three subgenera. There are about 42 species widely distributed in the Old World tropics; two species are eastern Palaearctic and another two extend into northern Australia. The genus is absent from the New World.

Ficalbia includes seven species, four of which are Afrotropical and three Oriental; the distribution of one of the latter (*F. minima*), extends into Australasia.

Larvae of most species of *Mimomyia* and *Ficalbia* are found in swamps and marshes. Those of a few species, such as *M. pallida*, *M. modesta* and *M. hybrida*, have a larval siphon that is inserted into plants to obtain oxygen (Fig. 5.11e), but the structural modifications are less elaborate than in *Mansonia* and *Coquillettidia*. Larvae of *Mimomyia (Ingramia)* are, however, plant-axil breeders; 16 species are unique to Madagascar – none occurs on mainland Africa – and three are Oriental. Many species have conspicuous stellate setae on the thorax (Fig. 5.22b) and abdomen while some, such as *M. levicastilloi*, have remarkable head setae (Fig. 5.22a). Neither *Mimomyia* nor *Ficalbia* has any medical importance, and only a few species are anthropophagic.

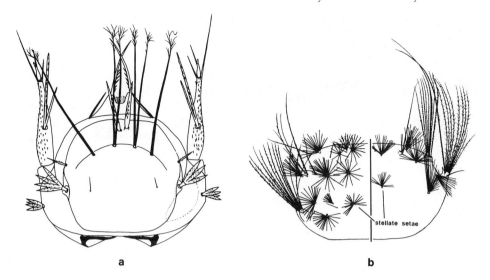

Figure 5.22 Larval setae in *Mimomyia*: (*a*) head setae of *M. levicastelloi*; (*b*) thoracic setae of *M. roubaudi*, dorsal surface (left of line) and ventral surface (right of line).

Hodgesia: There are 11 species restricted to the Old World tropics and all very small. A few species bite man but have no medical importance. In some species there is sexual dimorphism, adult males having peculiar modifications such as conspicuous tufts of setae on the legs which are absent in females. Larvae are mainly found in marshes.

Uranotaenia: The genus comprises some 193 species and has two subgenera, *Uranotaenia* s.str. and *Pseudoficalbia*. Most species are tropical, but several occur in Australia and *U. unguiculata* in Europe and the Middle East. The distribution of *U. anhydor* and *U. lowii* extends from the Neotropical region into the southern USA, *U. sapphirina* is found in the West Indies and from Mexico to Canada. Larvae of most species live in marshes and other ground collections of water-containing vegetation, but several are found in plant axils, bamboo, tree-holes and rock-pools. Adults are small, and some have silvery or metallic bluish scales. A few species bite man but none is a disease vector. Several species feed on birds, but *U. lowii* is an exceptional mosquito which bites frogs and toads.

Aedeomyia: There are six *Aedeomyia* species, three of which are Afrotropical, one Australian, one Neotropical and one widespread in the Oriental and Australasian regions. Adults are small and very scaly mosquitoes with small tufts of semi-erect scales on the legs (Fig. 5.16*c*). Although adults bite man, and have been incriminated as vectors of a few arboviruses, they are not generally considered to be important disease vectors. Larvae are remarkably hairy, having the body covered with beautiful plumose setae, some of which are stellate (Fig. 5.10*a,b*); larvae also have exceptionally long lateral plumose thoracic setae. The antennae are unusual in being greatly swollen (Fig. 5.10*b*). The body is weakly sclerotized and tracheation greatly reduced. Larvae can stay submerged for long periods, perhaps because respiration is mainly cuticular.

4. Miscellaneous genera This assorted collection comprises *Culex*, *Culiseta*, *Deinocerites*, *Galindomyia* and *Orthopodomyia*.

Culex: This is by far the largest, commonest and most important genus of the group. There are some 751 species arranged in 22 subgenera, 13 of which have ten or fewer species. *Culex* mosquitoes are found in all zoogeographical regions, their distribution ranging from hot humid tropics to cool temperate regions but not extending to the extreme northern latitudes where *Aedes* have so successfully established themselves.

While most *Culex* species form their eggs in a raft that floats on the water, some species are unusual in depositing the raft on aquatic vegetation. Even more unexpectedly, a very few species (e.g. *C. alogistus* and *C. pilosus*) lay their eggs singly, not in a raft, while the Central American *C. gaudeator* lays her eggs on the water surface in a gelatinous mass. *Culex* eggs, with few exceptions, cannot withstand desiccation and hatch within a few days. *Culex* mosquitoes breed in ground collections of water, such as ditches, small pools and ponds. Several *Culex* species, for example *C. tritaeniorhynchus* and *C. tarsalis*, are common in marshes and ricefields, while other species colonize domestic utensils, bamboo, tree-holes and other container-habitats. A few species are found in organically polluted waters; for example larvae of the important filariasis vector *C. quinquefasciatus* are commonly found in cesspits. (For many years, this mosquito was known as *C. fatigans* or *C. pipiens fatigans*, but most experts now agree that the name *quinquefasciatus* is correct.)

The larval siphon can be short, and even barrel-shaped, or extremely long and narrow, but there is always more than one pair of subventral tufts (1–S). Adults are usually dull-brownish mosquitoes with brown or black wing scales, but some species of *Culex* (*Culex*) have areas of pale and dark scales on the wings somewhat like *Anopheles*. In females the abdomen is blunt-tipped (Fig. 5.18*d*), and in males the palps rather hairy and often upturned (Fig. 5.23). Because *Culex* adults lack conspicuous ornamentation they can be difficult to identify, and females of some species are identifiable only by examination of the pharyngeal armature; in males identification often has to be made on the complex structures of the genitalia. It can be difficult to identify individual species, but the genus is easily recognized by the presence of tarsal pulvilli (Fig. 5.16*d*) and the absence of postspiracular setae.

Culex mosquitoes mainly bite at night. Many species feed predominantly or exclusively on birds, a few on amphibians and reptiles, and several on mammals (including man); some feed on both birds and mammals. Because eggs cannot tolerate desiccation there is continuous breeding throughout the year where this is environmentally possible. In cold climates inseminated females develop large fat reserves and overwinter in man-made or natural shelters such as cellars and caves. Some arboviruses can survive the winter in hibernating females, but whether or not this is important in the epidemiology of viral diseases generally remains obscure.

Some subgenera are restricted to certain regions. *Melanoconion* for example, is virtually confined to the Neotropical region, but eight species extend their ranges into the USA. On the other hand, *C. abominator* is found only in the USA. Several species of the subgenus *Melanoconion*, such as *C. taeniopus*, are important vectors of encephalitis and other arboviruses. Other subgenera, including *Culiciomyia*, *Eumelanomyia* and *Lophoceraomyia*, occur only in the Old World.

In contrast, the subgenus *Culex*, which has the most species, has an almost worldwide distribution and contains most of the medically important species, especially the vectors of the encephalitis viruses. *Culex tritaeniorhynchus*, *C. gelidus* and *C. vishnui* transmit Japanese encephalitis in the Oriental region and *C. tarsalis*, *C. nigripalpus*, *C. restuans* and *C. pipiens* are among the vectors of Eastern and

Figure 5.23 Head of male *Culex* showing the very bushy and, in this genus, characteristically upturned palps.

Western equine encephalomyelitis and St Louis equine encephalitis in North America. Murray Valley encephalitis is spread in Australia by *C. annulirostris*. *Culex univittatus* transmits West Nile fever in Africa. Rift Valley fever is spread in Egypt by *C. pipiens* and *C. quinquefasciatus* and in southern Africa by *C. theileri*. Many of these viruses are zoonotic in birds and/or mammals, and most of the vectors feed on these hosts in addition to man.

Culex quinquefasciatus is the most important urban vector of Bancroftian filariasis throughout most of the tropics. It belongs to the *C. pipiens* complex, which comprises several species and infraspecific forms.

Culiseta: The genus *Culiseta*, formerly known as *Theobaldia*, contains about 35 species. Most of these have a temperate distribution, and some, such as *C. alaskaensis*, *C. silvestris* and *C. incidens*, extend into northern or subarctic areas. Only about seven species occur in tropical areas. None occur in South America, but two are found in Central America. Although *Culiseta* is a small genus there is considerable morphological diversity and it is divided into seven subgenera, the most important being *Culiseta* and *Culicella*.

Adults are fairly large and superficially resemble those of *Culex*, but their tarsi lack pulvilli and the wings of several species (subgenus *Culiseta*) have one or two dark median spots of scales. In most species eggs are formed into rafts that float on the water, but in some, such as *C. (Culicella) morsitans* of North America, North Africa and the Palaearctic region east to Siberia, the rafts are also deposited on damp leaf litter just above the water-line. In England, *C. morsitans* has been seen laying eggs individually on leaf litter. *Culiseta* eggs cannot withstand desiccation, but at least in *C. morsitans* can remain viable for six months on damp surfaces.

Larval habitats are usually ground collections of water with submerged vegetation, such as ponds, ditches, swamps, and ricefields, but in Africa *C. fraseri* occurs in tree-holes, and in Europe *C. longiareolata* is often found in wells and rock-pools. In subtropical and tropical regions breeding can continue throughout the year, but the species of temperate regions overwinter as larvae or hibernating adults, or more rarely as eggs on damp substrates. A few species, such as the European *C. annulata*, are unusual in overwintering as larvae or as hibernating adults that periodically take blood-meals. Several species, such as *C. melanura*, feed on birds, others on various mammals, and a few bite reptiles. Some species can be annoying pests, and in North America *C. inornata*, *C. melanura* and *C. dyari* are vectors of both Eastern and Western equine encephalomyelitis.

Deinocerites: This is a primitive New World genus. There are only about 18 species, most of which are Neotropical, but a few, including *D. cancer*, also occur in the USA. In many respects adult *Deinocerites* resemble *Culex* mosquitoes but differ in that the antennae of both sexes are non-plumose and longer than the proboscis, and the palps are very short (Fig. 5.24a). The larval head capsule has a pair of peculiar small lateral pouch-like extensions the function of which is unknown.

The mating behaviour is reminiscent of that in *Opifex fuscus* as copulation takes place precociously. All species breed in crab-holes. The eggs are laid above the water-line on the tunnel walls and after hatching larvae crawl down to the water – an unusual tactic for mosquitoes. The eggs cannot withstand desiccation.

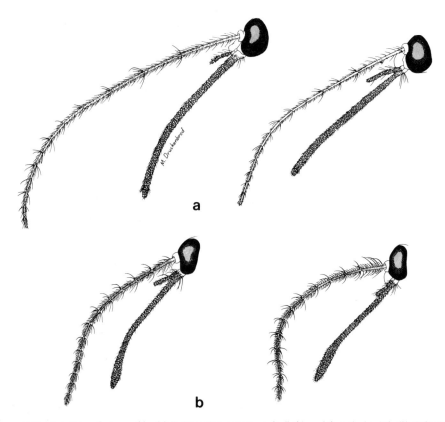

Figure 5.24 Adult heads in profile: (*a*) *Deinocerites cancer*, male (left) and female (right); (*b*) *Galindomyia leei*, male (left) and female (right).

Adults feed on a variety of mammals and birds and occasionally on man. *Deinocerites pseudes* can be a vector of St Louis encephalitis and Venezuelan equine encephalomyelitis viruses.

Galindomyia: This genus contains one species, *G. leei* from Colombia. Adults superficially resemble *Culex* but as in *Deinocerites* the antennae of both sexes are non-plumose and longer than the proboscis, and the palps are very short (Fig. 5.24b). *Galindomyia leei*, however, differs in that the proboscis is swollen apically in both sexes. The species has no medical importance.

Orthopodomyia: This is a relatively small genus with a basically Neotropical and Oriental distribution. *Orthopodomyia alba* and *O. signifera*, however, extend their range into the USA and Canada. A very few species occur in the Afrotropical region and only *O. pulcripalpis* occurs in the Palaearctic region. Adults are rather strikingly marked mosquitoes with bands and stripes of whitish, silvery or sometimes golden scales, or mottled with pale and dark scales. Larvae, like those of *Armigeres*, are unusual in lacking a pecten and in that the two main tracheae end in large thoracic air sacs.

Larvae occur almost exclusively in tree-holes and bamboo, although a few Neotropical species are found in bromeliads and in spathes of *Heliconia* (Heliconiaceae). Eggs are almost completely surrounded by a longitudinal gelatinous flange which helps them adhere just above the water-line to the inside surfaces of tree-holes. Eggs can probably tolerate some degree of desiccation. In Northern temperate areas species overwinter as eggs, larvae or adults, but as *Orthopodomyia* are rather rare and elusive relatively little is known about the biology. The feeding habits remain largely unknown, but birds are probably important hosts. *Orthopodomyia* mosquitoes are not regarded as disease vectors. Eastern equine encephalomyelitis has been isolated in Mexico from *O. signifera*, but as this species is almost entirely zoophagic it cannot be an important vector of the virus to man.

Species complexes

It is believed that most mosquito species differ sufficiently in their morphology to be readily recognized as distinct species – though this is arguable. However, several supposedly single species have been found to consist of several similar-looking species called sibling or isomorphic species. Groups of such morphologically similar species comprise species complexes, members of which are reproductively isolated. Crosses between sibling species of *Anopheles* often produce viable hybrids but the resulting F_1 adults are usually sterile, whereas in culicines many crosses between members of a complex produce viable fertile offspring with reduced fertility in successive hybrid generations (Barr, 1981; Rai, 1982).

Although crossing experiments can detect species complexes they are laborious and can be very difficult, so attention has mainly focused on biochemical methods and cytotaxonomy. The most commonly employed and useful biochemical method is enzyme electrophoresis (zymotaxonomy). Sympatric sibling species of the European *Aedes detritus* and *Aedes caspius* were first recognized by electrophoresis. In both instances proof of speciation rested on the lack of heterozygosity of the diagnostic isoenzymes in specimens breeding together in the same larval habitats. Electrophoresis has also proved useful in separating the sibling species *Aedes triseriatus* and *Aedes hendersoni*; the former is the primary vector in the USA of La Crosse encephalitis, while *A. hendersoni* does not appear to transmit any important arboviruses. Several other species complexes and groups have been investigated by zymotaxonomy (see Cianchi et al., 1985; Munstermann, 1988; Narang et al., 1990; Service, 1988).

The most widely used method for identifying species within *Anopheles* species complexes is cytotaxonomy, based on the banding patterns of the polytene chromosomes in the larval salivary glands and ovarian nurse cells of adults (Fig. 5.25). Mosquitoes have three pairs of chromosomes (2n = 6), except for *Chagasia bathana* which has four pairs (2n = 8). In the Anophelinae, but not the Culicinae, there is a pair of distinctly recognizable heterosomes, the so-called sex or XY chromosomes. The first new mosquito species to be described almost entirely on its chromosomes was *Anopheles beklemishevi* of the *A. maculipennis* complex. *Anopheles engarensis* of the *A. hyrcanus* group was also described mainly from its chromosomes. At the infraspecific level, chromosomal inversion polymorphism has sometimes been associated with behavioural or morphological characteristics. For

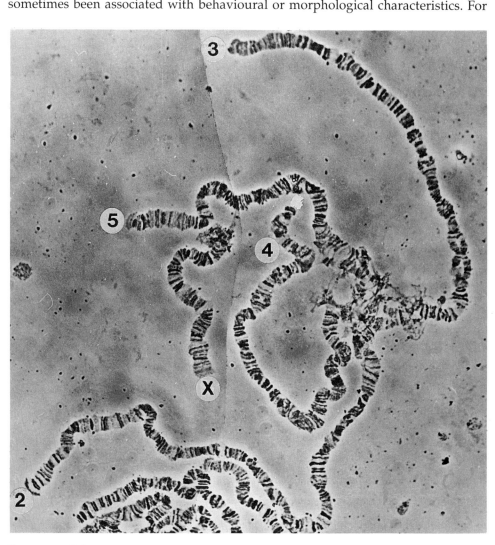

Figure 5.25 Polytene chromosomes from adult ovarian nurse cells of *Anopheles gambiae*. Landmark features labelled are the species-diagnostic X arm and the four autosomal arms (2–5).

example, in *Anopheles arabiensis*, an African malaria vector of the *A. gambiae* complex, certain inversions are believed to be associated with outdoor resting habits and degree of man-biting. In India the vectorial capacity of the malaria vector *Anopheles stephensi* differs between urban and rural populations: although there are no apparent pre- or postcopulatory barriers between these populations, urban populations are highly polymorphic in their polytene chromosomes, whereas rural populations are almost monomorphic (Coluzzi et al., 1985).

Although polytene chromosomes have been of enormous value in unravelling *Anopheles* species complexes, they have been of virtually no use with culicine complexes because the species do not have readable chromosomes. It is, for example, notoriously difficult to get good preparations of *Aedes* chromosomes, although sometimes reasonably good polytene chromosomes can be obtained from *Culex* and *Mansonia* mosquitoes. On the other hand, beautiful chromosome spreads have been made from the larval salivary glands of *Orthopodomyia pulcripalpis*, *Wyeomyia smithii* and *Toxorhynchites* and *Sabethes* species.

Good reviews on problems of mosquito identification and cytogenetics are given by Kitzmiller (1976), Pal et al. (1981), Steiner et al. (1982), White (1979, 1984) and Wright and Pal (1967). Service (1988) contains several papers dealing with mosquito systematics. Only a few of the better known mosquito complexes are discussed here.

Anopheles gambiae complex

The pioneering chromosomal work of Coluzzi and his colleagues showed that the African *Anopheles gambiae* morphospecies is a complex of six species, four of which are freshwater species. One of these is *A. gambiae* s.str. which is highly anthropophagic and rests predominantly in houses (endophilic). It is widespread in Africa, though better adapted to wetter regions than to savanna areas. The other common species of the complex is *A. arabiensis*, which although it often bites man and rests indoors also commonly bites cattle and rests out of doors (exophilic). It is widely distributed and often sympatric with *A. gambiae*, but is commoner in drier savanna areas; it is the only species of the complex known from the Sudan, and is common in Ethiopia where *A. gambiae* is unknown. The third freshwater member of the complex, *A. quadriannulatus*, is zoophagic and exophilic, and has a patchy distribution in eastern and southern Africa, Ethiopia and Zanzibar. The fourth species, *Anopheles bwambae*, is restricted to the Semliki forest in the Rift Valley near the Zaire border, where it breeds in geothermal mineral springs, and is markedly anthropophagic (White, 1985).

The two salt-water species of the complex are *A. melas*, which is found along the West African coast and is basically endophilic and anthropophagic, and the East African predominantly coastal *A. merus*, which is more exophilic than endophilic, but nevertheless frequently feeds on man. The species has recently been found breeding in freshwater habitats in South Africa.

Anopheles gambiae and *A. arabiensis* are sympatric in many parts of their range, and are also sympatric with *A. merus* in East Africa and *A. melas* in West Africa. In a few inland localities *A. gambiae*, *A. arabiensis* and *A. quadrinannulatus* are sympatric, while in some localities in southern Mozambique these three species coexist with *A. merus*.

Salinity tests can distinguish *A. merus* and *A. melas* from the freshwater species, but there are no reliable morphological methods for distinguishing species of the *A.*

gambiae complex and cytotaxonomy remains the cornerstone for species discrimination. Gillies and de Meillon (1968) and Gillies and Coetzee (1987) give useful accounts of the chromosomal taxonomy of species in the *A. gambiae* complex.

The most important malaria vectors are *A. gambiae* and *A. arabiensis*. However, because the former is more endophilic than *A. arabiensis*, insecticidal house-spraying, as practised in malaria control campaigns, is more effective against it than against *A. arabiensis*. *Anopheles merus* is not such an efficient malaria vector as *A. gambiae* or *A. arabiensis*; the West African *A. melas*, on the other hand, is a good vector. *Anopheles melas* commonly rests in houses and is therefore more easily controlled by house-spraying than *A. merus*. There is no need to control *A. quadriannulatus* as it is mainly zoophagic and hence not a malaria vector. Although *A. bwambae* bites man and can transmit malaria, its very restricted distribution makes it only an insignificant local vector compared to some other species. All species, except *A. quadriannulatus*, are also vectors of Bancroftian filariasis to a greater or lesser degree.

Anopheles maculipennis complex
Over 50 years ago it was realized that *A. maculipennis*, a common mosquito associated with malaria in Europe, existed in a variety of forms with differing egg morphology, so-called 'egg-types'. A little later it was discovered that *A. maculipennis* was common in areas where there was little or no malaria and this led to the idea of 'anophelism without malaria'. Careful classical morphological, biological and hybridization experiments showed that *A. maculipennis* was composed of sibling species, and that some of these, such as *A. maculipennis* s.str. and *A. messeae*, are basically zoophagic and thus played no part in malaria transmission, whereas others, such as *A. atroparvus*, *A. labranchiae* and *A. sacharovi*, were vectors. This at last explained the often focal nature of malaria transmission in Europe. The *A. maculipennis* complex was the first mosquito complex to be discovered and is therefore of historical interest. Initially species discrimination was in part based on egg morphology and wing length, but it was later shown that most species could be identified by their larval salivary gland chromosomes (*A. atroparvus* and *A. labranchiae* are homosequential) and many by isoenzyme electrophoresis.

There are at least 15 species within the complex. In the Palaearctic region the complex comprises *A. atroparvus*, *A. beklemishevi*, *A. labranchiae*, *A. maculipennis* s.str., *A. martinius*, *A. melanoon*, *A. messeae*, *A. sacharovi*, *A. sicaulti* and *A. subalpinus*. Some of these species are sympatric, others allopatric; *A. sicaulti* is found only in Morocco and *A. beklemishevi* and *A. martinius* are known only from the USSR. Some species overwinter in cold northern climates as hibernating adults (e.g. *A. maculipennis*, *A. messeae* – the most widespread species of the complex – and *A. atroparvus*), whereas others are restricted to warmer climates (e.g. *A. sacharovi*, *A. melanoon* and *A. labranchiae*). Although most species breed in fresh water, *A. atroparvus*, *A. labranchiae* and *A. sacharovi* are common in salt-water habitats and particularly abundant in coastal areas.

There are five Nearctic species, all of which are more or less allopatric. They comprise *A. aztecus* of Mexico, *A. freeborni* of western USA, *A. occidentalis* of northwestern USA, *A. earlei* ranging from northeastern USA across to almost the eastern limit of the *A. freeborni* distribution, and *A. hermsi* from southern California. Useful references are given by Kitzmiller (1977) and White (1978).

Other Anopheles species complexes
There are several more anopheline complexes besides those mentioned above. These include: *Anopheles quadrimaculatus* complex containing four North American species; *A. punctulatus* complex containing *A. punctulatus* s.str., *A. koliensis* and siblings 1, 2 and 3 of *A. farauti* in tropical Australasia (mainly New Guinea and Solomon Islands); *A. culicifacies* complex (four species), *A. dirus* complex (eight species), *A. leucosphyrus* complex (22 species) and *A. maculatus* complex (four species) of the Oriental region; and the *A. marshalli* complex containing four African species. *Anopheles albitarsis, A. darlingi* and *A. nuneztovari* of the Neotropical region are probably also distinct species complexes because each contains allopatric, and sometimes sympatric, populations which differ chromosomally and in their efficiency as malaria vectors.

Culex pipiens complex
Because of the poor quality of polytene chromosomes in *Culex* mosquitoes, cytotaxonomy is of little use for distinguishing species within the *C. pipiens* complex. Some species of the complex differ morphologically in minor details, and males can sometimes be distinguished by their genitalia; the same is true for the temperate *C. pipiens* from the tropical *C. quinquefasciatus*. However, species within the complex are mostly different in their behaviour and biology.

Laboratory crosses can indicate the degree of closeness between various members of the complex, but hybridization does not necessarily mean that this occurs in nature, where precopulatory isolating mechanisms can prevent interbreeding. In fact, neither morphological characters nor cross-mating experiments have resolved the complexity of the *C. pipiens* complex. One difficulty is that the complex is still in the process of speciation, so even within a distinct species, such as *C. pipiens* s.str., not all populations are interfertile. Infertility in this instance is due to cytoplasmic incompatibility caused by symbiotic rickettsia-like microorganisms (RLMOS) (*Wolbachia pipientis*) in the gonads of both sexes. Males lacking symbionts (i.e. aposymbiotic) are compatible with all populations of *C. pipiens*, whereas aposymbiotic females are compatible only with aposymbiotic males. Adults harbouring symbionts are fertile with the opposite sex only so long as their RLMOS are of the same type. Aposymbiotic strains can be produced by rearing larvae in water containing tetracycline (see Magnin and Pasteur, 1987 for references). The evolutionary role of *Wolbachia*-induced incompatibility remains a fascinating but unresolved issue; Miles and Paterson (1979) have reasoned that the symbiont-free condition is ancestral. *Wolbachia* symbionts also occur in the *Aedes scutellaris* complex.

The two principal species of the *C. pipiens* complex are *C. pipiens* s.str. and *C. quinquefasciatus*. *Culex pipiens* s.str. is essentially a temperate Holarctic mosquito, although it occurs in highland areas of eastern Africa, at lower altitudes in the cool climate of southern Africa, and in southern Argentina, Australia and Asia. It is essentially a bird-biting eurygamous species; in temperate areas female adults with well-developed fat reserves undergo winter diapause (hibernation). In autumn, diapause is initiated by shortening of daylength and low temperatures, factors which cause adults to switch from biting birds to feeding on sugar. After diapause ends in the spring females take blood-meals and develop eggs. Breeding occurs in water-butts, tanks, ponds and ditches. *Culex quinquefasciatus*, the southern and tropical counterpart of *C. pipiens* s.str., is the world's third most commonly distributed mosquito (*Aedes vexans* and *Aedes aegypti* being more cosmopolitan but usually less abundant). It differs from *C. pipiens* s.str. in that it bites man, is an important urban vector of Bancroftian filariasis, is stenogamous and has non-hibernating adults. The two species can be distinguished by their male genitalia. In some localities they are sympatric, and sometimes, as in parts of the USA and Japan, natural hybrids occur. Hybrids are recognized by having genitalia intermediate in structure between those of the two parent species. Because of such hybridization, some believe that *C. pipiens* and *C. quinquefasciatus* should be considered subspecies. Larvae of *C. quinquefasciatus* occur in organically polluted waters, including latrines, septic tanks and soakaway pits. It is autogenous at low frequency in Tanzania.

A form that has previously been called *C. molestus*, or *C. autogenicus*, is indistinguishable morphologically from *C. pipiens* s.str. and is now considered synonymous with it even though it differs biologically. The adults are autogenous, man-biting and stenogamous, and do not hibernate. Breeding is continuous through the winter months and larvae are unusual in colonizing underground, or otherwise covered and dark, collections of water. (The 'form' is responsible for transmitting Bancroftian filariasis in Egypt and Turkey.)

Culex pallens (Japan, China, Mexico, western and middle USA) is taxonomically very close to *C. pipiens* s.str. and variously regarded as an infraspecific form of it or as a distinct species. Biologically, *C. pallens* is intermediate between *C. pipiens* s.str. and *C. quinquefasciatus*. Adults are anautogenous, eurygamous and anthropophagic, and can be important vectors of Bancroftian filariasis in Japan and China. Lastly there are *C. globocoxitus* (Australia) and *C. australicus* (Australasia): the former can interbreed with *C. quinquefasciatus* (but hybrids die) and the latter has male genitalia that are intermediate between those of *C. pipiens* and *C. quinquefasciatus*.

The complete nomenclatural synonymy for *C. pipiens* and *C. quinquefasciatus* is set out on pp. 229–234 of Harbach (1988).

Aedes scutellaris complex
This complex includes about 35 forms scattered over the Pacific and parts of South East Asia and northern Australia. Nearly all species of the complex are container-breeders, larvae being found in coconut husks, coral rock-pools, tree-holes and sometimes plant axils and domestic containers, including water-storage pots. *Aedes kesseli*, however, is unusual as its larvae have been collected from ground waters, including borrow pits.

Speciation seems to have run riot with different species evolving on various islands or archipelagos. For example, *Aedes cooki* is found only on Niue and the Tonga Islands, *A. pseudoscutellaris* is restricted to Fiji, *A. kesseli*, *A. tongae* and *A. tabu* to Tonga, *A. upolensis* to Samoa, *A. futunae* to Îsle de Horn (Wallis and Futuna group), and *A. rotumae* to Rotuma (northern outlier island associated with Fiji). *Aedes hebrideus* and *A. pernotatus* occur on small islands of the Santa Cruz archipelago in Vanuatu. The namesake of the complex, *A. scutellaris*, is found in Papua New Guinea, the nearby islands of Seram, Ambon and Aru, and in Timor and Queensland, but the very closely related *A. katherinensis* is known only from northern Australia. A few species, however, are more widely distributed, e.g. *A. polynesiensis* found on numerous Pacific islands from Fiji through Polynesia to Pitcairn and Easter Island.

The intractable problem of preparing readable polytene chromosomes from *Aedes* has meant that the relationships of the members of this complex have had to be determined largely by time-consuming crossing experiments; consequently these have been made between few members of the complex, and the exact status of many entities remains undetermined. However, we know that some taxa are completely incompatible, including *A. polynesiensis* and *A. pseudoscutellaris*, whereas others show varying degrees of compatibility. Crosses between *A. scutellaris* from the Papua New Guinea area and *A. malayensis*, its counterpart in Malaysia, are fully compatible but those between *A. scutellaris* and *A. polynesiensis*, and between *A. pseudoscutellaris* and *A. polynesiensis* are sterile. As in *Culex pipiens* complex, rickettsia-like symbionts occur in the gonads of at least some species, and it seems that crossing relationships can sometimes be determined by maternally-inherited symbionts.

Some species, including *A. polynesiensis*, *A. pseudoscutellaris*, *A. tabu* and *A. cooki*, are vectors of subperiodic Bancroftian filariasis, and a few, including *A. polynesiensis*, are also dengue vectors. Others, *A. polynesiensis* and *A. cooki* among them, are experimentally susceptible to Brugian filariasis. In contrast, *A. malayensis* and *A. scutellaris* s.str. are refractory to filarial infections. Apparently there are differences between the isoenzymes of susceptible and refractory species (Townson et al., 1977). The complex is reviewed by Meek (1988), and Hartberg (1975) reviews reproductive isolation and phylogeny in the genus *Aedes*.

Other culicine complexes
Several other culicine mosquitoes exist as complexes of groups. For example, there are complexes in *Trichoprosopon* in the Neotropical region and in *Aedes communis*, *A. punctor* and *A. excrucians* of the Holarctic realm. *Aedes simpsoni* is a complex in Africa, and *Culex sitiens* and *C. vishnui* are complexes in the Oriental region. Knowledge is mostly limited to morphology and behavioural differences, extremely little being understood about the genetic affinities of the species comprising these complexes.

Identification key to adult females of principal anthropophagic genera

This key is to genera containing man-biting species that are involved, or might be involved, in transmitting diseases to humans. Species of *Toxorhynchites* do not suck blood, but this genus has been included because specimens attract attention by their large size and metallic colours. Figure 5.6 illustrates the parts of the thorax featuring in the key. Special care needs to be taken not to confuse prespiracular setae (which are sometimes difficult to see) with postpronotal setae.

1 Proboscis seen in profile straight or recurved near middle at an angle of not more than 40°, long or short (e.g. Figs 5.17*b*, 5.19*b*, left). Small or medium-sized mosquitoes with wing spread of 5–17 mm.. 2
— Proboscis seen in profile strongly bent, recurved near middle at approximately 90° (Fig. 5.13*b*), long. Exceptionally large mosquitoes with wing spread of 12–24 mm. [Abdomen metallic green, blue or violet and bearing tufts of hair-like scales on sides of apical segments (Fig. 5.13*c*): widespread tropical areas, also Japan and Nearctic region].. **Toxorhynchites**
2 Palps short, usually 0.20 to 0.33 times as long as proboscis (Figs 5.2*a*, 5.4, 5.19*b*, left). Scutellum usually trilobed (Fig. 5.15, right). Wing scales more or less

uniformly coloured, not usually in dark and pale blocks (but wing sometimes appearing mottled because of mixed dark and pale scales). Abdominal sternites and tergites with dense covering of scales 3

— Palps long, about equal in length to proboscis (Fig. 5.14, upper left). Scutellum evenly rounded (Fig. 5.15, left). Wing scales often arranged in discrete black and white blocks (Fig. 5.7*b*). Abdominal sternites, and usually tergites, entirely or mainly without scales .... **Anopheles**

3   Antennae short and with tuft of scales on first flagellomere. Thorax without prespiracular scales or setae. Mid and hind femora with large apical tufts of semi-erect scales (Fig. 5.16*c*). [Usually conspicuously scaly mosquitoes, absent from Nearctic and Palaearctic regions] .... **Aedeomyia**

— Without such combination of characters...4

4   Thorax without prespiracular and postspiracular setae. Wing veins with small broad scales (except sometimes on apical part of wing); alula bare or with broad decumbent scales (Fig. 5.26*a,d*). Tarsal claws toothless (simple). [Old World only and mainly tropical] .... **Mimomyia**

— Without such combination of characters.. 5

5   Antennae extending beyond end of proboscis by distance at least equal to length of its last four flagellomeres, and with first flagellomere at least three times as long as apical flagellomere (Fig. 5.24*a*). Thorax without postspiracular setae. [Nearctic and Neotropical regions] .... **Deinocerites**

— Without such combination of characters... 6

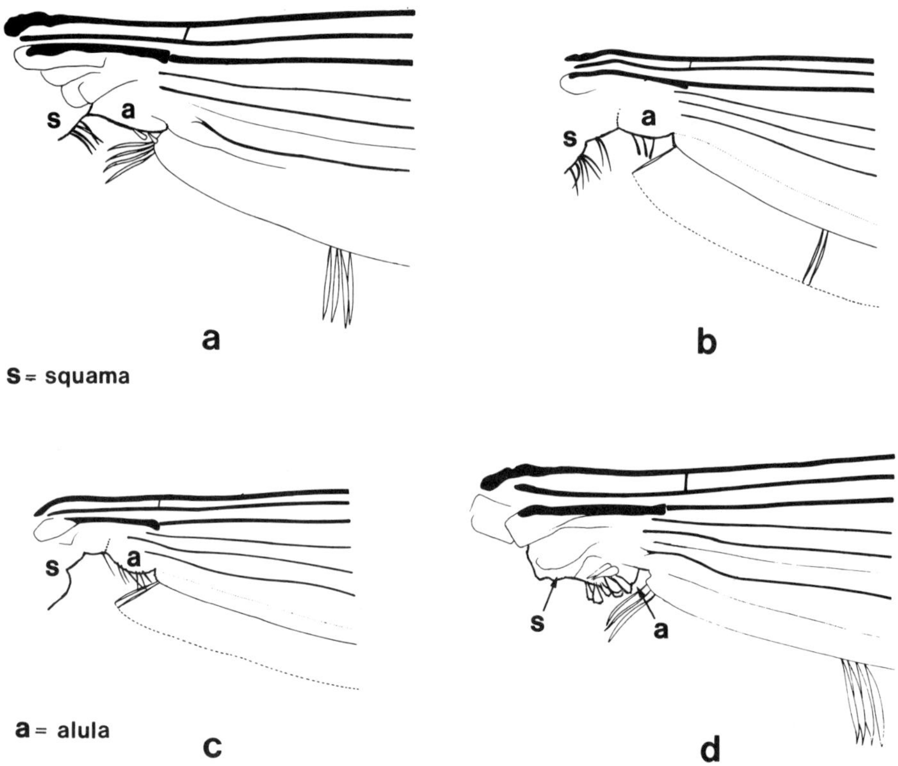

Figure 5.26 Adult wing base showing differences in vestiture of the squama and alula: (*a*) *Mimomyia luzonensis*; (*b*) *Trichoprosopon pallidiventer*; (*c*) *Wyeomyia moerbista*; (*d*) *Mimomyia deguzmanae*.

6   Thorax without postspiracular setae (except in *Mansonia* and occasionally *Heizmannia*). Abdomen with square or rounded tip and non-retractable apical segment (e.g. as *Culex*, Fig. 5.18d); cerci inconspicuous.................................... 7
—   Thorax usually with postspiracular setae (if these absent then palps at least half as long as proboscis). Abdomen with more or less pointed tip and retractable apical segment (e.g. as *Aedes*, Fig. 5.18c); cerci conspicuously protruding (Fig. 5.18c) 17
7   Thorax with following combination of characters: mesopostnotum usually without tuft of apical setae; antepronotal lobes small, with or without scales; prealar knob usually with numerous setae (prealar setae); mesanepimeron with at least one seta on its lower part; mesomeron long, its upper margin distinctly dorsal to base of hind coxa (Fig. 5.27a) 8
—   Thorax with following combination of characters: mesopostnotum with small tuft of apical setae (except in some *Tripteroides*) [setae sometimes difficult to see if obscured by halter]; antepronotal lobes usually large and conspicuous (Fig. 5.18a,b), covered with scales; prealar knob usually bare, if with setae then these usually four (but up to ten in some *Trichoprosopon*); mesanepimeron almost always bare on its lower part (but one or two setae in *Heizmannia*); mesomeron short, its upper margin in line with base of hind coxa (Fig. 5.27b) or slightly ventral to it (rarely slightly dorsal to it).. 11
8   Tarsi without pulvilli 9
—   Tarsi with well-developed pulvilli (Fig. 5.16d) .... **Culex**
9   Thorax without prespiracular setae, with or without postspiracular setae.........10
—   Thorax with prespiracular setae (usually numerous), without postspiracular setae .... **Culiseta**
10 Thorax with postspiracular setae. Wing veins covered with broad dark and pale scales giving the wings a speckled or mottled appearance (Fig. 5.21). Mosquitoes without yellow appearance, body and leg scaling not including any bright yellow scales.. **Mansonia**
—   Thorax without postspiracular setae. Wing veins with narrow pale or yellowish scales and wings not appearing mottled. Mosquitoes with conspicuously yellow appearance, or at least body and leg scaling including some bright yellow scales .... **Coquillettidia**

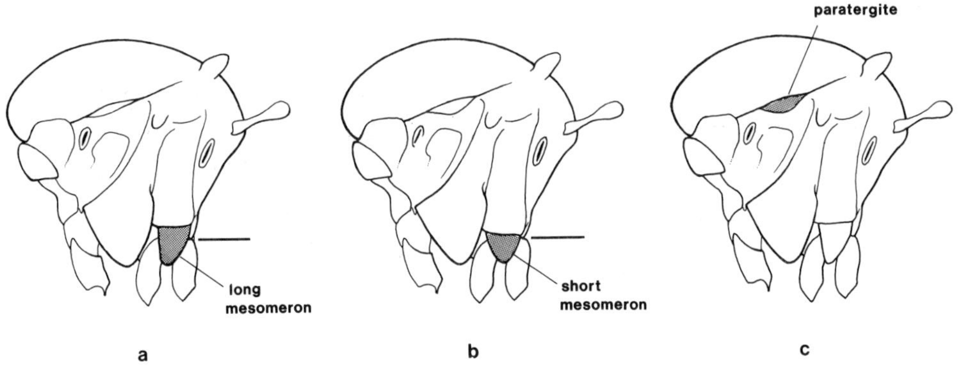

Figure 5.27 Left side of adult culicine thorax in diagrammatic outline showing: (a) long mesomeron, occurring in some genera (see identification key couplet 7); (b) short mesomeron, occurring in some genera (see identification key couplet 7); (c) broad paratergite occurring in *Eretmapodites*.

11 Thorax with spiracular scales *or* at least one seta on prespiracular area 12
— Thorax without prespiracular scales or setae. [South East Asia]....................
.... **Heizmannia**

12 Thorax with prealar knob bearing at most six setae. Wing with cell r2 (upper fork cell) usually longer, often much longer, than its stem (vein R2 + 3, Fig. 5.7*a*); without setae on vein R and base of subcosta. [Australasian, Oriental and far eastern Palaearctic (mainly China and Japan) regions]..........................
.... **Tripteroides**
— Without such combination of characters. [Neotropical and (rarely) Nearctic regions]... 13

13 Proboscis as long as or longer than fore femur. Scutal and pleural scale colour varied but scutum without golden scales. Thorax without prespiracular scales but with at least one prespiracular seta. Hind tarsi with two claws...................14
— Proboscis shorter than fore femur. Scutal scales golden, bronze or purplish and pleural scales abundantly silvery and golden. Thorax with prespiracular scales but no prespiracular setae. Hind tarsi with one claw. [Neotropical region] ....
.... **Limatus**

14 Scutum with broad and flat metallic scales. Thorax with antepronotal lobes very large and almost meeting in midline, in perfect specimens appearing to form a continuous collar in front of scutum because of very dense scale cover (Fig. 5.16*b*). Tarsi of one or more pairs of legs usually with inner and outer rows of long scales forming broad and conspicuous tarsal 'paddles' (Fig. 5.16*a*). [Neotropical region].. **Sabethes**
— Scutum usually without such metallic scales. Thorax with large or small antepronotal lobes. Tarsi without 'paddles' 15

15 Vertex of head (Fig. 5.2*a*) without erect scales. Clypeus without setae but sometimes with scales. Thorax with large antepronotal lobes almost meeting each other in midline of thorax (Fig. 5.18*a*). Squama bare or with one to three setae or hair-like scales on upper part near alula (Fig. 5.26*c*). [Nearctic and Neotropical regions] .... **Wyeomyia**
— Vertex of head with erect scales. Clypeus sometimes with setae and also scales. Thorax usually with small widely separated antepronotal lobes. Squama often fringed with small setae, especially numerous on lower part (Fig. 5.26*b*) ....
.... 16

16 Proboscis 1.2–1.4 times as long as fore femur. Thorax with setae of mesokatepisternal row not extending upwards as far as lower margin of mesanepimeron. [Neotropical region].. **Johnbelkinia**
— Proboscis 0.8–1.2 times as long as fore femur. Thorax with setae of mesokatepisternal row extending upwards as far as lower margin of mesanepimeron. [Neotropical region]... **Trichoprosopon**

17 Scutum not covered with broad metallic green, blue or bronze scales; antepronotal lobes not enlarged, widely separated from each other anteriorly. Legs often with pale bands, especially on tarsi. Small to medium-sized species with wing spread of 6–17 mm. [Day-biting and night-biting mosquitoes in varied habitats] 18
— Scutum covered with broad and flat metallic green, blue or bronze scales; antepronotal lobes very large, almost meeting anteriorly in midline of thorax (Fig. 5.18*b*). Legs usually uniformly dark. Small species with wing spread of 6–9 mm. [Day-biting Neotropical mosquitoes of forest habitats]..........................
.... **Haemagogus** (subgenus **Haemagogus**)

18 Head with broad silvery scales posteriorly. Thorax usually partly or largely yellowish or with orange tint, with broad silvery scales on pleural regions; paratergites broad and bare (Fig. 5.27c). Abdomen with broad silvery scales on posterior corners of tergites. [Afrotropical region] .... **Eretmapodites**
— Without such combination of characters.. 19
19 Thorax without prespiracular setae, with or without postspiracular setae.........20
— Thorax with both prespiracular and postspiracular setae. [Nearctic and Neotropical regions] .... **Psorophora**
20 Head with scales of vertex all broad and decumbent, erect scales absent. Thorax with or without postspiracular setae: if present, then proboscis slightly curved (Fig. 5.19b) and slightly compressed laterally, and one lower mesanepimeral seta present: if absent, then palps at least half as long as proboscis. Abdomen usually slightly more than twice as long as thorax; abdominal segment 8 partially retracted. [Oriental and Australasian regions] .... **Armigeres**
— Head with scales of vertex broad and decumbent laterally, often narrow and curved in midline, numerous erect scales present. Thorax with postspiracular setae (except subgenus *Ayurakitia* with two species in Thailand); proboscis not curved and compressed (except subgenus *Alanstonea* with two Oriental species), and lower mesanepisternal setae present (except in *Alanstonea*). Abdomen usually not more than twice as long as thorax; abdominal segment 8 almost entirely retractile (Fig. 5.18c) .... **Aedes**

Faunal and taxonomic literature

The faunal and taxonomic literature on mosquitoes is enormous, much larger than that of any other medically important group of insects. Fortunately, entrance to this vast literature is provided through the world catalogue of Knight and Stone (1977) and its updating supplements by Knight (1978), Ward (1984), Gaffigan and Ward (1985) and Ward (1992). This ongoing catalogue lists the countries and islands from which each mosquito species has been recorded and is a basic reference which should be consulted by anyone interested in the mosquitoes of a particular area.

Also outstanding in importance as a source of information is *Mosquito Systematics*, a specialist journal containing original articles on culicid taxonomy and faunistics. The *Journal of the American Mosquito Control Association* (called *Mosquito News* until 1983) has been particularly useful in providing bibliographic updates of the literature on all aspects of culicidology, but, regrettably, this valuable service ceased in 1991.

Inevitably, it is possible to give here only a critically selected list of the most important faunal and taxonomic works such as catalogues, major regional and country overviews and revisionary studies of particular genera, subgenera or groups. Entries are concentrated on works containing identification keys, and to economize on space details are not always provided (e.g. names of subgenera within a major genus such as *Aedes*) if these can be found immediately by reading the title of the paper where it is listed in the chapter bibliography. The many country checklists which are no more than uncritical inventories of names are omitted.

World: Generic key (Mattingly, 1971, 1973); *Aedeomyia* (Tyson, 1970a, world species key); *Aedes* (Reinert, 1970a, 1973a subg. *Diceromyia* and *Aedimorphus* zoogeography); *Culex* (Sirivanakarn, 1971, partial reclassification with keys);

Culiseta (Maslov, 1967, subgenera); *Orthopodomyia* (Zavortink, 1968); Sabethini (Mattingly, 1981, Old World genera); *Uranotaenia* (Peyton, 1972, classification).

Palaearctic: Catalogue (Minář, 1991); Afghanistan (Danilov, 1985a,b); Arabia (Mattingly and Knight, 1956); Britain (Cranston et al., 1987); China (Lu and Li, 1982, Lu et al., 1988, Lu and Su, 1987, aedines); Czechoslovakia (Kramář, 1958); Egypt and southwestern Asia (Harbach, 1985, 1988, generic key, *Culex* species key); France (Moussiegt, 1986, literature, distribution); Germany (Mohrig, 1969); Hungary (Mihályi and Gulyás, 1963); Japan and Korea (Tanaka et al., 1979, Toma and Miyagi, 1986, Ryukyu Islands); Iran (Zaim and Cranston, 1986, culicines, with checklist); Iraq (Ibrahim et al., 1983, culicines; Hudson and Abul-Hab, 1987, culicine adults); Mediterranean area (Rioux, 1958, Senevet and Andarelli, 1959); Morocco (Guy, 1959, *Anopheles*); Scandinavia (Natvig, 1948); Turkey (Postiglione et al., 1973, *Anopheles*); USSR (Gutsevich et al., 1970, 1974, Gutsevich and Dubitskiy, 1987, checklist, *Aedes* keys); *Anopheles maculipennis* complex (Deruaz et al., 1985, Korvenkonitio et al., 1979, biochemical taxonomy).

Afrotropical: Catalogue (White, 1980); regional anopheline fauna (Gillies and de Meillon, 1968; Gillies and Coetzee, 1987); regional culicine and toxorhynchitine fauna (Edwards, 1941, adults, pupae; Hopkins, 1952, larvae; Service, 1990, adults and larvae, keys for genera other than *Aedes* and *Culex*, largely supersedes Edwards and Hopkins for genera covered); Angola (Ribeiro and Cunha Ramos, 1973, 1980, *Aedes*, *Culex*); Cape Verde Islands (Ribeiro et al., 1980); Madagascar (Grjebine, 1966, *Anopheles*; Grjebine, 1986, *Ficalbia* and *Mimomyia*; Ravaonjanahary, 1978, *Aedes*); Seychelles (Mattingly and Brown, 1955); South Africa (Muspratt, 1955, adults; Muspratt, 1956, *Aedes* in part, adults, larvae). *Aedes*, associated yellow fever (Cordellier et al., 1977, Huang and Ward, 1981, species keys); *Aedes*, in part (Mattingly, 1952, 1953; Tyson, 1970c; McIntosh, 1971, 1973; Rodhain and Boutonnier, 1983; Huang, 1990); *Anopheles gambiae* complex (White, 1985, morphology and sibling relationships; Coluzzi, 1968, Coluzzi and Sabatini, 1969, Coluzzi et al., 1985 for cytology; Mahon et al., 1976, Miles, 1978, 1979 for isoenzymes; Collins et al., 1988, Gale and Crampton, 1987, 1988 for DNA probe identification); *Culex* (Vattier and Hamon, 1962, regional key *Culiciomyia* larvae).

Oriental: Catalogue (Stone and Delfinado, 1973); generic key (Mattingly, 1971); Indomalayan culicine fauna (Mattingly, 1957a, b, 1958, 1959, 1961, 1965); South East Asian fauna (see series of major works published by American Entomological Institute (1967–1973) with running title 'Contributions to the mosquito fauna of southeast Asia', some in following selected entries: Apiwathnasorn, 1986, list; Delfinado, 1967, 1968, *Aedes* in part; Dobrotworsky, 1971, *Culiseta*; Huang, 1972, *Aedes* in part; Mattingly, 1970, *Heizmannia*; Peyton, 1977, *Uranotaenia* in part; Reinert, 1970b, 1973a, b, *Aedes* in part; Sirivanakarn, 1971, *Culex* in part; Tyson, 1970a, *Aedeomyia*; Tyson, 1970b, *Aedes* in part; Zavortink, 1971, *Orthopodomyia*); Borneo (Reid, 1968, *Anopheles*); Burma (= Myanmar) (Barraud, 1934, non-anophelines); Cambodia (Klein, 1973, *Aedes* in part; Klein, 1977); India (Barraud, 1934, culicines and toxorhynchitines; Rao, 1984, *Anopheles*, taxonomy, biological and medical data; Das et al., 1989, *Anopheles* key); Malaya (Reid, 1968, *Anopheles*, major work with biology and vector importance data; Wharton, 1962, *Mansonia*); Nepal (Darsie and Pradhan, 1990); Philippines (Delfinado, 1966, Culicini; Cagampang-Ramos and Darsie, 1970, anophelines); Sri Lanka (Barraud, 1934, culicines and toxorhynchitines; Reinert, 1984, *Aedes* in part); Thailand (Harrison et al., 1990; Rattanarithikul, 1982, generic key; Bram, 1967a, *Culex* s.str.; Delfinado, 1967, *Aedes* in part; Harrison and

Scanlon, 1975, Harrison, 1980, *Anopheles*); Vietnam (Borel, 1930). *Aedes* (Huang, 1979, subg. *Stegomyia*; Reinert, 1970b, 1973a, 1974, 1981, 1990, some subgenera); *Anopheles* (Peyton, 1989, *leucosphyrus* group classification); *Culex* (Sirivanakarn, 1972, 1976, 1977, some subgenera); *Tripteroides* (Mattingly, 1981, some subgenera).

Australasian: Catalogue (Evenhuis and Gon III, 1989); regional fauna (12-volume taxonomic coverage available under general title 'The Culicidae of the Australasian region' and including biological and medical importance data, see references Lee et al., 1980, 1982 and 1984 (volumes 1–3), Debenham, 1987a,b, 1988a (volumes 4–6), 1989a,b (volumes 7 and 8), Debenham, 1988b,c (volumes 9 and 10), Hicks, 1989 (volume 11) and Debenham et al., 1989 (volume 12)), Guam (Nowell and Ward, 1989, literature entrée only); Maluku (Bonne-Wepster, 1954, with Lesser Sunda Islands); New Guinea (Assem and Bonne-Wepster, 1964; Bonne-Wepster, 1954; Huang, 1968a,b, *Aedes* in part; Sirivanakarn, 1968, *Culex* in part); New Zealand (Belkin, 1968); Polynesia (Huang, 1977); Samoa (Shivaji, 1976); South Pacific, general (Belkin, 1962); Tonga (Shivaji, 1976; Huang and Hitchcock, 1980). *Tripteroides* (Mattingly, 1980, reclassification).

Nearctic: Catalogue (Stone, 1965, America north of Mexico); regional fauna (Darsie and Ward, 1981, major coverage, amendments Ward and Darsie, 1982; Darsie, 1989, pupal keys); Canada (Wood et al., 1979, includes distribution and medical importance data). *Aedes* (Arnell, 1976, *scapularis* group); *Anopheles* (Floore et al., 1976, *crucians* subgroup; Zavortink, 1970, tree-hole species); *Culex* (Bram, 1967b, *Culex* s.str.).

Neotropical: Regional fauna (Lane, 1953, old but still useful; Forattini, 1962, anophelines; Forattini, 1965a,b, culicines; Vargas, 1972, 1974, generic keys, adults, larvae); South America, west (Goreham and Stojanovich, 1973); Argentina (Darsie, 1985, Mitchell and Darsie, 1985); Brazil (Xavier and Mattos, 1965, 1975, Xavier et al., 1983, culicine distribution); Caribbean (see 'Mosquitoes of Middle America' series of papers in *Mosquito Systematics*, starting Belkin and Heinemann, 1973, Dominican Republic and ending Heinemann et al., 1980, Trinidad and Tobago); Central America (Wilkerson and Strickman, 1990, female anophelines); Cuba (García Avila et al., 1981, adults except *Culex*); Guatemala (Clark-Gil and Darsie, 1983); Jamaica (Belkin et al., 1970); Mexico (Vargas and Martínez Palacios, 1956, anophelines; Wilkerson and Strickman, 1990, female anophelines); Venezuela (Cova-Garcia, 1961, anophelines; Cova-Garcia et al., 1966, culicines). *Aedes* (Arnell, 1976, *scapularis* group; Berlin, 1969, subg. *Howardina*; Zavortink, 1972, ex *Finlaya* species); *Anopheles* (Faran and Linthicum, 1981, subg. *Nyssorhynchus*, Amazonia; Linthicum, 1988, *Nyssorhynchus* in part; Zavortink, 1970, tree-hole *Anopheles*; Zavortink, 1973, subg. *Kerteszia*); *Culex* (Berlin and Belkin, 1980, three subgenera; Bram, 1967b, *Culex* s. str., Foote, 1954, subg. *Melanoconion, Mochlostyrax*, immatures; Sirivanakarn, 1982, subg. *Melanoconion*); *Deinocerites* (Adames, 1971); *Haemagogus* (Arnell, 1973); *Mansonia* and *Coquillettidia* (Ronderos and Bachmann, 1963); *Phoniomyia* (Correa and Ramalho, 1956); *Trichoprosopon* (Zavortink, 1979, classification); *Uranotaenia* (Galindo et al., 1954).

BIOLOGY

The book by Clements (1963) entitled *The Physiology of Mosquitoes* contains a much wider spectrum of biological information than the title might suggest and is an important work. However, it is now rather outdated and a two-volume work by Clements, retitled *The Biology of Mosquitoes*, is intended to replace it: volume 1 covers development, nutrition and reproduction (Clements, 1992) and volume 2 will cover senses and behaviour (Clements, in preparation). A book by Service (1993) covers ecology with the focus on sampling techniques.

Adults

Mating and swarming

After emergence from the pupa the teneral adult usually seeks shelter amongst vegetation until ready for mating, which in the case of the female usually occurs a few hours after emergence, but sometimes much sooner. In some species one or two days may elapse between emergence and mating. Males do not mate until their genitalia have rotated through 180°, a process that often takes 20–24 hours but in some species as little as 6–12 hours.

Mating usually occurs in flight. The tone of the female wing beat attracts males, their antennae being sound receptors which aid in mate location; odour seems mostly to play an insignificant role. The wing beat tone of females varies considerably between species, but is commonly about 250–320 Hz. The tone of the male wing beat is some 100–200 Hz higher. Sometimes males of several species swarm within a small area, but the swarms are usually separated by orientation to different visual markers, although occasionally what is apparently a single swarm comprises males of more than one species. How frequently close encounters with the wrong species occur is not known, but a surface pheromone, in at least some species, mediates species recognition at close range.

A male usually seizes a female with his hind legs, whereupon copulation occurs in flight, or the pair drops to the ground. In some species, including *Aedes aegypti*, the male is below the female and holds onto her as the two fly around 'face to face'. But in many, if not most, mosquitoes after the male genitalia have locked into the female he lets go with his legs and hangs head downwards while in flight. Copulation frequently lasts 5–60 seconds, but in some species it takes several minutes, and in a few such as *Deinocerites cancer* and *Culiseta inornata* it may last for 45 minutes or more. A single male can mate with several females, without necessarily inseminating them all.

The formation of male swarms in some species, particularly anophelines, is an essential behavioural prelude to mating. Yet in some species, mating can occur independently of swarms. For instance, almost all female *Coquillettidia fuscopennata* arriving at a male swarm at which copulation is taking place are sometimes already inseminated. Male swarms are most commonly encountered around sunset and over markers. These can be tall trees, buildings, fences, bushes, a branch or twig, the observer's head or less conspicuous markers such as a small clearing on the ground which contrasts with a paler or darker background. At low latitudes swarming usually lasts only 10–30 minutes or so. In a few species such as *Aedes varipalpus* and *Aedes triseriatus* males swarm over small mammals and seize females as they arrive to feed on them, while in *Aedes aegypti* and some *Coquillettidia* mating, though not as a rule swarming, also occurs as females arrive to feed on a host.

Opifex fuscus is found only in coastal rock-pools in New Zealand. Males have rotated their genitalia within only five to six hours after emergence, after which they fly low over the water surface in pursuit of mature female pupae, which they seize. As soon as the pupal skin begins to split the male thrusts his abdomen into the split and starts copulating with the female before she has completely emerged. He then drags her to the edge of the pool where mating lasts for a further 10–20 minutes. If a male selects a male pupa his mistake is not discovered until emergence of the adult is almost complete. Good accounts of this fascinating species are given by Haeger and Provost (1965) and Slooten and Lambert (1983).

A mating process almost as bizarre occurs in *Deinocerites cancer*, a Neotropical crab-hole-breeding mosquito. Newly emerged males remain within the tunnels of land-crabs, walking on the water surface in search of either newly emerged females with which they immediately mate, or female pupae. They grab pupae and 'fight off' male adults that approach closely (Conner and Itagaki, 1984).

After copulation males of at least some species of *Anopheles, Aedes* and *Psorophora* insert a gelatinous plug, formed from secretions of the male accessory glands, into the female's genital chamber. It presents a physical barrier preventing other males inseminating the mated female. Such a mating plug dissolves within 24–36 hours after copulation. It was formerly believed that females only mated once, but there is increasing evidence that in at least some species females mate more than once, and that this may become necessary because their stored sperm is depleted. Sperm remains viable in females for a long time – up to 9–12 months in species that overwinter as inseminated females.

The terms 'eurygamous' and 'stenogamous' are used to describe respectively the necessity for a large space for mating, and the ability to mate in a confined space, such as in the laboratory in a small cage or even a test tube.

Downes (1969) provides a good general review of the relevance of swarming to mating, whereas swarming in the *Anopheles gambiae* complex is described by Charlwood and Jones (1980). Young and Downe (1987) describe substances in the male accessory glands that induce sexual refractoriness in the females. Gynandromorphs have been recorded in the genera *Aedes, Armigeres, Coquillettidia, Culex, Culiseta, Haemagogus, Mansonia, Orthopodomyia, Toxorhynchites* and *Trichoprosopon*, notably in species such as *Aedes aegypti, Culex quinquefasciatus, Culex tritaeniorhynchus* and *Culex tarsalis*. Intersexes also sometimes occur (especially among Holarctic *Aedes*). Clements (1992) discusses the distinction between intersexes and gynandromorphs.

Sugar-feeding
Both sexes of *Toxorhynchites* feed exclusively on sweet secretions, such as nectar, extrafloral nectaries, honey-dew (including that secreted by aphids) and fruit. Males of anophelines and culicines are also restricted to sugar-feeding. Apparently females also commonly feed on sugary substances. Sugar-feeding has most often been recorded in temperate species, but it seems that it is more common in tropical mosquitoes than previously realized. In some temperate mosquitoes sugar-feeding occurs for several days or even weeks before females switch to blood-feeding. The reason for such protracted abstinence from blood remains unknown. Joseph (1970) reviews fruit-feeding.

A sugar-meal is diverted to the ventral diverticulum (crop) which acts as a kind of fuel tank; later some part, or all, is passed to the stomach, where it is digested. Sugar-meals provide energy necessary for flight, dispersal and most other biological activities, though not usually for egg development (but see below).

Autogeny

Most female mosquitoes are anautogenous, that is they require a blood-meal for maturation of the ovaries. There are exceptions, however, and *Toxorhynchites* females are incapable of biting and therefore autogenous. Females of the tropical genus *Malaya* never take blood-meals but feed on the regurgitated stomach contents of ants.

In addition to these two genera, autogeny has been recorded in about 60 species in numerous genera, including *Anopheles, Culex, Aedes, Eretmapodites, Culiseta, Uranotaenia, Orthopodomyia, Deinocerites, Opifex, Wyeomyia, Coquillettidia* and *Tripteroides*. Autogeny can be a survival mechanism when there is a shortage of suitable hosts, as in arctic areas, or serve as an alternative strategy to blood-feeding. Autogeny usually involves just the first ovarian cycle, that is during the few days following emergence females can lay eggs without a blood-meal, but thereafter blood-feeding is essential for subsequent egg batches. There are, however, some exceptions, and the incidence of autogeny in a species can vary considerably in different geographical populations, and also seasonally. For example, northern populations of *Wyeomyia smithii* never take blood-meals and are fully autogenous, females laying two, or rarely more, egg batches, while southern populations require blood after the first oviposition if further batches of eggs are to develop. Autogeny is usually obligatory for a species, but in a very few species it is facultative: for example, if females of the Canadian *Aedes nigripes* and *Aedes impiger* fail to get a blood-meal within ten days or so after emergence they will use their remaining food reserves to develop a small batch of eggs.

Usually nutrients obtained during larval growth are utilized by adults for autogenous development of their eggs, but some species use their protein reserves for autogeny, whereas others histolyse their indirect flight muscles. The number of eggs laid following autogenous ovarian development is nearly always smaller than the number resulting from haematophagy, but an exception is *Aedes atropalpus*, a mosquito which can develop some 200 eggs autogenously, comparable to the numbers developed after blood-feeding.

Autogeny has been reported in several important vectors with various levels of occurrence, for example *Culex tarsalis* (vector of western and St Louis encephalitis), certain species of the *Aedes scutellaris* group (vectors of Brugian filariasis) including *Aedes polynesiensis* (Rivière, 1983), and very rarely natural populations of *Aedes aegypti* (Trpis, 1977). Autogeny will of course reduce the efficiency of a species as a vector. See O'Meara (1984, 1985) for reviews.

Blood-feeding

Females of most mosquitoes need a blood-meal before they can develop their eggs. Blood-feeding can take place before or after mating; this seems to depend on the species and local circumstances. In *Anopheles stephensi* the first blood-meal is taken on the first or second night of a female's life regardless of whether or not she has been inseminated. Females of most species do not start host-seeking until about 24 hours old; some bite sooner than this, whereas others wait much longer. In Britain, for example, *Aedes cantans*, which feeds readily on mammals, will not bite until about three weeks old, and similar delays have been observed in *Culiseta morsitans*, a bird-biting mosquito.

Females are attracted to their hosts by various stimuli, including convection currents, carbon dioxide, moisture, lactic acid, possibly 1-octen-3 ol, and by vision, but mostly by an assortment of little known olfactory components conveniently labelled 'host odours' (see Gillies, 1980; Sutcliffe, 1987). These olfactory stimuli often act in combination with one another. No known single host attractant can equal the attractiveness of natural odours emanating from a host. Vision plays a not inconsiderable role in host location, especially by day-biting mosquitoes. Mosquitoes of the genus *Eretmapodites*, which is confined to Africa, are attracted in greater numbers to a clearly visible person standing in a forest than to one lying on the ground, and in even greater numbers to a person who moves around rather than to a stationary one. Somewhat surprisingly, vision can also play a role in host orientation by night-flying mosquitoes, seemingly even on dark moonless nights. Allan et al. (1987) present a brief review of vision in mosquitoes and other biting flies.

Hungry mosquitoes follow a chain of behavioural responses which eventually leads them to a suitable host. This type of host-seeking behaviour has often been called 'appential flight' or 'appetitive behaviour'. Separate mechanisms may be involved for long-, medium- and close-range orientation to a host. For example, long-range attraction usually involves host-specific odours, while a mixture of host odours and CO_2 is important in medium-range attraction. At close range, warm moist convection currents and visual cues are usually important. Gillies (1988) presents a short review of host-feeding and other behaviours in anophelines.

The distance of attraction of a host to a hungry mosquito varies according to species of both host and mosquito, topography and meteorological conditions, but seems to range from about 7 to 30 m; the larger or more numerous the bait the greater the distance an odour plume travels. Laboratory and field observations have demonstrated that mosquitoes seem to be particularly attracted to hosts on which mosquitoes are already feeding. There is some evidence to suggest that host-seeking mosquitoes respond to some chemical stimulus, possibly an 'invitation pheromone', produced by engorging mosquitoes. Although such an invitation to 'share dinner' is advantageous to the mosquito population as a whole, it would seem to be disadvantageous to the individual mosquitoes already feeding, because increased feeding success could lead to increased competition for larval habitats and resources.

Some people are undoubtedly more attractive to mosquitoes than others, but the relative attractiveness of an individual can differ from day to day. Adult humans are generally considerably more attractive than young children or infants, and there is some evidence to suggest that men are more attractive than women (Muirhead-Thomson, 1982; Port et al., 1980).

The amount of blood ingested during a single bout of feeding depends much on the size of the mosquito, but is generally between 2 and 5 mg, being about 4 mg in *Aedes aegypti*, but only 1–2.5 mg in many *Anopheles*. Some species, such as *Culiseta annulata, Psorophora cyanescens, Psorophora confinnis, Aedes sollicitans* and *Culex quinquefasciatus*, however, ingest 6–10 mg of blood. *Psorophora ciliata* is a particularly large mosquito and its blood-meal can be up to 25 mg of blood (see Gooding, 1972 for references).

A brief review of blood-feeding in haematophagous insects, including mosquitoes, is given by Friend and Smith (1977). Bowen (1991) reviews the physiology of host-seeking behaviour.

Host preferences
Most species do not bite man, but instead a variety of other hosts. Some species, such as *Culex pipiens* of northern temperate regions, the Holarctic *Culiseta morsitans*

and the African *Coquillettidia metallica*, feed exclusively on birds, whereas others, such as the North American *Culex tarsalis* and *Culex nigripalpus*, feed on mammals as well as on birds. *Deinocerites cancer* is even more catholic in its taste; although principally engorging on birds it also feeds to a small extent on mammals, reptiles and amphibians – that is, all four classes of land vertebrates. Several other *Deinocerites,* such as *D. dyari*, feed predominantly on lizards. In Florida armadillos are commonly fed upon by *Aedes infirmatus, Psorophora ferox* and *Aedes fulvus pallens*. Many *Ficalbia, Mimomyia* and several *Uranotaenia* feed on amphibians. A few species have been recorded feeding on bats. Fish, usually mud skippers that flop out of the water and lie half exposed in mangrove swamps, are fed upon by some salt-marsh species. Mosquitoes have even been observed feeding on insects such as caterpillars, and moreover developed eggs on such a diet.

Females of *Malaya*, a genus which occurs in Asia, the Pacific and Africa, fly up and down over trees or bamboo searching for ants of the genus *Crematogaster*. When an ant is descending a tree, and therefore likely to be gorged with honey-dew, the mosquito often settles in its path, inserts her proboscis into the ant's mouth and sucks out the honey-dew. The ant appears to gain nothing from this relationship, so why it permits itself to be robbed of its meal is not understood.

Mosquitoes feeding predominantly on humans are referred to as anthropophagic, those feeding on non-human hosts are called zoophagic, and those feeding on birds are sometimes called ornithophagic. Often these terms are used in a relative not absolute sense. Useful summaries are given for mosquito host preferences by Washino and Tempelis (1983) and for feeding on ectothermic animals by Heatwole and Shine (1976).

Hosts can exhibit pronounced anti-feeding tactics such as swishing of tails, stamping, quivering, and pecking, to dislodge or divert mosquitoes to less attractive but more quiescent animals. Ornithophagic species often bite nestlings because they are generally more passive than older birds, but nevertheless adult birds are fed upon, especially around the eyes and beak. Some of the most mosquito-ridden areas lie in northern temperate regions, and there are many accounts of reindeer and caribou climbing above the snow line, or moving to windy shores and exposed areas, to seek refuge from mosquito attacks. In Alaska, bears and reindeer are driven into the water to escape attacks, especially by *Aedes punctodes* and *Aedes cataphylla*. Such defensive reactions by the host can result in incomplete blood-meals, necessitating feeding again soon on other hosts; such multiple feeding can increase the chance of disease transmission.

Biting times and places
Some mosquitoes bite during daylight (e.g. species of *Wyeomyia, Psorophora ferox, Limatus flavisetosus* and many species of *Aedes*). Others bite at night (e.g. most *Anopheles, Mansonia* and *Culex*); some species bite only or mainly at dusk (e.g. *Aedes ingrami*), and a few species seem to bite more or less indiscriminately throughout the daytime (*Eretmapodites chrysogaster*), and others during the day and night (*Aedes cumminsii, Aedes domesticus*). Many species, however, have regular, highly synchronized diel biting patterns such as the very pronounced unimodal peak in biting exhibited around sunset by *Aedes africanus*. Some day-biting species, such as *Aedes aegypti*, have clearly defined biting peaks just after sunrise and just before sunset. Mosquitoes feeding on nocturnally active animals that hide in burrows

during the day presumably benefit from biting at night. Similarly, avian feeders are more likely to be successful in obtaining blood-meals from birds at night when they are roosting or sitting on their nests. Periodicity of biting times, commonly expressed as their biting cycle, is basically controlled by endogenous (circadian) rhythms, but with environmental factors such as wind, temperature and moonlight capable of modifying biting activity. In some species such as *Coquillettidia fuscopennata* and *Anopheles implexus* biting patterns are known to be different in different microhabitats. Biting periodicities of nulliparous and parous females are usually the same.

Rowland (1989) gives useful references on circadian activity patterns in mosquitoes and reports on circadian flight patterns in *Anopheles stephensi*.

A few mosquitoes enter houses to feed and are termed endophagic, whereas most species are exophagic, that is they feed out of doors. While many species are entirely exophagic, no species is exclusively endophagic. Some mosquitoes bite predominantly in forests or woods; others, although breeding in these areas, regularly fly outside to feed, usually returning to them afterwards. Other species bite mainly in and around farms, while many anthropophagic species bite in towns and villages. Within a forest or wood different species can be feeding at different heights. For example, in northern Europe the ornithophagic *Culex pipiens* mainly feeds on birds resting in trees, whereas most *Aedes* species feeding on mammals bite near ground level. A few species, such as *Coquillettidia fuscopennata* and *Aedes africanus* in East African forests, undergo daily vertical migration. Most mosquitoes, however, bite at ground level and this has sometimes led people to construct houses or rooms on stilts.

The times and places of biting can have important epidemiological consequences. *Anopheles albimanus*, an important malaria vector in Central America, bites out of doors early in the evening; consequently, more or less all members of a village community will be bitten. *Anopheles gambiae*, the principal malaria vector in Africa, bites predominantly indoors and after midnight, and again most of the population will be at risk unless protected by mosquito bed-nets. An exophagic mosquito feeding late in the evening, however, will seldom be biting young children as they will be in bed. There are occupational hazards too. Mosquitoes biting in forests will mainly be feeding on woodcutters, hunters, charcoal burners – usually men – whereas both men and women will get bitten by those species that bite in and around villages.

An engorged mosquito's first priority is to rest somewhere safe from predators and digest her blood-meal, develop her eggs and become gravid. A few species, including *Culex quinquefasciatus* and several important malaria vectors, rest inside houses after engorging and are termed endophilic. In contrast most species are exophilic, that is they rest in various outdoor shelters, including rodent burrows, ventilation holes in termite mounds, cracks in the ground, tree-holes, amongst grass and other vegetation, on tree trunks, as well as on man-made structures such as fences, and in road culverts and under bridges. The distribution and abundance of exophilic mosquitoes may vary seasonally. For example, in India *Culex vishnui*, a vector of Japanese encephalitis, rests both indoors and outdoors, but the proportion resting in houses decreases during the hot dry season.

Most mosquitoes are zoophagic, exophagic and exophilic, but several important vector species are largely endophagic and endophilic. Endophagic species can also be exophilic. Exophily poses problems in malaria control programmes which are based on spraying the interior surfaces of houses with persistent insecticides like DDT or malathion. This strategy only works as long as the vectors rest in houses, and several malaria vectors, such as *Anopheles dirus* of South East Asia, are exophilic.

Ovarian development

Ovarian development has been conveniently divided into five major stages (I–V), some stages having minor subdivisions. Unfed mosquitoes arriving at bait have the ovaries in a so-called resting stage (IIa–IIb), beyond which development will not occur without a blood-meal, except of course in autogenous species. When about half the blood-meal has been digested and the ovaries have advanced to stages IIIa or IIIb the mosquito is referred to as half-gravid, and when all blood has been digested and eggs (i.e. follicles in stage V) are ready to be laid the female is said to be gravid. Most tropical mosquitoes digest their blood-meal within two or three days, but species living in colder climates take longer. In Britain, for instance, digestion can take four to six days in the summer, and 13–26 days during cool months. Usually only a single blood-meal is needed for a mosquito to become gravid, but in some species a certain proportion require two blood-meals; and rarely, such as in *Anopheles caroni* and *Anopheles hamoni*, three to four blood-meals are required to develop the first batch of eggs. In laboratory experiments a few North American floodwater species (*Aedes* and *Psorophora*) sometimes need eight or even 14 blood-meals before the ovaries mature (Washino, 1977), but this is very atypical of mosquitoes. After completing the first cycle of blood-feeding and oviposition (the gonotrophic cycle), most mosquitoes will then lay subsequent batches of eggs after just a single blood-meal.

The frequency of blood-feeding is largely determined by the time needed to complete the gonotrophic cycle, and how soon afterwards hungry adults seek out hosts. In tropical countries the gonotrophic cycle often lasts just two to three days, and females can re-feed every two to three days; whereas in temperate countries re-feeding occurs at much longer intervals, such as one to three weeks. The greater the frequency of blood-feeding the higher the probability that a vector transmits diseases. In nature not uncommonly females may complete four or five ovipositions, and sometimes as many as eight to 12; and in the laboratory a female can rarely lay eggs up to 20 times. A mosquito which has not laid eggs is termed nulliparous, and one that has oviposited one or more times is called parous. Epidemiologically it is important to distinguish between these two categories. Before ovaries develop beyond the resting stage nulliparous females can be identified by the tightly coiled terminations of the tracheoles supplying the ovarioles. The number of ovipositions completed by a female can sometimes be determined by carefully counting the dilatations in the pedicels of the ovarioles. The presence of live aquatic mites of the families Arrenuridae and Limnesiidae of female adults almost always indicates nulliparity (Smith, 1988). Useful accounts or ovarian age-grading methods are given by Detinova (1962, 1968) and Tyndale-Biscoe (1984).

The process of complete ovarian development following a blood-meal is termed gonotrophic concordance, to distinguish it from gonotrophic dissociation, a term applied when the ovaries fail to develop after a blood-meal (e.g. as in hibernating *Anopheles atroparvus*). Gonotrophic dissociation is known mainly in *Anopheles* but has been recorded in *Culex tritaeniorhynchus* (Washino, 1977). Another condition, gonotrophic discordance, describes cases where blood-feeding leads to the maturation of eggs, whereupon the gravid female delays oviposition and may re-feed one or more times. This phenomenon is most commonly encountered in tropical *Anopheles*, e.g. *A. culicifacies* and *A. stephensi*.

Oviposition
Egg laying appears to be mainly under the control of a circadian rhythm. Most species oviposit during sunset or during the night, but there are several notable exceptions. For example, in nature *Aedes aegypti* has discrete oviposition peaks within two hours after sunrise and two hours before sunset, and both *Aedes africanus* and *Toxorhynchites noctezuma* have clear oviposition peaks in the mid-afternoon.

Anopheles usually alight on the water surface to lay their eggs, but a few scatter them while in flight. *Toxorhynchites* species oviposit in flight, dropping the eggs into natural or man-made containers. The South East Asian *Tripteroides bambusa* shoots her eggs into hollow bamboo stems, while the Neotropical *Sabethes chloropterus*, and possibly also certain *Toxorhynchites*, hover in front of small holes in the sides of bamboo stems and flick their eggs through the openings onto the water. Other bamboo-breeding mosquitoes, such as *Armigeres dolichocephalus* and *Armigeres angustus*, squeeze through small holes bored by beetles in bamboo and lay their eggs in ribbons. More remarkably, several species, including *Armigeres flavus* and *Armigeres annulitarsis*, lay their eggs in the form of an egg raft stuck to the hind legs which are then dipped into the water to float them off. *Trichoprosopon digitatum* is unusual in depositing her eggs in fruit husks, or bamboo, in a loose egg raft which she holds between her middle legs for some 30 hours until they hatch (Fig. 5.28). Such parental care in mosquitoes is unique. The selective advantage of this exceptional brooding behaviour remains obscure, but it is probably to prevent egg rafts from becoming washed out by rain or stranded at the edges of the cocoa husks and drying out; it could also be to guard the eggs against predators.

Opifex fuscus dips the abdomen under the water and glues her eggs onto the sides of rock-pools, while *Mansonia* (*Mansonioides*) sticks the eggs as a gelatinous mass on the undersurfaces of floating vegetation (Fig. 5.12b). Most *Culex* species form their eggs into a raft which floats on the water surface (Fig. 5.12b), but a few atypical *Culex* deposit egg rafts on floating vegetation, and a few do not form egg rafts at all. Other genera laying eggs in a raft include *Coquillettidia*, *Culiseta* and *Uranotaenia* (*Uranotaenia*). In contrast, *Aedes* and *Psorophora*, and some quasi-sabethine mosquitoes such as *Haemagogus*, *Heizmannia* and *Eretmapodites*, as well as some *Armigeres*, lay their eggs on damp mud and leaf litter of pools and ponds, wet walls of tree-holes, bamboo, rock-pools, leaf axils and on the damp interior surfaces of water-storage pots. However, a few *Aedes*, such as *A. caspius* and *A. mariae*, sometimes deposit their eggs directly on the water surface, as do most species of the subgenus *Mucidus*. Interestingly, the East African *Aedes pembaensis* lays her eggs on the legs of freshwater crabs.

Water chemistry is clearly important in the selection of oviposition sites. Some species, such as *Anopheles melas* and *Anopheles merus* in West and East Africa respectively, *Anopheles aquasalis* in Central and South America, *Anopheles sundaicus* in India and south-west Asia, *Aedes detritus* in Europe, *Aedes sollicitans* and *Aedes taeniorhynchus* in North America and *Aedes australis* in Australia and New Zealand oviposit in saline habitats. These species are mainly coastal in their distribution, but a few

Figure 5.28 Adult female of *Trichoprosopon* guarding her eggs.

halophilous species such as *Aedes natronius* in East Africa breed in inland saltwaters. Females of *Culex quinquefasciatus*, *Culex stigmatosoma* and *Culex nebulosus* nearly always oviposit in water contaminated with human or animal excreta, or household refuse. Sometimes volatile chemicals produced by decomposition of organic debris are the principal attractants, for example log-ponds are particularly attractive oviposition sites for *Culex tarsalis* and *C. quinquefasciatus* in North America, and waters contaminated with chicken manure or rice-straw infusions are very attractive to *Culex pallens*.

There is increasing evidence that in many instances pheromones are produced by eggs or larvae signalling that the habitat is suitable for survival of the aquatic stages. Oviposition pheromones have been associated with *Culex tarsalis*, *Culex quinquefasciatus*, *Culex pipiens*, *Aedes atropalpus*, *Aedes triseriatus*, and possibly *Aedes albopictus* and *Aedes polynesiensis*. A useful review of this important subject is presented by Bentley and Day (1989).

Longevity

In general, adults in temperate, and possibly subarctic and arctic, regions live on average about four to five weeks. In hot tropical areas life-expectancy is shorter, probably on average one to two weeks, and in some species in certain situations average life may be only three to five days. Females that hibernate or aestivate live much longer; for example in Europe some *Culex pipiens* survive from August until May.

Hibernation and aestivation
Only a few temperate and arctic species overwinter as hibernating adults; most survive the winter as unhatched eggs or as slowly growing larvae. Females that hibernate accumulate fat reserves by feeding on sugary secretions; usually only inseminated nulliparous females hibernate, the males die at the beginning of hibernation. Initiation of hibernation is caused by reduced photoperiod and to a lesser extent by low temperatures. Hibernating mosquitoes are found in various dark and usually humid quarters, such as caves, large tree-holes, rock screes, rodent-holes and an assortment of man-made shelters including cellars, abandoned buildings, bunkers, old mine shafts, pigsties, cattle sheds, outdoor latrines and barns.

Some species such as *Culex pipiens, Culex tarsalis, Anopheles maculipennis, Anopheles earlei, Anopheles punctipennis* and *Culiseta impatiens*, undergo complete hibernation, that is they never take blood-meals during hibernation and consequently their fat reserves become progressively depleted, and there is usually a large overwintering mortality. The few survivors emerging from hibernation many months later take a blood-meal and lay eggs. The duration of hibernation varies. *Culex pipiens* in Europe stays in hibernation for about eight months, in Central Washington and North Dakota *Culex tarsalis* adults remain in hibernation for six months or more, but in California for only two to three months.

Partial hibernation is exhibited by many *Anopheles*, such as *A. freeborni* and *A. franciscanus* in the USA and *A. atroparvus* in the Palaearctic region, and by other mosquitoes such as *Culiseta annulata*. As their reserves are utilized during the winter months they emerge on sunny days and take blood-meals which enable their fat-body to be built up again, this being an example of gonotrophic dissociation. It is assumed that an increase in daylength and rise in temperature terminate hibernation.

The importance of aestivation in mosquitoes as a survival mechanism during dry and usually hot periods has received comparatively little attention. There are, however, several records of females of temperate and tropical *Anopheles* sheltering in cracks and crevices during severe dry seasons, periodically taking blood-meals but deferring oviposition. With increasing humidities and re-establishment of breeding habitats adults leave their shelters and lay eggs. Aestivation over many months has been reported in *Anopheles arabiensis* from the Sudan, and in *Culiseta inornata*, which is found from Canada to northern Mexico. In northern latitudes *C. inornata* is gonoactive in the spring to early autumn and then overwinters as a non-blood-feeding hibernating adult, but further south the adult is active in late autumn through to early spring; it survives the extreme summer heat by aestivation.

Dispersal
Energy for flight is supplied by glycogen which is rapidly assimilated from a sugar-meal; females can also derive energy for flying from their blood-meals, but few blood-fed mosquitoes fly far. Most mosquitoes probably disperse only a few hundred metres or so from their emergence sites, for example *Aedes aegypti* usually probably flies only 25–100 m or so. Normally *Anopheles* do not usually fly more than 2 km, but in certain circumstances they can regularly fly 3–5 km. The distance mosquitoes fly is determined largely by the environment: if suitable hosts and breeding places are nearby mosquitoes do not have to disperse far, but if one or both are more distant greater dispersal will be necessary.

There are, however, several accounts of long distance migration, assisted in almost all instances by the wind. For instance in Egypt, *Anopheles pharoensis* has been recorded about 72 km from the nearest known larval habitats, and there is even the possibility that malaria-infected *A. pharoensis* have flown as far as 280 km in the deserts of North Africa and the Middle East. Adults of *Aedes sollicitans* have been caught on board a boat no closer than 177 km to the North Carolina coast: swarms of *Anopheles pulcherrimus* have invaded a ship 25 km off the Arabian coast. In Australia the salt-water *Aedes vigilax* has been caught 97 km inland and also on a boat 32 km out at sea, while *Aedes theobaldi* is known to disperse as far as 64 km in a few days. In Florida, and probably elsewhere, wind apparently commonly sweeps *Aedes taeniorhynchus* 30–60 km from its emergence sites. There is evidence that in Illinois *Aedes vexans* has been carried 145–370 km within one or two days, and as far as 741 km in association with a cold front.

Mosquitoes have also become long-distance international travellers, and are not particular which airline they use (Russell, 1987)! A review of mosquito dispersal is given by Johnson (1969).

Phoresy

Very rarely *Linognathus* species of lice, parasitic on antelopes, have been found clinging to the legs of mosquitoes (*Aedes, Eretmapodites*) in Africa. Pseudoscorpions (*Cheiridium*) have been found on *Culex tarsalis* and *Anopheles freeborni* in the USA

Eggs

Newly laid eggs are white but turn brown or black within one to two hours, and after a further 15–18 hours the serosal cuticle is formed beneath the outer chorion. In aedine eggs this makes them resistant to desiccation. Female mosquitoes usually lay 30–500 or so eggs at one oviposition – the number depending much on the species – but not necessarily all of them at one site. The number laid depends on the species, nutrition and size of the adults.

The tolerance of eggs to extremes of temperature is more or less related to their natural environment. For example, eggs of some *Psorophora* species in southern USA and Mexico, and of *Aedes dupreei* in USA, hatch only when temperature reaches 25°C. Eggs of tropical species such as *Anopheles stephensi* and *Anopheles culicifacies* die when temperatures fall below about 10°C. In contrast, eggs of many temperate *Aedes* (e.g. *A. stimulans, A. hexodontus*) hatch at temperatures as low as 5°C. Those of the Nearctic *Anopheles walkeri* can survive several days at temperatures as low as –21°C. Many temperate and subarctic *Aedes* and species of *Culiseta* (*Culicella*) lay eggs that can survive many weeks or months if habitats become frozen.

Anopheles eggs usually cannot tolerate dryness for more than a few hours or at most a day or so, but their survival depends on the severity of the desiccation. In tropical regions, where embryonic development is rapid, eggs usually hatch within about two to three days; but if they become stranded on wet substrates hatching can be delayed for a few days, or very occasionally for several weeks. In cooler temperate regions, embryonic development may take about a week or more and, although at the end of this time most eggs will have hatched, a few may remain unhatched for a further two to four weeks. *Anopheles walkeri* overwinters as eggs, an unusual tactic for *Anopheles*.

Most *Culex* species form their eggs into a raft (Fig. 5.12*b*). The eggs stand upright in staggered rows on the water surface with the narrower micropylar hydrofugic end uppermost; the broader 'head' (anterior) end is readily wetted by water. Eggs are held together as a raft entirely by surface forces, with the enormous surface area presented by the minute chorionic papillae covering the eggs assisting cohesion. Minute droplets of fluid, containing lipids and oviposition pheromones, are secreted at the upper micropylar end of the eggs of *Culex* and *Culiseta* mosquitoes, and presumably attract gravid females to the preferred oviposition sites.

Culex eggs hatch within a few days and cannot normally survive desiccation, but *C. alogistus* and *C. pilosus* lay single eggs above the water-line which can withstand partial desiccation for about a month, as can eggs of *C. aikenii* which are laid singly on aquatic vegetation. *Culex taeniopus* lays her peculiarly curved eggs in small clusters on damp surfaces, where they can withstand desiccation for a relatively long time – a very unusual characteristic for a *Culex* mosquito. The females of *C. gaudeator* are apparently unique amongst mosquitoes in laying their eggs in a gelatinous mass that floats on the water surface.

Most *Culiseta* species collect their eggs into rafts, but a few species, belonging to subgenera *Culicella* and *Neotheobaldia* deposit them singly.

Species of *Uranotaenia* of the subgenus *Pseudoficalbia* appear to lay eggs singly on the water surface, whereas those of the subgenus *Uranotaenia* form rafts, but this apparent subgeneric difference might not be invariable.

Normally all the eggs in a raft hatch within a few hours of each other, but sometimes a few remain unhatched until a day or so later. Eggs of *Culex pipiens* and *Culiseta annulata* have been observed that failed to hatch until eight and 13 days, respectively, after the first eggs in the rafts had hatched.

Almost all the species discussed so far lay eggs that cannot tolerate desiccation. Consequently the eggs hatch soon after completion of embryonic development, in as little as two to three days in tropical species but after a week or more in temperate and subarctic species. In contrast are the species having eggs that can withstand desiccation. Eggs of *Haemagogus equinus* can survive partial desiccation for seven months, and some species of *Heizmannia* can hatch from eggs that have been kept mainly dry for three weeks. Eggs of *Opifex fuscus* can remain dry for six months, and it seems that eggs of *Eretmapodites* and *Armigeres* can tolerate at least some degree of drying. It is, however, the eggs of *Aedes* and *Psorophora* that are best known for the ability to resist desiccation.

Before *Aedes* eggs can withstand drying they must be 'conditioned' to allow the protective serosal cuticle to be formed, and the endochorion to become tanned and more or less impermeable to water, and for the embryo to develop into a mature larva. It is this young larva encapsulated in the protective egg shell that resists desiccation. If eggs become dry before larvae have matured they collapse and die. How long dry eggs can remain alive depends largely on relative humidity. Eggs of *Aedes aegypti* kept at a relative humidity of 70% or more have remained viable for as long as 15.5 months. *Aedes cantans* lays eggs amongst wet leaf litter but they collapse and die unless kept at relative humidities above 85%; at such high humidities some eggs can survive for up to four and a half years (Service, 1977b). Generally, the longer the eggs are kept the fewer hatch.

When *Aedes* eggs are flooded, hatching can occur within minutes, but usually some eggs remain unhatched until later. For example, a staggered hatch over many days occurs after a single soaking, larvae appearing in instalments, up to three months with *A. cantans* and up to about 45 days in *A. aegypti*. Soon after oviposition the surfaces of eggs of *A. aegypti* and other *Aedes* become colonized by bacteria which apparently lower the oxygen in the immediate microenvironment of the eggs and stimulate hatching. Eggs that harbour more bacteria than others will hatch earlier, and on emergence the young larvae browse over the surface of unhatched eggs, reducing their bacterial colonies and thus delaying hatching. Inhibition of egg hatching when food is limited and when there are already numerous larvae in the habitat is clearly advantageous as it reduces competition among larvae for food (Livdahl and Edgerly, 1987). Erratic hatching also occurs in eggs of *Psorophora, Haemagogus* and *Heizmannia,* and probably in other aedine genera, but whether the microbial theory applies in all such cases is unknown.

Not only do eggs exhibit a staggered hatch after immersion in water, but repeated soaking and desiccation for short periods stimulates hatching in instalments. For example, eggs of *Aedes detritus* and *A. cantans* require 12 and 15 soakings respectively before all of them hatch (Service, 1968, 1977b). There can also be seasonal variations in the proportions hatching on different soakings. In Senegal almost all eggs of *Aedes luteocephalus* laid at the beginning of the rains hatched on their first soaking, whereas hatching was spread over several soakings in those deposited towards the end of the rainy season.

Hatching in instalments after repeated immersion is a common phenomenon in *Psorophora* mosquitoes and has also been recorded in *Haemagogus* and *Heizmannia.* In *Haemagogus equinus* as many as ten soakings are sometimes needed to induce all eggs to hatch. The combination of delayed hatching of aedine eggs on prolonged submersion in water, and the necessity for several soakings and desiccations to promote hatching of some eggs, clearly increases survival rates of mosquitoes living in temporary habitats liable suddenly to dry out.

Aedes eggs that fail to hatch in deoxygenated water are probably in diapause. Facultative diapause is induced by environmental cues, such as changes in daylength and/or temperature which are perceived by the stage preceding and/or including the diapausing stage. Certain stimuli are also needed to terminate diapause, such as prolonged exposure to high or low temperatures, or exposure to increased photoperiod. Many northern temperate *Aedes* mosquitoes overwinter as diapausing eggs. In the summer some northern *Aedes,* such as *A. vexans, A. campestris* and *A. canadensis,* lay non-diapausing eggs which hatch if flooded during the summer, and also diapausing eggs which are responsible for winter survival (Wood et al., 1979).

The onset of diapause in *Aedes triseriatus* living in tree-holes in New York is induced mainly by autumnal shortening of daylength; then in the following spring diapause is broken by longer photoperiods (16 hours or more), and eggs start hatching. In contrast, populations of *A. triseriatus* in Georgia do not overwinter as eggs; instead, short daylengths induce larval diapause. When *Aedes togoi* are reared at 27°C with 16 hours' daylength, diapause does not occur in any life-stages, but at 15°C and with 12 hours of light adults lay eggs that enter diapause. Maintaining them at 15°C but reducing the photoperiod to ten hours results in larvae diapausing in the fourth instar. Sometimes, as in *Aedes dorsalis,* diapause is broken only by a combination of long photoperiod and increased temperature.

Larvae

All mosquito larvae are aquatic and metapneustic and pass through four larval instars.

Larval development
Speed of larval development depends on food supply, water temperature and the species. In the tropics larval development of *Aedes* colonizing temporary habitats is often completed within a week. Pupae of *Aedes vittatus* have been collected from rock-pools three to four days after the pools were flooded with rainwater, and they can complete development from egg hatching to pupation in 53 hours! Predatory larvae of *Toxorhynchites*, commonly inhabiting tree-holes and water-filled bamboo, can need two to five weeks to complete their development, and when prey is limited this can be extended to three to five months.

With plentiful food *Wyeomyia vanduzeei* and *Wyeomyia medioalbipes* living in plant axils can complete larval development within about 13 days, but many weeks or months (up to 173 days for the latter species) can be required if there is a food shortage. Undoubtedly food shortages can greatly prolong mosquito larval development, and in small container-habitats like tree-holes, plant axils, pitcher plants and domestic water-storage pots food often seems to be an important limiting factor (Frank and Lounibos, 1983).

In contrast to the rapid growth of most tropical mosquitoes, larvae of temperate and subarctic species may live for many months. Temperate region *Aedes* hatching in the winter or early spring from overwintering eggs often take two to four months before they pupate. If extreme conditions affect temperate species these can overwinter for six, or more rarely eight, months as larvae. *Anopheles claviger* in ponds in Europe, *Anopheles plumbeus* and *Orthopodomyia pulcripalpis* in tree-holes in Europe, *Anopheles barberi*, *Aedes triseriatus* and *Orthopodomyia signifera* in tree-holes in North America, *Anopheles pulcherrimus* in shallow lakes and marshes in the USSR (although in warmer parts of its distribution breeding continues throughout the year), and *Wyeomyia smithii* in leaf axils of *Sarracenia* pitcher plants in northern latitudes of North America, are examples. *Coquillettidia* species such as the North American *C. perturbans* and European *C. richiardii* also overwinter as larvae.

There is often insufficient evidence to determine whether overwintering larvae are in diapause (initiated and terminated by environmental cues) or are quiescent, needing only warmer temperatures to speed up larval development. In some mosquitoes, however, such as *Aedes triseriatus*, *Anopheles plumbeus*, *Anopheles barberi* and *Orthopodomyia signifera*, a reduction in daylength is known to induce larval diapause; the first species also diapauses as eggs. In others, such as *Anopheles pulcherrimus*, *Anopheles claviger* and *Wyeomyia smithii*, cold temperatures as well as short daylengths are needed to maintain diapause. There can also be geographical variations in photoperiod response for inducing diapause, as known in *Aedes sierrensis* in the USA.

The European *Culiseta annulata* is unusual in that it overwinters as larvae as well as partially hibernating adults.

Feeding behaviour
Five functional feeding types can be recognized among larvae: (*a*) scavengers that ingest mainly dead food by scraping and nibbling at animal carcasses and detritus; (*b*) bottom-feeders that browse over bottom debris, feeding on living and non-living material (much of it scraped from submerged surfaces); (*c*) filter-feeders that strain phytoplankton and zooplankton from the water; (*d*) filter-feeders that feed from the surface film; (*e*) predators that consume other living organisms (usually mosquito larvae).

Many culicine larvae are browsers and scavengers, but some *Aedes* larvae which normally browse also filter-feed when they surface to breathe. Larvae of a few species, such as *Culex sinensis* and *Culex bitaeniorhynchus*, feed on filamentous algae.

Anopheles larvae are good examples of surface-feeders. Because their mouthbrushes and mouth are ventral, larvae have to swivel the head through 180° in order to sweep the undersurface of the water and waft very fine food particles towards the mouth. Many culicine larvae also live at the water surface, simultaneously feeding and breathing, but because they have siphons, which can sometimes be very long, they can feed well below the surface, and so are plankton-feeders. Other plankton-feeders, such as *Mansonia* and *Coquillettidia* and some *Mimomyia* have larvae whose siphon is inserted into plants. Surface and plankton-feeders ingest bacteria, yeasts, protozoans, diatoms, desmids and other microflora and microfauna as well as non-living very fine suspended matter.

In most species having predatory larvae the mandibles are used to catch prey. This consists of small aquatic organisms, particularly mosquito larvae. Predatory larvae are sometimes cannibalistic. Occasionally all, or nearly all, species in a genus such as *Toxorhynchites* and *Eretmapodites* or subgenera such as *Culex (Lutzia)*, *Aedes (Mucidus)* and *Psorophora (Psorophora)* are predaceous. Only a few species in genera such as *Trichoprosopon*, *Armigeres* and *Zeugnomyia* are predators. The best-known predators are the voracious *Toxorhynchites* larvae, which can devour up to 350–450 larvae during their life. *Toxorhynchites* larvae sometimes display compulsive killing behaviour (Corbet, 1985) a few days before pupation, killing far more larvae than they can eat. The selective advantage of this mass destruction remains uncertain.

Eretmapodites larvae breed in containers and many species are obligatory predators and facultative cannibals; others are facultative predators that can complete development on a non-living diet. *Tripteroides (Rachisoura)* species breed mainly in tree-holes and bamboo; they have predaceous larvae, but larvae will also readily feed on dead and decaying insects falling into the water. Their maxillae are modified into extremely long claw-like clasping organisms (Fig. 5.10c), believed to be adapted to this type of feeding.

The larval environment
Some species, like *Anopheles gambiae*, occur in sunlit pools lacking vegetation or in ricefields, whereas vegetation appears to be essential for species such as *Anopheles funestus*. Others, like *Anopheles umbrosus*, occur in heavily shaded forest pools. Several mosquitoes, including *Anopheles albimanus* and various *Culex* species, are found in waters thick with filamentous algae which probably afford them some protection from predators, whereas others such as *Anopheles philippinensis* prefer habitats without filamentous algae. Because of their respiratory adaptations, larvae of *Mansonia*, *Coquillettidia* and a few *Mimomyia* are restricted to waters having *Pistia*, *Eichhornia*, *Hydrocharis*, *Typha* or some other suitable floating or rooted vegetation. Larvae of *Eretmapodites*, *Trichoprosopon* and *Armigeres* are common in the viscous, foetid, anaerobic water collecting in coconut or cocoa pods. Larvae of *Eretmapodites subsimplicipes* seem to thrive in the thick mucilaginous slime left behind when most water has evaporated from fruit husks.

Temperature can be a limiting factor. Certain tropical mosquitoes can tolerate, at least for a short time, temperatures as high as 40–46°C; e.g. in the Seychelles larvae of *Aedes vigilax vansomerenae* tolerate rock-pool water as hot as 50°C. In contrast, larvae of north-temperate mosquitoes, such as *Culiseta inornata* in Canada, die if subjected to 29°C – as do *Aedes flavescens* larvae held at 25°C. In temperate and arctic regions water temperatures can drop to near freezing point for part of the year, forcing larvae to shelter in the slightly warmer water at the bottom of pools and ponds, and sometimes to bury themselves in mud to avoid being frozen. Larvae sometimes aggregate in parts of pools receiving the most incident solar radiation. Although larvae of most species are killed when frozen, a few can tolerate freezing; *Wyeomyia smithii* larvae, for example, can become frozen in solid blocks of ice in pitcher plants and yet suffer very little mortality.

Most *Anopheles* larvae require some degree of oxygenation of the water and a few, such as *Anopheles minimus*, seem to be restricted to highly oxygenated waters, such as stream edges. Many other species, including *Culex quinquefasciatus*, *Armigeres subalbatus* and *Trichoprosopon digitatum*, flourish in poorly oxygenated waters, such as stagnant swamps, cesspits, logging ponds, coconut pits and foetid water in fruit husks. Most mosquito larvae have rather a wide tolerance to pH, but only a few are found in waters having a pH of more than 10.5; one exception is the East African *Aedes natronius* which breeds in crater lakes having a pH of > 11. At the other extreme, larvae of *Culex quinquefasciatus* have been found in water having a pH of 1.6.

Most mosquito larvae are found in fresh water and are unable to tolerate even moderate salinities, but a few breed in salt-water habitats. The Palaearctic *Aedes detritus* can tolerate chlorine concentrations more than twice those of sea-water. Larvae of *Culex tenagius* seem to withstand environmental extremes, having been collected from a saline lake in Tanzania with > 80 g NaCl/litre, alkalinity of >0.32 normal sodium hydroxide, pH >11, and temperature as high as 42°C. The physiology of osmoregulation is reviewed by Bradley (1987).

Larvae are unable to survive desiccation, and when habitats slowly dry out they often seek temporary shelter under wet leaves, amongst damp debris or in mud, where they can survive for many days and sometimes even weeks. Larvae can also live many weeks in a thin film of water at the base of leaf axils.

Larval habitats
Most classifications of larval habitats make an important distinction between surface collections of water (ponds, pools, marshes etc.) and container-type habitats (tree-holes, bamboo, plant axils, domestic water-storage pots). This poses problems. For instance, rock-pools can be considered container-type habitats, but if they are large, contain much mud and vegetation, they are ecologically more similar to ground pools. Crab-holes can be regarded as ground collections of water, albeit subterranean, or as specialized container-habitats. The present classification could be considered too detailed because bamboo habitats are listed separately from tree-holes, and fallen leaves as distinct from fruit husks; this is intentional, to draw attention to the great diversity of mosquito breeding places. The word phytotelmata (Frank and Lounibos, 1983) is sometimes used to describe small collections of water held in various parts of terrestrial plants (e.g. tree-holes, bamboo, leaf axils, flower bracts, pitchers, fallen leaves and fruits). Among the more unusual breeding places are water-filled cavities formed in large cacti in Colombia colonized by *Haemagogus equinus* and *H. anastasionis*, holes in *Euphorbia* plants in Nigeria used by *Aedes* species (including *A. aegypti*), water-filled ants' nests in South East Asia used by *Malaya* larvae, and water accumulating on the concave tops of polyporaceous fungi in East Africa colonized by *Eretmapodites subsimplicies* and

E. grahami. Terrestrial cup fungi provide larval habitats for *Aedes palmarum* in Australia and for *Aedes dobodurus* in Papua New Guinea. In South East Asia larvae of *Aedes albopictus* are sometimes found in cups placed on rubber trees to collect latex.

A few species are restricted to a single type of habitat; for instance larvae of *Armigeres dolicocephalus* are found only in bored bamboo and *Wyeomyia smithii* only in *Sarracenia* pitcher plants. Many other species occur in a rather wide range of, usually similar, breeding places. For example, *Aedes aegypti* is found in numerous natural and artificial containers while *Trichoprosopon digitatum*, although mainly found in bamboo or fallen fruit husks, colonizes tree-holes, leaf axils, fallen leaves and artificial containers.

Useful accounts of larval ecology and habitats are given by Assem (1961), Dąbrowska-Prot (1979) and Laird (1988). The following is my classification of mosquito larval habitats.

1 **Ground habitats**
1.1 *Still waters*

1.1.1 Permanent or semi-permanent
    (a) Marshes
    (b) Swamps
    (c) Exposed ponds, borrow pits
    (d) Forest ponds
    (e) Ditches
    (f) Ricefields
    (g) Polluted waters (e.g. cesspits, log-ponds)
    (h) Wells
    (i) Salt-water marshes
    (j) Salt-water ponds
    (k) Subterranean (caves, flooded cellars)

1.1.2 Temporary
    (l) Exposed pools and puddles
    (m) Forest pools
    (n) Hoof-prints
    (o) Salt-water pools

1.2 *Flowing waters*
    (p) Exposed streams, ditches, irrigation channels
    (q) Forest streams
    (r) Gravel stream beds

2 **Container habitats**
2.1 *Natural*
    (s) Tree-holes
    (t) Bamboo
    (u) Leaf axils, including bromeliads and plant bracts
    (v) Pitcher plants
    (w) Fallen fruit husks
    (x) Fallen leaves

(y) Empty snail shells
(z) Rock-pools
(aa) Crab-holes

2.2 *Man-made*
(bb) Water tanks, cisterns
(cc) Latrines, septic tanks
(dd) Domestic water-storage pots
(ee) Discarded tin cans, bottles, tyres
(ff) Miscellaneous (cemetery vases, gutters, broken boats etc.)

Permanent ground habitats (a–e)   These habitats usually have vascular vegetation and are colonized by a great variety of mosquitoes, including species of *Anopheles, Hodgesia, Uranotaenia, Aedeomyia, Ficalbia, Mimomyia, Coquillettidia* and *Mansonia*. Several *Culex, Culiseta* and *Aedes* species also commonly breed in these more permanent collections of water.

Ricefields (f)   Most mosquitoes breeding in ricefields are *Anopheles* or *Culex* species; many are important disease vectors. Malaria vectors commonly found in ricefields include *Anopheles gambiae* and *A. arabiensis* in Africa, *A. culicifacies* and *A. subpictus* in the Indian subcontinent, *A. darlingi* in South America, *A. sinensis* in China and *A. aconitus* in much of South East Asia. *Anopheles freeborni* is commonly found in ricefields in the western USA, and was once an important malaria vector. The principal vector of Japanese encephalitis, *Culex tritaeniorhynchus*, breeds extensively in ricefields. Other common mosquitoes found in ricefields include *Culex tarsalis*, a vector of Western equine encephalomyelitis and St Louis encephalitis, various *Aedes* species (such as *A. melanimon* and *A. nigromaculis*) and *Psorophora confinnis*, all found in the USA.

Polluted waters (g)   Mosquito larvae mostly cannot withstand high degrees of organic pollution, but some species such as *Culex quinquefasciatus, C. stigmatosoma* and *Culiseta incidens* in the USA, *Culex cinereus* of East Africa, and *Armigeres subalbatus* and *A. kuchingensis* of South East Asia, are common in cesspits, septic tanks, latrines, log-ponds, as well as drains and ditches found in urban slums. Larvae of *Trichoprosopon digitatum* frequently occur in foul-smelling waters of bamboo and fruit husks.

Wells (h)   A few species, including some important malaria vectors such as *Anopheles stephensi* in India and *A. claviger* in the Middle East, are often found in the cool dark waters of wells.

Salt-water habitats (i, j, o)   In most areas of the world some mosquitoes have colonized salt-water estuaries and marshes, mangrove swamps, and saline small pools of water, including those formed in coral and coastal rock outcrops.
In West Africa the malaria vector *Anopheles melas* is common in *Avicennia* mangrove swamps; in East Africa its counterpart *A. merus* breeds not only in salt-water but sometimes in fresh water. In India and Malaysia *Anopheles sundaicus* breeds in coastal salt-waters, while in South America a local malaria vector along the coast is *Anopheles aquasalis*. In Europe *Aedes detritus* and *Aedes caspius* are

common coastal mosquitoes, while in North America *Aedes sollicitans* and *Aedes taeniorhynchus* are important coastal pest mosquitoes. Other salt-water species include the African *Culex thalassius*, the Australasian *Aedes vigilax*, the African and Australasian *Culex sitiens*, and *Opifex fuscus* which lives in brackish-water rockpools in New Zealand. A few species occur in inland brackish waters: *Aedes natronius*, for example, is found in saline crater lakes in Uganda and *Anopheles multicolor* in brackish desert oases in North Africa.

Subterranean waters (k) The *molestus* 'form' of *Culex pipiens* breeds in dark underground collections of water, such as flooded cellars and abandoned mine tunnels. It used to breed in pockets of water formed under platforms of the London underground (tube) stations. In Hungary, *C. pipiens* has been recorded breeding 300 m down a coal mine, and in India *Culex quinquefasciatus* has been found in coal mines at depths of 305 and 457 m. In Africa larvae of *Anopheles smithii* are sometimes found in caves, as are larvae of *Anopheles vanhoofi* and a few other *Anopheles* species.

Temporary waters (l–o) Small ponds, pools, roadside ruts, animal hoof-prints and other temporary habitats have generally been best exploited by aedine mosquitoes – mainly *Aedes* and *Psorophora* which have drought-resistant eggs and can complete their larval development within about a week. In the tropics, larval development of some *Anopheles* lasts only 7–11 days, and throughout most of Africa *Anopheles gambiae* and *Anopheles arabiensis* are very common in the muddy waters of rain pools, puddles and cattle hoof-prints, as are larvae of *Anopheles vagus* in the Indian subcontinent and *Anopheles punctulatus* in Papua New Guinea.

Flowing waters (p–r) Mosquitoes are not found in fast flowing waters, nevertheless a few species, principally *Anopheles* such as *A. pseudopunctipennis* of South America and *A. superpictus* of southern Europe, *A. nili* and *A. funestus* of Africa, *A. minimus* of India and *A. maculatus* in Malaysia, have successfully colonized the shallow edges of streams where there is moderately or gently flowing water. In Australasia *Culex starckeae* larvae are found in briskly flowing waters.

Natural container-habitats (s–aa) All *Toxorhynchites* and sabethines breed in containers, as do about 43% of culicine species, but only a few *Anopheles* occur in containers. Most species breeding in containers lay drought-resistant eggs, and some have rapid larval development – adaptations which help them cope with transient habitats. Basically mosquitoes living in containers are forest mosquitoes, although certain species have invaded domestic and peridomestic habitats.

Tree-holes (s) Tree-holes are the most widespread of natural container-habitats and are found throughout tropical and temperate regions. Virtually all water-filled tree-holes are formed in hardwood trees, although occasionally larvae occur in conifers and palm trees. There are different ecological types of tree-holes. Some arise as rot-holes after a branch has fallen or been cut off. Sometimes water collects in depressions, called pans, formed where the trunk divides, or between the buttress roots of trees.

Mosquitoes can show a marked preference for specific types of tree-holes; for example, *Sabethes chloropterus* and Australian species of *Aedes (Chaetocruiomyia)* are restricted to tree-holes having narrow apertures. In West Africa *Aedes pseudoafricanus* breeds only in holes in *Avicennia* mangrove trees, while *Aedes thibaulti* of South East Asia occurs predominantly in basal tree-holes (pans) in gum (*Nyssa*) trees.

Tree-holes support a rich variety of mosquitoes, including several species of *Toxorhynchites*, *Sabethes*, *Haemagogus*, *Heizmannia*, *Tripteroides*, *Topomyia*, *Orthopodomyia*, numerous species of *Aedes* (especially of subgenera *Finlaya* and *Stegomyia*), some *Anopheles* and a few *Culex*. Typical examples of well-known mosquitoes found in tree-holes are *Anopheles plumbeus* and *Aedes geniculatus* (Europe), *Toxorhynchites rutilus*, *Anopheles barberi*, *Orthopodomyia signifera*, *Aedes triseriatus* and *Aedes sierrensis* (USA), *Aedes africanus* and *Aedes luteocephalus* (Africa), *Aedes seoulensis*, *Aedes chemulpoensis* and *Culex brevipalpis* (China), *Aedes candidoscutellum* (Australia), *Anopheles omorii* and *Culex kyotoensis* (Japan) and *Anopheles barianensis* (Asia).

Bamboo (t) After bamboo has been cut the water-filled stumps are colonized by many *Aedes* species, several *Toxorhynchites*, *Eretmapodites*, *Trichoprosopon*, *Sabethes*, and a few *Anopheles* and *Uranotaenia* species. Many mosquitoes found in bamboo stumps also readily breed in tree-holes, but some, such as *Trichoprosopon* and *Armigeres*, rarely occur in them. The South East Asian *Armigeres dolichocephalus* and *Tripteroides bambusa*, however, colonize dark and enclosed habitats and can be found in bamboo where rain water has trickled through holes bored by beetles and accumulated within the stems. In Africa, *Uranotaenia garnhami* and sometimes *U. shillitonis* are found in holes that have been bored by caterpillars.

Leaf axils and pitcher plants (u, v) Several mosquitoes, including many *Aedes*, several *Trichoprosopon*, *Topomyia*, *Malaya*, *Wyeomyia* and *Mimomyia (Ingramia)*, oviposit in water accumulating in the leaf axils of plants, such as *Colocasia*, *Xanthosoma*, *Mauritia*, *Pandanus*, *Sansevieria*, *Ravenala*, *Dracaena*, *Lobelia* and *Strelitzia*.

Anopheles larvae are very rarely found in leaf axils (apart from bromeliads), an exception being *A. hackeri* breeding in leaf axils of *Nypa* palms. There are some 2000 species of bromeliad plants in the Americas, but mosquitoes breed only in the tank varieties, that is those that collect water (1–45 litres) in the axils of their leaves. About 214 mosquito species have been collected from bromeliads (Frank and Lounibos, 1983), including several *Toxorhynchites*, some *Trichoprosopon*, a few *Wyeomyia* and *Culex*, and about five species of *Anopheles (Kerteszia)* – including *A. bellator*, an important malaria vector in the West Indies.

Several species of *Wyeomyia* and *Trichoprosopon* colonize pitcher plants, the best known being the North American *Wyeomyia smithii* in *Sarracenia purpurea*. In South East Asia and Papua New Guinea at least 94 species in eight genera are found in *Nepenthes* pitcher plants, the most successful being *Tripteroides*. Several *Culex (Lophoceraomyia)*, a few *Aedes*, *Toxorhynchites* and *Armigeres*, and a single *Anopheles* species, also occur in these plants. Although *Nepenthes* are absent from mainland Africa, they grow in Madagascar where they are colonized by several *Uranotaenia (Pseudoficalbia)* species and *Mimomyia (Ingramia) jeansottei*.

Heliconia flower bracts usually hold less than 10 ml of water, but they are colonized by *Wyeomyia*, a few *Culex*, *Toxorhynchites* and *Trichoprosopon* species.

Ground containers (w–y) Numerous small container-type habitats, such as split and rotting cocoa pods and coconut husks which litter forest floors, provide ideal habitats for *Eretmapodites* in Africa and for *Armigeres* in Asia. When the decaying pulp diminishes and the water becomes less organically polluted, species of *Toxorhynchites, Trichoprosopon, Aedes* and *Uranotaenia* invade these pods and husks. Rachids and spadix sheaths (e.g. coconut palms), leaves of bananas and other large plants, bracts and palm fronds falling onto the forest floor can all collect small, but sufficient, quantities of water for colonization by *Aedes, Eretmapodites* and *Culex* mosquitoes.

Larvae of a few species of *Aedes (Finlaya)*, such as *A. palmarum* in Australia, can live in the small collections of water that accumulate in cup fungi, although as the name suggests this mosquito is more commonly found breeding in fallen palm fronds. In East Africa *Eretmapodites subsimplicipes* and *E. grahami* breed in the water collecting on top of polyporaceous fungi.

Various *Aedes* and *Eretmapodites* species breed in empty snail shells, such as the large *Achatina* shells found in Africa.

Rock-pools (z) Several mosquito species encountered in rock-holes are also found in small ground pools, and a few even in tree-holes, but others such as *Aedes vittatus, Aedes atropalpus* and *Aedes fluviatilis* are more or less restricted to rock-pools. Several *Uranotaenia, Toxorhynchites* and a few *Culex* and *Anopheles* also live in rock-pools.

In New Zealand, *Opifex fuscus* breeds in brackish rock-pools, while the Oriental coastal *Aedes togoi* is commonly found in fresh water or brackish rock-pools. In the Mediterranean, *Aedes mariae* is found in saline rock-pools.

Coastal rock-pools formed in coral support either brackish-water mosquitoes, or, if the pools are formed on elevated coral reefs, freshwater species, such as *Aedes aegypti* and the East African *Aedes bromeliae*. The latter species, however, is much more common in leaf axils.

Crab-holes (aa) In the New World, *Deinocerites* species have specialized in breeding in the long and tortuous tunnels made by land crabs. Several *Aedes* species, especially those of the subgenera *Cancraedes, Geoskusea* and *Levua,* as well as mosquitoes like *Aedes irritans* of West Africa, *Aedes pembaensis* of East Africa and a few *Uranotaenia, Culex* and *Culiseta,* are also found in crab-holes.

Man-made containers (bb–ff) Water-tanks and cisterns, often sited on the roofs of houses in the Indian subcontinent, are an important source of *Anopheles stephensi*, the principal vector of urban malaria. Other mosquitoes found in water-tanks include *Anopheles claviger*, various *Culex* such as *C. pipiens*, and a few *Culiseta* species. Domestic water-storage pots, gourds, tin cans, bottles, tyres, which are commonly scattered around villages and towns, support peridomestic breeding of several mosquitoes, including *Aedes aegypti*, the *A. scutellaris* complex, *A. albopictus,* other *Aedes* and, less frequently, various *Culex* and *Toxorhynchites* mosquitoes. Flower pots and cemetery urns provide unusual habitats. A survey in 1977 showed that in the major cemetery in Caracas, Venezuela, there were some 190 000 flower pots which supported large populations of *Culex quinquefasciatus* and *Aedes aegypti*. Rose Hill Memorial Park in an urban part of Los Angeles also contains an estimated 190 000 flower vases in which mosquitoes are breeding – a remarkable (but correctly cited) coincidence with the number in the Caracas cemetery!

In Sri Lanka illegal gem mining is responsible for creating numerous water-filled pits which become colonized by *Anopheles culicifacies*. In Asia, pits dug to soak coconut husks in the preparation of coir – the fibre used to make rope and door mats – provide ideal breeding places for *Culex quinquefasciatus* and *Culex gelidus*.

Larval mortalities
Although mosquito larvae can survive for three to five days or so on wet substrates, desiccation nevertheless causes appreciable mortality in many species colonizing temporary collections of water, such as small pools, rock-pools, tree-holes, snail shells and fallen fruit husks. On the other hand, torrential rain can flush out some habitats and lead to high larval mortality.

The most permanent collections of water frequently support a large variety of predators such as fish, tadpoles, turtles, aquatic birds, dragonflies, beetles, chaoborids, Hemiptera, predatory mosquito larvae, and even carnivorous plants like the bladderwort (*Utricularia*).

Adult flies of the Dolichopodidae, Scathophagidae, Ephydridae, Anthomyiidae and Muscidae prey on larvae and pupae when these are at the water surface, as well as on adult mosquitoes. Other aquatic predators that are frequently overlooked are non-aquatic spiders (e.g. lycosids) which dash nimbly across the water surface and seize emerging mosquitoes. Cockroaches, mites, ants and various other scavengers devour stranded mosquito larvae and eggs.

Mosquito larvae are also killed by pathogenic bacteria and viruses (iridescent, cytoplasmic, polyhedrosis and baculoviruses), protozoans of the Microspora (*Nosema algerae, Vavraia culicis, Amblyospora* and *Thelohania* species), fungi (*Coelomomyces, Lagenidium*) and nematodes (*Romanomermis culicivorax, Octomyomermis muspratti*). Larvae sometimes become covered with epibionts, such as *Vorticella* and *Epistylis*, which although not killing them can make them more sluggish and more vulnerable to predators.

In contrast to mosquitoes colonizing relatively large habitats, container-breeding mosquitoes, especially those living in plant axils, pitcher plants, *Heliconia* bracts and tree-holes, often seem remarkably free of predators. It appears that mosquito populations in these phytotelmata, and probably other natural containers, are mainly limited by competition for resources, in particular food. It also seems that lack of food can limit population size of *Aedes aegypti* breeding in water-pots (Subra, 1983).

Undoubtedly natural enemies cause large larval mortalities, but listing them would serve little purpose. Annotated lists are in Jenkins (1964), Roberts and Strand (1977) and Roberts and Castillo (1980). A bibliography of larvivorous fish is in Gerberich and Laird (1985).

Some believe that heat-stable metabolites are excreted by larvae when they are severely overcrowded, and that such so-called growth-retardants or overcrowding factors aid self-regulation of population size (see Mori, 1979, for references). Others have failed to find evidence for this, and assert that lack of food *per se* is responsible for larval population regulation (Dye, 1984).

Pupae

Pupation occurs at the water surface. The larval integument splits along the mid-dorsal line allowing the pupal thorax to emerge, and very soon the trumpets spring up and come into contact with the water surface; pupal respiration then comes into operation while the still-encased abdomen wriggles free of its larval integument.

The whole process takes about three to five minutes. At first the pupal abdomen is straight; it takes a little longer for it to curve underneath the cephalothorax and for the pupa to assume its characteristic comma shape.

Pupae are unable to feed. Being less dense than water, they normally spend most of their time at the water surface breathing through the paired respiratory trumpets; exceptions are pupae of *Mansonia*, *Coquillettidia* and a few *Mimomyia* which, like the larvae, pierce submerged plants to breathe (Fig. 5.20).

Buoyancy is due to air trapped in a ventral air-space formed by the developing mouthparts, legs and wings. When pupae are disturbed by shadows or vibrations they descend in a quick zigzagging fashion by alternately flexing and stretching the abdomen. When swimming stops, pupae slowly float to the water surface, where their position is held by the hydrophilic outer coating of the trumpets and by the large paired dendritic float hairs on the first abdominal segment. Mosquito pupae are the most active of any insect pupae.

Pupal duration is mainly determined by temperature. In hot tropical countries it is usually two to three days, but can be as short as 26 hours at temperatures of about 30°C. In temperate regions pupal life usually lasts about a week, but can be two to three weeks in cold weather. Duration also varies according to species, for example in *Mansonia* and *Coquillettidia* the pupal stage lasts six to nine days. No species overwinters as pupae. In dry conditions pupae tolerate partial desiccation better than larvae; they survive well on damp substrates.

A few hours before emergence the pupa noticeably darkens, then some five to ten minutes before the adult emerges the abdomen is straightened and the pupa lies almost parallel to the water surface. Air appears beneath the integument which then splits mid-dorsally; the thorax and head of the adult, followed by the antennae and mouthparts, and finally the legs and abdomen, begin to emerge. During emergence the adult swallows air which distends the stomach, and it is this that forces the adult from its pupal case. Emergence usually takes 12–15 minutes, and within minutes afterwards the newly emerged adult can make very short flights. Male larvae generally pupate before females and in most species male pupal life is shorter than female pupal life. Males usually emerge a day or so before females. Such protandry occurs in many, if not most, species; protogyny at emergence has been recorded in only a few. Most species emerge in the early evening or late at night.

The presence of pupae in a habitat reflects its productivity, but even so, not all pupae will give rise to sexually mature adults, many will be eaten by predators. Furthermore, emerging and newly emerged adults are also very susceptible to predation, and adults have to live sufficiently long to reproduce before there is any new 'input' into the habitat.

The sex ratio of emerging populations calculated over the entire emergence period is usually about unity, although in high arctic *Aedes impiger* and *A. nigripes* females predominate. There can be seasonal variations; *Aedes triseriatus*, a tree-hole species, produces predominantly males at the beginning of the season but females are commonest in the later part of the season (Scholl and DeFoliart, 1978).

MEDICAL IMPORTANCE

Mosquitoes transmit a variety of infections to man, malaria, filariasis and the arbovirus diseases of yellow fever and dengue being the best known. Some of these diseases, such as yellow fever and Brugian filariasis, are zoonotic. The *WHO Weekly Epidemiological Record* is an informative publication giving the latest data on the prevalence of vector-borne, and other, diseases. Certain chapters in Service (1989a) describe how human demography, migration, urbanization, irrigation and other development schemes can affect mosquito vector populations and the epidemiology of malaria, filariasis and various arbovirus infections. Service (1989b) briefly describes the impact of rice-growing on mosquito-borne diseases.

Malaria

Although since the 1940s there have been some spectacular anti-malaria campaigns, malaria unquestionably remains the most important vector-borne disease. In many areas such as India, Sri Lanka, Guyana and Mexico, there have been resurgences of malaria, whereas in tropical Africa there has been no real reduction in malaria. In contrast, malaria mortality and morbidity have been reduced in China, and malaria has been eradicated from Europe (except Turkey), USA, Chile, Israel, Australia, Brunei and 17 islands (including Cyprus, Taiwan, Mauritius, the Seychelles and several Caribbean islands). Currently, about 2100 million people live under the threat of malaria in 103 countries, and about 445 million of these are in areas where there is no control. It has been estimated that in Africa about one million infants die each year from *P. falciparum* malaria. All four human malarial parasites, namely *Plasmodium falciparum, P. vivax, P. malariae* and *P. ovale*, are transmitted by *Anopheles* mosquitoes.

Although there are some 422 *Anopheles* species, only about 70 are malaria vectors and of these probably only about 40 are important. Malaria vectors are often divided into primary and secondary vectors, but this is rather unsatisfactory because a species can be a so-called primary vector in one area and a secondary vector or even a non-vector in another. White (1982) preferred to classify *Anopheles* as main vectors (widespread, dominant vectors) and subsidiary vectors, i.e. incidental vectors which by themselves are incapable of sustaining transmission, or secondary vectors localized in their distribution but able to be the principal vectors within their areas. An example is the West African salt-water *A. melas*, which is restricted to coastal areas but can here be an important malaria vector. The most important vectors are listed in Table 5.2 according to the epidemiological zones (Fig. 5.29) of Macdonald (1957): categorization as main and subsidiary vectors is somewhat subjective.

The ability of *Anopheles* species to transmit malaria depends much on their physiological susceptibility. Vector competence is frequently higher with co-indigenous strains of *Plasmodium* than exotic ones (Warren and Collins, 1981). For example, *A. atroparvus* from Italy and England is refractory to *P. falciparum* from Africa, whereas it is readily infected with European *P. falciparum*. In addition to the susceptibility of a mosquito species to malaria, its capability as a vector (vectorial capacity) depends on a combination of biological factors, particularly the survival rate determining the proportion of adult mosquitoes living sufficiently long to become infective, the degree of anthropophagy, and the vector population size. One of the world's more efficient vectors is *A. gambiae* s.str., an African vector characterized by a high degree of mosquito–man contact and a relatively high survival rate; as a consequence, sporozoite infections of about 5% are common. In contrast, *A. culicifacies* in India is mainly zoophagic and has a lower survival rate; its

Table 5.2 *Anopheles* vectors of malaria in the epidemiological zones

Zone numbers and geographical naming follow Macdonald (1957) and zone limits are shown in Fig. 5.29. Abbreviations for subgenera given in parentheses are: A, *Anopheles* s.str.; C, *Cellia*; K, *Kerteszia*; N, *Nyssorhynchus*. Main (primary) vectors are distinguished from subsidiary (incidental and local) vectors by bold type. North America is included for completeness but is no longer a malarious area.

1. North American
A. (A.) freeborni
A. (A.) quadrimaculatus
A. (N.) albimanus

2. Central American
A. (A.) aztecus
A. (A.) punctimacula
A. (A.) pseudopunctipennis
A. (N.) albimanus
A. (N.) albitarsis
A. (N.) aquasalis
A. (N.) argyritarsis
A. (N.) darlingi

3. South American
A. (A.) pseudopunctipennis
A. (A.) punctimacula
A. (K.) bellator
A. (K.) cruzii
A. (K.) neivai
A. (N.) albimanus
A. (N.) albitarsis
A. (N.) aquasalis
A. (N.) argyritarsis
A. (N.) braziliensis
A. (N.) darlingi
A. (N.) nuneztovari
A. (N.) triannulatus

4. North Eurasian
A. (A.) atroparvus
A. (A.) messeae
A. (A.) sacharovi
A. (A.) sinensis
A. (C.) pattoni

5. Mediterranean
A. (A.) atroparvus
A. (A.) claviger
A. (A.) labranchiae
A. (A.) messeae
A. (A.) sacharovi
A. (C.) hispaniola
A. (C.) superpictus

6. Afro-Arabian
A. (C.) culicifacies
A. (C.) fluviatilis
A. (C.) hispaniola
A. (C.) multicolor
A. (C.) pharoensis
A. (C.) sergentii

7. Afrotropical
A. (C.) arabiensis
A. (C.) funestus
A. (C.) gambiae
A. (C.) melas
A. (C.) merus
A. (C.) moucheti
A. (C.) nili
A. (C.) pharoensis

8. Indo-Iranian
A. (A.) sacharovi
A. (C.) aconitus
A. (C.) annularis
A. (C.) culicifacies
A. (C.) fluviatilis
A. (C.) jeyporiensis
A. (C.) minimus
A. (C.) philippinensis
A. (C.) pulcherrimus
A. (C.) stephensi
A. (C.) sundaicus
A. (C.) superpictus
A. (C.) tessellatus
A. (C.) varuna

9. Indo-Chinese hills
A. (A.) nigerrimus
A. (C.) annularis
A. (C.) culicifacies
A. (C.) dirus
A. (C.) fluviatilis
A. (C.) jeyporiensis
A. (C.) maculatus
A. (C.) minimus

10. Malaysian
A. (A.) campestris
A. (A.) donaldi
A. (A.) letifer
A. (A.) nigerrimus
A. (A.) whartoni
A. (C.) aconitus
A. (C.) balabacensis
A. (C.) dirus
A. (C.) flavirostris
A. (C.) jeyporiensis
A. (C.) leucosphyrus
A. (C.) ludlowae
A. (C.) maculatus
A. (C.) mangyanus

Table 5.2 (continued)

A. (C.) minimus	12. Australasian
A. (C.) philippinensis	A. (A.) bancroftii
A. (C.) subpictus	A. (C.) farauti type 1
A. (C.) sundaicus	A. (C.) farauti type 2
	A. (C.) hilli
11. Chinese	A. (C.) karwari
A. (A.) anthropophagus	A. (C.) koliensis
A. (A.) sinensis	A. (C.) punctulatus
A. (C.) balabacensis	A. (C.) subpictus
A. (C.) jeyporiensis	
A. (C.) pattoni	

sporozoite rate is usually 0.1% or less. Nevertheless, because of the enormous biting populations, especially during the monsoon season, it is an important vector in many areas. White (1982) presents a concise account of the ecology and genetics of malaria vectors.

To become infective, an *Anopheles* must suck up male and female gametocytes with her blood-meal. Male gametocytes (microgametocytes) exflagellate and produce male gametes (microgametes) which fertilize the female gametes (macrogametes) that have arisen from the macrogametocytes. The resultant zygote elongates into an ookinete and penetrates the stomach of the mosquito and develops into a spherical oocyst on its outer wall. Its nucleus repeatedly divides to produce thousands of spindle-shaped sporozoites which are released into the haemocoel after about eight days. These are carried to all parts of the mosquito's body, but most concentrate in the salivary glands. Sporozoites are inoculated into a

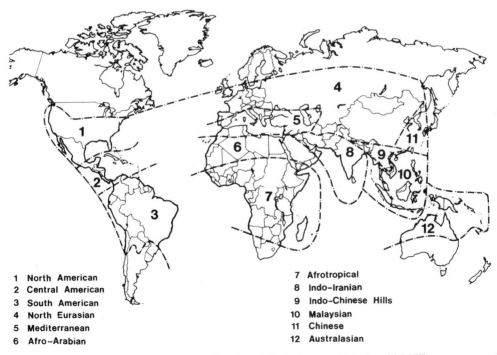

1 North American
2 Central American
3 South American
4 North Eurasian
5 Mediterranean
6 Afro-Arabian
7 Afrotropical
8 Indo-Iranian
9 Indo-Chinese Hills
10 Malaysian
11 Chinese
12 Australasian

Figure 5.29 Map of the malaria epidemiological zones of Macdonald (1957).

host with saliva when the mosquito feeds. The time for this extrinsic (exogenous) cycle to be completed depends on temperature and the malarial species. For example, at 24°C sporogony takes nine days in *P. vivax*, 11 days in *P. falciparum* and 24 days in *P. malariae*. Mosquitoes do not have to live so long to transmit *P. vivax* malaria. The salivary glands can contain as many as 60 000–70 000 sporozoites, but very few are injected into a person and a bite by an infective mosquito does not always result in malaria transmission. A mosquito once infective remains so throughout life. Accounts of pathological features induced by malarial infections are given by Maier et al. (1987) and Rossignol et al. (1986).

Salivary glands can be easily dissected out, crushed in saline and examined under a microscope for sporozoites (Ramsey et al., 1986). It is not possible, however, to distinguish sporozoites of the four human malarias by routine microscopy. Moreover, in some areas *Anopheles* can be transmitting non-human malarias whose sporozoites are indistinguishable from those of the human malarias. However, immunological tests involving detection of circumsporozite antigens, such as the enzyme-linked immunosorbent assay (ELISA), immunoradiometric assay (IRMA), indirect fluorescent antibody (IFA) and those using anti-sporozoite monoclonal antibodies, not only distinguish human and non-human malarias but also individually recognize the sporozoites of all four human malarias. These methods can also quantify infections (Burkot and Wirtz, 1986; Sucharit and Supavej, 1987). Unfortunately it has recently been shown that a positive immunological reaction cannot always distinguish between infective and infected mosquitoes (Beier et al., 1990).

It is normally necessary to dissect or process several hundreds or thousands of mosquitoes to get reliable sporozoite rates, since in most areas rates are below 0.1%. In Africa, however, the *A. gambiae* complex and *A. funestus* commonly have sporozoite rates of 5% or more. This, together with large biting densities, can result in people receiving several infective bites a night.

Several prophylactic and chemotherapeutic drugs are effective against malaria parasites, but *P. falciparum* malaria has in many areas become resistant to chloroquine and to other antimalarial drugs.

Most malaria vectors bite at night; some like *A. gambiae* bite predominantly indoors late at night, others such as *A. albimanus* in Central America out of doors early in the evenings. Clearly time of biting, and whether it occurs indoors or outside, are both epidemiologically important and relevant to control strategies.

Good overviews on malaria are given by Bruce-Chwatt (1985), Strickland (1986), Wernsdorfer and McGregor (1988), Kettle (1984), Harrison (1978, historical), Oaks et al. (1991) and Gilles and Warrell (1993). Reviews of the malaria literature are in Haworth (1987, 1989). A now classic analysis of malaria control in northern Nigeria is given by Molineaux and Gramiccia (1980). Much information on biology, vector bionomics and control is summarized for Afrotropical malaria vectors in Zahar (1984, 1985a, b, c) and for European and Mediterranean malaria vectors in Zahar (1988, 1990a, b, 1991).

Filariasis

Many malaria vectors also transmit the filarial worm *Wuchereria bancrofti* and some transmit the *Brugia malayi* filaria. The malaria vector *Anopheles barbirostris* is also a vector of *Brugia timori*. In addition, *Culex quinquefasciatus* is an important vector of *W. bancrofti*, while in the Pacific and some parts of Asia, various *Aedes* are vectors. *Mansonia* mosquitoes are the principal vectors of *B. malayi*. The distribution of the various types of filariasis according to vector zones is shown in Fig. 5.30 and the vectors are listed in Table 5.3. There are an estimated 751 million people at risk of lymphatic filariasis in 76 countries, and some 79 million people actually infected.

The development of these three filarial parasites in mosquitoes is essentially the same. Microfilariae in the host's blood are ingested by the vector, and in some mosquitoes, such as certain *Anopheles* species, which have a well-developed pharyngeal armature, many are physically destroyed during their passage to the mid-gut; others may be excreted through the anus. Surviving microfilariae commence exsheathment within a few minutes of entering the stomach and penetrate its wall to pass into the haemocoel. From here they migrate to the thoracic flight muscles, where the small larvae become more or less inactive, grow stumpier and after two days have developed into 'sausage-shaped' forms. These moult twice and the resultant third stage larvae migrate through the head and reach the fleshy labium of the proboscis after ten days or more. If very high numbers of microfilariae are ingested their development can cause high mosquito mortality or reduce flight capability. When an infective mosquito feeds, a few infective parasite larvae (1.2–1.6 mm long) rupture the labella and are deposited in a drop of haemolymph onto the skin; many die, but a few enter the skin through the mosquito's bite, or abrasions. This appears to be a very inefficient method of transmission, and it has been estimated that between 2700 and one million or more infective bites would be needed to produce a patent microfilaraemia (review by Southgate, 1984).

In man, the parasites pass to the lymphatic system. Adult female worms, after 8–12 months, release thousands of microfilariae which migrate to the blood, and are able to continue producing microfilariae over the next 15–18 years. Microfilariae can obstruct the lymph system and result in the legs and scrotum swelling to grotesque proportions, a condition known as 'elephantiasis'. Microfilariae of some

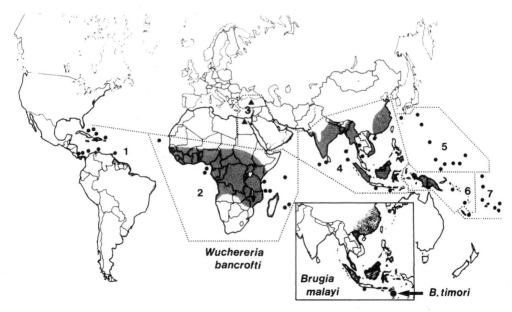

Figure 5.30 Map of the geographical distribution of lymphatic filariasis and distribution zones of its vectors. Black dots indicate islands where infection occurs; triangles indicate the localized nature of Bancroftian filariasis in Turkey and Egypt. Numbers correspond to the endemic zones listed and named in Table 5.3: 1, Tropical America; 2, Tropical Africa; 3, Middle East; 4, South Asia; 5, Far East; 6, New Guinea; 7, Polynesia. (Slightly modified from World Health Organization, 1987.)

Table 5.3 Mosquito-transmitted filarial parasites of man and their vectors in the endemic zones

The table is updated from World Health Organization (1987) and zone limits are shown on Fig. 5.30. Abbreviations for subgenera given in parentheses are: A, *Anopheles* s.str.; C, *Cellia*; Cx, *Culex*; Cq, *Coquillettidia* s.str.; F, *Finlaya*; K, *Kerteszia*; M, *Mansonioides*; N, *Nyssorhynchus*; O, *Ochlerotatus*; S, *Stegomyia*. Main (primary) vectors are distinguished from subsidiary (incidental and local) vectors by bold type.

Filaria species and form	Endemic zones	Vectors
Wuchereria bancrofti: nocturnal periodic		
Zone 1	Tropical America	*Anopheles (K.) bellator* *Anopheles (N.) albimanus* *Anopheles (N.) aquasalis* ***Anopheles (N.) darlingi*** *Aedes (O.) scapularis* *Aedes (O.) taeniorhynchus* ***Culex (Cx.) quinquefasciatus*** *Mansonia (M.) titillans*
Zone 2	Tropical Africa	***Anopheles (C.) arabiensis*** ***Anopheles (C.) funestus*** ***Anopheles (C.) gambiae*** *Anopheles (C.) hancocki* ***Anopheles (C.) melas*** ***Anopheles (C.) merus*** *Anopheles (C.) nili* *Anopheles (C.) pauliani* *Anopheles (C.) wellcomei* *Culex (Cx.) antennatus* ***Culex (Cx.) quinquefasciatus***
Zone 3	Middle East	*Culex (Cx.) antennatus* ***Culex (Cx.) pipiens*** form *molestus* ***Culex (Cx.) quinquefasciatus***
Zone 4	South Asia	*Anopheles (A.) barbirostris* ***Anopheles (A.) donaldi*** *Anopheles (A.) kweiyangensis* ***Anopheles (A.) letifer*** *Anopheles (A.) nigerrimus* ***Anopheles (A.) sinensis*** ***Anopheles (A.) whartoni*** *Anopheles (C.) aconitus* ***Anopheles (C.) anthropophagus*** ***Anopheles (C.) balabacensis*** ***Anopheles (C.) dirus*** ***Anopheles (C.) flavirostris*** ***Anopheles (C.) jeyporiensis*** ***Anopheles (C.) leucosphyrus*** ***Anopheles (C.) maculatus*** ***Anopheles (C.) minimus*** *Anopheles (C.) philippinensis* ***Anopheles (C.) subpictus*** ***Anopheles (C.) tessellatus*** ***Anopheles (C.) vagus***

Table 5.3 (*continued*)

		Aedes (F.) poicilius
		Aedes (F.) togoi
		Culex (Cx.) bitaeniorhynchus
		Culex (Cx.) pipiens form **pallens**
		Culex (Cx.) quinquefasciatus
		Culex (Cx.) sitiens
Zone 5	Far East	**Culex (Cx.) pipiens** form **pallens**
		Culex quinquefasciatus
Zone 6	New Guinea	*Anopheles (A.) bancroftii*
		Anopheles (C.) farauti
		Anopheles (C.) koliensis
		Anopheles (C.) punctulatus
		Aedes (F.) kochi
		Culex (Cx.) annulirostris
		Culex (Cx.) bitaeniorhynchus
		Culex (Cx.) quinquefasciatus
		Mansonia (M.) uniformis
Wuchereria bancrofti: nocturnal subperiodic		
Zone 4	South Asia	**Aedes (F.) harinasutai**
		Aedes (F.) niveus
Wuchereria bancrofti: diurnal subperiodic		
Zone 7	Polynesia	*Aedes (F.) fijiensis*
		Aedes (F.) oceanicus
		Aedes (F.) samoanus
		Aedes (F.) tutuilae
		Aedes (O.) vigilax
		Aedes (S.) cooki
		Aedes (S.) futunae
		Aedes (S.) kesseli
		Aedes (S.) polynesiensis
		Aedes (S.) pseudoscutellaris
		Aedes (S.) tongae
		Aedes (S.) upolensis
Brugia malayi: nocturnal periodic	South Asia	*Anopheles (A.) barbirostris*
		Anopheles (A.) campestris
		Anopheles (A.) donaldi
		Anopheles (A.) kweiyangensis
		Anopheles (A.) nigerrimus
		Anopheles (A.) sinensis
		Anopheles (C.) anthropophagus
		Aedes (F.) kiangsiensis
		Aedes (F.) togoi
		Mansonia (M.) annulata
		Mansonia (M.) annulifera
		Mansonia (M.) bonneae
		Mansonia (M.) dives
		Mansonia (M.) indiana
		Mansonia (M.) uniformis

Table 5.3 (continued)

Brugia malayi: nocturnal subperiodic	South Asia	*Mansonia (M.) annulata* *Mansonia (M.) bonneae* *Mansonia (M.) dives* *Mansonia (M.) uniformis* *Coquillettidia (Cq.) crassipes*
Brugia malayi: ? diurnal subperiodic	Thailand	?
Brugia timori: nocturnal periodic	Flores, Timor, Alor, Roti	*Anopheles (A.) barbirostris*

forms of both Bancroftian and Brugian filariasis exhibit pronounced nocturnal periodicity in their numbers in the peripheral blood, in response to daytime and night-time difference between the oxygen tension in arterial and alveolar blood.

Drugs that kill adult worms are highly toxic and not routinely used. Diethylcarbamazine (DEC), the drug most used in mass chemotherapy, kills the microfilariae but not usually adult worms so there must be repeated therapy, about every year, to kill new microfilariae if disease control is to be sustained. Chemotherapy has been successful in Malaysia, Japan and a few Pacific islands, and in some parts of India. Death of numerous microfilariae causes side-reactions which can sometimes be severe, and include headache, vomiting, urticaria and fever. Recently ivermectin has proved a valuable microfilaricide.

Mosquito dissections for filarial infections are relatively simple. The head is removed and placed in saline, and the proboscis pulled apart so that infective worms will emerge into the saline. The remainder of the head and thorax can be teased apart and examined for immature filarial worms. Large numbers of mosquitoes can be processed more quickly by crushing them in saline and placing them on a sieve immersed in normal saline. Motile larvae drop from the sieve through a funnel into a centrifuge tube of saline. It can then be necessary to concentrate the larvae by centrifugation. Infection rates in mosquitoes are about 0.1–5%, depending greatly on vector species and local epidemiology. Careful examination and staining are needed to determine whether infective larvae are those of *W. bancrofti*, *B. malayi*, *B. timori*, or non-medically important filarial worms of mammals or birds.

Useful references are Brengues (1975), Brunhes (1975), Denham and McGreevy (1977), Sasa (1976), World Health Organization (1987, 1992b) and Zahar et al. (1980).

Bancroftian filariasis
Wuchereria bancrofti occurs in most tropical regions from Latin America through Africa and Asia to the Pacific, extending into some subtropical regions. Over most of its range the microfilariae exhibit pronounced nocturnal periodicity (peak periodicity 2300–0300 h), and are ingested by night-biting mosquitoes. *Culex quinquefasciatus* is a common urban vector in most areas, but populations in West Africa are not very susceptible to infection, whereas in East Africa *C. quinquefasciatus* is an efficient vector. In rural areas transmission is maintained

mainly by *Anopheles* mosquitoes. There are about 63 known vectors of Bancroftian filariasis (Table 5.3).

In Papua New Guinea and West Irian *Mansonia uniformis* and night-biting *Culex* are vectors, while in Japan and China *Aedes togoi* is also a vector of nocturnal periodic Bancroftian filariasis. In the Philippines *Aedes poicilius* is an important vector. A diurnal subperiodic form of *W. bancrofti* occurs in Polynesia, where the nocturnal form is absent, and the most important vector is the day-biting *Aedes polynesiensis*. In Fiji, *Aedes pseudoscutellaris* is a vector, whereas in New Caledonia, where this species does not occur, transmission is by *Aedes vigilax*. (See Table 5.3 for other local vectors.) In Thailand there is a nocturnal subperiodic form of *W. bancrofti* transmitted by the *Aedes niveus* group of mosquitoes. There are no known animal reservoirs of Bancroftian filariasis.

Brugian filariasis
Brugia malayi is not found in the Americas or Africa, but occurs from the Indian subcontinent through Asia to Japan; it has been virtually eradicated, however, from Sri Lanka and Taiwan, and dramatically reduced in mainland China. It is generally considered to exist in two forms which differ in microfilarial periodicity in the vertebrate hosts. There are some 16 species of mosquito vector (Table 5.3). The nocturnal form occurs in southern India, West Malaysia, Thailand, Indonesia, Sarawak, South Korea, North Vietnam, China and parts of Japan. There do not appear to be very important reservoirs, although cats might play a minor role. Transmission is by night-biting *Mansonia* mosquitoes such as *M. indiana*, *M. annulifera* and *M. uniformis*, and also by *Anopheles*. In China, Korea and Japan, *Aedes togoi* is a vector of nocturnal periodic *B. malayi*.

The nocturnal subperiodic form is found in West Malaysia, Sumatra, North Borneo, Thailand, Indonesia and in the Palawan islands, and is transmitted by *Mansonia* species, mainly *M. annulata*, *M. dives*, *M. bonneae* and *M. uniformis*. This form of Brugian filariasis is a zoonosis. It is essentially a disease of swamp monkeys, especially the leaf monkeys (*Presbytis*), but *Macaca* monkeys, domestic and wild cats such as civets and also pangolins provide minor reservoirs of infection. In southern Thailand there is some evidence that a localized sympatric diurnal subperiodic form exists (see Denham and McGreevy, 1977, for references).

Timor filariasis
Brugia timori is known only from the Indonesian islands of Timor, Flores, Alor and Roti. Its microfilariae are nocturnally periodic and are transmitted to man by *Anopheles barbirostris*. There is no known animal reservoir.

Dirofilariasis
Dirofilaria immitis, the heartworm of dogs, occurs mainly in the tropics and subtropics, but also extends into southern Europe and North America. In addition to dogs, the worm infects other canids, rarely cats and very occasionally man. It differs from *Wuchereria* and *Brugia* parasites in undergoing development in the mosquito's Malpighian tubules. At least 72 species of *Anopheles*, *Aedes*, *Culex* and *Mansonia* are susceptible to the parasites. It is not considered an important human disease and only about 100 cases have been recorded, mainly in the USA.

Arboviruses

Of the 504 arboviruses recognized by Karabalsos (1985), more than 200 are known or suspected of being mosquito-borne and about 100 infect humans. The best known of the latter are those causing yellow fever and dengue. A host having virus in its blood is said to be viraemic. Viruses sucked up by a mosquito multiply in the stomach and within a few days pass across the stomach wall into the haemocoel. From here they migrate to the salivary glands and are inoculated into a host when the infective mosquito feeds. There are, however, various so-called barriers to be overcome. For example, a mosquito must ingest sufficient virus for it to overcome the 'gut barrier' and pass into the haemocoel, then the 'salivary gland infection barrier' must be surmounted before the salivary glands are infected, and more recently it has been shown that there is even a 'salivary gland escape barrier' to be overcome before viruses are inoculated into a victim (DeFoliart et al., 1987).

Detection of virus-infected mosquitoes relies on grinding up pools of several score mosquitoes, filtration and intracerebral inoculation into newborn mice, inoculation into chick embryos, into mosquito cell lines (*Aedes aegypti*, *A. albopictus*, *A. pseudoscutellaris*), or intrathoracic inoculation into adult *Toxorhynchites* mosquitoes, followed by immunological characterization. Viruses can also be detected in mosquitoes by immunofluorescent and complement-fixation tests, fluorescent antibody tests, use of type specific-monoclonal antibodies, enzyme-linked immunosorbent assay (ELISA), and enzyme-linked immunofluorescent assay (ELFA) (DeFoliart et al., 1987; World Health Organization, 1986a).

In 1973 it was discovered that there was trans-ovarial transmission of La Crosse virus in *Aedes triseriatus*; it was later found that other viruses, such as Rift Valley fever, St Louis encephalitis, Gamboa virus, yellow fever and dengue (DeFoliart et al., 1987), could similarly be transmitted. More rarely there is venereal transmission, that is to say virus-infected sperm passing to the female during mating. Some of the important biological considerations in the transmission of dengue, yellow fever and other arboviruses are discussed by Dégallier et al. (1988a,b).

There are no drugs that can be used either to prevent or cure arboviral infections and disease control is through vaccination and vector control. See Monath (1988) on the epidemiology and ecology of arboviruses in general and Reeves (1990) for a major account of their epidemiology and control in California.

Yellow fever

Yellow fever is essentially a zoonotic disease of forest monkeys in Africa and Central and South America that can spread to man, although in urban transmission there is direct transmission between people without involvement of monkeys. Formerly yellow fever occurred in the Americas as far north as Philadelphia and Baltimore. Yellow fever virus is found in mosquito salivary glands 12–15 days after an infective blood-meal, or at 30°C in as little as two to three days. In Panama it has been shown that yellow fever virus can be trans-ovarially transmitted to the eggs of *Haemagogus* species, and in West Africa field-caught male *Aedes* have been infected. The significance of trans-ovarial transmission in the epidemiology of the disease is not yet clear, but it could perhaps serve to maintain the virus during the dry season. In Africa yellow fever virus has also been isolated from ticks (*Amblyomma variegatum*), and it is possible these arthropods are an alternative reservoir from which infection passes to monkeys. Some argue that mosquitoes and not monkeys are the basic reservoir.

The African forest (sylvatic) cycle of transmission involves monkeys of the family Cercopithecidae (both cercopithecine and colobine), and more rarely the lesser bush-baby (*Galago senegalensis*), becoming infected by *Aedes africanus*, a forest mosquito that breeds in tree-holes and feeds mainly in the tree canopy around sunset. This sylvatic cycle maintains a reservoir of disease. More adventurous monkeys invade farms, where they can be bitten by *Aedes bromeliae* (formerly considered as *A. simpsoni*), a mosquito that breeds in leaf axils of banana and other crops and which bites man as well as monkeys. This so-called rural cycle bridges the sylvatic and urban cycles, the latter being maintained principally as man-to-man transmission by *Aedes aegypti*, a species breeding in peridomestic containers. In West Africa there is increasing evidence that other mosquitoes such as *Aedes taylori*, *A. furcifer*, *A. opok*, *A. neoafricanus* and *A. luteocephalus* spread the virus from monkeys to man in rural areas. There might also be man-to-man transmission by *A. africanus*. There have been recent epidemics of yellow fever in The Gambia (1978–79), Ghana (1978–79, 1983), Burkina Faso (1983), Mali (1987) and Cameroun (1990), epidemics occurred in Nigeria in 1986, 1987 and 1991. The Nigerian outbreak was the largest epidemic for three decades, producing at least 3361 cases and 631 deaths – with some estimates as high as 9800 cases and 5600 deaths.

In the Americas the reservoir is provided by forest monkeys, this time of the family Cebidae, especially howler (*Alouatta*) and spider monkeys (*Ateles*). The jungle cycle is maintained by arboreal forest mosquitoes that breed in tree-holes, such as various *Haemagogus* species including *H. albomaculatus*, *H. equinus*, *H. janthinomys*, *H. leucocelaenus*, *H. spegazzinii* and *H. capricornii*, and also by *Sabethes chloropterus* and sometimes *Aedes* species. When wood-cutters enter the jungle to fell trees mosquitoes which normally fly and bite in the tree canopy descend and bite them. The urban cycle is maintained by domestic *A. aegypti*. Although yellow fever is reported yearly from South America and a hundred or so cases have been reported in Colombia (1978), Bolivia (1981) and Peru (1986), there have been no recent epidemics on the scale of those of 1948–57; there seems, however, to have been an upward trend since 1983. The last known case of urban-transmitted yellow fever occurred in 1954 in Trinidad. Between 1965 and 1990 yellow fever was reported from 16 African countries and 12 Latin American countries.

Control of yellow fever is by immunization with 17D vaccine which protects for at least ten years, probably much longer: if a person recovers from yellow fever there is protective immunity for life. Insecticidal campaigns are directed against *A. aegypti*. The World Health Organization (1986a) reviews prevention and control of yellow fever, and Cordellier et al. (1977) is a practical guide to the study of African vectors.

Dengue
Dengue is widely distributed in the tropics, occurring throughout most of South East Asia, the Pacific, northern Australia, the Indian subcontinent, West and East Africa, from the USA through Central to northern South America, and in the Caribbean. There are at least four serological types of dengue virus. A more severe form, dengue haemorrhagic fever (DHF) (World Health Organization, 1986b) with an associated dengue shock syndrome (DSS), causes infant mortality; it spread from

South East Asia to India, and in 1981 to Cuba where an epidemic of 344 203 cases occurred (of which it seems that 1109 were DHF). Subsequently at least 12 other countries in the Americas reported DHF. In this region dengue transmission has recently been characterized by more frequent epidemic activity. Mexico in particular has experienced large epidemics of dengue, with the first case of DHF being reported in 1984. Other epidemics occurred in Brazil in 1986 (47 370) and 1987 (89 394), in Puerto Rico in 1986 (10 659), in Ecuador in 1988 (an estimated 420 000 individuals infected) and in Venezuela in 1990 (5416). In Asia, dengue is often a bigger menace; in 1987 there were over 600 000 cases of DHF in South East Asia. In Malaysia, the worst outbreak occurred in 1983 with 3005 detected cases, while in the same year Vietnam had its most severe epidemic of 149 519 cases and 1798 deaths. During 1987 there were over 7292 cases in Myanmar (Burma), 22 765 cases in Indonesia and 171 630 cases in Thailand.

Both the classical and haemorrhagic forms are transmitted principally by *Aedes aegypti*, and in South East Asia to a lesser extent by *Aedes albopictus*. Mosquitoes of the *Aedes scutellaris* group can also transmit dengue in the Pacific islands; in 1988–89 there were 20 220 cases in French Polynesia and in 1989 a reported 18 000 cases in New Caledonia. There are no known animal reservoirs, although in Malaysia there is a forest cycle of dengue amongst monkeys transmitted by species of the *Aedes niveus* group. There is some evidence of trans-ovarial transmission but its importance in the epidemiology of the disease remains unknown. No reliable vaccine is at present available and dengue control is based on control of the vectors.

Encephalitides
Three alphaviruses, namely Eastern equine encephalomyelitis (EEE), Western equine encephalomyelitis (WEE) and Venezuelan equine encephalomyelitis (VEE), and three flaviviruses, Japanese encephalitis (JE), St Louis encephalitis (SLE) and Murray Valley encephalitis (MVE), are all zoonotic, having wild animals as their main hosts. They all cause encephalitis in man. Horses are particularly susceptible to these viruses, and often die. Basically there is a maintenance cycle in wild hosts, an amplifying cycle in a susceptible host – often a domestic animal – and a cycle involving humans. In some cases the viraemia in humans (and also horses) is insufficient to infect mosquitoes and in these instances man is sometimes referred to as a 'dead-end host'. Other known mosquito-borne arboviruses include Ross River virus (Australia), Chikungunya (Africa, Asia), Sindbis (Africa, Asia, Australia), La Crosse (North America), Rift Valley fever (Africa), West Nile (Africa, India, Europe, Australia), and O'nyong-nyong virus (Africa). The last differs from the other virus infections mentioned in being transmitted by *Anopheles*, particularly *A. gambiae* complex and *A. funestus*. About another 18 arboviruses are transmitted to man by anophelines, but very few human cases are normally reported.

CONTROL

Mosquito control is aimed at reducing biting nuisances and preventing or stopping disease transmission. Greater efforts in controlling mosquitoes are made in the USA than anywhere else, often because mosquitoes there are vectors of encephalitides viruses, but also because they can constitute serious biting nuisances and money is available to support control. In the disease-endemic tropical countries, mosquito control is usually implemented to reduce vector transmission of malaria and other mosquito-borne diseases; nevertheless most communities are more concerned with controlling an immediate biting nuisance than disease transmission. Control can be directed at the pre-adult aquatic stages, adults or both stages. Several methods employing 'appropriate technology' in vector control have been described by Curtis (1989).

Larval control

There are three basic methods of larval control – physical control, insecticidal control and biological control. Brief descriptions of each are given here.

Physical control
This is sometimes referred to as mechanical or environmental control, or source reduction. A simple approach is to fill in with rubble, sand and earth larval habitats ranging in size from tree-holes to marshes. Filling in tree-holes can prove difficult because many could be at great heights and not easy to locate, or there could be just too many for the method to be practical. It is also impossible to fill in all the small, ephemeral and scattered pools, puddles and car ruts that are often important breeding places. More permanent habitats such as ponds and borrow pits might constitute important sources of water or provide local drainage and cannot be destroyed. Large breeding places such as marshes can be drained, but this can be a costly engineering exercise. There may sometimes also be environmental objections to drainage.

Alternatively, marshy areas having numerous scattered pools and small ponds can be excavated to create vertical-sided reservoirs or impoundments of permanent water. This usually prevents breeding by aedine mosquitoes, but can attract *Anopheles*, *Culex*, *Mansonia* or *Coquillettidia* species, although the likelihood of this can be minimized by stocking the water with fish and encouraging water-fowl. Maintaining fluctuating water levels in impoundments, as practised by the Tennessee Valley Authority in the USA, can strand and kill mosquito larvae and pupae. Large impoundments can sometimes be used for recreational purposes, such as angling and sailing.

Ditches and small streams can be realigned to increase water flow and prevent the build-up of static pockets of water. The periodical opening of sluice gates in India and Malaysia has been practised to flush out larvae (*Anopheles minimus* and *A. maculatus*) breeding in small isolated areas of still-water. Other environmental modifications include removing overhanging vegetation to reduce breeding by shade-loving mosquitoes such as *Anopheles dirus*. Deforestation can also eliminate this malaria vector. Conversely, planting vegetation along streams and reservoirs makes habitats inimical to sun-loving species such as *Anopheles gambiae*. Before the advent of DDT, shading small areas of water was successfully practised in Malaysia to create unfavourable habitats for *Anopheles maculatus*. Forestation can also decrease breeding of some mosquitoes, but may promote breeding of others.

In parts of China, intermittent irrigation of ricefields, to allow drying out every three to five days for 24–48 hours, substantially reduces populations of *Culex tritaeniorhynchus*, the principal vector of Japanese encephalitis, and the malaria vector *Anopheles sinensis*. There are claims this practice has also decreased malaria morbidity.

Theoretically, it should be easy to prevent *Aedes aegypti* and *A. albopictus* breeding in water-storage containers, either by ensuring they are covered or screened to exclude ovipositing mosquitoes yet let in rainwater, or by regularly changing their water. Moreover, simple sanitation can prevent these species breeding in discarded utensils, such as tin cans, jars and tyres. However, irregularity of piped-water and difficulties in persuading people to cooperate often makes this approach impractical. Estimates that Bangkok has over one million domestic water-containers capable of breeding *Aedes aegypti* shows the magnitude of the problem. There are, however, two examples that prove it is possible to control *A. aegypti*. In Cuba, after the 1981 dengue haemorrhagic fever (DHF) epidemic, government propaganda and an 'esprit de corps' resulted in a drastic clean-up of all standing waters and a dramatic reduction of *A. aegypti*, while in Singapore health education combined with law enforcement has over the past 30 years drastically reduced peridomestic breeding of *Aedes* dengue vectors (Chan, 1985). Both are excellent examples of control by source reduction, that is, when it is strictly enforced.

In the main cemetery in Caracas, Venezuela, there are some 190 000 flower vases, and mourners maintain water in 43% of them during the dry season, thus creating breeding places for *A. aegypti* when other larval habitats are mainly dry. One proposed solution has been to replace the flowers with artificial ones, so obviating the need for water.

Defective septic tanks, pit latrines, drains and ditches which become blocked allowing the accumulation of stagnant water are highly productive breeding places of *Culex quinquefasciatus*. Sanitary measures are the key to successful elimination of such breeding but are rarely implemented. An innovative form of control consists of pouring expanded polystyrene beads into pit latrines so that they form a 2–3 cm layer floating on the water surface, thus presenting a physical barrier against oviposition by *C. quinquefasciatus*.

An account of environmental management is presented by World Health Organization (1982) and Bruce-Chwatt (1985) gives a concise summary of drainage methods for *Anopheles* control.

Insecticidal control

One of the older control methods is to spray mineral oils on the water surface to kill mosquito larvae by suffocation and poisoning. An efficient oil must spread thinly over the water, but this is hindered by vegetation. Kerosene (paraffin), diesel oil, waste engine oil and various mixtures have been widely used, but proprietary oils, such as Malariol and Flit MLO, made specifically as larvicides, are more efficient. Addition of organophosphate or carbamate insecticides greatly enhances their toxicity. Monolayer films of lecithins and various non-petroleum oils such as aliphatic amines and fatty alcohols (nonionic surfactants) interfere with the properties of the water interface and cause larvae, pupae and even emerging or ovipositing adults to drown. This method is increasingly used, especially as petroleum oils become more costly.

Another old control method is to apply Paris Green (copper acetoarsenite) as a fine dust to the water surface. It acts as a simple stomach poison, killing surface-feeders such as *Anopheles*. Paris Green, when formulated as granules or pellets that sink, can sometimes be effective against culicines. There have been no instances of mosquitoes developing resistance to either oils or Paris Green.

With the arrival of DDT larvicide the use of oils and Paris Green was greatly reduced. However, because of its persistence in the environment and accumulation into animal and plant tissues, neither it nor other organochlorine insecticides should be used as mosquito larvicides. The recommended chemicals for larviciding are organophosphates such as malathion, pirimiphos-methyl and fenitrothion, carbamates such as propoxur, and pyrethroids such as deltamethrin and permethrin. In highly polluted waters the organophosphates fenthion or chlorpyrifos are recommended. A list of suitable larvicides is given by World Health Organization (1984). These insecticides, as well as oils and Paris Green, are usually applied to the water every 10–14 days, more frequently on polluted waters.

Temephos (Abate), an organophosphate of very low mammalian toxicity, is used to treat potable waters; briquettes or microencapsulated formulations which slowly release temephos over several days or weeks can control *Aedes aegypti* breeding in water-storage pots. In Rangoon, however, people refuse to have their water pots treated with temephos, because they regard any insecticide in drinking water as environmental contamination. Instead they readily accept the introduction of larvivorous fish in their water-storage containers to control *A. aegypti*. There will probably be increasing objections by communities to the introduction of chemicals for mosquito control into drinking water.

Insecticides formulated as slow-release granules (0.25–0.6 mm) or pellets (0.6–2 mm) can be scattered over marshy areas when these are relatively dry. When the water level rises it not only stimulates aedine eggs to hatch but causes the granules to release their toxicants which kill the newly emerged larvae. Pellets and granules are also more effective than liquid formulations in penetrating aquatic vegetation, and thus give better control in such situations.

Insect growth regulators (insect development inhibitors) such as methoprene (Altosid) which arrests larval development, and diflubenzuron (Dimilin) which inhibits chitin formation of the pre-adult stages, can also be applied to water to control mosquito breeding. Methoprene can be introduced into drinking water (see Laird and Miles, 1985).

A relatively easy way to control *Mansonia* and *Coquillettidia* species breeding in ponds, borrow pits and reservoirs is to apply herbicides, such as diquat, 2,4-D and MCPA, to kill aquatic plants such as *Pistia* and *Salvinia* on which the immatures depend for their respiration.

The bacterium *Bacillus thuringiensis* has long been commercially available for controlling agricultural and forest pests, but it was only in 1976 that a new isolate, called var. *israelensis* (or serotype H-14), was discovered that contains a crystalline endotoxin in its spores that is highly toxic to mosquito larvae. It is particularly effective against *Aedes*, *Psorophora* and *Culex*, but somewhat less so against *Anopheles* larvae. It is relatively easy to produce large quantities of *B. thuringiensis* var. *israelensis* as a powder of dead bacteria and spores that can be mixed with water and sprayed onto breeding places. There is no biological recycling in the water. *Bacillus thuringiensis* is a microbial insecticide which like other insecticides must be repeatedly applied to give long-lasting control. To date, mosquitoes have not developed resistance to it.

Various aspects of larvicidal control are outlined by World Health Organization (1973), Bruce-Chwatt (1985), Laird and Miles (1983) and Gilles and Warrell (1993).

Biological control

Biological control of mosquitoes dates back many hundreds of years and was much in vogue during the early part of this century; but, as with environmental control, it was largely replaced with the emergence of the DDT era. However, problems with insecticide resistance and greater awareness of environmental contamination have resulted in renewed interest in biological control. There are numerous records of mosquitoes being infected with pathogenic viruses, bacteria, protozoans, fungi and parasitic nematodes (Jenkins, 1964; Roberts and Castillo, 1980; Roberts and Strand, 1977; Roberts et al., 1983), but very few pathogens have been exploited for biological control, and even fewer considered commercially viable. The most promising biocontrol agents are probably *Bacillus sphaericus* and the nematode *Romanomermis culicivorax*, but again there are difficulties in using these organisms for effective long-term control (Service, 1985). Trials in Colombia with *R. culicivorax*, in addition to causing a substantial reduction in numbers of *Anopheles albimanus*, have been associated with a marked decline in malaria prevalence – providing apparently the first instance of a pathogen or parasite affecting malaria transmission.

Larvivorous fish are the most widely used biological control agents, the most famous being the top minnow or mosquito fish, *Gambusia affinis*, which exists as two subspecies: *G. affinis affinis* and *G. affinis holbrooki*. Originally a native of southern USA, this small voracious predator can produce 50–300 offspring a year and has been introduced into about 60 countries. In some parts of the world, including the USA, commercial companies supply *Gambusia* for mosquito control. *Gambusia* are surface feeders and therefore suited to preying on *Anopheles*. Although they have been introduced into many countries to control malaria, they sometimes appear to be more effective against culicines. The guppy, *Poecilia reticulata*, a native of South Africa, is also widely used, and can tolerate organic pollution better than *G. affinis*, but is not such a voracious predator. *Gambusia affinis* is an aggressive fish sometimes suspected of destroying eggs and small fry of native fish; some workers consequently do not recommend its introduction and place the emphasis on using indigenous fish for biological control.

Fish are more appropriately introduced into more or less permanent collections of water, such as ponds, borrow pits, wells, cisterns and small streams, but so-called annual fish (e.g. *Nothobranchius* and *Cynolebias*), which have eggs semi-resistant to drought, can be used in more ephemeral larval habitats. In ricefields, fish should be diverted to irrigation canals when fields are drained, and then used for restocking. In China, carp in paddy fields have reduced biting populations of both *Anopheles sinensis* and *Culex tritaenorhynchus*. In the Indian subcontinent fish kept in ornamental ponds, cisterns and wells in many towns are believed to reduce larval breeding of *A. stephensi*, the vector of urban malaria. Fish have been credited with reducing malaria transmission in Greece, Iran, Afghanistan, Somalia, China and the USSR. However, there have been very few convincing studies anywhere showing that fish can reduce malaria prevalence. In Myanmar (Burma) a local fish, *Trichogaster trichopterus*, is commonly placed in 200-litre water-storage drums and jars to keep them free of *Aedes* mosquitoes. Similarly in China, effective control of *Aedes aegypti* in household water-containers is achieved by the local edible catfish (*Clarias fuscus*).

There are many other predators of mosquitoes, including the large *Toxorhynchites* larvae used with limited success to reduce container-breeding *Aedes*. There are, however, several ecological reasons why *Toxorhynchites* is ill-suited to mosquito control (Service, 1985).

Good accounts of biological control of mosquitoes are given by Chapman (1985), Laird and Miles (1985) and Rishikesh et al. (in Wernsdorfer and McGregor, 1988). The role of mermithid nematodes in biological control is reviewed by Petersen (1985). The World Health Organization has issued mimeographed data sheets on biological control agents succinctly describing their biological characteristics, maintenance, rearing and use; other information is given in WHO/VBC and WHO/BCDS mimeographed documents. A *Biocontrol Newsletter* published by the Commonwealth Agricultural Bureaux contains review articles on biological control of pests of agricultural, medical and veterinary importance, and abstracts of published papers. See de Barjac and Sutherland (1990) for a comprehensive account of mosquito (and blackfly) control with *B. thuringiensis* var. *israelensis* and *B. sphaericus*, and Weiser (1991) for a manual on handling and identifying parasites and pathogens.

Genetic control
Genetic control can be regarded as a special category of biological control. There are several ways of introducing sterility into mosquito populations, such as by ionizing radiation and hybrid sterility, but most commonly chemosterilants such as alkylating chemicals are used. Other methods involve using cytoplasmic incompatibility, translocations, introducing lethal genes, or genes that make the mosquito refractory as a vector, and meiotic drive leading to sex distortion such as production of excessive males (Laird and Miles, 1985). Most of these ideas have not been adequately field-tested. The enthusiasm over genetic control during the late 1960s to early 1970s led to field trials against several species, including *Anopheles gambiae*, *Aedes aegypti*, *Culex quinquefasciatus* and *C. pipiens* (Asman et al., 1981). In El Salvador, the release of some 4.36 million chemosterilized male *Anopheles albimanus* over four and a half months in an isolated coastal area of about 15 km^2 caused a more than 97% reduction of the biting population, and 99% reduction of the larval population; a later expanded trial failed. In reality, although there have been some successes with agricultural pests and a few veterinary and medically important insects (Curtis, 1985, review), there have been no widescale successes with mosquitoes. Whereas trials in El Salvador and elsewhere may be intellectually stimulating, there are many difficult logistic, as well as scientific, problems to be overcome before genetic methods can offer a practical means of control.

Adult control

Control of adult mosquitoes can be based on personal protection measures or on more organized methods, often under the auspices of government agencies, directed more at the community. Some of these methods are outlined below.

Personal protection
Mosquito nets provide a simple means of protection against night-biting mosquitoes, especially malaria vectors. There is, however, little evidence that on a community basis untreated bed-nets reduce malaria prevalence, although in two

trials in The Gambia retrospective surveys suggested that bed-nets reduced splenomegaly.

During the 1980s it became fashionable to experiment with insecticide-impregnated bed-nets, the insecticide usually being permethrin, although in China the slightly more toxic deltamethrin was widely used. Such nets exert a repellent as well as an insecticidal effect on mosquitoes, and moreover still reduce biting even when nets become torn. Impregnated bed-net trials have been undertaken in many countries including Burkina Faso, Mali, Tanzania, The Gambia, Uganda, Ethiopia, Papua New Guinea, Pakistan, Malaysia and, on a much larger scale, in China, but their effectiveness against malaria remains questionable. For example, in Mali there was a reduction in parasite and spleen rates in children sleeping under deltamethrin nets, whereas in The Gambia although there was a significant reduction in clinical cases of malaria in children using permethrin-treated nets they had no effect on splenomegaly or parasitaemia. In China deltamethrin nets, and in Burkina Faso permethrin-treated bed-nets, have remained effective for more than a year, but reports from elsewhere suggest their insecticidal life is only about three to four months; much depends on whether the nets are washed after treatment. Nets can be reimpregnated by dipping them in a plastic dustbin containing insecticide. Curtis (1989) and Rozendaal (1989) review the use of impregnated nets.

Other personal protection methods include the application of insect repellents. The best is probably diethyltoluamide (DEET), which can sometimes remain effective for 6–13 hours, but dimethyl phthalate (DIMP) is more effective against certain *Anopheles*, while cyclohexamethylene carbamide is effective against some mosquitoes for 10–16 hours. Ethyl hexanediol is also effective against mosquitoes, but usually gives a shorter period of protection than DEET. Repellent or insecticide-impregnated (e.g. permethrin) clothing, such as wide-mesh netting jackets, hoods or anklets, give longer protection than when repellents are applied to the skin. When not in use clothing is best stored in plastic bags to prolong its active life. Garments have to be retreated after laundering although those treated with dibutyl phthalate remain effective after one or two washings.

Anti-mosquito coils which are ignited and smoulder to produce an insecticidal smoke are widely used in bedrooms at night in tropical countries. The better coils contain pyrethrins or pyrethroids such as permethrin or bioallethrin and burn for eight to ten hours; however, they rarely kill mosquitoes and function mainly by deterring them from entering rooms. Another means of producing repellent vapour consists of putting tiny insecticide-impregnated mats into a mains-operated electric heater; results can be good for about ten hours.

Simple hand-operated pump-action atomizers (e.g. 'flit guns') filled with 0.05–0.2% pyrethrum and a synergist (e.g. piperonyl butoxide) dissolved in kerosene, or aerosol canisters containing a variety of insecticides (often natural or synthetic pyrethrums, DDT, malathion or dichlorvos (DDVP)) are commonly used to spray bedrooms against mosquitoes – and other obnoxious pests.

Mosquito-proofing houses with screens in doorways, windows and other entry points can be an effective, although sometimes costly, control method. The strands should be 0.025–0.03 mm thick and there need to be about 16–18 holes per linear 2.5 cm, giving holes about 1.2–1.3 mm in diameter. This leaves about 70% free area for ventilation.

In some areas of Japan a decrease in Japanese encephalitis, and in parts of the USA of St Louis encephalitis, has been attributed to people staying indoors in air-conditioned houses in the evenings and watching television, instead of sitting on their porches exposed to mosquito bites (Service, 1989a). Television must therefore be included in the inventory of control measures!

Insecticidal fogs, mists and aerosols
Various hand-held (Swingfog, Dynafog) or vehicle-mounted (Leco, Micro-Gen, TIFA) machines generate insecticidal mists (51–100 µm) or aerosols (<50 µm) suitable for killing outdoor resting (exophilic) adult mosquitoes. Optimum droplet size depends on environmental and operational conditions, as well as the insecticide used. Fogs are produced when very fine aerosol droplets (<15 µm) are so plentiful that they reduce visibility. Although such methods can be spectacular and generate local interest, they give only temporary relief, because areas cleared of mosquitoes are soon reinvaded by others. If sustained control is needed repeated misting at about weekly intervals will be required, but the interval will depend on the species involved.

Aerial applications from fixed-wing aircraft or helicopters often give better and more rapid coverage and usually more effective control than ground-based applications. Ultra-low-volume (ULV) techniques which apply small amounts of concentrated insecticides, often just 225–500 ml ha^{-1}, are increasingly used. Commonly employed insecticides are malathion, fenitrothion, pirimiphos-methyl, propoxur, chlorpyrifos and the pyrethroids: a more complete list is given by World Health Organization (1984). Apart from reducing biting nuisances of exophilic and exophagic mosquitoes, aerial spraying is used in potential or actual epidemic situations to control disease outbreaks. In emergency situations aerial spraying is the most effective method of killing the vectors and stopping transmission. ULV spraying has controlled outbreaks of dengue haemorrhagic fever in South East Asia, Japanese encephalitis in Korea, Western equine encephalomyelitis in Canada, Venezuelan equine encephalomyelitis in South America, Murray Valley encephalitis and Ross River disease in the Pacific region, and malaria in Haiti. However, the effectiveness of ULV methods in stopping dengue epidemics has recently been questioned. ULV applications should be made in the evenings or early mornings, when there are temperature inversions, to allow aerosol droplets to drift downwards and not be swept into the upper air on thermals. These spraying times often coincide with peak mosquito activity, which increases the exposure of adults to the insecticides. Although usually aimed at exophilic adults, aerosol applications often kill endophilic species, and may also be larvicidal; but for more effective larval control a larger droplet size is usually needed.

Residual house-spraying
In the 1950s and 1960s it was generally believed that spraying the interior walls, roofs or ceilings of houses with residual insecticides like DDT would in most countries, but admittedly not in those of sub-Saharan Africa, result in malaria eradication. This approach hinges on the belief that malaria vectors rest indoors, and that in so doing pick up lethal doses of insecticide. Logistically this strategy is attractive because spraying 200 mg m^{-2} of DDT water-dispersible powder remains effective in killing mosquitoes for six months. Consequently houses need be sprayed just twice a year, or in areas where malaria transmission is restricted to the rainy season, only once – just before the monsoon season. House-spraying often

becomes popular because it kills bed-bugs, cockroaches and house-flies, at least initially. But many important mosquito vectors, especially those in South East Asia and the tropical Americas, are to a larger or lesser degree exophilic. In addition there is evidence that house-spraying may promote the selection of vector populations which are exophilic. Such apparent changes in behaviour, from endophily to exophily, have been documented in the *Anopheles gambiae* complex in Zimbabwe, *A. nuneztovari* in Venezuela, *A. farauti* in the Solomon Islands, *A. punctulatus* in Papua New Guinea, *A. minimus* in Thailand and *A. philippinensis* in India (see Pant in Wernsdorfer and McGregor, 1988, for summary).

Other problems include the development of DDT resistance necessitating spraying with organophosphate, carbamate or pyrethroid insecticides, which apart from being more expensive usually have to be applied every three or four months because they are less persistent. They are also more toxic than DDT, hence spraymen have to adopt better protection against insecticidal poisoning. Another difficulty is that householders may become disillusioned after several rounds of spraying and refuse spraymen entry to their houses. It has also become apparent that few countries can afford large-scale repetitive house-spraying malaria campaigns.

Problems encountered in the Indian malaria eradication campaign are discussed by Reuben (in Service, 1989a) and good accounts of house-spraying given by Bruce-Chwatt (1985), Laird and Miles (1983), Pant (in Wernsdorfer and McGregor, 1988) and Gilles and Warrell (1993). Details of spraying equipment used in vector control operations have been given by World Health Organization (1990) and approved insecticides are listed by World Health Organization (1984). Properties and specifications of pesticides are covered by Ware (1983) and World Health Organization (1985). Wernsdorfer and McGregor (1988, volume 2) contains important chapters on control of malaria vectors.

Insecticide resistance

Insecticide resistance can be a serious constraint to mosquito control. A list by Brown (in Laird and Miles, 1983) of mosquitoes resistant to various insecticides has been updated by Brown (1986) and World Health Organization (1992). Insecticidal spraying of agricultural crops such as cotton and rice has sometimes caused mosquito vectors to become resistant. The best-known example is the resistance to a broad range of insecticides developed by the malaria vector *Anopheles albimanus* in El Salvador, apparently in response to cotton spraying. Information on resistance and effects that agricultural development can have on vector-borne diseases has been given by the Food and Agriculture Organization (1987). Insecticide susceptibility kits for determining the susceptibility of mosquito larvae and adults to a variety of insecticides can be obtained from the World Health Organization.

COLLECTING, PRESERVING AND REARING MATERIAL

Comprehensive accounts of sampling methods for all life-history stages of mosquitoes are given in Service (1993), and a critical review of collecting methods for adults in Service (1977a). Information on collection and preservation of *Anopheles* is in World Health Organization (1975). Some commonly used collecting methods are outlined below, with notes on preservation of specimens.

Eggs

Collection

Eggs of *Culex* and *Coquillettidia* and others laid in rafts are easily seen and collected from the water surface. They are best transported to the laboratory on wet filter paper, otherwise they disintegrate. Gelatinous egg masses of *Mansonia* are found on the underside of aquatic vegetation. Aedine eggs are obtained by removing mud and leaf litter from breeding places and flushing it with water through a series of graded sieves. Eggs float to the surface when the residue is immersed in a solution of sodium chloride, magnesium sulphate or cane sugar with specific gravity about 1.2. Examination of the rough surfaces of rock-pools and clay pots can reveal *Aedes* eggs, and removing bottom debris or scraping the walls of tree-holes often yields eggs of *Aedes* and other tree-hole-breeding mosquitoes. Deposited eggs can often be obtained if the insides of water-filled pots, rock-pools, tree-holes or bamboo are lined with filter-paper or blotting-paper with its lower edge dipping in the water; these strips of paper are easily removed, and are most effective if coloured pink, green or grey. A method for collecting eggs of domestic and peridomestic *Aedes* mosquitoes is the use of oviposition traps. These consist of a black glass jar, tin or plastic beaker containing water and a strip of hardboard with its lower part in the water as an oviposition paddle. Eggs are laid on the rough side of these strips, which are numbered and taken weekly to the laboratory in small self-sealing plastic bags. Such 'ovitraps' are widely used to monitor *Aedes aegypti* populations. Cut sections of bamboo lined with filter-paper or blotting-paper, or paper towelling, can also be used. Ovitraps can be placed in various types of habitat and at different heights.

Various types of water receptacle filled with water can be used to attract ovipositing females and so obtain eggs. Organically polluted water is attractive to *Culex quinquefasciatus*.

Preservation

Drought-resistant eggs of *Aedes*, *Psorophora* and other aedine mosquitoes can be kept alive for months or years on damp, or sometimes even dry, filter-paper. When immersed in water having a hatching stimulus (e.g. yeast) many eggs hatch within a few minutes, others after an hour or so.

Aedine eggs laid on filter-paper are preserved by placing them in 70% ethanol, or in a small tube with a small plug of cotton wool at the bottom soaked in 7% formalin. Eggs of other genera can also be preserved in this way. Slide preparations can be made of eggs using conventional mounting media, but to aid examination of any chorionic sculpture they should first be bleached on a microscope slide using either commercial hair bleach or concentrated hydrochloric acid dropped onto potassium chlorate crystals (Kalpage and Brust, 1968). Alternatively, eggs can be bleached in small tubes in a solution of sodium chlorite, glacial acetic acid and distilled water (Trpis, 1970); this bleaches black eggs to a brown colour.

Larvae and pupae

Collection

Numerous methods are used to collect mosquito larvae, the choice depending considerably on the larval habitat. Larvae can sometimes be pipetted directly from the water surface into small specimen tubes, but those colonizing ground collections of water (ranging from small puddles to ricefields and swamps) are collected with various utensils of different shapes and sizes, e.g. soup ladles, dippers, bowls and plastic photographic trays. Aquatic nets and large collecting utensils are useful for detecting breeding when larval densities are low. Larvae in wells can be caught by carefully lowering a bucket or net into the water and pulling it up against the sides of the well. Long pipettes attached to large capacity bulbs (c. 50 ml) can be inserted into tree-holes to remove water containing larvae. Larvae can also be siphoned or pumped out of tree-holes, bamboo and inaccessible ground collections of water such as crab-holes, and tipped into a bowl. Thin pipettes are needed to collect larvae from the axils of bananas, pineapples and bromeliad plants. Ovitraps, with or without paddles, sections of bamboo (termed bamboo pots or cups), clay pots, motor vehicle tyres etc. are often used as artificial breeding sites to monitor container-breeding mosquitoes.

Special procedures (Walker and Crans, 1986) are needed for the *Mansonia*, *Coquillettidia* and *Mimomyia* immatures which insert their larval siphons and pupal respiratory trumpets into submerged plants. Occasionally traps containing a light-source and having one-way funnels or slits have been used to collect mosquito larvae (Service et al., 1983).

Larvae can be transported to the laboratory in glass or plastic specimen tubes (75 × 25 mm), in larger containers such as Mason (Kilner) jars, or in plastic self-sealing plastic bags. Collections must be protected against high temperatures and direct sunlight. To prevent larvae from drowning through being continuously shaken on a long journey they can be stranded on wet filter paper overlying cotton wool soaked in water, or placed on wet mud contained in Petri dishes or bowls.

Preservation
Immature stages can be killed and preserved in 70% ethanol or any other standard preservative (Chapter 1). For taxonomic examination larvae must be mounted dorsal side uppermost on microscope slides. To examine the pecten and anal segment of anophelines, and the comb, anal segment and siphon of culicines, it is best to cut off the abdomen after segment 7 and turn the terminal segments sideways. Slide preparations can also be made of larval skins, but care is needed as setae are easily detached. The pupal abdomen can be cut from the cephalothorax and mounted dorsally while the cephalothorax is placed on its side. Pupal preparations are best made from pupal exuviae; the abdomen and thoracic metanotum are detached and mounted with their dorsal side uppermost, while the cephalothorax is opened out and mounted with the ventral and lateral surfaces uppermost.

Adults

Collection
Both females and males can be collected by sweep-netting vegetation or using motor-powered aspirators. They can also be aspirated from under tree buttresses, bridges and culverts, and from rodent holes, crab-holes, tree-holes, granaries and other shaded niches. Alternatively, artificial resting sites, such as earthen pits and red wooden boxes, can be made to attract exophilic adults.

Several species, especially malaria vectors and *Culex quinquefasciatus*, rest in houses, and can be collected with aspirators. Alternatively houses or bedrooms can be space-sprayed with 0.05–0.1% pyrethrum in paraffin (kerosene), and the knocked-down adults collected from white sheets spread over the floor and furniture. Simple 30-cm cube exit traps fitted to doors and windows collect samples of mosquitoes leaving houses. The presence of blood-fed females in these collections indicates the species is predominantly or partially exophilic.

Anthropophagic species are caught in human bait catches. These can be performed in houses, on porches, in village compounds, on farms and in forests, during the entire, or parts of the, day and night; the place and time depends much on the ecology of the mosquitoes. Collections are normally made at ground level, but they are sometimes performed at various heights. Many species which bite man also feed on animals and consequently changes in their population size can be monitored by using animals such as cows as bait. Adults biting these animals can be collected periodically with aspirators or small hand-nets. Alternatively, the animal bait is placed under a large mosquito net raised a few centimetres above the ground, which then acts as a simple trap. When species attracted to different animals need to be compared, animal baits can be placed in a variety of cages and traps, which are usually provided with one-way entrances (cones, slits, baffles) allowing hungry mosquitoes to enter but hindering them from escaping.

Some species, including certain important vectors, can be caught in traps baited with carbon dioxide, either emanating from 1–2 kg of dry ice or from a gas cylinder. Other species are caught in light-traps, some of which like the American or New Jersey trap operate from mains electricity and use 40-W or more powerful light-bulbs. The Monks Wood light-trap employs a 23-cm 6-W daylight or ultraviolet fluorescent light-tube and is powered by a 12-V car battery. The most widely used light-trap is the CDC trap, or one of its many modifications, which has a torch bulb and operates from four 1.5-V dry-cell torch batteries connected in series. All light-traps have a fan to suck the catch down into a collecting bag. Apart from placing light-traps in various out-of-door situations, the CDC trap, and to a lesser extent the Monks Wood trap, are very useful in catching endophilic and endophagic species when placed in bedrooms. A few traps such as the EVS trap (Rohe and Fall, 1979) employ a combination of carbon dioxide and light.

Traps have also been designed to catch newly emerged adults. For example, simple cages can be fitted over cracked septic tanks, ventilation pipes of pit latrines and tree-holes, or mosquito bed-nets placed over ground collections of water. Numerous other traps, such as Malaise traps and suction traps are also used for sampling.

Preservation

Live adults can be taken from the field to the laboratory in test tubes plugged with cotton wool, or in paper or plastic beakers covered with mosquito netting. They should not be exposed to direct sunlight, and in hot weather are best placed in a cool box.

Adult mosquitoes should not be preserved in ethanol or other liquids. Whenever possible freshly killed specimens should be pinned through the thorax with a micropin staged on a piece of polyporus, plastazote (i.e. polyethylene foam plastic), card, cork or celluloid-type plastic. Great care is needed not to denude the mosquitoes of scales as in many species these are crucial for identification. The male genitalia, necessary for identification of some species, have to be removed and, after treatment with 10% KOH or NaOH, mounted on a microscope slide. Sometimes slides need also to be made of other parts, e.g. tarsal claws.

If pinning is not immediately possible dead adults can be carefully layered between soft tissue or toilet paper, which is then packed amongst cotton wool. After relaxing them in a humid atmosphere (water with 5% phenol or 0.1% merthiolate to prevent fungal growth) they can be pinned.

Rearing

Simple rearing procedures

Unfed female mosquitoes collected by various sampling methods can be fed on a variety of animals, and humans if they are not infective disease vectors. Blood-fed mosquitoes, including wild-caught blood-fed individuals, and half-gravid females, should be supplied with cotton wool soaked in water (or 10% sugar solution) and kept in paper or plastic beakers, or small cages. When they become gravid they can either be caged and provided with appropriate oviposition substrates, or individually aspirated into 75 × 25 mm glass vials and covered with mosquito netting. A few grass strands can be stood vertically in the vials, or a strip of filter-paper or blotting-paper placed down one side, to provide more suitable resting surfaces. The type of oviposition substrate placed in the tubes depends on how the eggs will be laid. For example, for *Culex* species and others that form their eggs into a raft, a small quantity of water is added. *Anopheles* will also lay their eggs on water, but to prevent adults drowning before they have oviposited it is sometimes better to replace the water with a wad of cotton wool covered with a disc of filter-paper. Water is then added until a very thin layer forms on the filter paper. Most *Aedes* and *Psorophora* species will lay their eggs on very damp cotton wool, mud or leaf litter. Oviposition is usually followed by high mortality, but if females remain alive they can be removed, re-fed and later returned to tubes for oviposition.

After oviposition, eggs of *Aedes*, *Psorophora*, *Haemagogus* and other mosquitoes having drought-resistant eggs must be kept moist for a few days or weeks (depending on the species) to allow the embryos to develop into mature larvae. Eggs can then be stored in a humid atmosphere for many months before being immersed in water for hatching.

Eggs of anophelines and most culicines, except those of *Aedes*, *Psorophora* and a few others, usually hatch within a few days after being laid, the time depending much on temperature. *Aedes* eggs, however, usually require a hatching stimulus, such as a reduction of the oxygen content of the water produced by adding yeast. Eggs of some *Aedes* species could be in diapause and fail to hatch unless diapause is broken by suitable photoperiod and/or temperature regimes.

Field-collected larvae, or those hatching from eggs, can be placed in small plastic, glass, or porcelain containers, domestic bowls, chemistry beakers, or plastic photographic trays, and kept either in field-collected water – so long as all predators have been removed – or in tap water if this is not highly chlorinated. *Anopheles* larvae are filter-feeders and are best fed daily on foods that float on the water surface, such as finely ground proprietary dried 'baby-food' cereals (e.g. Farex), dried yeast, liver powder, dog biscuits or food sold for aquaria fish. Care should be taken not to overfeed, and scum formation must be prevented. Culicine larvae, many of which are bottom-feeders, can be fed by putting into the water a yeast tablet, uncooked porridge oats, coarsely ground dog biscuits or rabbit food pellets.

Water will probably have to be changed once or twice a week to remove excess food and larval faeces. Aeration of the water is not usually necessary although aquatic vegetation or ordinary grass is sometimes added to bowls containing *Anopheles* larvae. Larval development takes about a week or several weeks according to the species, temperature and amount of available food. When pupae are formed they are removed with a pipette and placed in a bowl of clean water in a cage, or individually in 75 x 25 mm glass vials. The pupal period lasts two or more days. Adults that emerge should not be killed for taxonomic studies until at least 24 hours old to ensure they have 'hardened' (i.e. the cuticle has tanned).

Adults of both sexes can be maintained alive for many weeks by supplying them with sugar solutions or sugar cubes. However, unless the species is autogenous, females will require one, and sometimes two, blood-meals before they can develop eggs. Some species will readily feed on an arm thrust in the cage, or on anaesthetized laboratory animals such as guinea pigs, rabbits or mice, but others will feed only on specific mammals, birds or (rarely) amphibians. Nocturnal feeders will feed best at night, unless the room is darkened and the mosquitoes have been conditioned to a reversal of night and day periods. Small dishes of water, or strips of filter or blotting paper with the lower ends in water, can be placed in cages for egg laying; but unless insemination has occurred the eggs will be sterile. Stenogamous species will mate in 30-cm cube cages, whereas others (eurygamous species) will mate only in much larger cages. Some species require a gradual decrease in light, mimicking twilight, to stimulate mating, while many fail to mate under almost any laboratory conditions however carefully provided.

Colonization

Only very few mosquito species have been colonized. Some such as *Aedes aegypti*, *Culex quinquefasciatus* and even a few *Anopheles* (including *A. stephensi* and *A. albimanus*) are easily colonized, but with other species, including many important disease vectors, colonization is difficult or virtually impossible. Colonization can sometimes be undertaken in a corner of the laboratory, but usually an insectary is needed. With skill and ingenuity very simple insectaries lacking electricity are often sufficient. Usually, however, insectaries need to have controlled temperature, humidity and lighting.

Useful tips on colonization are given by Ellis and Brust (1973) and Sucharit and Supavej (1987). Gerberg (1970) has summarized many colonization procedures and gives many references. The World Health Organization (1975) presents information on blood-feeding *Anopheles* and simple colonization procedures. Methods for establishing bioclimatic chambers and rearing rooms for insects are described by Leppla and Ashley (1978).

REFERENCES

Adames, A. J. 1971. Mosquito studies (Diptera, Culicidae) XXIV. A revision of the crabhole mosquitoes of the genus *Deinocerites*. *Contributions of the American Entomological Institute* **7**(2): 1–154.

Allan, S. A., Day, J. F. and Edman, J. D. 1987. Visual ecology of biting flies. *Annual Review of Entomology* 32: 297–316.

Apiwathnasorn, C. 1986. *A list of mosquito species in southeast Asia.* 73 pp. Faculty of Tropical Medicine, Mahidol University, Bangkok.

Arnell, J. H. 1973. Mosquito studies (Diptera, Culicidae) XXXII. A revision of the genus *Haemagogus. Contributions of the American Entomological Institute* 10 (2): ii + 1–174.

Arnell, J. H. 1976. Mosquito studies (Diptera, Culicidae) XXXIII. A revision of the *scapularis* group of *Aedes (Ochlerotatus). Contributions of the American Entomological Institute* 13 (3): 1–144.

Asman, S. M., McDonald, P. T. and Prout, T. 1981. Field studies of genetic control systems for mosquitoes. *Annual Review of Entomology* 26: 289–318.

Assem, J. van den 1961. Mosquitoes collected in the Hollandia area, Netherlands New Guinea, with notes on the ecology of the larvae. *Tijdschrift voor Entomologie* 104: 17–30.

Assem, J. van den and Bonne-Wepster, J. 1964. New Guinea Culicidae, a synopsis of vectors, pests and common species. *Zoologische Bijdragen* 6: 1–136.

Barr, A. R. 1981. The *Culex pipiens* complex. Pp. 123–136 in Pal, R., Kitzmiller, J. B. and Kanda, T. (eds), *Cytogenetics and genetics of vectors. Proceedings of a Symposium of the XVth International Congress of Entomology.* Tokyo, Japan. x + 265 pp. Elsevier Biomedical, Oxford.

Barraud, P. J. 1934. *The fauna of British India, including Ceylon and Burma.* Diptera vol. 5, Family Culicidae. Tribes Megarhinini and Culicini. xxvi + 463 pp. Taylor & Francis, London.

Bates, M. 1949. *The natural history of mosquitoes.* xv + 378 pp. Macmillan, New York. [Facsimile by Harper Torchbooks, 1965.]

Beier, J. C., Perkins, P. V., Koros, J. K., Onyango, F. K., Gargan, T. P., Wirtz, R. A., Koech, D. K. and Roberts, C. R. 1990. Malaria sporozoite detection by dissection and ELISA to assess infectivity of Afrotropical *Anopheles* (Diptera: Culicidae). *Journal of Medical Entomology* 27: 377–384.

Belkin, J. N. 1962. *The mosquitoes of the South Pacific (Diptera, Culicidae)*: vol. 1, xii + 608 pp.; vol. 2, 412 pp. University of California Press, Berkeley and Los Angeles.

Belkin, J. N. 1968. Mosquito studies (Diptera, Culicidae) VII. The Culicidae of New Zealand. *Contributions of the American Entomological Institute* 3 (1): 1–182.

Belkin, J. N. and Heinemann, S. J. 1973. Collection records of the project 'Mosquitoes of Middle America' I. Introduction; Dominican Republic (RDO). *Mosquito Systematics* 5: 201–220.

Belkin, J. N., Heinemann, S. J. and Page, W. A. 1970. Mosquito studies (Diptera, Culicidae) XXI. The Culicidae of Jamaica. *Contributions of the American Entomological Institute* 6 (1): 1–458.

Belton, P. 1989. The structure and probable function of the internal cuticular parts of Johnston's organ in mosquitoes (*Aedes aegypti*). *Canadian Journal of Zoology* 67: 2625–2632.

Bentley, M. D. and Day, J. F. 1989. Chemical ecology and behavioral aspects of mosquito oviposition. *Annual Review of Entomology* 34: 401–421.

Berlin, O. G. W. 1969. Mosquito studies (Diptera, Culicidae) XII. A revision of the Neotropical subgenus *Howardina* of *Aedes. Contributions of the American Entomological Institute* 4 (2): 1–190.

Berlin, O. G. W. and Belkin, J. N. 1980. Mosquito studies (Diptera, Culicidae) XXXVI. Subgenera *Aedinus, Tinolestes* and *Anoedioporpa* of *Culex. Contributions of the American Entomological Institute* 17 (2): ii + 1–104.

Bonne-Wepster, J. 1954. Synopsis of a hundred common non-anopheline mosquitoes of the Greater and Lesser Sundas, the Moluccas and New Guinea. *Publications of the Royal Tropical Institute of Amsterdam, Special Publication* 111: 1–147.

Borel, E. 1930. Les moustiques de la Cochinchine et du Sud-Annam. *Collection de la Société de Pathologie exotique*, Monographie 3: 1–423.

Bowen, M. F. 1991. The sensory physiology of host-seeking behavior in mosquitoes. *Annual Review of Entomology* 36: 139–158.

Bradley, T. L. (1987) Physiology of osmoregulation in mosquitoes. *Annual Review of Entomology* 32: 439–462.

Bram, R. A. 1967a. Contributions to the mosquito fauna of southeast Asia II. The genus *Culex* in Thailand (Diptera: Culicidae). *Contributions of the American Entomological Institute* **2** (1): iii + 1–296.

Bram, R. A. 1967b. Classification of *Culex* subgenus *Culex* in the New World (Diptera: Culicidae). *Proceedings of the United States National Museum* **120** (3557): 1–120.

Brengues, J. 1975. La filariose de Bancroft en Afrique de l'ouest. *Mémoires de l'ORSTOM* **79**: 1–299.

Brown, A. W. A. 1986. Insecticide resistance in mosquitoes: pragmatic review. *Journal of the American Mosquito Control Association* **2**: 123–140.

Bruce-Chwatt, L. J. 1985. *Essential malariology*. Second edition. xii + 452 pp. Heinemann Medical Books, London.

Brunhes, J. 1975. La filariose de Bancroft dans la sous-région Malgache (Comores-Madagascar–Réunion). *Mémoires de l'ORSTOM* **81**: 1–212.

Burkot, T. R. and Wirtz, R. A. 1986. Immunoassays of malaria sporozoites in mosquitoes. *Parasitology Today* **2**: 155–157.

Cagampang-Ramos, A. and Darsie, R. F. 1970. Illustrated keys to the *Anopheles* mosquitoes of the Philippine islands. *USAF Fifth Epidemiological Flight, PACAF, Technical Report* **70–1**: 1–49. San Francisco.

Chan, K. L. 1985. *Singapore's dengue haemorrhagic fever programme: A case study on the successful control of Aedes aegypti and Aedes albopictus using mainly environmental measures as a part of integrated vector control*. Southeast Asian Medical Information Center Publication No. 5: 114 pp. Tokyo.

Chapman, H. C. (ed.) 1985. Biological control of mosquitoes. *American Mosquito Control Association Bulletin* **6**: 1–218.

Charlwood, J. D. and Jones, M. D. R. 1980. Mating in the mosquito *Anopheles gambiae* s. l. II. Swarming behaviour. *Physiological Entomology* **5**: 315–320.

Cianchi, R., Urbanelli, S., Villani, F., Sabatini, A. and Bullini, L. 1985. Electrophoretic studies in mosquitoes: Recent studies. *Parassitologia* **27**: 157–167.

Clark-Gil, S. and Darsie, R. F. 1983. The mosquitoes of Guatemala. Their identification, distribution and bionomics, with keys to adult females and larvae in English and Spanish. *Mosquito Systematics* **15**: 151–284.

Clements, A. N. 1963. *The physiology of mosquitoes*. ix + 393 pp. Pergamon Press, Oxford.

Clements, A. N. 1992. *The biology of mosquitoes*: vol. 1, *Development, nutrition and reproduction*. xii + 509 pp. Chapman & Hall, London.

Clements, A. N. (in preparation). *The biology of mosquitoes*: vol. 2, *Sensory reception and behaviour*. Chapman & Hall, London.

Cockburn, A. F., Tarrant, C. A. and Mitchell, S. 1988. Use of DNA probes to distinguish sibling species of the *Anopheles quadrimaculatus* complex. *Florida Entomologist* **71**: 299–302.

Collins, F. H., Petrarca, V., Mpofu, S., Brandling-Bennett, A. D., Were, J. B. O., Rasmussen, M. O. and Finnerty, V. 1988. Comparison of DNA probe and cytogenetic methods for identifying field collected *Anopheles gambiae* complex mosquitoes. *American Journal of Tropical Medicine and Hygiene* **39**: 545–550.

Coluzzi, M. 1968. Cromosomi politenici delle cellule nutrici ovariche nel complesso *gambiae* del genere *Anopheles*. *Parassitologia* **10**: 179–183.

Coluzzi, M., Petrarca, V. and Di Deco, M. 1985. Chromosomal inversion intergradation and incipient speciation in *Anopheles gambiae*. *Bollettino di Zoologia* **52**: 45–63.

Coluzzi, M. and Sabatini, A. 1969. Cytogenetic observations on the salt-water species *Anopheles merus* and *Anopheles melas* of the *gambiae* complex. *Parassitologia* **11**: 177–187.

Conner, W. E. and Itagaki, H. 1984. Pupal attendance in the crabhole mosquito *Deinocerites cancer*: the effects of pupal sex and age. *Physiological Entomology* **9**: 263–267.

Corbet, P. S. 1985. Prepupal killing behavior in *Toxorhynchites brevipalpis*. Pp. 407–417 in Lounibos, L. P., Rey, J. R. and Frank, J. H. (eds), *Mosquito ecology: proceedings of a workshop*. xix + 579 pp. Florida Medical Entomology Laboratory, Vero Beach, Florida.

Cordellier, R., Germain, M., Hervy, J.-P. and Mouchet, J. 1977. *Guide pratique pour l'étude des vecteurs de fièvre jaune en Afrique et méthodes de lutte*. 114 pp. Initiations – Documentations Techniques – No. 33. Office de la Récherche Scientifique et Technique Outre-mer. Paris.

Cornel, A. J. and **Hunt, R. H.** 1991. *Aedes albopictus* in Africa? First record of live specimens in imported tires in Cape Town. *Journal of the American Mosquito Control Association* **7**: 107–108.

Correa, R. R. and **Ramalho, G. R.** 1956. Revisão de *Phoniomyia* Theobald, 1903 (Diptera, Culicidae, Sabethini). *Folia Clinica et Biologica* **25**: 1–176.

Cova-Garcia, P. 1961. *Notas sobre los Anofelinos de Venezuela y su identificación*. Segunda edicion. 213 pp. Editora Grafos, Caracas.

Cova-Garcia, P., Sutil, E. and **Rausseo, J. A.** 1966. *Mosquitos (Culicinos) de Venezuela*: vol. 1, 410 pp.; vol. 2, 406 pp. [+ 5 pp. unnumbered index]. Ministerio de Sanidad y Asistencia Social, Caracas.

Cranston, P. S., Ramsdale, C. D., Snow, K. R. and **White, G. B.** 1987. Keys to the adults, male hypopygia, fourth-instar larvae and pupae of the British mosquitoes (Culicidae) with notes on their ecology and medical importance. *Freshwater Biological Association Scientific Publication* **48**: 1–152.

Curtis, C. F. 1985. Genetic control of insect pests: growth industry or lead balloon. *Biological Journal of the Linnean Society* **26**: 359–374.

Curtis, C. F. (ed.) 1989. *Appropriate technology in vector control*. 227 pp. CRC Press, Boca Raton, Florida.

Dąbrowska-Prot, E. 1979. Mosquitoes – the components of aquatic and terrestrial ecosystems. *Polish Ecological Studies* **5**: 5–88.

Danilov, V. N. 1985a. Mosquitoes (Diptera, Culicidae) of Afghanistan. Communication I. Identification table of the females. *Meditsinskaya Parazitologiya i Parazitarnye Bolezni* **1985** (2): 67–72. [In Russian with English summary.]

Danilov, V. N. 1985b. Mosquitoes of Afghanistan. 2. A key to fourth-stage larvae. *Meditsinskaya Parazitologiya i Parazitarnye Bolezni* **1985** (4): 51–55. [In Russian with English summary.]

Darsie, R. F. 1985. Mosquitoes of Argentina. Part I. Keys for identification of adult females and fourth stage larvae in English and Spanish (Diptera, Culicidae). *Mosquito Systematics* **17**: 153–253. [The references for this paper are in Mitchell and Darsie, 1985.]

Darsie, R. F. 1989. Keys to the genera, and to the species of five minor genera, of mosquito pupae occurring in the Nearctic region (Diptera, Culicidae). *Mosquito Systematics* **21**: 1–10.

Darsie, R. F. and **Pradhan, S. P.** 1990. The mosquitoes of Nepal: their identification, distribution and biology. *Mosquito Systematics* **22**: 69–130.

Darsie, R. F. and **Ward, R. A.** 1981. Identification and geographical distribution of the mosquitoes of North America, north of Mexico. *Mosquito Systematics Supplement* **1**: 1–313.

Das, B. P., Rajagopal, R. and **Akiyama, J.** 1989. Pictorial key to the species of Indian anopheline mosquitoes. *Journal of Pure and Applied Zoology* **2**: 131–162.

de Barjac, H. and **Sutherland, D. J.** (eds) 1990. *Bacterial control of mosquitoes and black flies. Biochemistry, genetics and applications of Bacillus thuringiensis and Bacillus sphaericus*. xix + 349 pp. Rutgers University Press, New Brunswick.

Debenham, M. L. (ed.) 1987a. *The Culicidae of the Australasian region*: vol. 4, Nomenclature, synonymy, literature, distribution, biology and relation to disease: genus *Aedes*, subgenera *Scutomyia, Stegomyia, Verrallina*. xiii + 324 pp. (Entomology Monograph No. 2 [part]). Australian Government Publishing Service, Canberra.

Debenham, M. L. (ed.) 1987b. *The Culicidae of the Australasian region*: vol. 5, Nomenclature, synonomy, literature, distribution, biology and relation to disease: genus *Anopheles* subgenera *Anopheles, Cellia*. ix + 315 pp. (Entomology Monograph No. 2 [part]). Australian Government Publishing Service, Canberra.

Debenham, M. L. (ed.) 1988a. *The Culicidae of the Australasian region*: vol. 6, Nomenclature, synonymy, literature, distribution, biology and relation to disease: genera *Armigeres, Bironella* and *Coquillettidia*. ix + 124 pp. (Entomology Monograph No. 2 [part]). Australian Government Publishing Service, Canberra.

Debenham, M. L. (ed.) 1988b. *The Culicidae of the Australasian region*: vol. 9, Nomenclature, synonymy, literature, distribution, biology and relation to disease: genera *Culex* (subgenera *Lutzia*, *Neoculex*, subgenus undecided), genera *Culiseta, Ficalbia, Heizmannia, Hodgesia, Malaya, Mansonia*. ix + 162 pp. (Entomology Monograph No. 2 [part]). Australian Government Publishing Service, Canberra.

Debenham, M. L. (ed.) 1988c. *The Culicidae of the Australasian region*: vol. 10, Nomenclature, synonymy, literature, distribution, biology and relation to disease: genera *Maorigoeldia, Mimomyia, Opifex, Orthopodomyia, Topomyia, Toxorhynchites*. ix + 105 pp. (Entomology Monograph No. 2 [part]). Australian Government Publishing Service, Canberra.

Debenham, M. L. (ed.) 1989a. *The Culicidae of the Australasian region*: vol. 7, Nomenclature, synonymy, literature, distribution, biology and relation to disease: genus *Culex*, subgenera *Acallyntrum, Culex*. ix + 281 pp. (Entomology Monograph No. 2 [part]). Australian Government Publishing Service, Canberra.

Debenham, M. L. (ed.) 1989b. *The Culicidae of the Australasian region*: vol: 8, Nomenclature, synonymy, literature, distribution, biology and relation to disease: genus *Culex*, subgenera *Culiciomyia, Eumelanomyia, Lophoceraomyia*. ix + 171 pp. (Entomology Monograph No. 2 [part]). Australian Government Publishing Service, Canberra.

Debenham, M. L., Hicks, M. M. and **Griffiths, M.** 1989. *The Culicidae of the Australasian region*: vol. 12, Summary of taxonomic changes, revised alphabetical list of species, supplementary bibliography, errata and addenda, geographic guide to species, synopsis of disease relationships, indexes. xxv + 217 pp. (Entomology Monograph No. 2 [part]). Australian Government Publishing Service, Canberra.

DeFoliart, G. R., Grimstad, P. R. and **Watts, D. M**. 1987. Advances in mosquito-borne arbovirus/vector research. *Annual Review of Entomology* **32**: 479–505.

Dégallier, N., Hervé, J.-P., da Rosa, A.P.A.T. and **Sa, G. C**. 1988a. *Aedes aegypti* (L.): Importance de la bioécologie dans la transmission de la dengue et des autres arbovirus. Première partie. *Bulletin de la Société de Pathologie exotique* **81**: 97–110.

Dégallier, N., Hervé, J.-P., da Rosa, A.P.A.T. and **Sa, G. C**. 1988b. *Aedes aegypti* (L.): Importance de la bioécologie dans la transmission de la dengue et des autres arbovirus. Deuxième partie: Bibliographie. *Bulletin de la Société de Pathologie exotique* **81**: 111–124.

Delfinado, M. D. 1966. The culicine mosquitoes of the Philippines, tribe Culicini (Diptera, Culicidae). *Memoirs of the American Entomological Institute* **7**: 1–252.

Delfinado, M. D. 1967. Contributions to the mosquito fauna of southeast – Asia I. The genus *Aedes*, subgenus *Neomacleaya* Theobald in Thailand. *Contributions of the American Entomological Institute* **1** (8): 1–37.

Delfinado, M. D. 1968. Contributions to the mosquito fauna of southeast – Asia III. The genus *Aedes*, subgenus *Neomacleaya* Theobald in southeast Asia. *Contributions of the American Entomological Institute* **2** (4): 1–74.

Denham, D. A. and **McGreevy, P. B**. 1977. Brugian filariasis: epidemiological and experimental studies. *Advances in Parasitology* **15**: 243–309.

Deruaz, D., Pichot, J. and **Petavy, A. F**. 1985. Identification of the Palearctic species of the *Anopheles maculipennis* complex (Diptera: Culicidae) by isoelectric focusing of proteins. *Rivista di Parassitologia* **46**: 107–114.

Detinova, T. S. 1962. Age-grouping methods in Diptera of medical importance. With special reference to some vectors of malaria. *World Health Organization Monograph* **47**: 1–216.

Detinova, T. S. 1968. Age structure of insect populations of medical importance. *Annual Review of Entomology* **13**: 427–450.

Dobrotworsky, N. V. 1971. Contributions to the mosquito fauna of southeast Asia X. The genus *Culiseta* Felt in southeast Asia. *Contributions of the American Entomological Institute* **7** (3): 38–61.

Downes, J. A. 1969. The swarming and mating flight of Diptera. *Annual Review of Entomology* **14**: 271–298.

Dye, C. 1984. Competition amongst larval *Aedes aegypti*: the role of interference. *Ecological Entomology* **9**: 355–357.

Edwards, F. W. 1941. *Mosquitoes of the Ethiopian region III. – Culicine adults and pupae.* vi + 499 pp. British Museum (Natural History), London.

Ellis, R. A. and **Brust, R. A.** 1973. Sibling species delimitation in the *Aedes communis* (DeGeer) aggregate (Diptera: Culicidae). *Canadian Journal of Zoology* **51**: 915–959.

Evenhuis, N. L. and **Gon III, S. M.** 1989. Family Culicidae. Pp. 191–218 in Evenhuis, N. L. (ed.), *Catalog of the Diptera of the Australasian and Oceanian regions.* 1155 pp. Bishop Museum Press, Honolulu, and E. J. Brill, Leiden.

Faran, M. E. 1980. Mosquito studies (Diptera, Culicidae) XXXIV. A revision of the *albimanus* section of the subgenus *Nyssorhynchus* of *Anopheles*. *Contributions of the American Entomological Institute* **15** (7): 1–215.

Faran, M. E. and **Linthicum, K. J.** 1981. A handbook of the Amazonian species of *Anopheles (Nyssorhynchus)* (Diptera: Culicidae). *Mosquito Systematics* **13**: 1–91.

Floore, T. G., Harrison, B. A. and **Eldridge, B. F.** 1976. The *Anopheles (Anopheles) crucians* subgroup in the United States (Diptera: Culicidae). *Mosquito Systematics* **8**: 1–109.

Food and Agriculture Organization 1987. *Effects of agricultural development on vector-borne diseases.* Mimeographed document AGL/MISC/12/87, vi + 144 pp. Food and Agriculture Organization, Rome.

Foote, R. H. 1954. The larvae and pupae of the mosquitoes belonging to the *Culex* subgenera *Melanoconion* and *Mochlostyrax*. *United States Department of Agriculture Technical Bulletin* **1091**: 1–126.

Forattini, O. P. 1962. *Entomologia Médica*: vol. 1, Parte Geral. Diptera, Anophelini. 662 pp. Faculdade de Higiene e Saúde Pública, S~ao Paulo.

Forattini, O. P. 1965a. *Entomologia Médica*: vol. 2, Culicini: *Culex, Aedes* e *Psorophora*. 506 pp. Universidade de S~ao Paulo.

Forattini, O. P. 1965b. *Entomologia Médica*: vol. 3, Culicini: *Haemagogus, Mansonia, Culiseta*, Sabethini. Toxorhynchitini. Arboviruses. Filariose bancroftiana. Genética, 416 pp. Universidade de S~ao Paulo.

Frank, J. H. and **Lounibos, L. P.** (eds) 1983. *Phytotelmata: terrestrial plants as hosts for aquatic insect communities.* vii + 293 pp. Plexus Publishing, New Jersey.

Friend, W. G. and **Smith, J. J. B.** 1977. Factors affecting feeding by bloodsucking insects. *Annual Review of Entomology* **22**: 309–331.

Gaffigan, T. V. and **Ward, R. A.** 1985. Index to the second supplement to 'A catalog of the mosquitoes of the world', with corrections and additions. *Mosquito Systematics* **17**: 52–63.

Gale, K. R. and **Crampton, J. M.** 1987. DNA probes for species identification of mosquitoes in the *Anopheles gambiae* complex. *Medical and Veterinary Entomology* **1**: 127–136.

Gale, K. R. and **Crampton, J. M.** 1988. Use of a male-specific DNA probe to distinguish female mosquitoes of the *Anopheles gambiae* species complex. *Medical and Veterinary Entomology* **2**: 77–79.

Galindo, P., Blanton, F. S. and **Peyton, E. L.** 1954. A revision of the *Uranotaenia* of Panama with notes on other American species of the genus. *Annals of the Entomological Society of America* **47**: 107–177.

García Avila, I., Gutsevich, A. V. and **Gonzalez Broche, R.** 1981. Determinación de las especies de mosquitos (Culicidae) de Cuba, según preparados microscópicos de la cabeza de las hembras (exceptuados las especies del género *Culex*). *Poeyana* **231**: 1–12.

Gerberg, E. J. 1970. Manual for mosquito rearing and experimental techniques. *Manual of the American Mosquito Control Association* **5**: 1–109.

Gerberich, J. B. and **Laird, M.** 1985. Larvivorous fish in the biocontrol of mosquitoes, with a selected bibliography of recent literature. Pp. 47–76 in Laird, M. and Miles, J. W. (eds), *Integrated mosquito control methodologies*: vol. 2, *Biocontrol and other innovative components and future directions*. xviii + 444 pp. Academic Press, London and New York.

Gilles, H. M. and **Warrell, D. A.** (1993). *Bruce-Chwatt's Essential Malariology.* Third edition. Edward Arnold, London.

Gillett, J. D. 1971. *Mosquitos* [sic]. xiii + 274 pp. Weidenfeld & Nicolson, London.

Gillies, M. T. 1980. The role of carbon dioxide in host-finding by mosquitoes (Diptera: Culicidae): a review. *Bulletin of Entomological Research* **70**: 525–532.

Gillies, M. T. 1988. Anopheline mosquitos [sic]: vector behaviour and bionomics. Pp. 453–485 in Wernsdorfer, W. H. and McGregor, I. (eds), *Malaria: principles and practice of malariology*: vol. 1, xv + pp. 1–912. Churchill Livingstone, Edinburgh. [See Wernsdorfer and McGregor entry for indexing details.]

Gillies, M. T. and **Coetzee, M**. 1987. A supplement to the Anophelinae of Africa south of the Sahara (Afrotropical region). *Publications of the South African Institute for Medical Research* **55**: 1–143.

Gillies, M. T. and **de Meillon, B**. 1968. The Anophelinae of Africa south of the Sahara (Ethiopian zoogeographical region). *Publications of the South African Institute for Medical Research* **54**: 1–343.

Gooding, R. H. 1972. Digestive processes of haematophagous insects. I. A literature review. *Quaestiones Entomologicae* **8**: 5–60.

Gordon, R. M. and **Lumsden, W. H. R**. 1939. A study of the behaviour of the mouthparts of mosquitoes when taking up blood from living tissue; together with some observations on the ingestion of microfilariae. *Annals of Tropical Medicine and Parasitology* **33**: 259–278.

Goreham, J. R. and **Stojanovich, C. J**. 1973. Clave illustrada para los mosquitos anofelinos de sudamerica occidental. Illustrated keys to the anopheline mosquitoes of western South America. *Mosquito Systematics* **5**: 97–156.

Grjebine, A. 1966. Insectes Diptères Culicidae Anophelinae. *Faune de Madagascar* **22**: 1–487.

Grjebine, A. 1986. Insectes Diptères Culicidae Culicinae Ficalbiini. *Faune de Madagascar* **68**: 1–441.

Gutsevich, A. V. and **Dubitskiy, A. M**. 1987. New species of mosquitoes in the fauna of the USSR. *Mosquito Systematics* **19**: 1–92. [English translation from Russian; original publication in *Parazitologichesky Sbornik* **30**: 97–165 (1981).]

Gutsevich, A. V., Monchadskii, A. S. and **Shtakel'berg, A. A**. 1970. Mosquitoes: family Culicidae. *Fauna of the USSR*. New Series. No. 100, *Insects, Diptera* **3** (4). 384 pp. [In Russian: see Gutsevich et al. (1974).]

Gutsevich, A. V., Monchadskii, A. S. and **Shtakel'berg, A. A**. 1974. *Mosquitoes Family Culicidae*. iii + 408 pp. Israel Program for Scientific Translations, Jerusalem. [English translation of Gutsevich et al. (1970).]

Guy, Y. 1959. Les *Anophèles* du Maroc. *Mémoires de la Société des Sciences naturelles et Physiques du Maroc* (n.s.) (Zoologie) **7**: 1–235.

Haeger, J. S. and **Provost, M. W**. 1965. Colonization and biology of *Opifex fuscus*. *Transactions of the Royal Society of New Zealand* (Zoology) **6**: 21–31.

Harbach, R. E. 1977. Comparative and functional morphology of the mandibles of some fourth stage mosquito larvae (Diptera: Culicidae). *Zoomorphologie* **87**: 217–236.

Harbach, R. E. 1985. Pictorial keys to the genera of mosquitoes, subgenera of *Culex* and species of *Culex (Culex)* occurring in southwestern Asia and Egypt, with a note on the subgeneric placement of *Culex deserticola* (Diptera: Culicidae). *Mosquito Systematics* **17**: 83–107.

Harbach, R. E. 1988. The mosquitoes of the subgenus *Culex* in southwestern Asia and Egypt (Diptera: Culicidae). *Contributions of the American Entomological Institute* **24** (1): vi + 1–240.

Harbach, R. E. and **Knight, K. L**. 1980. *Taxonomists' glossary of mosquito anatomy*. xi + 415 pp. Plexus Publishing, New Jersey.

Harbach, R. E. and **Knight, K. L**. 1981. Corrections and additions to *Taxonomists' Glossary of Mosquito Anatomy*. *Mosquito Systematics* **13**: 201–217.

Harrison, B. A. 1980. Medical Entomology Studies – XIII. The *Myzomyia* series of *Anopheles* (*Cellia*) in Thailand, with emphasis on intra-interspecific variations (Diptera: Culicidae). *Contributions of the American Entomological Society* **17** (4): iv + 1–195.

Harrison, B. A., Rattanarithikul, R., Peyton, E. L. and **Mongkolpanya, K**. 1990. Taxonomic changes, revised occurrence records and notes on the Culicidae of Thailand and neighboring countries. *Mosquito Systematics* **22**: 196–227.

Harrison, B. A. and Scanlon, J. E. 1975. Medical entomological studies – II. The subgenus *Anopheles* in Thailand (Diptera: Culicidae). *Contributions of the American Entomological Society* **12** (1): iv + 1–307.

Harrison, G. 1978. *Mosquitoes, malaria and man: a history of the hostilities since 1880.* viii + 314 pp. John Murray, London.

Hartberg, W. K. 1975. Comments on reproductive isolation and phylogeny of mosquitoes of the genus *Aedes*. *Mosquito Systematics* **7**: 193–206.

Haworth, J. 1987. Malaria in man: its epidemiology, clinical aspects and control. A review of recent abstracts from *Tropical Diseases Bulletin* and *Abstracts on Hygiene and Communicable Diseases* January 1984–June 1986. *Tropical Diseases Bulletin* **84** (4): R1–R51.

Haworth, J. 1989. Malaria in man: its epidemiology, clinical aspects and control. A review of recent abstracts from *Tropical Diseases Bulletin* and *Abstracts on Hygiene and Communicable Diseases* July 1986–June 1988. *Tropical Diseases Bulletin* **86** (10): R1–R66.

Heatwole, H. and Shine, R. 1976. Mosquitoes feeding on ectothermic vertebrates: a review and new data. *Australian Zoologist* **19**: 69–75.

Heinemann, S. J., Aitken, T. H. G. and Belkin, J. N. 1980. Collection records of the project 'Mosquitoes of Middle America'. 14. Trinidad and Tobago. *Mosquito Systematics* **12**: 179–284.

Hicks, M. M. (ed.) 1989. *The Culicidae of the Australasian region*: vol. 11, Nomenclature, synonymy, literature, distribution, biology and relation to disease: genera *Tripteroides, Uranotaenia, Wyeomyia, Zeugnomyia*. ix + 306 pp. (Entomology Monograph No. 2 [part]). Australian Government Publishing Service, Canberra.

Hopkins, G. H. E. 1952. *Mosquitoes of the Ethiopian region I. – Larval bionomics of mosquitoes and taxonomy of culicine larvae.* Second edition. viii + 355 pp. British Museum (Natural History), London.

Horsfall, W. R. 1972. *Mosquitoes. Their bionomics and relation to disease.* x + 723 pp. Hafner Publishing, New York.

Huang, Y.-M. 1968a. A new subgenus of *Aedes* (Diptera, Culicidae) with illustrated key to the subgenera of the Papuan subregion (Diptera: Culicidae). *Journal of Medical Entomology* **5**: 169–188.

Huang, Y.-M. 1968b. New Guinea mosquitoes, I. *Aedes* (*Verrallina*) of the Papuan subregion (Diptera: Culicidae). *Pacific Insects Monograph* **17**: 1–73.

Huang, Y.-M. 1972. Contributions to the mosquito fauna of southeast Asia. XIV. The subgenus *Stegomyia* of *Aedes* in southeast Asia I – The *scutellaris* group of species. *Contributions of the American Entomological Institute* **9** (1): 1–109.

Huang, Y.-M. 1977. The mosquitoes of Polynesia with a pictorial key to some species associated with filariasis and/ or dengue fever. *Mosquito Systematics* **9**: 289–322.

Huang, Y.-M. 1979. Medical entomology studies – XI. The subgenus *Stegomyia* of *Aedes* in the Oriental region with keys to the species (Diptera: Culicidae). *Contributions of the American Entomological Institute* **15** (6): ii + 1–79.

Huang, Y.-M.. 1990. The subgenus *Stegomyia* of *Aedes* in the Afrotropical region. 1. The *africanus* group of species (Diptera: Culicidae). *Contributions of the American Entomological Institute* **26** (1): 1–90.

Huang, Y.-M. and Hitchcock, J. C. 1980. Medical entomology studies – XII. A revision of the *Aedes scutellaris* group of Tonga (Diptera: Culicidae). *Contributions of the American Entomological Institute* **17** (3): 1–107.

Huang, Y.-M. and Ward, R. A. 1981. A pictorial key for the identification of the mosquitoes associated with yellow fever in Africa. *Mosquito Systematics* **13**: 138–149.

Hudson, J. E. and Abul-Hab, J. 1987. Keys to the species of adult female culicine (Diptera, Culicidae) mosquitoes of Iraq. *Bulletin of Endemic Diseases* **28**: 53–59.

Ibrahim, I. K., Al-Samarae, T. Y. M., Zaini, M. A. and Kasal, S. 1983. Identification key for Iraqi culicine mosquitoe larvae (Culicinae: Diptera). *Bulletin of Endemic Diseases* **22/23**: 89–113.

Jenkins, D. W. 1964. Pathogens, parasites and predators of medically important arthropods. Annotated list and bibliography. *Bulletin of the World Health Organization* **30** (Supplement): 1–150.
Jobling, B. 1987. *Anatomical drawings of biting flies.* 119 pp. British Museum (Natural History) and Wellcome Trust, London.
Johnson, C. G. 1969. *Migration and dispersal of insects by flight.* xvi + 763 pp. Methuen, London.
Joseph, S. R. 1970. Fruit feeding of mosquitoes in nature. *Proceedings of the New Jersey Mosquito Extermination Association* **57**: 125–131.
Kalpage, K. S. and **Brust, R. A.** 1968. Mosquitoes of Manitoba. I. Descriptions and a key to *Aedes* eggs (Diptera: Culicidae). *Canadian Journal of Zoology* **46**: 699–718.
Karabalsos, N. (ed.) 1985. *International catalogue of arboviruses including certain other viruses of vertebrates.* Third edition. 1147 pp. American Society of Tropical Medicine and Hygiene, San Antonio, Texas.
Kettle, D. S. 1984. *Medical and veterinary entomology.* 659 pp. Croom Helm, London and Sydney.
Kitzmiller, J. B. 1976. Genetics, cytogenetics, and evolution of mosquitoes. *Advances in Genetics* **18**: 315–433.
Kitzmiller, J. B. 1977. Chromosomal differences among species of *Anopheles* mosquitoes. *Mosquito Systematics* **9**: 112–122.
Klein, J.-M. 1973. Contributions to the mosquito fauna of southeast Asia XVII. The Cambodian *Aedes* (*Neomacleaya*) species with some new species descriptions (Diptera: Culicidae). *Contributions of the American Entomological Institute* **10** (1): 1–21.
Klein, J.-M. 1977. La faune des moustiques du Cambodge I. Anophelinae (Diptera, Culicidae). *Cahiers ORSTOM* (Entomologie médicale et Parasitologie) **15**: 107–122.
Knight, K. L. 1978. *Supplement to a catalog of the mosquitoes of the world* (Diptera: Culicidae). 107 pp. Entomological Society of America, College Park, Maryland.
Knight, K. L. and **Stone, A.** 1977. *A catalog of the mosquitoes of the world* (Diptera: Culicidae). Second edition. xi + 611 pp. Entomological Society of America, College Park, Maryland.
Korvenkonitio, P., Loki, J., Saura, A. and **Ulmanen, I.** 1979. *Anopheles maculipennis* complex (Diptera: Culicidae) in northern Europe: species diagnosis by egg structure and enzyme polymorphism. *Journal of Medical Entomology* **16**: 169–176.
Kramář, J. 1958. Komáři bodaví – Culicinae (Rád: Dvoukřídlí – Diptera). *Fauna CSR* **13**: 1–286. [In Czech with Russian and German summaries.]
Laird, M. 1988. *The natural history of larval mosquito habitats.* xxvii + 555 pp. Academic Press, London.
Laird, M. and **Miles, J. W.** (eds) 1983. *Integrated mosquito control methodologies*: vol. 1, *Experience and components from conventional chemical control.* xiii + 369 pp. Academic Press, London and New York.
Laird, M. and **Miles, J. W.** (eds) 1985. *Integrated mosquito control methodologies*: vol. 2, *Biocontrol and other innovative components, and future directions.* xviii + 444 pp. Academic Press, London and New York.
Lane, J. 1953. *Neotropical Culicidae*: vol. 1, Dixinae, Chaoborinae and Culicinae, tribes Anophelini, Toxorhynchitini and Culicini (genus *Culex* only); vol. 2, Tribe Culicini, *Deinocerites, Uranotaenia, Mansonia, Orthopodomyia, Aedomyia, Aedes, Psorophora, Haemagogus,* tribe Sabethini, *Trichoprosopon, Wyeomyia, Phoniomyia, Limatus* and *Sabethes*. 1112 pp. University of São Paulo, São Paulo.
Lee, D. J., Hicks, M. M., Griffiths, M., Russell, R. C. and **Marks, E. N.** 1980. *The Culicidae of the Australasian region*: vol. 1, lxix + 248 pp. (Entomology Monograph No. 2 [part]). Australian Government Publishing Service, Canberra.
Lee, D. J., Hicks, M. M., Griffiths, M., Russell, R. C. and **Marks, E. N.** 1982. *The Culicidae of the Australasian region*: vol. 2, Nomenclature, synonymy, literature, distribution, biology and relation to disease: genus *Aedeomyia,* genus *Aedes* (Subgenera [*Aedes*], Aedimorphus,

Chaetocruiomyia, Christophersiomyia, Edwardsaedes and *Finlaya*). v + 286 pp. (Entomology Monograph No. 2 [part]). Australian Government Publishing Service, Canberra.

Lee, D. J., Hicks, M. M., Griffiths, M., Russell, R. C. and **Marks, E. N.** 1984. *The Culicidae of the Australasian region*: vol. 3, Nomenclature, synonymy, literature, distribution, biology and relation to disease: genus *Aedes*, subgenera *Geoskusea, Halaedes, Huaedes, Leptosomatomyia, Levua, Lorrainea, Macleaya, Mucidus, Neomelaniconion, Nothoskusea, Ochlerotatus, Paraedes, Pseudoskusea, Rhinoskusea*. v + 257 pp. (Entomology Monograph No. 2 [part]). Australian Government Publishing Service, Canberra.

Leppla, N. C. and **Ashley, T. R.** (eds) 1978. Facilities for insect research and production. *United States Department of Agriculture Technical Bulletin* **1576**: 1–86.

Linthicum, K. J. 1988. A revision of the *argyritarsis* section of the subgenus *Nyssorhynchus* of *Anopheles* (Diptera: Culicidae). *Mosquito Systematics* **20**: 98–271.

Livdahl, T. P. and **Edgerly, J. S.** 1987. Egg hatching inhibition: field evidence for population regulation in a treehole mosquito. *Ecological Entomology* **12**: 395–399.

Lu, B. L. and **Li, B. S.** 1982. Identification of Chinese mosquitoes. Pp. 1–159 in Lu, B. S. (ed.), *Identification handbook for medically important animals in China*. 956 pp. People's Health Publication Company, Beijing. [In Chinese.]

Lu, B. L. and **Su, L.** 1987. *A handbook for the identification of Chinese aedine mosquitoes*. 160 pp. Science Press, Beijing. [In Chinese.]

Lu, B. L., Chen, B. H., Xu, R. and **Ji, S.** 1988. *A checklist of Chinese mosquitoes (Diptera: Culicidae)*. 164 pp. Guizhou People's Publishing House, Beijing. [In Chinese with English title page and introduction.]

Lutz, M. 1985. Eine fossile Stechmucke aus dem Unter-Oligozen von Cereste, Frankreich (Diptera, Culicidae). *Palaeontologische Zeitschrift* **59** (3–4): 269–276.

Macdonald, G. 1957. *The epidemiology and control of malaria*. xiv + 201 + xl + 11 pp. Oxford University Press, London.

Magnin, M. and **Pasteur, N.** 1987. Incompatibilités cytoplasmiques dans le complex *Culex pipiens*. Un revue. *Cahiers ORSTOM* (Entomologie médicale et Parasitologie) **25**: 45–53.

Mahon, R. J., Green, C. A. and **Hunt, R. H.** 1976. Diagnostic allozymes for routine identification of adults of the *Anopheles gambiae* complex (Diptera: Culicidae). *Bulletin of Entomological Research* **66**: 25–31.

Maier, W. A., Becker-Feldman, H. and **Seitz, H. M.** 1987. Pathology of malaria-infected mosquitoes. *Parasitology Today* **3**: 216–218.

Manning, D. L., Evenhuis, N. L. and **Steffan, W. A.** 1982. Supplement to annotated bibliography of *Toxorhynchites* (Diptera: Culicidae). *Journal of Medical Entomology* **19**: 429–486.

Maslov, A. V. 1967. Blood-sucking mosquitoes of the subtribe Culisetina (Diptera, Culicidae) of the world fauna. *Opredeliteli po Faune SSSR* **93**: 1–182. [In Russian: English translation titled *Blood-sucking mosquitoes of the subtribe Culisetina (Diptera, Culicidae) in world fauna*, xv + 248 pp., Smithsonian Institution Libraries and The National Science Foundation, Washington, D. C., 1989.]

Mattingly, P. F. 1952. The sub-genus *Stegomyia* (Diptera: Culicidae) in the Ethiopian region I. A preliminary study of the distribution of species occurring in the West African sub-region with notes on taxonomy and bionomics. *Bulletin of the British Museum (Natural History)* (Entomology) **2**: 235–304.

Mattingly, P. F. 1953. The sub-genus *Stegomyia* (Diptera: Culicidae) in the Ethiopian region II. Distribution of species confined to the East and South African sub-region. *Bulletin of the British Museum (Natural History)* (Entomology) **3**: 1–65.

Mattingly, P. F. 1957a. *The culicine mosquitoes of the Indomalayan area. Part I: Genus Ficalbia Theobald*. 61 pp. British Museum (Natural History), London.

Mattingly, P. F. 1957b. *The culicine mosquitoes of the Indomalayan area. Part II: Genus Heizmannia Ludlow*. 57 pp. British Museum (Natural History), London.

Mattingly, P. F. 1958. *The culicine mosquitoes of the Indomalayan area.* Part III: *Genus Aëdes Meigen, subgenera Paraëdes Edwards, Rhinoskusea Edwards and Cancraëdes Edwards.* 61 pp. British Museum (Natural History), London.

Mattingly, P. F. 1959. *The culicine mosquitoes of the Indomalayan area.* Part IV: *Genus Aëdes Meigen, subgenera Skusea Theobald, Diceromyia Theobald, Geoskusea Edwards and Christophersiomyia Barraud.* 61 pp. British Museum (Natural History), London.

Mattingly, P. F. 1961. *The culicine mosquitoes of the Indomalayan area.* Part V: *Genus Aëdes Meigen, subgenera Mucidus Theobald, Ochlerotatus Lynch Arribalzaga and Neomelaniconion Newstead.* 62 pp. British Museum (Natural History), London.

Mattingly, P. F. 1965. *The culicine mosquitoes of the Indomalayan area.* Part VI: *Genus Aëdes Meigen, subgenus Stegomyia Theobald (Groups A, B and D).* 67 pp. British Museum (Natural History), London.

Mattingly, P. F. 1969. *The biology of mosquito-borne disease.* xii + 184 pp. George Allen & Unwin, London.

Mattingly, P. F. 1970. Contributions to the mosquito fauna of southeast Asia. – VI. The genus *Heizmannia* Ludlow in Southeast Asia. *Contributions of the American Entomological Institute* **5** (7): 1–104.

Mattingly, P. F. 1971. Contributions to the mosquito fauna of southeast Asia. XII. Illustrated keys to the genera of mosquitoes (Diptera, Culicidae). *Contributions of the American Entomological Institute* **7** (4): 1–84.

Mattingly, P. F. 1973. Culicidae (Mosquitoes). Pp. 37–107 in Smith, K. G. V. (ed.), *Insects and other arthropods of medical importance.* xiv + 561 pp. British Museum (Natural History), London.

Mattingly, P. F. 1980. An interim reclassification of the genus *Tripteroides* with particular reference to the Australasian subgenera. *Mosquito Systematics* **12**: 164–171.

Mattingly, P. F. 1981. Medical entomology studies – XIV. The subgenera *Rachionotomyia, Tricholeptomyia* and *Tripteroides* in the Oriental region (Diptera: Culicidae). *Contributions of the American Entomological Institute* **17** (5): ii + 1–147.

Mattingly, P. F. and **Brown, E. S.** 1955. The mosquitos (Diptera, Culicidae) of the Seychelles. *Bulletin of Entomological Research* **46**: 69–110.

Mattingly, P. F. and **Knight, K. L.** 1956. The mosquitoes of Arabia I. *Bulletin of the British Museum (Natural History)* (Entomology) **4**: 91–141.

McAlpine, J. F., Peterson, B. V., Shewell, G. E., Teskey, H. J., Vockeroth, J. R. and **Wood, D. M.** 1981. *Manual of Nearctic Diptera*: vol. 1, vi + 674 pp. Research Branch, Agriculture Canada (Monograph No. 27).

McIntosh, B. M. 1971. The aedine subgenus *Neomelanoconion* Newstead (Culicidae, Diptera) in southern Africa with descriptions of two new species. *Journal of the Entomological Society of Southern Africa* **34**: 319–333.

McIntosh, B. M. 1973. A taxonomic re-assessment of *Aedes (Ochlerotatus) caballus* (Theobald) (Diptera: Culicidae) including a description of a new species of *Ochlerotatus. Journal of the Entomologial Society of Southern Africa* **36**: 261–269.

McIver, S. B. 1982. Sensilla of mosquitoes (Diptera: Culicidae). *Journal of Medical Entomology* **19**: 489–535.

Meek, S. R. 1988. Compatibility of members of the *Aedes (Stegomyia) scutellaris* subgroup of mosquitoes (Diptera: Culicidae) and its relevance to the control of filariasis. Pp. 115–132 in Service, M. W. (ed.), *Biosystematics of haematophagous insects.* xi + 363 pp. Clarendon Press, Oxford.

Mihályi, F. and **Gulyás, M.** 1963. *Magyarország csípő szúnyogjai* [= Biting gnats of Hungary]. *Leírásuk, életmódjuk és az ellenük való védekezés.* 229 pp. Akadémiai Kiadó, Budapest. [In Hungarian.]

Miles, S. J. 1978. Enzyme variation in the *Anopheles gambiae* Giles group of species (Diptera: Culicidae). *Bulletin of Entomological Research* **68**: 85–96.

Miles, S. J. 1979. A biochemical key to adult members of the *Anopheles gambiae* group of species (Diptera: Culicidae). *Journal of Medical Entomology* **15**: 297–299.

Miles, S. J. and **Paterson, H. E.** 1979. Protein variation and systematics in the *Culex pipiens* group of species. *Mosquito Systematics* **11**: 187–202.

Minář, J. 1991. Family Culicidae. Pp. 74–113 in Soós, A. and Papp, L. (eds), *Catalogue of Palaearctic Diptera*: vol. 2, Psychodidae – Chironomidae. 499 pp. Elsevier, Amsterdam.

Mitchell, C. J and **Darsie, R. F.** 1985. Mosquitoes of Argentina. Part II: Geographic distribution and bibliography (Diptera, Culicidae). *Mosquito Systematics* **17**: 279–360.

Mohrig, W. 1969. Die Culiciden Deutschlands. Untersuchungen zur Taxonomie, Biologie und Ökologie der einheimischen Stechmücken. *Parasitologische Schriftenreihe* **18**: 1–260.

Molineaux, L. and **Gramiccia, G.** 1980. *The Garki project: research on the epidemiology and control of malaria in the Sudan savanna of West Africa*. 311 pp. World Health Organization, Geneva.

Monath, T. P. 1988. *Arboviruses: Epidemiology and ecology*: vol.1, *General principles*, 329 pp; vol. 2, *African horse sickness to dengue*, 292 pp; vol. 3, *Eastern equine encephalomyelitis to O'nyong nyong*, 234 pp; vol. 4, *Oropouche fever to Venezuelan equine encephalomyelitis*, 243 pp; vol. 5, *Vesicular stomatitis to yellow fever*, 241 pp. CRC Press, Boca Raton, Florida.

Mori, A. 1979. Effects of larval density and nutrition on some attributes of immature and adult *Aedes albopictus*. *Tropical Medicine* **21**: 85–103.

Moussiegt, O. 1986. *Moustiques de France. Bibliographie et répartition. Inventaires de faune et de flore* **30**: vii + 184 pp. Museum national d'Histoire naturelle, Paris.

Muirhead-Thomson, R. [as 'E'] C. 1982. *Behaviour patterns of blood-sucking flies*. vii + 224 pp. Pergamon Press, Oxford.

Munstermann, L. E. 1988. Biochemical systematics of nine Nearctic *Aedes* mosquitoes (subgenus *Ochlerotatus, annulipes* group B). Pp. 133–147 in Service, M. W. (ed.), *Biosystematics of haematophagous insects*. xi + 363 pp. Clarendon Press, Oxford.

Muspratt, J. 1955. Research on South African Culicini (Diptera, Culicidae) III. – Check-list of the species and their distribution, with notes on taxonomy, bionomics and identification. *Journal of the Entomological Society of Southern Africa* **18**: 149–207.

Muspratt, J. 1956. The *Stegomyia* mosquitoes of South Africa and some neighbouring territories. Including chapters on the mosquito-borne virus diseases of the Ethiopian zoogeographical region of Africa. *Memoirs of the Entomological Society of Southern Africa* **4**: 1–138.

Narang, S. K., Seawright, J. A. and **Kaiser, P. E.** 1990. Evidence for microgeographic genetic subdivision of *Anopheles quadrimaculatus* species C. *Journal of the American Mosquito Control Association* **6**: 179–187.

Natvig, L. R. 1948. Contributions to the knowledge of the Danish and Fennoscandian mosquitoes. Culicini. *Norsk Entomologisk Tidsskrift* (Supplement I): xxii + 1–567.

Nowell, W. R. and **Ward, R. A.** 1989. Literature pertaining to the mosquito fauna and the mosquito-borne diseases on Guam. Addendum. *Mosquito Systematics* **21**: 25–39.

Oaks, S. C., Mitchell, V. S., Pearson, G. W. and **Carpenter, C. J.** (eds) 1991. *Malaria: obstacles and opportunities*. xv + 309 pp. National Academy Press, Washington D.C.

O'Meara, G. F. 1984. Gonotrophic interactions in mosquitoes: kicking the blood-feeding habit. *Florida Entomologist* **68**: 122–133.

O'Meara, G. F. 1985. Ecology and autogeny in mosquitoes. Pp. 459–471 in Lounibos, L. P., Rey, J. R. and Frank, J. H. (eds), *Mosquito ecology: proceedings of a workshop*. xix + 579 pp. Florida Medical Entomology Laboratory, Vero Beach, Florida

Pal, R., Kitzmiller, J. B. and **Kanda, T.** 1981. *Cytogenetics and genetics of vectors. Proceedings of a Symposium of the XVth International Congress of Entomology*. Tokyo, Japan. x + 265 pp. Elsevier Biomedical, Oxford.

Panyim, S., Yasothornsrikul, S. and **Baimai, V.** 1988a. Species-specific DNA sequences from the *Anopheles dirus* complex – a potential for efficient identification of isomorphic species. Pp. 193–202 in Service, M. W. (ed.), *The biosystematics of haematophagous insects*. xi + 363 pp. Clarendon Press, Oxford.

Panyim, S. Yasothornsrikul, S., Tungpradubkul, S., Baimai, V., Rosenberg, R., Andre, R. G. and **Green, C. A.** 1988b. Identification of isomorphic malaria vectors using a DNA probe. *American Journal of Tropical Medicine and Hygiene* **38**: 47–49.

Pappas, L. G., Pappas, C. D. and **Grossman, G. L.** 1986. Hemodynamics of human skin during mosquito (Diptera: Culicidae) blood feeding. *Journal of Medical Entomology* **23**: 581–587.

Petersen, J. J. 1985. Nematodes as biological control agents: Part I. Mermithidae. *Advances in Parasitology* **24**: 307–344.

Peyton, E. L. 1972. A subgeneric classification of the genus *Uranotaenia* Lynch Arribalzaga, with a historical review and notes on other categories. *Mosquito Systematics* **4**: 16–40.

Peyton, E. L. 1977. Medical entomology studies – X. A revision of the subgenus *Pseudoficalbia* of the genus *Uranotaenia* in southeast Asia (Diptera: Culicidae). *Contributions of the American Entomological Institute* **14** (3), iv + 1–273.

Peyton, E. L. 1989. A new classification for the *leucosphyrus* group of *Anopheles (Cellia)*. *Mosquito Systematics* **21**: 197–205.

Phillips, A., Milligan, P. J. M., Broomfield, G. and **Molyneux, D. H.** 1988. Identification of medically important Diptera by analysis of cuticular hydrocarbons. Pp. 39–59 in Service, M. W. (ed.), *Biosystematics of haematophagous insects*. xi + 363 pp. Clarendon Press, Oxford.

Port, G. R., Boreham, P. F. L. and **Bryan, J. H.** 1980. The relationship of host size to feeding by mosquitoes of the *Anopheles gambiae* Giles complex (Diptera, Culicidae). *Bulletin of Entomological Research* **70**: 133–144.

Postiglione, M., Tabanli, B. and **Ramsdale, C. D.** 1973. The *Anopheles* of Turkey. *Rivista di Parassitologia* **34**: 127–157.

Rai, K. S., Pashley, D. P. and **Munstermann, L. E.** 1982. Genetics of speciation in aedine mosquitoes. Pp. 84–129 in Steiner, W. W. M., Tabachnick, W. J., Rai, K. S. and Narang, S. (eds), *Recent developments in the genetics of insect disease vectors*. 665 pp. Stipes Publishing, Champaign, Illinois.

Ramsey, J. M., Bown, D. N., Aron, J. L., Beaudoin, R. L. and **Mendez, J. F.** 1986. Field trial in Chiapas, Mexico, of a rapid detection method for malaria in anopheline vectors with low infection rates. *American Journal of Tropical Medicine and Hygiene* **35**: 234–238.

Rao, T. R. 1984. *The anophelines of India*. Revised edition. xvi + 518 pp. Malaria Research Council, Delhi.

Rashed, S. S. and **Mulla, M. S.** 1990. Comparative functional morphology of the mouth brushes of mosquito larvae (Diptera: Culicidae). *Journal of Medical Entomology* **27**: 429–439.

Rattanarithikul, R. 1982. A guide to the genera of mosquitoes (Diptera: Culicidae) of Thailand with illustrated keys, biological notes and preservation and mounting techniques. *Mosquito Systematics* **14**: 139–208.

Ravaonjanahary, C. 1978. Les *Aedes* de Madagascar (Diptera – Culicidae). 1. – Étude monographique du genre. 2. Biologie d'*Aedes (Diceromyia) tiptoni*. *Travaux et Documents de l'ORSTOM* **87**: 1–210.

Reeves, W. C. (ed.) 1990. *Epidemiology and control of mosquito-borne arboviruses in California, 1943–1987*. xiv + 508 pp. California Mosquito and Vector Control Association, Sacramento, California.

Reid, J. A. 1968. Anopheline mosquitoes of Malaya and Borneo. *Studies from the Institute for Medical Research, Malaysia* **31**: xiii + 1–520.

Reinert, J. F. 1970a. The zoogeography of *Aedes (Diceromyia)* Theobald (Diptera: Culicidae). *Journal of the Entomological Society of Southern Africa* **33**: 129–141.

Reinert, J. F. 1970b. Contributions to the mosquito fauna of southeast Asia. – V Genus *Aedes*, subgenus *Diceromyia* Theobald in southeast Asia. *Contributions of the American Entomological Institute* **5** (4): 1–43.

Reinert, J. F. 1973a. Contributions to the mosquito fauna of southeast Asia. XVI. Genus *Aedes* Meigen, subgenus *Aedimorphus* Theobald in southeast Asia. *Contributions of the American Entomological Institute* **9** (5): 1–218.

Reinert, J. F. 1973b. Contributions to the mosquito fauna of southeast Asia. – XIX. *Bothaella*, a new subgenus of *Aedes* Meigen. *Contributions of the American Entomological Institute* **10** (3): 1–51.

Reinert, J. F. 1974. Medical entomology studies – I. A new interpretation of the subgenus *Verrallina* of the genus *Aedes* (Diptera: Culicidae). *Contributions of the American Entomological Institute* **11** (1): iv + 1–249.

Reinert, J. F. 1981. Medical entomology studies – XV. A revision of the subgenus *Paraedes* of the genus *Aedes* (Diptera: Culicidae). *Contributions of the American Entomological Institute* **18** (4): 1–91.

Reinert, J. F. 1984. Medical entomology studies – XVI. A review of the species of subgenus *Verrallina*, genus *Aedes*, from Sri Lanka and a revised description of the subgenus (Diptera: Culicidae). *Mosquito Systematics* **16**: 1–130.

Reinert, J. F. 1990. Medical entomology studies – XVII. Biosystematics of *Kenknightia*, a new subgenus of the mosquito genus *Aedes* Meigen from the Oriental region (Diptera: Culicidae). *Contributions of the American Entomological Institute* **26** (2): iv + 1–119.

Ribeiro, H. and **da Cunha Ramos, H.** 1973. Research on the mosquitoes of Angola VIII. – The genus *Aedes* Meigen, 1818 (Diptera: Culicidae). Check-list with new records, keys to females and larvae, distribution and taxonomic and bioecological notes. *Anais do Instituto de Higiene e Medicina Tropical* **1**: 107–138.

Ribeiro, H. and **da Cunha Ramos, H.** 1980. Research on the mosquitoes of Angola (Diptera, Culicidae). X. – The genus *Culex* L., 1758. Check-list with new records, keys to females and larvae, distribution, and taxonomic and bioecological notes. *Junta de Investigações Científicas do Ultramar, Estudos, Ensaios et Documentos* **134**: 1–175.

Ribeiro, H., da Cunha Ramos, H., Capela, R. A. and **Pires, C. A.** 1980. Os mosquitos de Cabo Verde (Diptera: Culicidae). Sistemática, distribuição, bioecologia e importância médica. *Junta de Investigações Científicas do Ultramar, Estudos, Ensaios et Documentos* **135**: 1–141.

Ribeiro, J. M. C. 1987. Role of saliva in blood-feeding by arthropods. *Annual Review of Entomology* **32**: 463–478.

Rioux, J. A. 1958. Les Culicides du 'Midi' méditerranéen. Étude systématique et écologique. *Encyclopédie Entomologique* (A) **35**: v + 1–303.

Rivière, F. 1983. Mise en évidence de l'autogenèse chez *Aedes (Stegomyia) polynesiensis* Marks, 1951 en Polynésie Française. *Cahiers ORSTOM* (Entomologie médicale et Parasitologie) **21**: 77–81.

Roberts, D. W. and **Castillo, J. M.** 1980. Bibliography on pathogens of medically important arthropods: 1980. *Bulletin of the World Health Organization* **58** (Supplement) : 1–197.

Roberts, D. W. and **Strand, M. A.** 1977. Pathogens of medically important arthropods. *Bulletin of the World Health Organization* **55** (Supplement): 1–419.

Roberts, D. W., Daoust, R. A. and **Wraight, S. P.** 1983. Bibliography of pathogens of medically important arthropods: 1981. *VBC/83.1*: 324 pp. World Health Organization, Geneva.

Rodhain, F. and **Boutonnier, A.** 1983. Description d'un nouvel *Aedes* du souse-genre *Ochlerotatus* (Diptera: Culicidae) de Madagascar: *Aedes ambreensis* nova species et considérations générales sur les femelles du sous-genre *Ochlerotatus* dans la région afrotropicale. *Bulletin de la Société de Pathologie exotique* **76**: 825–833.

Rohe, D. L. and **Fall, R. P.** 1979. A miniature battery powered CO_2 baited light trap for mosquito borne encephalitis surveillance. *Bulletin of the Society of Vector Ecologists* **4**: 24–27.

Ronderos, R. A. and **Bachmann, A. O.** 1963. Mansoniini neotropicales. I. (Diptera-Culicidae). *Revista de la Sociedad Entomólogica Argentina* **26**: 57–65.

Rossignol, P. A., Ribeiro, J. M. C. and **Spielman, A.** 1986. Increased biting rate and reduced fertility in sporozoite-infected mosquitoes. *American Journal of Tropical Medicine and Hygiene* **35**: 277–279.

Rowland, M. 1989. Changes in the circadian flight activity of the mosquito *Anopheles stephensi* associated with insemination, blood-feeding, oviposition and nocturnal light intensity. *Physiological Entomology* **14**: 77–84.

Rozendaal, J. A. 1989. Impregnated mosquito nets and curtains for self-protection and vector control. *Tropical Diseases Bulletin* **86** (7): R1–R41.

Russell, C. 1987. Survival of insects in the wheel bays of Boeing 747B aircraft on flights to tropical and temperate airports. *Bulletin of the World Health Organization* **65**:659–662.

Sabatini, A., Raineri, V., Trovato, G. and Coluzzi, M. 1990. *Aedes albopictus* in Italia e possible diffusione della specie nell'area mediterranea. *Parassitologia* **32**: 301–304.

Sasa, M. 1976. *Human filariasis. A global survey of epidemiology and control.* vii + 819 pp. University of Tokyo Press, Tokyo.

Savage, H. M. and Miller, B. R. 1991. First confirmation of breeding populations of *Aedes albopictus* in continental Africa. *Vector Ecology Newsletter* **22** (4): 5–6.

Scholl, P. J. and DeFoliart, G. R. 1978. The influence of seasonal sex ratio on the number of annual generations of *Aedes triseriatus*. *Annals of the Entomological Society of America* **71**: 677–679.

Senevet, G. and Andarelli, L. 1959. Les moustiques de l'Afrique du Nord et du bassin méditerranéen. Les genres *Culex, Uranotaenia, Theobaldia, Orthopodomyia* et *Mansonia*. *Encyclopédie entomologique* (A) **37**: 1–383.

Service, M. W. 1968. The ecology of the immature stages of *Aedes detritus* (Diptera: Culicidae). *Journal of Applied Ecology* **5**: 513–630.

Service, M. W. 1977a. A critical review of procedures for sampling populations of adult mosquitoes. *Bulletin of Entomological Research* **67**: 343–382.

Service, M. W. 1977b. Ecological and biological studies on *Aedes cantans* (Meig.) (Diptera: Culicidae) in southern England. *Journal of Applied Ecology* **14**: 159–196.

Service, M. W. 1978. Patrick Manson and the story of bancroftian [sic] filariasis. Pp. 11–14, in Willmott, S. (ed.), *Medical Entomology Centenary 23rd to 25th November 1977. Symposium Proceedings.* 144 pp. Royal Society of Tropical Medicine and Hygiene, London.

Service, M. W. 1985. Some ecological considerations basic to the biocontrol of Culicidae and other medically important athropods. Pp. 9–30 and 429–431 in Laird, M. and Miles, J. W. (eds), *Integrated mosquito control methodologies:* vol. 2, *Biocontrol and other innovative components, and future directions.* xviii + 444 pp. Academic Press, London and New York.

Service, M. W. 1988. New tools for old taxonomic problems in bloodsucking insects. Pp. 325–345 in Service, M. W. (ed.), *Biosystematics of haematophagous insects.* xi + 363 pp. Clarendon Press, Oxford.

Service, M. W. (ed.) 1989a. *Demography and vector-borne diseases.* 402 pp. CRC Press, Boca Raton, Florida.

Service, M. W. 1989b. Rice, a challenge to health. *Parasitology Today* **5**: 162–165.

Service, M. W. 1990. *Handbook to the Afrotropical toxorhynchitine and culicine mosquitoes, excepting Aedes and Culex.* British Museum (Natural History), London.

Service, M. W. 1993. *Mosquito ecology: field sampling methods.* Second edition. xiii + 988 pp. Elsevier Applied Science, London and New York.

Service, M. W., Sulaiman, S. and Esena, R. 1983. A chemical aquatic light trap for mosquito larvae (Diptera: Culicidae). *Journal of Medical Entomology* **20**: 659–663.

Shivaji, R. 1976. An annotated checklist and keys to the mosquitoes of Samoa and Tonga. *Mosquito Systematics* **8**: 298–318.

Sirivanakarn, S. 1968. New Guinea mosquitoes. I. The *Culex* subgenus *Lophoceraomyia* in New Guinea and Bismarck Archipelago (Diptera: Culicidae). *Pacific Insects Monograph* **17**: 75–186.

Sirivanakarn, S. 1971. Contributions to the mosquito fauna of southeast Asia. XI. A proposed reclassification of *Neoculex* Dyar based principally on the male terminalia. *Contributions of the American Entomological Institute* **7** (3): 62–85.

Sirivanakarn, S. 1972. Contributions to the mosquito fauna of southeast Asia. XIII. The genus *Cule x*, subgenus *Eumelanomyia* Theobald in southeast Asia and adjacent areas. *Contributions of the American Entomological Institute* **8** (6): i + 1–86.

Sirivanakarn, S. 1976. Medical entomology studies – III. A revision of the subgenus *Culex* in the Oriental region (Diptera: Culicidae). *Contributions of the American Entomological Institute* **12** (2): iii + 1–272.

Sirivanakarn, S. 1977. Medical entomology studies – VI. A revision of the subgenus *Lophoceraomyia* of the genus *Culex* in the Oriental region (Diptera: Culicidae). *Contributions of the American Entomological Institute* **13** (4): iii + 1–245.

Sirivanakarn, S. 1982. A review of the systematics and a proposed scheme of internal classification of the New World subgenus *Melanoconion* of *Culex* (Diptera, Culicidae). *Mosquito Systematics* **14**: 265–333.

Slooten, E. and **Lambert, D. M.** 1983. Evolutionary studies of the New Zealand coastal mosquito *Opifex fuscus* (Hutton). I. Mating behaviour. *Behaviour* **84**: 157–172.

Smith, B. P. 1988. Host–parasite interaction and impact of larval water mites on insects. *Annual Review of Entomology* **33**: 487–507.

Southgate, B. A. 1984. Recent advances in the epidemiology and control of filarial infections including entomological aspects of transmission. *Transactions of the Royal Society of Tropical Medicine and Hygiene* **78** (Supplement): 19–28.

Steffan, W. A. and **Evenhuis, N. L.** 1981. Biology of *Toxorhynchites*. *Annual Review of Entomology* **26**: 159–181.

Steffan, W. A. and **Evenhuis, N. L.** 1985. Classification of the subgenus *Toxorhynchites* (Diptera: Culicidae) I. Australasian, eastern Palearctic, and Oriental species-groups. *Journal of Medical Entomology* **22**: 421–446.

Steffan, W. A., Evenhuis, N. L. and **Manning, D. L.** 1980. Annotated bibliography of *Toxorhynchites* (Diptera: Culicidae). *Journal of Medical Entomology* (Supplement) **3**: 1–140.

Steiner, W. W. M., Tabachnick, W. J., Rai, K. S. and **Narang, S.** 1982. *Recent developments in the genetics of insect disease vectors*. A Symposium Proceedings. 665 pp. Stipes, Champaign, Illinois.

Stone, A. 1965. Family Culicidae. Pp. 98–120 in Stone, A., Sabrosky, C. W., Wirth, W. W., Foote, R. H. and Coulson, J. R. (eds), *A catalog of the Diptera of the Americas North of Mexico*. Agriculture Handbook 276, iv + 1696 pp. U.S. Department of Agriculture, Washington, D.C.

Stone, A. and **Delfinado, M. D.** 1973. Family Culicidae. Pp. 266–343 in Delfinado, M. D. and Hardy, D. E. (eds), *A catalog of the Diptera of the Oriental region*: vol. 1, *Suborder Nematocera*. 618 pp. University Press of Hawaii, Honolulu.

Strickland, G. T. 1986. *Clinics in tropical medicine and communicable diseases*: vol. 1 (1), *Malaria*. x + 279 pp. Saunders, London.

Subra, R. 1983. The regulation of preimaginal populations of *Aedes aegypti* L. (Diptera: Culicidae) on the Kenya coast. I. Preimaginal population dynamics and the role of human behaviour. *Annals of Tropical Medicine and Parasitology* **77**: 195–201.

Sucharit, S. and **Supavej, S.** 1987. *Practical entomology: malaria and filariasis*. iv + 139 pp. Mahidol University, Bangkok.

Sutcliffe, J. F. 1987. Distance orientation of biting flies to their hosts. *Insect Science and its Application* **8**: 611–616.

Tanaka, K., Mizusawa, K. and **Saugstad, E. S.** 1979. A revision of the adult and larval mosquitoes of Japan (including the Ryukyu archipelago and the Ogasawara Islands) and Korea (Diptera: Culicidae). *Contributions of the American Entomological Institute* **16**: vii + 1–987.

Tenorio, J. A. 1977. Revision of the genus *Bironella* (Diptera: Culicidae). *Journal of Medical Entomology* **14**: 317–361.

Thurman, E. B. 1959. A contribution to a revision of the Culicidae of northern Thailand. *Bulletin of the University of Maryland Agricultural Experimental Station* **A-100**: 1–180.

Titus, R. G. and **Ribeiro, J. M. C.** 1990. The role of vector saliva in transmission of arthropod-borne disease. *Parasitology Today* **6**: 157–160.

Toma, T. and Miyagi, I. 1986. The mosquito fauna of the Ryukyu archipelago with identification keys, pupal descriptions and notes on biology, medical importance and distribution. *Mosquito Systematics* **18**: 1–109.

Townson, H., Meredith, S. E. O. and Thomas, K. 1977. Studies of enzymes in the *Aedes scutellaris* group. *Transactions of the Royal Society of Tropical Medicine and Hygiene* **71**: 110.

Trpis, M. 1970. A new bleaching and decalcifying method for general use in zoology. *Canadian Journal of Zoology* **48**: 892–893.

Trpis, M. 1977. Autogeny in diverse populations of *Aedes aegypti* from East Africa. *Tropenmedizin und Parasitologie* **28**: 77–82.

Tyndale-Biscoe, M. 1984. Age-grading methods in adult insects: a review. *Bulletin of Entomological Research* **74**: 341–377.

Tyson, W. H. 1970a. Contributions to the mosquito fauna of southeast Asia. VII. Genus *Aedeomyia* Theobald in southeast Asia. *Contributions of the American Entomological Institute* **6** (2): 1–27.

Tyson, W. H. 1970b. Contributions to the mosquito fauna of southeast Asia. VIII. Genus *Aedes*, subgenus *Mucidus* Theobald in southeast Asia. *Contributions of the American Entomological Institute* **6** (2): 28–80.

Tyson, W. H. 1970c. Notes on African *Aedes*, subgenus *Mucidus* (Diptera: Culicidae). *Journal of the Entomological Society of Southern Africa* **33**: 81–88.

Vargas, L. 1972. Clave para identificar géneros de mosquitos de las Américas usando caracteres de las hembras. *Boletín Informativo de la Dirección de Malariologia y Saneamiento Ambiental* **12**: 204–206.

Vargas, L. 1974. Bilingual key to the New World genera of mosquitoes (Diptera: Culicidae) based on the fourth stage larvae. *California Vector Views* **21**: 15–18.

Vargas, L. and Martínez Palacios, A. 1956. *Anofelinos Mexicanos taxonomia y distribución*. 181 pp. Secretaria de Salubridad y Asistencia, Mexico, D. F.

Vattier, G. and Hamon, J. 1962. Description de la larve et de la nymphe de *Culex* (*Culiciomyia*) *gilliesi* Hamon et van Someren, 1961. Clef des larves du sous-genre *Culiciomyia* connues en Afrique au sud du Sahara. *Bulletin de la Société de Pathologie exotique* **55**: 246–252.

Walker, E. D. and Crans, W. R. 1986. A simple method for sampling *Coquillettidia perturbans* larvae. *Journal of the American Mosquito Control Association* **2**: 239–240.

Ward, R. A. 1984. Second supplement to 'A catalog of the mosquitoes of the world: (Diptera: Culicidae)'. *Mosquito Systematics* **16**: 227–270.

Ward, R. A. 1992. Third supplement to 'A catalog of the mosquitoes of the world: (Diptera: Culicidae)'. *Mosquito Systematics* **24**: 177–230.

Ward, R. A. and Darsie, R. F. 1982. Corrections and additions to the publication, *Identification and Geographical Distribution of the Mosquitoes of North America, North of Mexico*. *Mosquito Systematics* **14**: 209–219.

Ware, G. W. 1983. *Pesticides: theory and application*. x + 308 pp. Freeman, San Francisco.

Warren, M. and Collins, W. E. 1981. Vector–parasite interactions and the epidemiology of malaria. Pp. 266–274 in Canning, E. U. (ed.), *Parasitology topics: a presentation volume to P. C. C. Garnham, F. R. S. on the occasion of his 80th birthday 1981*. 289 pp. Society for Protozoologists, Special Publication No. 1., Allen Press, Lawrence, Kansas.

Washino, R. K. 1977. Review article. The physiological ecology of gonotrophic dissociation and related phenomena in mosquitoes. *Journal of Medical Entomology* **13**: 381–388.

Washino, R. K. and Tempelis, C. H. 1983. Mosquito host bloodmeal identification: Methodology and data analysis. *Annual Review of Entomology* **28**: 179–201.

Weiser, J. 1991. *Biological control of vectors: manual for collecting, field determination and handling of biofactors for control of vectors*. 189 pp. John Wiley, Chichester.

Wernsdorfer, W. H. and McGregor, I. (eds) 1988. *Malaria: principles and practice of malariology*: vol. 1, xv + pp. 1–912; vol. 2, xv + pp. 913–1818. Churchill Livingstone, Edinburgh. [Each volume has identical indexes separately paginated from text as A1–A37 (Author) and S1–S70 (Subject).]

Wharton, R. H. 1962. The biology of *Mansonia* mosquitoes in relation to the transmission of filariasis in Malaya. *Bulletin of the Institute of Medical Research of Malaya* **11**: iii + 1–114.

White, G. B. 1978. Systematic reappraisal of the *Anopheles maculipennis* complex. *Mosquito Systematics* **10**: 14–44.

White, G. B. 1979. Identification of mosquitoes as vectors of malaria and filariasis. Pp. 103–143 in Taylor, A. E. R. and Muller, R. (eds), *Problems in the identification of parasites and their vectors*. Symposia of the British Society for Parasitology, **17**, vii + 221 pp. Blackwell Scientific Publications, Oxford.

White, G. B. 1980. Family Culicidae. Pp. 114–148 in Crosskey, R. W. (ed.), *Catalogue of the Diptera of the Afrotropical region*. 1437 pp. British Museum (Natural History), London.

White, G. B. 1982. Malaria vector ecology and genetics. Pp. 207–212 in Cohen, S. (ed.), Malaria. *British Medical Bulletin* **38**: 115–218.

White, G. B. 1984. Needs and progress in the application of new techniques to mosquito identification. Pp. 293–332 in Newton, B. N. and Michal, F. (eds), *New approaches to the identification of parasites and their vectors*. 466 pp. Schwabe, Basel.

White, G. B. 1985. *Anopheles bwambae* sp. n., a malaria vector in the Semliki Valley, Uganda, and its relationships with other sibling species of the *An. gambiae* complex (Diptera: Culicidae). *Systematic Entomology* **10**: 501–522.

Wilkerson, R. C. and Peyton, E. L. 1990. Standardized nomenclature for the costal wing spots of the genus *Anopheles* and other spotted-wing mosquitoes (Diptera: Culicidae). *Journal of Medical Entomology* **27**: 207–224.

Wilkerson, R. C. and Strickman, D. 1990. Illustrated key of the female anopheline mosquitoes of Central America and Mexico. *Journal of the American Mosquito Control Association* **6**: 7–34.

Wood, D. M., Dang, P. T. and Ellis, R. A. 1979. The insects and arachnids of Canada. Part 6. The mosquitoes of Canada: Diptera: Culicidae. *Research Branch Agriculture Canada Publication* **1686**: 1–390.

World Health Organization 1973. Manual on larval control operations in malaria programmes. *WHO Offset Publication* **1**: 1–199.

World Health Organization 1975. Manual on practical entomology in malaria. Prepared by the WHO Division of Malaria and other Parasitic Diseases. Part II. Methods and techniques. *WHO Offset Publication* **13**: 1–191.

World Health Organization 1982. Manual on environmental management for mosquito control with special emphasis on malaria vectors. *WHO Offset Publication* **66**: 1–283.

World Health Organization 1984. *Chemical methods for the control of arthropod vectors and pests of public health importance*. vi + 109 pp. World Health Organization, Geneva.

World Health Organization 1985. *Specifications for pesticides used in public health. Insecticides – molluscicides – repellents – methods*. Sixth edition. 384 pp. World Health Organization, Geneva.

World Health Organization 1986a. *Prevention and control of yellow fever in Africa*. v + 94 pp. World Health Organization, Geneva.

World Health Organization 1986b. *Dengue haemorrhagic fever: diagnosis, treatment and control*. vii + 58 pp. World Health Organization, Geneva.

World Health Organization 1987. Control of lymphatic filariasis. A manual for health personnel. 89 pp. World Health Organization, Geneva.

World Health Organization 1990. *Equipment for vector control*. Third edition. 310 pp. World Health Organization, Geneva.

World Health Organization 1992a. Vector resistance to pesticides. Fifteenth Report of the WHO Expert Committee on Vector Biology and Control. *World Health Organization Technical Report Series* **818**: 1–62.

World Health Organization 1992b. Lymphatic filariasis: the disease and its control. *World Health Organization Technical Report Series* **821**: vi + 1–71.

Wright, J. W. and Pal, R. (eds) 1967. *Genetics of insect vectors of disease*. xix + 794 pp. Elsevier, Amsterdam.

Xavier, S. H. and **Mattos, S. S**. 1965. Distribuição geográfica dos culicineos no Brasil (Diptera, Culicidae). I. Estado de Goias. *Revista Brasileira de Malariologia e Doenças Tropicais* **17**: 269–291.

Xavier, S. H. and **Mattos, S. S.** 1975. Geographical distribution of the Culicinae in Brazil. III. State of Pará (Diptera, Culicidae). *Mosquito Systematics* **7**: 234–268.

Xavier, S. H., Mattos, S. S., Calábria, P. V. and **Cerqueira, E.** 1983. Geographical distribution of Culicinae in Brazil VII. State of Ceará. (Diptera, Culicidae). *Mosquito Systematics* **15**: 127–140.

Young, A. D. M. and **Downe, A. E. R.** 1987. Male accessory gland substances and the control of sexual receptivity in female *Culex tarsalis*. *Physiological Entomology* **12**: 233–239.

Zahar, A. R. 1984. Vector bionomics in the epidemiology and control of malaria. Part I. The WHO African region and the southern WHO eastern Mediterranean region. Section I: Malaria vectors of the Afrotropical region – general information. Section II: An overview of malaria control measures and the recent malaria situation. Mimeographed document VBC/84.6 and MAP/84.3, 109 pp. World Health Organization, Geneva.

Zahar, A. R. 1985a. Vector bionomics in the epidemiology and control of malaria. Part I. The WHO African region and the southern WHO eastern Mediterranean region. Section III: Vector bionomics, malaria epidemiology and controls by geographic areas. (A) West Africa. Mimeographed document VBC/85.1 and MAP/85.1, 225 pp. World Health Organization, Geneva.

Zahar, A. R. 1985b. Vector bionomics in the epidemiology and control of malaria. Part I. The WHO African region and the southern WHO eastern Mediterranean region. Section III: Vector bionomics, malaria epidemiology and control by geographic areas. (B) Equatorial Africa. (C) Southern Africa. Mimeographed document VBC/85.2 and MAP/ 85.2, 136 pp. World Health Organization, Geneva.

Zahar, A. R. 1985c. Vector bionomics in the epidemiology and control of malaria. Part I. The WHO African region and the southern WHO eastern Mediterranean region. Section III. Vector bionomics, malaria epidemiology and control by geographic areas. (D) East Africa, (E) Eastern outer islands, (F) Southwestern Arabia. Mimeographed document VBC/85.3 and MAP/85.3, 244 pp. World Health Organization, Geneva.

Zahar, A. R. 1988. Vector bionomics in the epidemiology and control of malaria. Part II. The WHO European region and the WHO eastern Mediterranean region. Volume I. Vector laboratory studies. Mimeographed document VBC/88.5 and MAP/88.2, 228 pp. World Health Organization, Geneva.

Zahar, A. R. 1990a. Vector bionomics in the epidemiology and control of malaria. Part II. The WHO European region and the WHO eastern Mediterranean region. Volume II. Applied field studies. Section I. An overview of the recent malaria situation and current problems. Section II. Vector distribution. Mimeographed document VBC/90.1 and MAL/90.1, 90 pp. World Health Organization, Geneva.

Zahar, A. R. 1990b. Vector bionomics in the epidemiology and control of malaria. Part II. The WHO European region and the WHO eastern Mediterranean region. Volume II. Applied field studies. Section III(A). The Mediterranean basin. Mimeographed document VBC/90.2 and MAL/90.2, 226 pp. World Health Organization, Geneva.

Zahar, A. R. 1991. Vector bionomics in the epidemiology and control of malaria. Part II. The WHO European region and the WHO eastern Mediterranean region. Volume II. Applied field studies. Section III: Vector bionomics, malaria epidemiology and control by geographical areas, (B) Asia west of India. Mimeographed document VBC/90.3 and MAL/90.3, 352 pp. World Health Organization, Geneva. [Document bears '90' [= 1990] year date but was not issued until 1991.]

Zahar, A. R., King, M. and **Chow, C. Y.** 1980. A review and an annotated bibliography on subperiodic Bancroftian filariasis with special reference to its vectors in Polynesia, South Pacific. Mimeographed document. iii + 492 pp. WHO Regional Office for the Western Pacific, World Health Organization, Manila.

Zaim, M. and **Cranston, P. S.** 1986. Checklist and keys to the Culicinae of Iran (Diptera: Culicidae). *Mosquito Systematics* **18**: 233–245.

Zavortink, T. L. 1968. Mosquito studies (Diptera, Culicidae) VIII. A prodome of the genus *Orthopodomyia*. *Contributions of the American Entomological Institute* **3** (2): 1–121.

Zavortink, T. L. 1970. Mosquito studies (Diptera, Culicidae) – XIX. The tree hole *Anopheles* of the New World. *Contributions of the American Entomological Institute* **5** (2): 1–35.

Zavortink, T. L. 1971. Contributions to the mosquito fauna of southeast Asia IX. The genus *Orthopodomyia* Theobald in southeast Asia. *Contributions of the American Entomological Institute* **7** (3): 1–37.

Zavortink, T. L. 1972. ['1964' error]. Mosquito studies (Diptera, Culicidae) XXVIII. The New World species formerly placed in *Aedes (Finlaya)*. *Contributions of the American Entomological Institute* **8** (3): 1–206.

Zavortink, T. L. 1973. Mosquito studies (Diptera, Culicidae) XXIX. A review of the subgenus *Kerteszia* of *Anopheles*. *Contributions of the American Entomological Institute* **9** (3): 1–54.

Zavortink, T. L. 1979. Mosquito studies (Diptera, Culicidae) XXXV. The new sabethine genus *Johnbelkinia* and a preliminary reclassification of the composite genus *Trichoprosopon*. *Contributions of the American Entomological Institute* **17** (1): 1–61.

CHAPTER SIX

Blackflies (Simuliidae)

R. W. Crosskey

The significance of Simuliidae among medically important insects is twofold: first, that this family includes all the carriers of human onchocerciasis ('river blindness'), and secondly that the species include some of the world's most persistent and demoralizing man-biting insect pests. There are several areas of the world, such as rural eastern Canada and southern New Zealand, where simuliids are the most dreaded noxious arthropods even though they transmit no human disease. In tropical medicine, however, it is the role of simuliids as vectors of human onchocerciasis that makes headlines in the fast-growing literature about this disease (Muller and Horsburgh, 1987: bibliography). All vectors belong to *Simulium*. In this genus, as in other Simuliidae and other biting Nematocera, only the females suck blood and males are harmless. The hosts are birds and mammals (only warm-blooded vertebrates are bitten).

Among birds, blackflies cyclically transmit most of the protozoan blood parasites of the genus *Leucocytozoon* (Apicomplexa: Plasmodiidae) for which the life cycle is known: *L. smithi* is the cause of 'turkey malaria', a virulent infection with economic consequences for turkey breeders in the United States, and *L. simondi* is a pathogenic parasite among domestic ducks. They also transmit a filarial worm among ducks and are involved in the transmission of certain avian trypanosomes. Among mammals, they cyclically transmit various species of parasitic nematode worms of the genera *Dirofilaria* (in bears), *Mansonella* (in man) and *Onchocerca* (in ungulate mammals and man). These genera all belong to the filarioid family Onchocercidae, but 'onchocerciasis' refers strictly to infections with *Onchocerca* and is often qualified as bovine or human according to whether the infection is in cattle or man. Bovine onchocerciasis (produced by several species different from that responsible for infection in man) is very common and widespread but non-pathogenic. It is not mentioned again, and 'onchocerciasis' as used hereafter in this chapter refers to human onchocerciasis, the disease produced in man by infection with *Onchocerca volvulus*. Another filarial parasite of humans, *Mansonella ozzardi*, sometimes has *Simulium* vectors, but is not pathogenic.

The Simuliidae are unique in medically important insects because of the habitat requirements of the immature stages. These develop in the flowing water of rivers and streams and are closely adapted only for life in this kind of aquatic environment: in ecological terminology, they are lotic organisms occupying lotic habitat (i.e. running waters in contrast to lentic or still waters). Fine-tuning of geographical

Medical Insects and Arachnids Edited by Richard P. Lane and Roger W. Crosskey.
Published in 1993 by Chapman & Hall ISBN 0 412 40000 6

distribution is determined by where lotic habitat is available, and blackflies are absent from streamless areas where other biting flies often thrive, e.g. sandflies in desert regions or mosquitoes on coral islands. Nevertheless, simuliids are broadly speaking cosmopolitan and abundant in all zoogeographical regions from arctic to equatorial latitudes. Altitudinally, in the Andes and on high mountains of East Africa, they occur at up to 4000–4500 metres (although at such heights they do not suck blood).

Adult simuliids are small and stout-bodied flies whose popular name is partly a misnomer; though black predominates, by no means all species are dark, and many Neotropical 'blackflies' (such as the aptly named *Simulium ochraceum*) are mainly orange or yellow. Fossils show that blackflies have existed at least since Middle Jurassic times, about 170 million years BP. At the end of 1991 the number of named species was about 1570, but so-called 'new' species are still being discovered and described. Species complexes (Chapter 1) are unusually prevalent in the Simuliidae, and undoubtedly many hundreds (perhaps thousands) of sibling species await recognition.

RECOGNITION AND ELEMENTS OF STRUCTURE

Peterson (1981) and Crosskey (1990) provide detailed accounts of morphology, the latter including also a detailed diagnosis for the family Simuliidae. The following abbreviated account notes the chief family recognition features and the structural characteristics which vary most within the family and are therefore of taxonomic use.

The Simuliidae form one of the most uniform families of Diptera, and blackflies are easily recognized as larvae, pupae and adults. Only the simple egg has no distinctive features, but even this stage can usually be recognized among aquatic eggs of comparable size (length 0.15–0.30 mm) by their shape: instead of being perfectly ovoid or ellipsoid, the egg has a distinct bulge on one side (Fig. 6.1b).

The larva is immediately recognizable by its possession of two pseudopods (usually called prolegs) and a pair of large cephalic fans on the head (Fig. 6.1a). The thoracic proleg extends from the underside of the thorax and the abdominal proleg forms the whole posterior end of the abdomen; each is crowned with a terminal circlet formed of numerous minute hooks arranged in parallel rows around a bare centre. Both prolegs are present in all species in all their developmental instars. Cephalic fans are less strictly diagnostic for the family, as they are not universally present: they are absent in the arctic genus *Gymnopais*, absent or much reduced in a few other aberrant simuliids, and undeveloped in the first instar larva of *Prosimulium*. The pupa (Fig. 6.1c) is recognized by its enclosure in a case (cocoon), the paired spiracular gills on the thorax, and the unique arrangement of hooks on the abdomen. The hook groundplan (onchotaxy) comprises a transverse row of four tergal hooks per side on segments 3 and 4, and two sternal hooks per side on segments 5–7, and is completely diagnostic for Simuliidae – even though (rarely) it can be somewhat masked by the development of extra hooks.

The adult blackfly, apart from its stocky form (Fig. 6.2), is best recognized by its wings, eyes, antennae and abdominal base. The wings are short and broad, with a large anal lobe, and characterized by crowding of the true (tubular) veins towards the leading edge and by sinuous curvature of vein CuA2 (except in *Gigantodax* in which this vein is straight). The head of the female is dichoptic (eyes separated) and

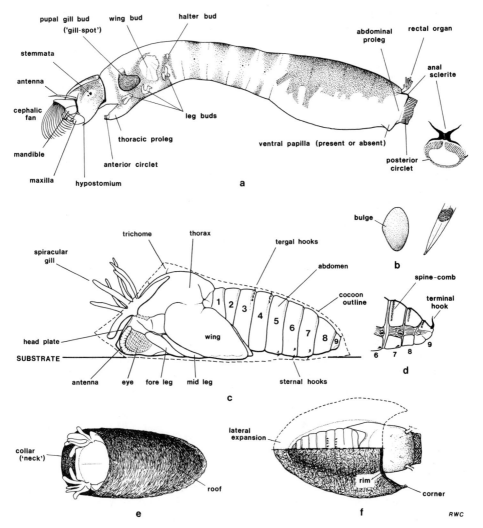

Figure 6.1 Basic morphology of the immature stages of Simuliidae: (*a*) mature larva in left side view (with apex of the abdomen in dorsal view); (*b*) egg, showing typical bulged shape (with small group of layered eggs on leaf tip); (*c*) pupa in left side view with outline of cocoon shown by pecked line (drawn from *Simulium lineatum*); (*d*) tip of pupal abdomen of *Prosimulium*, showing long terminal hooks; (*e*) cocoon of shoe-shaped type, dorsal view (collar present); (*f*) cocoon of slipper-shaped type, dorsal view (collar absent).

its eyes have uniformly small lenses (Fig. 6.3); the male head is nearly always holoptic (eyes meeting) and has eyes which are abruptly distinguished into an upper part with very large lenses (ommatidia) and a lower part with small lenses (Fig. 6.4); in a few unusual species the male head is narrowly dichoptic and the eye lenses not much differentiated in size. The antennae are short, more or less cylindrical and (unlike many Nematocera) short-haired in both sexes; there are nearly always 11 segments, but only 10 in *Austrosimulium* and very rarely 9. A diagnostic feature is a curious modification of the first abdominal tergite to form a prominent flange ('basal scale', Fig. 6.2) fringed with fine hairs (extremely long in the male).

244 *Blackflies (Simuliidae)*

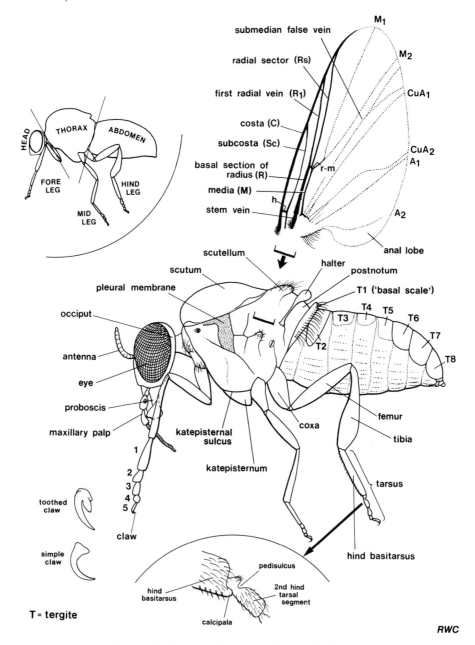

Figure 6.2 Basic morphology of adult female *Simulium*.

Larva

The head capsule floor is rarely fully sclerotized and shows a membranous area of varied shape and size extending forwards from the hind margin (the postgenal cleft, important in taxonomy). The head usually has pigmented head-spots, 'positive' when they occur at points of muscle attachment to the cranium and 'negative' when pigmentation surrounds these points. A mid-ventral forward extension of the head capsule, the hypostomium, has terminal teeth of varied shape;

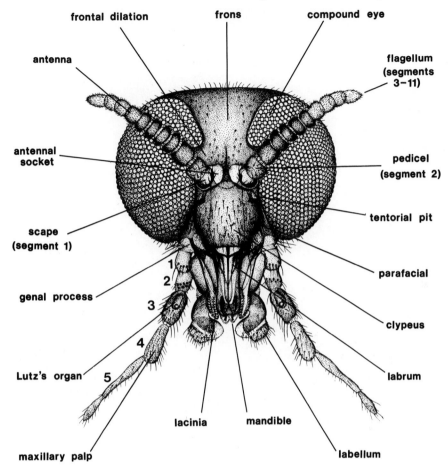

Figure 6.3 Head of female *Simulium* in facial view. (Modified from Jobling (1987) and reproduced courtesy of the Trustees of the Wellcome Trust.)

typically there is a transverse row of nine conical teeth in which the outside pair and the middle tooth are prominent, but mature larvae of *Simulium neavei* group have 13 hypostomial teeth of about equal size, and in *Gymnopais* the teeth are flat and chisel-like in association with loss of cephalic fans and the need to scrape food from the substrate instead of filtering it from the current. A pair of black stemmata or 'eye-spots' (absent only in the blind larva of *Parasimulium*) occurs on each side of the head. Apart from cephalic fans, the mouthparts comprise a bushy labrum overhanging the mouth entrance (cibarium), paired mandibles and maxillae laterally, and a very complex moustached structure (labiohypopharynx) beneath the mouth opening through which the silk (used for various purposes, see later) is extruded. The mandibles bear teeth and brushes whose diverse form can be useful in taxonomy. The thorax and abdomen are not sharply differentiated and their segmentation is poorly marked, but body shape differs slightly in relation to lotic habitat type: in larvae of slow waters the last segments of the abdomen have an evenly expanded profile (Fig. 6.1*a*), whereas in larvae of cascades the profile of the

Figure 6.4 Holoptic head of a typical male *Simulium* showing the large (upper) and small (lower) eye facets.

abdomen widens gradually towards the end before sharply contracting to the abdominal proleg. Many larvae of the *Simulium damnosum* complex have curious paired dorsal protuberances ('tubercles') on the first few abdominal segments (function unknown). The body cuticle is bare (but for scattered ultramicroscopic sensilla) or (occasionally) has a vestiture or setae or scales of varied shapes. Larvae of many species possess a pair of conical protuberances (ventral papillae) on the underside of the abdomen just before the posterior circlet (Fig. 6.1a). The rectal organ, a thin-walled trilobed structure in which each lobe is often subdivided into finger-like lobules, is often visible extruded through the anus. An X-like (in *Ectemnia* Y-like) anal sclerite lies between the anus and the posterior circlet. Buds of the adult's wings and legs, and of the pupal gill, are visible on older larvae, the gill bud forming a very large 'gill-spot' which blackens as the larva matures. Fully grown larvae measure about 4–12 mm in length, according to species and the nutritional success of the individual.

Pupa

The pupa is of the obtect type in which the head and thoracic appendages are sealed to the body. It shows four features notable for structural variation and associated use in taxonomy – gills, thoracic trichomes, abdominal onchotaxy and cocoon shape. The Simuliidae are unique among insects for the extraordinarily diverse (often quite bizarre) form of the specialized spiracular pupal gills, made of lifeless cuticle, which enable the developing (pharate) adult to breathe while under water; coming in seemingly endless variety, these structures are as yet unexplained evolutionary 'experiments' with the gill plastron – a plastron being an air-holding cuticular device on an insect body surface which permits gas exchange at an air–water interface (see Crosskey, 1990, Chapter 11). Trichomes – sensilla on the upper surface of the thorax – are usually tiny and sparse simple hairs, but are occasionally bifid or trifid, and can even (as in some South American species) be multibranched sensilla like the stellate setae of certain mosquito larvae. The abdominal onchotaxy shows some variation, such as development of hooks extra to the groundplan (e.g. in some species pupating on mayfly nymphs), presence of long terminal hooks in place of small terminal tubercles, and the presence of backwardly directed dorsal spine-combs (Fig. 6.1*d*), the last being characteristic of pupae in which the cocoon lacks a collar. In *Parasimulium*, which pupates in stream-bed sediments, *Gymnopais* and *Crozetia*, the pupa is naked, but in all other blackflies it is partly or wholly encased in a cocoon built by the larva from secreted silk. Cocoon shape varies considerably but shows two basic types, slipper-shaped and shoe-shaped, which differ by presence or absence of an anterior collar (Fig. 6.1*e,f*) and are respectively associated with slow and fast waters. Pupae of species living in tumultuous cascades are often deeply recessed in highly protective cocoons with long forward boot-like extensions. The cocoons of some *Simulium* species have large anterolateral holes that influence the water flow patterns around the pupae (Eymann, 1991).

Adult

General features which vary greatly are size and colour. Wing length is the best measure of size and ranges from 1.4 to 6.0 mm in different species. Legs are often all black in males but variegated yellow and black in females, and the abdomen of females sometimes has a spotted pattern. Some body areas (especially the scutum) often have reflectant patterns which reverse from silver to black as light falls on ultramicroscopic pile (pollinosity) inclined at different angles. Unlike some mosquitoes (Chapter 5), the body and wings are not ornamented with large scales.

The proboscis is shorter than the height of the head, directed downwards (Fig. 6.3), and composed of two basic elements – posterior labium and anterior syntrophium. The labium consists mainly of two large fleshy labella which are abundantly provided with sensilla and enclose the syntrophium when at rest but part company when the fly needs to feed. The syntrophium comprises six stylets: labrum, paired mandibles, paired laciniae (parts of the maxillae) and hypopharynx. The mandibles together form a transverse screen, and the space in front of them and behind the labrum is the food channel through which all the adult's liquid food enters the alimentary canal; a comparable space behind the mandibles, in front of the hypopharynx and between the lacinia, is the salivary channel down which (in the female) secreted saliva passes when the fly bites. All stylets in female flies are tipped with teeth used during bloodsucking, but in males (and females of certain non-bloodsucking species) the stylet tips are weak and ragged – unadapted for biting. A cross-section of the proboscis is shown in Fig. 6.10a. The maxillary palps have five segments, of which the last is typically long and sinuous but occasionally short and sausage-like; the third segment contains a large CO_2-sensitive sensory pit (Lutz's organ).

The thorax shows two significantly varying features. The katepisternum, which is usually demarcated above by a deep and continuous groove (katepisternal sulcus, Fig. 6.2) running between the mid coxae, is sometimes differentiated from the rest of the thorax by a shallow and incomplete furrow; and the pleural (anepisternal) membranes – very large in Simuliidae compared to other Diptera – are usually bare but are hairy in both sexes of some species. The legs always have five tarsal segments (tarsomeres) and paired tarsal claws of equal size on all legs, but the profile of the fore tarsi and/or the hind basitarsi can be narrow or conspicuously widened. The claws differ sexually, those of males being over-hung basally by a large hood-like structure absent in females. The talon-like claws of females (always alike on each leg) are either toothless (simple) or armed with one small or large basal tooth. Two leg structures occurring in both sexes – the calcipala and pedisulcus – are unique to Simuliidae but not present in all species; the calcipala is a semicircular inner flange on the apex of each hind basitarsus, and the pedisulcus is a groove across the basal part of each second hind tarsomere (Fig. 6.2). The wings vary to a minor extent in vestiture of the veins, and vein Rs is sometimes forked (Fig. 6.5a).

The abdomen of most females, apart from the tergites, is largely membranous (Fig. 6.2), in association with the need to accommodate a large blood-meal, and there are usually no evident sternites before segment 8. Sternites, however, are present in males and often in females of non-bloodsucking species. Genitalia of males form a compact hypopygium in which configuration of the styles, parameral spines, ventral plate and median sclerite vary greatly. Terminalia of females are fairly simple and vary mainly in the shape of the ovipositor lobes (= hypogynial valves), which are usually rounded but sometimes drawn out into tapering processes; all species possess a Y-shaped genital fork (modification of sternite 9). There is one spermatheca, which is usually hard and spherical.

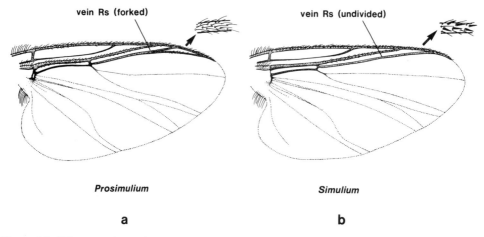

Figure 6.5 Wing venation and wing vestiture: (a) *Prosimulium*, with vein Rs forked; (b) *Simulium*, with vein Rs undivided.

CLASSIFICATION AND IDENTIFICATION

The morphological features used in taxonomy are listed in Peterson and Dang (1981). At present, classification relies more or less entirely on morphology and non-morphological approaches to taxonomy have been little researched from the perspective of their possible use above the species level. However, data obtained from mitochondrial DNA sequencing (Xiong and Kocher, 1991) and from larval polytene chromosomes (Rothfels, 1979) broadly support the higher taxonomic groups based on morphology, whatever rank these are given by different specialists.

The classification of Simuliidae is not fully agreed and two main 'rival' systems exist, drawn largely on geographical lines. An increasingly elaborated scheme developed in Russia has resulted in four subfamilies, the Parasimuliinae, Gymnopaidinae, Prosimuliinae and Simuliinae, of which one (Simuliinae) contains five tribes (Rubtsov, 1974). This means a total of eight suprageneric taxa to classify a family so homogeneous that in reality it is hard to find fully satisfactory diagnostic characters for subgenera or genera, let alone higher categories. A corollary of this is that assemblages of species which most workers rank at the highest as species-groups or subgenera (usually parts of *Simulium*) are treated as full genera by Rubtsov and those who follow his system (mainly in eastern Europe, Germany and Italy).

The Rubtsov approach to classification has not found favour in North and South America, Britain, France, Iberia or Japan (areas of recent high taxonomic activity on the family) and here preference is given to a simpler classification recognizing two subfamilies, Parasimuliinae and Simuliinae, of which the latter contains two tribes, the Prosimuliini and Simuliini (Table 6.1). This system was proposed by Crosskey (1969) and used later for a checklist of world species (Crosskey, 1988) and as a framework in which to describe simuliid biology (Crosskey, 1990). Even this simple system, however, can be considered top-heavy with higher taxa for flies so generally uniform as the Simuliidae: there is difficulty in unequivocally placing certain genera in either the Prosimuliini or the Simuliini. The aberrant *Parasimulium* – the only genus constituting the subfamily Parasimuliinae – has been found, since discovery of its long-unknown immatures, to have many characteristics of Prosimuliini. Still, the subfamily is cladistically warranted.

The principal diagnostic characters of Parasimuliinae by which the subfamily differs from all other blackflies (= Simuliinae) are:

Eyes of male almost meeting *below* antennae and with only some central facets
  ‣larger (though indistinctly) than other facets.
Wing with vein R1 short, joining costa about halfway along wing length, and with
    branches of Rs fork widely separated.
Katepisternum much reduced, with ventrally pointed profile and without furrow
    demarcating it from rest of thorax.
Larval head without stemmata.

In the Simuliinae, three main characters distinguish typical prosimuliines and simuliines. In Prosimuliini the pupal cocoon (except in *Ectemnia*) is ragged and shapeless or nearly absent, the adult thorax has an incomplete and shallow groove between the katepisternum and the rest of the thorax, and the hind leg lacks the pedisulcus. In Simuliini, on the other hand, the cocoon has a well-formed shape

250   Blackflies (Simuliidae)

Table 6.1 Synopsis of genera and *Simulium* subgenera

The table includes all genera recognized in the world checklist of Crosskey (1988). Subgenera of *Simulium* are listed (prefixed 's.') which contain man-biting species or (if they do not) are important subgenera with 20 or more species. Genera and subgenera that include man-biting species are marked with an asterisk (*). Numbers of species refer only to those formally named (some genera/subgenera also contain undescribed sibling species); the world total and some individual generic/subgeneric numbers of species are higher than those in Crosskey (1990, pp. 32–34) because of recent description of new species. (Data to end of 1991.)

Genus	No. of spp.	Regional or country distribution
PARASIMULIINAE		
Parasimulium	4	Northwestern USA
SIMULIINAE		
Prosimuliini		
Araucnephia	1	Chile
Araucnephioides	1	Chile
*Cnephia**	10	Holarctic
Cnesia	3	Argentina, Chile
Cnesiamima	1	Argentina, Chile
Crozetia	2	Crozet Islands
Ectemnia	2	Canada, USA
Gigantodax	65	Central and South America (Andean)
Greniera	12	Palaearctic, northeastern Nearctic
Gymnopais	12	Northern Holarctic (mainly arctic)
Levitinia	2	Central Asia, Middle East
Lutzsimulium	4	Northern Argentina, southern Brazil
Mayacnephia	11	Western Canada to Mexico/Guatemala
Metacnephia	51	Holarctic
Paraustrosimulium	1	Southern Chile, Tierra del Fuego
Piezosimulium	1	USA (Colorado)
*Prosimulium**	110	Holarctic, eastern and southern Africa
Stegopterna	9	Northern Holarctic (including Japan)
Sulcicnephia	21	Palaearctic (mainly Central Asia)
Tlalocomyia	1	Central Mexico
Twinnia	10	Northern Holarctic (including Japan)
Genus uncertain	9[a]	Australia
Simuliini		
*Austrosimulium**	25	Australia and New Zealand
*Simulium**	1203	Worldwide (except New Zealand)
s. *Anasolen**	(11)	Tropical and southern Africa
s. *Boophthora**	(5)	Palaearctic (including Japan)
s. *Byssodon**	(13)	Holarctic, tropical Africa, Sri Lanka
s. *Edwardsellum***[b]	(22)	Tropical and southern Africa, Yemen
s. *Eusimulium*	(33)	Holarctic (also Guatemala, Taiwan)
s. *Gomphostilbia*	(61)	Austro-Oriental, eastern Palaearctic
s. *Hearlea*	(20)	Western Nearctic, Central America
s. *Hebridosimulium**	(3)	Fiji, Vanuatu, Tahiti
s. *Hellichiella*	(26)	Holarctic
s. *Hemicnetha*	(21)	Western USA to central Neotropics
s. *Himalayum**	(2)	Himalayas (Afghanistan to Burma)
s. *Inseliellum**	(24)	Polynesia and Micronesia
s. *Lewisellum***[c]	(9)	Tropical Africa
s. *Meilloniellum**	(5)	Afrotropical (including Arabia)
s. *Metomphalus***[d]	(33)	Afrotropical (including Arabia)
s. *Montisimulium*	(37)	Palaearctic (mainly Central Asia)
s. *Morops**	(55)	Australasian (mainly New Guinea)

Table 6.1 (cont.)

s. *Nevermannia*	(169)	All regions except Neotropical
s. *Notolepria**e	(5)	Central and South America
s. *Parabyssodon**	(3)	Northern Holarctic
s. *Pomeroyellum*	(42)	Afrotropical (including Madagascar)
s. *Psaroniocompsa**	(38)	South America and Panama
s. *Psilopelmia**f	(50)	Western Nearctic, Neotropical
s. *Psilozia**	(3)	Nearctic and Iceland
s. *Pternaspatha*	(31)	Western South America (Andean)
s. *Schoenbaueria**	(17)	Northern Holarctic
s. *Simulium* s. str.*g	(292)	Holarctic, Oriental and northern Neotropical
s. *Trichodagmia**	(9)	South America
s. *Wilhelmia**	(17)	Palaearctic (including Japan)
Unlisted subgenera	(106)	
Unplaced species	(41)	
Total world species	1571	

[a] Species wrongly assigned to *Cnephia* in the literature.
[b] Includes the *Simulium damnosum* complex (onchocerciasis vectors).
[c] Includes the *Simulium neavei* group (onchocerciasis vectors).
[d] Includes *Simulium albivirgulatum* (onchocerciasis vector).
[e] Includes the *Simulium exiguum* complex (onchocerciasis vectors).
[f] Includes the *Simulium ochraceum* complex (onchocerciasis vectors).
[g] Includes the *Simulium metallicum* complex (onchocerciasis vectors).

(even if it covers only the pupal abdomen), the adult thorax has a complete and deep groove between the katepisternum and the rest of the thorax, and the hind leg possesses the pedisulcus. Other, less sharply diagnostic, differences are:

Calcipala usually absent in prosimuliines (virtually always present in simuliines).
Wing vein Rs often forked and costa often without spinules in prosimuliines (Rs undivided and costa with spinules among the hair in simuliines) (Fig. 6.5a,b).
Fore tarsi narrow in prosimuliines (often with dilated profile in simuliines).
Pupa often with long terminal hooks and with spine-combs in prosimuliines (Fig. 6.1d) (without such hooks and often without spine-combs in simuliines).
Larval cuticle bare in prosimuliines (often with spines or scales in simuliines).
Larval rectal organ lobes undivided in prosimuliines (often secondarily divided into lobules in simuliines).

Genera and subgenera

A few genera (or subgenera) have distinctive chromosomal features (e.g. the Holarctic genus *Metacnephia* has a whole-arm chromosomal interchange not present in its near relatives) but in practice genera and subgenera are all defined only on the basis of morphology. In the Rubtsov system, mentioned earlier, about 60 genera are recognized, mainly because assemblies of species which most specialists rank at no higher level than subgenera within a broad *Simulium* are ranked as full genera – despite the impossibility in almost all cases of finding fully satisfactory characters which cover all the included species in all their life stages. The Rubtsov system is considered very undesirable by the present author, and the classification here used is that in which (apart from the genus *Austrosimulium*) all elements of Simuliini are ranked as subgenera within the one large genus, *Simulium*. (This is analogous to the classification of other major genera of bloodsucking Diptera such as *Phlebotomus*, *Aedes*, *Culicoides* and *Tabanus*.) The genera recognized, and the principal subgenera of *Simulium*, are listed alphabetically under each tribe in Table 6.1.

Species groups and complexes

In simuliid taxonomy it is usual (but still not universal) to use the term 'complex' for an assemblage of sibling species which can only be reliably distinguished by non-morphological criteria, and 'group' for an assemblage of closely similar but morphologically recognized species (each of which might prove to be a complex when better known). Example: the little-studied *Simulium neavei* group contains members which are extremely alike but still identified only on morphology in the absence of other data; the much-studied *S. damnosum* complex contains many members virtually identical in their morphology and identified by non-morphological criteria.

In practice, nearly all species are based on features of the hard-parts morphology (e.g. male genital or pupal gill structure) since nothing is yet known about the genetic, chromosomal or biochemical attributes of most of the 1570 or so described and named morphospecies. However, as sibling speciation is extremely prevalent in Simuliidae it is certain that – besides the complexes already known – many other named species (perhaps most) are simply morphospecies (Chapter 1) of the 'museum' taxonomist, and will prove to be complexes of genetically distinct (reproductively isolated) sibling species whenever they are examined with non-morphological techniques.

All simuliid larvae have, in the nuclei of their silk gland cells, large polytene chromosomes which show banding patterns and other features of great value as specific diagnostic characters, and for determining cytophylogeny (evolutionary relationships shown by stepwise rearrangements in chromosomal structure). The gametic haploid complement (n) of the karyotype is three chromosomes and the diploid six (very rarely two and four by reduction), but pairing is so intimate that ostensibly there are only three chromosomes. By convention, these are numbered I–III in decreasing order of size, and their arms (parts either side of the centromere) called L (long) and S (short) according to their relative lengths (Fig. 6.6).

While the mainframe of taxonomy remains morphology, chromosomal studies have a profound impact at species level. Chromosomes often provide the only reliable means of identifying the individual members of a complex which contains pests and disease vectors, and this gives their study paramount importance. Many complexes contain half a dozen or more cytospecies (sibling species diagnosed and identified by chromosomal criteria). In *Simulium damnosum* – the largest known complex and containing the major vectors of African onchocerciasis – about 40 cytologically different entities have been recognized (see Crosskey, 1987: review) some of which are certainly distinct species in nature and others provisionally called 'cytotypes' or 'cytoforms' because evidence for their reproductive isolation is inconclusive. Other medically important complexes are *S. ochraceum*, *S. exiguum* and *S. metallicum*, all including onchocerciasis vectors in Latin America; the last contains at least 11 chromosomal segregates, of which six appear to be sibling species (Conn, 1988). Some important man-biting pests in temperate regions belong to complexes, e.g. *S. venustum*.

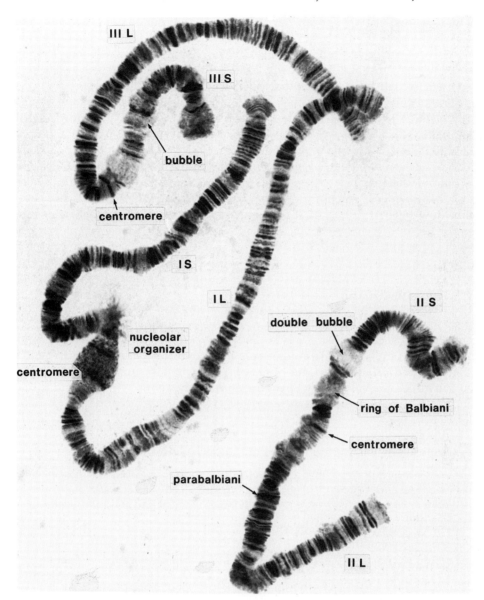

Figure 6.6 Structural features of the complete polytene chromosome complement from the nucleus of a blackfly larval silk gland cell. (Illustrated from a preparation of *Simulium sirbanum* from Rio Corubal in Guinea Bissau, West Africa. Photograph courtesy of Magda Charalambous.)

Polytene chromosomes can sometimes be obtained from adult fly Malpighian tubules, but in routine practice cytology can yield data only for the larval stage. Members of complexes are virtually isomorphic, incapable of being recognized reliably if at all by their morphology, and identification of adult females of man-biting simuliids to the sibling species level – which is much needed in epidemiological study and control of onchocerciasis – is so far impossible (at least for routine purposes) in the absence of cytological and morphological criteria. Some morphometric studies have been made of adult flies, but these have not yielded criteria for distinguishing siblings that can be readily used in practice. To circumvent these difficulties, biochemical (molecular) approaches are in use involving enzyme systems, body surface hydrocarbons and the genetic DNA (Townson et al., 1988: review). The value of DNA sequencing within the *S. damnosum* complex is under study (e.g. Post and Crampton, 1988; Post and Flook, 1992). Little has so far been achieved, however, by these methods and isoenzymes in particular have not lived up to expectation (their usefulness being very limited even in the much-studied *S. damnosum* complex). Chromosomes continue to provide the keystone characters. The now-massive chromosomal literature is scattered and overview works are much needed: Rothfels (1979, 1981, 1988) and Chubareva and Petrova (1979) provide starting points. The cytological literature on the *S. damnosum* complex has grown piecemeal, and there is now outstanding need of a comprehensive treatment to make cytotaxonomic findings comprehensible to a wider audience in medical entomology – with chromosomal criteria linked to morphological/morphometric and biochemical data where possible.

Identification

Generic identification

Man-biting behaviour, and thus potential medical importance, is essentially restricted to three genera: *Prosimulium*, *Austrosimulium* and *Simulium*. The onchocerciasis vectors belong only to *Simulium*, a genus distinguishable from all others in the family by this simple rule-of-thumb: **any adult blackfly with a pedisulcus and calcipala on the hind leg and with 11-segmented antenna is a *Simulium*.** In some continents *Simulium* is the only genus which includes man-biting species and geography is thus the easiest approach to identifying genera.

Geographical key to simuliid genera containing man-biting species

Areas shown are those in which the genera named are concerned in man-biting, not necessarily the entire generic range. For example, *Simulium* occurs in Australia and *Prosimulium* in Africa but the species present in these continents are not anthropophilic.

1 Specimens from Australasian region 2
— Specimens from other areas 3
2 Specimens from New Zealand or Australia. [Antennae with ten segments] .... **Austrosimulium**
— Specimens from Melanesia or Polynesia. [Antennae with 11 segments] .... **Simulium**
3 Specimens from Europe, Japan, North America or USSR (Holarctic realm*) 4
— Specimens from Africa, Central America (including Mexico) or South America... .... **Simulium**
4 Wing costa with intermixture of stout spinules and fine hairs (Fig. 6.5b); vein Rs undivided. Hind leg with pedisulcus and calcipala (Fig. 6.2) .... **Simulium**
— Wing costa with uniformly fine hairs, no spinules (Fig. 6.5a); vein Rs forked near its end (but branches sometimes closely parallel and fork then inconspicuous). Hind leg without pedisulcus and calcipala .... **Prosimulium**

*Some simuliids in this realm are troublesome swarming irritants even though they bite man hardly at all. *Simulium vittatum* s.l. is particularly pestilent in Iceland.

Identification of complexes and species
It is impossible to provide easy means of identifying medically important blackfly species. Nearly every one important for its role as an onchocerciasis vector, or simply as a biting pest, belongs to a different subgenus or species group, i.e. there is no relationship in *Simulium* between the capacity to transmit *Onchocerca volvulus*, or the propensity to feed on humans, and the phylogenetic position of species in the classification. Here only a few notable features are mentioned, followed by a key to man-biters in tropical Africa.

Simulium neavei group   This is a tropical African group with eight named species of which *S. neavei* and *S. woodi* are vectors in East Africa. Larvae and pupae are found only on the bodies of *Potamonautes* river-crabs (Fig. 6.9) in an ecological relationship (obligate phoresy) unique in the Simuliidae. This habit enables immatures to be instantly recognized. The group is very uniform and species identification difficult in all life stages. Microscopic differences in surface sculpture of the larval cuticle have been found useful for species recognition (e.g. Lewis and Raybould, 1974).

Simulium damnosum complex   By far the largest known in the Simuliidae (approximately 40 cytospecies and cytotypes), this complex occurs over almost the whole Afrotropical region (including Yemen but excluding Madagascar). Fortunately, from the viewpoint of medical entomology, the complex has several clear-cut diagnostic characters which permit it to be quite easily recognized. These are:

Fore tarsi enlarged and with a conspicuous hair-crest (Fig. 6.7a).
Legs all black except for a sharply contrasting creamy-white band on each hind basitarsus (Fig. 6.7a).
Pupa gill with somewhat the form of a hand of bananas (branches stoutly tubular, Fig. 6.7b).
Larva with 'unshaved' appearance (Fig. 6.7c,d) caused by cuticular covering of small upstanding black spines and scales, and (often) with very prominent dorsal abdominal tubercles (Fig. 6.7d). (Setae of the larval cuticle occur even on the thoracic proleg, and in this the complex is unique in the Simuliidae.)

An enormous amount of morphological research has gone into the complex in an effort to find external features by which individual members can be reliably identified to species (especially biting females). Some identification is possible by experienced specialists, but even then more on a high-probability rather than absolute reliability basis. Differences exist between siblings in, for example, larval scaling and tubercular development, and adult female antennal length/colour/segment shape and wing stem-vein tuft colour, but all are intangible and difficult to use reliably in practice. Some literature cannot be uncritically trusted. The most careful studies of morphology (larval as well as adult) in populations of known chromosomal/enzymatic identity have revealed considerable intraspecific variability in most sibling species. No up-to-date overview paper is available. Vajime and Dunbar (1975) published a major cytological work establishing the basic composition of the complex, with a key to West African species based on polytene chromosomes ('cytospecies') but their paper is now partly superseded and identification of chromosomal material should be entrusted to a cytotaxonomic specialist. See Dunbar and Vajime (1981) for a basic idea of cytophylogeny and Boakye et al. (1993) for a revision of the *S. sanctipauli* subcomplex.

256 *Blackflies (Simuliidae)*

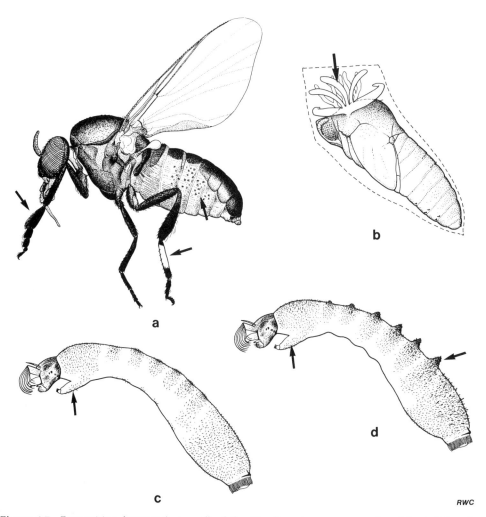

Figure 6.7 Recognition features (arrowed) of the *Simulium damnosum* complex: (*a*) adult female, showing the broad and hair-crested fore tarsi, white-banded hind basitarsi and clumped lateral vestiture of the abdomen; (*b*) pupa, showing the banana-like gill; (*c* and *d*) larva, showing the general covering of small setae and scales which give the body an 'unshaved' appearance, setae on the thoracic proleg (a feature unique in Simuliidae), and large abdominal tubercles often present in larvae of sibling species from forest areas. Tubercles are typically absent in siblings of savanna areas (as *c*).

Other complexes Problems such as those just mentioned are likely to occur with the Latin American *S. ochraceum* complex (Central America), *S. metallicum* complex (Central and northern South America) and *S. exiguum* complex (northwestern South America). Each of these comprises several cytospecies and/or cytotypes, but at this stage of knowledge the constituent taxa are not critically diagnosed or formally named. A morphological key to larvae of Mexican cytotypes of *S. metallicum* complex is in Millest (1990). *Simulium exiguum* is currently under research by A. J. Shelley and colleagues and some clarification of the complex can be expected shortly.

Key to *Simulium* females attracted to humans in tropical Africa

The following is a simple key to aid identification of adult females often or sometimes caught while biting or landing on humans in tropical Africa. The localities noted are those where attraction to man is usually reported, not complete geographical ranges of the species. Leg coloration refers to ground colour and not vestiture colour (which is often different). Slide preparations are useful for seeing whether or not the claws are toothed.

1   Wing with basal section of radial vein (Fig. 6.2) finely haired along its length 2
—   Wing with basal section of radial vein bare. [Small silvery grey species with toothed claws: mainly Sudan and northern Nigeria] .... **S. griseicolle**
2   Pleural membrane haired. Legs conspicuously bicolorous, partly reddish yellow and partly blackish brown 3
—   Pleural membrane bare. Legs either bicolorous or mainly or entirely black 4
3   Claws with large basal tooth (easily seen at low magnifications). Hind tibia with a dark band near base in addition to dark apex. [Small silvery grey species, wing length 1.4–2.2 mm: mainly northern savannas of West Africa] .... **S. adersi**
—   Claws with very small (inconspicuous) basal tooth. Hind tibia without dark band near base. [Larger grey-black species, wing length 3.0–3.75 mm: Ethiopia, eastern Africa] .... **S. dentulosum**
4   Fore tarsi narrow, basitarsi in profile 5.5–7.0 times as long as greatest breadth; without definite hair-crest. Abdomen uniformly covered with recumbent hair, terminal tergites not bare and shining 5
—   Fore tarsi broad, basitarsi in profile 3.6–4.2 times as long as greatest breadth; with long and conspicuous hair-crest (Fig. 6.7a). Abdomen not uniformly covered with recumbent hair, posterior tergites shining and bare (except for some tiny erect hairs) and sides with small clumps of silver-yellow or silver hair. [Medium-sized flies (wing length 2.0–2.4 mm) with black body and black legs, except for creamy-white band on hind basitarsus (Fig. 6.7a): widespread tropical Africa, also Yemen] .... **S. damnosum** complex
5   Large blackish or dark brown species, wing length 2.5–3.6 mm. Claws with small basal tooth (sometimes hardly detectable). Abdomen with hair either entirely silvery yellow to golden or partly coppery or black. Scutum without pattern 6
—   Very small greyish species (blackish in *wellmanni*), wing length 1.3–2.3 mm. Claws without trace of basal tooth (simple and talon-like). Abdomen with uniformly silvery hair. Scutum with three fine dark longitudinal lines (concealed by hair vestiture in perfect specimens and best seen as fly is turned) 10
6   Abdomen with hair mostly black, yellow only at base and/or on mid-lateral parts. Legs entirely black (not even a pale band on hind basitarsi). [Cameroun] 7
—   Abdomen with hair mostly silvery yellow or golden, sometimes coppery or bronze on middle segments. Legs black or very dark brown with basal parts of hind basitarsi dingy or clear reddish yellow (mid parts of tibiae also sometimes faintly paler). [Eastern Africa] 8

7   Very dark species with hair of frons, clypeus, scutellum and hind part of scutum black or bronze-black. Abdomen with yellow hair only on first two segments. [Forest areas].. **S. dukei**
— Paler species with hair of frons, clypeus, much of scutellum and all of scutum silvery yellow to pale golden. Abdomen with yellow hair on first two segments and mid-laterally (sides of segments 3–6). [Mainly savanna areas] ... . **S. ovazzae**
8   Abdominal hair unicolorous silvery yellow or golden.. 9
— Abdominal hair not unicolorous silvery yellow or golden, darker coppery or bronze on middle segments. [Tanzania] .... **S. woodi**
9   Legs black except for indistinctly dingy yellow bases to hind basitarsi (these not definitely pale-banded). [Uganda and eastern Zaire]............................ **S. neavei**
— Legs not almost uniformly black, mid part of each tibia and hind femoral/tibial 'knees' dingy reddish orange or brownish; hind basitarsi distinctly reddish yellow on basal two-thirds or so (fairly definitely pale-banded). [Southwestern Ethiopia] .... **S. ethiopiense**
10 Antennae with first two segments dingy yellow or reddish, paler than black remainder. Legs not entirely black, at least hind basitarsi with pale band ... 11
— Antennae entirely black. Legs entirely black. [Angola] .... **S. wellmanni**
11 Legs mainly pale reddish yellow, only tarsi, apices of tibiae and (sometimes) mid femora, brown or blackish brown. [Mainly northern Nigeria] .... **S. bovis**
— Legs dark brown or black-brown, yellowish only on basal two-thirds of hind basitarsi (basal parts of tibiae also sometimes rather pale). [Zaire basin, Zambia]... **S. albivirgulatum**

Faunal and taxonomic literature

General: Crosskey (1988) gives a world checklist of 1461 species described up to 1986 and considered valid, but more than one hundred species have since been described. Catalogues exist for all faunal regions but not all use the same classification and most are somewhat outdated, especially in nomenclature. Works cited below under the country names include keys to adults, pupae and larvae unless life stages are specified: then there are keys only to the stage(s) stated. Works relating to particular North American states and provinces are useful beyond their title-specified coverage: e.g. Currie on the Alberta fauna is an important work relevant for all Canadian prairie provinces.

Palaearctic: Catalogue (Rubtsov and Yankovsky, 1988); regional fauna (Rubtsov, 1959–1964, monograph in German); Britain (Davies, 1968); Czechoslovakia (Knoz, 1965); Denmark (Jensen, 1984, pupae and larvae); France (Grenier, 1953); Italy (Rivosecchi, 1978); Japan (Uemoto, 1985, pupae and larvae); Pakistan (Lewis, 1973, pupae); USSR (Rubtsov, 1956, 1961, bloodsucking species; Terteryan, 1968, Armenia; Patrusheva, 1982, Siberia, keys to genera but not species).

Afrotropical: Catalogue (Crosskey, 1980); regional fauna (Freeman and de Meillon, 1953, outdated but essential work, pupae and adults); West Africa (Crosskey, 1960, larvae); Arabia (Crosskey and Büttiker, 1982); *Simulium neavei* group (Lewis and Raybould, 1974, adults).

Oriental: Catalogue (Crosskey, 1973); Philippines (Takaoka, 1983); Sabah (Smart and Clifford, 1969); Sulawesi (Takaoka and Roberts, 1988); Taiwan (Takaoka, 1979);

Thailand (Takaoka and Suzuki, 1984). No comprehensive works for India and Sri Lanka: for India see papers by I. M. Puri (*Indian Journal of Medical Research*, 1932–1933) and by M. Datta (mostly in *Oriental Insects*, 1973–present) (trace for both authors through *Zoological Record*); for Sri Lanka see papers by D. M. Davies and H. Györkös in *Canadian Journal of Zoology* (1987–1992).

Australasian: Catalogue (Crosskey, 1989); Australia (Colbo, 1976, *Simulium*; Dumbleton, 1973, *Austrosimulium*); New Guinea (Smart and Clifford, 1965); New Zealand (Dumbleton, 1973; Crosby, 1990, pupae and larvae); Society Islands (Craig, 1987); *Austrosimulium* (Dumbleton, 1973, revision); *Simulium* (Crosskey, 1967, classification).

Nearctic: Catalogue (Stone, 1965); regional fauna (Peterson, 1981, key to genera/subgenera; Adler et al., complete monograph in preparation); Canada (Currie, 1986, Alberta, pupae and larvae; Davies et al., 1962, Ontario, adults and pupae; Wood et al., 1963, Ontario, larvae); USA (Adler and Kim, 1986, Pennsylvania, pupae and larvae; Cupp and Gordon, 1983, northeastern states, faunal overview, no keys; Stone and Jamnback, 1955, New York; Stone, 1964, Connecticut; Merritt et al., 1978, mid-western states, pupae and larvae; Snoddy and Noblet, 1976, southeastern states, pupae and larvae; Stone and Snoddy, 1969, Alabama; Wirth and Stone, 1956, California, pupae and larvae). *Prosimulium* (Peterson, 1970, Canada and Alaska).

Neotropical: Catalogue (Vulcano, 1967), regional prosimuliine fauna (Wygodzinsky and Coscarón, 1973), Guatemala (Dalmat, 1955); Mexico (Vargas et al., 1946, larvae, males by genitalia only; Vargas and Diaz Najera, 1957, adults and pupae). *Gigantodax* (Wygodzinsky and Coscarón, 1989, full revision); *Simulium amazonicum* group (Shelley et al., 1982); *Simulium* subgenus *Pternaspatha* (Wygodzinsky and Coscarón, 1967, Coscarón and Wygodzinsky, 1972).

BIOLOGY

The biology of blackflies is comprehensively covered by Crosskey (1990) and that book (containing a bibliography of about 1200 works published up to mid-1989) should be consulted by anyone wanting more detail than it has been possible to provide here. Also useful are edited books by Laird (1981) and Kim and Merritt (1988), although their focus is largely on ecology and control.

Life history

The basic egg–larval–pupal–adult life cycle (Fig. 6.8a) is complicated more than that of other Diptera by the high and unstable number of larval instars and the importance of pharate phases in development. In mosquitoes, and most other Nematocera, the number of larval instars is fixed at four, in higher flies at three, but in simuliids the number is six to nine (commonly seven) – and recent evidence has shown up to 11 in *Simulium vittatum* complex (Colbo, 1989). Moreover, the number can vary within a species. In warm tropical environments, where duration of larval life is usually six to ten days (even as little as four days in *S. sirbanum* developing in West African waters above 30°C), moulting must occur about once a day during the rapid growth.

260 *Blackflies (Simuliidae)*

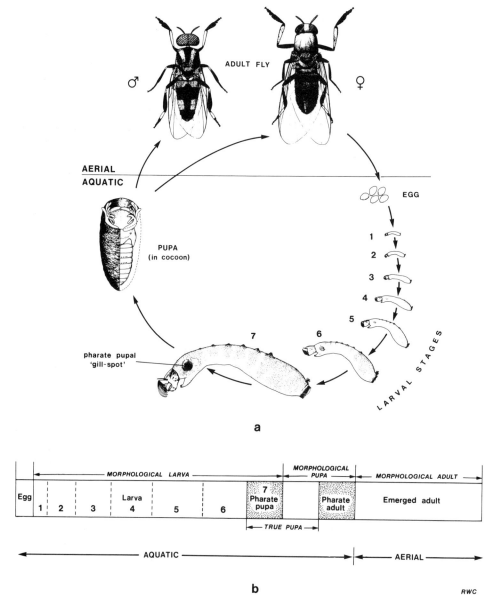

Figure 6.8 Life cycle of the Simuliidae: (*a*) instars of the basic life cycle illustrated from *Simulium damnosum* complex; (*b*) diagram of the pharate (hidden) phases of the blackfly life cycle (see text explanation).

The terms 'pharate pupa' and 'pharate adult' are often used in the blackfly literature and the following few words are needed to clarify their meaning. An insect is in a pharate phase when the new cuticle of a life stage (instar) has separated from the old cuticle of the immediately previous instar, but this old cuticle has not been discarded (moulted). Such pharate (= hidden) phases are usually very brief, but the Simuliidae are unusual because at the metamorphosis from larva-to-pupa and pupa-to-adult the two processes (parting and shedding of cuticles) are greatly out of step. The physiological separation of new from old cuticles occurs quite a long time before the old cuticle of the larva is shed (and the larva transforms into the pupa) and before the old cuticle of the pupa is shed (and the pupa transforms morphologically into the adult). Consequently, the apparent last larval instar (identified by its black thoracic 'gill-spots', Fig. 6.8a) is actually the pharate pupa – although it is nevertheless usually regarded as the 'mature larva' and is the cuticular larva used for taxonomic descriptions of the larval stage. Transformation of pharate into morphological pupa – pupation in the ordinary sense – is followed soon afterwards by separation of pupal and adult cuticles, and for most of its time (usually three to ten days) the ostensible pupa is the pharate adult living an underwater existence in a shell provided by dead pupal cuticle. At emergence from this shell, the adult shoots to the water surface enclosed in an air bubble expelled from its respiratory system. The relationship in development between physiological and morphological stages is shown diagrammatically in Fig. 6.8b.

Eggs often hatch within one to two days in warm tropical regions and the total time taken from egg to adult can be as little as two weeks, so allowing – as in the *S. damnosum* complex – for the completion of as many as 15–20 generations each year. In cool temperate and arctic areas, on the other hand, many species have only one annual generation (they are univoltine), the eggs either not hatching for several months after being laid or hatching quite quickly but larvae then developing very slowly through the long winter. From two to four annual generations is usual in temperate latitudes where winter is not too severe. This number obtains with many bloodsucking pest species. Oddly, the *S. posticatum* man-biting pest of southern England is univoltine despite the mild winters prevailing there.

In many species the life history strategy depends on eggs having a resting phase (diapause) which outlasts unfavourable environmental conditions such as extended drought or winter cold. Larvae of some *Prosimulium*, for example, hatch irregularly through the winter from eggs laid late the previous spring in summer-drying streams. Simuliid larvae which overwinter in cold northern latitudes are chemically protected from supercooling and can lie dormant under ice of frozen rivers, feeding and growing again when the water warms up in spring. In *Austrosimulium pestilens* the eggs can stay viable in dry creek beds for as much as two years and then hatch when floods eventually disturb the sediments and release them into flowing water. Even in this species the eggs must stay moist, deep in the interstices of the sediment, for the simuliid egg is never truly drought-resistant.

The larva in the lotic habitat

The aquatic larvae of blackflies can be found in virtually every kind of running water – fast or slow, placid or turbulent, warm or cold, acid or alkaline, spring-fed or run-off, and rich or poor in food or oxygen. Some species are adaptable and occupy a diversity of lotic niches, others are specialist inhabitants of restricted niches such as springs or lake outfalls. The larvae are normally found anchored to immersed inorganic or vegetational substrates – pebbles, boulders or fixed smooth bedrock, trailing grass, bed-rooted waterplants – but those of a few species attach to the bodies of other arthropods. The latter habit occurs in about 2% of species and for the blackfly is an ecological association (obligate phoresy) essential for survival. The species concerned are called for convenience 'phoretics' and occur in tropical Africa on crabs (e.g. *S. neavei*, Fig. 6.9), prawns or mayfly nymphs, and in Central

262 *Blackflies (Simuliidae)*

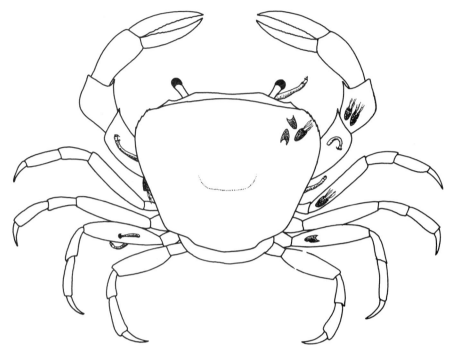

Figure 6.9 African river-crab of the genus *Potamonautes* carrying larvae, pupae and empty cocoons of *Simulium neavei*. Attachment occurs mainly on the bases of the walking limbs and chelipeds (pincers), margins of the carapace, and on the eye stalks (as representative positions shown).

Asia on mayfly nymphs. The larvae settle on the carrier when very young (even in the first instar) and always pupate on the carrier's body.

Anchorage in the current is achieved by the larva inserting the hooks of the abdominal proleg into a pad formed of silk copiously produced from its huge silk-glands and applied to a firm submerged surface. From this fixed point the larva can move in two ways, either by unlatching the hooks and drifting freely downstream until it is snared on a new substrate by a reeled out thread of sticky silk, or by looping along on the substrate semi-sideways with alternate attachment and detachment of the prolegs. Larvae of nearly all species are suspension-feeders which open their cephalic fans in the current to intercept material passing downstream. Mechanical sieving of particles in suspension is aided by secretion of sticky mucosubstance spread upon the fan-ray filter-apparatus from glands in the labrum. The filtrate is pushed into the cibarium and passes thence through the gut for digestion of the organic component. The larvae of *Twinnia* and *Gymnopais* have lost the cephalic fans and feed by scraping the substrate with their chisel-like hypostomial teeth. *Gymnopais* larvae occur in food-poor glacial streams where they subsist on scant algae before transforming into naked pupae secured to the clean-washed stones merely by a pad of silk (Wood, 1978).

Mating

Prosimulium ursinum and some *Gymnopais* of the arctic do not mate as they have abandoned the male sex and reproduce parthenogenetically from unfertilized eggs.

Males and females of normal bisexual species mate soon after emergence, either crawling about randomly to do so on some damp surface near the stream (as in *Cnephia dacotensis* and *Crozetia*) or meeting when flying females enter a precopulatory swarm of males formed at some 'swarm-marker' such as a tree branch or the corner of a building. Formation of such precopulatory male swarms is capricious and often influenced by weather. They occur mainly in species of temperate regions, where the adult flight season is cut short by winters. As in mosquitoes, they are rare in the tropics: for example, mating swarms occur in the *Simulium damnosum* complex but are so unusual that hardly any of the entomologists who have worked many years on these vectors have ever seen one.

The specialized eyes of the male (Fig. 6.4) provide this sex with excellent vision for a mating role, and a swarm-entering female is quickly seen and grabbed for attempted mating. Actual copulation, however, does not occur in the air, the non-rotated male genitalia obliging the coupling pair to fall out of the swarm onto a solid surface before they can engage the terminalia. The male usually packages the sperm before its transfer in a bag-like spermatophore which is placed into the genital recess of the female and can remain there safely while she flies or feeds. After the spermatophore is opened (presumably by enzyme action) the sperm travel along the spermathecal duct to be stored in the spermatheca until required to fertilize the eggs. Females are assumed to copulate only once and to fertilize all the eggs, however many batches are produced, from the one insemination. Copulation takes from a few seconds up to half an hour. In the *S. decorum* complex the ejaculate has been found to contain 600–7000 sperm.

Biting behaviour and hosts

Blackflies bite by day and in the open, and these are the two cardinal features of bloodsucking with most significance for the man–fly contact involved in the epidemiology of onchocerciasis. Biting activity is markedly seasonal in temperate regions (most intense in spring and early summer and absent in winter) but typically occurs year-round in warm tropical areas where there is continuous production of adults. Some species show peaks of biting activity around noon, others early and late in the day, but whatever the innate pattern it can be influenced by weather conditions. Some species are prone to land on the host without necessarily biting (landing rate is distinct from the biting rate) but in onchocerciasis vectors the biting tends to be so determined that the biting rate and the landing rate are virtually the same. Precise assessment of biting rates is very difficult on animal hosts, but on humans the FMH (number of flies settling to bite a man in one hour) is a useful practical measure of biting rate and widely used in onchocerciasis research for assessment of transmission potentials and monitoring fly populations before and after control. Among vectors, the FMH of *S. oyapockense* year-round in the Amazon basin is about 400–500, while that of *S. damnosum* complex in West Africa rarely exceeds 200 and is usually nearer 30–60 at least in the savanna biome; lowest vector biting populations are shown by *S. woodi*, in which the FMH rarely exceeds 10 and was calculated on a year-round basis at Amani (Tanzania) as only 3.2. Much higher biting rates occur in some pest species of temperate areas; *Austrosimulium ungulatum* in New Zealand, for example, often registers an FMH figure as high as 1000.

Not all blackfly species suck blood, even among those with fully formed biting mouthparts, and nearly 40 species are known in which the stylets of the proboscis are too weak and poorly shaped for them to be able to cut and pierce the skin of an animal host. Ordinarily, however, the mechanism of the proboscis is perfectly adapted for bloodsucking on the vertebrate hosts. During engorgement, strong apical teeth on the tip of the labrum stretch the host's skin, while the finely serrate tips of the mandibles incise it, the hypopharynx pierces the wound and injects it with saliva, and the laciniae drive deeply home into the dermis to act as anchors which hold the head tightly down to the skin (Fig. 6.10b). Presence in the saliva of a factor inhibiting coagulation of ingested blood has recently been confirmed (Jacobs et al., 1990). Engorgement usually takes 3–6 minutes, and a fly typically imbibes about its own weight in blood during the feed. Within a few hours the stomach wall secretes around the blood-meal a peritrophic membrane of a strength and thickness which varies between species (and thus could be related to the ease of escape of ingested microfilarial parasites from the fly's stomach into the haemocoel before subsequent development).

Some species feed willingly both on birds and mammals, but most have a strong and sometimes exclusive preference for 'feathers over fur' and have claws adapted to the vestiture of the host. Those that are ornithophilic have a large tooth at the base of the claw which helps them crawl among the feathers and down of their avian hosts, whereas those that are mammalophilic have simple talon-like claws. Intermediates occur, however, and in the *S. damnosum* complex (whose members bite birds *and* mammals) the claws have a tiny peg-like tooth. Thirteen orders of birds and eight orders of mammals (including man for the primates) have been recorded as providing hosts. On mammals there is a strong tendency for different species to feed on particular areas of the host's body. On man, for example, the *S. damnosum* complex mainly chooses the legs, whereas *S. ochraceum* complex prefers the head and torso; on its preferred horse host, *S. equinum* has a strong predilection for the ears, whereas *S. ornatum* on its (usual) cattle host mainly bites the belly. Host location by Simuliidae is reviewed by Sutcliffe (1986).

Oviposition

Females of some species (mainly at high latitudes or high altitudes) do not suck blood and nourish their developing eggs entirely from fat-body reserves accumulated during larval life (autogeny). Most species, however, appear to be anautogenous, using blood proteins obtained by biting host animals for egg maturation – although in some there is a half-way house (primiparous autogeny) in which the first egg batch is produced autogenously. All anautogenous simuliids show gonotrophic concordance, i.e. each egg batch is nourished with the blood ingested during the immediately previous blood-meal. The cycle from blood-meal ingestion to egg laying (gonotrophic or ovarian cycle) is usually completed in 3–7 days, depending to some extent on temperature. The cycle is shortest in warm regions, and a blood-meal to gravidity time as little as 24 hours can occur in *Austrosimulium pestilens* in Queensland. Female blackflies appear to live in nature for up to three or four weeks, so there is potentially ample time for the production of several successive egg batches.

The relatively primitive prosimuliines produce on average small batches of rather large eggs (no more than about 30–90 eggs in arctic *Gymnopais*) but in *Simulium* the

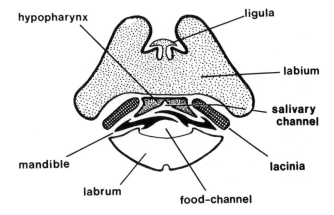

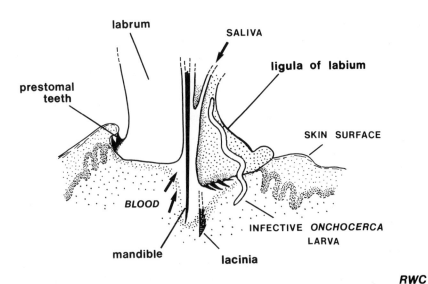

Figure 6.10 The biting mechanism of adult female Simuliidae: (*a*) cross-section of proboscis slightly below its mid-point (semi-diagrammatic); (*b*) diagram of the action of the mouthparts, seen in side view, during bloodsucking (with an infective filarial larva shown escaping into the skin).

smaller eggs are produced at about 200–800 per batch (sometimes more than a thousand) but there is an age-related decline in fecundity. The species, day-length associated with season and latitude, and local weather, all have an influence on oviposition, but in favourable conditions most species lay their eggs late in the day – sometimes aggregating towards sunset on riverside vegetation awaiting an environmental cue before starting to lay their eggs together. Submerged objects can sometimes be thickly encrusted with huge conglomerations of eggs piled indiscriminately, especially in complexes such as *S. ornatum* and *S. damnosum* in which the gravid females seem to be lured to places where other females have laid by some chemical substance given off by the earlier-laid eggs. However, research has still to prove the existence of egg pheromones in Simuliidae.

Most species adopt one of four basic egg-laying methods, namely scattering, dabbing, stringing and layering. Scattering involves laying the eggs on the water surface so that they can sink to the bottom and diapause in the bed sediments. This method is used, for instance, by *Prosimulium* and *Austrosimulium pestilens* to outlast times of drought, and by *S. arcticum* IIS-10.11 (the cattle-biting pest sibling of *S. arcticum* complex) whose eggs survive under the ice of its winter-frozen riverine habitat. Dabbing involves intermittently sticking small groups of eggs on firm substrate, usually the solid bedrock of cascades (as in *S. pictipes*) or large surface-floating leaves (as with *S. ochraceum* complex). Stringing is used by a few *Simulium* species – notably *S. damnosum* and *S. vittatum* complexes – and results from the fly walking about on just-submerged substrate while voiding the eggs; the eggs lie on their sides in looped chains or are heaped up here and there where the fly halted briefly in its perambulations. In the last of the four methods, layering, the fly remains on the substrate while she deposits the whole complement of eggs in one neat layer packed closely together and (often) standing vertically on end. In *S. posticatum* the ovipositional habit is very unusual and not classifiable into one of the basic categories: small groups of eggs are laid above the water-line deep in riverbank cracks and (presumably) hatch when earth is later eroded into the river. Females of some species, even though they normally string or layer the eggs, will crawl down emergent objects into the water and lay well below the surface – even at a depth of several centimetres.

The oviposition habit of the *S. neavei* vector group and other phoretics remains unknown, but it is certain that the eggs are not laid directly on the carriers.

Dispersal

Some primitive blackflies merely flit about on streamside stones, but most *Simulium* (particularly bloodsucking species) are capable of considerable dispersal and some of long-range migrations. Physiological experiments have shown that the innate flying ability (range, duration, speed) is about equal in male and female flies, but males nevertheless disperse little compared to females and their precopulatory swarms usually form well within half a kilometre of the emergence sites and often close to them. On the other hand, the females (energized for flight mainly by plant nectar) often move long distances seeking hosts and oviposition sites – both radially across country and along river courses. Dispersal of bird-biting species has been little studied, but mark-and-recapture methods used on several mammalophilic *Simulium* species have proved them capable of 15–35 km flight ranges. Some species make enormous wind-assisted migrations from the emergence sites, up to 150–225 km in outbreaks of *S. arcticum* complex and *S. luggeri* (the Canadian prairie livestock pests) and as much as 400–600 km in West African savanna siblings of the *S. damnosum* complex which transmit ocular onchocerciasis. Migrations in this complex are linked to seasonal movements of the Intertropical Convergence Zone (ITCZ), and are potentially a threat to vector control because they enable the females to recolonize remote breeding sites – and because the mostly old and parous flies transport *O. volvulus* parasites back into controlled areas. Garms and

Walsh (1988) describe the sequence of events involved in such long-range migration.

Some blackflies occur on remotely isolated volcanic islands never connected to any landmass, presumably as the result of long-range dispersal on the winds.

MEDICAL IMPORTANCE

That humans can suffer serious reactions to blackfly bites began to be documented last century, but the Simuliidae were not suspected of involvement in human disease transmission until 1905. In that year, Louis Sambon first promulgated the '*Simulium* theory of pellagra' – as it came to be called – in which pellagra was supposed to be caused by a *Simulium*-transmitted protozoan blood parasite. Attempts were made between 1906 and 1915 to verify Sambon's theory, especially by the Pellagra Commissions of the central United States (where pellagra was very serious), but no evidence was obtained and by 1916 the theory was finally abandoned. In reality, the wasted body and insanity of pellagra are caused by a specific dietary deficiency. (See Crosskey, 1990, pp. 590–592, for an extended account.)

The only disease confirmed as transmitted among humans by the Simuliidae is onchocerciasis, caused by the filarial worm *Onchocerca volvulus* specific to man. Historically, Rodolfo Robles was first to hypothesize this association (in 1917), observing close coincidence in Guatemala between distribution of the disease and of man-biting *Simulium*: after the relationship was proved, the disease became known in Central American Spanish as the 'Enfermedad de Robles'. Blacklock (1926) obtained first proof that onchocerciasis is transmitted by *Simulium*, working with *S. damnosum* in Sierra Leone.

At different times, *Simulium* has been suspected of involvement with Kaposi's sarcoma (a cancer) in Africa, and with Brazilian pemphigus (an auto-immune disease) but evidence is wanting and a relationship unlikely. There is also no evidence that simuliids transmit viruses biologically, and transmission of virus to man in any manner appears improbable – though simuliids can mechanically transmit myxomatosis virus of rabbits. *Simulium* has been suspected of a minor role in Venezuelan equine encephalomyelitis (VEE) arbovirus transmission during major Caribbean epidemics but this is now thought 'highly unlikely' (Homan et al., 1985).

On the proved side is the transmission (along with *Culicoides* midges, Chapter 7) of *Mansonella ozzardi*, a filarial parasite of man in which the developmental cycle in the vector exactly parallels that of *O. volvulus* and which also has no animal reservoir. Some simuliids of the *S. amazonicum* group are vectors in western Amazonian parts of Colombia and Brazil (Shelley, 1988).

Effects of blackfly bites on humans

The Simuliidae occur worldwide but the man-biting habit has a more limited distribution. It is especially prevalent in eastern North America, Fennoscandia and northern USSR (especially Siberia), the Amazon basin, and in the human onchocerciasis areas of Africa and Latin America, but can be locally intense elsewhere (e.g. New Zealand, Belize, Panama). Major areas notably free from man-biting are

268　*Blackflies (Simuliidae)*

the Mediterranean basin, Middle East, central and southern Asia, and Australia (except sometimes during massive cattle-pest outbreaks) and Andean and temperate South America. Species that can be said to attain pest status are listed in Table 6.2. Genera involved are *Simulium* alone in tropical regions, *Simulium* and (more restrictedly) *Prosimulium* in the northern Holarctic, and *Austrosimulium* in New Zealand and Australia.

Table 6.2　Onchocerciasis vectors and other important man-biting blackflies

Zoogeographical region and species	Principal area as pest or vector	Importance to humans
Afrotropical		
Simulium albivirgulatum	Zaire river basin	Only vector in Cuvette Centrale focus
Simulium damnosum s.str.*	West Africa to southern Sudan, Uganda	Vector (mainly in savannas)
Simulium ethiopiense	Southwestern Ethiopia	Vector (Jimma focus)
*Simulium kilibanum**	Eastern Africa	Vector in some foci (including Ruzizi)
*Simulium mengense**	Cameroun	Vector (in forest–savanna mosaic areas)
Simulium neavei	Uganda and eastern Zaire (formerly also Kenya)	Localized vector (eradicated in Kenya)
*Simulium rasyani**	Yemen	Only vector
*Simulium sanctipauli**	West Africa	Vector (mainly in forest areas)
*Simulium sirbanum**	Trans-Africa from Senegal to southern Sudan, also northern Sudan (Nile)	Vector (in dry savanna areas, also in Abu Hamed focus)
*Simulium soubrense**	West Africa	Vector (mainly in forest areas)
*Simulium squamosum**	West Africa, ?Zaire	Vector (locally in forest–savanna mosaic areas)
Simulium woodi	Tanzania	Vector (in Amani and Uluguru foci)
*Simulium yahense**	West Africa	Vector (locally and mainly in upland forest areas)
'Jimma' sibling*	Ethiopia	Presumed vector (in southwestern foci)
'Kapere' sibling*	Eastern Zaire	Likely vector (Kivu)
'Ketaketa' sibling*	Tanzania	Likely vector (Kilosa)
'Nkusi' sibling*	Tanzania	Likely vector (some small foci)
Australasian		
Austrosimulium australense	New Zealand	Minor to major biting pest (North Island)
Austrosimulium pestilens	Queensland	Biting pest (at times of major outbreaks)
Austrosimulium ungulatum	New Zealand	Major biting pest (South Island)
Simulium buissoni ['no-no' or 'nau-nau']	Marquesas Islands	Major biting pest (only Nuka-Hira Island)
Simulium jolyi	Vanuatu	Localized biting pest
Simulium laciniatum	Fiji	Localized biting pest
Nearctic		
Cnephia pecuarum ['buffalo gnat']	USA (Mississippi basin)	Biting pest in 19th century (human death cause reported), some resurgence in 1980s

Table 6.2 (*cont.*)

Prosimulium mixtum	Eastern Canada and northeastern USA	Major biting pest (spring–summer)
Simulium jenningsi (complex)	Eastern USA	Minor to major biting pest (mainly spring)
Simulium parnassum	Canada and northern USA	Minor to major biting pest
Simulium venustum (complex)	Canada and eastern USA	Major biting pest (spring–summer)
Simulium vittatum (complex)	Iceland, North America	Very severe annoyance
Neotropical		
Simulium amazonicum	Brazilian Amazon and Colombia	Major biting pest (also vector of mansonelliasis)
Simulium argentiscutum	Brazilian Amazon and Colombia	Minor to major biting pest (also vector of mansonelliasis)
Simulium callidum	Guatemala and Mexico	? Secondary vector
Simulium exiguum (complex)	Colombia and Ecuador	Primary vector
	Northern Venezuela	Secondary vector
Simulium guianense	Venezuela–Brazil border region (Amazonia)	Primary vector in highlands, secondary in lowlands
Simulium horacioi	Guatemala	? Secondary vector
Simulium metallicum (complex)	Guatemala and Mexico	Secondary vector
	Northern Venezuela	Primary vector
Simulium ochraceum (complex)	Guatemala and Mexico	Primary vector
Simulium oyapockense	Brazil–Venezuela border region (Amazonia)	Vector in lowlands
	Guyana, eastern Colombia, southern Venezuela	Mansonelliasis vector
Simulium pertinax	Southern Brazil	Locally major biting pest
Simulium quadrivittatum	Belize to Panama	Major biting pest
	Ecuador	Secondary vector
Simulium sanguineum	Northwestern South America	Minor to major biting pest
Oriental		
Simulium indicum	Himalayas	Localized biting pest
Palaearctic		
Simulium arakawae	Japan	Localized biting pest
Simulium cholodkovskii	Siberia	Major biting pest
Simulium colombaschense ['Golubatz fly']	Europe (Danube basin)	Historical: pest during pre-1950s outbreaks
Simulium decimatum	Russia (mainly Urals)	Major biting pest
Simulium erythrocephalum	Europe and USSR	Minor biting pest (occasional important outbreaks in middle Danube basin)
Simulium maculatum	Across mid-USSR	Minor to major biting pest
Simulium ornatum (complex)	Europe and USSR	Minor but widespread biting pest
Simulium posticatum ['Blandford fly']	Southern England	Biting pest considered locally important
Simulium reptans	Northern Europe	Minor but widespread biting pest
Simulium transiens	Across northern USSR	Major biting pest
Simulium truncatum	Fennoscandia	Minor to major biting pest
Simulium tuberosum (complex)	Northern Europe (Scotland included)	Minor to major biting pest

*Members of the *Simulium damnosum* complex.

Some species that hardly ever bite man can be almost unendurable because the females swarm about the body in clouds, intermittently landing and crawling on any exposed skin, or darting into the eyes and ears, mouth and nostrils. The arch-villain with this habit is *Simulium vittatum* complex, superabundant in North America and Iceland. More important, however, are pest species whose bite can produce severe pathological reactions.

The human victim is often unaware at the time of being bitten, since blood-hungry females settle silently on the skin and cutting the wound is often not instantly painful. Unsensitized people are often heavily bitten without being troubled, but in sensitized individuals there can be very severe allergic reactions to a single bite; children, especially, can suffer severely. Most reactions are acute, but severe reactions can be chronic and prolonged for several months. The range of clinical, pathological and histological responses is described in medical literature: see particularly Stokes (1914) and Gudgel and Grauer (1954). Palpable purpuric lesions are typical of blackfly-induced dermatitis (e.g. De Villiers, 1987).

After flies have fed, the site of the bite is usually evidenced by a small haemorrhage. Blood usually oozes from bites of larger temperate-region species, but often not from the bites of small tropical species. Typically the puncture becomes surrounded by a large flat wheal, and in those who do not suffer much from bites this is the only reaction – wheals vanishing without itching; scratching as relief from itching, however, often leads to suppurating pustules and other lesions. Sensitized persons can suffer painful gross swelling of bitten areas (e.g. oedema of the legs, or eyelids so swollen that the sufferer is temporarily blind). Asthmatic symptoms are not unusual.

Repeated biting by blackflies such as *S. venustum* complex in North America, its close relative *S. posticatum* in England, *S. erythrocephalum* in central Europe, and *S. aokii* in Japan, can cause a medically recognized syndrome of headache, feverish sweating and shivering, nausea, swollen and tender lymph glands, acutely aching joints, lassitude and psychological depression. In rural eastern Canada and north-eastern United States, where massive blackfly attack is usual each spring, this is well known as 'blackfly fever'. Here and elsewhere the syndrome can have economic impact on the community, leading to loss of work, disability benefits, extended treatment, and (in a few cases) hospitalization.

Some pre-twentieth century reports exist of human deaths directly caused by blackfly bites, some possibly valid but verification in all cases lacking. In recent times, *Simulium* has been suspected of causing haemorrhagic syndrome of Altamira (HSA) – a cutaneous sickness that has become evident since the early 1970s in settlers along the Trans-Amazon highway in Brazil and has resulted in a few fatalities (Pinheiro et al., 1977). A pyodermatitis-like condition called haemorrhagic exanthem of Bolivia (sometimes fatal) has also been suggestively associated with *Simulium* (Noble et al., 1974), but evidence is no more concrete than for HSA.

Onchocerciasis

Onchocerciasis is a non-fatal dermal and ocular disease caused by the filarial nematode worm *Onchocerca volvulus* and transmitted only by *Simulium*. Man is the definitive host in which the parasite multiplies, *Simulium* the intermediate host in which it undergoes part of its immature life cycle (Fig. 6.11). Larvae of the worm

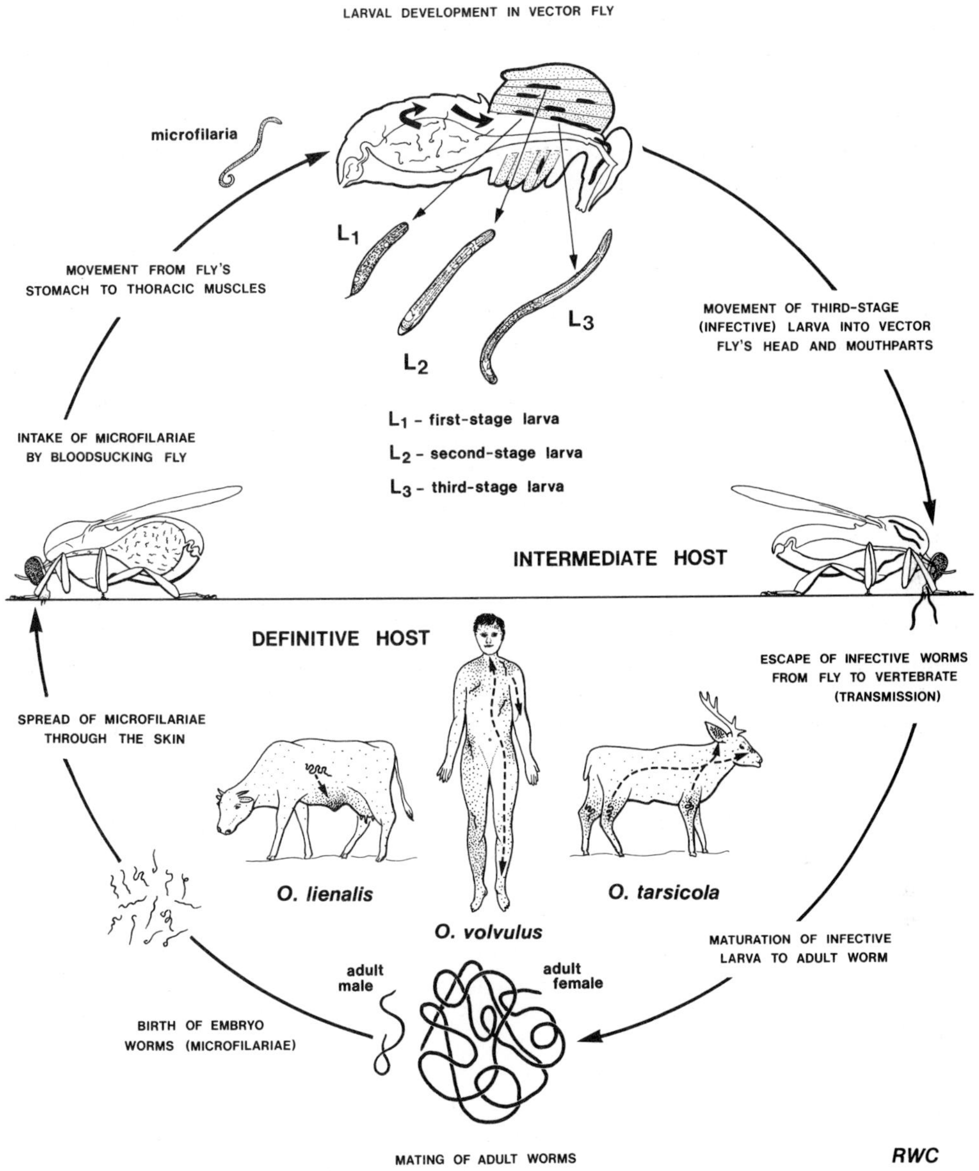

Medical importance 271

Figure 6.11 The life cycle of *Onchocerca volvulus* and other *Onchocerca* species known to have *Simulium* vectors.

occur in the human skin as microfilariae, are ingested by the vector flies when these suck blood, transform within the flies into worms at the infective (L3) stage, and re-enter the human definitive host when the flies take subsequent blood-meals. The cycle is maintained simply by inter-human transmission via the fly: there is no animal reservoir of *O. volvulus* (onchocerciasis is not a zoonosis).

The disease produces much morbidity in the most heavily infected communities and great suffering to the infected individual. It is manifested in three main symptoms: skin lesions associated with the presence of microfilariae in the dermis of the skin, usually accompanied by itching (often very violent); painless nodules produced mainly where the tissues are thin over the bones (knees, ribs, pelvis, scalp); eye lesions which can lead to blindness. Adult *O. volvulus* of both sexes live largely in the nodules, and these are the main sites where (after mating) females give birth to highly active microfilariae (so-called 'embryo worms'). Nodules are thus reproductive centres from which microfilariae disseminate through the skin and (sometimes) ultimately reach the eyes; the small ones can be felt beneath the skin and large ones are outwardly obvious as domed swellings reaching as much as four centimetres in length. The precise cause of eye lesions is uncertain, but they seem to be mainly associated with the death of microfilariae which have penetrated into the eyes from the skin. Once a person is blind the condition cannot be cured. The disease affects only poor rural communities and often goes undetected in areas where it causes little or no blindness.

Onchocercal blindness can be devastating to badly hit communities such as those in the savanna area of West Africa. Here blindness attains an economic level in many villages, where 10% of the people can be blind and up to 30% of the men of productive age. About 95% of all cases worldwide are in Africa, where the disease is endemic over vast areas of tropical western and equatorial Africa between 13°N and 10°S and in many localized foci scattered through eastern and central Africa from Ethiopia to Tanzania; outlier pockets of infection occur also in Malawi, on the Nile in northern Sudan, and in Yemen (Fig. 6.12). The only other area of the world in which the disease occurs is tropical Latin America, where it was presumably historically introduced through transportation from Africa of infected slaves. About a dozen small foci have been discovered in the New World, five in southern Mexico and Guatemala and the rest in northern South America (Venezuela and one focus each in Brazil, Colombia and Ecuador) (Fig. 6.13).

Understandably – since according to the World Health Organization (1987), 85.5 million people in total are at risk from the disease, 17.5 million infected and 340 000 blind – an enormous research effort has been devoted to onchocerciasis and its vectors in modern times and the literature is now vast. Muller and Horsburgh (1987) provide a bibliography citing approximately 2000 works published up to 1985, but unfortunately there is still no comprehensive book on the disease comparable to those available on malaria and filariasis. Manson-Bahr and Bell (1987) and World Health Organization (1987) provide general starting points and Buck (1974) deals with clinical and pathological aspects.

The parasite in vector flies
The biting habits of simuliids make them well adapted for the accidental uptake of the skin-borne microfilariae of *O. volvulus* and for transmission of disease caused by the parasite. Some *Simulium* species are efficient vectors, largely because of six aspects of female fly biology applying to bloodsucking simuliids in general:

1. Pool-feeding habit, which causes laceration of the skin and release of microfilariae into the blood pool.
2. Unwillingness of the fly to be dislodged while feeding.
3. Long engorgement time, which increases the chance of microfilarial uptake.
4. The need for blood-meals every few days.
5. Average fly life span much longer than the time taken for parasite development from ingested microfilaria to infective worm, allowing infective worms to develop from two or more microfilarial intakes.
6. Toleration of considerable levels of parasitization without seriously impaired function.

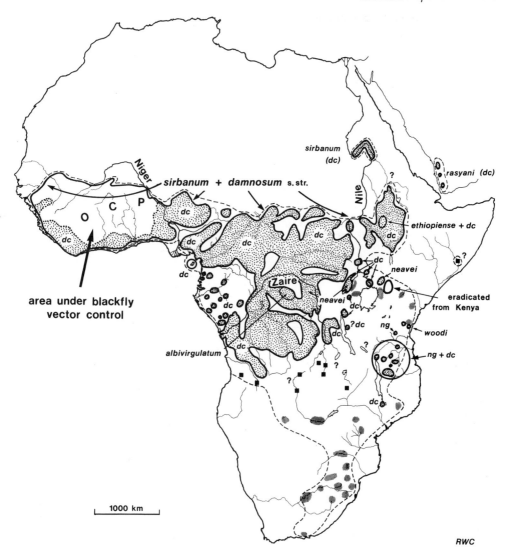

Figure 6.12 The approximate geographical distribution of human onchocerciasis in Africa (shaded areas with heavy outlines). The *Simulium* vectors are named for the principal endemic areas and foci (*dc* = *damnosum* complex, *ng* = unidentified members of *neavei* group). The distribution limits of the *Simulium damnosum* complex are shown by the pecked line: small black squares within this line represent sites where females have been found man-biting but there is no onchocerciasis and shaded patches in southern and eastern Africa indicate areas where the complex occurs but is non-anthropophilic. OCP = area of the Onchocerciasis Control Programme in West Africa.

Another important factor is that each vector species feeds for preference on areas of the body where skin-borne microfilariae are most abundant.

Certain constraints offset the factors just mentioned and lower the chances of disease transmission. The most important are zoophily (taking of blood-meals from animals) and the peritrophic membrane (the hardened sheath secreted around a freshly ingested blood-meal). Although vectors such as the *S. damnosum* and *S. ochraceum* complexes bite humans determinedly, others such as *S. metallicum* sometimes more hesitantly, all vectors sometimes feed on mammals other than man and some of them on birds – reducing the likelihood of *O. volvulus* being imbibed because its microfilariae occur only in the skin of man. Any microfilariae ingested in

Figure 6.13 The geographical distribution of human onchocerciasis in Latin America. The *Simulium* primary vector complex or species is indicated for each focus with the name of any additional (secondary) vector given in parentheses.

human blood immediately attempt to escape into the fly's haemocoel, but most, failing to penetrate the peritrophic membrane as it hardens, simply perish. The few microfilariae that succeed in breaking through the gut wall migrate to the flight muscles of the thorax where (as in other *Onchocerca* parasites) they undergo metamorphosis. In six to nine days (depending partly on temperature) the parasites grow enormously and transform into active third-stage (infective) larval worms which move into the fly's head ready for transmission when it next bites. Infective worms usually escape from the fly via the soft parts of the labium (Fig. 6.10*b*).

Distribution and importance of vectors
From the medical point of view the *Simulium damnosum* complex contains by far the most important blackflies anywhere in the world, being responsible for over 90% of onchocerciasis cases worldwide and more than 95% of cases in Africa. The complex is restricted to Africa (plus Yemen) but is distributed much more widely

than the disease and is entirely zoophilic over much of its vast range (Fig. 6.12). The distribution of anthropophilic members, depending on their identity, is linked either to the forest or to the savanna biome, and certain siblings are therefore associated mainly with transmission of a forest type of onchocerciasis (in which there is usually little blindness despite heavy infection) or with a savanna type (in which there are gross skin changes and much more blindness). *Simulium sirbanum* and *S. damnosum* s.str. are dangerous vectors of ocular onchocerciasis across the northern savannas from Senegal to southern Sudan and Uganda, and are the primary target species for vector control in the OCP area; the first also apparently maintains the isolated focus at Abu Hamed on the middle Nile. Both are long-range migrants able annually to recolonize their breeding sites in large open rivers (e.g. Fig. 6.14b), many of which stop flowing for a prolonged time in the dry season. *Simulium sanctipauli* and its kin (forming a cytological subcomplex) occur widely in forest areas of West Africa, where they are the principal vectors. The more localized *S. squamosum*, however, also plays some part in transmission where it occurs patchily through forest–savanna mosaic areas from Ivory Coast to Zaire. All these members disperse relatively little compared to their savanna counterparts, and usually breed in permanently flowing, smaller and more shaded rivers. They do not occur in eastern Africa, where some isolated foci, mainly in highland areas, are maintained by siblings different from those in West Africa (Table 6.2) – most of them not formally named and all essentially undiagnosed; *S. kilibanum* is probably responsible for the Ruzizi focus in Burundi, but nothing is known of which members maintain the large endemic areas in central and eastern Zaire. In Yemen, *S. rasyani* is the vectorial (and apparently only) member of the complex; its breeding sites are sun-warmed westward-flowing wadis subject to wide variation in flow.

The other African vectors, collectively responsible for probably no more than 5% of cases in Africa, are *S. albivirgulatum*, *S. neavei*, *S. woodi* and *S. ethiopiense*. *Simulium albivirgulatum* is a species of the Zaire basin which breeds in slow to swift but smooth-flowing rivers lined by gallery forest and is a vector only in the Cuvette Centrale focus situated between the Tshuapa and Zaire (Congo) rivers (an area lacking other potential vectors). The other three species belong to the *S. neavei* group and their larvae and pupae develop in obligate phoretic association with river-crabs – thereby collectively providing the only instance in which onchocerciasis is part of an ecosystem involving another organism besides *Simulium*. The crabs occur mainly in turbid forest streams (Fig. 6.14a). Research on the crabs is essential for understanding the biological basis of onchocerciasis transmitted by species of the *S. neavei* group, and for their control (since successful larviciding depends upon not disturbing the crabs). Disease foci with which the *S. neavei* group is associated are localized compared to the large endemic areas with which the *S. damnosum* complex is often associated, and occur in east-central Africa and Zaire (Raybould and White, 1979: review). In some of them, including West Nile District of Uganda and the central highlands of Tanzania, the *S. damnosum* complex plays a combined part in transmission, but in other foci a species of the *S. neavei* group is alone responsible: *Simulium neavei*, for example, maintains the Mount Elgon and Bugoma foci in Uganda, *S. woodi* is the vector at Amani in northeastern Tanzania. The little-studied *S. ethiopiense* is believed to be partly responsible for the Jimma focus in southwestern Ethiopia. *Simulium neavei* was formerly the vector in Kenya but was eradicated from the country in the 1950s by McMahon et al. (1958). A remarkable feature of the *S. neavei* group is the very low fly densities sufficient to

276 *Blackflies (Simuliidae)*

Figure 6.14 Larval habitats of *Simulium* vectors of human onchocerciasis in Africa: (*a*) typical turbid small-stream forest habitat of *S. neavei*; (*b*) typical open 'white water' river habitat of *S. damnosum* s.str. and *S. sirbanum*, principal vectors of ocular onchocerciasis in savanna areas.

maintain disease foci. High biting densities are found at very few places (the well-known *S. neavei* site of Bufumbo on Mount Elgon being one) and at Amani the *S. woodi* biting density hardly ever exceeds 10 FMH even at times of peak activity.

The disease foci in eastern Africa where both *S. damnosum* complex and *S. neavei* group participate in transmission have been too little studied to determine which, if either, is merely a secondary vector whose transmission potential is so low that without the other vector the disease would die out. In some of the small Latin American foci the more complete transmission potential data indicate a distinction in two-vector foci between species/complexes which would keep a focus 'alive' (primary vectors) if the other was eliminated, and those which almost certainly could not (secondary vectors). In the southern Mexican and Guatemalan foci – where onchocerciasis is associated with the coffee-growing communities on forested Pacific-facing slopes with an abundance of tiny streams – *S. ochraceum* complex is the primary and *S. metallicum* complex the secondary vector. On the other hand, in open country of northern Venezuela *S. metallicum* complex has the primary role and *S. exiguum* complex is secondary. *Simulium exiguum* s.l., though, is the only vector in the one small Colombian focus and is the primary vector (with *S. quadrivittatum* secondary) in the focus in Ecuador. Another two-vector situation exists in Amazonia, where the focus straddles the Brazil–Venezuela frontier. Here the disease affects very small, widely scattered and remotely isolated communities of Yanomami Indians. It is transmitted at higher altitudes along the mountainous border by *S. guianense* but at lower elevations on the Brazilian side by *S. oyapockense*. The primary vector breeding sites are quite different in South America from those in Central America: whereas the early stages of *S. ochraceum* complex are found in the most minute watercourses, sometimes in unchannelled water running across the forest floor, those of *S. exiguum* occur mainly at banks of shingle in rivers, and those of *S. oyapockense* in turbid, deep and turbulent water of large Amazonian rivers.

A very rare feature for the Simuliidae is the presence in *S. ochraceum* and *S. quadrivittatum* females of a long jagged cibarial armature where the cibarium joins the pharynx. It has several times been observed that *O. volvulus* microfilariae can become impaled and killed on this device, but it is questionable whether this has any measurable effect on total transmission levels in these vectors.

CONTROL

The large body of information on simuliid control has not been collated and at present the control-orientated chapters in Laird (1981) and Kim and Merritt (1988) provide the most useful general entrée. The following is an outline sketch of control problems and practice.

Control is aimed at eliminating or reducing contact between bloodsucking female blackflies and their hosts. However, this can rarely be done by attacking the adult directly and is usually approached through control of the larval stage in its lotic habitat. Adult flies are susceptible to insecticides, but are usually too dispersed for insecticidal spraying or fogging to achieve more than temporary or very local palliative relief from biting.

Control of the livestock pest blackflies *S. arcticum* and *S. luggeri* (especially the former) has been practised in prairie Canada for many years, but almost all other control measures have been or are directed either at species or complexes important as vectors of onchocerciasis (namely at the *S. damnosum* complex and *S. neavei* in Africa and the *S. ochraceum* complex in Mexico and Guatemala) or at species so distressing to human communities that relief is demanded from an intolerable but seasonal biting problem (e.g. *Prosimulium mixtum* and *S. venustum* complex in eastern Canada and northeastern USA and various *Simulium* species in Siberia). Historically, control falls into three eras: pre-DDT, DDT and post-DDT. In the first, control was thought hopeless and was limited to experimental use of substances (such as kerosene) which might make larvae release their hold. About *S. damnosum* it was said that breeding in river rapids 'proved the futility of attempting to attack the fly in its early stages' (Gibbins and Loewenthal, 1933). The earliest trial with DDT against *Simulium* (in Guatemala in 1944) showed its extraordinary effectiveness when used as a larvicide at very low dosages, and in the 1946–70 DDT era it was used quite successfully in almost all control schemes of the time. Though a cheap and effective means of chemical control, the use of DDT was ruled out for blackfly control after the risk of its accumulation in the food chain became clear and other insecticides are now used in the post-DDT era. The most general is temephos, but chlorphoxim and *Bacillus thuringiensis* H-14 are also in current use. Trials continue with various other compounds.

Control methods

The lotic larval habitat is both a help and a hindrance. On the positive side, linear confinement of larvae in streams and rivers restricts the target area for the insecticide and flow ensures that it will be carried considerable distances downstream to the larval attachment sites when applied at only a few places. On the negative side, the vulnerability of stream and river discharge to changes in weather and climate can pose many practical problems. Nevertheless, all vector control is currently based on insecticides applied to the lotic breeding sites as larvicidal formulations. Little practical use is possible of biological or physical methods of control, though in theory these could be helpful.

The early vector control schemes, of which there were about 40 in 15 countries before the OCP began in the early 1970s, almost all depended on ground application of larvicide, but aerial spraying is the current methodology. It is particularly effective against 'linear' vectors such as *S. sirbanum* which breed in large and open rivers but much less so against 'area' vectors breeding in networks of small streams – and almost useless against *S. ochraceum* complex in which the larvae occur in innumerable trickles protected from aerially applied insecticide by the close forest canopy. In the latter situation, the sheer difficulty of finding and reaching every site makes ground application almost equally useless and *S. ochraceum* is still effectively uncontrolled.

Larvae are susceptible to very low dosages of larvicide (often under 0.1 parts per million) applied for 30 minutes or even less, but applications need to be frequently repeated because larval life is short (usually seven to ten days but a month or more in *S. neavei* on crabs) and chemical insecticides have no effect on eggs or pupae. Use is beginning to be made of *Bacillus thuringiensis* H-14 as a larvicide, since this sporeforming bacterium produces protein crystals which when ingested by filtering larvae release a toxin which so damages the gut wall that death is induced within quite a short time. The filter-feeding habit of larvae is of the utmost importance in relation to the mode of action of all materials used as larvicides (Walsh, 1985: review).

Larval and adult blackflies have many parasites and predators (Crosskey, 1990: review and tables) but no natural enemy is yet in use as a biocontrol agent. Some mermithid worms (notably *Mesomermis flumenalis*) have seemed promising in laboratory experiments but have failed when released into natural rivers and streams. The potential of bacteria, fungi, nematodes, protozoans and viruses for biological control is reviewed by Lacey and Undeen (1988), who stress that even if promising organisms are found their application to rivers used for drinking water supply or recreation might not be publicly acceptable.

Physical elimination of larvae is sometimes possible by removing substrate or by altering river discharge or configuration. Cutting water-weed in spring, for example, can deprive larvae of substrate and produce self-help palliative control. Larval populations of *S. chutteri*, a South African cattle pest, have been temporarily reduced by lowering the river level to induce larval release and then raising it again to flush the larvae away. Deliberate impoundment of rivers to destroy larval sites permanently is rarely possible, but fortuitous elimination of larval sites has resulted from some major civil engineering works: damming of the Danube has permanently drowned the rapids where the notorious *S. colombaschense* pest formerly bred, and dams on the River Niger in Nigeria have flooded the 300-km stretch of rapids upstream of Jebba where *S. damnosum* complex used to breed.

The Onchocerciasis Control Programme in West Africa (OCP)

The only major current vector control scheme is the Onchocerciasis Control Programme in West Africa (OCP, originally suffixed 'in the Volta River Basin area') whose chief targets are *S. sirbanum* and *S. damnosum* s.str. The basic aim is to recover for resettlement and productive agriculture the vast area of savanna West Africa (to the west of Nigeria) in which villages have been deserted and farming abandoned in potentially fertile river valleys because of 'river blindness'. The Programme, sponsored by World Bank and executed by the World Health Organization, started in 1974 with involvement of seven countries, and by 1987 temephos larviciding was taking place on 23 000 km of river over an area of 764 000 km^2. Problems arose quite early on when it was found that parous flies can be long-range migrants able to recolonize cleared rivers from distant permanent reservoir sources. This, combined with some developing resistance to temephos, obliged the Programme to enlarge its area of operations and adapt its strategy, and since 1987 it has expanded westwards and southwards to cover 50 000 km of river in 11 countries over an area of 1.3 million km^2; chlorphoxim and *Bacillus thuringiensis* H-14 are now used as larvicides in addition to temephos (Guillet at al., 1990). The results of vector control on the onchocerciasis have been little short of dramatic: transmission of *Onchocerca volvulus* has ceased over most of the OCP area and children are growing up free from the risk of infection. A full description of the first ten years of the Programme can be found in World Health Organization (1985).

COLLECTING, PRESERVING AND REARING MATERIAL

Being specialized insects, the Simuliidae require some special collecting and handling techniques. These are briefly described here, but readers should also consult Chapter 1 for general information that is relevant, e.g. on storing pinned collections, data labelling, voucher material and specimen mailing.

Larvae and pupae are best removed from their substrates with fine forceps, care being taken with pupae to apply a light side-to-side pressure which will loosen the cocoons and allow them to be picked off without being crushed. For perfect specimens, larvae and pupae should be taken straight from the stream and put immediately into ready-prepared vials of preserving fluid (they deteriorate fast if left for any length of time in water). Shape and pigmentation are often important in larval identification and correct preservation is essential. The best preservative is ethyl alcohol at 80% strength: 70% alcohol is unsatisfactory because water taken in with larvae can produce over-dilution to the point where specimens can partially decompose. Polyvinyl lactophenol and chloral-gum-based media must never be used as they have a disastrous shrivelling effect; formalin should not be used as it has a strong bleaching effect. Larvae collected for chromosome study should be placed briefly on absorbent tissue to blot off stream water and then put into a vial containing freshly mixed Carnoy's fixative (1 part glacial acetic acid eq. : 2–3 parts of absolute ethanol). For the best results, fluid preservative (whether alcohol or Carnoy's) should be changed soon after initial specimen collection and within at most 24 hours. Specimens in Carnoy's should be refrigerated or kept as cool as practicable.

Collection of immatures of the *S. neavei* group is only possible by first collecting the carrier crabs on which they live: McMahon et al. (1958) have described bait-trap and box-trap techniques for this.

For long-term storage of alcohol-preserved material it is essential to submerge the vials or tubes in alcohol-filled jars and obviate the risk of specimens drying out (Fig. 6.15a). Wide-necked jam jars with plastic-lined twist-grip caps, easily available and cheap, are excellent for the purpose. Glass tubes of 50×12.5 mm are the most suited to accommodate typical specimen sample and label sizes.

Adults can be collected in tubes or aspirators from animal hosts, swept from vegetation, or caught in various kinds of sampling trap (Service, 1977, 1981). They can be preserved in alcohol, but this should not be done if identifications are needed: fluid preservation makes certain features difficult to see and gradually decolorizes specimens (thereby destroying important taxonomic features). Adults intended for identification or as voucher material should be micro-pinned through the thorax (in either position indicated in Fig. 6.15e), secured centrally on micro-pins ideally 12.5 mm in length, and staged as shown in Fig. 6.15c. Flies should not be gummed onto a card point, stuck with shellac to the shaft of a large pin, or impaled on the *tip* of a micro-pin, as these often-used methods make flies very vulnerable to loss or damage (especially when in the mail).

For many purposes, especially taxonomy of little-known faunas, it is necessary to rear adults from pupae. This is usually the only practical way to obtain males. A simple procedure which yields high quality flies individually associated with their pupal skins is as follows:

1. Remove the blackest-looking pupae from substrate.
2. Blot off surplus water so pupae are only just damp.
3. Place pupae on a pad of just damp cotton wool tightly packed into bottom of a tube large enough for insertion of a finger (corked glass tubes size 7.5×2.5 cm are recommended) (Fig. 6.15b).
4. Place tubes in a cool and preferably dark place to await fly emergence.
5. Move flies into dry tubes as soon as possible after emergence and keep them alive to harden for about 24 hours.
6. Kill and micro-pin fly.

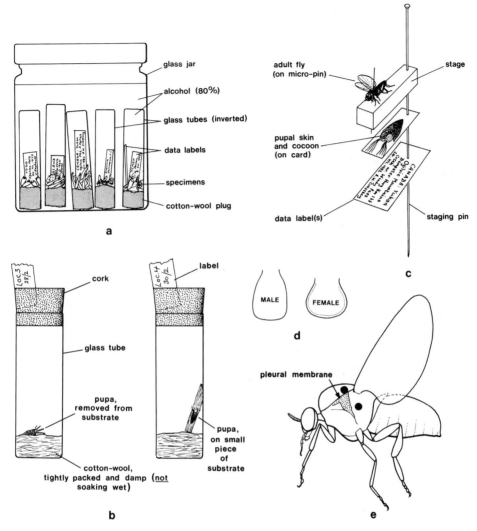

Figure 6.15 Best techniques for preserving and rearing blackflies: (*a*) storage method for alcohol-preserved immature stages; (*b*) rearing method to obtain individual adult flies associated with their pupal skins; (*c*) recommended method for preserving pinned adult flies and their pupal skins; (*d*) head plate shape in pupal skins useful for sex recognition; (*e*) adult fly showing (black circles) recommended alternative positions for micro-pinning – just behind the pleural membrane or just above it (through the scutum).

If results are poor this is usually due to excess dampness in the emergence tubes. These should be inspected frequently and any visible water vapour wiped away; pupae can be tipped out and momentarily surface-dried on tissue- or filter-paper if they appear unduly wet. If pupae are on vegetation this should be trimmed to the minimum to prevent 'sweating' in the tubes (as this spoils emerged flies by causing them to become stuck down by the wings). Pupae should not be placed simply on damp or wet filter-paper in emergence tubes as this dries too quickly in warm conditions unless repeatedly wetted, and emerged flies spoil if they crawl underneath it.

The pupal skin of the emerged fly can be impaled temporarily on the same micropin, but for permanency should be gummed on a piece of card fixed to the staging pin beneath the adult (Fig. 6.15c). Glycerine in microvials is not recommended for preserving pupal skins as it ultimately hardens and makes recovery of skins very difficult. Skin and adult must not be associated if there is any doubt that they belong together. The pupal head plate is differently shaped in males and females (Fig. 6.15d) and this can help in correct association.

If association of pupal skins with adults is not needed, then flies can simply be reared *en masse* by putting pupae on their damp natural substrate (e.g. sticks or grass) into a plastic bag and aspirating out the flies that emerge. Flies can be killed with the usual killing agents such as ethyl acetate and chloroform, or by exposure to heat.

Slide preparations of whole larvae and adults are useless, and specimens should never be sent to specialists as slide material.

Colonization

Flowing or circulating water facsimiles of larval habitat can be established in laboratories fairly easily and several species have been reared through their egg-to-adult life cycles for research connected with chemical insecticides, potential biological control agents and surrogate hosts for *Onchocerca* parasites. Nevertheless, blackflies are recalcitrant laboratory insects and self-perpetuating colonies have not been successfully established – mainly for reasons associated with the failure of most species to mate or blood-feed in captivity. Edman and Simmons (1985) provide a well-referenced review of large-scale rearing systems and the colonization problem.

REFERENCES

Adler, P. H. and **Kim, K. C.** 1986. The black flies of Pennsylvania (Simuliidae, Diptera). Bionomics, taxonomy, and distribution. *Pennsylvania State University College of Agriculture Bulletin* **856**: 1–88.

Blacklock, D. B. 1926. The development of *Onchocerca volvulus* in *Simulium damnosum*. *Annals of Tropical Medicine and Parasitology* **20**: 1–48.

Boakye, D.A., Post, R.J., Mosha, F.W., Surtees, D.P., and **Baker R.H.A.** (1993) Cytotaxonomic revision of the *Simulium sanctipauli* subcomplex (Diptera: Simuliidae) in Guinea and the adjacent countries including descriptions of two new species. *Bulletin of Entomological Research* **83**.

Buck, A. A. (ed.) 1974. *Onchocerciasis: symptomatology, pathology, diagnosis.* 80 pp. World Health Organization, Geneva.

Chubareva, L. A. and **Petrova, N. A.** 1979. Basic characteristics of blackfly karyotypes (Diptera, Simuliidae) in the world fauna. Pp. 58–95 in Skarlato, O. A. (ed.), *Karyosystematics of the invertebrate animals.* 118 pp. Akademiya Nauk SSSR, Leningrad. [In Russian.]

Colbo, M. H. 1976. Four new species of *Simulium* Latreille (Diptera: Simuliidae) from Australia. *Journal of the Australian Entomological Society* **15**: 253–269. [Contains keys to Australian *Simulium*.]

Colbo, M. H. 1989. *Simulium vittatum* (Simuliidae: Diptera), a black fly with a variable instar number. *Canadian Journal of Zoology* **67**: 1730–1732.

Conn, J. 1988. A cytological study of the *Simulium metallicum* complex (Diptera: Simuliidae) from Central and South America. Pp. 221–243 in Service, M. W. (ed.), *Biosystematics of haematophagous insects.* xi + 363 pp. Clarendon Press, Oxford.

Coscarón, S. and **Wygodzinsky, P.** 1972. Taxonomy and distribution of the black fly subgenus *Simulium (Pternaspatha)* Enderlein (Simuliidae, Diptera, Insecta). *Bulletin of the American Museum of Natural History* **147**: 199–240.

Craig, D. A. 1987. A taxonomic account of the black flies (Diptera: Simuliidae) of the Society Islands – Tahiti, Moorea and Raiatea. *Quaestiones Entomologicae* **23**: 372–429.

Crosby, T. K. 1990. Simuliidae (black flies or sandflies). Pp. 67–72 in Winterbourn, M. J. and Gregson, K. L. D., Guide to the aquatic insects of New Zealand. Revised edition. *Bulletin of the Entomological Society of New Zealand* **9**: 1–95.

Crosskey, R. W. 1960. A taxonomic study of the larvae of West African Simuliidae (Diptera: Nematocera) with comments on the morphology of the larval black-fly head. *Bulletin of the British Museum (Natural History) (Entomology)* **10**: 1–74.

Crosskey, R. W. 1967. The classification of *Simulium* Latreille (Diptera, Simuliidae) from Australia, New Guinea and the Western Pacific. *Journal of Natural History* **1**: 23–51.

Crosskey, R. W. 1969. A re-classification of the Simuliidae (Diptera) of Africa and its islands. *Bulletin of the British Museum (Natural History) (Entomology)* Supplement **14**: 1–195.

Crosskey, R. W. 1973. Family Simuliidae. Pp. 423–430 in Delfinado, M. D. and Hardy, D. E. (eds), *A catalog of the Diptera of the Oriental region*: vol. 1, *Suborder Nematocera*. 618 pp. University Press of Hawaii, Honolulu.

Crosskey, R. W. 1980. Family Simuliidae. Pp. 203–210 in Crosskey, R. W. (ed.), *Catalogue of the Diptera of the Afrotropical region*. 1437 pp. British Museum (Natural History), London.

Crosskey, R. W. 1987. A taxa summary for the *Simulium damnosum* complex, with special reference to distribution outside the control areas of West Africa. *Annals of Tropical Medicine and Parasitology* **81**: 181–192.

Crosskey, R. W. 1988. An annotated checklist of the world black flies (Diptera: Simuliidae). Pp. 425–520 in Kim and Merritt [1988].

Crosskey, R. W. 1989. Family Simuliidae. Pp. 221–225 in Evenhuis, N. L (ed.), *Catalog of the Diptera of the Australasian and Oceanian regions*. 1155 pp. Bishop Museum Press, Honolulu, and E. J. Brill, Leiden.

Crosskey, R. W. 1990. *The natural history of blackflies*. ix + 711 pp. John Wiley, Chichester.

Crosskey, R. W. and **Büttiker, W.** 1982. Insects of Saudi Arabia. Diptera: Fam. Simuliidae. *Fauna of Saudi Arabia* **4**: 398–446.

Cupp, E. W. and **Gordon, A. E.** (eds) 1983. Notes on the systematics, distribution, and bionomics of black flies (Diptera: Simuliidae) in the northeastern United States. *Search: Agriculture* **25**: 1–74.

Currie, D. C. 1986. An annotated list of and keys to the immature black flies of Alberta (Diptera: Simuliidae). *Memoirs of the Entomological Society of Canada* **134**: 1–90.

Dalmat, H. T. 1955. The black flies (Diptera, Simuliidae) of Guatemala and their role as vectors of onchocerciasis. *Smithsonian Miscellaneous Publications* **125** (1): vii + 1–425.

Davies, D. M., Peterson, B. V. and **Wood, D. M.** 1962. The black flies (Diptera: Simuliidae) of Ontario. Part I. Adult identification and distribution with descriptions of six new species. *Proceedings of the Entomological Society of Ontario* **92** (1961): 70–154.

Davies, L. 1968. A key to the British species of Simuliidae (Diptera) in the larval, pupal and adult stages. *Freshwater Biological Association Scientific Publication* **24**: 1–126.

De Villiers, P. C. 1987. *Simulium* dermatitis in man – clinical and biological features in South Africa. *South African Medical Journal* **71**: 523–525.

Dumbleton, L. J. 1973. The genus *Austrosimulium* Tonnoir (Diptera: Simuliidae) with particular reference to the New Zealand fauna. *New Zealand Journal of Science* **15** (1972): 480–584.

Dunbar, R.W. and **Vajime, C.G.** 1981. Cytotaxonomy of the *Simulium damnosum* complex. Pp. 31–43 in Laird, M. (ed.), *Blackflies: the future for biological methods in integrated control*. xii + 399 pp. Academic Press, London.

Edman, J. D. and **Simmons, K. R.** 1985. Rearing and colonization of black flies (Diptera: Simuliidae). *Journal of Medical Entomology* **22**: 1–17.

Eymann, M. 1991. Flow patterns around cocoons and pupae of blackflies of the genus *Simulium* (Diptera: Simuliidae). *Hydrobiologia* **215**: 223–229.

Freeman, P. and de Meillon, B. 1953. *Simuliidae of the Ethiopian region*. vii + 224 pp. British Museum (Natural History), London.

Garms, R. and Walsh, J. F. 1988. The migration and dispersal of black flies: *Simulium damnosum* s.l., the main vector of human onchocerciasis. Pp. 201–214 in Kim and Merritt [1988].

Gibbins, E. G. and Loewenthal, L. J. A. 1933. Cutaneous onchocerciasis in a *Simulium damnosum*-infested region of Uganda. *Annals of Tropical Medicine and Parasitology* **27**: 489–496.

Grenier, P. 1953. Simuliidae de France et d'Afrique du Nord (systématique, biologie, importance médicale). *Encyclopédie entomologique* (A) **24**: 1–170.

Gudgel, E. F. and Grauer, F. H. 1954. Acute and chronic reactions to black fly bites (*Simulium* fly). *Archives of Dermatology and Syphilology* **70**: 609–615.

Guillet, P., Kurtak, D. C., Philippon, B. and Meyer, R. 1990. Use of *Bacillus thuringiensis israelensis* for onchocerciasis control in West Africa. Pp. 187–201 in de Barjac, H. and Sutherland, D. J. (eds), *Bacterial control of mosquitoes and black flies: biochemistry, genetics & applications of Bacillus thuringiensis israelensis and Bacillus sphaericus*. xix + 349 pp. Rutgers University Press, New Brunswick.

Homan, E. J., Zuluaga, F. N., Yuill, T. M. and Lorbacher de R. H. 1985. Studies on the transmission of Venezuelan Equine Encephalitis virus by Colombian Simuliidae (Diptera). *American Journal of Tropical Medicine and Hygiene* **34**: 799–804.

Jacobs, J. W., Cupp, E. W., Sardana, M. and Friedman, P. A. 1990. Isolation and characterization of a coagulation factor Xa inhibitor from black fly salivary glands. *Thrombosis and Haemostasis* **64**: 235–238.

Jensen, F. 1984. A revision of the taxonomy and distribution of the Danish black-flies (Diptera: Simuliidae), with keys to the larval and pupal stages. *Natura Jutlandica* **21**: 69–116.

Jobling, B. 1987. *Anatomical drawings of biting flies*. 119 pp. British Museum (Natural History) and Wellcome Trust, London.

Kim, K. C. and Merritt, R. W. (eds) [1988]. *Black flies: ecology, population management, and annotated world list*. xi + 528 pp., '1987'. Pennsylvania State University, University Park and London. [Book dated 1987 on verso of title page: published 1 March 1988.]

Knoz, J. 1965. To [sic] identification of Czechoslovakian black-flies (Diptera, Simuliidae). *Folia Přírodovědecké Fakulty University J. E. Purkyně v Brně* **6** (5): 1–52. [In English with Czech summary: additional to the numbered pages are 90 unnumbered pages of figures, index and contents list.]

Lacey, L. A. and Undeen, A. H. 1988. The biological control potential of pathogens and parasites of black flies. Pp. 327–340 in Kim and Merritt [1988].

Laird, M. (ed.) 1981. *Blackflies: the future for biological methods in integrated control*. xii + 399 pp. Academic Press, London.

Lewis, D. J. 1973. The Simuliidae (Diptera) of Pakistan. *Bulletin of Entomological Research* **62**: 453–470.

Lewis, D. J. and Raybould, J. N. 1974. The subgenus *Lewisellum* of *Simulium* in Tanzania (Diptera: Simuliidae). *Revue de Zoologie africaine* **88**: 225–240.

Manson-Bahr, P. E. C. and Bell, D. R. (eds) 1987. *Manson's tropical diseases*. Nineteenth edition. xvii + 1557 pp. Baillière Tindall, London.

McMahon, J. P., Highton, R. B. and Goiny, H. 1958. The eradication of *Simulium neavei* from Kenya. *Bulletin of the World Health Organization* **19**: 75–107.

Merritt, R. W., Ross, D. H. and Peterson, B. V. 1978. Larval ecology of some lower Michigan black flies (Diptera: Simuliidae) with keys to the immature stages. *Great Lakes Entomologist* **11**: 177–208.

Millest, A. L. 1990. Differences in the larval head patterns and body coloration of members of the *Simulium metallicum* species complex (Diptera: Simuliidae) from Mexico. *Bulletin of Entomological Research* **80**: 191–194.

Muller, R. and Horsburgh, R. C. R. 1987. *Bibliography of onchocerciasis* (1841–1985). Part I – Author Index, vi + 143 pp; Part II – Keyword Index, 292 pp. C.A.B. International Institute of Parasitology. [Parts are separately paginated but bound as one work.]

Noble, J., Valverde, L., Eguia, O. E., Serrate, O. and Antezana, E. 1974. Hemorrhagic exanthem of Bolivia. Studies of an unusual hemorrhagic disease in high altitude dwellers at sea level. *American Journal of Epidemiology* **99**: 123–130.

Patrusheva, V. D. 1982. *Blackflies of Siberia and the Far East (annotated catalogue and handbook of species)*. 321 pp. Izdatel'stvo 'Nauka' Sibirskoe-Otdelenie, Novosibirsk. [In Russian.]

Peterson, B. V. 1970. The *Prosimulium* of Canada and Alaska (Diptera: Simuliidae). *Memoirs of the Entomological Society of Canada* **69**: 1–216.

Peterson, B. V. 1981. Simuliidae. Pp. 355–391 in McAlpine, J. F., Peterson, B. V., Shewell, G. E., Teskey, H. J., Vockeroth, J. R. and Wod, D. M. *Manual of Nearctic Diptera*: vol. 1, vi + 674 pp. Research Branch, Canada Agriculture (Monograph No. 27).

Peterson, B. V. and Dang, P. T. 1981. Morphological means of separating siblings of the *Simulium damnosum* complex (Diptera: Simuliidae). Pp. 45–56 in Laird, M. (ed.), *Blackflies: the future for biological methods in integrated control*. xii + 399 pp. Academic Press, London.

Pinheiro, F. P., Bensabath, G., Rosa, A. P. A. T., Lainson, R., Shaw, J. J., Ward, R., Fraiha, H., Moraes, M. A. P., Gueiros, Z. M., Lins, Z. C. and Mendes, R. 1977. Public health hazards among workers along the Trans-Amazon highway. *Journal of Occupational Medicine* **19**: 490–497.

Post, R. J. and Crampton, J. M. 1988. The taxonomic use of variation in repetitive DNA sequences in the *Simulium damnosum* complex. Pp. 245–256 in Service, M. (ed.), *Biosystematics of haematophagous insects*. xi + 363 pp. Clarendon Press, Oxford.

Post, R. J. and Flook, P. 1992. DNA probes for the identification of members of the *Simulium damnosum* complex (Diptera: Simuliidae). *Medical and Veterinary Entomology* **6**: 379–384.

Raybould, J. N. and White, G. B. 1979. The distribution, bionomics and control of onchocerciasis vectors (Diptera : Simuliidae) in eastern Africa and the Yemen. *Tropenmedizin und Parasitologie* **30**: 505–547.

Rivosecchi, L. 1978. Simuliidae Diptera Nematocera. *Fauna d'Italia* **13**: viii + 1–533.

Robles, R. 1917. Enfermedad nueva en Guatemala. *La Juventud Médica* **17**: 97–115.

Rothfels, K. H. 1979. Cytotaxonomy of black flies (Simuliidae). *Annual Review of Entomology* **24**: 507–539.

Rothfels, K. H. 1981. Cytotaxonomy: principles and their application to some northern species-complexes in *Simulium*. Pp. 19–29 in Laird, M. (ed.), *Blackflies: the future for biological methods in integrated control*. xii + 399 pp. Academic Press, London.

Rothfels, K. H. 1988. Cytological approaches to black fly taxonomy. Pp. 39–52 in Kim and Merritt [1988].

Rubtsov, I. A. 1956. Blackflies (Fam. Simuliidae). *Fauna of the USSR*. New Series. No. 64, *Insects, Diptera* **6** (6). 859 pp. Akademii Nauk SSSR, Moscow and Leningrad. [In Russian: English translation titled *Blackflies (Simuliidae) [Moshki (sem. Simuliidae)]*, xxviii + 1042 pp., Amerind Publishing Co., New Delhi, 1989.]

Rubtsov, I. A. 1959–1964. Simuliidae (Melusinidae). In Lindner, E. (ed.), *Die Fliegen der palaearktischen Region* **14**: 1–689. [In German: work issued irregularly in non-sequentially numbered Lieferungen between 3 March 1959 and 15 May 1964.]

Rubtsov, I. A. 1961. A short guide to bloodsucking blackflies in the fauna of the USSR. *Opredeliteli po Faune SSSR* **77**: 1–227. [In Russian: English translation titled *Short keys to the bloodsucking Simuliidae of the USSR*, vi + 228 pp., Israel Program for Scientific Translations, Jerusalem, 1969.]

Rubtsov, I. A. 1974. On the evolution, phylogeny and classification of the blackfly family (Simuliidae, Diptera). *Trudy Zoologicheskogo Instituta* **53**: 230–281. [In Russian.]

Rubtsov [Rubzov], I. A. and Yankovsky, A. V. 1988. Family Simuliidae. Pp. 114–186 in Soós, A. and Papp, L. (eds), *Catalogue of Palaearctic Diptera*: vol. 3, Ceratopogonidae – Mycetophilidae. 448 pp. Elsevier, Amsterdam.

Service, M. W. 1977. Methods for sampling adult Simuliidae, with special reference to the *Simulium damnosum* complex. *Tropical Pest Bulletin* **5**: 1–48.

Service, M. W. 1981. Sampling methods for adults. Pp 287–296 in Laird, M. (ed.), *Blackflies: the future for biological method in integrated control*. xii + 399pp. Academic Press, London.

Shelley, A. J. 1988. Biosystematics and medical importance of the *Simulium amazonicum* group and the *S. exiguum* complex in Latin America. Pp. 203–220 in Service, M. (ed.), *Biosystematics of haematophagous insects*. xi + 363 pp. Clarendon Press, Oxford.

Shelley, A. J., Pinger, R. R. and Moraes, M. A. P. 1982. The taxonomy, biology and medical importance of *Simulium amazonicum* Goeldi (Diptera: Simuliidae), with a review of related species. *Bulletin of the British Museum (Natural History)* (Entomology) **44**: 1–29.

Smart, J. and Clifford, E. A. 1965. Simuliidae (Diptera) of the Territory of Papua and New Guinea. *Pacific Insects* **7**: 505–619.

Smart, J. and Clifford, E. A. 1969. Simuliidae (Diptera) of Sabah (British North Borneo). *Zoological Journal of the Linnean Society* **48**: 9–47.

Snoddy, E. L. and Noblet, R. 1976. Identification of the immature black flies (Diptera: Simuliidae) of the southeastern U.S. With some aspects of the adult role in transmission of *Leucocytozoon smithi* to turkeys. *South Carolina Agricultural Experiment Station Technical Bulletin* **1057**: 1–58.

Stokes, J. H. 1914. A clinical, pathological and experimental study of the lesions produced by the bites of the 'black fly' (*Simulium venustum*). *Journal of Cutaneous Diseases* **32**: 751–769, 830–856.

Stone, A. 1964. Guide to the insects of Connecticut Part VI. The Diptera or true flies of Connecticut. Ninth fascicle. Simuliidae and Thaumaleidae. *State Geological and Natural History Survey of Connecticut Bulletin* **97**: vii + 1–126.

Stone, A. 1965. Family Simuliidae. Pp. 181–189 in Stone, A., Sabrosky, C. W., Wirth, W. W., Foote, R. H. and Coulson, J. R. (eds), *A catalog of the Diptera of America North of Mexico*. Agriculture Handbook 276, iv + 1696 pp. U.S. Department of Agriculture, Washington, D.C.

Stone, A. and Jamnback, H. A. 1955. The black flies of New York State (Diptera: Simuliidae). *New York State Museum Bulletin* **349**: 1–144.

Stone, A. and Snoddy, E. L. 1969. The black flies of Alabama (Diptera: Simuliidae). *Auburn University Agricultural Experiment Station Bulletin* **390**: 1–93.

Sutcliffe, J. F. 1986. Black fly host location: a review. *Canadian Journal of Zoology* **64**: 1041–1053.

Takaoka, H. 1979. The black flies of Taiwan (Diptera: Simuliidae). *Pacific Insects* **20**: 365–403.

Takaoka, H. 1983. *The blackflies (Diptera: Simuliidae) of the Philippines*. xi + 199 pp. Japan Society for the Promotion of Science, Tokyo. [In English with Japanese summary.]

Takaoka, H. and Roberts, D. M. 1988. Notes on blackflies (Diptera: Simuliidae) from Sulawesi, Indonesia. *Japanese Journal of Tropical Medicine and Hygiene* **16**: 191–219.

Takaoka, H. and Suzuki, H. 1984. The blackflies (Diptera: Simuliidae) from Thailand. *Japanese Journal of Sanitary Zoology* **35**: 7–45. [In English with Japanese summary.]

Terteryan, A. E. 1968. Diptera: blackflies (Simuliidae). *Fauna of the Armenian SSR*. 271 pp. Izdatel'stvo Akademii Nauk Armyanskoi SSR, Erevan. [In Russian.]

Townson, H., Post, R. J. and Phillips, A. 1988. Biochemical approaches to black fly taxonomy. Pp. 24–38 in Kim and Merritt [1988].

Uemoto, K. 1985. Simuliidae. Pp. 323–336 in Kawai, T. (ed.), *An illustrated book of aquatic insects of Japan*. viii + 409 pp. Tokai University Press, Tokyo. [In Japanese.]

Vajime, C. G. and Dunbar, R. W. 1975. Chromosomal identification of eight species of the subgenus *Edwardsellum* near and including *Simulium (Edwardsellum) damnosum* Theobald (Diptera: Simuliidae). *Tropenmedizin und Parasitologie* **26**: 111–138.

Vargas, L. and Díaz Nájera, A. 1957. Simúlidos Mexicanos. *Revista del Instituto de Salubridad y Enfermedades Tropicales* **17**: 143–399.

Vargas, L., Martínez Palacios, A. and Díaz Nájera, A. 1946. Simúlidos de Mexico. Dados sobre sistemática y morfologiá. Descripción de neuvos subgéneros y especies. *Revista del Instituto de Salubridad y Enfermedades Tropicales* **7**: 95–192.

Vulcano, M. A. 1967. *A catalogue of the Diptera of the Americas South of the United States.* 16. Family Simuliidae. 44 pp. Departamento de Zoologia, Secretaria da Agricultura, São Paulo.

Walsh, J. F. 1985. The feeding behaviour of *Simulium* larvae, and the development, testing and monitoring of the use of larvicides, with special reference to the control of *Simulium damnosum* Theobald s. l. (Diptera: Simuliidae): a review. *Bulletin of Entomological Research* **75**: 549–594.

Wirth, W. W. and **Stone, A.** 1956. Aquatic Diptera. Pp. 372–482 in Usinger, R. L. (ed.), *Aquatic insects of California with keys to the North American genera and California species.* ix + 508 pp. University of California Press, Berkeley and Los Angeles.

Wood, D. M. 1978. Taxonomy of the Nearctic species of *Twinnia* and *Gymnopais* (Diptera: Simuliidae) and a discussion of the ancestry of the Simuliidae. *Canadian Entomologist* **110**: 1297–1337.

Wood, D. M., Peterson, B. V., Davies, D. M. and **Györkös, H.** 1963. The black flies (Diptera: Simuliidae) of Ontario. Part II. Larval identification. With descriptions and illustrations. *Proceedings of the Entomological Society of Ontario* **93** (1962): 99–129.

World Health Organization 1985. *Ten years of onchocerciasis control in West Africa. Review of the work of the Onchocerciasis Control Programme in the Volta River Basin area from 1974 to 1984.* Document OCP/GVA/85.1B, 113 pp. World Health Organization, Geneva.

World Health Organization 1987. WHO Expert Committee on Onchocerciasis. Third report. *World Health Organization Technical Report Series* **752**: 1–167.

Wygodzinsky, P. and **Coscarón, S.** 1967. A review of *Simulium (Pternaspatha)* Enderlein (Simuliidae, Diptera). *Bulletin of the American Museum of Natural History* **136**: 47–116.

Wygodzinsky, P. and **Coscarón, S.** 1973. A review of the Mesoamerican and South American black flies of the tribe Prosimuliini (Simuliinae, Simuliidae). *Bulletin of the American Museum of Natural History* **151**: 129–199.

Wygodzinsky, P. and **Coscarón, S.** 1989. Revision of the black fly genus *Gigantodax* (Diptera: Simuliidae). *Bulletin of the American Museum of Natural History* **189**: 1–269.

Xiong, B. and **Kocher, T. D.** 1991. Comparison of mitochondrial DNA sequences of seven morphospecies of black flies (Diptera: Simuliidae). *Genome* **34**: 306–311.

CHAPTER SEVEN

Biting midges (Ceratopogonidae)

John Boorman

The Ceratopogonidae form a family of small nematocerous flies, usually less than 3 or 4 mm in length. They are closely related to the Chironomidae, the non-biting midges, but can be distinguished from them by the presence of biting mouthparts in the female, the wing venation and by the short fore legs (Fig. 7.1). They are best known as 'biting midges' (often, as here, abbreviated to 'midges'), but are also called 'sandflies' (usually, and best, reserved for Phlebotominae), 'no-see-ums', 'punkies' or simply 'biting gnats'. Some 5000 species in 60 or more genera have been described. They are distributed worldwide, with the exception of the Arctic and Antarctic.

The females of most species suck blood, either of vertebrates or other insects (Downes, 1978). Some nectar-feeding species of *Forcipomyia* are of great economic importance as pollinators of tropical crop plants, particularly cocoa (e.g. Winder, 1978).

Members of four genera are known to suck the blood of vertebrate animals. The most important genus is *Culicoides* containing over 1000 species of which about 50 are implicated in the spread of pathogens and parasites to man and other animals. Most species are tiny (1–2 mm, the largest less than 4 mm) and their wings speckled with light and dark patterns; some species, though, have unmarked wings (Fig. 7.4b). In general, those involved in disease transmission have fairly distinctive wing patterns (Figs 7.4, 7.5). The genus *Culicoides* is found from subarctic areas to the south of America, Africa and Australia, but seems to be absent from the extreme south of South America and from New Zealand. *Leptoconops* is a genus with about 120 species of small black flies with milky-white wings found in the tropics and subtropics. Although they are not known to be involved in the spread of disease, they occur in such numbers that they are important biting pests, especially in coastal areas where they can rapidly colonize sandy infilled areas in land development; their numbers, coupled with their habit of biting during the day, can be a major problem to tourism (e.g. Linley and Davies, 1971). *Leptoconops* bites are painful and reactions to them last for several days in sensitive people. *Forcipomyia* contains the subgenus *Lasiohelea* with about 150 species which can be vicious biters of man and livestock; one species transmits cattle filariae in Australia. Finally, *Austroconops* from Western Australia includes only one species, a day-biting midge not known to be of any medical or veterinary significance. Some fossil species and extinct genera are known.

Medical Insects and Arachnids Edited by Richard P. Lane and Roger W. Crosskey.
Published in 1993 by Chapman & Hall ISBN 0 412 40000 6

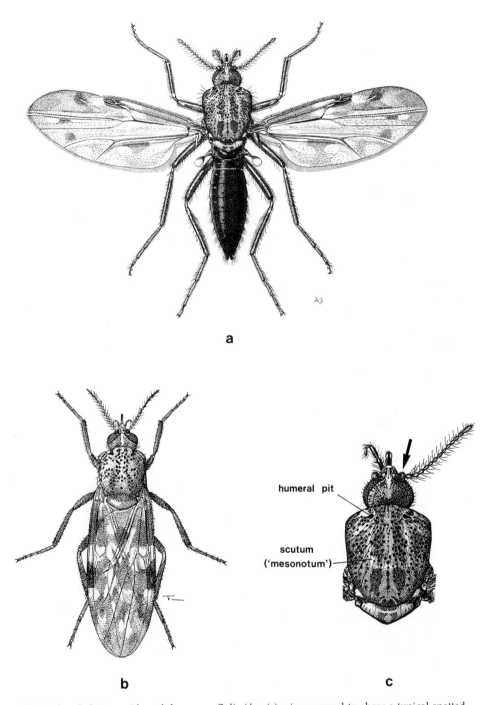

Figure 7.1 A female biting midge of the genus *Culicoides*: (*a*) wings spread to show a typical spotted wing pattern and the abdominal shape; (*b*) resting attitude of the living fly in which the wings are fully closed over the abdomen; (*c*) upper surface of the thorax showing the well-developed humeral pits characteristic of the genus and a representative 'mesonotal' pattern. The head included in (*c*) has one of its antennae to show (arrowed) the swollen pedicel (second segment) visible from above and the long hairy flagellum; one of the maxillary palps is also shown (left).

Biting midges (Ceratopogonidae)

RECOGNITION AND ELEMENTS OF STRUCTURE

The Ceratopogonidae can be distinguished from other families of Nematocera by the following combination of characters:

> Antennae usually with 14 visible segments (occasionally 12, 13 or 15).
> Male antenna usually with prominent plumes.
> Female mouthparts short, adapted for piercing and sucking.
> Legs usually short and stout.
> Wings with the median vein forked and usually with two radial cells (except in the Leptoconopinae).
> When at rest, wings closed flat over the abdomen (Fig. 7.1b).
> Most species brown or black, some orange or yellow.

Midges are identified to species by morphological characters which can often only be seen on slide-mounted specimens (Kremer et al., 1988). The head (Fig. 7.2a) bears a pair of large compound eyes which can be separated or contiguous, and hairy or bare. The mouthparts consist of a rigid labrum-epipharynx, a pair of mandibles and laciniae (= 'maxillae'), both of which usually bear teeth, and a hypopharynx, also with apical teeth. The head/proboscis ratio (H/P ratio) (interocular seta to the tormae/tormae to the tip of the labrum), is often quoted but has limited diagnostic value. The cibarium bears teeth which differ between species in *Lasiohelea* and also occur in some *Culicoides*. The maxillary palps usually have five segments, of which the third is often swollen with a large pit containing sensory setae. The shape of this pit, or sometimes pits, varies between species. The palpal ratio, i.e. length/breadth of the third segment, provides a useful measure of the segment shape. The antenna consists of a basal scape, an enlarged pedicel and 10–13 flagellar segments. In descriptions, all these segments are usually given Roman numerals. In the male, the flagellar segments bear whorls of long setae and the antennae therefore appear

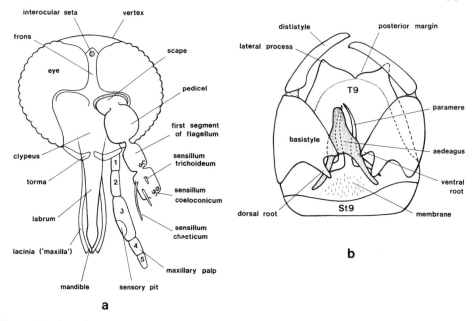

Figure 7.2 Some basic morphological features of *Culicoides* shown semi-diagrammatically: (*a*) head of the female in facial view, with only one maxillary palp and with only the first four segments of one antenna shown; (*b*) genitalia of the male, showing the parts most useful in species identification.

plumose. In the female the distal five segments are elongated; the ratio of the combined length of the apical five to the basal eight segments is termed the antennal ratio and is very useful in identification of species. The position and number of antennal sensilla (Fig. 7.2a) has great taxonomic importance, particularly the sensilla coeloconica and sensilla trichodea ('soies transparentes' of French authors) (Wirth and Navai, 1978: terminology).

The wings have great taxonomic importance. Their useful features are shown in Fig. 7.3b–d. Kremer (1966, p. 28) gives a comparative table of various naming systems that have been used for the veins and cells, but all of these have been superseded by the terminology of McAlpine et al. (1981). The venation in *Leptoconops* is considerably reduced (Fig. 7.3b) and the wings milky-white in life. In *Lasiohelea* the first radial cell is occluded but the second is considerably elongated (Fig. 7.3c) and the wings are densely covered with macrotrichia. In *Culicoides* two subequal radial cells are usually present and there is often a pattern of pale and dark spots (Figs 7.4, 7.5). The first and second costal spots are two convenient reference points on the *Culicoides* wing (Fig. 7.3d). In *Austroconops* the wings are

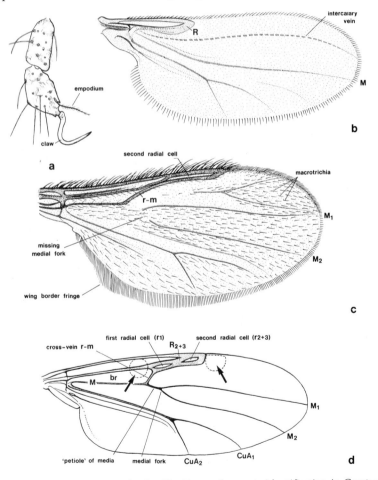

Figure 7.3 Some features important in classification and generic identification in Ceratopogonidae: (a) tip of tarsus in midges of subfamily Forcipomyiinae, showing well-developed empodium and strongly curved claws (only one claw illustrated); (b) wing of typical Leptoconopinae, illustrated by *Leptoconops rhodesiensis* from Zambia; (c) wing of *Forcipomyia* subgenus *Lasiohelea*, illustrated by the Afrotropical species *F. lefanui*; (d) wing venation of *Culicoides*. Notation for the wing veins and cells is that of McAlpine et al. (1981). Arrows on (d) indicate the usual positions of the costal spots.

Figure 7.4 Wings of some important species of *Forcipomyia* and *Culicoides*, female: (a) *F. (Lasiohelea) lefanui*, tropical Africa; (b) *C. inornatipennis*, tropical Africa, species unusual for the genus in lacking wing spots; (c) *C. imicola*, Africa, Mediterranean and Middle East to India and Laos; (d) *C. fulvithorax*, tropical Africa; (e) *C. milnei*, tropical Africa; (f) *C. grahamii*, tropical Africa; (g) *C. puncticollis*, North Africa, Europe and Middle East; (h) *C. pulicaris*, Eurasia. All wings are of female midges.

clear and have two elongate radial cells. Where the wing markings are difficult to see (in slide mounts) they can be enhanced by examination under dark-field illumination. The humeral pits are oval areas of shining cuticle on the 'shoulders' of the thorax which are present in some genera, notably *Culicoides* (Fig. 7.1c). The mesonotum of *Culicoides* often has light and dark areas, and these patterns (more easily seen in dry specimens) are used in species identification (Fig. 7.1c). The legs have little diagnostic value in the bloodsucking species.

Recognition and elements of structure 293

Figure 7.5 Wings of some important species of *Culicoides*: (*a*) *C. oxystoma*, Middle East, Oriental region and Australia; (*b*) *C. brevitarsis*, Oriental and Australasian regions; (*c*) *C. arakawae*, Japan, China, Korea and eastern Russia; (*d*) *C. variipennis*, North America including northern Mexico; (*e*) *C. furens*, eastern seaboard of the USA, Caribbean Islands and Brazil; (*f*) *C. phlebotomus*, Caribbean islands and northern South America; (*g*) *C. insignis*, southeastern USA to Argentina; (*h*) *C. paraensis*, southern USA to South America. All wings are of female midges.

In *Culicoides*, the number of spermathecae differs between subgenera, e.g. one in *Monoculicoides* and *Beltranmyia*, two in *Culicoides* s.str. and *Oecacta*, or three in *Trithecoides* and *Pontoculicoides*. The spermathecae are joined to a common duct and at this point there is often a small sclerotized ring. Taxonomic details of the male genitalia (Fig. 7.2*b*) include the form of the basistyle and dististyle (gonocoxite and gonostyle of some authors), together with their dorsal and ventral roots; the aedeagus and parameres; the posterior margin of the ninth tergite (T9); and the form of the ninth sternite (St9) and the presence or absence of spicules on the ninth sternite membrane. The female genitalia are increasingly recognized as having taxonomic value.

The early stages of Ceratopogonidae are poorly known, but Blanton and Wirth (1979) and Glukhova (1989) describe the chaetotaxy (distribution and size of setae) of both larvae and pupae, and Glukhova (1979) the larvae of many genera of Ceratopogonidae, including *Culicoides*. The eggs of *Culicoides* are banana-shaped and bear minute mushroom-shaped projections called ansulae by which they adhere to the substrate. The larvae of *Culicoides* are slender and nematode-like, without appendages (Fig. 7.6b), and are usually whitish with occasional thoracic pigment patterns. They are aquatic and swim with a characteristic eel-like motion. The head capsule is sclerotized and contains the pharyngeal skeleton, a very useful feature for distinguishing between species. The larvae of *Leptoconops* are semi-aquatic and differ from those of other members of the family by their unsclerotized head capsule. In contrast, the terrestrial larvae of *Forcipomyia* subgenus *Lasiohelea* are found in sandy soil, under rotting bark or in damp moss. They possess prothoracic pseudopods and long, often branched, abdominal hairs which in life have tiny drops of liquid at their tip (Fig. 7.6a). The early stages of most species of this subgenus are still unknown. The pupae are usually hidden in litter or mud, among floating vegetation or, in the case of *Lasiohelea*, under moss or rotting bark. Those of *Leptoconops* species found on beaches occur in the sand at or near the high-water mark.

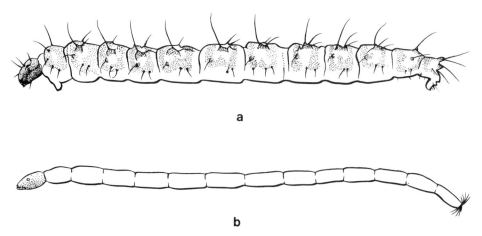

Figure 7.6 Larvae of Ceratopogonidae: (*a*) typical *Forcipomyia* larva, lateral view; (*b*) typical *Culicoides* larva, dorsal view.

CLASSIFICATION AND IDENTIFICATION

The classification and identification of biting midges is based largely on morphological characters. There have not been any detailed cytotaxonomic studies and to date the gross morphology of the chromosomes of few species has been described (e.g. 2n = 6 for *Culicoides*, Hagan and Hartberg, 1986). Enzyme electrophoresis is still in its infancy in this family (e.g. Nunamaker and

McKinnon, 1989) and DNA techniques and cuticular hydrocarbon analysis have yet to be applied.

Higher classification

In the generally accepted higher classification of the family there are four subfamilies: Leptoconopinae, Forcipomyiinae, Dasyheleinae and Ceratopogoninae. Recently the Austroconopinae has been described to accommodate the aberrant genus *Austroconops* (Borkent et al., 1987). The Ceratopogoninae is the largest subfamily and is divided into several tribes, namely Ceratopogonini, Culicoidini, Heteromyiini, Palpomyiini, Sphaeromyiini, Stenoxenini and Stilobezziini.

The established view of ceratopogonid phylogeny is that predaceous habits within the Ceratopogoninae are primitive and that they have led directly or indirectly to the more specialized bloodsucking habits of certain genera. Remm (1975) considered the ancestor as *Palpomyia*-like with progression from entomophagy (insect eating) through bloodfeeding to aphagia (no feeding in the adult stage). In contrast, Downes (1977, 1978) regarded the varied food sources utilized by the Forcipomyiinae as primitive (plesiotypic) for the Ceratopogonidae, progressing through the bloodfeeding habits of the Culicoidini to the hunting entomophagous behaviour of the Palpomyiini and Stenoxenini. A systematic arrangement reflecting this phylogeny and summarized by Wirth et al. (1974), with some changes and additions, is now adopted by most authors. However, the aberrant Leptoconopinae are still a matter of controversy. Krivosheina (1962) treated the Leptoconopinae as a separate family, a stance with which Remm (1975) and Downes (1977) were inclined to agree. Most authors, however, still treat them as members of the Ceratopogonidae, a view adopted here.

Many attempts have been made to erect subgenera within *Culicoides*, and more than 25 subgeneric names have been proposed. Some of these undoubtedly represent natural divisions but many, erected primarily for New World species (see Wirth et al., 1985, 1988), are difficult to apply to the fauna of other parts of the world. For this reason, many authors prefer to arrange the species loosely into groups, or to adopt an alphabetical arrangement. Glukhova (1977) has emphasized that any subgeneric classification of *Culicoides* must take account of larvae as well as adult characters, as has been done in *Forcipomyia* (Saunders, 1956; Wirth and Ratanaworabhan, 1978).

Six subgenera are recognized for *Leptoconops*: *Brachyconops*, *Holoconops*, *Leptoconops*, *Megaconops*, *Proleptoconops* and *Styloconops* (Boorman, 1989a).

Important genera and their identification

There are four genera within the family containing members which suck the blood of vertebrates and so are potential vectors. The key to world genera of Wirth et al. (1974) is somewhat out of date but is convenient and easy to use. Arnaud and Wirth (1964) give worldwide name lists of *Culicoides* species and Boorman (1989a) gives a name list of world species of *Austroconops*, *Leptoconops* and the subgenera *Lasiohelea* and *Dacnoforcipomyia* of *Forcipomyia*. *Dacnoforcipomyia* contains only one species, *F. anabaenae*, which has been recorded biting man in Singapore and Thailand. The species is also known from Queensland, Papua New Guinea and Japan (Debenham, 1983, 1989).

296 *Biting midges (Ceratopogonidae)*

Key to adult flies of ceratopogonid genera with species bloodsucking on vertebrates

It is not practicable to provide an identification key to the many man-biting species of Ceratopogonidae but the following key should assist in recognition of genera and subgenera to which they belong. The subgenus *Dacnoforcipomyia* of *Forcipomyia* cannot be distinguished as adults from *Lasiohelea*, and its only species (*F. anabaenae*, see text) therefore runs out to *Lasiohelea* in the key.

1 Wings with two medial veins (M1 and M2), either united at their bases at fully formed medial fork (Fig. 7.3*d*) or independent of each other because venation missing where fork would be (Fig. 7.3*c*); cross-vein r–m present (Fig. 7.3*c,d*). Antennae of female with 14 or 15 segments.. 2
— Wings with one medial vein running straight to wing margin without forking (Fig. 7.3*b*); cross-vein r–m absent. Antennae of female with 11–13 segments. [Subfamily LEPTOCONOPINAE]... **Leptoconops**
2 Last tarsal segment with large empodium (at least in female) and with strongly curved claws (Fig. 7.3*a*). [Wings with long hairs (macrotrichia) on much of their surface (Fig. 7.3*c*)] [Subfamily FORCIPOMYIINAE] 3
— Last tarsal segment with small or minute empodium and with gently curved or almost straight claws 5
3 Wings with minute microtrichia, these much smaller than abundant macrotrichia covering most of wing (Fig. 7.3*c*); macrotrichia lying flat on wing surface (recumbent) and often scale-like; hairs of hind border fringe in more than one row; costa long or short; first radial cell very narrow (Fig. 7.3*c*) and often obliterated... **Forcipomyia** 4
— Wings with large and conspicuous microtrichia; macrotrichia when present scattered, upright (erect) and never scale-like; hairs of hind border fringe forming one row of alternately long and short straight hairs; costa reaching far beyond middle of wing; second radial cell longer than first and usually open. [Non-bloodsucking midges]... [*Atrichopogon*]
4 Wings very hairy and with very long second radial cell (Fig. 7.3*c*). Cibarium of female with sharp teeth. Female with one spermatheca .... **Forcipomyia** (subgenus **Lasiohelea**)
— Midges without these characters all present together. [Non-bloodsucking midges].. [Other subgenera of *Forcipomyia*]
5 Antennae with smooth (unsculptured) segments. Wings with or without first radial cell; second radial cell present, not markedly square-ended and closed beyond middle of wing. Eyes usually bare. [Subfamilies CERATOPOGONINAE and AUSTROCONOPINAE].................................. 6
— Antennae with sculptured segments. Wings with first radial cell nearly or completely obliterated; second radial cell either obliterated or developed and square-ended, if present closed before middle of wing. Eyes covered with very short hairs (pubescent). [Subfamily DASYHELEINAE, non-bloodsucking midges] .... [*Dasyhelea*]
6 Wings with media dividing into M1 and M2 beyond cross-vein r–m (media therefore with 'petiole', Fig. 7.3*d*); cell m2 sometimes open at base; macrotrichia usually abundant; radial cells usually two, of more or less equal size (as in Fig. 7.3*d*). Legs all with claws of small and equal size in both sexes. Thorax with large and obvious humeral pits (Fig. 7.1*c*) 7

— Wings with media dividing into M1 and M2 at a point basal to cross-vein r–m; cell m2 nearly always complete; macrotrichia scanty or absent; radial cells one or two. Humeral pits small or absent. At least one pair of legs in female with large claws of equal or unequal size, legs of male with small claws of equal size. [Non-bloodsucking] .... [Other genera]

7 Maxillary palps with four segments, only one segment present beyond that which contains sensory pit. Wings without pattern; with two broad radial cells and cross-vein r–m so strongly oblique as to form an almost straight line between base of media and posterior side of radial cells; macrotrichia absent. [Subfamily AUSTROCONOPINAE] [one species only, *A. mcmillani*, south-western Australia].. **Austroconops**

— Palps with five segments, two present beyond third segment containing sensory pit (Fig. 7.2*a*). Wings with pattern (very few species without); with two minute radial cells and cross-vein r–m strongly angled in relation to media (Fig. 7.3*d*); macrotrichia present. [Subfamily CERATOPOGONINAE, tribe CULICOIDINI] [worldwide].. **Culicoides**

Faunal and taxonomic literature

The key-word bibliography of the Ceratopogonidae by Atchley et al. (1981) is of great value as an introduction to the literature. Unfortunately it only covers the years to 1978, but is augmented by Blanton and Wirth (1979) and Wirth and Hubert (1989).

Palaearctic: Catalogue (Remm, 1988); Europe (Havelka, 1978); Arabian peninsula (Boorman, 1989b). Algeria (Szadziewski, 1984); Britain (Boorman, 1986; Campbell and Pelham-Clinton, 1960); Cyprus (Boorman, 1974b); Czechoslovakia (Orszagh, 1980); France (Clastrier, 1973; Kremer, 1966); Israel (Braverman et al., 1976); Japan (Kitaoka, 1984a, b); Morocco (Chaker et al., 1980); Poland (Szadziewski, 1985); Spain and Portugal (Gil Collado and Sahuquillo Herraiz, 1986); Tibet (Lee, 1979); USSR (Glukhova, 1989).

Afrotropical: Catalogue (Wirth et al., 1980); West Africa (Cornet, 1970); East Africa (Glick, 1990). Angola (Callot et al., 1967); Central Africa (Itoua et al., 1989); Madagascar (Callot et al., 1968); Nigeria (Boorman and Dipeolu, 1979); Seychelles (Wirth and Messersmith, 1977); South Africa (Fiedler, 1951); Sudan (Boorman and Mellor, 1982). *Leptoconops* (Clastrier, 1984).

Oriental: Catalogue (Wirth, 1973); South East Asia (Chanthawanich and Delfinado, 1967, *Leptoconops*; Wirth and Hubert, 1989); China (Chu, 1981; Lee, 1980); Taiwan (Lien and Chen, 1983).

Australasian: Catalogue and bibliography (Debenham, 1979, 1989); Polynesia (Wirth and Arnaud, 1969).

Nearctic: Catalogue (Wirth, 1965); review and relevant bibliography (Downes and Wirth, 1981); atlas of wing patterns (Wirth et al., 1985).

Neotropical: Catalogue (Wirth, 1974); South America (Forattini, 1957; Spinelli and Wirth, 1986; Wirth et al., 1988); Haiti (Raccurt, 1984); Trinidad (Aitken et al., 1975); Venezuela (Ramirez Perez, 1984); West Indies (Wirth and Blanton, 1974).

There is no general work on immature stages, but the following will give access to the literature: Palaearctic (Glukhova, 1979, 1989, USSR; Kettle and Lawson, 1952, Britain); Afrotropical (Nevill, 1969, South Africa); Australasian (Kettle and Elson, 1978); Oriental (Wirth and Hubert, 1989, South East Asia). *Forcipomyia* subgenus *Lasiohelea* (Gornostaeva and Gachegova, 1972; Jeu and Rong, 1980); *Leptoconops* (Laurence and Mathias, 1972).

BIOLOGY

Life history

The larvae of *Culicoides* are found in a variety of habitats ranging from moist compost or leaf litter to mud at the margins of ponds and lakes or floating weed; those of veterinary importance are most often found in mud contaminated with animal waste products. Species of *Forcipomyia* (*Lasiohelea*) breed in moss or sandy soil or under bark, and those of *Leptoconops* in sandy soil or on beaches.

There are four larval instars. Many species, especially in more temperate climates, have only one annual generation (are univoltine) but elsewhere there are often several generations each year. Among *Culicoides variipennis* in Colorado the development time can be as little as two weeks and seven generations are produced during the year. Dormancy, either as diapause or aestivation, occurs in either eggs or larvae.

Mermithid nematodes commonly occur in various species. Worms of the genus *Helidomermis*, for example, parasitize *C. variipennis* larvae and up to 51% of a sample can be infected. As in other insects, parasitism can result in the production of intersexed adults that are usually sterile.

Mating

Mating usually takes place while in flight, and the males of most species form 'mating swarms' (Downes, 1978). Some species mate only once, but others several times (Linley and Adams, 1972). In *Culicoides nubeculosus*, sex-attractant pheromones are secreted from a porous ventral area of the abdomen during swarming (Ismail, 1982), whereas a contact mating pheromone is emitted by *C. melleus*. Formation of a packet of sperm (spermatophore), the mode of sperm transfer and activity of the spermatozoa have been described for *C. melleus* by Linley (1981).

Biting behaviour and hosts

Autogeny, the production of eggs without first taking blood, is known in at least 30 species of *Culicoides* and six species of *Leptoconops* (Linley, 1983) and might play a part in population maintenance in the absence of suitable hosts for blood-meals. In *C. impunctatus* for instance, a first batch of eggs is matured autogenously, and the females seek a blood-meal after oviposition. Some midges display a preference for biting on certain areas of the host, especially on large mammals such as cattle. In general, midges bite in still, humid and warm conditions.

The anatomy and histology of the gut of biting midges was described by Megahed (1956). Although most *Culicoides* and *Leptoconops* need a blood-meal in order to mature a batch of eggs, carbohydrates form an important part of the diet (Magnarelli, 1981), certainly in such major vectors as *C. brevitarsis* and *C. variipennis*. The size of the blood-meal in *Leptoconops* is around 2×10^{-4} ml and about twice this in *C. arakawae*; apparently the blood is concentrated by the insect during feeding by rapid diuresis. Several tests have been adapted specifically for determining the minute blood source of midges, e.g. the latex agglutination test.

In many species of midges, parous adults, those that have laid eggs at least once, are distinguished by the development of a dark red pigment in epidermal or subepidermal layers of the abdomen (Dyce, 1969). Survival rates in midges based on parous rates have been studied in Kenya, Britain, Israel and Zimbabwe (references in Braverman et al., 1985).

Oviposition

Up to a couple of hundred eggs are laid at each oviposition. Usually they are deposited on wet mud or leaf litter at the breeding sites. They are unable to withstand desiccation and so must remain damp to be viable. The blood-meal to oviposition (gonotrophic) cycle in some *Leptoconops* species has been described by Whitsel and Schoeppner (1970), in *Culicoides* species by many authors (including Linley, 1966 and Auriault, 1977), and in *Lasiohelea* species by Gornostaeva (1965).

Dispersal and movement

Local dispersal of midges plays an important part in their biology, but it is often over only a short distance (e.g. Murray, 1987). However, *C. variipennis* disperses at least as much as 2.8 km, and *Leptoconops* more than 2 km from a breeding site. The salt-marsh midge *C. mississippiensis* is known to travel more than 3 km in 24 hr without wind assistance. There is increasing circumstantial evidence that infected midges can be borne on the wind over considerable distances, and that the start of some disease outbreaks in animals may be initiated in this way. Bluetongue and Akabane viruses might have been spread to western Turkey by midges from Cyprus, and to eastern Turkey from Syria, respectively (Sellers and Pedgley, 1985) and from Cuba to Florida (Sellers and Maarouf, 1989). The distances which midges are postulated to cover are considerable. Reports exist of 700 km covered in 20 hr at 20°C during an outbreak of African horse-sickness in the Cape Verde Islands in 1944, and of some 110 km covered in 8.5 hr at 12°C and at a height of 1 km in the case of *Culicoides* involved in the spread of bluetongue and epizootic haemorrhagic disease of deer from Washington State to British Columbia in August and October 1987. Midges have been captured both in the upper air and far from land, e.g. *C. schultzei* group midges with an aircraft in Kenya at 1950 m and *C. schultzei* (or more probably *C. oxystoma*) on a ship in the East China Sea.

MEDICAL IMPORTANCE

Role as biting pests

Midges can occur in such numbers that they affect tourism (Linley and Davies, 1971) or forestry. In such circumstances the density of larvae can exceed 10 000 per square metre. Sensitivity to the bites of midges is thought to be responsible for an allergic condition in horses known in Britain as 'sweet itch', in Australia as 'Queensland itch', in Japan as 'kasen' and in the Philippines as 'dhobie itch'. Similar conditions have been reported in Israel, Florida, British Columbia and Iceland. Several species of *Culicoides* are involved: *C. pulicaris* in Britain, *C. robertsi* in Australia, *C. imicola* in Israel and *C. insignis*, *C. stellifer* and *C. venustus* in the USA. Sensitivity of domestic animals to the bites of *Leptoconops* is known in Japan and in South Africa. Allergic dermatitis in sheep in Britain is probably caused by the bites of *C. obsoletus*. Such attacks by pest species can be controlled by the use of insecticides.

Role as disease vectors

Investigations into the role of Ceratopogonidae as vectors of pathogens and parasites have been hampered by their small size and difficulties in their laboratory colonization. In midges biologically capable of acting as vectors of particular pathogens three factors affect their vectorial efficiency: the size of the blood-meal, the frequency of feeding or length of the gonotrophic cycle, and their survival in the wild. The very small blood-meal implies a low infection rate to midges for most pathogens. For example, a common viraemia peak in sheep infected with bluetongue virus is 10^4 to 10^5 infectious units per ml of blood; in a midge taking between 10^{-4} and 10^{-5} ml of blood, the chances of ingesting sufficient virus to infect a midge is small. The same argument applies for filariae; in Sierra Leone infection rates in three species of *Culicoides* infected with presumed *Onchocerca gutturosa* varied from 0.3 down to 0.06%.

The role of ceratopogonids in disease transmission has recently been reviewed by Linley et al. (1983), Linley (1985) and Wirth and Hubert (1989).

Midges of the genus *Culicoides* transmit several protozoans, notably species of *Haemoproteus*, *Hepatocystis* and trypanosomes. Most of these are parasites of birds, though *Hepatocystis kochi* (transmitted by *C. adersi* in eastern Africa) is a parasite of monkeys. Oddly, since *Leucocytozoon* otherwise has only simuliid vectors, *Culicoides* transmits one species of this genus, the *L. caulleryi* parasite of domestic fowl found especially in Japan. Some *Culicoides* also transmit a variety of filarial worms, some of which are parasites of birds and others of monkeys, raccoons, cattle, horses and buffaloes. The filarial parasites include at least four species of *Onchocerca*, namely *O. cervicalis*, *O. gibsoni*, *O. gutturosa* and *O. cebei* (= *O. sweetae*). *Forcipomyia* (*Lasiohelea*) can transmit *Onchocerca gibsoni* to cattle and *Icosiella neglecta* to frogs.

No protozoan blood parasites are transmitted to man by *Culicoides* but some of these midges are vectors of three species of human filariae, namely *Mansonella perstans*, *M. streptocerca* and *M. ozzardi*. *Mansonella perstans* occurs in central and West Africa, South America, Mexico, Trinidad and the Caribbean area but it is not usually pathogenic. In Africa it is transmitted by *Culicoides* of the *C. milnei* group, and by *C. grahamii* and *C. inornatipennis*; the vectors in the New World have not been discovered. There is some doubt whether *M. streptocerca* is pathogenic in man, although occasional skin disorders associated with infection occur. The worm is restricted to tropical Africa where it is probably transmitted by *C. grahamii*.

Mansonella ozzardi is widespread in South America and the Caribbean, where in some areas infection rates over 96% have been recorded. Much of the recent work on mansonelliasis has been carried out in Haiti (Lowrie and Raccurt, 1981) and Trinidad, where *Culicoides furens* and *C. phlebotomus* are the principal vectors.

Culicoides insinuatus is probably a vector in Colombia (Tidwell and Tidwell, 1982). *Culicoides barbosai* and *Leptoconops bequaerti* support the development of the parasite, but are apparently unimportant vectors.

A wide variety of viruses has been isolated from *Culicoides*, but only one, Oropouche virus, is known to be transmitted to man. It is recorded from Trinidad, Colombia and particularly Brazil, where numerous outbreaks have been recorded. It belongs to the Simbu group of viruses. No fatalities from the virus have been recorded, and most patients present with a variety of mild to severe symptoms; meningitis is an occasional complication. The principal vector is *Culicoides paraensis*, but mosquitoes (species of *Aedes* and *Coquillettidia* and *Culex quinquefasciatus*) are also involved.

Rift Valley fever is a serious and often fatal virus disease of animals and occasionally of man in Africa and neighbouring areas. It is known to be mosquito-borne but even though isolations of virus have been made from pools of *Culicoides* species in Kenya and Nigeria there remains some doubt whether *Culicoides* is really a vector. Similarly, the Congo and Dugbe viruses have been isolated from pools of *Culicoides* in Nigeria. In the New World, isolations of Venezuelan and Eastern equine encephalomyelitis have been made from *Culicoides*, and of Western equine encephalomyelitis from *Forcipomyia* (*Lasiohelea*).

Several virus diseases of animals, particularly bluetongue (BT), African horse sickness (AHS), bovine ephemeral fever (BEF) and Akabane (AKA), are transmitted by *Culicoides* and are of great economic significance.

Bluetongue virus occurs, or has occurred, in the United States, Canada, the Caribbean, South America, much of tropical and southern Africa, Spain, Portugal, the Greek Islands, Turkey, the Middle East, through India and Asia to Australasia. It causes acute to chronic symptoms in ruminants, often with high mortality. The epidemiology is complicated by the existence of over 20 serotypes of virus. Wild ruminants or indigenous breeds often show little sign of infection, but imported breeds can rapidly succumb. The major vectors are *Culicoides variipennis* and possibly *C. insignis* in the New World; *C. imicola* and possibly *C. obsoletus* in the Mediterranean and Middle East; and *C. imicola* and possibly also species of the *C. milnei* and *C. schultzei* groups, particularly *C. oxystoma*, in Africa. In Asia and Australasia, species of the subgenus *Avaritia* are known vectors, principally *C. imicola*, *C. brevitarsis*, *C. actoni*, *C. wadai* and *C. fulvus* (Barber and Jochim, 1985; Taylor, 1987).

Epizootic haemorrhagic disease (EHD) is an often fatal disease of deer first recognized in the United States, where it is transmitted by *Culicoides variipennis*. Antibodies to the virus have since been found in livestock in the Middle East, Africa and Asia, and in the Sudan EHD virus has been isolated from *C. oxystoma*. Viruses closely related to EHD have been isolated from *C. brevitarsis* in Australia.

African horse-sickness (AHS) is known from Africa, the Mediterranean basin, the Middle East and the Indian subcontinent. Infection with the virus produces moderate to severe disease in horses, in which it is often fatal. Vaccines are available for its control but (as with bluetongue virus) their use is limited by the existence of several virus serotypes. *Culicoides imicola* is a proven vector.

Bovine ephemeral fever, or three-day fever, is a debilitating disease of cattle which occurs in Africa, the Middle East, Australia (where widespread outbreaks can occur) and Japan. The vectors are probably species of the *Culicoides* subgenus *Avaritia*.

Another virus possibly transmitted by *Culicoides* is Akabane. This sometimes causes severe congenital abnormalities (arthrogryposis and hydranencephaly) in sheep and cattle but is not known to cause disease in humans. Antibodies to it have been found in the Middle East and neighbouring countries, and in Australia. The virus has been isolated from *C. brevitarsis* in Australia and from *C. oxystoma* in Japan.

A number of viruses other than those mentioned above have been isolated from *Culicoides* but have as yet no clear link with clinical disease even though in some instances antibodies have been found in animals. Some of these viruses are: Main Drain, Buttonwillow, Lokern (USA, isolations from *C. variipennis*); Kotonkan, Sabo, Sango, Sathuperi, Shgamonda, Shuni (Nigeria, from pools of mixed *Culicoides* including species of the *C. imicola* and *C. schultzei* groups); and D'Aguilar, Palyam, Warrego, Mitchell River (Australia, Australasia, from various species of *Culicoides* including *C. dycei* and *C. marksi*).

Midges which are not normally susceptible to oral infection with virus become infected if a mixed blood-meal of virus and microfilariae is taken (Mellor and Boorman, 1980). This could increase the number of vectors in areas where filariae of cattle are very common. Trans-ovarian transmission of virus, such as commonly occurs in *Phlebotomus* and in some mosquitoes, has not been demonstrated in *Culicoides*.

CONTROL

'One midge is an entomological curiosity, a thousand can be hell!' (Kettle, 1962). Where they occur in huge numbers, there is no doubt of the impact of midges as biting pests and many control efforts have been directed at them (Blanton and Wirth, 1979; Holbrook, 1985; Linley and Davies, 1971). The earliest attempts at control, before modern insecticides became generally available, involved the use of pyrethrins and oil applied to the breeding sites or land drainage by the construction of ditches and dykes. Environmental management has been used in Florida, Panama, the USSR and in Brazil, but the methods tend to be expensive and depend upon continued support over long periods. Artificial fluctuation in water levels in ponds can reduce the production of *C. variipennis* substantially.

Pest species of the genus *Culicoides* are generally susceptible to a wide range of insecticides and most of the common methods of application have been used, with varied results, in attempts to control both adult and larval populations. Difficulties arise not only in recognition of the (often) widely scattered breeding sites but because of windborne reintroduction of adults into treated areas. The treatment of animals with permethrin or with ivermectin has shown promise for use during outbreaks of animal disease. Other approaches have included the use of repellent-treated screens or netting, or clothing impregnated with repellents, and the use of cattle (the preferred host for *Culicoides imicola*) to protect sheep from infection with bluetongue.

Numerous attempts to control *Leptoconops* have been reported, e.g. by Reynolds and Vidot (1978) against *L. (Styloconops) spinosifrons* in the Seychelles. *Bacillus thuringiensis* seems to be ineffective against *Culicoides* and *Leptoconops*.

Wirth (1977) has reviewed the various parasites and pathogens of biting midges. As yet there is no effective biological control of these pests. Two viruses, an iridovirus and a birnavirus (both from *Culicoides* larvae in France) have been isolated since Wirth's review was published.

COLLECTING, PRESERVING AND REARING MATERIAL

The method employed for collecting biting midges depends upon the type of study in hand. Because of their small size the adult midges cannot be conveniently caught in hand-nets, although with a sharp eye to spot them, females can be collected from man or other hosts by aspirator.

Trapping

For the discovery and study of breeding sites, emergence traps such as those described by Davies (1966) are easy both to make and to use. In the tropics care should be taken to shade the traps to avoid the temperature inside rising to the point where the substrate is effectively sterilized. Larvae and pupae can be separated from their substrates by flotation, either with 30% magnesium sulphate or, preferably, with strong sugar solution which does not damage larvae or pupae.

In general, midges for studies on transmission of pathogens, or surveys for faunistic or taxonomic studies, are best sampled with light-traps. The use of carbon dioxide as an attractant with sticky traps, suction traps or light-traps will increase catches of certain species of both *Culicoides* and *Leptoconops*. Light-traps near animal pens produce numerous recently blood-fed insects, providing useful data on feeding habits if combined with blood-meal analysis. Traps near breeding sites yield useful information on the age structure of the population, and will usually trap greater numbers of males. In general, the more powerful the light source, the greater will be the number of midges caught; black-light lamps will generally attract greater numbers than tungsten lamps. In the tropics, traps can often catch more insects than can be dealt with.

Insects can be collected dry, in a fine-mesh cage, blown directly into a killing jar, or (preferably) blown into a jar containing water or normal (0.9%) saline with a little (0.01%) detergent to wet the insects. To a great extent the choice depends upon the purpose of the catch. For survey purposes, insects collected in detergent are ideal. The catch can either be preserved whole or sorted and the midges stored in 1–2% formalin or 70–80% alcohol. Insects in detergent decompose very rapidly in tropical daytime temperatures and can become useless for detailed study. For virus isolation studies, they should be collected dry, killed by freezing in a −70°C freezer and sorted rapidly over ice, or collected in detergent-saline and sorted as soon as possible after collection, preferably within an hour or two. It should be remembered that many arboviruses are detergent-sensitive, and these will be lost if the insects are collected in this way; but bluetongue and African horse-sickness viruses are detergent-resistant and this method can be used for these viruses. Both of these agents are rapidly destroyed at −20°C, and domestic-type freezers should never be used for storing midges intended for virus isolation studies. Dry ice is also best avoided, since the low pH induced by carbon dioxide can kill any virus in the insects. Liquid nitrogen is ideal for preserving insects for virus studies but supply, storage and transport of liquid gas may present problems under field conditions.

A useful and inexpensive method for catching biting midges for behavioural studies is the 'paddle trap', consisting of a square of mosquito screening greased with castor oil and mounted on a wooden stick.

Preservation

Midges for taxonomic purposes are best slide-mounted in phenol-balsam after dissection, preferably with the wings, head, abdomen and thorax (legs attached) under separate small cover slips. The mesonotum should be sliced off horizontally and mounted dorsal side up with the rest of the thorax. The abdomen should be orientated ventral side uppermost to display the genitalia. Other mountants, such as gum-chloral, are useful for temporary preparations but should not be used for permanent storage. Unmounted midges should be preserved in 70% or 80% alcohol or 1% or 2% formalin (i.e. a 1:100 dilution of concentrated formaldehyde solution). They should be stored in the dark because they are rapidly bleached by exposure to light. Larvae can be mounted in phenol-balsam after clearing in lactophenol (Chaker, 1982), and preserved in alcohol or formalin.

Rearing

Eggs are obtained from gravid females maintained in a damp chamber with sugar solution as a carbohydrate source. Females can be induced to lay eggs on damp filter paper by decapitating them after light anaesthesia with carbon dioxide.

Several species of *Culicoides* have been successfully maintained as laboratory colonies, e.g. *C. furens*, *C. arakawae*, *C. schultzei*, *C. riethi*, *C. nubeculosus* and *C. variipennis* (Boorman, 1974a). Attempts to colonize *Leptoconops kerteszi* have been unsuccessful but *Forcipomyia* (*Lasiohelea*) *taiwana* has been colonized. As yet none of the important vector species of the subgenus *Avaritia* has been colonized, and this remains a considerable barrier to assessing their relative importance as virus vectors. In the laboratory the later instars of many species can be reared on cultures of small nematodes on an agar substrate (e.g. Kettle et al., 1975).

REFERENCES

Aitken, T. H. G., Wirth, W. W., Williams, R. W., Davies, J. B. and **Tikasingh, E. S.** 1975. A review of the bloodsucking midges of Trinidad and Tobago, West Indies (Diptera: Ceratopogonidae). *Journal of Entomology* (B) **44**: 101–144.

Arnaud, P. H. and **Wirth, W. W.** 1964. A name list of world *Culicoides*, 1956–1962 (Diptera, Ceratopogonidae). *Proceedings of the Entomological Society of Washington* **66**: 19–32.

Atchley, W. R., Wirth, W. W., Gaskins, C. T. and **Strauss, S. L.** 1981. A bibliography and keyword index of the biting midges (Diptera: Ceratopogonidae). *United States Department of Agriculture Bibliographies and Literature of Agriculture* **13**: iv + 1–544.

Auriault, M. 1977. Contribution à l'étude biologique et écologique de *Culicoides grahamii* (Austen), 1909, II – Cycle gonotrophique (Diptera, Ceratopogonidae). *Cahiers ORSTOM* (Entomologie médicale et Parasitologie) **15**: 177–184.

Barber, T. L. and **Jochim, M. M.** (eds) 1985. *Bluetongue and related orbiviruses.* Progress in Clinical and Biological Research 178, 746 pp. Alan J. Liss, New York.

Blanton, F. S. and **Wirth, W. W.** 1979. *Arthropods of Florida and neighbouring land areas. 10. The sandflies (Culicoides) of Florida (Diptera: Ceratopogonidae).* 204 pp. Florida Department of Agriculture and Consumer Services, Gainesville, Florida.

Boorman, J. 1974a. The maintenance of laboratory colonies of *Culicoides variipennis* (Coq.), *C. nubeculosus* (Mg.) and *C. riethi* Kieff. (Diptera, Ceratopogonidae). *Bulletin of Entomological Research* **64**: 371–377.

Boorman, J. 1974b. *Culicoides* (Diptera, Ceratopogonidae) from Cyprus. *Cahiers ORSTOM* (Entomologie médicale et Parasitologie) **12**: 7–13.

Boorman, J. 1986. British *Culicoides* (Diptera: Ceratopogonidae): notes on distribution and biology. *Entomologist's Gazette* **37**: 253–266.

Boorman, J. 1989a. A name list of world *Austroconops, Leptoconops* and *Forcipomyia* (subgenera *Lasiohelea* and *Dacnoforcipomyia*) to 1985 (Diptera: Ceratopogonidae). *Cahiers ORSTOM* (Entomologie médicale et Parasitologie) **25** (1987: No. spécial): 53–62.

Boorman, J. 1989b. *Culicoides* (Diptera: Ceratopogonidae) of the Arabian peninsula with notes on their medical and veterinary importance. *Fauna of Saudi Arabia* **10**: 160–224.

Boorman, J. and **Dipeolu, O. O.** 1979. A taxonomic study of adult Nigerian *Culicoides* Latreille (Diptera: Ceratopogonidae) species. *Entomological Society of Nigeria Occasional Publication* **22**: 1–121.

Boorman, J. and **Mellor, P. S.** 1982. Notes on *Culicoides* (Diptera, Ceratopogonidae) from the Sudan in relation to the epidemiology of bluetongue virus disease. *Revue d'Élevage et de Médecine Vétérinaire des Pays Tropicaux* **35**: 173–178. [In this work the initials of the first author are given as J. P. T. instead of the usual J.]

Borkent, A., Wirth, W. W. and **Dyce, A. L.** 1987. The newly discovered male of *Austroconops* (Ceratopogonidae: Diptera) with a discussion of the phylogeny of the basal lineages of the Ceratopogonidae. *Proceedings of the Entomological Society of Washington* **89**: 587–606.

Braverman, Y., Boorman, J., Kremer, M. and **Delecolle, J. C.** 1976. Faunistic list of *Culicoides* (Diptera, Ceratopogonidae) from Israel. *Cahiers ORSTOM* (Entomologie médicale et Parasitologie) **14**: 179–185. [In this work the name of the first author is mispelled Bravermann on the title page.]

Braverman, Y., Linley, J. R., Marcus, R. and **Frish, K.** 1985. Seasonal survival and expectation of infective life of *Culicoides* spp. (Diptera: Ceratopogonidae) in Israel, with implications for bluetongue virus transmission and a comparison of the parous rate of C. *imicola* from Israel and Zimbabwe. *Journal of Medical Entomology* **22**: 476–484.

Callot, J., Kremer, M. and **Brunhes, J.** 1968. Étude de *Styloconops spinosifrons* et de *Culicoides* entomophages (Diptères Cératopogonidés) dont certains sont nouveaux pour la faune de Madagascar. *Cahiers ORSTOM* (Entomologie médicale et Parasitologie) **6**: 103–112.

Callot, J., Kremer, M. and **Molet, B.** 1967. Cératopogonidés (Diptères) de la région éthiopienne et particulièrement d'Angola (description d'espèces et de formes nouvelles). *Publicações Culturais da Companhia de Diamantes de Angola* **71**: 37–44.

Campbell, J. A. and **Pelham-Clinton, E. C.** 1960. A taxonomic review of the British species of *Culicoides* Latreille (Diptera, Ceratopogonidae). *Proceedings of the Royal Society of Edinburgh* (B) **67**: 181–302.

Chaker, E. 1982. *Culicoides* larvae; methods of study and mounting techniques. *Mosquito News* **42**: 517.

Chaker, E., Bailly-Choumara, H. and **Kremer, M.** 1980. Sixième contribution à l'étude faunistique des *Culicoides* du Maroc (Diptera, Ceratopogonidae). *Bulletin de l'Institut Scientifique* (Rabat) **4** (1979–1980): 81–86.

Chanthawanich, N. and **Delfinado, M. D.** 1967. Some species of *Leptoconops* of the Oriental and Pacific regions (Diptera, Ceratopogonidae). *Journal of Medical Entomology* **4**: 294–303.

Chu, F.-i. 1981. On the blood-sucking midges (Diptera, Ceratopogonidae) from the coastal regions of south-eastern China. *Acta Entomologica Sinica* **24**: 307–313. [In Chinese with English summary.]

Clastrier, J. 1973. Le genre *Leptoconops*, sous-genre *Holoconops* dans le Midi de la France (Dipt. Ceratopogonidae). *Annales de la Société Entomologique de France* (n.s.) **9**: 895–920.

Clastrier, J. 1984. Révision des espèces afrotropicales du sous-genre *Leptoconops* (s. str.) (Diptera, Ceratopogonidae). *Annales de Parasitologie humaine et comparée* **59**: 297–316.

Cornet, M. 1970. Les *Culicoides* (Diptera Ceratopogonidae) de l'Ouest africain (1ère note). *Cahiers ORSTOM* (Entomologie médicale et Parasitologie) **7** (1969): 341–364.

Davies, J. B. 1966. An evaluation of the emergence or box trap for estimating sand fly (*Culicoides* spp.: Heleidae) populations. *Mosquito News* **26**: 69–72.

Debenham, M. L. 1979. An annotated checklist and bibliography of Australasian region Ceratopogonidae (Diptera, Nematocera). *University of Sydney School of Public Health and Tropical Medicine Monograph Series* (Entomology) **1** (1978): xiv + 1–671.

Debenham, M. L. 1983. Australian species of the blood-feeding *Forcipomyia* subgenera, *Lasiohelea* and *Dacnoforcipomyia* (Diptera: Ceratopogonidae). *Australian Journal of Zoology* (Supplementary Series) **95**: 1–61.

Debenham, M. L. 1989. Family Ceratopogonidae. Pp. 226–251 in Evenhuis, N. L. (ed.), *Catalog of the Diptera of the Australasian and Oceanian regions*. 1155 pp. Bishop Museum Press, Honolulu, and E. J. Brill, Leiden.

Downes, J. A. 1977. *Leptoconops* – Leptoconopidae? *Mosquito News* **37**: 277.

Downes, J. A. 1978. Feeding and mating in the insectivorous Ceratopogonidae (Diptera). *Memoirs of the Entomological Society of Canada* **104**: 1–62.

Downes, J. A. and Wirth, W. W. 1981. Ceratopogonidae. Pp. 383–421 in McAlpine, J. F., Peterson, B. V., Shewell, G. E., Teskey, H. J., Vockeroth, J. R. and Wood, D. M., *Manual of Nearctic Diptera*: vol. 1, vi + 674 pp. Research Branch, Agriculture Canada (Monograph No. 27).

Dyce, A. L. 1969. The recognition of nulliparous and parous *Culicoides* (Diptera: Ceratopogonidae) without dissection. *Journal of the Australian Entomological Society* **8**: 11–15.

Fiedler, O. G. H. 1951. The South African biting midges of the genus *Culicoides* (Ceratopogonid., Dipt.). *Onderstepoort Journal of Veterinary Research* **25** (2): 3–33.

Forattini, O. P. 1957. *Culicoides* da Região Neotropical (Diptera. Ceratopogonidae). *Arquivos da Faculdade de Higiene e Saúde Pública da Universidade de São Paulo* **11**: 161–526.

Gil Collado, J. and Sahuquillo Herraiz, C. 1986. Claves de identificación de Ceratopogonidae de España peninsular II. Subfamilia Ceratopogoninae (Dip., Nematocera). *Graellsia* **41** (1985): 43–63.

Glick J. I. 1990. *Culicoides* biting midges (Diptera: Ceratopogonidae) of Kenya. *Journal of Medical Entomology* **27**: 87–195.

Glukhova, V. M. 1977. On the subgeneric classification of the genus *Culicoides* Latreille, 1809 (Diptera, Ceratopogonidae) considering morphological structures of the larval stage. *Parazitologichesky Sbornik* **27**: 112–118. [In Russian.]

Glukhova, V. M. 1979. *Biting midge larvae of the subfamilies Palpomyiinae and Ceratopogoninae in the fauna of the USSR (Diptera, Ceratopogonidae = Heleidae)*. 230 pp. 'Nauka', Leningrad. [In Russian.]

Glukhova, V. M. 1989. Bloodsucking midges of the genera *Culicoides* and *Forcipomyia* (Ceratopogonidae). *Fauna of the USSR*. New Series. No. 139, *Insects, Diptera* **3** (5). 406 [+ 2] pp. 'Nauka', Leningrad. [In Russian, unnumbered p. 407 provides English summary.]

Gornostaeva, R. M. 1965. The gonotrophic cycle of the biting midge *Lasiohelea sibirica* Bujan (Diptera, Heleidae). *Entomologicheskoe Obozrenie* **44**: 770–784. [In Russian: English translation in *Entomological Review*, Washington **44**: 454–461.]

Gornostaeva, R. M. and Gachegova, T. A. 1972. Preimaginal stages of *Lasiohelea sibirica* Bujanova (Ceratopogonidae). *Parazitologiya* **6**: 107–117. [In Russian.]

Hagan, D. V. and Hartberg, W. K. 1986. Preliminary observations on the mitotic chromosomes of *Culicoides variipennis* (Diptera: Ceratopogonidae). *Journal of Medical Entomology* **23**: 334–335.

Havelka, P. 1978. Ceratopogonidae. Pp. 441–458 in Illies, J. (ed.), *Limnofauna Europaea*. Second edition. xvii + 532 pp. Fischer, Stuttgart and New York, Swets and Zeitlinger, Amsterdam.

Holbrook, F. R. 1985. An overview of *Culicoides* control. Pp. 607–609 in Barber, T. L. and Jochim, M. M. (eds), *Bluetongue and related orbiviruses*. Progress in Clinical and Biological Research 178, 746 pp. Alan J. Liss, New York.

Ismail, M.-T. 1982. Factors affecting pheromone secretion and oviposition by fertilized females of *Culicoides nubeculosus* (Diptera: Ceratopogonidae). *Journal of Insect Physiology* **28**: 835–840.

Itoua, A., Cornet, M., Vattier-Bernard, G. and Trouillet, J. 1989. Les *Culicoides* (Diptera, Ceratopogonidae) d'Afrique Centrale. *Cahiers ORSTOM* (Entomologie médicale et Parasitologie) **25** (1987, No. spécial): 127–134.

Jeu, M.-h. and Rong, Y.-l. 1980. Morphological descriptions of the immature stages of *Forcipomyia (Lasiohelea) taiwana* (Shiraki) (Diptera, Ceratopogonidae). *Acta Entomologica Sinica* **23**: 66–75. [In Chinese with English summary.]

Kettle, D. S. 1962. The bionomics and control of *Culicoides* and *Leptoconops* (Diptera, Ceratopogonidae – Heleidae). *Annual Review of Entomology* **7**: 401–418.

Kettle, D. S. and Elson, M. M. 1978. Immature stages of more Australian *Culicoides* Latreille (Diptera: Ceratopogonidae). *Journal of the Australian Entomological Society* **17**: 171–187.

Kettle, D. S. and Lawson, J. W. H. 1952. The early stages of British biting midges *Culicoides* Latreille (Diptera: Ceratopogonidae) and allied genera. *Bulletin of Entomological Research* **43**: 421–467.

Kettle, D. S., Wild, C. H. and Elson, M. M. 1975. A new technique for rearing individual *Culicoides* larvae (Diptera: Ceratopogonidae). *Journal of Medical Entomology* **12**: 263–264.

Kitaoka, S. 1984a. Japanese *Culicoides* (Diptera: Ceratopogonidae) and keys for the species. I. *Bulletin of the National Institute of Animal Health* **87**: 73–89. [In Japanese.]

Kitaoka, S. 1984b. Japanese *Culicoides* (Diptera: Ceratopogonidae) and keys for the species. II. *Bulletin of the National Institute of Animal Health* **87**: 91–108. [In Japanese.]

Kremer, M. 1966. Contribution à l'étude du genre *Culicoides* Latreille particulièrement en France. *Encyclopédie Entomologique* (A) **39** (1965): 3–299.

Kremer, M., Waller, J. and Delecolle, J. C. 1988. Systématique des *Culicoides* (Diptères, Ceratopogonides). Critères actuels. *Bulletin de la Société Française de Parasitologie* **5** (1987): 123–132.

Krivosheina, N. P. 1962. Preimaginal stages of *Leptoconops (Holoconops) borealis* Gutz. and the systematic position of the group *Leptoconops* (Diptera, Nematocera). *Zoologichesky Zhurnal* **41**: 247–251. [In Russian.]

Laurence, B. R. and Mathias, P. L. 1972. The biology of *Leptoconops (Styloconops) spinosifrons* (Carter) (Diptera, Ceratopogonidae) in the Seychelles islands, with descriptions of the immature stages. *Journal of Medical Entomology* **9**: 51–59.

Lee, T.-s. 1979. Biting midges of Tibet, China (Diptera: Ceratopogonidae). *Acta Entomologica Sinica* **22**: 98–107. [In Chinese with English summary.]

Lee, T.-s. 1980. *Culicoides* of Yunnan Province, China (Diptera, Ceratopogonidae). *Acta Zootaxonomica Sinica* **5**: 85–88. [In Chinese with English summary.]

Lien, J.-c. and Chen, C.-s. 1983. Seasonal succession of some common species of the genus *Culicoides* (Diptera, Ceratopogonidae) in eastern Taiwan. *Journal of the Formosan Medical Association* **82**: 399–409.

Linley, J. R. 1966. The ovarian cycle in *Culicoides barbosai* Wirth & Blanton and *C. furens* (Poey) (Diptera: Ceratopogonidae). *Bulletin of Entomological Research* **57**: 1–17.

Linley, J. R. 1981. Emptying of the spermatophore and spermathecal filling in *Culicoides melleus* (Coq.) (Diptera: Ceratopogonidae). *Canadian Journal of Zoology* **59**: 347–356.

Linley, J. R. 1983. Autogeny in the Ceratopogonidae: literature and notes. *Florida Entomologist* **66**: 228–234.

Linley, J. R. 1985. Biting midges (Diptera: Ceratopogonidae) as vectors of nonviral animal pathogens. *Journal of Medical Entomology* **22**: 589–599.

Linley, J. R. and Adams, G. M. 1972. A study of the mating behaviour of *Culicoides melleus* (Coquillett) (Diptera: Ceratopogonidae). *Transactions of the Royal Entomological Society of London* **124**: 81–121.

Linley, J. R. and Davies, J. B. 1971. Sandflies and tourism in Florida and the Bahamas and Caribbean area. *Journal of Economic Entomology* **64**: 264–278.

Linley, J. R., Hoch, A. L. and Pinheiro, F. P. 1983. Biting midges (Diptera: Ceratopogonidae) and human health. *Journal of Medical Entomology* **20**: 347–364.

Lowrie, R. C. and Raccurt, C. 1981. *Mansonella ozzardi* in Haiti. II. Arthropod vector studies. *American Journal of Tropical Medicine and Hygiene* **30**: 598–603.

Magnarelli, L. A. 1981. Parity, follicular development, and sugar feeding in *Culicoides melleus* and *C. hollensis* (Diptera, Ceratopogonidae). *Environmental Entomology* **10**: 807–811.

McAlpine, J. F., Peterson, B. V., Shewell, G. E., Teskey, H. J., Vockeroth, J. R. and **Wood, D. M.** 1981. *Manual of Nearctic Diptera*: vol. 1, vi + 674 pp. Research Branch, Agriculture Canada (Monograph No. 27).

Megahed, M. M. 1956. Anatomy and histology of the alimentary tract of the female of the biting midge *Culicoides nubeculosus* Meigen (Diptera: Heleidae = Ceratopogonidae). *Parasitology* **46**: 22–47.

Mellor, P. S. and **Boorman, J.** 1980. Multiplication of bluetongue virus in *Culicoides nubeculosus* (Meigen) simultaneously infected with the virus and the microfilariae of *Onchocerca cervicalis* (Railliet & Henry). *Annals of Tropical Medicine and Parasitology* **74**: 463–469.

Murray, M. D. 1987. Local dispersal of the biting-midge *Culicoides brevitarsis* Kieffer (Diptera: Ceratopogonidae) in south-eastern Australia. *Australian Journal of Zoology* **35**: 559–573.

Nevill, E. M. 1969. The morphology of the immature stages of some South African *Culicoides* species (Diptera: Ceratopogonidae). *Onderstepoort Journal of Veterinary Research* **36**: 265–284.

Nunamaker, R. A. and **McKinnon, C. N.** 1989. Electrophoretic analyses of proteins and enzymes in *Culicoides variipennis* (Diptera: Ceratopogonidae). *Comparative Biochemistry and Physiology* **92B**: 9–16.

Orszagh, I. 1980. Family Ceratopogonidae. Pp. 20–143 in Chvála, M. (ed.), Krevasjící mouchy a střečci – Diptera [= bloodsucking flies and – Diptera]. Čeledi Ceratopogonidae, Simuliidae, Tabanidae, Hypodermatidae, Oestridae, Gasterophilidae, Hippoboscidae and Nycteribiidae. *Fauna CSSR* **22**, 538 pp. Československa Akademie Věd, Praha. [In Czech.]

Raccurt, C. 1984. Contribution à l'étude des *Culicoides* (Diptera: Ceratopogonidae) de la République d'Haiti (Grandes Antilles). *Récherches Haitiennes* **3**: 1–101.

Ramirez Perez, J. 1984. Revisión de los dipteros hematófagos del género *Culicoides* en Venezuela. *Boletin de la Direccion de Malariologia y Saneamiento Ambiental* **24**: 49–70.

Remm, H. 1975. On the classification of the biting midges (Diptera, Ceratopogonidae). *Parazitologiya* **9**: 393–397. [In Russian with English summary.]

Remm, H. 1988. Family Ceratopogonidae. Pp. 11–110 in Soós, A. and Papp, L. (eds), *Catalogue of Palaearctic Diptera*: vol. 3, Ceratopogonidae-Mycetophilidae. 448 pp. Elsevier, Amsterdam.

Reynolds, D. G. and **Vidot, A.** 1978. Chemical control of *Leptoconops spinosifrons* in the Seychelles. *PANS* [= Pest Articles and News Summaries] **24**: 19–26.

Saunders, L. G. 1956. Revision of the genus *Forcipomyia* based on characters of all stages (Diptera, Ceratopogonidae). *Canadian Journal of Zoology* **34**: 657–705.

Sellers, R. F. and **Maarouf, A. R.** 1989. Trajectory analysis and bluetongue virus serotype 2 in Florida 1982. *Canadian Journal of Veterinary Research* **53**: 100–102.

Sellers, R. F. and **Pedgley, D. E.** 1985. Possible windborne spread to western Turkey of bluetongue virus in 1977 and of Akabane virus in 1979. *Journal of Hygiene* **95**: 149–158.

Spinelli, G. R. and **Wirth, W. W.** 1986. Clave para la identificación de las espécies del género *Culicoides* Latreille presentes al sur de la Cuenca Amazónica. Nuevas citas y notas sinonimicas (Diptera: Ceratopogonidae). *Revista de la Sociedad Entomologica Argentina* **44** (1985): 49–73.

Szadziewski, R. 1984. Ceratopogonidae (Diptera) from Algeria. VI. *Culicoides* Latr. *Polskie Pismo Entomologiczne* **54**: 163–182.

Szadziewski, R. 1985. Przeglad faunistyczny krajowych kuczmanów z rodzaju *Culicoides* (Diptera, Ceratopogonidae) [A faunistic review of the Polish biting midges of the genus *Culicoides*]. *Polskie Pismo Entomologiczne* **55**: 283–341. [In Polish with English summary and subtitle.]

Taylor, W. P. (ed.) 1987. *Bluetongue in the Mediterranean region*. Proceedings of a meeting in the Community Programme for Coordination of Agricultural Research, Istituto Profilactico Sperimentale dell'Abruzzo e del Molise, Teramo, Italy, 3 and 4 October 1985. Commission of the European Communities Report EUR 10237 EN. 119 pp. Office for Official Publications of the European Communities, Luxemburg.

Tidwell, M. A. and Tidwell, M. A. 1982. Development of *Mansonella ozzardi* in *Simulium amazonicum, S. argentiscutum,* and *Culicoides insinuatus* from Amazonas, Colombia. *American Journal of Tropical Medicine and Hygiene* **31**: 1137–1141.

Whitsel, R. H. and Schoeppner, R. F. 1970. Observations on follicle development, and egg production in *Leptoconops torrens* (Diptera: Ceratopogonidae) with methods for obtaining viable eggs. *Annals of the Entomological Society of America* **63**: 1498–1502.

Winder, J. A. 1978. Cocoa flower Diptera; their identity, pollinating activity and breeding sites. *PANS* [= Pest Articles and News Summaries] **24**: 5–18.

Wirth, W. W. 1965. Family Ceratopogonidae. Pp. 121–142 in Stone, A., Sabrosky, C. W., Wirth, W. W., Foote, R. H. and Coulson, J. R. (eds), *A catalog of the Diptera of America North of Mexico*. Agriculture Handbook 276, iv + 1696 pp. U.S. Department of Agriculture, Washington, D.C.

Wirth, W. W. 1973. Family Ceratopogonidae. Pp. 346–388 in Delfinado, M. D. and Hardy, D. E. (eds), *A catalog of the Diptera of the Oriental region*: vol. 1, *Suborder Nematocera*. 618 pp. University Press of Hawaii, Honolulu.

Wirth, W. W. 1974. *A catalogue of the Diptera of the Americas South of the United States. 14. Family Ceratopogonidae*. 89 pp. Museu de Zoologia, Universidade de São Paulo.

Wirth, W. W. 1977. A review of the pathogens and parasites of the biting midges (Diptera: Ceratopogonidae). *Journal of the Washington Academy of Sciences* **67**: 60–75.

Wirth, W. W. and Arnaud, P. H. 1969. Polynesian biting midges of the genus *Culicoides* (Diptera: Ceratopogonidae). *Pacific Insects* **11**: 507–520.

Wirth, W. W. and Blanton, F. S. 1974. The West Indian sandflies of the genus *Culicoides* (Diptera: Ceratopogonidae). *United States Department of Agriculture Technical Bulletin* **1474**: iv + 1–98.

Wirth, W. W. and Hubert A. A. 1989. The *Culicoides* of Southeast Asia (Diptera: Ceratopogonidae). *Memoirs of the American Entomological Institute* **44**: iii + 1–508.

Wirth, W. W. and Messersmith, D. H. 1977. Notes on the biting midges of the Seychelles (Diptera: Ceratopogonidae). *Proceedings of the Entomological Society of Washington* **79**: 293–309.

Wirth, W. W. and Navai, S. 1978. Terminology of some antennal sensory organs of *Culicoides* biting midges (Diptera: Ceratopogonidae). *Journal of Medical Entomology* **15**: 43–49.

Wirth, W. W. and Ratanaworabhan, N. C. 1978. Studies on the genus *Forcipomyia*. V. Key to subgenera and description of a new subgenus related to *Euprojoannisia* Brèthes (Diptera: Ceratopogonidae). *Proceedings of the Entomological Society of Washington* **80**: 493–507.

Wirth, W. W., de Meillon, B. and Haeselbarth, E. 1980. Family Ceratopogonidae. Pp. 150–174 in Crosskey, R. W. (ed.), *Catalogue of the Diptera of the Afrotropical region*. 1437 pp. British Museum (Natural History), London.

Wirth, W. W., Dyce, A. L. and Peterson, B. V. 1985. An atlas of wing photographs, with a summary of the numerical characters of the Nearctic species of *Culicoides* (Diptera: Ceratopogonidae). *Contributions of the American Entomological Institute* **22** (4): 1–46.

Wirth, W. W., Dyce, A. L. and Spinelli, G. R. 1988. An atlas of wing photographs, with a summary of the numerical characters of the Neotropical species of *Culicoides* (Diptera: Ceratopogonidae). *Contributions of the American Entomological Institute* **25** (1): 1–72.

Wirth, W. W., Ratanaworabhan, N. C. and Blanton, F. S. 1974. Synopsis of the genera of Ceratopogonidae (Diptera). *Annales de Parasitologie humaine et comparée* **49**: 595–613.

CHAPTER EIGHT

Horse-flies, deer-flies and clegs (Tabanidae)

John E. Chainey

The Tabanidae form a large family of about 4000 described species found throughout the world. Although they are generally called horse-flies, several other vernacular names are used, including gadflies, stouts, elephant-flies, buffalo-flies, mooseflies, clegs (genus *Haematopota*), deer-flies (genus *Chrysops*) and greenheads (for Nearctic species of salt-marsh *Tabanus*). Many species bite man, but few are proven vectors of human disease and the only parasite cyclically transmitted among humans by tabanids is the filarial worm *Loa loa* carried by *Chrysops*. As mechanical vectors of pathogens, the flies are primarily of veterinary importance, but in certain circumstances they can apparently transmit to man diseases such as anthrax, tularaemia and (possibly) Lyme disease.

RECOGNITION AND ELEMENTS OF STRUCTURE

Although adult Tabanidae show variation between species in their general appearance, the family is one of the easiest to recognize among the Diptera (Figs 8.1–8.3). They are generally rather large flies, with a length range of 6–30 mm. Their bodies are generally brown, black or grey, usually with a pattern of pale areas, but can be yellow, green or metallic blue. Characteristic of the family are the large head and proboscis (Fig. 8.4*a,b*), the wing venation (Fig. 8.5*g,h*), large calypteres, pulvilliform empodium between large pulvilli (Fig. 8.5*f*), and the lack of bristles on the body. In many species, particularly of *Chrysops* and *Haematopota*, the eyes in life are brilliantly coloured and often marked with conspicuous patterns; the colour soon fades after death, however, and usually is not detectable on preserved specimens. Many species of Pangoniinae have an enormously elongated and stiletto-like proboscis (e.g. *Philoliche magrettii*, Fig. 8.1*a*).

The specific identification of adult Tabanidae is based largely on head structures (particularly the shape and proportions of the frons, antennae and maxillary palps) and the colour and patterns of the body and/or wings. Most body patterns are produced by a subtle combination of hairs overlying minute surface structures (usually referred to as pollinosity or tomentum) and give an effect that is often more easily appreciated by the unaided eye than through a microscope. They are often somewhat variable, however, even in the same species, and this, combined with the

Medical Insects and Arachnids Edited by Richard P. Lane and Roger W. Crosskey.
Published in 1993 by Chapman & Hall ISBN 0 412 40000 6

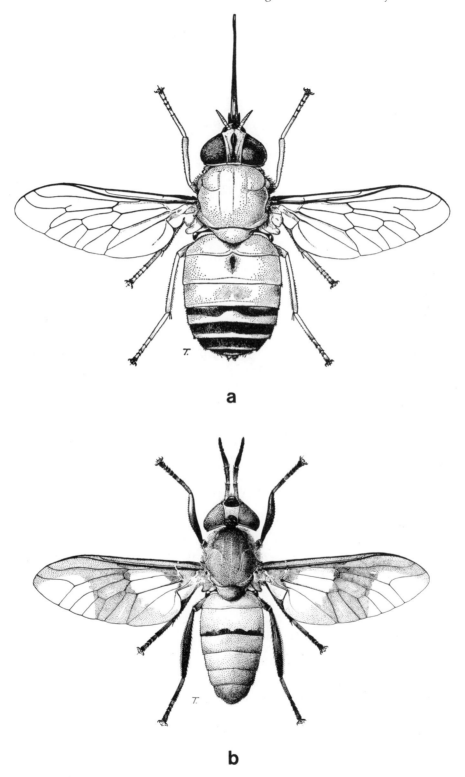

Figure 8.1 Female Tabanidae representative of the subfamilies Pangoniinae (top) and Chrysopsinae (bottom): (*a*) *Philoliche magrettii*, showing horizontally projecting stiletto-like proboscis usual in the Afrotropical tribe Philolichini; (*b*) *Chrysops fixissimus*, showing the long antennae and cross-banded wings of typical deer-flies.

312 *Horse-flies, deer-flies and clegs (Tabanidae)*

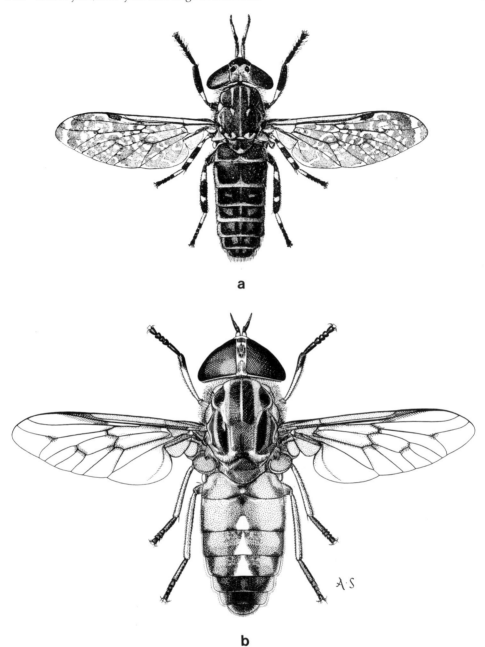

Figure 8.2 Female Tabanidae representative of the Tabaninae: (*a*) the cleg *Haematopota maculosifacies*, showing the widely separated eyes, rosette-patterned wings and parallel-sided abdomen of the tribe Haematopotini; (*b*) the horse-fly *Tabanus fraternus*, showing the narrowly separated eyes and oval abdomen of the tribe Tabanini. Both species occur in eastern and southern Africa.

general lack of structural characters, is the commonest cause of difficulty with specific identification. The genitalia are very simple in both sexes, only occasionally offering reliable features for separation of species. At subfamily and tribal level, however, differences in male styles and female ninth abdominal tergite (T9) provide important taxonomic characters; males of the South American *Mycteromyia*, a genus

Recognition and elements of structure 313

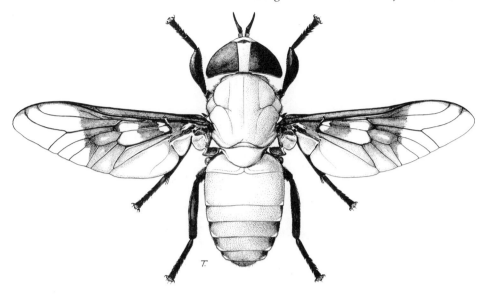

Figure 8.3 *Ancala africana* (female) showing the swollen tibiae of the fore legs and dark-banded wings characteristic of the Afrotropical genus *Ancala* (Tabanini).

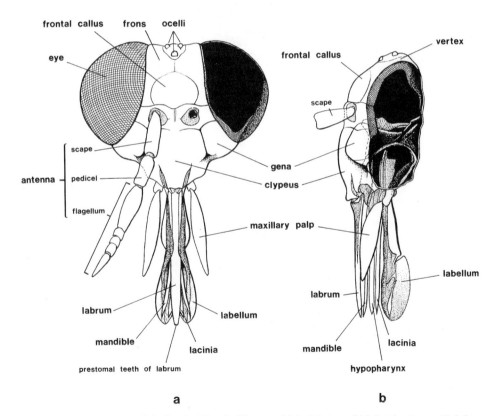

Figure 8.4 Basic structure of the head of female *Chrysops*: (*a*) facial view; (*b*) left side view, with left eye surface opened into head interior. (Modified from Jobling (1987) and reproduced courtesy of the Trustees of the Wellcome Trust.)

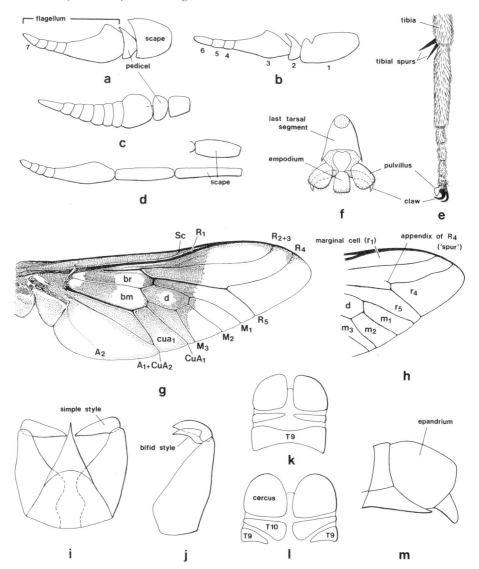

Figure 8.5 Structural features of Tabanidae important in family recognition and identification: (a–d) antenna in side view of (a) *Tabanus*, (b) *Haematopota*, (c) *Philoliche*, (d) *Chrysops* (with two shapes of scape shown); (e) part of leg showing tibial spurs present in subfamilies other than Tabaninae; (f) last segment of tarsus showing large pulvillus-like empodium of Tabanidae; (g and h) venation of wing and wing tip of Tabanidae with main veins and cells labelled according to the notation of McAlpine et al. (1981); (i) male genital style of simple (undivided) shape occurring in Pangoniinae other than Pangoniini; (j) male genital style of bifid shape occurring in Pangoniini; (k) female abdominal tip of Pangoniinae, showing undivided ninth tergite (T9); (l) female abdominal tip of Chrysopsinae, showing ninth tergite (T9) divided into two lateral parts; (m) massive male epandrium (T9) of the South American genus *Mycteromyia* (Pangoniinae).

sometimes put in a tribe distinct from other Pangoniinae, have this tergite strangely modified to form a bulbous epandrium (Fig. 8.5*m*).

Adult internal anatomy has been comprehensively illustrated by Jobling (1987) with *Chrysops* as the representative genus.

Larvae of Chrysopsinae and Tabaninae (Fig. 8.6*a*) are generally cylindrical, tapering at each end, and range 12–60 mm in length when fully grown. They are

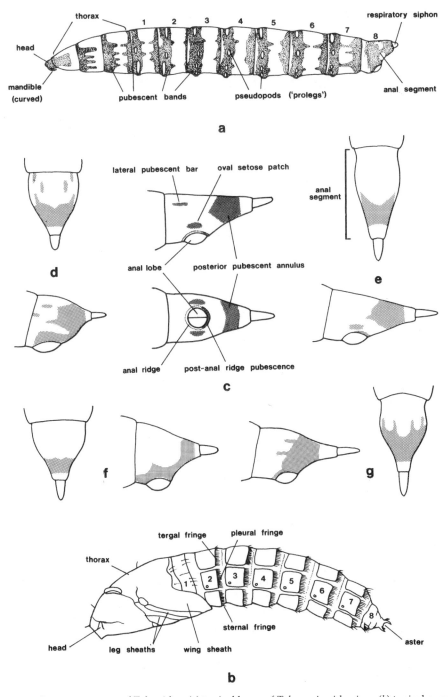

Figure 8.6 Immature stages of Tabanidae: (*a*) typical larva of *Tabanus* in side view; (*b*) typical pupa of *Tabanus* in side view; (*c*) diagram of the side and under-surface of the anal segment of a *Chrysops* larva showing the terminology of pubescent areas used in identification; (*d–g*) diagrams showing taxonomic differences in anal segment pubescence (dorsal and side views) between larvae of (*d*) *C. distinctipennis*, isolated bars present; (*e*) *C. silaceus*, annulus and post-anal ridge pubescence not continuous and oval setose patches present; (*f*) *C. griseicollis*, annulus and post-anal ridge pubescence continuous but no bars or patches; (*g*) *C. longicornis*, annulus and post-anal ridge pubescence continuous and bars extending forwards from the annulus. (Diagrams for *Chrysops* based on figures and key of Inaoka et al., 1988.)

white or cream coloured, sometimes brownish or greenish, often with dark rings near the borders of the segments. All larvae have retractable head capsule, three thoracic and eight abdominal segments, and a terminal respiratory siphon. The first seven abdominal segments bear three or four pairs of variably elongated pseudopods (so-called 'prolegs'). The larvae of subfamily Pangoniinae are poorly known but can be pear-shaped or cylindrical; they often bear pseudopods on the first five abdominal segments.

Pupae are usually brown and range 9–35 mm in length. More or less cylindrical, they typically have a dorsally arched body shape and large head with prominent leg and wing sheaths fused to the body (Fig. 8.6b). Two thoracic and eight abdominal segments are visible. The abdominal segments, except the first and the anal segment, are usually armed with fringes of short spines. The anal segment ends in two pointed and sclerotized tubercles forming the aster.

CLASSIFICATION AND IDENTIFICATION

Most authors broadly adopt the classification of Mackerras (1954, 1955a, 1955b) and recognize the following subfamilies and tribes: Pangoniinae (Pangoniini, Philolichini, Scionini), Chrysopsinae (Bouvieromyiini, Chrysopsini, Rhinomyzini) and Tabaninae (Diachlorini, Haematopotini, Tabanini). More controversial has been the treatment of Mackerras's Scepsidinae, a small group containing only non-biting species (see Fairchild, 1969). The distinction between the tribes Tabanini and Diachlorini is not very clear (especially with regard to a few genera in the Neotropical region), and Chainey (1987) has questioned the placement of *Pseudotabanus* and its allies in the Bouvieromyiini. A fourth tribe, the Mycteromyiini, is recognized by Coscarón and Philip (1979) within the Pangoniinae, i.e. the subfamily considered to be the most ancestral in the Tabanidae. The Tabaninae and Chrysopsinae are undoubtedly more closely related to each other than either is to the Pangoniinae, and the Tabaninae are considered to be the most derived subfamily. (Note: Chrysopsinae is the correct form of the subfamily name given in older literature as Chrysopinae.)

The subfamilies and tribes cannot be so perfectly defined on a world basis that it is possible to construct a simple key to the many genera assigned to these higher level taxa which happen to contain man-biting species. The only key which it is specially relevant and practicable to provide here is one to the man-biting *Chrysops* species involved, or potentially so, in loiasis transmission. However, the following characters important for identification of tribes and subfamilies should be noted: in males of Pangoniini the styles of the genitalia are bifid (Fig. 8.5j), whereas in other Pangoniinae they are undivided (Fig. 8.5i) and in females of Chrysopsinae the ninth abdominal tergite (T9) is split into two lateral parts (Fig. 8.5l) instead of being undivided as in Pangoniinae (Fig. 8.5k). Females of Tabaninae also have T9 divided, but in this subfamily (unlike Chrysopsinae) the caudal ends of the spermathecal ducts have characteristic 'mushroom-shaped expansions' (Oldroyd, 1973, Fig. 98B).

Taxonomy depends at present almost entirely on morphology. Electrophoresis has, however, been used successfully to distinguish two Nearctic pest species of *Tabanus* (Jacobson et al., 1981), and two forms of the common Palaearctic deer-fly

Chrysops caecutiens have been differentiated on larval chromosomes which might be sibling species (Ivanishchuk, 1983). Preliminary studies on cuticular hydrocarbons suggest some taxonomic usefulness (Hoppe et al., 1990).

Pupae are identified on their colour, size, antennal ridge shape, thoracic chaetotaxy, arrangement and number of spine fringes, and the number of tubercles on the aster. Larval identification relies mainly on body colour, number of pairs of pseudopods, shape of anal segment and respiratory siphon, and the arrangement and extent of cuticular striation and pubescence patterns (Teskey, 1969). Pubescent areas of the anal segment (Fig. 8.6c) are important for identifying larvae of African *Chrysops* (Inaoka et al., 1988), and patterns found in the loiasis vector species are shown in Fig. 8.6d–g. (It needs to be kept in mind that larvae are unknown for most of the 43 Afrotropical species of *Chrysops*, including *C. dimidiatus*, and therefore that immatures should always be reared to adults to confirm species identity.)

Key to African *Chrysops* species and other Tabanidae associated with human loiasis

The key covers the eight *Chrysops* species named in relation to human loiasis by the World Health Organization (1989), together with *C. centurionis*, a vector of *Loa* in forest monkeys which can be confused with *C. silaceus* or *C. dimidiatus*. Other species might very rarely bite man, so any doubtful specimens should be sent to a specialist for identification. *Haematopota* and *Hippocentrum* have been included because old records exist of some development of *Loa loa* in at least one species of each of these genera (see Krinsky, 1976). The key covers only bloodsucking female flies, and the word 'pollinosity' is used for the velvety coating seen on areas of the body which are not obviously bare and shining; these areas are usually paler than the underlying ground colour (which can be extensively revealed if the pollinosity should become abraded in older or damaged specimens).

1   Antennae with long cylindrical or barrel-like second segment (pedicel) which widely separates first (scape) from third segment and is often equal in length to one or other of these, sometimes both (Fig. 8.5d). Head with ocelli (Fig. 8.4a,b). Wings with conspicuous brown cross-band or unevenly dark on more than inner or outer half (Fig. 8.7). Tibiae with pair of small spurs at tip of undersurface. Flies with general appearance as in Fig. 8.1b. ....
....**Chrysops** 2
—   Antennae with small cup-like pedicel which does not widely separate first from third segment and is very much shorter than either of these (Fig. 8.5a,b). Head without ocelli. Wings without such dark pattern. Tibiae without spurs at tip of undersurface. Flies with general appearance as in Fig. 8.2a or 8.2b.............10
2   Wings with broad dark median cross-band (Fig. 8.7c,e,f,h). Scutellum black or mainly grey. Abdomen dark, with brown or almost black centre line or large transverse bands easily seen by naked eye (except mainly or partly reddish yellow in *longicornis*). Thorax dorsally without yellow stripes (if stripes present then grey, brown or black).. 3
—   Wings without cross-band but darkened to a varying extent on apical half (Fig. 8.7a,b,d,g). Scutellum bicolored, dark brown or black basally and yellow apically. Abdomen largely reddish yellow, usually with two blackish brown stripes easily seen by naked eye. Thorax dorsally often with yellow stripes against darker background 6

318 *Horse-flies, deer-flies and clegs (Tabanidae)*

Figure 8.7 Wings of females of some tropical African species of *Chrysops*: (*a*) *C. silaceus*; (*b*) *C. dimidiatus*; (*c*) *C. distinctipennis*; (*d*) *C. zahrai*; (*e*) *C. streptobalius*; (*f*) *C. griseicollis*; (*g*) *C. langi*; (*h*) *C. longicornis*. Note that the wing of *C. centurionis* is almost identical to that of *C. silaceus* and that in *C. streptobalius* the preapical pale spot visible in the wing shown here is often continuous with the apical pale area. The eight species for which wings are illustrated are those cited by World Health Organization (1989) as significant in transmission or suspected transmission of human loiasis.

3   Head bare and shining below eyes, on clypeus except for vertical pollinose median stripe, and on frontal callus. Thorax black, partly bare and shining on upper surface; scutellum black; thoracic hair yellowish white to golden or coppery red. Wings with dark colour along fore margin continuous through marginal cell (r1). Legs largely yellow or red, especially femora. Clypeus and genae swollen, facial profile strongly convex. Antennae with long and stick-like scape (except in *streptobalius*) 4
— Head with bare and shining areas confined to frontal callus, otherwise thickly coated with white pollinosity (and with long shaggy white hair). Thorax thickly and evenly coated with white pollinosity making it appear very pale

ashy grey to naked eye, only scutellum faintly brown centrally; thoracic hair pure white, very long and soft. Wings with dark colour along fore margin interrupted by colourless area in middle of marginal cell (r1) (Fig. 8.7f). Legs brownish black. Face not strongly swollen, its profile very gently curved. Antennae with distinctly swollen scape. [Antennae black. Abdomen pale ashy grey on sides and segment hind margins, otherwise dark brown to black.] [Cameroun, Nigeria, Uganda, Zaire] .... **C. griseicollis**

4 Abdomen bluish grey with a midline of large dark brown triangular marks of diminishing size (one on each segment). Antennae entirely black. Wings with cross-band slightly to strongly oblique and fading out indefinitely before hind margin (Fig. 8.7c,e). Thorax without transverse band of golden hair in front of scutellum 5

— Abdomen partly or mostly reddish yellow, first segment brown or black, second reddish yellow with triangular or arrow-shaped dark brown median mark, remainder varying from reddish yellow to black. Antennae with scape dingy yellow to tawny red, conspicuously paler than dark pedicel. Wings with cross-band not at all oblique and more or less reaching hind margin (Fig. 8.7h). Thorax with transverse band of thick silky golden hair in front of scutellum. [Widespread tropical Africa] .... **C. longicornis**

5 Thorax dorsally with bold pattern of five stripes, wide median brown stripe bordered by two pale grey stripes and these by shining black (ground colour) stripes. Wings with strongly oblique cross-band (Fig. 8.7e); cells br and bm dark brown except for milky white stripe across them; cell r3 with brown colour reaching at least to fork of veins R4 and R5, pale area of this cell not covering its full width but restricted to small milky white spot which either stands against middle of vein R4 or is isolated centrally in dark surrounding colour (as in Fig. 8.7e); extreme wing base dark brown. Antennae with slightly bulbous scape stouter than pedicel. Thoracic hair tufts mainly coppery red. [Ethiopia, but known also in Somalia] .... **C. streptobalius**

— Thorax dorsally with inconspicuous pattern of two narrow grey stripes on shining black ground colour. Wings with less oblique cross-band (Fig. 8.7c); cells br and bm colourless except for dark colour at extreme tip and trace of smokiness at base; cell r3 with brown colour not reaching fork of veins R4 and R5, colourless area much larger and covering full width of cell (meeting vein R2+3); extreme wing base clear. Antennae with long stick-like scape even narrower than pedicel. Thoracic hair tufts mainly yellowish white to yellow. [West to East and Central Africa] .... **C. distinctipennis**

6 Thorax dorsally with stripe pattern indefinite to naked eye. Legs mainly reddish yellow. Maxillary palps and proboscis reddish yellow. Wings with basal part of fore margin faintly yellow or pale brown; discal cell (d) darkest brown at base (Fig. 8.7a,b) or round a central pale area. Clypeus shining yellow or tawny yellow with central yellow pollinose stripe. Antennae dingy yellow or reddish yellow at least at base or apex and with long stick-like scape. Scutellum with dark part reaching sides and not strongly convex posteriorly.. 7

— Thorax dorsally with very bold black and golden-yellow striped pattern, easily seen by naked eye and comprising central black stripe, pair of submedian yellow stripes, pair of black sublateral stripes and yellow side margins. Legs black. Maxillary palps and proboscis black. Wings conspicuously dark brown along whole fore margin and rather evenly suffused with brown colour on apical half (Fig. 8.7d), discal cell (d) uniformly dark brown. Clypeus shining black and all bare. Antennae black and with bulbous scape. Scutellum with dark part not reaching sides and strongly convex behind. [Known from Cameroun and Nigeria] .... **C. zahrai**

7 Abdomen with brown marks of first segment completely separated by yellow or orange ground colour (paired dark stripes not united at base). Antennae with flagellum blackish brown, at most a little reddish on third segment. Wings near middle often distinctly yellow to naked eye... 8

— Abdomen with brown marks of first segment joined by continuous brown colour along centre of hind margin (paired dark stripes therefore united at base). Antennae with flagellum partly or wholly orange-red. Wings not at all yellow, darkened apical part rather evenly smoky brown (Fig 8.7g). [Known from Cameroun and Zaire]... **C. langi**

8 Abdomen with brown stripes poorly developed, narrow and usually extending back only as far as base of third segment, sometimes represented only as pair of basal dark spots on first segment or as spots and streaks on first and second segments (some *centurionis* specimens with bold stripes reaching fourth segment); yellow or orange ground colour at most darkening only to reddish tan on terminal segments. Wings distinctly yellow in middle beyond dark base of discal cell, to naked eye this yellowness contrasting with pale smoky brown wing tips (in Fig. 8.7a visible as paler middle area of wing).* Thorax dorsally with light or dark brown ground colour. Hind tibiae with yellow or dark hair 9

— Abdomen with paired black–brown stripes bold and obvious, broad and fully extending from first segment to hind margin of fourth segment; ground colour usually more or less yellow basally and becoming much darker orange–red to reddish brown on terminal segments (these sometimes with brown hind margins). Wings more or less evenly darkened beyond base of discal cell (Fig. 8.7b), not contrastingly yellow and pale smoky brown to naked eye. Thorax dorsally with brownish black ground colour. Hind tibiae with dark hair. [Guinea to Congo basin and northern Angola] .... **C. dimidiatus**

9 Thorax mid-dorsally with dark brown ground colour (nearly black to naked eye). Scutellum with dark area covering half its length. Hind tibiae with pale yellow hair. Abdomen usually not noticeably tapering and with almost uniformly bright orange ground colour. Length about 9 mm. [Ghana to Congo basin and Sudan]... **C. silaceus**

— Thorax mid-dorsally with tawny brown or light reddish brown ground colour (not dark to naked eye). Scutellum with dark area usually covering less than half its length. Hind tibiae with dark brown hair. Abdomen in most specimens evenly tapering and with basal ground colour usually orange–yellow rather than bright orange. Length about 11 mm. [Nigeria to Congo basin and Uganda]... **C. centurionis**

10 Antennae with six segments, first (scape) much longer than wide seen in side view (Fig. 8.5b). Wings with dappled pattern (e.g. as in Fig. 8.2a); in life held in rooflike manner which conceals abdomen. Frons broad, square or slightly higher than wide, and with pair of velvety spots positioned above frontal callus and close to eyes (Fig. 8.2a) (velvety spots absent in *Hippocentrum*) 11

---

* The wing of *Chrysops centurionis* is not figured here as it is almost identical with that of *C. silaceus*. Illustrations can be found in Oldroyd (1957, p. 103 and Plate VIII Fig. 5).

— Antennae with seven segments, scape subtriangular and not or hardly longer than wide seen in side view (Fig. 8.5a). Wings without dappled pattern, nearly always clear or mainly so (e.g. as in Fig. 8.2b); in life held almost flat when at rest and diverging behind so that abdomen largely exposed. Frons narrow, several times higher than wide (Fig. 8.2b); without paired velvety spots but with narrow bare central ridge above frontal callus (and usually connected to it). [*Tabanus* and allied Tabanini].................. Flies not associated with loiasis

11 Wings with variegated spots and streaks forming rosette-like patterning (Fig. 8.2a). Head mainly pollinose and dull. Maxillary palps dull and usually pale pollinose, not greatly swollen... **Haematopota**

— Wings without rosette-like patterning, darkened apical half crossed by one or two narrow white bands (easily seen by naked eye). Head bare and shining dark brown or black. Maxillary palps shining dark brown or black, large and conspicuously swollen... **Hippocentrum**

Geographical distribution

The Tabanidae are distributed worldwide from temperate to tropical climes (except for some oceanic islands) and range altitudinally from sea level to about 5000 metres. They are thought to have evolved first in southern Gondwana and then to have radiated northwards. This supposition is supported by the current distribution of both ancestral and derived groups. The distribution of the tribes is summarized below (with the numbers of species shown in parentheses for the genera mentioned under each tribe).

Subfamily Pangoniinae
Tribe Mycteromyiini: confined to southern Neotropical region (Chile and Argentina). The three genera contain non-bloodsucking flies.
Tribe Pangoniini: mostly Neotropical and Palaearctic, but represented in all regions. The 25 genera include *Esenbeckia* (75). Many species are non-bloodsucking.
Tribe Philolichini: chiefly Afrotropical (not Madagascar) but represented in the Australasian (three species in New Caledonia), Oriental and Palaearctic regions (one species in Morocco). The two genera include *Philoliche* (100).
Tribe Scionini: Australasian and Neotropical regions, one species in Nearctic. The five genera include *Fidena* (100), *Scaptia* (116) and *Scione* (40).

Subfamily Scepsidinae
Not divided into tribes and considered an artificial assemblage of species, all of them non-bloodsucking: Afrotropical region, genera *Adersia* (6) and *Braunsiomyia* (1) and Neotropical region, the genus *Scepsis* (1).

Subfamily Chrysopsinae
Tribe Bouvieromyiini: largely Afrotropical, but with some Oriental and Australasian representatives and a few Palaearctic and Nearctic species; absent from Neotropical region. The nine genera include *Aegophagamyia* (34), *Lilaea* (14) and *Rhigioglossa* (50).
Tribe Chrysopsini: worldwide, but very few species in Australasia and southern Neotropical region. The seven genera include *Chrysops* (250) and *Silvius* (33).

Tribe Rhinomyzini: Afrotropical and Oriental regions (one possible representative in Neotropics). The 12 genera include *Orgizomyia* (1) and *Tabanocella* (30).

Position uncertain: *Pseudotabanus* (45) and allies, confined to the Neotropical (Chile and Argentina) and Australasian regions, are of uncertain tribal placement.

Subfamily Tabaninae

Tribe Diachlorini: represented in all regions but mainly Australasian and Neotropical. The 45 genera include *Catachlorops* (60), *Chlorotabanus* (6), *Cydistomyia* (110), *Dasybasis* (150), *Dasychela* (7), *Diachlorus* (28), *Dichelacera* (70), *Dicladocera* (30), *Lepiselaga* (4), *Philipomyia* (3), *Philipotabanus* (26), *Selasoma* (1), *Stenotabanus* (70), *Stibasoma* (19) and *Stypommisa* (28).

Tribe Haematopotini: mainly Afrotropical (except Malagasy subregion), Oriental and Palaearctic, but also five Nearctic species. Absent from Australasian and Neotropical regions. The four genera include *Haematopota* (450) and *Hippocentrum* (5).

Tribe Tabanini: worldwide (except New Caledonia, New Zealand, Tasmania and southern tip of South America), but few species in Australasian region, southern tip of Afrotropical region (including Malagasy subregion) and temperate parts of Neotropical region. The 15 genera include *Ancala* (7), *Atylotus* (70), *Hybomitra* (200) and *Tabanus* (1500).

Faunal and taxonomic literature

Keys cited are to adults unless otherwise indicated.

General: Classification and distribution (Mackerras, 1954, 1955a, 1955b); world catalogue (Moucha, 1976).

Palaearctic: Catalogue (Chvála, 1988); Chrysopsinae and Pangoniinae (Leclercq, 1960, keys); Tabaninae (Leclercq, 1966, keys); British Isles (Oldroyd, 1969, keys); China (Xu, 1982, key; Wang, 1983, keys; Murdoch and Takahasi, 1969, keys to Manchurian species); Europe (Chvála et al., 1972, keys); Iran (Abbassian-Lintzen, 1964, keys); Israel (Theodor, 1965, keys); Japan (Murdoch and Takahasi, 1969, keys; Hayakawa, 1980, biology, *Tabanus iyoensis* group keys to immatures; Hayakawa, 1985, keys); Korea (Murdoch and Takahasi, 1969, keys); Saudi Arabia (Amoudi and Leclercq, 1988, checklist); USSR (Andreeva, 1990, keys to larvae; Olsuf'ev, 1977, keys).

Afrotropical: Catalogue (Chainey and Oldroyd, 1980); monograph (Oldroyd, 1952, 1954, 1957); Angola (Dias, 1960, key); Guinea-Bissau (Tendeiro, 1965, keys); Mali (Goodwin, 1982, keys to adults and immatures); Mozambique (Dias, 1966, keys); Nigeria (Inaoka et al., 1988, key to *Chrysops* larvae); South Africa (Usher, 1972, annotated list); *Rhigioglossa* (Chainey, 1987, keys).

Oriental: Catalogue (Stone, 1975); Borneo (Philip, 1960, keys); Indonesia (Schuurmans Stekhoven, 1926, keys); India (Senior-White, 1927, keys); Malaysia (Philip, 1960, keys); Philippines (Philip, 1959, keys); Sri Lanka (Burger, 1981, keys); Thailand (Philip, 1960, keys; Burton, 1978, keys to Tabanini); *Haematopota* (Stone and Philip, 1974, keys).

Australasian: Catalogue (Daniels, 1989); Australia (Mackerras, 1956a, general review; Mackerras, 1956b, revision of Pangoniini; Mackerras, 1959, annotated catalogue of Tabaninae; Mackerras, 1960, revision of Scionini; Mackerras, 1961, revision of Chrysopsinae; Mackerras, 1971, revision of *Tabanus*); New Guinea (Mackerras, 1964, keys); New Zealand (Mackerras, 1957, keys); South Pacific islands (Mackerras and Rageau, 1958, keys).

Nearctic: Catalogue (Philip, 1965); larvae (Teskey, 1990); genera (Pechuman and Teskey, 1981, keys to adults and immatures); Chrysopsinae (Philip, 1954, 1955, keys); Pangoniinae (Philip, 1954, keys); Tabaninae (Stone, 1938, keys); Alaska (Teskey, 1990); Arizona (Burger, 1974a,b,c, 1975, keys to adults; Burger, 1977, keys to immatures); California (Middlekauff and Lane, 1980, keys); Canada (Teskey, 1990); Florida (Jones and Anthony, 1964, keys); Illinois (Pechuman et al., 1983, keys); Louisiana (Tidwell, 1973, keys to adults and immatures); New York (Pechuman, 1981, keys); Ontario (Pechuman et al., 1961, keys); Tennessee (Goodwin et al., 1985, keys); Virginia (Pechuman, 1973, keys). *Haematopota* (Burger and Pechuman, 1986, key).

Neotropical: Catalogue (Fairchild, 1971); genera (Fairchild, 1969, keys); Mycteromyiini (Coscarón and Philip, 1979, revision); Antilles (Bequaert, 1940, keys); Chile (Coscarón and González, 1991, key and species list); Colombia (Wilkerson, 1979, keys); Cuba (Cruz and Garcia Avila, 1974, keys); Panama (Fairchild, 1986, keys); *Chlorotabanus* (Philip and Fairchild, 1956, key). *Dasybasis* (Coscarón and Philip, 1967, revision); *Diachlorus* (Wilkerson and Fairchild, 1982, key); *Dichelacera* (Fairchild and Philip, 1960, revision); *Esenbeckia* (Philip, 1978, key to Central American species; Wilkerson and Fairchild, 1983, key to South American species of subgenus *Esenbeckia); Lepiselaga* (Fairchild, 1966, key); *Scaptia* (Wilkerson and Coscarón, 1984, key to subgenus *Pseudoscione* species); *Tabanus* (Fairchild, 1976, key to *trivittatus* complex; Fairchild, 1983, key to *lineola* complex; Fairchild, 1984, key to large South American species).

BIOLOGY

Life history

For most species of Tabanidae, particularly tropical species, the immature stages are unknown; only in the Nearctic region are they described for more than half the fauna (Teskey, 1969). Of the nine species of *Chrysops* associated with *Loa loa* transmission the immatures of only five have been described (Fig. 8.6c; Inaoka et al., 1988).

Many tabanids breed in mud or damp soil, but the larvae of some species are either aquatic, occur in dry soil, mosses or in rot-holes in trees. Breeding sites include pasture, meadows, forest floor, pools, streams, boggy depressions, salt-marsh and, occasionally, beaches. The larvae have six to 13 instars (but this is variable even in the same species) and most require a long maturation period of several months or even up to several years. Most temperate species are univoltine and some have an obligatory diapause. In tropical zones two or three generations per year is typical. The larvae of most species are actively carnivorous and many, at least in captivity, are cannibalistic. *Chrysops* larvae reportedly feed on decaying organic matter, but the morphology of the mouthparts suggests that they are predators of small invertebrates. Some Tabanini species will attempt to feed on any living flesh that is available, even including humans in paddyfields. One Nearctic species attacks young spadefoot toads emerging *en masse* from desert pools. Some species construct characteristic mud cylinders to avoid desiccation in dry conditions. Pupation takes place close to, or within, the drier parts of the medium inhabited by the larva. A few species pupate in aquatic vegetation or under semi-submerged rocks. The pupal stage generally lasts from one to three weeks, and adults live for two to four weeks.

Natural enemies of adult tabanids include birds, spiders, dragonflies, wasps (Bembecidae) and robber-flies (Asilidae). The main predators of larvae are other tabanid larvae, but they are also preyed upon by larger invertebrates (e.g. Odonata nymphs) and vertebrates (e.g. wading birds, mice). Larvae can be parasitized by nematodes, Diptera (Bombyliidae and Tachinidae) and Hymenoptera (Pteromalidae) and eggs can be heavily parasitized by Hymenoptera (Scelionidae and Trichogrammatidae). However, pathogens infecting larvae, e.g. entomophthogenic fungi and microsporidia, might prove to be more important than predators and parasites in the regulation of tabanid populations.

Mating

Little is known about mating mechanisms (see Wilkerson et al., 1985), but the males of some species hover individually or in groups prior to pairing. One Afrotropical species forms fast-moving swarms into which females fly before copulation; this begins in the air and is usually completed at rest on nearby vegetation. The males of some species wait on vegetation and intercept passing females.

Biting behaviour and hosts

Only female tabanids bite and suck blood. Blood is needed, as in other biting flies, as a protein source for the production of eggs. Some species develop at least the first egg batch without blood (autogenously). The hosts are located by sight (colour and movement), odour, carbon dioxide emission and (possibly) body heat. Flies are attracted to large objects such as cars, and traps can be used to exploit this tendency (e.g. the Manitoba trap). *Chrysops silaceus* (a vector of loiasis) is attracted to smoke from wood fires. Apart from man, hosts include monkeys, equines, bovines, ruminants, rodents, some reptiles (e.g. turtles) and occasionally birds. Biting is achieved by slicing the mandibles through the skin tissues while using the laciniae ('maxillae') and fore legs as anchors. Saliva, which contains an anticoagulant, is pumped into the lesion before the pool of blood is sucked up between the large labella into the food canal between the labrum and hypopharynx (Fig. 8.4b).

In most bloodsucking species the females, and males, will also feed on sugars at flowers, and are sometimes found at the sap exuding from a wounded tree or at aphid honeydew. Some species are non-haematophagous, e.g. those of *Adersia* and *Mycteromyia*. The adults also visit pools, drinking while in flight.

Most tabanids are active diurnally, though some are crepuscular and a few, notably *Chlorotabanus,* are nocturnal. Biting activity varies between species, peaking once or twice during the day depending on environmental conditions (particularly humidity, light levels and temperature).

Oviposition

Eggs are usually laid in a large mass (up to 25 mm long) from one to four layers deep, each mass including up to a thousand eggs. A few species lay their eggs singly or in scattered groups. They are usually laid low down on vegetation overhanging a suitable breeding site, but sometimes also directly on rocks by pools and streams. In a few species the female covers the egg-mass with a whitish secretion that soon darkens. The function of this secretion is unknown. In Nearctic *Goniops* the female guards the eggs for several days until they hatch.

Tabanids are strong fliers, though there is little evidence of regular migration. The Afrotropical species *Tabanus taeniola* has been recorded on boats a mile from shore and on the islands of Madagascar and Aldabra suggesting that it is capable of flying for long distances. Similarly, the Nearctic *Tabanus atratus* has occurred on oil rigs over five miles offshore.

MEDICAL IMPORTANCE

Many tabanids are persistent biters. Their large size and painful bites attract attention and they are frequently disturbed whilst trying to blood-feed. This gives them considerable potential as mechanical vectors. Tabanids are of veterinary rather than medical importance (see reviews by Foil, 1989; Krinsky, 1976; Minter, 1987). However, the transmission of equine infectious anaemia virus (EIAV) to horses by tabanids is of interest in a medical context because EIAV has been suggested as a model for the morphologically similar human immunodeficiency virus (HIV). Diseases that can be mechanically transmitted to man by tabanids include anthrax (*Bacillus anthracis*) and, in parts of North America and the former USSR, tularaemia (*Francisella tularensis*). Recently, Foil (1989) has suggested that tabanids are involved in the transmission of Lyme disease (*Borrelia burgdorferi*).

The only known cyclical transmission of parasites to man by tabanids is of the filarial nematode worm *Loa loa*, the cause of human loiasis in forested areas of Central and West Africa (Fig. 8.8). This disease affects about one million people and is characterized by the temporary appearance of large swellings (known as 'Calabar swellings' after a town in southern Nigeria) at the sites where migrating adult worms occur; these swellings mainly affect the limbs and can be crippling while they last. The microfilarial worms have a daytime periodicity in man and can cause some irritation as they move about the body. Adult worms frequently move under the conjunctiva of the eye but rarely cause any permanent ocular damage. Loiasis is primarily transmitted in man by *Chrysops dimidiatus* and *C. silaceus*, but there remains some doubt about how many species have been included under these names (the immatures and adult male of *C. dimidiatus* are still undescribed). Where the disease occurs outside the range of these species, *C. distinctipennis* and *C. longicornis* are the most likely vectors. Loiasis is probably a zoonosis that involves man and monkey hosts and *Chrysops* vectors (Rodhain, 1980). The related *Loa loa*

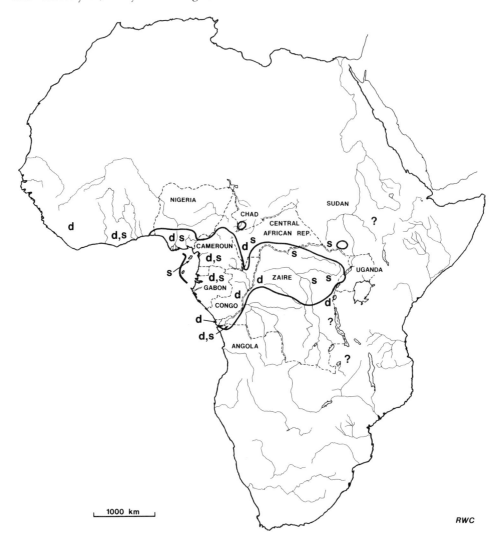

Figure 8.8 Approximate geographical distribution of human loiasis (thick unbroken line) and its chief vectors, *Chrysops dimidiatus* (**d**) and *Chrysops silaceus* (**s**). Query marks indicate countries (Ethiopia, Malawi and Zambia) from which *Loa* infection has been reported but where the existence of foci needs confirmation.

papionis has a nocturnal microfilarial periodicity and is apparently transmitted among monkeys by *C. langi* and *C. centurionis* (Duke, 1972). Other *Chrysops* species have the potential to transmit the disease and still need to be closely investigated. In the 1950s, much important and still relevant research was undertaken on *Chrysops* in the rain forest at Kumba in Cameroun, particularly on the biology in relation to *Loa* transmission, and the associated literature should be consulted by anyone interested in this topic: it is found in the *Annals of Tropical Medicine and Parasitology*, volumes 45–55, 1951–1961.

Tabanids often occur in such large numbers as to cause economic losses by affecting labourers and tourists. Some people are very sensitive to tabanid bites and require anti-histamines to combat the resultant swellings. The bites of certain species, such as *Diachlorus ferrugatus* of southern North America, and the widespread Neotropical species *Lepiselaga crassipes*, can be particularly irritating.

Carnivorous tabanid larvae sometimes bite man (e.g. workers in paddyfields) while hunting their invertebrate prey. The larva usually injects an immobilizing poison into its prey, and as a result the bites can cause severe skin reactions lasting several days.

CONTROL

Tabanids are generally difficult to control (see Anderson, 1985). The application of pesticides to breeding sites on a wide scale has not proved to be very effective and poses an environmental risk. Sudden changes engineered in the water levels of breeding sites by flooding and draining can be effective against certain species, but there is a risk that they will benefit other tabanid species; like pesticide use, such methods could also be environmentally damaging.

In limited areas the trapping of adult flies is probably the most effective means of control. Where localized breeding sites are known, some control can be achieved by isolating the oviposition sites, e.g. by clearing most of the vegetation around the edge of a pool habitat and physically destroying the egg-masses as they are laid.

Natural predators and parasites of adults and larvae are probably too generalized in their feeding habits to be effective control agents, but the use of parasites attacking the egg stage merits further investigation.

Repellents can be of some use for the protection of livestock from biting but require frequent application. On humans, most repellents seem to be of limited value. DEET (diethyltoluamide) gives some protection (Minter, 1987) but has been shown to have no effect against Nearctic species of *Chrysops*.

COLLECTING, PRESERVING AND REARING MATERIAL

The most effective way to collect large numbers of Tabanidae is to use canopy or Manitoba traps (optionally baited with octenol or CO_2). Large numbers can also be caught on or around livestock. It is important that tabanids are collected dry and pinned while still fresh. Specimens collected in alcohol can be dried by passing them through Cellosolve overnight and then xylene for about one hour, but the results are generally unsatisfactory. Any eye patterns should be recorded while the specimen is fresh, though they can usually be temporarily recovered by placing a dry specimen in a moist chamber. Most specimens die with their wings held down on the body and it is desirable that one wing be lifted up so that the abdominal pattern can be seen. Specimens that become greasy can usually be cleaned by immersing them in ethyl acetate for a few hours.

Larvae can be sieved from soil or mud. As many are cannibalistic, they must be held in individual containers if reared adults are required. However, *Chrysops* species can be reared *en masse*, keeping them in the substrate taken from their natural habitat or in damp tissues. Most larvae will thrive on a diet of house-fly larvae or pieces of earthworm. Those to be preserved should be killed in near-boiling water and then stored in 70–80% ethanol. Larval exuviae (moulted skins) can be similarly stored, but should be inflated with ethanol while fresh to prevent permanent wrinkling. Collected pupae should be embedded near the surface of their rearing media and any excess water removed. After emergence the empty exuviae can be kept dry with the adult specimen or stored in ethanol. Newly emerged adults should be kept in a cool dark place for 24 hours before being killed to allow for full development of the taxonomically important colours and patterns.

REFERENCES

Abbassian-Lintzen, R. 1964. Tabanidae (Diptera) of Iran X. List, keys and distribution of species occurring in Iran. *Annales de Parasitologie humaine et comparée* **39**: 285–327.

Amoudi, M. A. and **Leclercq, M.** 1988. *Tabanus riyadhae* (Diptera: Tabanidae), a new species from Saudi Arabia. *Journal of Medical Entomology* **25**: 399–401.

Anderson, J. F. 1985. The control of horse flies and deer flies (Diptera: Tabanidae). *Myia* **3**: 547–598.

Andreeva, R. V. 1990. *Identification of the larvae of horse-flies from the European part of the USSR, the Caucasus and Central Asia.* 171 pp. Naukova Dumka, Kiev. [In Russian.]

Bequaert, J. 1940. The Tabanidae of the Antilles (Dipt.). *Revista de Entomologia* (Rio de Janeiro) **11**: 253–369.

Burger, J. F. 1974a. The horse flies of Arizona. I. Introduction and zoogeography (Diptera: Tabanidae). *Proceedings of the Entomological Society of Washington* **76**: 99–118.

Burger, J. F. 1974b. Horse flies of Arizona.II. Notes on and keys to the adult Tabanidae of Arizona, subfamilies Pangoniinae and Chrysopsinae (Diptera). *Proceedings of the Entomological Society of Washington* **76**: 247–269.

Burger, J. F. 1974c. The horse flies of Arizona. III. Notes on and keys to the adult Tabanidae of Arizona, subfamily Tabaninae, except *Tabanus*. *Proceedings of the Entomological Society of Washington* **76**: 428–443.

Burger, J. F. 1975. Horse flies of Arizona IV. Notes on and keys to the adult Tabanidae of Arizona, subfamily Tabaninae, genus *Tabanus*. *Proceedings of the Entomological Society of Washington* **77**: 15–33.

Burger, J. F. 1977. The biosystematics of immature Arizona Tabanidae (Diptera). *Transactions of the American Entomological Society* **103**: 145–258.

Burger, J. F. 1981. A review of the horse flies (Diptera: Tabanidae) of Sri Lanka (Ceylon). *Entomologica Scandinavica* (Supplement) **11**: 81–123.

Burger, J. F. and **Pechuman, L. L.** 1986. A review of the genus *Haematopota* (Diptera: Tabanidae) in North America. *Journal of Medical Entomology* **23**: 345–352.

Burton, J. J. S. 1978. *Tabanini of Thailand above the Isthmus of Kra (Diptera: Tabanidae).* 165 pp. Entomological Reprint Specialists, Los Angeles.

Chainey, J. E. 1987. Afrotropical Tabanidae (Diptera): the genus *Rhigioglossa* Wiedemann, 1828 (including *Mesomyia* Macquart, 1850, as a subgenus). *Annals of the Natal Museum* **28**: 137–159.

Chainey, J. E. and **Oldroyd, H.** 1980. Family Tabanidae. Pp. 275–308 in Crosskey, R. W. (ed.), *Catalogue of the Diptera of the Afrotropical region.* 1437 pp. British Museum (Natural History), London.

Chvála, M. 1988. Family Tabanidae. Pp. 97–171 in Soós, A. and Papp, L. (eds), *Catalogue of Palaearctic Diptera*: vol. 5, Athericidae-Asilidae. 446 pp. Elsevier, Amsterdam.

Chvála, M., Lyneborg, L. and **Moucha, J.** 1972. *The horse flies of Europe (Diptera, Tabanidae).* 500 pp. Entomological Society of Copenhagen, Copenhagen.

Coscarón, S. and **González, C. R.** 1991. Tabanidae de Chile: lista de especies y clave para los generos conocidos de Chile (Diptera: Tabanidae). *Acta Entomologica Chilena* **16**: 125–150.

Coscarón, S. and **Philip, C. B.** 1967. Revision del genero 'Dasybasis' Macquart en la region neotropical (Diptera-Tabanidae).*Revista del Museo Argentino de Ciencias Naturales "Bernardino Rivadavia"* (Entomologia) **2**: 15–266.

Coscarón, S. and **Philip, C. B.** 1979. A revision of Mycteromyiini ("genus *Mycteromyia*" of authors), a new tribe of Neotropical horse flies (Diptera, Tabanidae). *Proceedings of the California Academy of Sciences* (4) **41**: 427–452.

Cruz, J. de la and **Garcia Avila, I.** 1974. Los Tábanos (Diptera: Tabanidae) de Cuba. *Poeyana* **125**: 1–91.

Dias, J. A. Travassos Santos 1960. Nova contribução ao estudo dos tabanídeos (Diptera: Tabanidae) de Angola. *Publicações Culturais de Companhia de Diamantes de Angola* **53**: 1–125.

Dias, J. A. Travassos Santos 1966. *Tabanídeos (Diptera-Tabanidae) de Moçambique. Contribução para o seu conhecimento.* xvi + 1283 pp. Lourenço, Marques [= Maputo].

Daniels, G. 1989. Family Tabanidae. Pp. 277–294 in Evenhuis, N. L. (ed.), *Catalog of the Diptera of the Australasian and Oceanian regions.* 1155 pp. Bishop Museum Press, Honolulu, and E. J. Brill, Leiden.

Duke, B. O. L. 1972. Behavioural aspects of the life cycle of *Loa*. Pp. 97–107 in Canning, E. U. and Wright, C. A. (eds), *Behavioural aspects of parasite transmission.* xi + 219 pp. Academic Press, London.

Fairchild, G. B. 1966. Notes on Neotropical Tabanidae VI. A new species of *Lepiselaga* Macq. with remarks on related genera. *Psyche* **72** (1965): 210–217.

Fairchild, G. B. 1969. Notes on Neotropical Tabanidae XII. Classification and distribution, with keys to genera and subgenera. *Arquivos de Zoologia,* São Paulo **17**: 199–255.

Fairchild, G. B. 1971. *A catalogue of the Diptera of the Americas South of the United States. 28. Family Tabanidae.* 163 pp. Museu de Zoologia, Universidade de São Paulo.

Fairchild, G. B. 1976. Notes on Neotropical Tabanidae XVI. The *Tabanus trivittatus* complex. *Studia Entomologica* **19**: 237–261.

Fairchild, G. B. 1983. Notes on Neotropical Tabanidae (Diptera) XIX. The *Tabanus lineola* complex. *Entomological Society of America Miscellaneous Publications* **57**: 1–51.

Fairchild, G. B. 1984. Notes on Neotropical Tabanidae (Diptera) XX. The larger species of *Tabanus* of eastern and South America. *Contributions of the American Entomological Institute* **21** (3): 1–50.

Fairchild, G. B. 1986. The Tabanidae of Panama. *Contributions of the American Entomological Institute* **22** (3): 1–139.

Fairchild, G. B. and **Philip, C. B.** 1960. A revision of the Neotropical genus *Dichelacera*, subgenus *Dichelacera*, Macquart (Diptera, Tabanidae). *Studia Entomologica* **3**: 1–96.

Foil, L. D. 1989. Tabanids as vectors of disease agents. *Parasitology Today* **5**: 88–96.

Goodwin, J. T. 1982. The Tabanidae (Diptera) of Mali. *Entomological Society of America Miscellaneous Publications* **13**: 1–141.

Goodwin, J. T., Mullens, B. A. and **Gerhardt, R. R.** 1985. The Tabanidae of Tennessee. *University of Tennessee Agricultural Experiment Station Bulletin* **642**: ix + 1–73.

Hayakawa, H. 1980. Biological studies on *Tabanus iyoensis* group of Japan, with special reference to their blood-sucking habits (Diptera, Tabanidae). *Tohoku National Agricultural Experiment Station Bulletin* **62**: 131–321.

Hayakawa, H. 1985. A key to the females of Japanese tabanid flies with a checklist of all species and subspecies (Diptera, Tabanidae). *Japanese Journal of Sanitary Zoology* **36**: 15–23. [In English with Japanese summary.]

Hoppe, K. L., Dillwith, J. W., Wright, R. E. and **Szumlas, D. E.** 1990. Identification of horse flies (Diptera: Tabanidae) by analysis of cuticular hydrocarbons. *Journal of Medical Entomology* **27**: 480–486.

Inaoka, T., Hori, E., Yamaguchi, K., Watanabe, M., Yoneyama, Y. and **Ogunba, E. O.** 1988. Morphology and identification of *Chrysops* larvae from Nigeria. *Medical and Veterinary Entomology* **2**: 141–152.

Ivanishchuk, P. P. 1983. On the taxonomic status of two forms of *Chrysops caecutiens* (Tabanidae). *Parazitologiya* **17**: 223–228. [In Russian with English summary.]

Jacobson, N. R., Hansens, E. J., Vrijenhoek, R. C., Swofford, D. L. and **Berlocher, S. H.** 1981. Electrophoretic detection of a sibling species of the salt marsh greenhead, *Tabanus nigrovittatus. Annals of the Entomological Society of America* **74**: 602–605.

Jobling, B. 1987. *Anatomical drawings of biting flies.* 119 pp. British Museum (Natural History) and Wellcome Trust, London.

Jones, C. M. and **Anthony, D. W.** 1964. The Tabanidae (Diptera) of Florida. *United States Department of Agriculture Technical Bulletin* **1295**: 1–85.

Krinsky, W. L. 1976. Animal disease agents transmitted by horse flies and deer flies (Diptera: Tabanidae). *Journal of Medical Entomology* **13**: 225–275.

Leclercq, M. 1960. Révision systématique et biogéographique des Tabanidae (Diptera) paléarctiques. I. Pangoniinae et Chrysopinae. *Mémoires. Institut royal des Sciences naturelles de Belgique* (2) **63**: 1–77.

Leclercq, M. 1966. Révision systématique et biogéographique des Tabanidae (Diptera) paléarctiques. II. Tabaninae. *Mémoires. Institut royal des Sciences naturelles de Belgique* (2) **80**: 1–237.

Mackerras, I. M. 1954. The classification and distribution of Tabanidae (Diptera). *Australian Journal of Zoology* **2**: 431–454.

Mackerras, I. M. 1955a. The classification and distribution of Tabanidae (Diptera). II. History: morphology: classification: subfamily Pangoniinae. *Australian Journal of Zoology* **3**: 439–511.

Mackerras, I. M. 1955b. The classification and distribution of Tabanidae (Diptera). III. Subfamilies Scepsidinae and Chrysopinae. *Australian Journal of Zoology* **3**: 583–633.

Mackerras, I. M. 1956a. The Tabanidae (Diptera) of Australia. I. General review. *Australian Journal of Zoology* **4**: 376–407.

Mackerras, I. M. 1956b. The Tabanidae (Diptera) of Australia. II. Subfamily Pangoniinae, tribe Pangoniini. *Australian Journal of Zoology* **4**: 408–443.

Mackerras, I. M. 1957. Tabanidae (Diptera) of New Zealand. *Transactions of the Royal Society of New Zealand* **84**: 581–610.

Mackerras, I. M. 1959. An annotated catalogue of described Australian Tabaninae (Diptera, Tabanidae). *Proceedings of the Linnean Society of New South Wales* **84**: 160–185.

Mackerras, I. M. 1960. The Tabanidae (Diptera) of Australia. III. Subfamily Pangoniinae, tribe Scionini and supplement to Pangoniini. *Australian Journal of Zoology* **8**: 1–152.

Mackerras, I. M. 1961. The Tabanidae (Diptera) of Australia. IV. Subfamily Chrysopinae. *Australian Journal of Zoology* **9**: 827–905.

Mackerras, I. M. 1964. The Tabanidae (Diptera) of New Guinea. *Pacific Insects* **6**: 69–210.

Mackerras, I. M. 1971. The Tabanidae (Diptera) of Australia V. Subfamily Tabaninae, tribe Tabanini. *Australian Journal of Zoology* (Supplementary Series) **4**: 1–54.

Mackerras, I. M. and **Rageau, J.** 1958. Tabanidae (Diptera) du Pacifique Sud. *Annales de Parasitologie humaine et comparée* **33**: 671–742.

McAlpine, J.F., Peterson, B.V., Shewell, G.E., Teskey H.J., Vockeroth, J.R. and **Wood, D.M.** 1981. *Manual of Nearctic Diptera*: vol.1, vi + 674 pp. Research Branch, Agriculture Canada (Monograph No.27).

Middlekauff, W. W. and **Lane, R. S.** 1980. Adult and immature Tabanidae (Diptera) of California. *Bulletin of the California Insect Survey* **22**: 1–99.

Minter, D. M. 1987. Tabanidae. Horse flies, clegs, deer flies. Pp. 1437–1447 in Manson-Bahr, P. E. C. and Bell, D. R. (eds), *Manson's tropical diseases.* Nineteenth edition. xvii + 1557 pp. Baillière Tindall, London.

Moucha, J. 1976. Horse-flies (Diptera: Tabanidae) of the world. Synoptic catalogue. *Acta entomologica Musei nationalis Pragae* (Supplementum) **7**: 1–319.

Murdoch, W. P. and **Takahasi, H.** 1969. The female Tabanidae of Japan, Korea and Manchuria. The life history, morphology, classification, systematics, distribution, evolution and geologic history of the family Tabanidae (Diptera). *Memoirs of the Entomological Society of Washington* **6**: 1–230.

Oldroyd, H. 1952. *The horse-flies (Diptera: Tabanidae) of the Ethiopian region.* I. *Haematopota and Hippocentrum.* ix + 226 pp. British Museum (Natural History), London.

Oldroyd, H. 1954. *The horse-flies (Diptera: Tabanidae) of the Ethiopian region.* II. *Tabanus and related genera.* x + 341 pp. British Museum (Natural History), London.

Oldroyd, H. 1957. *The horse-flies (Diptera: Tabanidae) of the Ethiopian region*. III. *Subfamilies Chrysopinae, Scepsidinae and Pangoniinae and a revised classification.* xii + 489 pp. British Museum (Natural History), London.

Oldroyd, H. 1969. Diptera Brachycera Section (a) Tabanoidea and Asiloidea. *Handbooks for the Identification of British Insects* **9** (4): 1–132.

Oldroyd, H. 1973. Tabanidae (horse-flies, clegs, deer-flies, etc.). Pp. 195–202 in Smith, K. G. V. (ed.), *Insects and other arthropods of medical importance.* xiv + 561 pp. British Museum (Natural History, London.

Olsuf'ev, N. G. 1977. Horse-flies. Family Tabanidae. *Fauna of the USSR.* New Series. No. 113, *Insects, Diptera* **7** (2). 434 [+2] pp. Izdatel'stvo 'Nauka', Leningrad. [In Russian.]

Pechuman, L. L. 1973. The insects of Virginia: No. 6. Horse flies and deer flies of Virginia (Diptera: Tabanidae). *Virginia Polytechnic Institute and State University Research Division Bulletin* **81**: 1–92.

Pechuman, L. L. 1981. The horse flies and deer flies of New York (Diptera, Tabanidae). Second edition. *Search: Agriculture* **18**: 1–68.

Pechuman, L. L. and **Teskey, H. J.** 1981. Tabanidae. Pp. 463–478 in McAlpine, J. F., Peterson, B. V., Shewell, G. E., Teskey, H. J., Vockeroth, J. R. and Wood, D. M. *Manual of Nearctic Diptera*: vol. 1, vi + 674 pp. Research Branch, Agriculture Canada (Monograph No. 27).

Pechuman, L. L., Teskey, H. J. and **Davies, D. M.** 1961. The Tabanidae (Diptera) of Ontario. *Proceedings of the Entomological Society of Ontario* **91** (1960): 77–121.

Pechuman, L. L., Webb, D. W. and **Teskey, H. J.** 1983. The Diptera, or true flies, of Illinois. I. Tabanidae. *Illinois Natural History Survey Bulletin* **33**: 1–121.

Philip, C. B. 1954. New North American Tabanidae. VIII. Notes on and keys to the genera and species of Pangoniinae exclusive of *Chrysops*. *Revista Brasileira de Entomologia* **2**: 13–60.

Philip, C. B. 1955. New North American Tabanidae. IX. Notes on and keys to the genus *Chrysops* Meigen. *Revista Brasileira de Entomologia* **3**: 47–128.

Philip, C. B. 1959. Philippine Zoological Expedition 1946–1947. Tabanidae (Diptera). *Fieldiana: Zoology* **33**: 543–625.

Philip, C. B. 1960. Malaysian parasites XXXVI. A summary review and records of Tabanidae from Malaya, Borneo, and Thailand. *Studies from the Institute for Medical Research, Federation of Malaya* **29**: 33–78.

Philip, C. B. 1965. Family Tabanidae. Pp. 319–342 in Stone, A., Sabrosky, C. W., Wirth, W. W., Foote, R. H. and Coulson, J. R. (eds), *A catalog of the Diptera of America North of Mexico*. Agriculture Handbook 276, iv + 1696 pp. U.S. Department of Agriculture, Washington, D.C.

Philip, C. B. 1978. New North American Tabanidae (Insecta, Diptera) XXIV. Further comments on certain Pangoniinae in Mexico with special reference to *Esenbeckia*. *Proceedings of the California Academy of Sciences* (4) **41**: 345–356.

Philip, C. B. and **Fairchild, G. B.** 1956. American biting flies of the genera *Chlorotabanus* Lutz and *Cryptotylus* Lutz (Diptera, Tabanidae). *Annals of the Entomological Society of America* **49**: 313–324.

Rodhain, F. 1980. Hypothèses concernant l'écologie dynamique des infections à *Loa*. *Bulletin de la Société de Pathologie exotique* **73**: 182–191.

Schuurmans Stekhoven, J. H. 1926. The bloodsucking arthropods of the Dutch East Indian archipelago VII. The tabanids of the Dutch East Indian archipelago (including those of some neighbouring countries). *Treubia* **6** (Supplement): 1–552.

Senior-White, R. 1927. *Catalogue of Indian insects*: Part 12, *Tabanidae*. 70 pp. Government of India, Calcutta.

Stone, A. 1938. The horseflies of the subfamily Tabaninae of the Nearctic region. *United States Department of Agriculture Miscellaneous Publication* **305**: 1–171.

Stone, A. 1975. Family Tabanidae. Pp. 43–81 in Delfinado, M. D. and Hardy, D. E. (eds), *A catalog of the Diptera of the Oriental region*: vol. 2, *Suborder Brachycera through Division Aschiza, suborder Cyclorrhapha*. 459 pp. University Press of Hawaii, Honolulu.

Stone, A. and **Philip, C. B.** 1974. The Oriental species of the tribe Haemotopotini (Diptera, Tabanidae). *United States Department of Agriculture Technical Bulletin* **1489**: 1–240.

Tendeiro, J. 1965. Novas observações sobre tabanídeos da Guiné Portuguesa. *Revista dos Estudos Gerais Universitários de Moçambique* (4, Ciências Veterinarias) **1** (1964): 1–256.

Teskey, H. J. 1969. Larvae and pupae of some eastern North American Tabanidae (Diptera). *Memoirs of the Entomological Society of Canada* **63**: 1–147.

Teskey, H. J. 1990. The insects and arachnids of Canada. Part 16. The horse flies and deer flies of Canada and Alaska: Diptera: Tabanidae. *Research Branch Agriculture Canada Publication* **1838**: 1–381.

Theodor, O. 1965. Tabanidae of Israel. *Israel Journal of Zoology* **14**: 241–257.

Tidwell, M. A. 1973. The Tabanidae (Diptera) of Louisiana. *Tulane Studies in Zoology and Botany* **18**: 1–95.

Usher, P. J. 1972. A review of the South African horsefly fauna (Diptera: Tabanidae). *Annals of the Natal Museum* **21**: 459–507.

Wang, Z.-m. 1983. *Economic insect fauna of China. 26. Diptera Tabanidae.* vi + 128 pp. Science Press, Beijing. [In Chinese.]

Wilkerson, R. C. 1979. Horse flies (Diptera: Tabanidae) of the Colombian Departments of Choco, Valle, and Cauca. *Cespedesia* **8**: 89–433. [Pp. 89–98 comprise a title page and keys in Spanish: a repeated title and the remainder of the work, pp. 99–433, are in English.]

Wilkerson, R. C. and **Coscarón, S.** 1984. A review of South American *Scaptia* (*Pseudoscione*) (Diptera: Tabanidae). *Journal of Medical Entomology* **21**: 213–236.

Wilkerson, R. C. and **Fairchild, G. B.** 1982. Five new species of *Diachlorus* (Diptera: Tabanidae) from South America with a revised key to species and new locality records. *Proceedings of the Entomological Society of Washington* **84**: 636–650.

Wilkerson, R. C. and **Fairchild, G. B.** 1983. A review of the South American species of *Esenbeckia* subgenus *Esenbeckia* (Diptera: Tabanidae). *Journal of Natural History* **17**: 519–567.

Wilkerson, R. C., Butler, J. F. and **Pechuman, L. L.** 1985. Swarming, hovering, and mating behavior of male horse flies and deer flies (Diptera: Tabanidae). *Myia* **3**: 515–546.

World Health Organization 1989. *Geographical distribution of arthropod-borne diseases and their principal vectors.* WHO/VBC/89.967, 134 pp. World Health Organization, Geneva.

Xu, R.-m. 1982. Identification of important Tabanidae in China. Pp. 237–342 in Lu, B. S. (ed.), *Identification handbook for medically important animals in China.* 956 pp. People's Health Publishing Company, Beijing. [In Chinese.]

CHAPTER NINE

Tsetse-flies (Glossinidae)

A. M. Jordan

The Glossinidae, or tsetse-flies, form a monogeneric family of the Diptera. The adults range in length from 6 to 14 mm and in all the 23 known species are various shades of brown – ranging from light yellowish brown to dark blackish brown. In some species the abdomen has alternate darker and lighter bands. Female flies give birth, at intervals of about nine days, to a single third-instar larva which rapidly burrows into the soil and transforms into a black puparium; according to the species, this varies in length from 3 to 8 mm.

Apart from two known localities in the Arabian peninsula, tsetse-flies occur only in Africa south of the Sahara desert and north of the temperate climes of the south of the continent. Within these limits, high ground where winter temperatures are too low, and areas which are devoid of woody vegetation (either naturally or because of the activities of man), are free of these insects. Some 11 million km^2 of Africa are infested. Habitats range from lowland rain forest, in West Africa and the Congo Basin, through the vast and varied savanna woodlands, to arid thicket vegetation on the margins of Africa's deserts. Fossil *Glossina*, attributed to four species, have been reported from sedimentary shales in Colorado, USA (Cockerell, 1918); there is some doubt as to the age of the deposits, but they are generally assigned to the Oligocene geological epoch (38–26 million years BP).

All tsetse-flies feed exclusively on blood, mainly from mammals but some species take meals from reptiles and birds. During feeding, tsetse-flies can transmit protozoan parasites of the genus *Trypanosoma* whose normal vertebrate hosts are the wild large mammals of Africa; the latter, unless stressed, generally do not suffer any ill-effects from the infections. Trypanosomes undergo a cycle of development within the fly, and once a fly has matured an infection it can remain infected for life. The circulation of trypanosomes between tsetse-flies and wild mammals has no practical significance until an infected fly feeds on a human or a domestic animal in which trypanosomes are pathogenic. Not all *Trypanosoma* species are equally infective to all mammals. Thus, only *T. brucei gambiense* and *T. brucei rhodesiense* infect man; *T. vivax* and *T. congolense* are important parasites of domestic animals (particularly cattle) and *T. simiae* is an important parasite of pigs.

Disease caused by *Trypanosoma* in man is commonly referred to as sleeping sickness. Only a few tsetse species have been incriminated as vectors, and even these do not transmit the disease throughout their ranges. However, all *Glossina* species are potential vectors of animal trypanosomiasis, although some are more important than others, partly depending on the type of habitat that they occupy.

Medical Insects and Arachnids Edited by Richard P. Lane and Roger W. Crosskey.
Published in 1993 by Chapman & Hall ISBN 0 412 40000 6

In the early years of the twentieth century many thousands of people died from sleeping sickness, particularly around Lake Victoria, in the Congo Basin and in the countries within the wide arc of the River Niger. Today, only some 25 000 new cases are reported each year (TDR, 1990) but many other sufferers from the disease undoubtedly go undiagnosed and uncounted in the, often remote, rural areas where the disease occurs.

Early attempts at vector control were mainly directed at those species of tsetse-flies of medical importance, but with the regression of the major epidemics, attention has shifted to the species primarily of veterinary importance. Vector control remains an important option in epidemic foci of the human disease.

RECOGNITION AND ELEMENTS OF STRUCTURE

The adult is the life stage of tsetse-flies most obvious to the casual observer and the stage most studied. Adults can be recognized in life by their habit of resting with the wings closed over the abdomen like the closed blades of a pair of scissors, with the tips extending slightly beyond the end of the abdomen, and by their long piercing proboscis, sheathed by two maxillary palps and projecting forward underneath the head (Fig. 9.1). Two other features distinguish the genus *Glossina* from all other Diptera: the discal medial cell (dm) of the wing, which is shaped like a cleaver and referred to as the 'hatchet' cell (Fig. 9.2*a*); and the presence of secondary branches on the hairs located on the upper surface of the arista of the antenna (Fig. 9.2*b*). The compound eyes are large, reddish brown and, unlike those of some other Diptera, widely separated in both sexes (Fig. 9.3). There are three ocelli located between the compound eyes. The antennae have two small basal segments and a third much longer one. Some features of the third segment are of taxonomic significance in some species. The arista, with its branched hairs, arises from near the base of the third segment. Chemoreceptors, thermoreceptors and hygroreceptors (Glasgow, 1970) are all located on the antennae and it has been suggested that other receptors may act as airspeed indicators (Vanderplank, 1950). The mouthparts (Fig. 9.4) of the sexes are identical, as both feed on blood. The proboscis consists of three parts, the labrum, labium, and hypopharynx, and is sheathed by two modified maxillary palps equal to it in length. The proboscis arises from a bulb-shaped base located directly under the head. Detailed descriptions of the mouthparts are given by Newstead et al. (1924) and Jobling (1933). The labium is U-shaped in cross-section and the most rigid of the components of the proboscis. The labrum, which encloses a tubular space, the food channel, along which imbibed blood passes, fits into the U of the labium. Between these two structures lies the delicate hypopharynx, which is a tube continuous with the common duct of the salivary glands and down which saliva, containing an anticoagulant, is pumped into the wound made by the proboscis. The tip of the labium, the labellum, is armed with teeth for piercing the skin of the host and either enters a capillary directly or breaks up capillaries to form a blood-pool from which fluid is sucked into the food channel by the action of the muscular pharyngeal pump. The mid-dorsal surface of the thorax is crossed by a groove (transverse suture). The scutellum is a flattened triangular hump, slightly elevated apically. Some features of the thorax (Fig. 9.5) have taxonomic significance, including the anepimeron (pteropleuron) and katepisternum (sternopleuron), the lower calypter (thoracic squama, a membranous

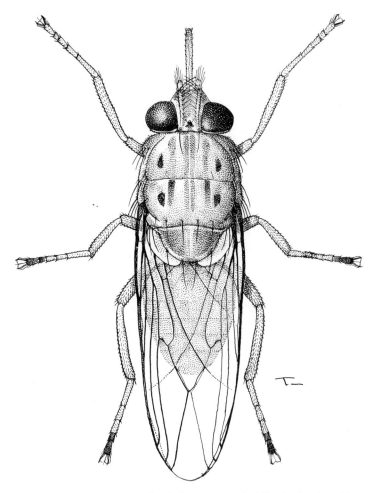

Figure 9.1 Resting attitude of a living tsetse-fly (*Glossina longipennis*).

lobe attached above the halter) and the greater ampulla (infra-alar bulla, a protuberance just in front of the wing base). The thoracic markings vary in intensity from species to species. The coloration of some of the tarsal segments is used to distinguish between species. The venation of the wings is common to all species (Fig. 9.3); supernumerary stub-veins occur in some individuals and appear to develop when the puparium is exposed to unduly high temperatures (Glasgow, 1960).

The abdomen has eight segments. It is able to distend greatly after feeding and (in the female) to accommodate the fully grown third-instar larva. The hectors of males are two lobes of a thickened plate, covered with short stiff hairs, on the sternum of the fifth abdominal segment. The sexes can be readily distinguished by the presence in the male of the button-like hypopygium, hinged under the posterior end of the abdomen (Fig. 9.6). The form of the male and female genitalia is of major taxonomic significance and it is impossible to distinguish between some closely related species on any morphological criteria other than features of the genitalia.

The male hypopygium is composed of a series of structures, the most conspicuous of which are the superior claspers hinged at their base along the posterior margin of the hypopygium. These clasp the end of the female abdomen

336 *Tsetse-flies (Glossinidae)*

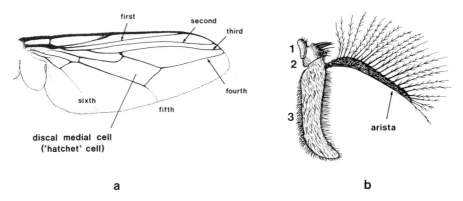

Figure 9.2 Two principal diagnostic features of the genus *Glossina*: (*a*) wing venation, showing the characteristic hatchet shape of the discal medial cell (dm) and the convenience notation of the long veins; (*b*) antenna, showing the three segments and the arista with its characteristically branched hairs. The antenna illustrated is that of *G. pallicera pallicera*.

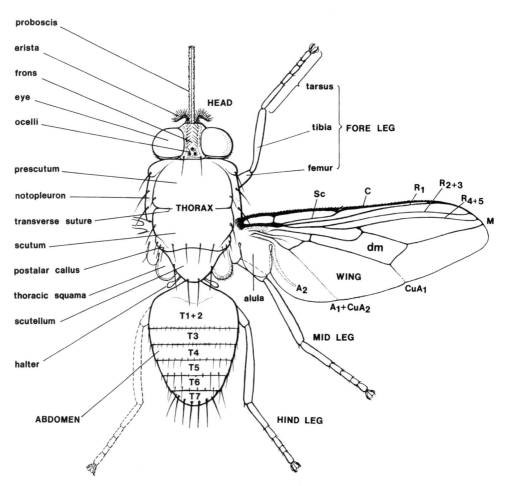

Figure 9.3 Basic morphology of an adult tsetse-fly. T = abdominal tergite. Wing vein and cell notation are those adopted for the higher Diptera by McAlpine et al. (1981).

Recognition and elements of structure 337

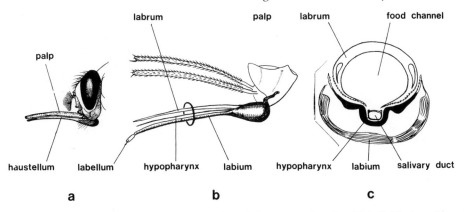

Figure 9.4 Proboscis of *Glossina*: (*a*) side view of head showing proboscis with bulb-like base beneath paired palps (latter slightly raised and only near one shown); (*b*) detail of proboscis and palps with latter separated from haustellum; (*c*) diagrammatic cross-section of the haustellum at position circled in (*b*). (Illustrations (*b*) and (*c*) modified from Newstead et al., 1924.)

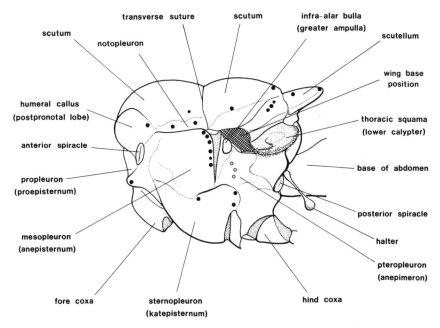

Figure 9.5 Left side of thorax of *Glossina* showing landmark features. Solid black dots show base positions of the major bristles; open circles on pteropleuron indicate positions where large bristles are present in *G. fusca* group (not in other groups).

during mating. The inferior claspers are smaller than the superior claspers but are more important for taxonomic purposes as, in many instances, there are striking morphological differences between closely allied species. The aedeagus, the distal part of the phallosome, is composed of a number of different structures which show much variation within the genus. These structures include the sclerotized harpes which, in the *fusca* group, often terminate in characteristic free points of great taxonomic significance (Fig. 9.11).

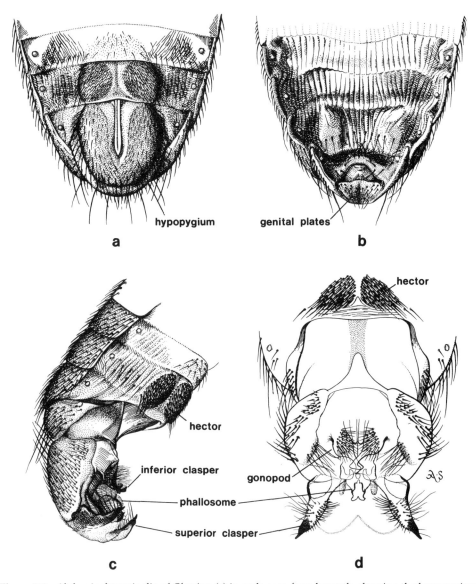

Figure 9.6 Abdominal terminalia of *Glossina*: (*a*) in male seen from beneath, showing the hypopygium drawn up into the abdomen; (*b*) in female seen from beneath, showing absence of hypopygium; (*c*) in male seen from ventrolateral position with hypopygium extended; (*d*) male hypopygium seen after maceration and flattening.

The external genital armature of the female comprises sclerotized plates surrounding the anus and vulva. The maximum number of plates is six: paired dorsal plates and anal plates and a single sternal plate and mediodorsal plate (Fig. 9.12). In some species groups one or more of these plates may be fused or missing. In 12 of the 13 species of the *fusca* group, a sclerotized plate, the signum, is present on the inner surface of the dorsal wall of the uterus near its anterior end. The form of the signum is the only reliable morphological criterion for distinguishing between females of some species of the *fusca* group (Fig. 9.15). The free-living third-

Recognition and elements of structure 339

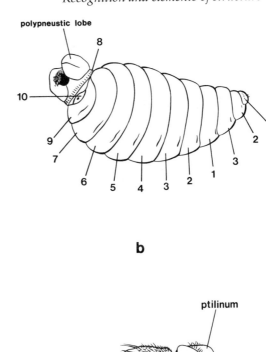
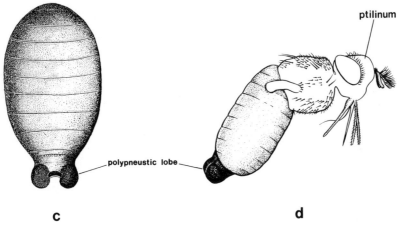

Figure 9.7 Larva, puparium and adult emergence of *Glossina*: (*a*) third-instar larva of *G. swynnertoni* during peristaltic movement; (*b*) third-instar larva of *G. swynnertoni* from ventrolateral aspect, diagrammatic, showing segmentation; (*c*) typical tsetse puparium, showing posterior spiracles developed into massive polypneustic lobes, (*d*) emerging adult of *G. morsitans*, showing the ptilinum with which the fly opens the puparium and effects its escape from the soil (see text). (Illustrations (*a*) and (*b*) based on Burtt and Jackson, 1951; (*d*) based on Newstead et al., 1924.)

instar larva lasts for only an hour or so. It is segmented, white in colour and characterized by black polypneustic lobes at the posterior end; these have a respiratory function during intra-uterine life and possibly during the short active extra-uterine life. There is an antenno-maxillary mass at the anterior end. Because of the brief existence of the free-living larva possible morphological differences between the species have not been investigated. The puparium (Fig. 9.7c) is also characterized by the prominent polypneustic lobes but during the development of the pupa and the subsequent metamorphosis of the imago these perform no respiratory function (Bursell, 1955). The form of the two lobes and the cavity between them can be used to distinguish between certain species (Fig. 9.16).

340 Tsetse-flies (Glossinidae)

CLASSIFICATION AND IDENTIFICATION

Higher classification

Many early authorities (e.g. Newstead et al., 1924) included *Glossina* in the family Muscidae, either within the Stomoxyinae, or in a separate subfamily, the Glossininae. However, it is now generally accepted that the Glossinidae is a monogeneric family (Brues et al., 1954; Haeselbarth et al., 1966) since *Stomoxys* is a true muscid (Zumpt, 1973) with little phylogenetic relationship to *Glossina*.

The relationships between the Glossinidae and other families of Diptera remain somewhat uncertain but it is widely accepted that the nearest living relatives of the Glossinidae are the louse-flies of the family Hippoboscidae (Bequaert, 1954; Bezzi, 1911; Griffiths, 1976; Hennig, 1965). There are two contrasting views on the origin of the tsetse-flies. The first (Pollock, 1971, 1973), postulates that the common ancestor of Hippoboscidae and Glossinidae closely resembled extant Gasterophilidae in terms of wing venation, thoracic musculature, abdominal segmentation and male genitalia. The three families are united in the superfamily Gasterophiloidea and can be defined in terms of their modification from the postulated basic gasterophiloid type as follows: Gasterophilidae, generalized gasterophiloid flies having atrophied mouthparts in the adult; Glossinidae, free-living gasterophiloid flies modified for adenotrophic viviparity; Hippoboscidae, gasterophiloid flies modified for adenotrophic viviparity and ectoparasitism.

The second hypothesis (Griffiths, 1976) disputes the gasterophiloid origin of *Glossina* and reverts to the conventional view that Gasterophilidae are closely related to oestroid flies, and that the Hippoboscidae are clearly members of the 'Pupipara' with the Streblidae and Nycteribiidae. The Glossinidae are regarded as a sister group of the 'Pupipara' (which Griffiths terms Hippoboscidae s.l.). The relationship between Glossinidae and 'Pupipara' (= Hippoboscidae s.l.) is not endorsed by Pollock despite a statement to the contrary by Griffiths.*

Subgenera

The genus *Glossina* can be divided into three distinct species groups, based upon features of the genitalia of both sexes (Newstead et al., 1924). Some authorities (Haeselbarth et al., 1966; Zumpt, 1936) have given these the status of subgenera.

Males of the *fusca* group (subgenus *Austenina*) are characterized by free superior claspers, without a membrane between them (Fig. 9.8b). The forms of the harpes vary markedly between species but often have free points. There are five genital plates in the female, one dorsal pair, one anal pair and a single median sternal plate (Figs 9.8a, 9.14). A signum is present in all species except *G. brevipalpis*.

In males of the *palpalis* group (subgenus *Nemorhina*) the superior claspers are connected by a thin membrane, deeply divided medially (Fig. 9.8d). There are six

---

* Editors' note: In this book we have followed the classification of Wood (1987) and McAlpine (1989) under which the 'gasterophilids' are treated as a subfamily of Oestridae (see Chapter 12) and the Glossinidae treated as a family of the Hippoboscoidea (= erstwhile 'Pupipara' + Glossinidae, see Chapter 3). We consider the question of the relationships of Glossinidae essentially settled. See discussion and character analysis in McAlpine (1989).

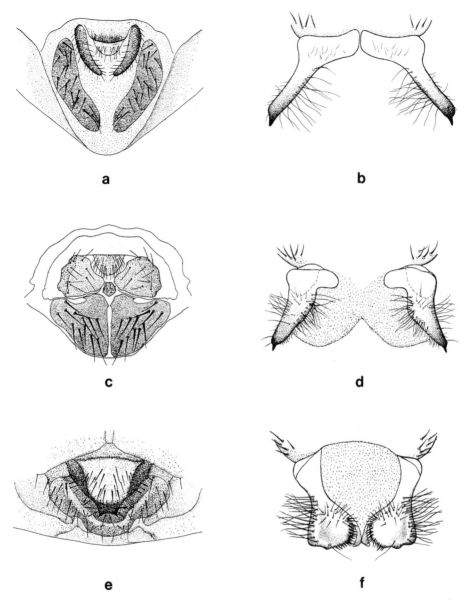

Figure 9.8 Typical form of female external genital armature (left) and male genital superior claspers (right) in the three species-groups of *Glossina*: (*a* and *b*) *G. fusca* group; (*c* and *d*) *G. palpalis* group; (*e* and *f*) *G. morsitans* group. (After Newstead et al., 1924.)

genital plates in the female (Fig. 9.8*c*). In addition to those present in the *fusca* group, there is a small mediodorsal plate.

The superior claspers of the males of the *morsitans* group (subgenus *Glossina* s.str.) are completely joined by a membrane and are fused distally (Fig. 9.8*f*). Their shape resembles that of the mammalian scapula. In the female there is a pair of fused anal plates (possibly including, medially, the mediodorsal plate of the *palpalis* group) and a median sternal plate (Fig. 9.8*e*). Dorsal plates are generally absent but occur in a reduced form in *G. austeni*.

Dias (1987) considered that differences between *G. austeni* and the other species of the *morsitans* group are sufficient to warrant the establishment of a fourth subgenus, with the proposed name of *Machadomyia*, to contain the single species, *G. austeni*. Whereas *G. austeni* does display some unusual features, the male genital armature, in particular, is clearly of the same form as the other species of the *morsitans* group, consequently the erection of a new subgenus is here considered not justified.

The division of the genus into three groups is in the main supported by the types of habitat occupied by the various species. The species of the *fusca* group typically occur in the lowland rain forests of West and West–Central Africa or the somewhat drier forest outliers in savanna woodlands around the margins of the forest. *Glossina longipennis* is an exception as it occurs in arid habitats in East Africa. The species of the *palpalis* group are also basically forest-dwellers although, additionally, the economically important species, *G. palpalis*, *G. fuscipes* and *G. tachinoides*, extend linearly far out into the African savannas along rivers, streams and lake shores. The species of the *morsitans* group typically occur in savannas ranging from moist savannas on the margins of the forest to dry savannas near the margins of the African deserts.

Most authors (e.g. Evens, 1953; Newstead et al., 1924) have considered that the series *fusca* group–*palpalis* group–*morsitans* group represents the course of evolution in *Glossina*. This interpretation is based on Newstead's view of the form of the male superior claspers alone (Machado, 1959). If the structure of the aedeagus and the female genital armature are considered, it is possible to come to a different conclusion. Thus the aedeagus is most complex in the *fusca* group and least so in the *morsitans* group. Alternatively, taken in isolation, the arrangement and full complement of six genital plates in the *palpalis* group might suggest that this group is the most primitive.

An ecological argument (Bursell, 1958) has been advanced to support the view that the *fusca* group contains the most primitive *Glossina* as all but one of these species are associated with lowland rain forest or forest relics assumed to represent the ancestral habitat. On the basis of the ability of their puparia to resist desiccation *G. brevipalpis* and *G. fuscipleuris* (of the *fusca* group) were considered closest to the ancestral type and a number of species of the *morsitans* group (and the exceptional *G. longipennis* of the *fusca* group) the most advanced – as the latter are able to survive in dry environments.

The assumption that the ancestral habitat of *Glossina* was humid forest has been challenged because *G. longipennis* has the most primitive features in the *fusca* group and is also the most xerophytic of all tsetse (Machado, 1959). It has been concluded, on the basis of various features of comparative anatomy, especially the elaborate harpes of the male and the presence of a signum in the female, that the *fusca* group contains the most specialized *Glossina* species and the *morsitans* group the most primitive (Pollock, 1974). The form of the superior claspers in the three groups involves the progressive deletion of the median area (the median lobes of the *morsitans* group and the membrane of the *morsitans* and *palpalis* groups) of the superior claspers. Thus Pollock envisaged the ancestral *Glossina* as the open savanna woodland *morsitans* group which gave rise to the forest/riverine *palpalis* group from which in turn the forest-dwelling *fusca* group was derived. *Glossina longipennis* was an early offshoot of the *fusca* group which reverted to the original dry habitat of the genus.

Species and subspecies

The genus *Glossina* is generally considered to comprise 23 species (Table 9.1), the exact number depending on the rank accorded to taxa by different authorities.

The species of the *fusca* group have been studied less than those of the *palpalis* and *morsitans* groups, perhaps because they are of limited economic importance. Based on features of the male and female genitalia, the *fusca* group can be divided into two subgroups. The first comprises *G. fusca*, *G. haningtoni*, *G. nashi*, *G. tabaniformis*, *G. schwetzi*, *G. vanhoofi* and *G. fuscipleuris* and has a generally western or central African distribution; the second comprises the less closely related *G. brevipalpis* and *G. longipennis* with an eastern African distribution (Machado, 1959); *Glossina nigrofusca* is intermediate between the two subgroups (Machado, 1959) and possesses a number of primitive features (Pollock, 1974). *Glossina medicorum* is more closely related to the first subgroup and *G. severini*, showing a mixture of features, is difficult to place (Machado, 1959). The most recently described tsetse species, *G. frezili*, is very closely related to *G. medicorum* (Gouteux, 1987). *Glossina frezili* has been recorded from Gabon (Maillot, 1957, as *medicorum*) and the Republic of Congo. The two species are separated geographically and occur in different habitats, *G. medicorum* in dry forest in West Africa and *G. frezili* in mangrove swamps (together with *G. caliginea* of the *palpalis* group).

Two species of the *fusca* group have been split into subspecies on the basis of their geographical isolation and morphological differences. *Glossina nigrofusca nigrofusca*, found in West Africa, is distinguishable from *G. n. hopkinsi*, which occurs from Chad to Uganda, by differences in the shape and pilosity of the antennae. However, in both the male and female genitalia the two subspecies are identical (Machado, 1959; Van Emden, 1944). Two subspecies of *G. fusca* are recognized (Jordan, 1965; Le Berre and Itard, 1960; Machado, 1959). *Glossina f. fusca* is a macrophallic form and occurs in lowland rain forest west of the Togo–Benin savanna gap and *G. f. congolensis* is a microphallic form occurring eastwards of this gap and continuing southwards around the edge of the central African rain forest to southern Zaire; however a few specimens of the latter subspecies have been found in Ghana. In addition to the subdivision of *G. fusca*, Machado (1959) recognized a geographical cline in the length of the first pair of harpes within *G. f. congolensis*, which are shorter in southern populations than in northern populations.

The *palpalis* group is a much more homogeneous assemblage of species than the *fusca* group. The rank of the taxa within the group proposed by Machado (1954) is now generally accepted. Three of the species in the group have been divided into subspecies on features of both the male and female genitalia. Within *G. palpalis*, *G. p. gambiensis* occurs to the west of the Togo–Benin savanna gap and *G. p. palpalis* mainly to the east (Challier et al., 1984), its distribution extending east to Cameroun and thence south, parallel with the Atlantic seaboard of Africa, as far as Angola. There is an extensive zone of overlap of the two subspecies in which intermediate forms frequently occur. Within the three subspecies of *G. fuscipes* some variation occurs but no incontrovertible transitional forms have been described in the limited zones of overlap (Machado, 1954). The three subspecies together occupy the block of lowland rain forest in the Congo Basin, extending outwards along watercourses into the surrounding savannas. Their distribution is extended to the east to include the shores and river systems associated with some of the great lakes of East Africa,

Table 9.1 Geographical distribution of tsetse-fly (*Glossina*) species and subspecies

The African countries with tsetse infestations are grouped as western, central, eastern and southern Africa, although some larger countries could be considered as belonging to two group areas. Species and subspecies within species groups are tabulated in a sequence approximating to that in which they appear across Africa directionally from western parts to central, eastern and southern parts of the continent.

	G. fusca group															G. palpalis group									G. morsitans group						
	fusca fusca	nigrofusca nigrofusca	tabaniformis	medicorum	fusca congolensis	haningtoni	fuscipleuris	nkashi	schwetzi	frezili	severini	vanhoofi	nigrofusca hopkinsi	longipennis	brevipalpis	palpalis gambiensis	tachinoides	palpalis palpalis	pallicera pallicera	caliginea	pallicera newsteadi	fuscipes fuscipes	fuscipes quanzensis	fuscipes martinii	morsitans submorsitans	longipalpis	pallidipes	morsitans centralis	austeni	swynnertoni	morsitans morsitans
Western Africa																															
Senegal																x									x						
The Gambia																x									x						
Guinea-Bissau	x															x									x	x					
Guinea	x	x	x													x	x	x	x						x	x					
Sierra Leone	x	x														x		x	x						x	x					
Mali				x												x	x								x	x					
Liberia	x	x	x																												
Ivory Coast	x	x	x	x												x	x	x	x												
Burkina Faso				x												x	x	x	x						x	x					
Ghana	x	x	x	x	x											x	x	x	x	x					x	x					
Togo	x															x	x	x							x	x					
Benin				x	x											x	x								x	x					
Niger																	x								x						
Nigeria		x		x	x	x										x	x	x	x	x					x	x					
Chad																	x								x						
Central Africa																															
Cameroun	x	x	x		x	x	x	x	x							x	x	x	x	x	x	x			x	x					
Cent. Afr. Rep.	x	x	x		x	x	x	x	x								x	x			x	x			x						
Eq. Guinea		x	x		x	x		x	x									x				x									
Gabon		x	x		x	x		x	x	x								x		x	x	x									

Table 9.1 (cont.)

G. fusca group

	fusca fusca	nigrofusca nigrofusca	tabaniformis	medicorum	fusca congolensis	haningtoni	fuscipleuris	nashi	schwetzi	frezili	severini	vanhoofi	nigrofusca hopkinsi	longipennis	brevipalpis
Congo			×		×	×		×	×	×					
Zaire			×		×	×	×		×		×	×	×		×
Eastern Africa					×		×								
Sudan														×	
Ethiopia															×
Somalia														×	×
Kenya				×			×						×	×	×
Uganda					×		×							×	×
Rwanda															
Burundi															
Tanzania						×								×	×
Southern Africa							×	×	×					×	
Angola			×												×
Zambia															×
Malawi															×
Mozambique															
Zimbabwe															
Botswana															
Namibia															
South Africa															×

G. palpalis group

	palpalis gambiensis	tachinoides	palpalis palpalis	pallicera pallicera	caliginea	pallicera newsteadi	fuscipes fuscipes	fuscipes quanzensis	fuscipes martinii
Congo			×			×	×	×	
Zaire			×			×	×	×	×
Eastern Africa		×							
Sudan		×					×		
Ethiopia							×		
Somalia									
Kenya							×		×
Uganda							×		×
Rwanda									×
Burundi									×
Tanzania									
Southern Africa			×			×	×	×	
Angola							×		
Zambia							×		
Malawi									
Mozambique									
Zimbabwe									
Botswana									
Namibia									
South Africa									

G. morsitans group

	morsitans submorsitans	longipalpis	pallidipes	morsitans centralis	austeni	swynnertoni	morsitans morsitans
Congo							
Zaire	×	×	×	×			
Eastern Africa	×		×				
Sudan	×		×				
Ethiopia			×				
Somalia	×		×		×		
Kenya	×		×	×	×	×	
Uganda	×		×	×			
Rwanda				×			
Burundi				×			
Tanzania			×	×	×	×	×
Southern Africa				×		×	
Angola			×	×			
Zambia			×	×			×
Malawi			×		×		×
Mozambique			×				×
Zimbabwe							×
Botswana				×			
Namibia				×			
South Africa					×		

particularly Lakes Victoria and Tanganyika. *Glossina f. fuscipes* occurs in the north and east of this area, *G. f. quanzensis* to the southwest and *G. f. martinii* to the southeast. The two, geographically separated, subspecies of *G. pallicera* (*G. p. pallicera* and *G. p. newsteadi*) are distinguished by features of the genitalia and the degree of pilosity of the third segment of the antenna.

The five species of the *morsitans* group can be aggregated on morphological criteria into two pairs of closely related species and one somewhat aberrant species, *G. austeni*. Although the two pairs of species, following now generally accepted practice, are considered distinct species (e.g. Machado, 1966) some consider them conspecific (e.g. Haeselbarth et al., 1966). Thus *G. longipalpis* and *G. pallidipes* can only be separated by the coloration of the last tarsal segment of the fore leg and minor differences of the male genitalia. The two species are separated geographically by about 1000 km. *Glossina morsitans* and *G. swynnertoni* are considered separate species (Machado, 1966) but laboratory crosses between the subspecies of *G. morsitans* and *G. swynnertoni* indicate that *G. swynnertoni* is no more distinct genetically from the subspecies of *G. morsitans* than these are from each other (Curtis, 1972). There is some confusion in the literature concerning the minor taxa of *G. morsitans* but the revision of Machado (1970) is now generally accepted. He recognized three 'races géographique majeures' which are now generally regarded as subspecies. The three subspecies of *G. morsitans* and *G. swynnertoni* can be distinguished most readily by features of the male genitalia; they are generally geographically isolated from one another but there is believed to be a zone of overlap of *G. m. morsitans* and *G. m. centralis* in Zambia and there is an overlap of *G. m. centralis* and *G. swynnertoni* in Tanzania. Dias (1956, 1987) considered that *G. austeni* should be divided into two subspecies, *G. a. austeni* and *G. a. mossurizensis* on the grounds that the latter is larger and darker than the former and that the two are geographically isolated from one another, with *G. a. mossurizensis* being restricted to dense wet forest in the Sitatonga Hills of west Mozambique. This ranking is not generally accepted.

The classification of *Glossina* is generally uncontroversial, and there is wide agreement as to which tsetse should be ranked as species or subspecies. Nevertheless, a few outstanding questions remain to be satisfactorily answered: should *G. pallidipes* be a subspecies of *G. longipalpis*? Should *G. swynnertoni* be a subspecies of *G. morsitans*? Is *G. austeni mossurizensis* a valid subspecies?

Because of the generally well-accepted taxonomy there has been little incentive to investigate methods other than morphology for distinguishing between species. However, the karyotypes of some *Glossina* species have been determined (Gooding, 1984; Itard, 1973; Southern, 1980) and in general found to confirm the division of the genus into three species-groups on morphological grounds. Only two species of the *fusca* group have been examined: in *G. fusca congolensis* $2n = 22$ and in *G. brevipalpis* $2n = 16$. It seems that relatively large numbers of chromosomes may prove to be typical of the *fusca* group. In the *palpalis* group, *G. fuscipes fuscipes*, *G. palpalis palpalis* and *G. tachinoides* all have $2n = 6$. All five species of the *morsitans* group also have $2n = 6$ large chromosomes but, usually, there are an additional but variable number (up to $2n = 8$) of small supernumerary or B chromosomes. The number of supernumeraries can vary both between species and between individuals of a single species or subspecies. Polytene chromosome maps have been constructed for *G. m. morsitans* and *G. austeni* (Southern and Pell, 1974; Southern et al., 1973) but further data of this type are required before interrelationships between the various taxa of *Glossina* can be determined on cytological criteria.

Some use has been made of electrophoretic techniques to examine the genetic diversity of a few natural populations of *Glossina* but although the data available are too scanty to significantly influence interpretation of speciation within the genus, Gooding (1982) placed *G. austeni* in the *palpalis* group rather than in the *morsitans* group. A phylogenetic tree has been constructed for various taxa of *Glossina* based on studies at the molecular level which largely confirms the accepted morphology-based phylogeny and, in particular, indicates that *G. austeni* is clearly more closely related to the species of the *morsitans* group than to the *palpalis* group (Cross and Dover, 1987).

Species and subspecies of *Glossina* can also be distinguished by qualitative and quantitative differences in cuticular hydrocarbons. There are striking differences, in both males and females, between the *morsitans* and *palpalis* groups but while detailed examination of taxa within these groups indicates the existence of variation in different quantities and patterns of cuticular components between species there is much less variation between subspecies (Carlson, 1981). However, there is still uncertainty whether species and subspecies show consistent patterns over their often large range.

The *fusca* group have more varied compounds than those of the other two groups, making gas chromatography–mass spectrometry (GC–MS) necessary for detailed study. Most of the species of the *fusca* group from central and West Africa contain long methylene bridges between methyl branches, and the two East African species *G. brevipalpis* and *G. longipennis* contain trimethyl or tetramethyl alkanes, as also found in the *palpalis* group. *Glossina medicorum* and *G. nigrofusca* appear to be intermediate between the two clusters as they display characters of each (Carlson, personal communication).

Long chain branched hydrocarbons in the cuticle of female tsetse are responsible for sex recognition when contacted by the male, and these compounds appear to be species-specific. The three major contact-stimulant pheromone compounds in *G. m. morsitans* females are 3-methylene-interrupted dimethyl and trimethyl alkanes of 37 carbon backbones (Nelson and Carlson, 1986). These compounds are present in similar proportions in *G. m. submorsitans* and *G. m. centralis* females as well as the closely related *G. swynnertoni*. However, chemically different major compounds having 9-methylene-interrupted dimethyl alkanes are responsible for sex stimulant activity in *G. pallidipes* (Carlson et al., 1984) and are present across the range of the species. The same compounds, plus larger quantities of homologous compounds modified by addition of two carbons to the end of the chain, occur in *G. longipalpis* females (Carlson, personal communication).

In females of the *palpalis* group, *G. p. palpalis* and *G. p. gambiensis* share major 3-methylene-interrupted trimethyl alkanes of 35 carbon backbones with *G. f. fuscipes*, whereas *G. tachinoides* contains a major 11-methylene-interrupted dimethyl alkane (Nelson et al., 1988).

Identification keys

Four keys are provided here, three (Keys A–C) for identification of the 23 species of *Glossina* as adult tsetse-flies, and a fourth (Key D) for recognition of the puparia so far as these are known. The keys are essentially those of Potts (1973) but have been considerably modified editorially in their presentation and (for the adult) now include one extra species (*frezili*) more recently described. Key A uses external features of the adult by which specimens can be identified with only the unaided eye or a pocket lens. Certain species of the *fusca* group cannot be recognized in this way, but these can be readily identified using features of the male and female genitalia. Key B uses male genital features on which all species are distinguishable, and Key C uses features of the genital plates and signa (the latter occurring *only* in the *fusca* group) for identification of females. Females of species of the *palpalis* and *fusca* groups can be distinguished from one another, but females of the *morsitans* group show no clear specific differences in their genitalia and species need to be identified from external features using Key A. As differences between closely related species (and especially between subspecies) can be quite minor, doubtful specimens may need to be compared with authoritatively identified specimens in a reference collection or sent to a specialist.

Key A: Identification key to adults based on external characters

1   Anepimeron (pteropleuron) of thorax (Fig. 9.5) bearing a few strong bristles of equal size to those on katepisternum (sternopleuron), these bristles clearly differentiated from general pleural vestiture. Lower calypter (thoracic squama) with marginal hairs curly and numerous, giving a woolly appearance. Medium-sized to large flies (length 9.5–14 mm) .... **Glossina fusca** group 3

    [External features do not satisfactorily separate some species of this group and specimens are more reliably identified by using Keys B and C.]

—   Anepimeron (pteropleuron) of thorax bearing only setulose hairs, some of which may be longer than others but none clearly differentiated as bristles like those of katepisternum (sternopleuron). Lower calypter (thoracic squama) with marginal hairs not curly, forming a neat and simple fringe. Small to medium-sized flies (length 6.5–11 mm) 2

2   Hind tarsi uniformly dark brown or blackish seen from above. Abdominal dorsum usually uniformly brown (generally dark), not showing distinct transverse dark bands on a paler background.. ....**Glossina palpalis** group (also some forms of **G. austeni**) 15

—   Hind tarsi dark brown or blackish only on distal segments, tips contrasting strongly with paler proximal segments. Abdominal dorsum generally with distinct dark transverse bands showing against a paler background .... **Glossina morsitans** group 23

3   Palps shorter than width of head or not exceeding it by more than a ninth of their length 4

—   Palps longer than width of head by one-sixth to one-third of their length......... 8

4   Wings dusky. Antenna with fringe length one-quarter to one-third of antennal width (Fig. 9.9c).. **G. tabaniformis**

—   Wings pale. Antenna with fringe length one-fifth of antennal width or less.......... 5

5   Thoracic dorsum with a conspicuous dark brown spot towards each corner. Proboscis with bulb darkened apically on underside. [Pale species, generally light yellowish brown].. **G. longipennis**

—   Thoracic dorsum without conspicuous dark brown spots. Proboscis with bulb uniformly coloured... 6

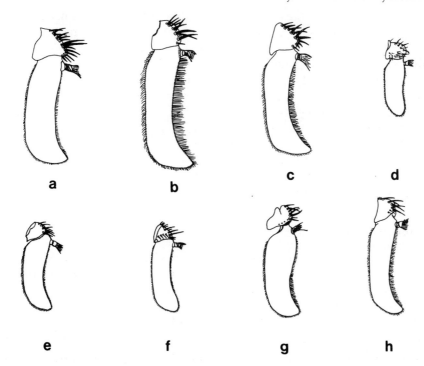

Figure 9.9 Antennae of *Glossina* species showing in profile the fringe of the last (third) segment: (*a*) *G. fusca*; (*b*) *G. nigrofusca nigrofusca*; (*c*) *G. tabaniformis*; (*d*) *G. tachinoides*; (*e*) *G. palpalis*; (*f*) *G. morsitans*; (*g*) *G. longipalpis*; (*h*) *G. pallidipes*. The first antennal segment is omitted. The antenna of *G. pallicera pallicera* is shown in Fig. 9.2.

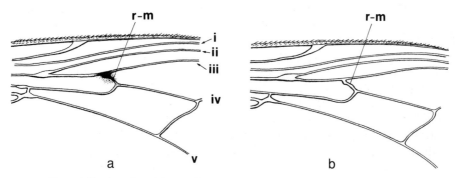

Figure 9.10 Part of *Glossina* wing showing the anterior cross-vein (r–m) in: (*a*) *G. brevipalpis*, both sexes; (*b*) *G. schwetzi*, male. (After Buxton, 1955.)

6   Wing with dark spot on anterior cross-vein (r–m) where it is thickened and strongly sclerotized 7
—   Wing without dark spot on anterior cross-vein (r–m) .... **G. schwetzi** (male, Fig. 9.10*b*), **G. medicorum** or **G. frezili**
7   Wing dark spot present in male and female (Fig. 9.10*a*)................ **G. brevipalpis**
—   Wing dark spot present only in female.. **G. schwetzi**
8   Hind tarsi uniformly dark dorsally 9
—   Hind tarsi dark dorsally on last two segments, which contrast in colour with pale basal segments 10

9 Fore and mid tarsi with last two segments pale, these at most only slightly darkened at tips. Antenna with third segment strongly and gradually recurved at tip (as in *nigrofusca*, Fig. 9.9b). [Thoracic pleura and hind coxae brownish grey] .... **G. severini**
— Fore and mid tarsi with last two segments largely dark, apical segment entirely dark dorsally (strongly contrasting with other segments) and penultimate segment at least with dark apical band. Antenna with third segment bluntly tipped, only slightly and abruptly recurved (as in *fusca*, Fig. 9.9a) .... **G. nashi**

[The two species running out at couplet 9 are better separated by examination of the genitalia.]

10 Antenna with fringe length less than one-quarter of antennal width. Hind tibiae with or without dark suffusions 11
— Antenna with fringe length half to three-quarters of antennal width (Fig. 9.9b). Hind tibiae with broad dark suffusions near middle and (much less distinctly) at their tips .... **G. nigrofusca nigrofusca**
11 Infra-alar bulla of thorax (Fig. 9.5) dark brownish and without a pale central vertical streak .... **G. fuscipleuris**
— Infra-alar bulla of thorax reddish and often with a pale central vertical streak 12
12 Antenna with fringe length about one-fifth of antennal width............................... .... **G. haningtoni**
— Antenna with fringe length less than one-sixth of antennal width (Fig. 9.9a) 13
13 Hind tibiae with dark suffusions near middle and (much less distinctly) at their tips.. **G. nigrofusca hopkinsi**
— Hind tibiae with or without dark suffusions, if present then scarcely less evident at apex than nearer base 14
14 Palps grey–black. Hind tarsi brown on first three segments .... **G. vanhoofi**
— Palps buff or dusky brown to grey–brown. Hind tarsi yellowish (but sometimes tending to brownish) on first three segments .... **G. fusca**

[The two species running out at couplet 14 are better separated by examination of the genitalia.]

15 Abdomen dorsally with interrupted dark segmental bands on a pale yellowish background. [Antenna as Fig. 9.9d]... **G. tachinoides**
— Abdomen dorsally without distinct segmental banding on a pale yellowish background.. 16
16 Abdomen reddish-ochraceus on dorsal surface. Small flies (length 7.5–8.5 mm).. .... **G. austeni**
— Abdomen brown to dark brown (sepia or clove) on dorsal surface. Larger flies (length 8.5–11 mm) 17
17 Antenna with fringe length one-quarter or more of antennal width .... **G. pallicera** s.l. 18
— Antenna with fringe length one-sixth or less of antennal width (Fig. 9.9e) 19
18 Antenna with third segment usually narrow relative to its length and strongly curved at apex; fringe length about three-fifths of width of third segment (Fig. 9.2b) ....**G. pallicera pallicera**

— Antenna with third segment not so shaped, more like a pea-pod (as in Fig. 9.9e); fringe length about a quarter of antennal width .... **G. pallicera newsteadi**
19 Abdomen dark brown to sepia brown on dorsal surface, without pale segmental hind margins; median pale area of second segment broad and more or less square .... **G. caliginea**
— Abdomen not so, segments narrowly pale along their hind margins; median pale area of second segment narrow and elongated 20

> [The species and subspecies running out at couplets 20–23 are better separated by examination of the genitalia.]

20 Abdomen variably coloured on dorsal surface but generally very dark. Hectors with hind margins not deeply cleft (e.g. as Fig. 9.6d) .... **G. fuscipes** s.l.
— Abdomen variably coloured but general tendency to be less dark. Hectors with hind margins deeply cleft by a forwardly pointed triangular area .... **G. palpalis** s.l.
21 Hectors with straight hind margins and uninterrupted hair covering .... **G. fuscipes quanzensis**
— Hectors with shallowly concave hind margins and with median interruptions in their hair covering 22
22 Hectors with hairless interruption in hind margin forming a narrow line. Body colour generally pale .... **G. fuscipes martinii**
— Hectors with hairless interruption in hind margin forming a forwardly pointing triangle. Body colour darker .... **G. fuscipes fuscipes**
23 Antenna with fringe length one-fifth to one-third of antennal width 24
— Antenna with fringe length not more than one-sixth of antennal width (e.g. Fig. 9.9f) 25
24 Fore and mid tarsi uniformly yellowish brown. Antenna with third segment about five times as long as its width, tip strongly recurved and tapering; fringe length one-third of antennal width (Fig. 9.9h) .... **G. pallidipes**
— Fore and mid tarsi with strongly defined dark tips. Antenna with third segment about 3.5 times as long as its width, tip only slightly and abruptly recurved and not tapering; fringe length one-fifth to one-quarter of antennal width (Fig. 9.9g) .... **G. longipalpis**
25 Abdomen reddish ochraceus to yellowish buff dorsally, with rather indistinctly darker transverse bands. Hind tarsi with last two segments not very much darker than basal segments .... **G. austeni**
— Abdomen yellowish or greyish yellow dorsally, with distinct medially interrupted dark brown to black transverse bands. Hind tarsi with last two segments dark and strongly contrasting with pale yellowish brown basal segments 26
26 Abdomen with hind margins of dark bands generally not clearly defined and their inner corners rounded (occasionally somewhat truncate, e.g. sometimes in ssp. *submorsitans*), median pale line not very sharply defined .... **G. morsitans** s.l.
— Abdomen with hind margins of dark bands clearly defined, inner corners squarely truncate and narrow median pale line thus very distinct .... **G. swynnertoni**

352 *Tsetse-flies (Glossinidae)*

Key B: Identification key to adult males based on genitalia

1 Superior claspers joined by a membrane (Fig. 9.8*d,f*)... 2
— Superior claspers free, not joined by membrane (Fig. 9.8*b*). [Superior claspers with long and narrow shape, terminating in claw-like tooth.] .... **Glossina fusca** group 3
2 Superior claspers with claw-like apical tooth (Fig. 9.8*d*).. **Glossina palpalis** group 16
— Superior claspers broadly dilated and club-like apically (Fig. 9.8*f*) .... **Glossina morsitans** group 24
3 Median process not prominent, in profile projecting at most only slightly beyond inferior claspers and never for more than their length 4
— Median process very prominent, projecting from between inferior claspers for twice their length or more (Fig. 9.11*j*)................................. **G. brevipalpis**
4 Harpes well developed, with one or more freely projecting processes (Fig. 9.11*a–f, h–m*).. 5
— Harpes poorly developed and not clearly differentiated from sclerotization of aedeagus, without projecting processes (Fig. 9.11*g*) .... **G. severini**
5 Harpes with one pair of freely projecting processes... 6
— Harpes with three pairs of freely projecting processes..................................... 9
6 Harpes processes simple, undivided 7
— Harpes processes bifid apically (Fig. 9.11*l*) .... **G. medicorum**
7 Harpes simple, without a triangular basal part.. 8
— Harpes with a basal triangular part from which the bottom corner is drawn out to form a strongly sclerotized blunt-ended process (Fig. 9.11*f*); harpes partly covered by a membrane studded with short scale-like spines .... **G. nigrofusca**

[The two subspecies do not differ in their male genitalia but can be differentiated by characters in couplets 10–13 of Key A.]

8 Harpes formed into a pair of long and slender processes curving upwards and tapering (Fig. 9.11*i*)... **G. longipennis**
— Harpes formed into a pair of stout processes with hull-shaped profile and fine terminal serrations (Fig. 9.11*m*) .... **G. frezili**
9 Harpes with no members of the processes bifid.. 10
— Harpes with members of distal pair of processes bifid.................................... 12
10 Harpes processes not dilated apically and not all of equal length.................. 11
— Harpes with proximal and middle pairs of processes dilated towards their tips (only distal processes tapering to a point); all three pairs of processes approximately of equal length (Fig. 9.11*c*) .... **G. vanhoofi**
11 Harpes with proximal pair of processes shorter than other two pairs (Fig. 9.11*k*) .... **G. schwetzi**
— Harpes with middle pair of processes shorter than other two pairs (Fig. 9.11*d*). [Appearance of middle pair of processes characteristic: members dark at base and clear and somewhat transparent on remainder]................. **G. fuscipleuris**
12 Harpes with all three pairs of processes tapering to a point............................ 13
— Harpes with one or other pairs of processes dilated apically or forming blunt protuberances.. 14
13 Harpes with proximal pair of processes much longer than other two pairs, form and disposition of processes as Fig. 9.11*b* .... **G. nashi**

Classification and identification 353

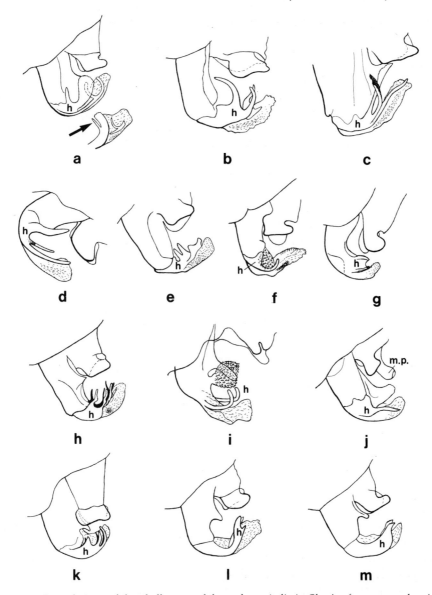

Figure 9.11 Lateral views of the phallosome of the male genitalia in *Glossina fusca* group showing the shapes of the harpes (h) used as identification characters (Key B): (*a*) *G. fusca fusca* (with phallosome tip in *G. fusca congolensis*); (*b*) *G. nashi*; (*c*) *G. vanhoofi*; (*d*) *G. fuscipleuris*; (*e*) *G. haningtoni*; (*f*) *G. nigrofusca*; (*g*) *G. severini*; (*h*) *G. tabaniformis*; (*i*) *G. longipennis*; (*j*) *G. brevipalpis* (m.p. = median process); (*k*) *G. schwetzi*; (*l*) *G. medicorum*; (*m*) *G. frezili*. Drawings slightly simplified and hairing omitted.

— Harpes with proximal pair of processes not longer than other two pairs, form and disposition of processes as Fig. 9.11*h*.................................. **G. tabaniformis**
14 Harpes with middle pair of processes broadly dilated; proximal processes and (bifid) distal processes with pointed tips (Fig. 9.11*a*)......................... 15
— Harpes with proximal and middle pairs of processes peg-like, distal pair formed into two blunt protuberances (Fig. 9.11*e*).................................... **G. haningtoni**
15 Harpes with proximal processes long, tapering to a point and about equal in length to middle pair of processes (Fig. 9.11*a*, arrowed); middle processes

with their serrate apical margins straight in profile; distal bifid processes with outer arm conspicuously shorter than inner arm; plates bearing proximal and middle harpes about 0.58–0.65 mm long ....
.... **G. fusca congolensis**
— Harpes with proximal processes short and peg-like, not as long as middle pair of processes (Fig. 9.11a); middle processes more broadly dilated and with their serrate apical margins convex in profile; distal bifid processes with outer arm as long as inner arm: plates bearing proximal and middle harpes about 0.68–0.77 mm long... **G. fusca fusca**
16 Superior claspers with free tooth much less than one-third of length of clasper. Inferior claspers with more or less foot-like head (e.g. Fig. 9.12a–g) 17
— Superior claspers with free tooth very long, almost one-third of length of clasper. Inferior claspers with swollen and notched (somewhat bilobed) terminal dilation (Fig. 9.12h)... **G. caliginea**
17 Inferior claspers with terminal dilation bifurcated 18
— Inferior claspers with terminal dilation not bifurcated 19
18 Inferior clasper internal lobe with flattened outline (Fig. 9.12f).............................
.... **G. pallicera pallicera**
— Inferior clasper internal lobe with pointed outline (Fig. 9.12g)..............................
.... **G. pallicera newsteadi**
19 Inferior claspers with 'neck' plainly longer than broad 20
— Inferior claspers with short 'neck' about as broad as long (Fig. 9.12i) ....
.... **G. tachinoides**
20 Inferior claspers with strongly developed external lobe (prominent and projecting at least slightly upwards) and with internal lobe (Fig. 9.12c–e).........
.... **G. fuscipes** s.l. 21
— Inferior claspers with feebly developed external lobe (without trace of upward projection) and without internal lobe (Fig. 9.12a,b)............... **G. palpalis** s.l. 23
21 Inferior claspers with terminal dilation more or less foot-like (Fig. 9.12d,e)...........
.... 22
— Inferior claspers with terminal dilation in form of curved pointed hook which evenly prolongs clasper neck (Fig. 9. 12c) .... **G. fuscipes fuscipes**
22 Inferior clasper terminal dilation markedly foot-like, with pronounced 'heel' and concave 'sole' (Fig. 9.12d); clasper internal lobe not very strongly projecting ...
.... **G. fuscipes martinii**
— Inferior clasper terminal dilation less strongly foot-like, 'heel' not pronounced and 'sole' more or less flat; clasper internal lobe strongly projecting (Fig. 9.12e)
.... **G. fuscipes quanzensis**
23 Inferior clasper terminal dilation large, its width greater than length of clasper 'neck', latter narrowing gradually from body of clasper (Fig. 9.12a)....................
.... **G. palpalis gambiensis**
— Inferior clasper terminal dilation relatively small, its width less than length of clasper 'neck', latter narrowed abruptly from body of clasper (Fig. 9.12b) ....
.... **G. palpalis palpalis**
24 Superior claspers with outer lateral angle forming a blunt tooth (Fig. 9.13a,b) 25
— Superior claspers with outer lateral angle either rounded or much developed, not forming a tapering tooth (Fig. 9.13c–e)... 26
25 Superior claspers with subterminal tooth; length of line of union between inner extensions of claspers about equal to greatest width of clasper (Fig. 9.13b).......
.... **G. pallidipes**

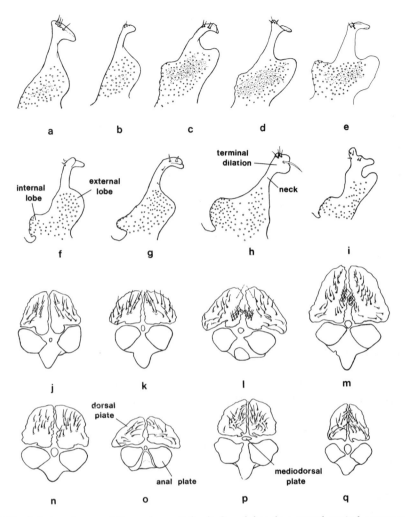

Figure 9.12 Inferior clasper of the male genitalia (*a–i*) and female external genital armature (*j–q*) in *Glossina palpalis* group: (*a*) *G. palpalis gambiensis*; (*b*) *G. palpalis palpalis*; (*c*) *G. fuscipes fuscipes*; (*d*) *G. fuscipes martinii*; (*e*) *G. fuscipes quanzensis*; (*f*) *G. pallicera pallicera*; (*g*) *G. pallicera newsteadi*; (*h*) *G. caliginea*; (*i*) *G. tachinoides*; (*j*) *G. palpalis gambiensis*; (*k*) *G. palpalis palpalis*; (*l*) *G. fuscipes fuscipes*; (*m*) *G. fuscipes quanzensis*; (*n*) *G. fuscipes martinii*; (*o*) *G. pallicera pallicera*; (*p*) *G. caliginea*; (*q*) *G. tachinoides*. (After Machado, 1954.)

— Superior claspers with terminal tooth; length of line of union between inner extensions of claspers plainly less than greatest width of clasper (Fig. 9.13*a*) .... **G. longipalpis**

26 Superior claspers with outer lateral angle rounded and not strongly produced 27

— Superior claspers with outer lateral angle narrowly rounded and strongly produced (Fig. 9.13*c*) .... **G. austeni**

27 Superior clasper median lobes with broad outwardly turned tips which project as far as hind margins of claspers or slightly beyond (Fig. 9.13*d*) .... **G. morsitans** s.l.

[The three subspecies can be distinguished by minor differences in genitalia (Machado, 1970) but specimens should be referred to a specialist for reliable subspecific identification. Geographical origin of a specimen is usually the surest guide to its identity.]

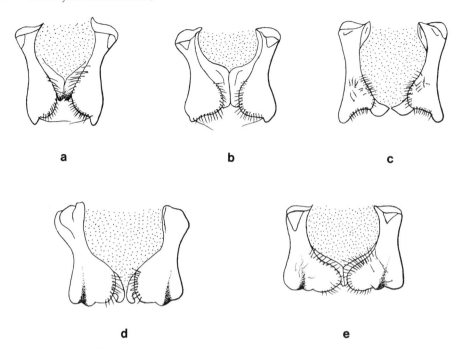

Figure 9.13 Superior claspers of the male genitalia in *Glossina morsitans* group: (*a*) *G. longipalpis*; (*b*) *G. pallidipes*; (*c*) *G. austeni*; (*d*) *G. morsitans*; (*e*) *G. swynnertoni*. (After Newstead et al., 1924; Machado, 1954.)

— Superior clasper median lobes with small pointed tips which do not generally project as far as hind margins of claspers (Fig. 9.13*e*).............. **G. swynnertoni**

Key C: Identification key to adult females based on genitalia

1 External genital armature formed of five or six well-defined sclerotized plates (terms as in Fig. 9.14). Uterus with or without sclerotized plate (signum) at its anterior end 2

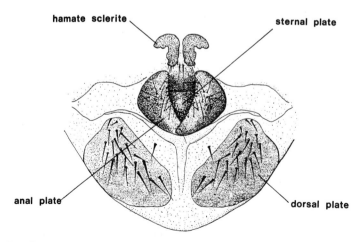

Figure 9.14 Female external genital armature of *Glossina brevipalpis*, showing the hamate sclerites unique to this species. (After Newstead et al., 1924.)

— External genital armature much reduced, well-defined sclerotized plates usually absent and not more than three if present (Fig. 9.8e). Signum absent .... **Glossina morsitans** group

[Females of this group need to be identified using Key A.]

2 External genital armature formed of five plates (Fig. 9.8a). Signum present (except in *brevipalpis*) .... **Glossina fusca** group 3
— External genital armature formed of six plates (Fig. 9.8c). Signum absent .... **Glossina palpalis** group 16
3 Signum absent. Hamate sclerites (paired hook-like or comma-shaped sclerotizations associated with bases of anal plates) present (Fig. 9.14) .... **G. brevipalpis**
— Signum present (weakly sclerotized in some species such as *longipennis*). Hamate sclerites absent 4
4 Signum with pair of separated sclerotized strips 5
— Signum wholly sclerotized 6
5 Signum not reduced, with two sharply defined dark-sclerotized strips expanded at their tips and close together at base (Fig. 9.15e) .... **G. severini**
— Signum much reduced, with two short and well-separated submedian pale-sclerotized strips (Fig. 9.15g) .... **G. nigrofusca**

[The two subspecies are alike in the signum but can be distinguished by characters in couplets 10–13 of Key A.]

6 Signum elongated vertically, narrowly constricted near bottom and constricted also near middle (its lateral outlines wavy as in Fig. 9.15a,h) 7
— Signum of various shapes but without two constrictions 8
7 Signum strongly flexed or even bent double in middle of its length, sides subparallel, with two transverse constrictions (towards bottom and near top) (Fig. 9.15h) .... **G. fuscipleuris**
— Signum not flexed, sides not subparallel, divided into two unequal parts by a transverse constriction, bottom with diverging lateral 'horns' and top more or less deeply bifurcate (Fig. 9.15a) .... **G. tabaniformis**
8 Signum cordiform in outline shape 9
— Signum not cordiform in outline shape 10
9 Signum usually shaped as Fig. 9.15d .... **G. medicorum**
— Signum shaped as Fig. 9.15l .... **G. frezili**

[Signum shape varies in *medicorum* (e.g. Nash and Jordan, 1959) and females are not necessarily so reliably distinguishable from those of *frezili* as Gouteux (1987) suggests. Care is needed in separation of these species on the signum character.]

10 Signum with pairs of curved dark-sclerotized thickenings 11
— Signum without such thickenings 14
11 Signum of subrotund and lobate form (Fig. 9.15k) 12
— Signum not of subrotund form 13
12 Signum lobes more than 0.55 mm wide .... **G. fusca fusca**
— Signum lobes less than 0.55 mm wide .... **G. fusca congolensis**
13 Signum dark thickenings positioned anteriorly and not continuous in midline (Fig. 9.15c) .... **G. schwetzi**
— Signum dark thickenings positioned posteriorly and meeting in midline to form a crescent (Fig. 9.15i) .... **G. haningtoni**

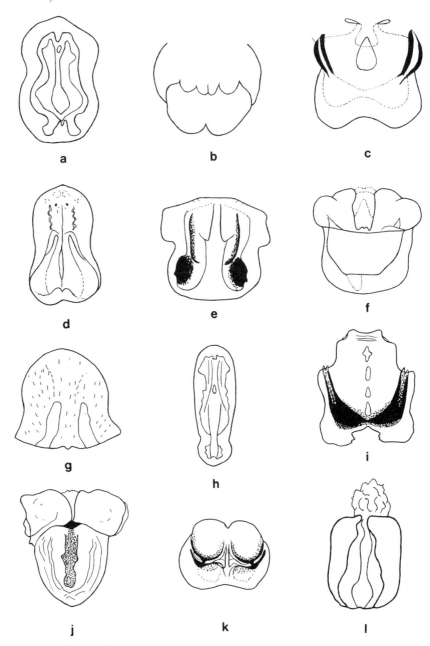

Figure 9.15 Female genital signa of *Glossina fusca* group species: (*a*) *G. tabaniformis*; (*b*) *G. longipennis*; (*c*) *G. schwetzi*; (*d*) *G. medicorum*; (*e*) *G. severini*; (*f*) *G. nashi*; (*g*) *G. nigrofusca*; (*h*) *G. fuscipleuris*; (*i*) *G. haningtoni*; (*j*) *G. vanhoofi*; (*k*) *G. fusca*; (*l*) *G. frezili*.

14 Signum in form of a backwardly directed semicircle incised in midline with a V-shaped notch between two lobes (Fig. 9.15*b*), usually very weakly sclerotized ....**G. longipennis**
— Signum not of this form 15
15 Signum with a pair of anterior bilobed tubercles tapering behind into hollow stalks which lead into a sporran-shaped pouch (Fig. 9.15*f*) .... **G. nashi**

— Signum comprising two parts, an upper hollow lobe capping a lower truncated cone containing a sclerotized plate shaped like a spear-head (Fig. 9.15j) .... **G. vanhoofi**

16 External genitalia with mediodorsal plate as tall as it is broad or taller (subcircular or suboval) 17
— External genitalia with mediodorsal plate broader than tall (transversely elongate) (Fig. 9.12p) .... **G. caliginea**

17 Dorsal plates taller than broad 18
— Dorsal plates as tall as broad or nearly so 21

18 Dorsal plates not nearly twice as tall as broad 19
— Dorsal plates nearly twice as tall as broad (Fig. 9.12q) .... **G. tachinoides**

19 Dorsal plates not extending laterally beyond width of anal plates 20
— Dorsal plates extending laterally well beyond width of anal plates (Fig. 9.12m) .... **G. fuscipes quanzensis**

20 Dorsal plates with inner angle extending markedly downwards below baseline of plate; mediodorsal plate very small (Fig. 9.12j) ...... **G. palpalis gambiensis**
— Dorsal plates with inner angle less strongly projecting below baseline of plate; mediodorsal plate large (Fig. 9.12n) .... **G. fuscipes martinii**

21 Dorsal plates about as broad as tall or only slightly broader 22
— Dorsal plates markedly broader than tall (Fig. 9.12o) .... **G. pallicera** s.l.

[Females of subspecies *pallicera pallicera* and *pallicera newsteadi* are not separable on their genitalia but can be distinguished by external characters used in Key A.]

22 Dorsal plates very close together and bare on inner angles (Fig. 9.12k); median space between plates bare .... **G. palpalis palpalis**
— Dorsal plates more widely separated and almost always haired on inner angles (Fig. 9.12l); median space between plates sometimes haired .... **G. fuscipes fuscipes**

Key D: Identification key to puparia

The following key is given to aid identification of the 14 tsetse species for which the puparium is known. It is likely to be most useful for distinguishing between puparia found in an area where the local *Glossina* species are known from prior identification of adult flies. The features used are shown labelled in Fig. 9.16n.

1 Polypneustic lobes separated by a conspicuous interlobe recess 2
— Polypneustic lobes appearing fused into a single mass, only a slight and barely noticeable hollow between them representing the interlobe recess (Fig. 9.16b). .... **G. longipennis**

2 Interlobe recess not V-shaped 3
— Interlobe recess V-shaped (Fig. 9.12e). [Large puparium, length 7–8 mm] .... **G. brevipalpis**

3 Interlobe recess with constriction giving it a 'key-hole' shape. Puparium small to medium size (length 3–6 mm) 4
— Interlobe recess without constriction. Puparium large (length 7–8.5 mm) 11

4 Interlobe recess with very marked constriction and well-formed 'key-hole' shape 5

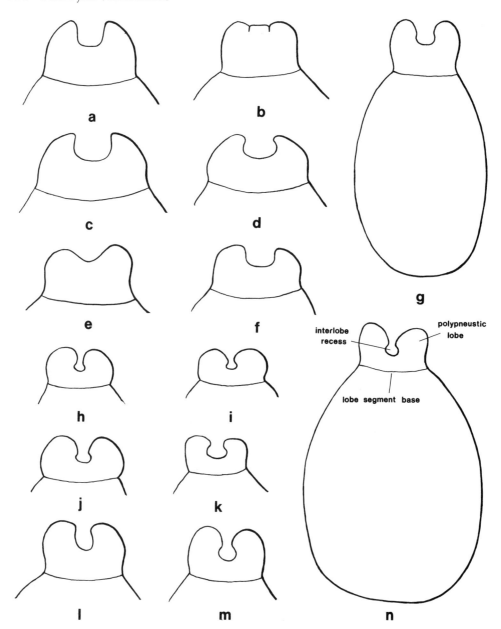

Figure 9.16 Outline shape of the polypneustic lobes of *Glossina* puparia (semi-diagrammatic): (*a*) *G. fuscipleuris*; (*b*) *G. longipennis*; (*c*) *G. tabaniformis*; (*d*) *G. fusca*; (*e*) *G. brevipalpis*; (*f*) *G. schwetzi*; (*g*) *G. austeni*; (*h*) *G. tachinoides*; (*i*) *G. palpalis*; (*j*) *G. fuscipes*; (*k*) *G. swynnertoni*; (*l*) *G. pallidipes*; (*m*) *G. longipalpis*; (*n*) *G. morsitans*.

— Interlobe recess with less pronounced constriction, nevertheless with quite obvious 'key-hole' shape 7

5 Puparium very small, length 3–5.4 mm. Polypneustic lobes close together, interlobe recess at narrowest point about one-fifth of width of each lobe or less; inner gap of 'key-hole' shape rounded, its depth nearly equal to its width (Fig. 9.16*h*) .... **G. tachinoides**

— Puparium small to medium size, length 5–5.6 mm. Interlobe recess with inner gap of 'key-hole' shallower, its depth less than its width 6
6 Polypneustic lobes close together, narrowest distance between them about a quarter of width of each lobe (Fig. 9.16*i*) .... **G. palpalis**
— Polypneustic lobes more widely separated, narrowest distance between them nearly half of width of each lobe (Fig. 9.16*j*) .... **G. fuscipes**
7 Polypneustic lobes appearing large in comparison to size of puparium, their combined breadth sometimes nearly approaching width of puparium (Fig. 9.16*g*). Puparium small, length about 5.5 mm .... **G. austeni**
— Polypneustic lobes not markedly large in comparison to size of puparium. Puparium of small to medium size, length about 5.6–6 mm............................ 8
8 Interlobe recess with bottom edge positioned about midway between lobe segment base and ends of lobes.. 9
— Interlobe recess with bottom edge positioned nearer to lobe segment base than to ends of lobes.. 10
9 Interlobe recess with very slight constriction and with narrowest part of 'key-hole' near ends of lobes (Fig. 9.16*l*) .... **G. pallidipes**
— Interlobe recess with more pronounced constriction and with narrowest part of 'key-hole' midway between bottom of recess and ends of lobes (Fig. 9.16*k*)...... .... **G. swynnertoni**
10 Interlobe recess with length of waist of 'key-hole' about half total depth of recess; width of inner part of 'key-hole' at least half of maximum width of each lobe (Fig. 9.16*m*).. **G. longipalpis**
— Interlobe recess with length of waist of 'key-hole' about two-thirds of depth of recess; width of inner part of 'key-hole' markedly less than half maximum width of each lobe (Fig. 9.16*n*).. **G. morsitans**
11 Interlobe recess in form of straight-sided U, polypneustic lobes not at all incurved at their ends... 12
— Interlobe recess not so, slightly narrowed at mouth and polypneustic lobes incurved at their ends... 13
12 Polypneustic lobe ends distinctly tapering; interlobe recess deep, its length nearly equal to distance between bottom of recess and lobe segment base (Fig. 9.16*a*) .... **G. fuscipleuris**
— Polypneustic lobe ends more rounded and hardly at all tapering; interlobe recess comparatively shallow, its length obviously less than distance between bottom of recess and lobe segment base (Fig. 9.16*f*)...................... **G. schwetzi**
13 Polypneustic lobes quite strongly incurved, distance between their tips markedly less than maximum width of the interlobe recess (Fig. 9.16*d*)............. .... **G. fusca**
— Polypneustic lobes only slightly incurved, distance between their tips only very little less than maximum width of interlobe recess (Fig. 9.16*c*).......................... .... **G. tabaniformis**

Geographical distribution and faunal works

Just as the taxonomy of *Glossina* is, except for a few minor details, generally accepted, so the distribution of the recognized species and subspecies of the genus is relatively well known – certainly much more so than is the case for the other groups of insects included in this book. Today *Glossina* species are essentially restricted to Africa although there are records for the genus from the Arabian peninsula. There is an old and since unconfirmed report of *G. tachinoides* in southern Yemen (Carter, 1906) and there are recent records of *G. f. fuscipes* and *G. m. submorsitans* from near Gizan in southwestern Saudi Arabia (Elsen et al., 1990). Because of this distribution of the genus it is inappropriate to list 'faunal works' in the sense that they are considered in other chapters.

The northern limit of the genus extends across the African continent from Senegal in the west to southern Somalia in the east. Much of this limit is at about 10–14°N but is only about 4°N in Somalia; the limit approximates to the southern edges of the Sahara and Somali deserts. The southern limit is less uniform, varying from about 10°S to 20°S in the south west, following the northern edges of the Kalahari and Namibian deserts, to about 20°S in the south east, but extending as far as 29°S adjacent to the eastern coast of Africa.

Within these general limits, infestations of *Glossina* are not continuous. Extensive areas naturally devoid of woody vegetation – or areas rendered so by the activities of man – are generally free of tsetse. Areas of higher ground are also devoid of tsetse, the upper altitudinal limit varying from about 1800 m near the equator to about 1300 m in, for example, Zimbabwe (Jack, 1927).

Simplistically, the tsetse-infested tropical and subtropical zones of Africa can be considered as comprising two climax vegetation types, lowland rain forest and savanna woodland. The two large zones of lowland rain forest, along the western African coast from Guinea to Ghana and in the basin of the Zaire river extending westwards to Nigeria, are surrounded, and separated in Togo–Benin, by savanna woodlands varying from regions of good tree growth and seasonally lush grasslands, adjacent to the forests, to regions of scanty trees and poor grassland, adjacent to the deserts.

Each tsetse species has its own relatively narrow ecological requirements within the range of habitats afforded by the forests and the savannas of Africa. Today the picture is much complicated by the activities of man, destroying habitats in some places and creating new habitats in other places. Much of the climax forest vegetation, particularly in West Africa, has been removed and many of the savanna woodlands have been modified by arable farmers, firewood collection or through overgrazing by domestic livestock (Jordan, 1986, 1989).

Numerous publications refer to the distribution of *Glossina* species in Africa and most countries have produced their own maps of tsetse distribution; many are in need of revision, however, particularly in the light of modifications of the environment by man. The most valuable overall picture is that of Ford and Katondo (1977), who illustrated the distribution of the species in the three groups of *Glossina* on three separate maps of Africa on a scale of 1:5 000 000. Figures 9.17–9.19 are based on these maps. Twelve of the 13 species of the *fusca* group (Fig. 9.17) inhabit various types of forest, ranging from true lowland rain forest, to forest-edge, to forest islands far from the main blocks of lowland rain forest. *Glossina longipennis* is exceptional amongst species of the *fusca* group, as it occurs only in arid habitats in East Africa. The *palpalis* group (Fig. 9.18) is also centred on the two blocks of lowland rain forest, but some species extend far out through the humid savannas and into the drier savannas along rivers and streams draining into the Atlantic Ocean, the Mediterranean Sea (upper reaches of the River Nile) and the inland drainage systems of some of the great African lakes – but not river systems draining into the Indian Ocean. Unlike the other two groups, the species of the *morsitans* group (Fig. 9.19) are restricted to the savanna woodlands surrounding the two blocks of lowland rain forest. Whereas in the wetter areas the flies are distributed widely throughout the woodland, in drier parts of their range, such as the Sudan savanna of West Africa, they are restricted to the mesophytic vegetation of the watercourses, particularly during the months of the severe dry season.

It is considered inappropriate to enter here into details of the distribution and habitat requirements of each of the species and subspecies of *Glossina*. Much information is available. Ford and Katondo (1977), Jordan (1986) and Potts (1973) give overviews with references to other publications and maps describing the situation in individual countries. Table 9.1 p.344 gives an indication of the distribution of each of the species and recognized subspecies; the various African countries with tsetse infestations are grouped as western, central, eastern or southern although some larger countries could be considered as belonging to two regions. For example, Angola is treated as being in southern Africa but the northern part of that country, especially Cabinda north of the Zaire river, clearly would be more appropriately placed in central Africa. The species and subspecies of *Glossina* within each of the three well-defined species-groups are listed in an order which approximates to that in which they appear in the African fauna as one moves from western to central to eastern to southern parts of the continent. The table therefore illustrates the way in which the tsetse fauna changes and highlights interesting areas of overlap.

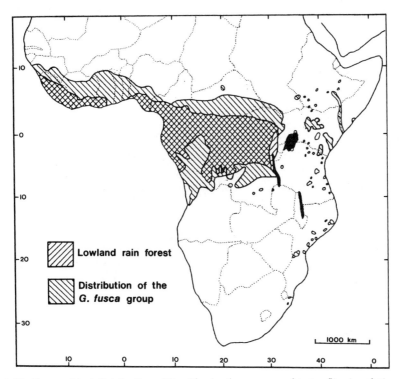

Figure 9.17 Geographical distribution of the *Glossina fusca* group of tsetse-flies in relation to the two zones of lowland rain forest. (After Jordan, 1986.)

364 Tsetse-flies (Glossinidae)

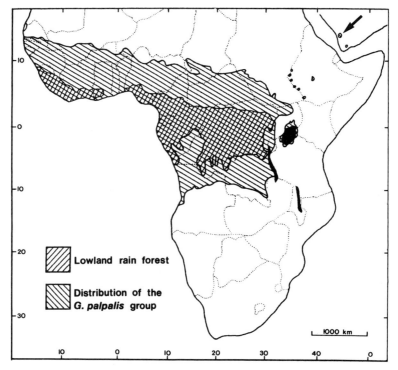

Figure 9.18 Geographical distribution of the *Glossina palpalis* group of tsetse-flies in relation to the two zones of lowland rain forest. (After Jordan, 1986.) The arrow marks the position of the locality recently discovered in the Arabian peninsula (Elsen et al., 1990).

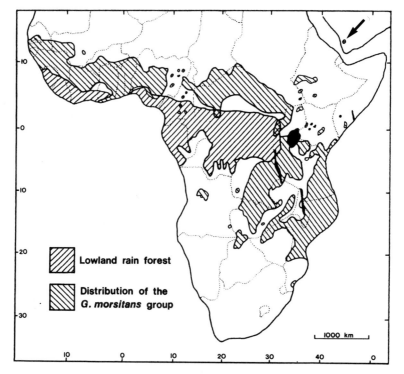

Figure 9.19 Geographical distribution of the *Glossina morsitans* group of tsetse-flies in relation to the two zones of lowland rain forest. (After Jordan, 1986.) The arrow marks the position of the locality recently discovered in the Arabian peninsula (Elsen et al., 1990).

BIOLOGY

Life history

The most distinctive feature of the life history of *Glossina* species, shared with only a few other small families of Diptera, is the retention of the single mature egg in the uterus of the female, where it hatches into a larva and is nourished by the products of a pair of modified accessory glands. This method of reproduction is referred to as adenotrophic viviparity. *Glossina* species breed throughout the year and thus there are no environmental cues that trigger particular events in the life cycle, although the duration of all stages is temperature-dependent. In the following account the duration of events is given for a mean temperature of 25°C.

The female fly has two ovaries, each of which has two ovarioles (Fig. 9.20). Eggs develop sequentially in the four ovarioles (Saunders, 1970) and the first is ovulated from the right ovary into the uterus when the fly is about nine days old and is followed 9–10 days later by the second ovulation, from the left ovary, and so on. In the uterus the egg is fertilized by one of a number of sperms released from the spermathecae. The first-instar larva develops within the egg shell and hatches after about three and a half days. The larva grows rapidly and moults twice. By the ninth day, the third-instar larva, with its two conspicuous black polypneustic lobes at the posterior end, is some 5 mm long and weighs about 35 mg in an average-size species such as *G. morsitans*. Larviposition now occurs, after which the free-living fully fed larva excretes the waste products accumulated during the period of intra-uterine life and burrows into the soil where the white third-instar skin is transformed within a few hours into a hard, almost black shell, the puparium,

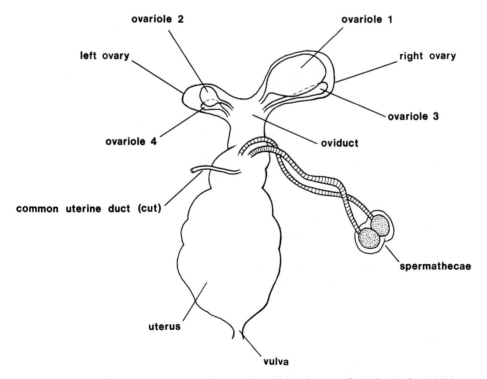

Figure 9.20 Reproductive system of a one-day-old female tsetse-fly. (After Jordan, 1986.)

within which pupation and metamorphosis to the adult stage takes place. The term puparium is now in general use to describe the puparium (strictly, the larval cuticle) and its contents (the pupa or pharate adult); in the earlier literature these were simply referred to as the 'pupa'.

Some 30 days later a single fly emerges from each puparium; the sex ratio on emergence is 1:1. At a mean temperature of 20°C the duration of the puparial period is about 47 days and at 30°C about 20 days. Males emerge about two days later than females. Below about 17°C and above about 32°C there are insufficient fat reserves within the puparium and development cannot be successfully completed (Bursell, 1960). At temperatures above this, adult flies exhibit negative phototaxis which causes them to concentrate in the darker, and hence cooler, parts of their immediate environment. They will, therefore, concentrate in thickets and also enter ant-bear holes, tree crevices and other well-shaded situations (e.g. Pilson and Pilson, 1967). This negative response to light at high temperatures is reflected in the sites in which gravid females deposit their larvae, rather loosely referred to as 'breeding sites'.

During the wet season puparia are deposited in scattered situations within the wide ambits of the flies; under such circumstances they are difficult to find. This is true throughout the year for species occupying more humid areas, such as lowland rain forest, where there are no marked seasonal variations in temperatures. This probably partly accounts for the lack of breeding site records for most members of the *fusca* group. In areas with a long, severe dry season, *Glossina* species select the best-shaded spots in which to larviposit and puparia can consequently occur in large 'aggregations' and can be readily found by the skilled collector under fallen logs, between the buttresses of trees, in large hollow trees and in animal burrows. The seasonal shift in breeding sites has been described for a number of species of the *morsitans* and *palpalis* groups, including *G. m. submorsitans* and *G. tachinoides* in northern Nigeria (Nash, 1942) and *G. swynnertoni* in Tanzania (Burtt, 1952).

Eclosion in *Glossina* is a similar process to that in other Muscomorpha, the young adult breaking out of the puparium and then pushing itself out of the soil by means of the ptilinum (Fig. 9.7d). The newly emerged fly is referred to as 'teneral' until it has taken its first blood-meal. A teneral fly can be recognized by its soft thorax, the eversibility of the ptilinum and the pale appearance of the abdomen when held up to the light.

Mating

Little is known about the precise way in which the two sexes of *Glossina* species locate one another; there is no evidence of males swarming, in the way that occurs with many Nematocera. It is generally assumed that mating usually occurs on or in the vicinity of the host. In some species of the *morsitans* group a 'following swarm' of mate-seeking males will follow a moving object, be it a host animal or vehicle, with the objective not to feed but to mate with any female attracted by the movement of the object (Bursell, 1961). Not only hungry females are associated with the following-swarm, as mate-seeking, replete females of *G. m. morsitans* and *G. pallidipes* are also in the swarm for brief periods (Vale, 1974).

However, this explanation, even for species of the *morsitans* group, is not all-embracing as there are clear differences between *G. m. morsitans* and *G. pallidipes*. Mating pairs of *G. m. morsitans* have been commonly observed on host animals and nearby vegetation but this is rarely, if ever, so with *G. pallidipes* despite an abundance of non-inseminated tenerals in female swarms near baits. Thus mating may normally occur away from hosts in this species (Vale, 1974; Vale et al., 1976). There is also laboratory evidence that female *G. pallidipes* mate later in life than other species (Rogers, 1972) and behavioural studies in the laboratory also suggest that the two species have very different, as yet not understood, mating systems (Wall, 1988).

Following-swarms have not been reported for species of the *palpalis* and *fusca* groups. It is assumed that the sexes of these species also meet on or near the host but there is no convincing explanation of why, even in very low density and often apparently dispersed tsetse populations, practically all females are inseminated.

As far as is known there are no volatile pheromones produced by *Glossina* species which influence the behaviour of other members of the same species over long distances. However, the females of some (Carlson et al., 1984; Huyton et al., 1980; Langley et al., 1975; McDowell et al., 1981; Offor et al., 1981) and probably all *Glossina* species possess a contact sex-recognition pheromone as one of their cuticular hydrocarbons which induces a copulatory response in males of the same species but not of other species. The existence of these contact sex-recognition pheromones explains the absence of interspecific pairing even in dense following-swarms.

The sexes remain in copula for an hour or two, during which time a spermatophore is formed from secretions of the male within the uterus of the female. Just before the end of copulation, sperms are ejaculated into the spermatophore (Pollock, 1970). Within a few hours, sperms are slowly released from the spermatophore and move up the spermathecal ducts into the pair of spermathecae. Active and viable sperms can remain in the spermathecae, nourished by secretions from the layer of cells which surrounds the cuticular lining of the lumen of each spermatheca (Jordan, 1972), throughout the life of the female. Thus there is no necessity for female *Glossina* to mate more than once. In the laboratory, females become less willing to mate as they age but the frequency of multiple mating by females in the field is not known. Male flies are fully potent when a few days old and have the capability of pairing on many occasions although, in the field, an individual male is unlikely to pair more than a few times. No spermatogenesis occurs after a male fly has emerged from its puparium (Curtis, 1968).

Biting behaviour and hosts

In all *Glossina* species both sexes feed only on vertebrate blood and, because the brief free-living third-instar larva does not feed, all life cycle stages are thus dependent on blood. It is particularly crucial that the adult fly obtains its first blood-meal quickly before the limited quantity of fat remaining from the pupal stage is exhausted (Bursell, 1960); for this reason, tsetse are less discriminatory at this time in the species of hosts from which meals are taken than they are later in life.

It has long been known that the senses of sight and smell are both involved in host location, but research since the early 1970s which has led to the development of improved methods for trapping *Glossina* species has greatly increased knowledge on the subject. The compound eyes of *Glossina* can perceive shape, colour and movement. The visual stimuli from a stationary host are perceived only at a short distance, perhaps about 10 m (Vale, 1974). A moving host, or especially a group of moving hosts, provides a much stronger visual stimulus; for instance, the limit at which *G. swynnertoni* responded to the visual stimuli of a moving herd of cattle was 130–180 m (Bax,1937). *Glossina* species also seem capable of genuine colour discrimination rather than simply an ability to respond to intensity contrasts (Green, 1986). Phthalogen blue is particularly attractive, at least to those species investigated, whereas black and ultra-violet-reflecting white are colours which stimulate landing. Yellow is particularly unattractive. Whereas advantage has been taken of these findings in the design of traps, satisfactory ecological explanation of the responses induced by colours have yet to be advanced.

The relative importance of sight and smell in host location is uncertain but is known to vary between species. Whereas the visual stimulus of a stationary host covers a limited area, but is increased when the animal moves, host odours have the potential of attracting flies over a much greater distance – though only from downwind of the odour source. From extensive studies in several African countries it can be concluded that species of the *morsitans* group are highly responsive to host odours, whereas species of the *palpalis* group respond little or not at all to those components of host odour known to be attractive to the *morsitans* group, other than to carbon dioxide. Data are scanty but so far it appears that species of the *fusca* group may be intermediate between those of the other two groups, although no true forest-dwelling species have yet been investigated.

For species of the *morsitans* group, certainly for *G. m. morsitans* and *G. pallidipes*, the most effective olfactory attractant is a large quantity of natural host odour, the effectiveness of which is maintained even if the animals giving off the odour are not visible to the flies (Vale, 1974). Components of both host breath and host urine have been identified as either initiators of fly activity or as substances which will induce them to fly upwind. In addition to carbon dioxide, acetone (Vale, 1980), for which butanone may be substituted, 1-octen-3-ol (Bursell, 1984; Hall et al., 1984) and 3-n-propyl phenol and 4-methyl phenol (Hassanali et al., 1986; Vale et al., 1988) have already been identified as attractants and it seems likely that other active compounds may also exist in host emanations.

Once a hungry fly has arrived in the vicinity of a host the actual site of attack varies with the species of *Glossina*. For example, subspecies of *G. morsitans* may bite man at any point (Jackson, 1933), *G. palpalis* attacks mainly above the waist and *G. tachinoides* mainly below the knee (Nash, 1948). Similarly, feeding *G. m. morsitans* have a broad spatial distribution over an ox whereas more than 80% of *G. pallidipes* land on the legs (Thomson, 1987). The warmth of the host's skin is probably the prime stimulus to probing. Although there are occasional references to tsetse being active, and even feeding, at night, these are exceptional and *Glossina* species are all essentially diurnal. However, not all are equally active throughout the hours of daylight. Measures of flies attracted to stationary hosts can be considered as giving the best estimate of differences in diurnal feeding activity, especially when more than one tsetse species is present. For example, Harley (1965) caught three species attracted to a stationary ox in Uganda and found different diurnal activity patterns, which themselves sometimes varied from one season to another. *Glossina pallidipes*, especially males, showed a gradual increase in feeding activity over the day until early evening, followed by a rapid decline. Rain depressed activity but there were no marked seasonal differences. *Glossina f. fuscipes* was most active at midday and *G. brevipalpis* near sunrise and sunset. In the Zambezi valley in Zimbabwe there are much more marked variations in climate than in Uganda, and in the hot season both *G. m. morsitans* and *G. pallidipes* begin feeding at dawn, activity rising to an early morning peak, ceasing when the ambient temperature exceeds about 32°C and then resuming in the late afternoon as temperatures fall. Some *fusca* group species – such as *G. tabaniformis* and *G. f. congolensis* (Jordan, 1962) and *G. longipennis* (Power, 1964) – have similar patterns of activity throughout the year. Conversely, a midday peak of biting activity is probably typical of *G. p. palpalis* at all times of the year, at least in the more humid parts of its distribution range (Page, 1959). *Glossina tachinoides* in Chad shows peak feeding activity at midday in the cool season but is most active in the early morning and especially at the end of the day in the hot dry season (Gruvel, 1970). If all these and other data are considered together, it is clear that many of the observed changes in feeding activity can be explained by changes in temperature and other environmental factors but some, such as the midday depression in activity in *G. m. morsitans*, are at least in part probably of an endogenous nature (Brady, 1972). An endogenous activity rhythm of this type may have adaptive significance, such as the similarity shown by times of activity of *G. m. morsitans* and its widely used host, the warthog (*Phacochoerus aethiopicus*).

Much is known about the hosts utilized by most *Glossina* species and often data for particular species are available from a number of localities. The subject has been well reviewed (Weitz, 1963) and although more data have accumulated recently, they have, in general, confirmed the conclusions reached. Perhaps the most significant discovery has been the realization that there are major differences in the feeding habits of some species of the *palpalis* group according to whether they occupy riverine habitats in drier zones or whether they roam widely in modified forest or peridomestic man-made habitats in humid zones of Africa. In drier areas species such as *G. palpalis* and *G. tachinoides* feed on a range of hosts including man but in the absence or scarcity of wild animal hosts in some of the more densely populated humid regions of Africa they feed extensively on domestic animals – often the domestic pig (*G. tachinoides*, Baldry, 1964; *G. palpalis*, Laveissière et al., 1986) – and rarely on man despite his apparent ready availability.

The feeding habits of the various species of *Glossina* are in part determined by the availability of hosts but some species show marked tendencies to feed on particular hosts. Many of the *palpalis* group are opportunist feeders; in some localities many meals are taken from reptiles and man. On the other hand, species of the *morsitans* and *fusca* groups have a relatively limited range of hosts. Man is not commonly fed upon (rarely, if ever, by some species of the *fusca* group) and has been shown to be repellent to a number of species (Ford, 1970; Vale, 1974). Nevertheless, really hungry individuals of some species, such as *G. morsitans*, will more readily attack man, and probably other rarely used hosts, than those which are well-fed. Such underdeveloped discriminative faculties may have survival value, as it has already been indicated that it is essential for the teneral fly, particularly under extreme climatic conditions, to obtain its first blood-meal before its supply of residual fat is exhausted.

Species of the *morsitans* group feed extensively on warthog and bushpig (*Potamochoerus porcus*) and also on certain species of Bovidae such as bushbuck (*Tragelaphus scriptus*), greater kudu (*Tragelaphus strepsiceros*) and buffalo (*Syncerus caffer*). The Suidae are, certainly outside national parks and game reserves, today more readily available to savanna tsetse flies than are the more obvious, and hence more readily hunted, Bovidae. Species like the warthog and the bushbuck are 'reliable' (Nash, 1969) hosts of tsetse, having relatively restricted home ranges in favoured tsetse habitat and being most active in the early morning and late afternoon when savanna species of tsetse are actively searching for food. Other hosts, such as buffalo, are readily fed upon by these species but are less reliable as they are here today and gone tomorrow. Still other species such as the red hartebeest (*Alcelaphus buselaphus*), waterbuck and allies (*Kobus* species) and impala (*Aepyceros melampus*) are apparently readily available to tsetse but are rarely fed upon. Why such mammals should not be fed upon by tsetse is not understood, especially as *G. m. morsitans* can digest impala blood as efficiently as that of other vertebrates (Langley, 1968). Still other species of potential host, such as the common zebra (*Equus burchellii*) (*quagga*) and the black wildebeest (*Connochaetes gnou*) may not be fed upon for a different reason: they are primarily animals of the open plains and not in such intimate contact with the fly.

No species of the *fusca* group and only some species of the *morsitans* and *palpalis* groups have been incriminated as vectors of human disease and the data given in Table 9.2 are restricted to the medically important species. The opportunist feeding of the *palpalis* group and the difference between the feeding habits of some species in riverine habitats in dry savanna areas and peridomestic habitats in humid areas is clear. The importance of certain savanna species of wild Bovidae and of wild Suidae to *G. m. centralis* (*G. m. morsitans* has similar feeding habits) and *G. pallidipes* is also apparent from Table 9.2.

Dispersal and movement

Glossina species are capable of flying at speeds of up to about 25 km per hour although normal flight speed is almost certainly less than this. Tsetse are incapable of sustained flight and activity takes the form of short bursts, probably of about five seconds duration. Female *G. m. morsitans* probably spend no more than a few minutes a day in flight, young males about 15 minutes and mature males 30–50 minutes (Bursell and Taylor, 1980). Periods of activity can be more prolonged in cold seasons than in hot seasons of the year. Spontaneous movement of this type implies that quite long distances can be flown but because of the short steps and, in areas of uniform habitat, essentially random direction, the rate of displacement is generally low. Bursell (1970) has suggested this might be of the order of 200 m per week in *G. morsitans* subspecies; advances of this species into new habitat also seem to occur at about this rate. If a fly is in a non-uniform environment and can relate to some major feature, such as a band of riparian vegetation, then much higher rates of dispersal are possible. For example, *G. p. palpalis* can travel 1000 m in a day and even over 2000 m (Nash and Page, 1953). Towards the end of the hunger cycle, if a species of the *morsitans* group encounters the odour given off by a host animal, an upwind anemotaxis (Vale, 1974) introduces a brief directional component to their movement but this will eventually be offset by the tendency of flies to move downwind after host encounter (Vale, 1977). The study of movement is difficult as it is complicated by the fact that many species of tsetse, and not just flies at the end of the hunger cycle, will readily follow or ride on men, animals and vehicles. Such 'movement' can even occur over extensive tracts of open land and has often been responsible for the breakdown of man-made 'barriers' designed to protect areas rendered free of tsetse from nearby infested areas.

Table 9.2 Sources of blood food (percéntages) of some medically important species of tsetse-flies (*Glossina*)

Lettered columns show data from the literature for species/subspecies, their associated habitats and geographical areas, as follows:
A – *G. fuscipes fuscipes*: lake-shore habitats, East Africa (Weitz, 1963).
B – *G. palpalis palpalis*: riverine habitats, northern Nigeria (Jordan et al., 1962).
C – *G. palpalis palpalis*: forest and peridomestic habitats near Vavoua, Ivory Coast (Laveissière et al., 1986).
D – *G. tachinoides*: riverine habitats, northern Nigeria (Jordan et al., 1962).
E – *G. tachinoides*: forest and peridomestic habitats near Nsukka, Nigeria (Baldry, 1964).
F – *G. morsitans centralis*: savanna woodland habitats, Daga-Iloi, Tanzania (Weitz and Glasgow, 1956).
G – *G. pallidipes*: dense thicket habitats, Lambwe Valley, Kenya (England and Baldry, 1972).
Percentage totals are shown in parentheses for clarity.

Vertebrate hosts	A	B	C	D	E	F	G
Reptiles	(34)	(38)	(2)	(8)	—	—	—
Birds	(1)	(6)	—	(1)	—	(2)	—
Mammals							
Primates							
Man	17	26	9	30	1	10	*
Unidentified/others	1	2	6	13	—	5	1
(Total)	(18)	(28)	(15)	(43)	(1)	(15)	(1)
Suidae							
Warthog	—	1	—	1	—	55	—
Bushpig	3	—	—	—	—	6	18
Domestic pig	—	—	56	—	64	—	—
Unidentified	—	4	—	1	—	—	—
(Total)	(3)	(5)	(56)	(2)	(64)	(61)	(18)
Bovidae							
Bushbuck	17	7	6	1	—	—	32
Buffalo	2	1	1	—	—	11	3
Duiker	—	—	—	1	—	1	3
Ox	1	5	—	8	—	—	—
Roan antelope	—	—	—	—	2	6	—
Reedbuck	—	—	—	—	—	3	—
Unidentified/others	17	8	18	21	7	—	43
(Total)	(37)	(21)	(25)	(31)	(9)	(21)	(81)
Other mammals							
Carnivore	3	1	—	1	—	2	—
Porcupine	—	—	—	4	2	—	—
Rhinoceros	—	—	—	—	—	1	—
Unidentified/others	2	2	1	11	25	—	2
(Total)	(5)	(3)	(1)	(16)	(27)	(3)	(2)
Total meals	590	266	694	426	152	281	922

* Two feeds only.

Tsetse-flies are exceptional among the Diptera in that they do not require carbohydrate from plant sugars for their flight energy. Instead they exploit the partial oxidation of proline, an amino acid derived from the blood-meal, and when the limited reserves of proline are exhausted, flight must cease. During rest following flight, proline reserves are reconstituted, probably involving alanine, bicarbonate and lipid carbon as precursors (Bursell et al., 1974).

Tsetse-flies spend more than 23 hours each day at rest and studies have been undertaken on the resting sites of a number of *Glossina* species, of all three species-groups, in many parts of Africa. The extensive literature on both day and night resting sites was reviewed by Langridge et al. (1963), Hadaway (1977) and Challier (1982). True resting sites, on which flies spend most of their time during the day or night, have to be distinguished from so-called 'watching sites' on which flies rest only temporarily when disturbed or when following a man, animal or vehicle. Watching sites are often on fallen logs, grass and the ground – sites never used as true resting sites – as well as on trees and shrubs.

Day resting sites are usually on woody parts of the vegetation and night resting sites are usually on the upper surfaces of leaves, often in tree canopies. As temperatures increase during the day, so flies tend to rest nearer to the ground. In areas where one or more species of the *palpalis* group coexist with one or more species of the *morsitans* group, the former tend to rest on lower sites than the latter.

Some observations on *G. m. morsitans* and *G. p. palpalis* in different parts of Nigeria illustrate some of these generalities. At the northern limit of *G. m. submorsitans*, in the hot, dry season, at temperatures below about 30°C the flies rest on the underside of horizontal or near-horizontal branches of 3–30 cm diameter and 1–4 m above the ground. At higher temperatures the flies move from branches on to tree trunks and above about 32°C they are commonly to be found on larger tree trunks, in shade, no more than 2 m above the ground. Further south, in the less extreme dry season of the northern Guinea savanna, *G. m. submorsitans* is less restricted in its resting sites, the flies spending less time on tree trunks and more on horizontal or near-horizontal branches (Fig. 9.21). Further south still, in the southern Guinea zone, resting sites are much more diffuse. Unlike *G. m. submorsitans*, *G. p. palpalis* seldom rests on tree trunks. During the day, in the savanna areas of West Africa, this species rests mainly on the underside of thin branches and stems of climbing plants up to 3 cm in diameter, most resting between ground level and 2 m. Roots of trees exposed by erosion and shaded by an overhanging river bank are often used as resting sites.

Examples of the various vegetational habitats mentioned above can be seen in Figs 9.22 and 9.23.

MEDICAL IMPORTANCE

Role as biting pests

Glossina species are not important biting pests of man or domestic animals as the density of populations is generally low, unlike the seasonally enormous numbers that can occur with some other species of biting insect. Local reaction to a tsetse bite is generally minimal, but occasionally severe weals and pronounced swelling can occur (Gordon and Crewe, 1948).

Role as disease vectors

As far as is known, the only pathogenic parasites transmitted by *Glossina* species are certain species of the genus *Trypanosoma*. These protozoan blood parasites generally

Figure 9.21 Commonly used resting sites of *Glossina morsitans submorsitans* at the hottest time of day in the dry season in Nigeria: (*a*) on shaded tree trunks in the Sudan savanna vegetation zone; (*b*) on the underside of tree branches in the northern Guinea savanna vegetation zone. (After Jordan, 1986.)

have no pathogenic effects on a range of African vertebrates but some are pathogenic to man and others to some domestic animals.

Flies become infected by ingesting vertebrate blood containing trypanosomes from infected hosts. Following ingestion the trypanosomes undergo a cycle of development in the fly the complexity and duration of which varies according to the subgenus of *Trypanosoma* concerned. In the subgenus *Duttonella* (which includes the important cattle parasite *T. vivax*) development takes place only in the proboscis of the fly, in *Nannomonas* (which includes the important cattle parasite *T. congolense* and the pig parasite *T. simiae*) in the mid-gut and in the proboscis, and in *Trypanozoon* (which includes the relatively non-pathogenic livestock parasite *T. brucei brucei* and the human parasites *T. b. gambiense* and *T. b. rhodesiense*) in the mid-gut and salivary glands. Once metacyclic forms of the parasite have been produced, the fly is potentially capable of infecting any susceptible host on which it subsequently feeds.

Trypanosomes can also be transmitted mechanically from one mammalian host to another on the mouthparts of tsetse-flies (and other biting flies) but, as trypanosomes can only survive outside the insect or mammalian host for a short time, successive hosts have to be probed by the fly in quick succession. Although it may be important at some times of the year (such as during seasonal peaks of biting fly activity), there is no evidence that normal cyclically transmitted African trypanosomiasis is sustainable in the absence of *Glossina*.

In economic terms, *Glossina* species are more important today as vectors of animal trypanosomiasis than as vectors of human trypanosomiasis. The animal disease causes mortality and morbidity of a number of species of domestic animal, particularly cattle. Even more important, the disease prevents the keeping of cattle and other domestic animals over extensive areas of Africa. Not only has this limited the amount of meat and milk available but it has also restricted the development of mixed arable and livestock farming and animals have not been readily available for draught purposes. Overviews of the significance of animal trypanosomiasis in African rural economies are given in Ford (1971) and Jordan (1986).

Figure 9.22 Two tsetse habitats of differing vegetational type in West Africa: (*a*) a habitat in a dry savanna area of Nigeria where *Glossina palpalis palpalis* is restricted to riparian vegetation; (*b*) a forest 'island' habitat of *Glossina fusca fusca* and *G. medicorum* in Ivory Coast.

Figure 9.23 Two tsetse habitats of conspicuously different type in West Africa: (*a*) savanna woodland habitat of *Glossina morsitans submorsitans* in The Gambia in which trypanotolerant N'Dama cattle are grazing; (*b*) peridomestic village habitat of *G. palpalis palpalis* in an area of southern Ivory Coast where the flies feed primarily on domestic pigs and man and where *Trypanosoma brucei gambiense* sleeping sickness occurs: in the left foreground is an example of the biconical trap used with much success in tsetse surveillance and control.

376 *Tsetse-flies (Glossinidae)*

There are two forms of human trypanosomiasis. *Trypanosoma brucei gambiense* is the causative organism of a classically chronic disease which occurs in West and Central Africa from Senegal to Sudan in the north and southwards to Angola and Zaire. The vectors of *T. b. gambiense* are species of the *palpalis* group, namely *G. palpalis*, *G. fuscipes* and *G. tachinoides*. *Glossina caliginea* is suspected of being a vector but as it has a very restricted distribution it could only be of little importance. The epidemiology of the disease is far from being fully understood. Disease prevalence is not directly related to the density of *Glossina* populations and epidemics can occur where tsetse are relatively scarce. Conversely, there are extensive areas where one or more of the three important vector species are widespread and abundant, but where sleeping sickness has never been recorded (Fig. 9.24). This is quite unlike animal trypanosomiasis, which occurs wherever there are tsetse-flies. *Trypanosoma brucei rhodesiense* is the causative organism of a typically more acute form of human trypanosomiasis. The epidemiology and the tsetse vectors of the two forms of the disease are different. Whereas man is probably the usual mammalian host of *T. b. gambiense*, wild and domestic animals can be

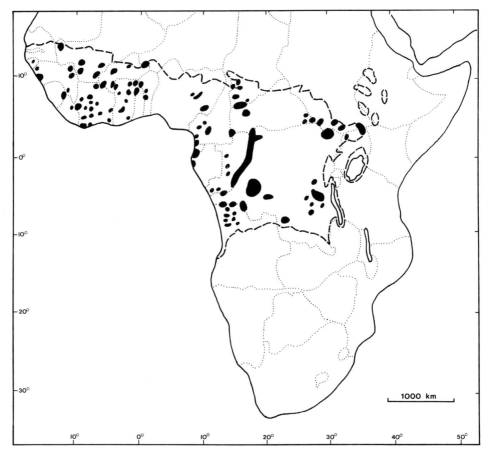

Figure 9.24 Geographical distribution of *Trypanosoma brucei gambiense* sleeping sickness and of its vectors. Disease foci are shown as solid black areas (after World Health Organization, 1986) and distribution limits of *Glossina palpalis*, *G. fuscipes* and *G. tachinoides* by the broken lines. (*T. b. gambiense* is considered to have occurred in former times in Busoga, Uganda, north of Lake Victoria, but the causative organism of sleeping sickness there now is *T. b. rhodesiense*.)

reservoir hosts but are generally outside the typical cycle of transmission; for *T. b rhodesiense* the usual mammalian hosts are wild ungulates and man is usually an adventitious host. The latter form of sleeping sickness is thus a true zoonosis and is generally associated with areas inhabited by wild animals in which man is essentially an intruder. The savanna woodlands of east and central southern Africa are the main areas where *T. b. rhodesiense* sleeping sickness occurs and where its vectors are species of the *morsitans* group inhabiting these woodlands, namely *G. m. morsitans*, *G. m. centralis* and *G. pallidipes*. *Glossina swynnertoni* has also been incriminated in the restricted area in which it occurs. As with *T. b. gambiense* sleeping sickness, there are extensive areas where the vectors occur but where the disease has never been recorded (Fig. 9.25). The situation around Lake Victoria is unusual. In the early years of the twentieth century there were major epidemics of sleeping sickness which originated in Uganda but spread around both the eastern and western shores of the lake to northern Tanzania. The parasite was *T. b. gambiense* and the vector *G. f. fuscipes*. After a series of well-documented biological events (summarized in Jordan, 1986), *T. b. rhodesiense* replaced *T. b gambiense* but

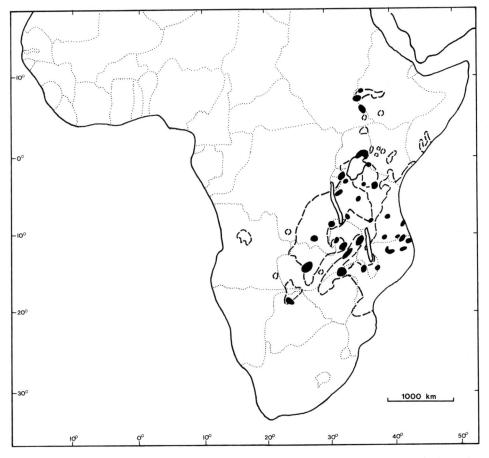

Figure 9.25 Geographical distribution of *Trypanosoma brucei rhodesiense* sleeping sickness and of its vectors of the *Glossina morsitans* group. Disease foci are shown as solid black areas (after World Health Organization, 1986) and distribution limits of *G. morsitans morsitans*, *G. morsitans centralis*, *G. swynnertoni* and *G. pallidipes* by the broken lines. (The vector of *T. b. rhodesiense* in Busoga, Uganda, north of Lake Victoria, is *G. fuscipes fuscipes*.)

G. f. fuscipes continued to be the vector. Epidemics of the disease occurred in 1964 in Kenya and in the Busoga District of Uganda in the mid-1970s, where an epidemic continues at the present time (1992). Cattle are an important mammalian reservoir of *T. b. rhodesiense* in some areas.

The classification of the course of the disease into chronic *T. b. gambiense* and acute *T. b rhodesiense* is convenient but not always accurate, as acute forms of *T. b. gambiense* sleeping sickness (Scott, 1970) and chronic forms of the *T. b. rhodesiense* disease (Apted, 1970) are known to occur. Both forms of the disease were much greater problems in the past than they are today. An account of the history of the disease is given by Duggan (1970). About 25 000 new cases of the disease are reported each year (TDR, 1990), although many other cases must occur in remote rural areas. Far more people contract the *T. b. gambiense* disease than the *T. b. rhodesiense* disease. Most cases have been reported for many years from Zaire (World Health Organization, 1986), but there were also numerous cases in Angola, southern Sudan and Uganda at the end of the 1980s. If surveillance and treatment of the disease is relaxed, as happened at the sudden end of the colonial era in Zaire and (for different reasons) in Uganda, a serious recrudescence of the disease can soon occur.

Even in areas of epidemic sleeping sickness the prevalence of infection of wild *Glossina* with trypanosomes infective to man seldom exceeds 0.1%, although much higher infection rates of trypanosomes infective to livestock occur in all *Glossina* species (reviews by Buxton, 1955, and Jordan, 1974). Rates of 10–15% in the *morsitans* group and 5% in the *palpalis* group are typical. The extremely low infection rates of *T. b. gambiense* and *T. b. rhodesiense* (and of the non-human parasite *T. b. brucei*) in *Glossina* has never been satisfactorily explained. A variety of factors influences the establishment of trypanosomes in tsetse. Three types of factor are involved (Molyneux, 1977):

1. Endogenous factors associated with the fly: fly age at infective feed; sex of fly; genetic differences between fly species or within species; behaviour (e.g. host preference); physiological and biochemical state of fly; concurrent infections of fly (viruses, bacteria, fungi).
2. Ecological factors: climatic factors; availability of infected hosts; hosts available for subsequent feeds.
3. Parasite and host: parasite numbers available to fly; type of parasite and its infectivity to fly including the immune state of host; subspecies or strain; susceptibility; intercurrent infection; behaviour and attractiveness to fly.

Clearly, equal weight cannot be given to all these factors but, equally clearly, the situation is complex. As far as *T. b. gambiense* and *T. b. rhodesiense* are concerned, a particularly crucial factor was the finding that tsetse generally only become infected with these parasites when they are very young, especially when taking their first blood-meal (Wijers, 1958). Even in epidemic situations, the chances of a fly taking its first blood-meal from an infected man are not high, in view of the range of hosts utilized (Table 9.2). It has been proposed that this susceptibility to infection may be related to the activity of the rickettsial symbionts in the larval and pupal stages which cause the build-up of D-glucosamine in the newly emerged fly (Maudlin and Welburn, 1988). D-glucosamine inhibits lectins in the tsetse mid-gut (Ibrahim et al., 1984), rendering them susceptible to infection. It is further postulated that older flies may not produce sufficient D-glucosamine to inhibit lectin production and that lectins in the fly are also involved in controlling maturation of trypanosomes from mid-gut forms to metacyclics, infective to the mammalian host, in the salivary glands (Welburn et al., 1989). Although this story is, as yet, incomplete, there is firm evidence that some species and, of particular relevance to the epidemiology of human trypanosomiasis, that some individual flies, have an intrinsic susceptibility or refractoriness to infection (Maudlin and Dukes, 1985). The existence of 'pockets' of susceptible tsetse may be at least a partial explanation of why human trypanosomiasis occurs in long-standing 'hot spots', whereas other apparently ecologically similar zones are free of the disease.

CONTROL

Because of the mode of reproduction of the genus, a tsetse population cannot be subjected to regular high mortality at any life cycle stage if the population is to survive. Much is known in qualitative terms of the causes of mortality in populations of *Glossina* but very little is known of the numbers of deaths that can be attributed to each cause at different times of the year. Known predators of adult tsetse range from monkeys to spiders, and of tsetse puparia from rodents to ants (Gruvel, 1977). Deaths from insect parasitoids attacking the puparia have been quantified and these can be seasonally high in some localities but negligible (and often virtually non-existent) elsewhere (Greathead, 1980; Simmonds et al., 1977). There is little information on the nature of density-dependent mortality factors but it has been suggested that high-density tsetse populations may provoke increased reactions in host animals so that food is relatively less available than it is to lower density populations. Mortality of both adults and puparia caused by some predators may be of a density-dependent nature. Only a few, unsuccessful, attempts have been made to release biological control agents – insect parasitoids – into tsetse habitats.

It has been estimated that if an added mortality of 4% per day can be imposed and sustained on any female tsetse population then it must go extinct, regardless of the level of any density-dependent or density-independent natural mortality. Under field conditions it is likely that only 2–3% mortality per day is required (Hargrove, 1988). Some methods of control result in much higher daily mortalities than this but they are often not sustained for long enough to achieve eradication. Eradication of tsetse populations has been achieved by a number of techniques but it is necessary to distinguish between short-lived and sustainable 'eradication'. Campaigns in a number of countries have demonstrated that it is possible to remove *Glossina* from an area but that it is much more difficult to ensure that the area is not subsequently reinvaded from nearby untreated infestations once control measures have been relaxed. In practice, successful eradication campaigns have been aimed at *Glossina* as vectors of animal trypanosomiasis and have only been successful when an isolated entire fly belt has been treated (e.g. du Toit, 1954) or when removal of the fly over a large area has been followed by human settlement and development of the land to an extent that both tsetse habitats and hosts are removed (e.g. Maclennan and Na'isa, 1971). All forms of man-made 'barrier' to fly movement (whether of cleared vegetation, persistent insecticide on vegetation or on devices placed in the habitat by man) have proved inadequate, on anything other than a limited time-scale, to prevent reinfestation of areas cleared of the fly.

The concept of tsetse eradication will not be discussed further, because it has never seriously been applied to tsetse as vectors of human disease. It is not a realistic objective unless foci of the disease occur within the much wider area covered by a campaign to eliminate the vectors of animal trypanosomiasis. A number of sleeping sickness foci were removed in this way following extensive tsetse eradication operations in northern Nigeria (Maclennan and Na'isa, 1971) which were primarily aimed at the removal of animal trypanosomiasis.

Vector control (as distinct from eradication) is, however, an option as a component of strategies to combat human trypanosomiasis. It is generally not a cost-effective option for the control of endemic sleeping sickness over large areas and control under these circumstances will have to continue to rely, as at present, on the ability of rural medical services to diagnose and treat the disease. On the other hand, during epidemics, which are usually fairly circumscribed in area, vector control becomes an acceptable strategy. It is desirable to integrate vector control measures with surveillance of the human population and the use of trypanocidal drugs to treat diagnosed cases. The choice of vector control technique will depend on the resources available and the ecology of the flies, e.g. in drier areas of West Africa it is necessary to control only the riverine species of the *palpalis* group, which transmit the disease; the savanna species of the *morsitans* group, which do not, can be ignored. Costs can often be reduced, without any loss of efficiency, by involving the local people in control operations.

Populations of *Glossina* can be controlled by a variety of methods, not all of which are appropriate for control of the vectors of human trypanosomiasis. There have been many reviews of past and present methods of tsetse control (e.g. Jordan, 1986; Mulligan, 1970).

Glossina species can be controlled effectively by removing the vegetation on which they depend for their habitat and by killing wild animals on which they depend for their food. Whereas the latter has never been practised as a control measure for the human disease, the complete or partial clearance of shrub or woody vegetation was widely practised in the past as a method for controlling the vectors of sleeping sickness. It can be very effective but it is labour-intensive and cleared land does not remain cleared for long unless regular, preferably annual, reslashing of vegetation is carried out.

With the advent of modern insecticides, vegetation clearing as a method of tsetse control fell into disuse – although it is noteworthy that the same effect is being achieved unintentionally today in parts of Africa as human populations expand and remove the woody vegetation either for settlement or for use as firewood for the rapidly growing cities. Insecticide applications can be classified into two categories according to whether they are persistent and remain lethal to *Glossina* for at least two to three months, or are applied as a non-residual aerosol that drifts with the wind and kills any tsetse on which sufficient insecticide impinges but which must be repeated at regular intervals if individuals emerging from puparia in the soil are to be available to the insecticide. Both ground and aerial application techniques have been developed but aerial methods are expensive and have rarely been employed in human disease situations. The ground spraying of carefully selected sites with residual insecticide is a realistic option and, when competently carried out, can rapidly reduce a vector population and hence disease transmission.

Several studies were undertaken prior to the advent of modern insecticides which showed that, because of the low reproductive potential of *Glossina*, it was sometimes possible to catch and kill enough tsetse flies in an area to cause a population to decline or even disappear. Various types of traps were designed. This early work showed that, whereas widely dispersed populations of the *morsitans* group could not be sufficiently reduced to significantly affect disease transmission, circumscribed populations of the *palpalis* group (e.g. on an island or in a linear riverine habitat) could be greatly reduced by trapping. Since about 1973, the advent of the biconical trap (Fig. 9.23b) for sampling populations of *Glossina* (Challier and Laveissière, 1974), there have been major advances in the development of traps and related devices for use in tsetse control. Trials, and larger campaigns, have been carried out in many countries. At least 90% control of the important human disease vectors of the *palpalis* group can be relatively easily achieved, especially when they occur in circumscribed habitats, both by the use of biconical traps impregnated with insecticide (e.g. Laveissière et al., 1981) or by simpler blue cloth screens (120 × 90 cm), impregnated with insecticide and hung from vegetation in the fly habitat (e.g. Laveissière and Couret, 1982). Other types of trap have also been used with success, sometimes without impregnation with insecticide. Dispersed populations of the *morsitans* group, including the vectors of *T. b. rhodesiense*, are more difficult to trap-out but rather more complicated cloth targets, impregnated with insecticide and with odour attractants dispensed in their vicinity, can achieve control of these species. The identification and exploitation of the components of host odour that attract *Glossina* has been a major recent development. By the end of the 1980s, several compounds were used routinely to increase the effectiveness of targets in a number of control operations. The *palpalis* group as a whole is much less responsive to odours than the *morsitans* group (although some species have been shown to respond to some known odours in addition to CO_2) and natural or artificial odours are not routinely used for control of these vectors of sleeping sickness. Traps and targets cause virtually no contamination of the environment with insecticide. Cloth screens are cheap to operate and local communities can – and should be encouraged to – assist in placement and maintenance. However, like other methods of vector control (as distinct from sustainable eradication), traps and targets require a continuing commitment if population suppression is to be maintained. Screens need to be replaced when damaged or stolen (which has been a major problem in some localities) and insecticide must be replaced at regular intervals if its efficacy is to be more than short-lived.

The sterile insect technique, of which there have been successful trials against some tsetse vectors of animal trypanosomiasis, is inappropriate for inclusion in control operations against the vectors of human trypanosomiasis. It is expensive, and hence cannot be justified for localized control, and even a temporary man-induced increase in numbers of potential vectors of sleeping sickness would be unacceptable.

COLLECTING, PRESERVING AND REARING MATERIAL

Tsetse puparia can be collected by sieving sand or soil from their breeding sites. This is most easily accomplished if there is a seasonal concentration of larviposition sites but even under these circumstances it is a skilled and tedious process. Adult tsetse can be caught by a number of methods. Some species are attracted to man and can be caught with a hand-net but others (such as *G. pallicera*, *G. austeni* and the species of the *fusca* group) are rarely attracted to man and it is necessary to use a bait animal – generally and most successfully an ox – to attract the flies. It is also possible to search for resting flies and catch them with a hand-net. This is particularly useful with the large species of the *fusca* group, which do not feed on man, but it is, again, a difficult and tedious procedure.

Many *Glossina* species can be caught in traps, although whether all species (particularly the species of the *fusca* group that occur in lowland rain forest) can be caught in this way is unknown. Many different types of tsetse trap have been designed (Buxton, 1955; Swynnerton, 1933). The biconical trap (Challier and Laveissière, 1974) is portable, sturdy and easily assembled and is highly effective at catching species of the *palpalis* group. It will also catch other species but other designs, such as the F3 and the epsilon trap (Flint, 1985; Vale, personal communication) are more efficient for catching species of the *morsitans* group.

Adult flies and puparia can be preserved in alcohol but if identification by a specialist is required, at least some adults should be preserved as dry-pinned specimens.

Whereas *Glossina* species are not difficult to keep alive in the laboratory, the maintenance of a self-supporting colony is a specialized procedure, mainly because the insects are so slow-breeding and it is necessary to keep females alive, on average, for at least a month. Adults and puparia should be maintained at a temperature of about 25°C and relative humidity of 60–80%. Adults must be provided with regular blood-meals and this can most conveniently be achieved by feeding on rabbits' ears (Nash et al., 1966) or the shaven flanks of guinea pigs (Geigy, 1948). Colonies of *Glossina* species are maintained at a number of African and European laboratories; in some of these the flies are maintained on living animal hosts and in others on blood presented to them under an artificial membrane (Mews et al., 1977). The subject is reviewed by Langley (1985).

REFERENCES

Apted, F. I. C. 1970. The epidemiology of Rhodesian sleeping sickness. Pp. 645–660 in Mulligan, H. W. (ed.), *The African trypanosomiases.* lxxxviii + 950 pp. George Allen & Unwin, London.

Baldry, D. A. T. 1964. Observations on a close association between *Glossina tachinoides* and domestic pigs near Nsukka, Eastern Nigeria. II. – Ecology and trypanosome infection rates in *G. tachinoides. Annals of Tropical Medicine and Parasitology* **58**: 32–44.

Bax, S. N. 1937. The senses of smell and sight in *Glossina swynnertoni. Bulletin of Entomological Research* **28**: 539–582.

Bequaert, J. 1954. The Hippoboscidae or louse-flies (Diptera) of mammals and birds. Part II. Taxonomy, evolution and revision of American genera and species. *Entomologica Americana* (n.s.) **34**: 1–232.

Bezzi, M. 1911. Études systématiques sur les Muscides hématophages du genre *Lyperosia. Archives de Parasitologie* **15**: 110–143.

Brady, J. 1972. Spontaneous, circadian components of tsetse fly activity. *Journal of Insect Physiology* **18**: 471–484.

Brues, C. T., Melander, A. L. and **Carpenter, F. M.** 1954. Keys to the living and extinct families of insects, and to the living families of other terrestrial arthropods. *Bulletin of the Museum of Comparative Zoology* (Harvard) **108**: v + 1–917.

Bursell, E. 1955. The polypneustic lobes of the tsetse larva (*Glossina*, Diptera). *Proceedings of the Royal Society* (B) **144**: 275–286.

Bursell, E. 1958. The water balance of tsetse pupae. *Philosophical Transactions of the Royal Society of London* (B) **241**: 179–210.

Bursell, E. 1960. The effect of temperature on the consumption of fat during pupal development in *Glossina*. *Bulletin of Entomological Research* **51**: 583–598.

Bursell, E. 1961. The behaviour of tsetse flies (*Glossina swynnertoni* Austen) in relation to problems of sampling. *Proceedings of the Royal Entomological Society of London* (A) **36**: 9–20.

Bursell, E. 1970. Dispersal and concentration of *Glossina*. Pp. 382–394 in Mulligan, H. W. (ed.), *The African trypanosomiases*. 1xxxviii + 950 pp. George Allen & Unwin, London.

Bursell, E. 1984. Effects of host odour on the behaviour of tsetse. *Insect Science and its Application* **5**: 345–349.

Bursell, E. and **Taylor, P.** 1980. An energy budget for *Glossina* (Diptera: Glossinidae). *Bulletin of Entomological Research* **70**: 187–196.

Bursell, E., Billing, K. C., Hargrove, J. W., McCabe, C. T. and **Slack, E.** 1974. Metabolism of the bloodmeal in tsetse flies (a review). *Acta Tropica* **31**: 297–320.

Burtt, E. 1952. The occurrence in nature of tsetse pupae (*Glossina swynnertoni* Austen). *Acta Tropica* **9**: 304–344.

Burtt, E. and **Jackson, C. H. N.** 1951. Illustrations of tsetse larvae. *Bulletin of Entomological Research* **41**: 523–527.

Buxton, P. A. 1955. The natural history of tsetse flies. An account of the biology of the genus *Glossina* (Diptera). *London School of Hygiene and Tropical Medicine Memoir* **10**, xviii + 816 pp. H. K. Lewis, London.

Carlson, D. A. 1981. Chemical taxonomy in tsetse flies (*Glossina* spp.) by analysis of cuticular components. Pp. 449–457 in International Scientific Council for Trypanosomiasis Research and Control, Seventeenth Meeting, Arusha, 1981. Publication No. 112, 664 pp. Organization of African Unity Scientific and Technical Research Commission.

Carlson, D. A., Nelson, D. R., Langley, P. A., Coates, T. W., Davis, T. L. and **Leegwater-van der Linden, M. E.** 1984. Contact sex pheromone in the tsetse fly *Glossina pallidipes* (Austen): identification and synthesis. *Journal of Chemical Ecology* **10**: 429–450.

Carter, R. M. 1906. Tsetse fly in Arabia. *British Medical Journal* **1906** (2): 1393–1394.

Challier, A. 1982. The ecology of tsetse (*Glossina* spp.) (Diptera: Glossinidae): a review (1970–1981). *Insect Science and its Application* **3**: 97–143.

Challier, A. and **Laveissière, C.** 1974. Un nouveau piège pour la capture des glossines (*Glossina*: Diptera, Muscidae): description et essais sur le terrain. *Cahiers ORSTOM* (Entomologie médicale et Parasitologie) **11** (1973): 251–262.

Challier, A., Gouteux, J. -P. and **Coosemans, M.** 1984. La limite géographique entre les sous-espèces *Glossina palpalis palpalis* (Rob.-Desv.) et *G. palpalis gambiensis* Vanderplank (Diptera: Glossinidae) en Afrique occidentale. *Cahiers ORSTOM* (Entomologie médicale et Parasitologie) **21** (1983): 207–220.

Cockerell, T. D. A. 1918. New species of North American fossil beetles, cockroaches, and tsetse flies. *Proceedings of the United States National Museum* **54**: 301–311.

Cross, N. C. P. and **Dover, G. A.** 1987. A novel arrangement of sequence elements surrounding the rDNA promoter and its spacer duplications in tsetse species. *Journal of Molecular Biology* **195**: 63–74.

Curtis, C. F. 1968. Some observations on reproduction and the effects of radiation on *Glossina austeni*. *Transactions of the Royal Society of Tropical Medicine and Hygiene* **62**: 124.

Curtis, C. F. 1972. Sterility from crosses between sub-species of the tsetse fly *Glossina morsitans*. *Acta Tropica* **29**: 250–268.

Dias, J. A. Travassos Santos 1956. Descrição de uma nova subespécie de tsé-tsé de Moçambique, *Glossina austeni mossurizensis* n. spp. *Moçambique* **88**: 51–78.

Dias, J. A. Travassos Santos 1987. Contribução para o estudo da sistemática do género *Glossina* Wiedemann, 1830 (Insecta, Brachycera, Cyclorrhapha, Glossinidae). Proposta para a criação de um novo subgénero? *Garcia de Orta* (Zoologia) **14**: 67–78.

Duggan, A. J. 1970. An historical perspective. Pp. xli–lxxxviii in Mulligan, H. W. (ed.), *The African trypanosomiases*. lxxxviii + 950 pp. George Allen & Unwin, London.

Du Toit, R. 1954. Trypanosomiasis in Zululand and the control of tsetse flies by chemical means. *Onderstepoort Journal of Veterinary Research* **26**: 317–387.

Elsen, P., Amoudi, M. A. and **Leclercq, M.** 1990. First record of *Glossina fuscipes fuscipes* Newstead, 1910 and *Glossina morsitans submorsitans* Newstead, 1910 in southwestern Saudi Arabia. *Annales de la Société Belge de Médecine Tropicale* **70**: 281–287.

England, E. C. and **Baldry, D. A. T.** 1972. The hosts and trypanosome infection rates of *Glossina pallidipes* in the Lambwe and Roo valleys. *Bulletin of the World Health Organization* **47**: 785–788.

Evens, F. M. J. C. 1953. Dispersion géographique des Glossines au Congo belge. *Mémoires. Institut royal des Sciences naturelles de Belgique* (2) **48**: 1–70.

Flint, S. 1985. A comparison of various traps for *Glossina* spp. (Glossinidae) and other Diptera. *Bulletin of Entomological Research* **75**: 529–534.

Ford, J. 1970. The search for food. Pp. 298–304 in Mulligan, H. W. (ed.), *The African trypanosomiases*. lxxxviii + 950 pp. George Allen & Unwin, London.

Ford, J. 1971. *The role of the trypanosomiases in African ecology. A study of the tsetse fly problem.* xiv + 568 pp. Clarendon Press, Oxford.

Ford, J. and **Katondo, K. M.** 1977. Maps of tsetse fly (*Glossina*) distribution in Africa 1973, according to sub-generic groups on scale 1:5,000,000. *Bulletin of Animal Health and Production in Africa* **15**: 187–193.

Geigy, R. 1948. Élevage de *Glossina palpalis*. *Acta Tropica* **5**: 201–218.

Glasgow, J. P. 1960. Variations in the venation of *Glossina* Wiedemann (Diptera: Muscidae). *Proceedings of the Royal Entomological Society of London* (A) **35**: 49–57.

Glasgow, J. P. 1970. The genus *Glossina*: introduction. Pp. 225–242 in Mulligan, H. W. (ed.), *The African trypanosomiases*. lxxxviii + 950 pp. George Allen & Unwin, London.

Gooding, R. H. 1982. Classification of nine species and subspecies of tsetse flies (Diptera: Glossinidae: *Glossina* Wiedemann) based on molecular genetics and breeding data. *Canadian Journal of Zoology* **60**: 2737–2744.

Gooding, R. H. 1984. Tsetse genetics: a review. *Quaestiones Entomologicae* **20**: 89–128.

Gordon, R. M. and **Crewe, W.** 1948. The mechanisms by which mosquitoes and tsetse-flies obtain their blood-meal, the histology of the lesions produced, and the subsequent reactions of the mammalian host; together with some observations on the feeding of *Chrysops* and *Cimex*. *Annals of Tropical Medicine and Parasitology* **42**: 334–356.

Gouteux, J.-P. 1987. Une nouvelle glossine du Congo: *Glossina* (*Austenina*) *frezili* sp. nov. (Diptera: Glossinidae). *Tropical Medicine and Parasitology* **38**: 97–100.

Greathead, D. J. 1980. Biological control of tsetse flies: an assessment of insect parasitoids as control agents. *Biocontrol News and Information* **1**: 111–123.

Green, C. H. 1986. Effects of colours and synthetic odours on the attraction of *Glossina pallidipes* and *G. morsitans morsitans* to traps and screens. *Physiological Entomology* **11**: 411–421.

Griffiths, G. C. D. 1976. Comments on some recent studies of tsetse-fly phylogeny and structure. *Systematic Entomology* **1**: 15–18.

Gruvel, J. 1970. Observations écologiques concernant *Glossina tachinoides* dans la région du Bas-Chari (Tchad). Pp. 445–454 in Azevedo, J. Fraga de (ed.), *Criação da mosca tsé-tsé em laboratorio e sua aplicação prática. Tsetse fly breeding under laboratory conditions and its practical application*. 1st International Symposium, 22nd and 23rd April 1969, Lisbon. 524 pp. Junta da Investigações do Ultramar, Lisboa.

Gruvel, J. 1977. Predators. Pp. 45–55 in Laird, M. (ed.), *Tsetse: the future for biological methods in integrated control*. 220 pp. International Development Research Centre, Ottawa.

Hadaway, A. B. 1977. Resting behaviour of tsetse flies, and its relevance to their control with residual insecticides. *Miscellaneous Report No. 36*. 11 pp. Centre for Overseas Pest Research, London.

Haeselbarth, E., Segerman, J. and Zumpt, F. 1966. The arthropod parasites of vertebrates in Africa south of the Sahara (Ethiopian region). 3 (Insecta excl. Phthiraptera). *Publications of the South African Institute for Medical Research* **13** (52): 1–283.

Hall, D. R., Beevor, P. S., Cork, A., Nesbitt, B. F. and Vale, G. A. 1984. 1-Octen-3-ol: a potent olfactory stimulant and attractant for tsetse isolated from cattle odours. *Insect Science and its Application* **5**: 335–339.

Hargrove, J. W. 1988. Tsetse: the limits to population growth. *Medical and Veterinary Entomology* **2**: 203–217.

Harley, J. M. B. 1965. Activity cycles of *Glossina pallidipes* Aust., *G. palpalis fuscipes* Newst. and *G. brevipalpis* Newst. *Bulletin of Entomological Research* **56**: 141–160.

Hassanali, A., McDowell, P. G., Owaga, M. L. A. and Saini, R. K. 1986. Identification of tsetse attractants from excretory products of a wild host animal, *Syncerus caffer*. *Insect Science and its Application* **7**: 5–9.

Hennig, W. 1965. Vorarbeiten zu einem phylogenetischen System der Muscidae (Diptera: Cyclorrhapha). *Stuttgarter Beiträge zur Naturkunde* **141**: 1–100.

Huyton, P. M., Langley, P. A., Carlson, D. A. and Schwarz, M. 1980. Specificity of contact sex pheromones in tsetse flies, *Glossina* spp. *Physiological Entomology* **5**: 253–264.

Ibrahim, E. A. R., Ingram, G. A. and Molyneux, D. H. 1984. Haemagglutinins and parasite agglutinins in haemolymph and gut of *Glossina*. *Tropenmedizin und Parasitologie* **35**: 151–156.

Itard, J. 1973. Revue des connaissances actuelles sur la cytogénétique des Glossines (Diptera). *Revue d'Élevage et de Médecine Vétérinaire des Pays Tropicaux* **26**: 151–167.

Jack, R. W. 1927. Some environmental factors relating to the distribution of *Glossina morsitans* Westw. in Southern Rhodesia. *South African Journal of Science* **24**: 457–475.

Jackson, C. H. N. 1933. On an advance of tsetse-fly in central Tanganyika. *Transactions of the Royal Entomological Society of London* **81**: 205–221.

Jobling, B. 1933. A revision of the structure of the head, mouth-part [sic] and salivary glands of *Glossina palpalis* Rob.-Desv. *Parasitology* **24**: 449–490.

Jordan, A. M. 1962. The ecology of the *fusca* group of tsetse flies (*Glossina*) in southern Nigeria. *Bulletin of Entomological Research* **53**: 355–385.

Jordan, A. M. 1965. The status of *Glossina fusca* Walker (Diptera, Muscidae) in West Africa. *Annals of Tropical Medicine and Parasitology* **59**: 219–225.

Jordan, A. M. 1972. Extracellular ducts within the wall of the spermatheca of tsetse flies (*Glossina* spp.) (Dipt., Glossinidae). *Bulletin of Entomological Research* **61**: 669–672.

Jordan, A. M. 1974. Recent developments in the ecology and methods of control of tsetse flies (*Glossina* spp.) (Dipt., Glossinidae) – a review. *Bulletin of Entomological Research* **63**: 361–399.

Jordan, A. M. 1986. *Trypanosomiasis control and African rural development*. x + 357 pp. Longman, London and New York.

Jordan, A. M. 1989. Man and changing patterns of the African trypanosomiases. Pp. 47–58, in Service, M. W. (ed.), *Demography and vector-borne diseases*. 402 pp. CRC Press Inc., Boca Raton, Florida.

Jordan, A. M., Lee-Jones, F. and Weitz, B. 1962. The natural hosts of tsetse flies in Northern Nigeria. *Annals of Tropical Medicine and Parasitology* **56**: 430–442.

Langley, P. A. 1968. The effect of feeding the tsetse fly *Glossina morsitans* Westw. on impala blood. *Bulletin of Entomological Research* **58**: 295–298.

Langley, P. A. 1985. *Glossina* spp. Pp. 97–112 in Singh, P. and Moore, R. F. (eds), *Handbook of insect rearing*: vol. 2, viii + 514 pp. Elsevier, Amsterdam.

Langley, P. A., Pimley, R. W. and Carlson, D. A. 1975. Sex recognition pheromone in tsetse fly *Glossina morsitans*. *Nature* **254**: 51–53.

Langridge, W. P., Kernaghan, R. J. and Glover, P. E. 1963. A review of recent knowledge of the ecology of the main vectors of trypanosomiasis. *Bulletin of the World Health Organization* **28**: 671–701.

Laveissière, C. and **Couret, D.** 1982. Essai de lutte contre les glossines riveraines à l'aide d'écrans imprégnés d'insecticide. *Cahiers ORSTOM* (Entomologie médicale et Parasitologie) **19** (1981): 271–283.

Laveissière, C., Couret, D. and **Kienon, J.-P** 1981. Lutte contre les glossines riveraines à l'aide de pièges biconiques imprégnés d'insecticide, en zone de savane humide. 5. Note de synthèse. *Cahiers ORSTOM* (Entomologie médicale et Parasitologie) **19**: 49–54.

Laveissière, C., Couret, D., Staak, C. and **Hervouet, J.-P.** 1986. *Glossina palpalis* et ses hôtes en secteur forestier de Côte d'Ivoire. *Cahiers ORSTOM* (Entomologie médicale et Parasitologie) **23** (1985): 297–303.

Le Berre, R. and **Itard, J.** 1960. Validité des sous-espèces *Glossina fusca fusca* Walker, 1879 et *Glossina fusca congolensis* Newstead et Evans, 1921. Diptera, Muscidae. *Bulletin de la Société de Pathologie exotique* **53**: 542–550.

Machado, A. de Barros 1954. Révision systématique des Glossines du groupe *palpalis* (Diptera). *Publicações Culturais da Companhia de Diamantes de Angola* **22**: 1–189.

Machado, A. de Barros 1959. Nouvelles contributions à l'étude systématique et biogéographique des Glossines (Diptera). *Publicações Culturais da Companhia de Diamantes de Angola* **46**: 13–90.

Machado, A. de Barros 1966. Remarques sur la systématique des Glossines du groupe *morsitans*. *Proceedings of the First International Congress of Parasitology*, Rome (1964) **2**:981–982.

Machado, A. de Barros 1970. Les races géographiques de *Glossina morsitans*. Pp. 471–486 in Azevedo, J. Fraga de (ed.), *Criaçõa da mosca tsé-tsé em laboratorio e sua aplicação prática. Tsetse fly breeding under laboratory conditions and its practical application.* 1st International Symposium, 22nd and 23rd April 1969, Lisbon. 524 pp. Junta de Investigações do Ultramar, Lisboa.

Maclennan, K. J. R. and **Na'isa, B. K.** 1971. The current status and future prospects regarding tsetse extermination in Nigeria. Pp. 303–309 in International Council for Trypanosomiasis Research, Thirteenth Meeting, Lagos, 1971 [Report]. 328 pp. Organisation of African Unity Scientific, Technical and Research Commission Publications Bureau, Niamey.

Maillot, L. 1957. Présence de *Glossina medicorum* Austen, 1911 au Gabon (Afrique Équatoriale Française). *Bulletin de la Société de Pathologie exotique* **49** (1956): 823–827.

Maudlin, I. and **Dukes, P.** 1985. Extrachromosomal inheritance of susceptibility to trypanosome infection in tsetse flies. I. Selection of susceptible and refractory lines of *Glossina morsitans morsitans*. *Annals of Tropical Medicine and Parasitology* **79**: 317–324.

Maudlin, I. and **Welburn, S. C.** 1988. Tsetse immunity and the transmission of trypanosomiasis. *Parasitology Today* **4**: 109–111.

McAlpine, J. F. 1989. Phylogeny and classification of the Muscomorpha. Pp. 1397–1518 in McAlpine, J. F. (ed.), *Manual of Nearctic Diptera*: vol. 3, vi + pp. 1333–1581. Research Branch, Agriculture Canada (Monograph No. 32).

McAlpine, J. F., Peterson, B. V., Shewell, G. E., Teskey, H. J., Vockeroth, J. R. and **Wood, D. M.** 1981. *Manual of Nearctic Diptera*: vol. 1, vi + 674 pp. Research Branch, Agriculture Canada (Monograph No. 27).

McDowell, P. G., Whitehead, D. L., Chaudhury, M. F. B. and **Snow, W. F.** 1981. The isolation and identification of the cuticular sex-stimulant pheromone of the tsetse *Glossina pallidipes* Austen (Diptera: Glossinidae). *Insect Science and its Application* **2**: 181–187.

Mews, A. R., Langley, P. A., Pimley, R. W. and **Flood, M. E. T.** 1977. Large-scale rearing of tsetse flies (*Glossina* spp.) in the absence of a living host. *Bulletin of Entomological Research* **67**: 119–128.

Molyneux, D. H. 1977. Vector relationships in the Trypanosomatidae. *Advances in Parasitology* **15**: 1–82.

Mulligan, H. W. (ed.) 1970. *The African trypanosomiases.* lxxxviii + 950 pp. George Allen and Unwin, London.

Nash, T. A. M. 1942. A study of the causes leading to the seasonal evacuation of a tsetse breeding-ground. *Bulletin of Entomological Research* **32**: 327–339.

Nash, T. A. M. 1948. *Tsetse flies in British West Africa*. 77 pp. His Majesty's Stationery Office, London.

Nash, T. A. M. 1969. *Africa's bane. The tsetse fly.* 224 pp. Collins, London.
Nash, T. A. M. and **Jordan, A. M.** 1959. A guide to the identification of the West African species of the *fusca* group of tsetse-flies, by dissection of the genitalia. *Annals of Tropical Medicine and Parasitology* **53**: 72–88.
Nash, T. A. M. and **Page, W. A.** 1953. The ecology of *Glossina palpalis* in Northern Nigeria. *Transactions of the Royal Entomological Society of London* **104**: 71–169.
Nash, T. A. M., Jordan, A. M. and **Boyle, J. A.** 1966. A promising method for rearing *Glossina austeni* (Newst.) on a small scale, based on the use of rabbits' ears for feeding. *Transactions of the Royal Society of Tropical Medicine and Hygiene* **60**: 183–188.
Nelson, D. R. and **Carlson, D. A.** 1986. Cuticular hydrocarbons of the tsetse flies *Glossina morsitans morsitans*, *G. austeni* and *G. pallidipes*. *Insect Biochemistry* **16**: 403–416.
Nelson, D. R., Carlson, D. A. and **Fatland, C. L.** 1988. Cuticular hydrocarbons of the tsetse flies. II: *G. p. palpalis*, *G. p. gambiensis*, *G. fuscipes*, *G. tachinoides* and *G. brevipalpis*. *Journal of Chemical Ecology* **14**: 963–987.
Newstead, R., Evans, A. M. and **Potts, W. H.** 1924. Guide to the study of tsetse-flies. *Liverpool School of Tropical Medicine Memoirs* (n.s.) **1**: xi + 1–332. University of Liverpool Press, Liverpool, and Hodder & Stoughton, London.
Offor, I. I., Carlson, D. A., Gadzama, N. M. and **Bozimo, H. T.** 1981. Sex recognition pheromone in the West African tsetse fly, *Glossina palpalis palpalis* (Robineau-Desvoidy). *Insect Science and its Application* **1**: 417–420.
Page, W. A. 1959. The ecology of *Glossina palpalis* (R.-D.) in southern Nigeria. *Bulletin of Entomological Research* **50**: 617–631.
Pilson, R. D. and **Pilson, B. M.** 1967. Behaviour studies of *Glossina morsitans* Westw. in the field. *Bulletin of Entomological Research* **57**: 227–257.
Pollock, J. N. 1970. Sperm transfer by spermatophores in *Glossina austeni* Newstead. *Nature* **225**: 1063–1064.
Pollock, J. N. 1971. Origin of the tsetse flies: a new theory. *Journal of Entomology* (B) **40**: 101–109.
Pollock, J. N. 1973. A comparison of the male genitalia and abdominal segmentation in *Gasterophilus* and *Glossina* (Diptera), with notes on the gasterophiloid origin of the tsetse flies. *Transactions of the Royal Entomological Society of London* **125**: 107–124.
Pollock, J. N. 1974. Functional morphology of the phallosome in *Glossina* (Diptera, Glossinidae) and its evolutionary implications. *Zoologica Scripta* **3**: 185–192.
Potts, W. H. 1973. Glossinidae (tsetse-flies). Pp. 209–249 in Smith, K. G. V. (ed.), *Insects and other arthropods of medical importance.* xiv + 561 pp. British Museum (Natural History), London.
Power, R. J. B. 1964. The activity pattern of *Glossina longipennis* Corti (Diptera: Muscidae). *Proceedings of the Royal Entomological Society of London* (A) **39**: 5–14.
Rogers, A. 1972. Studies on the mating of *Glossina pallidipes* Austen. I: The age at mating. *Annals of Tropical Medicine and Parasitology* **66**: 515–523.
Saunders, D. S. 1970. Reproduction in *Glossina.* Pp. 327–344 in Mulligan, H. W. (ed.), *The African trypanosomiases.* lxxxviii + 950 pp. George Allen & Unwin, London.
Scott, D. 1970. The epidemiology of Gambian sleeping sickness. Pp. 614–644 in Mulligan, H. W. (ed.), *The African trypanosomiases.* lxxxviii + 950 pp. George Allen & Unwin, London.
Simmonds, F. J., Jordan, A. M. and **Touré, S. M.** 1977. Parasitoids. Pp. 57–74 in Laird, M. (ed.), *Tsetse: the future for biological methods in integrated control.* 220 pp. International Development Research Centre, Ottawa.
Southern, D. I. 1980. Chromosome diversity in tsetse flies. Pp. 225–243 in Blackman, R. L., Hewitt, G. M. and Ashburner, M. (eds), *Insect cytogenetics.* (Symposia of the Royal Entomological Society of London 10.) viii + 278 pp. Blackwell Scientific Publications, Oxford.
Southern, D. I. and **Pell, P. E.** 1974. Comparative analysis of the polytene chromosomes of *Glossina austeni* and *Glossina morsitans morsitans*. *Chromosoma* **47**: 213–226.
Southern, D. I., Pell, P. E. and **Craig-Cameron, T. A.** 1973. Polytene chromosomes of the tsetse fly *Glossina morsitans morsitans*. *Chromosoma* **40**: 107–120.

Swynnerton, C. F. M. 1933. Some traps for tsetse-flies. *Bulletin of Entomological Research* **24**: 69–102.

TDR 1990. Tropical diseases in media spotlight. *TDR News* **31** (March 1990): 3.

Thomson, M. C. 1987. The effect on tsetse flies (*Glossina* spp.) of deltamethrin applied to cattle either as a spray or incorporated into ear-tags. *Tropical Pest Management* **33**: 329–335.

Vale, G. A. 1974. The response of tsetse flies (Diptera, Glossinidae) to mobile and stationary baits. *Bulletin of Entomological Research* **64**: 545–588.

Vale, G. A. 1977. Feeding responses of tsetse flies (Diptera: Glossinidae) to stationary hosts. *Bulletin of Entomological Research* **67**: 635–649.

Vale, G. A. 1980. Field studies of the responses of tsetse flies (Glossinidae) and other Diptera to carbon dioxide, acetone and other chemicals. *Bulletin of Entomological Research* **70**: 563–570.

Vale, G. A., Hall, D. R. and **Gough, A. J. E.** 1988. The olfactory responses of tsetse flies, *Glossina* spp. (Diptera: Glossinidae) to phenols and urine in the field. *Bulletin of Entomological Research* **78**: 293–300.

Vale, G. A., Hargrove, J. W., Jordan, A. M., Langley, P. A. and **Mews, A. R.** 1976. Survival and behaviour of tsetse flies (Diptera: Glossinidae) released in the field: a comparison between wild flies and animal-fed and *in vitro*-fed laboratory-reared flies. *Bulletin of Entomological Research* **66**: 731–744.

Vanderplank, F. L. 1950. Air-speed/wing-tip speed ratios of insect flight. *Nature* **165**: 806–807.

Van Emden, F. 1944. A new sub-species of *Glossina* from Uganda (Diptera). *Bulletin of Entomological Research* **35**: 193–196.

Wall, R. 1988. Analysis of the mating activity of male tsetse flies *Glossina m. morsitans* and *G. pallidipes* in the laboratory. *Physiological Entomology* **13**: 103–110.

Weitz, B. 1963. The feeding habits of *Glossina*. *Bulletin of the World Health Organization* **28**: 711–729.

Weitz, B. and **Glasgow, J. P.** 1956. The natural hosts of some species of *Glossina* in East Africa. *Transactions of the Royal Society of Tropical Medicine and Hygiene* **50**: 593–612.

Welburn, S. C., Maudlin, I. and **Ellis, D. S.** 1989. Rate of trypanosome killing by lectins in midguts of different species and strains of *Glossina*. *Medical and Veterinary Entomology* **3**: 77–82.

Wijers, D. J. B. 1958. Factors that may influence the infection rate of *Glossina palpalis* with *Trypanosoma gambiense*. I – The age of the fly at the time of the infected feed. *Annals of Tropical Medicine and Parasitology* **52**: 385–390.

Wood, D. M. 1987. Oestridae. Pp. 1147–1158 in McAlpine, J. F. (ed.), *Manual of Nearctic Diptera*: vol. 2, vi + pp. 675–1332. Research Branch, Agriculture Canada (Monograph No. 28).

World Health Organization 1986. Epidemiology and control of African trypanosomiasis. Report of a WHO Expert Committee. *World Health Organization Technical Report Series* **739**: 1–127.

Zumpt, F. 1936. *Die Tsetsefliegen. Ihre Erkennungsmerkmale, Lebensweise und Bekampfung.* iv + 149 pp. Gustav Fischer, Jena.

Zumpt, F. 1973. *The stomoxyine biting flies of the world. Diptera: Muscidae. Taxonomy, biology, economic importance and control measures.* vii + 175 pp. Gustav Fischer, Stuttgart.

CHAPTER TEN

Stable-flies and horn-flies (bloodsucking Muscidae)

R. W. Crosskey

The Muscidae occur worldwide and include nearly 4000 species. The adults of most of them cannot bite and usually depend for their nutrition on a diet other than blood, but some among this huge number – about 50 species comprising the subfamily Stomoxyinae – are unusual muscids because they are able to bite with a specially adapted proboscis and feed by directly gorging on the blood of large mammals (usually ruminants). On account of this habit, which is common to both sexes of the flies, the stomoxyines can be annoying to humans, major irritant pests of cattle, and (rarely) transmitters of certain parasites and pathogens. They include the stable-fly (*Stomoxys calcitrans*) and the horn-flies (*Haematobia* species), and it is with these that this chapter is mainly concerned. Also covered, however, is *Musca crassirostris* (the Indian cattle-fly), as despite belonging to the (normally) non-biting Muscinae this species is able to bite a vertebrate host and suck its blood in a direct manner like that of a stomoxyine; the fly also resembles the Stomoxyinae in developing through the larval stages in the dung of cattle.

Stomoxyines are nondescript flies of small or medium size (length 2–8 mm), often superficially like house-flies. *Stomoxys calcitrans* is such a look-alike that it is often called the biting house-fly – and in Germany there is even a saying that 'winter will be early when the house-flies start biting'. The practical importance of these flies, however, is veterinary rather than medical and a large literature upon them has developed which pertains mainly in applied entomology. Nearly all the important literature can easily be traced using bibliographies for the genus *Stomoxys* by Morgan et al. (1983) and for the genus *Haematobia* by Morgan and Thomas (1974, 1977).

RECOGNITION AND ELEMENTS OF STRUCTURE

Characteristics for recognition of Muscidae are given in Chapter 11. Within the family, adult stomoxyines are best recognized by their rather slender, stiff and 'horny' proboscis (Fig. 10.2*a*–*e*) which cannot be retracted, projects forwards more than downwards, and is often partly visible when the fly is seen from above. The flies are easily distinguished from tsetse-flies (Chapter 9) by the palps, arista and wing veins: the palps of stomoxyines do not sheath the proboscis, the hairs of the arista are not branched (Fig. 10.3*a*,*b*), and the median discal wing cell (dm) is not

Medical Insects and Arachnids Edited by Richard P. Lane and Roger W. Crosskey.
Published in 1993 by Chapman & Hall ISBN 0 412 40000 6

390 *Stable-flies and horn-flies (bloodsucking Muscidae)*

hatchet-shaped; also the tips of the wings *usually* diverge when the fly is at rest instead of closing like scissors over the abdomen. The general structure is like that of other calyptrate flies (Chapter 3), and as in other muscids the frons is narrower in male than female flies.

There are three larval instars and a puparium. The larvae (Fig. 10.1b) are whitish maggots very much like house-fly maggots but recognized by oblique ridge-like ornamentation across the abdominal pseudopods (Fig. 10.1d,e). The length when mature ranges from 6 to 12 mm. The cephalopharyngeal skeleton has mouth-hooks of unequal size (the left being smaller than the right) and a narrowly jagged dorsal wing which is much smaller than the ventral wing (Fig. 10.1c). The anterior spiracles protrude as divided hand-like structures with three to seven lobes (Fig. 10.1b). The posterior spiracles are positioned far above the midline of the last abdominal segment and are not raised above the general surface; the three spiracular slits are weakly sinuous to strongly serpentine (Fig. 10.1f,g). The

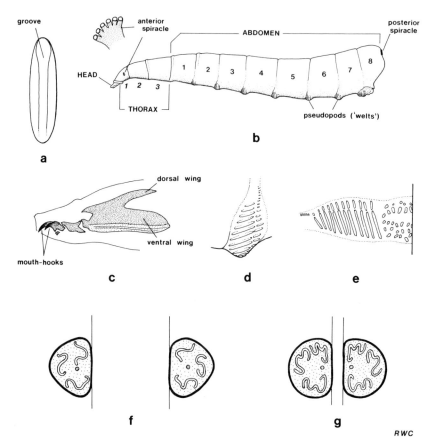

Figure 10.1 Features of early stages in horn-flies and stable-flies (Stomoxyinae): (a) egg shape in *Haematobia irritans*, typical for stomoxyine eggs; (b) *Stomoxys calcitrans*, third-stage larva in side view; (c) *S. calcitrans*, cephalopharyngeal skeleton of third-stage larva; (d and e) *S. calcitrans*, cuticular ornamentation of the abdominal pseudopods of third-stage larva in lateral and (one side only) ventral views; (f) *S. calcitrans*, posterior spiracles of third-stage larva (terminal view, semi-diagrammatic); (g) *Haematobia irritans* and *H. exigua*, posterior spiracles of third-stage larva (terminal view, semi-diagrammatic). Diagonal ridge-like thickenings on the pseudopod cuticle are characteristic of stomoxyine larvae and can be seen in (d) and (e); vertical lines in (f) and (g) call attention to the marked difference in posterior spiracle spacing between *Stomoxys* and *Haematobia* larvae.

puparium has the usual barrel shape of calyptrate fly puparia; it is recognized by faint diagonal ridges on the pseudopods inherited from the third-stage larva.

The egg is of the long and ovoid *Musca*-type (Skidmore, 1985). It is slightly concave on the upper surface and grooved along its length (Fig. 10.1a). Depending on species, it measures 0.85–1.70 mm in length. The shell (chorion) has numerous air-spaces (aeropyles) which show minor differences in form between genera (Hinton, 1960).

Descriptions and illustrations of early stages of the common species are given in the old but still good papers of Thomsen (1935) and Muirhead Thomson (1937) and in the more modern works of Schumann (1963) and Skidmore (1985).

CLASSIFICATION AND IDENTIFICATION

The subfamily Stomoxyinae is classified into ten genera (Zumpt, 1973), of which only three, *Haematobosca*, *Stomoxys* and *Haematobia*, include more than half a dozen species; the last is still sometimes called *Lyperosia*, but this name is a junior synonym of *Haematobia* and its use an error. Only morphological characters are employed in taxonomy. For generic identification the usable characters are the size and shape of the palps, the disposition of hairs on the arista, the number of katepisternal bristles ('sternopleural setae'), and the presence or absence of hairs on the proepisternum ('propleuron'), notopleuron and postalar part of the thorax. Species identification depends mainly on body colour and pattern, leg colour, frons width proportions, curvature and setation of certain wing veins, occurrence or form of various bristles and hairs on parts of the legs, and male genital structure; female terminalia have little diagnostic value.

Early stages of most species are unknown and larval taxonomy in its infancy. Specific differences between larvae, especially in the cephalopharyngeal skeleton, cuticular ornamentation and posterior spiracles (slit shape and distance between spiracles) have some use in identification. Eggs vary in size, colour and width of the dorsal groove, and can sometimes be identified to species by these features.

Identification key to adults of man-biting species of Muscidae

[Note that *Musca crassirostris* of the subfamily Muscinae is included to assist its recognition and differentiate it from Stomoxyinae (see also Chapter 11). Parts of the thorax mentioned in the key are illustrated in Fig. 11.3.]

1   Proboscis in profile narrow and shaft-like except at swollen base (Fig. 10.2a,e). Arista with hairs confined to upper side (Fig. 10.3a,b). Wing edge with ends of veins R4+5 and M separated by a distance greater than length of r–m cross-vein; vein M with gentle and widely obtuse bend (Fig. 10.3c,d). Katepisternum with one or two bristles. Lower calypter seen in side view not reaching back nearly to first abdominal segment (T1+2); its outline almost semicircular and inner part of margin not adjacent to scutellum... 2

—   Proboscis in profile stout and boat-like (Fig. 10.4b). Arista with hairs on upper and lower sides (Fig. 10.4b). Wing edge with ends of veins R4+5 and M near together, separated by a distance less than length of r–m cross-vein; vein M with abrupt bend changing direction almost at a right-angle (Fig. 10.4a). Katepisternum with three bristles (arranged 1 + 2). Lower calypter seen in

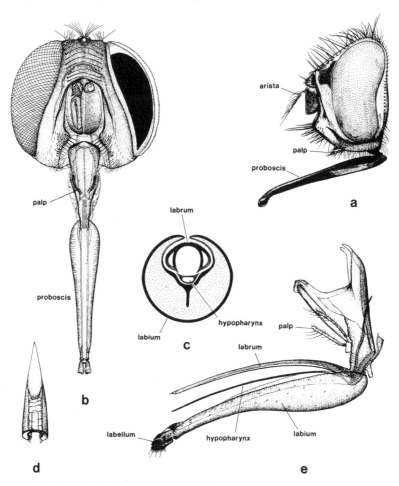

Figure 10.2 Head and proboscis of adult *Stomoxys calcitrans*, commonest biting muscid fly bloodsucking on humans: (*a*) side view; (*b*) facial view, with proboscis depressed and left eye surface removed; (*c*) cross-section of proboscis near the middle of its length (semi-diagrammatic); (*d*) tip of proboscis; (*e*) proboscis in left side view with stylets exposed. (Figures (*b*), (*d*) and (*e*) reproduced with relabelling from Jobling (1987) courtesy of the Trustees of the Wellcome Trust.)

    side view almost reaching to first abdominal segment (T1+2); its outline not semicircular, hind margin almost straight and inner part of margin adjacent to basal part of scutellum. [Appearance of fly as Fig. 10.4*a*: Oriental and Afrotropical regions, Middle East].................................... **Musca crassirostris**

2 Palps very large, extending far in front of head and distinctly dilated (Fig. 10.3*b*), as long as proboscis. Katepisternum with two bristles (near upper anterior and posterior corners). Head with posterior surface almost flat on upper part and strongly swollen on lower part (Fig. 10.3*b*). Abdomen widest near base and somewhat tapering (Fig.10.3*j*). Proepisternum, katepimeron and notopleuron bare. Body length 3–5 mm. [Posterior spiracles of third-stage larva very close together (Fig. 10.1*g*) and with very twisty slits]..........................
.... **Haematobia** 3

— Palps small and very slender, not extending in front of head and not dilated, less than half as long as proboscis (Fig. 10.3*a*). Katepisternum with one bristle

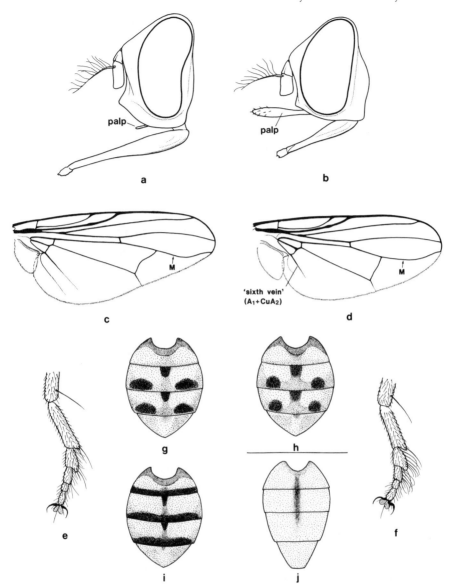

Figure 10.3 Important features used in identification of man-biting Stomoxyinae: (*a*) head outline in *Stomoxys*, hairing omitted; (*b*) head outline in *Haematobia*, hairing omitted; (*c*) wing of *S. calcitrans*; (*d*) wing of *H. irritans*; (*e*) hind tarsus of male *H. irritans*; (*f*) hind tarsus of male *H. exigua*; (*g–i*) typical abdominal pattern (drawn from males) of (*g*) *Stomoxys sitiens*, (*h*) *S. calcitrans* and (*i*) *S. niger*; (*j*) abdomen of male of *H. irritans* (*H. exigua* similar). Figures (*c*) and (*d*) show (arrowed) the gentle curve of vein M characteristic of the stomoxyine wing (compare the abrupt bend of this vein found in *Musca*, Fig. 10.4*a*).

(near upper posterior corner). Head with posterior surface not strongly swollen on lower part (Fig. 10.3*a*). Abdomen widest near middle and with more rounded outline shape (Fig. 10.3*g–i*). Proepisternum, katepimeron and notopleuron hairy. Body length 5–7 mm. [Posterior spiracles of third-stage larva very far apart (Fig. 10.1*f*) and with less twisty slits] .... **Stomoxys** 5

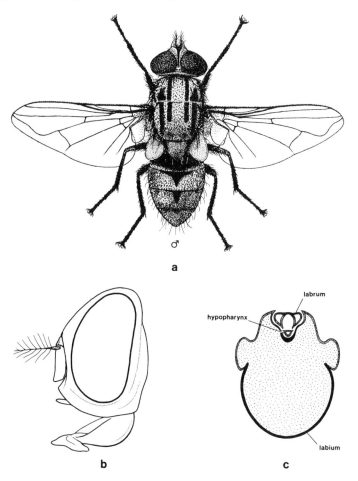

Figure 10.4 *Musca crassirostris* (Indian cattle-fly): (*a*) male fly, with wings in 'set' position; (*b*) head outline showing the bulbous proboscis, hairing omitted; (*c*) cross-section of proboscis near the middle of its length (semi-diagrammatic).

3   Wing with 'sixth vein' (Fig. 10.3*d*) extending half way or more towards wing edge. Abdomen with at least a short and narrow dark centre line (Fig. 10.3*j*). Male hind tarsi asymmetrical, basal segments expanded to form outer dorsal flange bearing some or many very long fine hairs (Fig. 10.3*e, f*) 4
— Wing with 'sixth vein' not extending even half way towards wing edge. Abdomen unicolorous, without trace of dark central line. Male hind tarsi normal, symmetrical, undilated dorsally and without such very long hairs. [Arabian peninsula, Afrotropical and Oriental regions] .... **H. minuta**
4   Hind tarsus of male without continuous row of long fine hairs on upper internal edge of second and third segments (Fig. 10.2*e*). Hind femur near base with an obvious isolated ventral bristle. Flies mostly dark, palps and proboscis usually dark brown, legs often mainly blackish brown, bristles and hairs of upper surface of thorax usually brown or black. [Palaearctic region including Japan, New World, Hawaii] .... **H. irritans**
— Hind tarsus of male with continuous row of very long fine apically curled hairs forming a conspicuous fringe along upper internal edge of second and third

segments (Fig. 10.3f). Hind femur near base without an obvious ventral bristle. Flies usually rather pallid, palpi and proboscis usually yellow or reddish, legs usually extensively reddish yellow, bristles and hairs of upper surface of thorax mainly dingy yellow to coppery red. [Oriental region, Seychelles, Australia and western Pacific islands]..............................**H. exigua***

[Individuals of *H. irritans* and *H. exigua*, especially females, cannot always be reliably distinguished and geographical origin is often the best guide to identity. Southern specimens of *irritans* tend to be paler than northern specimens and hardly distinguishable from *exigua*. See Kano et al. (1972) for photographs of the hind tarsal hair character used for males.]

5   Abdomen with upper surface of first segment (T1+2) uniformly grey or yellowish grey (not darkened along hind margin); middle segments (T3 and T4) each with a pair of widely separated sublateral brown spots near hind margin (Fig. 10.3g, h) 6

—   Abdomen with upper surface of first segment margined by obvious black-brown band; middle segments without spots but with whole of hind margins black-brown and shiny (Fig. 10.3i). [Male frons narrow, at most twice as wide as antenna. Abdomen with central dark line of middle segments narrow, often merging with dark hind margins.] [Afrotropical region including Madagascar] ....**S. niger***

6   Abdomen with sublateral spots of middle segments (T3 and T4) more or less circular (Fig. 10.3h). Frons very wide, in male about three times and in female about four times as wide as antenna seen in profile (width of vertex seen from above 0.28–0.33 of head width in male, 0.38–0.41 of head width in female). [Worldwide] ....**S. calcitrans**

[Some specimens show ill-defined or wider than usual abdominal spots. In doubtful cases check frons width. The wide frons is specifically diagnostic among *Stomoxys* species and especially reliable for male flies.]

—   Abdomen with sublateral spots of middle segments wider and more oval (Fig. 10.3g). Frons narrower, in male about twice as wide and in female about three times as wide as antenna seen in profile (width of vertex seen from above 0.19–0.21 of head width in male, 0.32–0.35 of head width in female). [Afrotropical and Oriental regions, Egypt] ....**S. sitiens**

*   Zumpt (1973) considered *H. exigua* to be a subspecies of *H. irritans* but specialists now treat it as a valid species: *S. nigra* in Zumpt is now correctly spelt *niger* in accordance with masculine gender of the generic name (Steyskal, 1975).

Faunal and taxonomic literature

General: A monograph by Zumpt (1973) provides taxonomic coverage for the world fauna and includes a checklist and keys to adults of described species. It is the standard work and provides data on geographical distribution and biology. Earlier works and post-1973 taxonomic catalogues and country-based faunal papers which usefully amplify Zumpt are listed below. A starting-point key to larvae on a worldwide basis is in Skidmore's (1985) book on muscid biology.

Palaearctic: Catalogue (Pont, 1986); regional taxonomy (Hennig, 1964, key; Schumann, 1963, larvae); Britain (Fonseca, 1968, adult key; Hinton, 1960, eggs);

Japan (Hayashi and Shinonaga, 1979, adult key; Ishijima, 1967, larval key; Kano, 1953, egg key).

Afrotropical: Catalogue (Pont, 1980); regional taxonomy (Zumpt, 1950, adult key); Arabian peninsula (Pont, 1991, adult key).

Oriental: Catalogue (Pont, 1977); regional taxonomy (Van Emden, 1965, adult, larval, egg keys); Taiwan (Shinonaga and Kano, 1987, adult key).

Australasian: Catalogue (Pont, 1989).

Nearctic: Genera (Huckett and Vockeroth, 1987).

Neotropical: Catalogue (Pont, 1972).

BIOLOGY

Adult stomoxyines live in close association with the large mammals which provide their blood diet. Man is never the primary blood source, however, and flies feed mainly on large quadrupeds. These include all forms of domestic livestock (particularly cattle and water-buffalo) and, in Africa, notably large wild mammals such as giraffe, buffalo, zebra, antelopes and rhinoceros. In the horn-flies the association with the animal host is so complete that the flies are virtually obligate parasites, hardly leaving the host day or night; *Haematobia irritans* spends 98% of its adult life on cattle (Hillerton, 1985). Even the *Haematobia* larva is cow-pat dependent. In *Stomoxys*, on the other hand, the association is far less intimate both for adults and early stages, and this makes it convenient to discuss each genus separately here. (See Zumpt, 1973, for a lead to the present negligible knowledge of biology in other genera.)

Haematobia (horn-flies)

The biology of *H. irritans* and *H. exigua* is essentially similar. Eggs are laid on freshly dropped dung, and females leave the host to lay when it voids its faeces. Their responsiveness to sloppy bovine excrement (which crusts over as cow-pats) is such that faeces cease to be attractive to them within ten minutes of being dropped – and a reason why *Haematobia* seldom uses animals such as sheep or horses as hosts is certainly that its larvae develop better in dung of the cow-pat type than in hardened pellet-like faeces with low moisture content. As droppings dry out, larvae move to the moister parts and when mature transform into puparia within the cow-pat or in the soil underneath.

Temperature greatly influences the time for development from egg to adult. This can be under two weeks in very warm conditions (when eggs hatch in as little as 16 hours and larvae mature in about a week) but up to six weeks under cool conditions – and even several months in colder climates where there is sometimes puparial diapause. Emerging flies waste little time in seeking an animal as a blood source, both sexes settling on the host and (usually) mating there a day or two after emergence. The females mate only once and can produce their first eggs within three days of emergence. On the host animal, flies mainly congregate on shoulders and flanks where least disturbed by the swishing tail; they adopt a characteristic head-downwards posture on non-horizontal surfaces, and frequently cluster round the bases of the horns (hence 'horn-fly'). Blood is taken two or three times daily and the adult is thought to live about four to eight weeks.

Stomoxys (stable-flies)

The aptly named and universally distributed stable-fly, *S. calcitrans*, differs from horn-flies by spending comparatively little time on the hosts and by its females being rarely directly attracted to animal faeces. The eggs are laid, and the larvae develop, in almost any kind of decaying organic matter that contains plenty of moisture. In stables and byres, horse or cow dung mixed with the urine-soaked straw is the common development medium, whereas outdoors larvae can be found in heaps of rotting vegetables, in fermenting grass cuttings, lakeshore and seashore debris, or humus-rich soil; development in accumulations of dead mayflies is on record. The egg stage lasts from under a day to five days, the larval stage usually for about six to ten days, and the puparial stage usually five to seven days, but total development time from oviposition to emergence varies from twelve days to six weeks depending on the temperature. Several batches of eggs are laid, and one female can lay 600 or even more eggs in her lifetime; each batch is small in cool months but can contain as many as 100–200 eggs at warmer times.

Adults mate in the air and not (as in *Haematobia*) on a host. Females copulate only once but males will mate several females. The association with hosts is much less positive than in *Haematobia*, and the adults of *S. calcitrans* can be found as much sunning themselves on walls, vehicles and implements around stables and farmyards as they can be found in open pastures where there are grazing stock. Flies will gorge blood from their livestock hosts for as long as ten or 15 minutes if undisturbed, but biting is inhibited by low temperature and usually does not occur much below about 14°C. The hosts are mainly cattle, horses, donkeys, mules and dogs, but other mammals (including humans) are sometimes attacked.

The biology of the other species which sometimes bite man, *S. niger* of the Afrotropical region and *S. sitiens* in Africa and the Orient, though little studied, are evidently similar to that of *S. calcitrans*. The former species, however, is more closely associated with bovines and apparently more willing to lay eggs directly on their dung-pats. *Stomoxys sitiens* seems to be less attracted to bovine hosts than to camels and equines.

For completeness, *S. ochrosoma* deserves mention. The biology of this yellow-bodied East African fly is evidently different from that of other *Stomoxys* for it drops its eggs into columns of driver-ants (Dorylinae) to be carried off to the ant bivouacs (Thorpe, 1942). What happens to the larvae unfortunately remains unknown.

Musca crassirostris (Indian cattle-fly)

This species, although not a stomoxyine (see earlier), is so similar to *Haematobia* in general biology that it should be mentioned at this point. It is widely distributed from Africa through the Middle East to South East Asia and is a true bloodsucker which bites particularly cattle but also horses, donkeys and occasionally man. Both sexes suck blood. In size and appearance (Fig. 10.4a) it is very like the house-fly, but differs by its bulbous proboscis (Fig.10.4b) in which the labella can be parted to expose very large prestomal teeth for rasping at the host skin and producing a bleeding wound – while the fly simultaneously introduces a powerful anticoagulant. The cross-section of this proboscis is shown in Fig. 10.4c. Biting occurs mainly on the legs and belly, and flies are not easily disturbed whilst bloodsucking.

Oviposition takes place on patches of freshly dropped cow dung, the female crawling about on the dung to find a suitable crevice in which to thrust her ovipositor quite deeply and lay her 40–50 rather large eggs (length about 2 mm). When many flies are about they tend to lay gregariously, crowding together with their heads outwards and ovipositors adjacent – with the result that eggs become heaped in irregular masses. Eggs hatch in a few hours, and the lemon-yellow larvae develop for a few days within the dung before crawling out together and burying themselves in the earth or beneath leaves to pupate. Larva and puparium are typical for *Musca*. Old accounts by Patton and Cragg of the life history (1912), and of structure and functioning of the proboscis (1913), both using the synonymous name *Philaematomyia insignis*, are essentially not superseded and should still be consulted. The egg and larval stage have been described by Muirhead Thomson (1947).

MEDICAL IMPORTANCE

The stomoxyine flies do not carry any cyclically transmitted parasites of man, and have only very minor medical importance. Two aspects of their relationship to man should nevertheless be briefly considered: the distress caused by biting and the possibility of mechanical transmission of pathogenic organisms to or between humans.

Species of *Haematobia*, and also *Musca crassirostris*, bite man relatively little and are rarely a serious irritant. However, the African horn-fly (*H. minuta*) can be a troublesome biter at times in the Arabian peninsula, and *M. crassirostris* an annoyance in the dry grassland areas of India – where the flies will persistently alight on the skin even though few of them actually bite. More significant are *Stomoxys calcitrans* and *S. niger*. The former, though cosmopolitan in its range, is the chief stomoxyine pest of man in temperate latitudes and relatively unimportant in the equatorial regions; in tropical Africa its place is taken by the aggravating *S. niger*. Both species deliver a sharp stinging bite and have a propensity to feed on the ankles, readily stabbing through socks and stockings. Such attacks, when endlessly repeated, can be very unpleasant and disturbing – to animals as much as humans. Dogs everywhere are very viciously bitten, and in the United States *Stomoxys calcitrans* is widely known as the 'dog-fly'. On the Atlantic shoreline from New Jersey to Florida the flies often breed in accumulated organic flotsam and in summer attain the status of serious pests; inland, too, recreations such as fishing and camping sometimes have to be abandoned as unendurable because of the intensity of biting attacks. The stomoxyine bite, however, is not toxic, and even though humans can suffer severe immediate discomfort there are no consequent pathological reactions or lasting effects.

Evidence is mainly negative on whether contamination of the proboscis in stomoxyines can lead to transmission of organisms to man, but experiments have shown that these flies can in some circumstances mechanically transmit bacteria and viruses between animals, and a cultured species of *Leishmania* (a protozoan causing human cutaneous leishmaniasis) has been experimentally transmitted to humans by *Stomoxys calcitrans* (Berberian, 1939, reported a *L. tropica*). Fortunately, stomoxyines do not transmit human faecal pathogens because they are not attracted by human food or excrement. Early this century it was thought in the United States that these flies were to blame for poliomyelitis, but confirmation of early intangible evidence was not forthcoming and by the mid-1920s it was concluded that 'after an exhaustive enquiry the stable fly has now left the court without this stain upon its character' (Dick, 1925).

On the other hand, it is thought possible that the spirochaete causing louse-borne relapsing fever (*Borrelia recurrentis*) might in rare circumstances be stomoxyine-carried, and there is some dubious evidence for a stomoxyine-induced death from anthrax (*Bacillus anthracis*): a man died of this cause after being bitten by *S. calcitrans* which flew to him from the carcase of a cow which had died from the disease. This case is cited by Zumpt (1973), whose book should be consulted for a full lead into the literature on pathogen transmission. On the trypanosomes causing human sleeping sickness, Zumpt concludes unequivocally that for this disease 'Stomoxyinae definitely play no role in the epidemiology'.

A few vague reports exist of stomoxyine larvae causing myiasis in man – see Zumpt (1973) and Chapter 12. The same is true for *Musca crassirostris*, whose larvae occur in rare cases of human intestinal myiasis – usually brought about by the accidental swallowing of larvae during religious rites which involve coprophagy.

Stomoxyines are more important from the veterinary than the medical viewpoint and can be very serious biting pests of economic concern when the health of domestic animals is adversely affected and the milk output of dairy herds reduced. The flies remain under suspicion for occasional mechanical transmission of bovine anthrax, brucellosis and other infections of herd animals, and include proved vectors of two cyclically transmitted nematode parasites: *Stomoxys calcitrans* is intermediate host of *Habronema microstoma*, a worm whose adult infests the stomach of horses, and *Haematobia irritans* intermediate host of *Stephanofilaria stilesi* which produces bloody sore-like lesions on the bellies of cattle; the latter infection is known as stephanofilariasis and is fairly well documented (Hibler, 1966). See Zumpt (1973) for an introduction to the literature of all these veterinary relationships of stomoxyines.

CONTROL

The control of stomoxyines is motivated only by their veterinary importance, since the flies are never more than nuisance pests to humans. Control of cattle-biting pest species can become necessary if their biting populations exceed a tolerated threshold level. Cattle become distressed by horn-fly attack when infestations exceed 400–500 flies per head, and in the United States (where horn-fly can be a major veterinary problem) control measures are then considered advisable (Bruce, 1964). Control mainly relies upon insecticides applied to infested animals. Apart from insecticides, limited control by cattle-fly traps is sometimes possible; these are large screened traps suitably positioned where herd animals must pass daily, e.g. to obtain their drinking water or enter a barn. Many other types of trap have been tried experimentally, but none with outstanding success. Consideration has been given to the sterile male control technique and to possible use of parasites or predators for biological control but such ideas have not to date borne fruit. Literature on control can be traced through the bibliographies of Morgan and Thomas (1974, 1977) and Morgan et al. (1983); much American literature can be found in the *Journal of Economic Entomology*.

COLLECTING, PRESERVING AND REARING MATERIAL

In general, no special techniques are required beyond those applicable to most insects and described in Chapter 1. Flies swarming in stables and around potential hosts can be netted, and those on the skin of a living or freshly killed animal sucked up by aspirator. Taxonomic features of adults are best seen when these are preserved as dry flies, and pinned specimens are much preferable to fluid-preserved specimens if identifications are needed. Zumpt (1973) provides more details, and stresses that flies should never be preserved in formalin. Larvae, like those of other Diptera, preserve well in 80% ethanol, but are best killed first in hot water as this helps to retain well their natural shape. References on laboratory-rearing techniques can also be found in Zumpt's monograph.

REFERENCES

Berberian, D. A. 1939. A second note on successful transmission of Oriental sore by the bites of *Stomoxys calcitrans*. *Annals of Tropical Medicine and Parasitology* **33**: 95–96.

Bruce, W. G. 1964. The history and biology of the horn fly, *Haematobia irritans* (Linnaeus); with comments on control. *North Carolina Agricultural Experiment Station Technical Bulletin* **157**: 1–32.

Dick, R. 1925. The epidemiology and administrative control of anterior poliomyelitis. *Medical Journal of Australia* **1**: 536–541.

Fonseca, E. C. M. d'A. 1968. Diptera Cyclorrhapha Calyptrata, Section (*b*) Muscidae. *Handbooks for the Identification of British Insects* **10** (4) (b): 1–119.

Hayashi, A. and **Shinonaga, S.** 1979. *Flies: ecology and control*. 210 pp. Bun-Eido, Tokyo. [In Japanese.]

Hennig, W. 1964. Muscidae [part]. In Lindner, E. (ed.), *Die Fliegen der palaearktischen Region* **63b**: 1009–1056, 1057–1110 (separately dated Lieferungen).

Hibler, C. P. 1966. Development of *Stephanofilaria stilesi* in the horn fly. *Journal of Parasitology* **52**: 890–898.

Hillerton, J. E 1985. Sexing of *Haematobia irritans* (L.) (Dipt., Muscidae). *Entomologist's Monthly Magazine* **121**: 211–212.

Hinton, H. E. 1960. The chorionic plastron and its role in the eggs of the Muscinae (Diptera). *Quarterly Journal of Microscopical Science* **101**: 313–332.

Huckett, H. C. and **Vockeroth, J. R.** 1987. Muscidae. Pp. 1115–1131 in McAlpine, J. F. (ed.), *Manual of Nearctic Diptera*: vol. 2, vi + pp. 675–1332. Research Branch, Agriculture Canada (Monograph No. 28).

Ishijima, H. 1967. Revision of the third stage larvae of synanthropic flies of Japan (Diptera: Anthomyiidae, Muscidae, Calliphoridae and Sarcophagidae). *Japanese Journal of Sanitary Zoology* **18**: 47–100.

Jobling, B. 1987. *Anatomical drawings of biting flies*. 119 pp. British Museum (Natural History), London, and Wellcome Trust, London.

Kano, R. 1953. Notes on the flies of medical importance in Japan – Part VII. Eggs and larvae of Stomoxydinae in Japan. *Japanese Journal of Experimental Medicine* **23**: 187–195.

Kano, R., Shinonaga, S. and **Hasegawa, T.** 1972. On the specific name of *Haematobia* (Diptera, Muscidae) from Japan. *Japanese Journal of Sanitary Zoology* **23**: 49–56.

Morgan, C. E. and **Thomas, G. D.** 1974. Annotated bibliography of the horn fly, *Haematobia irritans* (L.), including references on the buffalo fly, *H. exigua* (de Meijere), and other species belonging to the genus *Haematobia*. *United States Department of Agriculture Miscellaneous Publication* **1278**: 1–134.

Morgan, C. E. and **Thomas, G. D.** 1977. Supplement I: annotated bibliography of the horn fly, *Haematobia irritans irritans* (L.) including references on the buffalo fly, *H. irritans exigua* (de Meijere), and other species belonging to the genus *Haematobia*. *United States Department of Agriculture Miscellaneous Publication* **1278** (Supplement I): 1–38.

Morgan, C. E., Thomas, G. D. and **Hall, R. D.** 1983. Annotated bibliography of the stable fly, *Stomoxys calcitrans* (L.), including references on other species belonging to the genus *Stomoxys*. *University of Missouri–Colombia Agricultural Experiment Station Research Bulletin* **1049**: 1–190.

Muirhead Thomson, R. C. 1937. Observations on the biology and larvae of the Anthomyidae. *Parasitology* **39**: 273–358.

Muirhead Thomson, R. C. 1947. Notes on the breeding habits and early stages of some muscids associated with cattle in Assam. *Proceedings of the Royal Entomological Society of London* (A) **22**: 89–100.

Patton, W. S. and **Cragg, F. W.** 1912. The life history of *Philaematomyia insignis*, Austen. *Annals of Tropical Medicine and Parasitology* **5**: 515–520.

Patton, W. S. and **Cragg, F. W.** 1913. *A textbook of medical entomology*. xxxiii + 766 pp. Christian Literature Society for India, London, Madras and Calcutta.

Pont, A. C. 1972. *A catalogue of the Diptera of the Americas South of the United States*. 97. *Family Muscidae*. 111 pp. Museu de Zoologia, Universidade de São Paulo.

Pont, A. C. 1977. Family Muscidae. Pp. 451–523 in Delfinado, M. D. and Hardy, D. E. (eds), *A catalog of the Diptera of the Oriental region*: vol. 3, *Suborder Cyclorrhapha (excluding Division Aschiza)*. x + 854 pp. University Press of Hawaii, Honolulu.

Pont, A. C. 1980. Family Muscidae. Pp. 721–761 in Crosskey, R. W. (ed.), *Catalogue of the Diptera of the Afrotropical region*. 1437 pp. British Museum (Natural History), London.

Pont, A. C. 1986. Family Muscidae. Pp. 57–215 in Soós, A. and Papp, L. (eds), *Catalogue of Palaearctic Diptera*: vol. 11, Scathophagidae–Hypodermatidae. 346 pp. Elsevier, Amsterdam.

Pont, A. C. 1989. Family Muscidae. Pp. 675–699 in Evenhuis, N. L. (ed.), *Catalog of the Diptera of the Australasian and Oceanian regions*. 1155 pp. Bishop Museum Press, Honolulu, and E. J. Brill, Leiden.

Pont, A. C. 1991. A review of the Fanniidae and Muscidae (Diptera) of the Arabian peninsula. *Fauna of Saudi Arabia* **12**: 312–365.

Schumann, H. 1963. Zur Larvalsystematik der Muscinae nebst Beschreibung einiger Musciden- und Anthomyidenlarven. *Deutsche Entomologische Zeitschrift* (n.s) **10**: 134–151.

Shinonaga, S. and **Kano, R.** 1987. Studies on Muscidae from Taiwan (Diptera) Part 1. Muscinae and Stomoxyinae. *Sieboldia* (Supplement): 31–46.

Skidmore, P. 1985. *The biology of the Muscidae of the world*. [xii] + 550 pp. W. Junk Publishers, Dordrecht.

Steyskal, G. C. 1975. The gender of the genus-name *Stomoxys* Geoffroy, 1762 (Diptera, Muscidae). *Proceedings of the Entomological Society of Washington* **77**: 163.

Thomsen, M. 1935. A comparative study of the development of the Stomoxydinae [sic] (especially *Haematobia stimulans* Meigen) with remarks on other coprophagous muscids. *Proceedings of the Zoological Society of London* **1935**: 531–550.

Thorpe, W. H. 1942. Observations on *Stomoxys ochrosoma* Speiser (Diptera Muscidae) as an associate of army ants (Dorylinae) in East Africa. *Proceedings of the Royal Entomological Society of London* (A) **17**: 38–41.

Van Emden, F. I. 1965. *The fauna of India and adjacent countries. Diptera* vol. 7: *Muscidae*, Part 1. xiv + 647 pp. Government of India, Delhi.

Zumpt, F. 1950. Key to the Stomoxydinae of the Ethiopian region, with description of a new *Haematobia* and a new *Rhinomusca* species from Zululand. *Anais do Instituto de Medicina Tropical* **7**: 397–426.

Zumpt, F. 1973. *The stomoxyine biting flies of the world. Diptera: Muscidae. Taxonomy, biology, economic importance and control measures*. viii + 175 pp. Gustav Fischer, Stuttgart.

CHAPTER ELEVEN

House-flies, blow-flies and their allies (calyptrate Diptera)

R. W. Crosskey and R. P. Lane

The calyptrate Diptera are relatively homogenous in their appearance compared to the Nematocera, being mostly robust flies with conspicuous bristles – in fact the epitome of the popular idea of 'flies'. The 18 000 or so described species form a natural group ('subsection') within the higher Diptera (i.e. the infraorder Muscomorpha). Their classification is described in brief in Chapter 3.

The calyptrates as a whole have medical significance in three ways: first, biting and bloodsucking occurs in two families, the Glossinidae (tsetse-flies, Chapter 9) and Muscidae (of the subfamily Stomoxyinae, Chapter 10); secondly, larvae of some species in several families are a cause of human myiasis; and thirdly, the adults of some species – referred to as synanthropic flies from the Greek *syn* (with) and *anthropos* (man) – are not only common and annoying around human habitations but have the potential for mechanical transmission of pathogenic organisms because of their habit of feeding on human food as well as rubbish and excrement. Some of these synanthropic species are so common they have colloquial names, at least in English, e.g. *Musca domestica* is the house-fly (Fig. 11.1a), *Fannia canicularis* is the lesser house-fly (Fig. 11.1b), and *Fannia scalaris* is the latrine-fly.

This chapter is concerned with the *adults* of medically important synanthropic and myiasis-producing calyptrate flies, the latter including the Old World screw-worm flies (genus *Chrysomya*, Fig. 11.2a) and the Tumbu fly of Africa (Fig. 11.2b). Identification and biology of the larvae of the myiasis-producing Diptera are considered in detail in Chapter 12.

RECOGNITION OF CALYPTRATES AND ELEMENTS OF STRUCTURE

The flies concerned in this chapter are 'calyptrate' Diptera (Chapter 3) and the essential starting point for recognition of the various families is to understand the recognition features of the Calyptratae (a huge subsection of the infraorder

Medical Insects and Arachnids Edited by Richard P. Lane and Roger W. Crosskey.
Published in 1993 by Chapman & Hall ISBN 0 412 40000 6

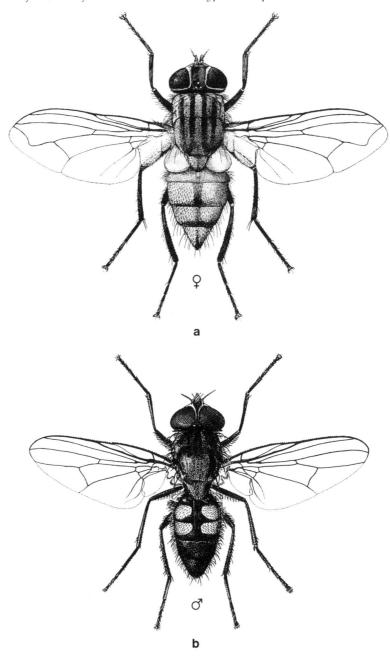

Figure 11.1 Two species of muscoid flies of medical importance: (*a*) female of the house-fly, *Musca domestica* (Muscidae); (*b*) male of the lesser house-fly, *Fannia canicularis* (Fanniidae). Major differences between the two genera can be seen in the wings and calypteres (lobes behind the wing bases): the central long vein (M) describes a strong forward bend in *Musca* and runs straight to the wing edge in *Fannia*, the calypteres are very large and obvious in *Musca*, small and inconspicuous in *Fannia*.

Muscomorpha). In brief, adult calyptrate flies simultaneously possess the following chief diagnostic characters:

1. Head with a fine inverted U-like or horseshoe-like line (ptilinal fissure) arching over the bases of the antennae and running downwards towards the epistome

Recognition of calyptrates and elements of structure 405

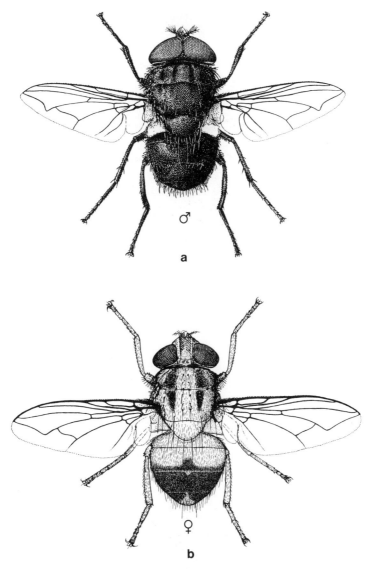

Figure 11.2 Two species of calliphorid flies of medical importance: (*a*) male of an Old World screwworm fly, *Chrysomya megacephala*; (*b*) female of the Tumbu fly of Africa, *Cordylobia anthropophaga*.

(Fig. 3.2*a*), and with a small sclerite (lunule) between the fissure and the antennae.
2. Antenna with a fine longitudinal groove (antennal seam) along the outer edge of the second segment (Fig. 3.2*c*).
3. Side wall of the thorax with a prominent boss or knob (greater ampulla) slightly below and in front of the wing insertion (Figs 3.4 and 11.3).
4. Two membranous lobes (calypteres) interposed between the membranous wing base and the wall of the thorax (Fig. 3.5*b*) (called upper and lower calypter or alar and thoracic squama).

Besides these principal groundplan features, adult calyptrates show some other structural elements which help towards recognition. The body, especially the thorax (Fig. 11.3), usually has many strong bristles set in a fixed arrangement (chaetotaxy). The head is typically holoptic or almost so in males (Fig. 3.2b) and dichoptic in females (but sometimes the male frons is as wide as that of the female). The third antennal segment has an arista (Fig. 3.2c) which can be bare, pubescent (with short hairs about as long as its basal width) or plumose (bearing very long fine hairs on upper and lower edges, Fig. 3.2d, or along only the upper edge). The head (usually) has a pair of strong crossed or converging bristles (vibrissae) at the epistome (Fig. 3.2a). The thoracic scutum (often miscalled the 'mesonotum') is divided into anterior and posterior parts by a complete transverse suture (Fig. 11.3). The wing has a well-developed lobe (alula) on the hind margin near the base (Fig. 3.5b). Wing venation comprises few veins and these are nearly always configured such that the discal medial cell (dm) is a large and obvious feature (Fig. 11.4c–h); there is only one medial vein (M), and in many calyptrates this kinks abruptly forwards and meets the wing edge near the end of vein R4+5 (occasionally, but not in any medically important genera, the two veins join towards the wing tip and run to the edge as a common vein or 'petiole'). The abdomen is differentiated into a large preabdomen which encloses and largely conceals the postabdomen and genitalia (T5 is usually the last fully visible tergite, Fig. 11.6b).

Sexing of flies

Sexing of adult calyptrate flies can be difficult, especially when male and female do not differ noticeably in head structure or when the genitalia are recessed in the

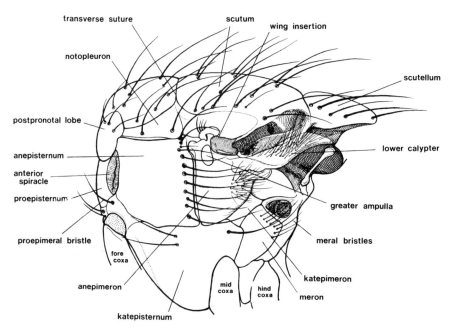

Figure 11.3 Side view of the thorax of a typical bristly calyptrate fly to show the main landmarks. The thorax illustrated is representative of Calliphoridae (one of the families in which the meron has a row of bristles) and is that of *Calliphora* (a genus in which the lower calypter is hairy on its upper surface).

abdomen. Sexing is easy in many medically important calyptrate genera, e.g. *Fannia, Haematobia, Calliphora, Chrysomya, Cochliomyia* and *Cordylobia*, because the head of the male is completely or nearly holoptic (eyes meeting, Fig. 3.2b) and that of the female dichoptic (eyes widely separated, Fig. 3.2a); in such flies the eyes of the male are nearly touching and the frons reduced to a narrow strip in which the parafrontals meet or nearly meet each other in the midline of the head; the frontal vitta is more or less obliterated.

This difference, however, is not universal in Calyptratae. In *Glossina*, the head is equally dichoptic in both sexes, and in many genera the frons width of the male is less than that of the female but not conspicuously so. Sex recognition in species with sexually similar heads necessitates examining the underside of the tip of the abdomen; male sex is shown by the genital hypopygium being at least partly visible within the terminal abdominal tergites (cf. Fig 9.6a and 9.6b of male and female terminalia in *Glossina*).

IDENTIFICATION OF MEDICALLY IMPORTANT CALYPTRATE FLIES

Identification of calyptrate flies is largely a matter for Diptera specialists because of the complex taxonomy of the several families and the enormous numbers of genera and species – of which relatively very few have medical importance. Even the families are not always clearly distinguishable because of the existence of aberrant genera not fitting textbook diagnoses based only on typical taxa. Hence the keys given below must be used with due care, bearing in mind that they cannot be expected to cover all the unimportant genera of which specimens might be obtained by general collecting.

Identification keys

The following keys are aimed at providing 'self-help' starting points for the identification of adults of the most important calyptrate flies of interest in medical entomology. They can be used for recognition of flies which have been obtained in direct relationship with man, i.e. caught while biting or reared from myiasis-inducing larvae, or for which there is strong circumstantial evidence of pathogen transmission among humans because of close association with humans in their food or habitations. The first key (Key A) to calyptrate families provides the lead to succeeding lettered keys (B–H) to genera and species within the families.

Parts of the head and thorax mentioned in the keys are labelled in Figs 3.2 and 11.3, respectively, the wing veins in Fig. 3.5b, and the numbering of abdominal tergites in Fig. 11.6b. Sclerites of the thorax are named on the modern McAlpine (1981a) system, which is now superseding old and more familiar taxonomic terms as follows:

Old taxonomic term	McAlpine term (Fig. 11.3)
humeral callus	postpronotal lobe
hypopleuron	meron
mesonotum	scutum

mesopleuron — anepisternum
notopleuron — notopleuron
prescutum — scutum (part before transverse suture)
propleuron (propleural depression) — proepisternum
pteropleuron — anepimeron
scutum — scutum (part behind transverse suture)
sternopleuron — katepisternum

Key A: Identification key to families of calyptrate flies

The families Nycteribiidae and Streblidae (parasites of bats), Rhinophoridae (parasites of woodlice) and Mystacinobiidae (one species of wingless fly associated with bat guano) are omitted; they have no medical importance and their specialized habits make it extremely unlikely they will be collected by persons who use this book. On the other hand, families without medical importance but for which there is some practical possibility of medical entomologists collecting specimens are included (their names in square brackets).

1   Wings with normal calyptrate venation (Fig. 11.4c–h), veins not crowded together towards leading edge. Thorax normal, not dorsoventrally flattened, with coxal attachments of mid and hind legs positioned close together near midline of ventral surface. Claws not toothed. Typical flies (e.g. appearance as Figs 11.1, 11.2, 11.5a and 11.6a).. 2
— Wings with abnormal venation in which veins mainly crowded into leading half of wing (Fig. 11.4a,b), or wings absent (*Melophagus ovinus*, Fig. 17.3f). Thorax abnormal, broad and dorsoventrally flattened, with coxal attachments of mid and hind legs widely separated and positioned far from midline of ventral surface. Claws toothed, very strong and recurved. Atypical flies (e.g. appearance as Fig. 17.3d,f). [Ectoparasites of mammals other than bats and of birds]... HIPPOBOSCIDAE

2   Wings with discal medial cell (dm) widening gradually and more or less regularly from its base, cell br not bulging backwards into basal half of cell dm (Fig. 11.4c–g). Arista bare (Fig. 3.2c) or haired (Figs 3.2d and 10.3a,b) but never with feathery branched hairs. Proboscis usually downwardly directed and with large and fleshy labella, if of forwardly directed piercing type (Stomoxyinae, Fig. 10.2a) then not embraced by palps................................. 3
— Wings with discal medial cell (dm) of characteristic cleaver-like shape ('hatchet cell'), caused by bulging of cell br into basal half of dm and backward deflection of M before its junction with cross-vein r–m (Figs 9.3 and 11.4h). Arista with feathery short-branched hairs (upper side only) (Fig. 9.2b). Proboscis long and stiletto-like, projecting forwards and completely embraced by long palps when fly at rest (Fig. 9.3). [One genus, *Glossina*]..........................
.... GLOSSINIDAE (see Chapter 9)

3   Flies with strong bristles on body and usually also legs, bristles often very strong on thorax (Fig. 11.3). Mouthparts fully developed as a downwardly directed or forwardly projecting proboscis easily visible in side view (Figs 3.2d and 10.2a). Vibrissae present (Fig. 3.2a,d). Head not unusually bulbous. Antennae usually neither very small nor sunken... 4

Identification of medically important calyptrate flies 409

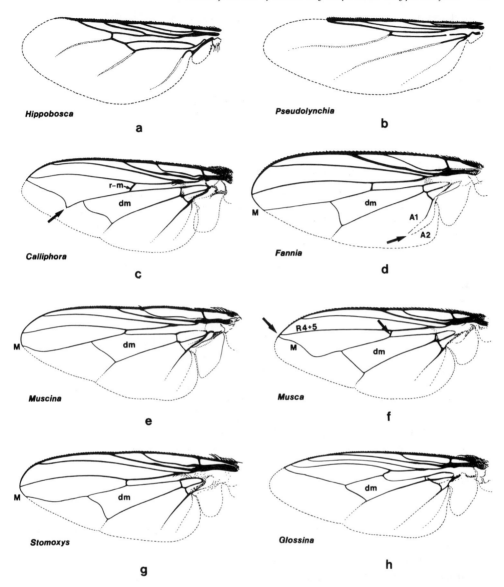

Figure 11.4 Wing venation of representative calyptrate flies: (*a* and *b*) *Hippobosca* and *Pseudolynchia*, showing the crowding of the veins into the leading half of the wing typical of Hippoboscidae; (*c*) *Calliphora*, showing the sharp angle of vein M (arrowed) usual in Calliphoridae, Sarcophagidae and Tachinidae; (*d*) *Fannia*, showing the convergence (arrowed) of the anal veins characteristic of Fanniidae; (*e* and *g*) *Muscina* and *Stomoxys*, showing slight apical forward curve of vein M seen in some Muscidae; (*f*) *Musca*, with strongly bent vein M ending close to R4+5 (arrowed), interval no more than the length of the r–m cross-vein (arrowed); (*h*) *Glossina*, showing the characteristic 'hatchet' shape of cell dm.

— Flies without definite bristles (Fig. 11.5*a*). Proboscis almost absent, mouthparts atrophied and invisible in side view of fly. Vibrissae absent. Head distinctly bulbous and with very small and sunken antennae (Fig. 11.5*b*,*c*). [Obligate larval parasites of vertebrates]...
.... OESTRIDAE

410 House-flies, blow-flies and their allies (calyptrate Diptera)

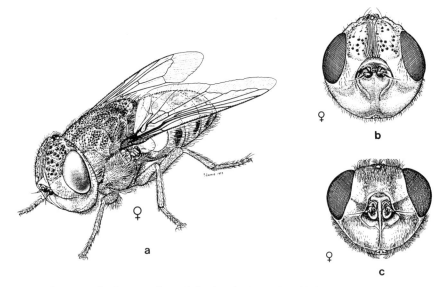

Figure 11.5 Features of calyptrate flies of the family Oestridae: (*a*) female of the sheep nostril-fly, *Oestrus ovis*; (*b* and *c*) head of a female of *Oestrus ovis*, facial view; (*c*) head of a female of *Gasterophilus intestinalis* (horse bot-fly), facial view. The illustrations are representative of oestrids in showing the non-bristly body and legs, bulbous head, tiny and sunken antennae, and virtual absence of mouthparts.

[Following Wood (1987), this family includes cuterebrines, gasterophilines and hypodermatines, groups often treated in the past as families. In practice, members are seen and identified in their myiasis-producing larval stage (Chapter 12), adults being very difficult to rear and rarely obtained. Refer to Zumpt (1965, Old World), Wood (1987, Nearctic) and Guimarães et al. (1983, Neotropical) for help in adult identification.]

4   Thorax with a more or less vertical row of strongly developed bristles on meron (meral bristles, Fig. 11.3) 5
—   Thorax without bristles on meron, this area either bare or with some soft and haphazardly placed hairs 7
5   Thorax not bulbously convex beneath scutellum (Fig. 11.3). 6
—   Thorax bulbously convex beneath scutellum (Fig. 3.4). [Larval parasitoids of insects: worldwide] .... [TACHINIDAE]
6   Flies of generally dull grey appearance, thorax with three black scutal stripes and abdomen with chequered, striped or spotted brown or black pattern (Fig. 11.6*a*, *b*). Notopleuron of thorax usually with three or four bristles (two in *Wohlfahrtia*). Head dichoptic in both sexes .... SARCOPHAGIDAE (Key F)
—   Flies of other appearance, usually with blue–black, violet–blue, blue–green or green metallic lustre (thorax without three bold black stripes and abdomen not chequered or variegated); sometimes yellowish brown or partly so. Notopleuron of thorax with two bristles (Fig. 11.3). Head usually holoptic in male, always dichoptic in female .... CALLIPHORIDAE (Key B)
7   Wings with vein A1* incomplete, not nearly reaching wing edge, membrane uncreased beyond vein end (Fig. 11.4*d*–*g*); vein M either running straight to wing edge *or* deflected forwards to meet costa near end of vein R4+5 (Fig. 11.4*f*). Scutellum bare on lower surface 8

— Wings with vein A1 usually complete, reaching wing edge even if only faintly as crease-like fold in membrane; vein M running straight to wing edge without forward deflection. Scutellum usually with a patch of hairs near tip of ventral surface. [Breeding associated mainly with living or decaying plant material, sometimes dung: mainly Holarctic realm].. [ANTHOMYIIDAE and SCATHOPHAGIDAE]

8   Wings with strong curved A2 vein whose tip approaches vein A1* (Fig. 11.4d) in such a way that imagined extensions of these veins would intersect before they reached wing edge; vein Sc almost straight on apical two-thirds (Fig. 11.4d); vein M running straight to wing edge (Fig. 11.4d). Calypteres not conspicuous, small and subcircular, lower calypter not much larger than upper calypter... FANNIIDAE (Key G)

— Wings with vein A2 not strongly curved (Fig. 11.4e–g), an imagined extension not intersecting vein A1 before that vein reached wing edge; vein Sc curved fowards in its apical third or quarter (Fig. 11.4e–g); vein M running straight to wing edge *or* deflected weakly forwards apically (Fig. 11.4e) *or* with an abrupt forward bend (Fig. 11.4f). Calypteres conspicuous, lower calypter much larger than upper calypter and in form of a broad lobe abutting scutellum ('*Musca*' type as visible in Fig. 11.1a) or a subtriangular outwardly pointing lobe not abutting scutellum... MUSCIDAE (Key H)

Key B: Identification key to medically important calliphorid genera

1   Flies of predominantly green, blue, violet–blue or bluish black body colour, conspicuously metallic (except in *Calliphora*). Legs black or brownish black, sometimes rather paler on tibiae. Thorax with hairy proepisternum (depressed area in front of and below anterior spiracle, Fig. 11.3).................... 2

— Flies of predominantly reddish yellow or reddish brown to mid-brown body colour (abdomen sometimes almost black), not metallic. Legs mainly or entirely reddish yellow. Thorax with bare proepisternum. [Tropical Africa] .... **Cordylobia** and **Auchmeromyia** (Key C)

2   Wing with stem-vein (base of R) entirely bare (Fig. 3.5b) 3
— Wing with stem-vein finely haired along its hind margin.. 4

3   Flies with glossy green or coppery green thorax and abdomen. Lower calypter bare. Body length 6–9 mm. ['Greenbottles'].. **Lucilia**

   [Occasionally involved in human myiasis, notably the cosmopolitan species *Lucilia sericata*.]

— Flies with black thorax and steely dark blue, bluish-violet or blue–black slightly metallic abdomen. Lower calypter with long dark hair on upper surface (Fig. 11.3). Body length 10–14 mm. ['Bluebottles'] .... **Calliphora** (Key D)

4   Head with almost entirely black ground colour and black hair. Lower calypter bare; *upper* calypter hairy on outer half of dorsal surface. [Not tropical, Palaearctic and Nearctic regions only].. 5

---

*   McAlpine (1981a) calls the sixth vein of the calyptrate wing A1+CuA2 (see Fig. 3.5b) but for this key it is simply called A1 in accord with McAlpine's (1981b) key to dipterous families and with labelling of muscoid wing figures in McAlpine (1987).

— Head with ground colour of at least lower half entirely or mainly bright orange or dirty orange–red and with white, yellow or golden–orange hair. Lower calypter haired on whole dorsal surface (*Chrysomya*) or at base concealed beneath upper calypter (*Cochliomyia*); upper calypter bare on dorsal surface. [Mainly tropical, neither genus solely Holarctic] .... **Cochliomyia** and **Chrysomya** (Key E)

5 Thorax with anterior spiracle black or blackish brown (not obvious against thoracic ground colour). Upper calypter smoky and with dark dorsal hair (latter very obvious). Male: head distinctly dichoptic, frons at narrowest wider than antenna and parafrontals fully separated by complete frontal vitta .... **Protophormia**

[One included species, *Protophormia terraenovae*.]

— Thorax with anterior spiracle yellow or orange (very obvious against dark thoracic ground colour). Upper calypter white with white or yellowish white hair (latter not obvious at first sight). Male: head holoptic, frons at narrowest *very* much narrower than antenna, parafrontals meeting in midline and eliminating upper part of frontal vitta .... **Phormia**

[One included species, *Phormia regina*.]

Key C: Identification key to species of *Auchmeromyia* and *Cordylobia* (Calliphoridae)

1 Thorax with only one bristle (presutural bristle) immediately above notopleuron (this bristle conspicuously isolated and no bristles present between it and groove delimiting postpronotal lobe). Male: genital cerci fused to form a long and narrow recurved hook-like process (outwardly visible) .... **Auchmeromyia** 2

— Thorax with two or three bristles immediately above notopleuron (one or two definite bristles present between presutural bristle and groove delimiting postpronotal lobe. Male: genital cerci paired and free as normal (not always outwardly visible) .... **Cordylobia** 3

2 Abdomen with apparent second segment (T3) elongate, about 1.5 times as long as T4 in male and twice as long in female. [Congo floor-maggot] .... **Auchmeromyia senegalensis**

— Abdomen with apparent second segment (T3) normal, about equal in length to T4 .... other species (not medically important)

3 Legs and sides of thorax entirely reddish yellow. Abdomen almost always not uniformly dark (at least basal part normally reddish yellow). Wings clear. Antennae small, third segment at most only twice as long as second segment. Smaller species, length 7–11 mm 4

— Legs and sides of thorax extensively darkened, reddish brown to brownish black. Abdomen entirely dark mahogany–brown to black, without pollinosity and very glossy to unaided eye. Wings smoky with orange–yellow base (evident to unaided eye). Antennae larger, third segment conspicuously more than twice as long as second segment. Large species, length 12–14 mm (of rather bulbous appearance). [Head of male widely dichoptic, frons about equal in width to length of third antennal segment] .... **Cordylobia rodhaini**

4 Abdomen shining, without pollinosity. Lower calypteres enlarged (reaching far beyond level of tip of scutellum when fly seen from above), dull and opaque. Thorax with hair of katepisternum and anepisternum black. Male: head widely dichoptic, frons as wide as length of antennae.... **Cordylobia ruandae**

— Abdomen dull, with thin white or ashy pollinosity (given a stippled appearance because hair bases on black areas forming irregular tiny dark islands among pale pollinosity). Lower calypteres normal (reaching about level with tip of scutellum), glassy and translucent. Thorax with hair of katepisternum and lower half of anepisternum yellowish white (that of upper part of latter black, stiff and unusually bristly). Male: head holoptic. [Tumbu fly, female in Fig. 11.2*b*] .... **Cordylobia anthropophaga**

Key D: Identification key to species of *Calliphora* affecting human health

Adults of the two species common in human habitations, potentially important in pathogen transmission, and reported in cases of human myiasis, are distinguishable by the following couplet:

1 Abdomen with patches of thick white pollinosity which shift conspicuous in appearance from white to steely dark blue or blue–black as light direction alters (pollinosity quite obvious to unaided eye). Genae of head with orange ground colour. Lower part of head with black hair. Thorax with anterior spiracle bright orange. [Mainly Holarctic and Australasian regions, southern Afrotropical region].. **Calliphora vicina**

[Species known in older literature by the preoccupied name *Calliphora erythrocephala*.]

— Abdomen uniformly and thinly white pollinose, appearance of pollinose covering not patchily shifting as light direction alters (pollinosity inconspicuous to unaided eye and abdomen appearing mainly shiny). Genae of head with black ground colour. Lower part of head with golden–orange hair. Thorax with anterior spiracle dingy yellow–brown. [Mainly Holarctic regions, not Afrotropical].. **Calliphora vomitoria**

Key E: Identification key to species of *Chrysomya* and *Cochliomyia* (Calliphoridae) associated with human myiasis

1 Scutum of thorax without bold black stripes (appearing uniformly green to violet–blue *to unaided eye*). Lower calypter hairy on whole upper surface (as in *Calliphora*, Fig. 11.3). Palpi normal, slightly dilated or clubbed, almost as long as third antennal segment. Greater ampulla (Fig. 11.3) covered with long and obvious hair. [Afrotropical, Oriental, Australasian and southern Palaearctic regions: also Neotropical region (introduced)]............................ **Chrysomya** 2

— Scutum of thorax with three bold black stripes (visible to unaided eye against general coppery green to deep blue colour). Lower calypter bare on most of its surface, haired only at base (hair usually partly hidden beneath upper calypter in specimens with unspread wings). Palpi tiny and filiform, only about half as long as third antennal segment. Greater ampulla with very short downy hair not at all obvious. [Nearctic and Neotropical regions: also North Africa (introduced but believed eradicated)]............................ **Cochliomyia** 5

2   Anterior spiracle (Fig. 11.3) pale, creamy white to pale yellow. General colour green and usually with little or no blue tinge 3
—   Anterior spiracle dark, tawny or orange–brown to smoky black. General colour greenish blue or blue (often with violet tinge) 4
3   Thorax with proepimeral bristle (strong upcurved bristle just below anterior spiracle, as in Fig. 11.3). [Widespread Oriental and Australasian regions (not Africa): also Japan, Central America and Argentina (introduced)] .... **Chrysomya rufifacies**
—   Thorax without proepimeral bristle. [Widespread Afrotropical (including Madagascar) and southwestern Palaearctic regions, northern Oriental region from India to China: also Neotropical region (introduced)] .... **Chrysomya albiceps**
4   Lower calypter smoky brown (obviously darker than creamy white base of upper calypter). Female: frontal vitta of head (feature labelled on Fig. 3.2*a*) with side margins bowed distinctly outwards. Male: eyes with facets conspicuously enlarged on upper three quarters and sharply demarcated from small facets on lowermost quarter. [Widespread Oriental and Australasian regions, eastern fringes of Palaearctic region, Malagasia: also West and South Africa, Middle East and Neotropical region (introduced)] .... **Chrysomya megacephala**
—   Lower calypter white, concolorous with upper calypter. Female: frontal vitta of head with parallel side margins. Male: eyes with facets not of conspicuously contrasting size, uppermost facets not much larger than lowermost facets. [Mainland Afrotropical region, eastern Arabian peninsula, Oriental region to New Guinea (not Australia)] .... **Chrysomya bezziana**
5   Abdomen with conspicuous white spot on each side of last visible tergite (T5, apparent fourth segment). Thorax with middle dorsal stripe extending whole length of scutum. Parafrontals (Fig. 3.2*a*) with white or yellow hair. Occiput uniformly black. Basicosta (small scale-like plate at base of wing, Fig. 3.5*b*) of female tawny–yellow or orange. [Widespread in New World from northern United States to Argentina, including Caribbean islands] .... **Cochliomyia macellaria**
—   Abdomen without white spots on last visible tergite. Thorax with middle dorsal stripe not extending whole length of scutum, absent on anterior part of scutum or developed only near transverse suture. Parafrontals with black hair. Occiput not uniformly black, upper part with central tawny–orange or brick-red patch below mid-vertex (sometimes indefinite in male). Basicosta of female dark brown or black. [South and Central America, Caribbean islands (formerly also United States and Mexico)] .... **Cochliomyia hominivorax**

Key F: Identification key to Sarcophagidae associated with human myiasis

1   Arista almost bare (bearing only microscopic hairs that are not longer than basal diameter of arista). Abdomen with pattern of black spots (sometimes very large) which contrasts with yellowish grey pollinose areas (Fig. 11.6*b*), pattern fixed in relation to changes in light direction. Notopleuron and katepisternum both with two setae .... **Wohlfahrtia** 2

Identification of medically important calyptrate flies 415

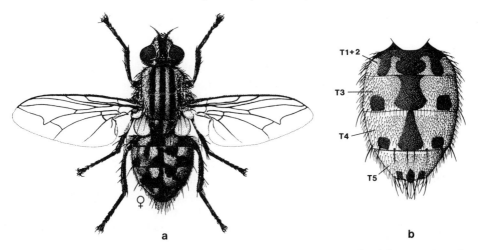

Figure 11.6 Features of Sarcophagidae associated with human myiasis: (*a*) a fly of the genus *Sarcophaga* showing the characteristically three-striped thorax and chequered abdomen; (*b*) abdominal pattern in *Wohlfahrtia magnifica*, principal Old World sarcophagid causing myiasis of man. Tergites (T) are numbered as for the calyptrate abdomen generally: note that the tergites of the four externally apparent segments are tergites 1–5, the first 'segment' of the adult calyptrate fly abdomen being formed by fused tergites of actual segments 1 and 2 (hence the last apparent segment is T5).

— Arista plumose (long and conspicuous hairs on upper and lower side along its basal half or more). Abdomen with dark and light chequered pattern (Fig. 11.6*a*) shifting in appearance with changes in light direction (silvery, ashy grey, or yellowish grey pollinose patches changing to black as fly is turned). Notopleuron and katepisternum both with three or four setae ....
.... Species of **Sarcophaga**

[This genus is understood here in its wide sense (see later text). *Sarcophaga crassipalpis* and *S. cruentata*, reported to cause human myiasis (Chapter 12), have the tip of the abdomen (including genitalia) orange–red instead of black (as in most species).]

2 Abdomen mainly pale yellowish-grey, thickly pale pollinose with three rows of small segmental spots in which at most only median spots of T3 and T4 extend whole length of segment, spots of outer two rows very small and almost circular (Fig. 11.6*b*). Palpi black or yellow. Male: mid and hind tibiae with short hair not or hardly longer than tibial width (not bushy). [Palaearctic and northern Afrotropical regions] 2
— Abdomen mainly black, with three rows of large black spots that extend full length of each segment and sometimes variably coalesce with spots of neighbouring rows, areas between spots ashy grey. Palpi yellow. Male: mid and hind tibiae very bushy, with long ventral hair two or three times as long as tibial width. [North America, Mongolia and eastern Siberia] ....
.... **Wohlfahrtia vigil**
2 Palpi yellow. Abdomen with very small spots, outer spots of basal segment (T1+2) tiny and isolated against hind margin, middle spots of T3 and T4 not reaching forwards to hind margins of previous segments (or only just on T3). Body, to unaided eye, with distinct yellow tinge both on thorax and abdomen. Antennae with third segment much longer than width of parafacial and more

than twice as long as second segment. Genitalia orange–red. [Middle East, Arabian peninsula, southwestern Asia to northwestern India, also northeastern Africa and Atlantic coastal Africa].................. **Wohlfahrtia nuba**

— Palpi black. Abdomen with larger spots, outer spots of basal segment extending forwards and merging at extreme abdominal base with middle spot, middle spots of T3 and T4 reaching forwards broadly to hind margins of previous segments (Fig. 11.6b). Body, to unaided eye, appearing greyer on thorax than abdomen. Antennae with third segment equal in length to or shorter than width of (very broad) parafacial and only twice as long as second segment. Genitalia black. [Palaearctic region from western Europe and North Africa through Middle East and Central Asia to China].. **Wohlfahrtia magnifica**

Key G: Identification key to species of Fanniidae closely associated with man

To see the scutal stripes the specimen should be illuminated from in front and tilted head downwards. To see the hind tibial bristles reliably the tibia should be observed in exact side view.

1. Arista bare. Head of male holoptic (Fig. 11.1b). Scutum in front of transverse suture with all four dorsocentral bristles strong (anterior one of each pair very well developed and not much smaller than posterior one)*............... **Fannia** 2
— Flies without such features present simultaneously .... Flies of no medical importance

   [The three other fanniid genera exit here. Specimens are unlikely to be collected in association with humans. The genera differ from *Fannia* by having pubescent to plumose arista, *or* dichoptic males, *or* the anterior one of each pair of presutural dorsocentral bristles much weaker than the posterior one.]

2. Palpi black. Antennae entirely black. [Worldwide]................................. 3
— Palpi yellow. Antennae not entirely black, first two segments orange–yellow. [North America] .... **Fannia benjamini**
3. Head widely dichoptic.. female flies 4
— Head holoptic .... male flies 5
4. Abdomen black. Thorax dorsally with two broad black central scutal stripes separated by dark grey area, general appearance very dark grey to black. Parafacials completely bare. Hind tibia on underside with a row of three widely and evenly spaced bristles .... **Fannia scalaris**
— Abdomen mainly black but usually with yellowish or tawny-reddish base. Thorax dorsally with a very narrow brownish middle stripe flanked by grey areas (in equivalent position to paired black stripes in *scalaris*). Parafacials haired on upper half. Hind tibia on underside with two widely spaced bristles .... **Fannia canicularis**
5. Abdomen all black in ground colour. Thorax dorsally somewhat shining black and without well-defined stripes. Mid tibia with conspicuous mid-ventral flange margined with tiny teeth (Fig. 11.7a) .... **Fannia scalaris**

* The bristles referred to here are the very large bristles standing on the scutum between the back of the head and the transverse suture. They are two, in line, on either side of the midline of the scutum (the largest bristles on the scutum in front of the transverse suture).

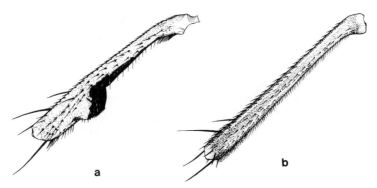

Figure 11.7 Tibia of the mid leg in males of the medically important species of *Fannia*: (*a*) latrine-fly, *Fannia scalaris*; (*b*) lesser house-fly, *Fannia canicularis*.

— Abdomen not all black in ground colour, first two or three segments (T1+2, T3 and sometimes T4) with large paired dingy yellow to bright reddish yellow lateral spots (these usually covering most of first two tergites but only anterior half of T4 when present on this tergite) (Fig. 11.1*b*). Thorax dorsally dull grey with variably evident median brown stripe. Mid tibia without specialization, evenly fringed with very fine short hairs (Fig. 11.7*b*).................. .... **Fannia canicularis**

Key H: Identification key to Muscidae for recognition of *Musca* species affecting man

1 Proboscis stout and directed downwards, with large and fleshy labella (of licking type) (Fig. 3.2*d*) 2
— Proboscis slender, directed forwards or more or less so, with small and inconspicuous labella (of piercing type) (Figs 10.2*a* and 10.3*a,b*)......................... .... Subfamily STOMOXYINAE (see Chapter 10)
2 Wing with vein M running straight to wing edge (as in *Fannia*, Fig. 11.4*d*) or with only slight forward deflection near its end (as in *Muscina*, Fig. 11.4*e*), ends of veins M and R4+5 separated by distance very much more than length of cross-vein r–m...................... Muscidae with little or no medical importance
— Wing with vein M deflected forwards at a strong bend and ending at wing edge near to end of vein R4+5, distance between ends of M and R4+5 not more than length of cross-vein r–m (Fig. 11.4*f*).. 3
3 Metallic bright green flies. Greater ampulla hairy .... **Neomyia**

   [Genus worldwide, known till recently by invalid synonymous name *Orthellia*. Species breeding in dung and rarely attracted to humans or dwellings, no medical importance.]

— Non-metallic flies, variously dull black, grey, brown or partly yellow–orange. Greater ampulla bare .... **Musca** 4
4 Palpi black or blackish brown. Mid tibia without a bristle near middle of underside. Proboscis normal, not enlarged and boat-shaped. [Normal non-biting species] 5
— Palpi yellow. Mid tibia with an isolated bristle near middle of underside. Proboscis much enlarged, its profile boat-shaped (Fig. 10.4*b*). [Oriental and

Afrotropical regions, Middle East: Indian cattle-fly, biting and bloodsucking species (see Chapter 10)]...... **Musca crassirostris**

5   Scutum with four black stripes in front of transverse suture and two very broad black stripes behind transverse suture. Thorax with proepisternum bare. [Old World only]...... 6

—   Scutum with four black stripes continuous along its whole length (Fig. 11.1a), stripes not merging behind transverse suture into single pair. Thorax with hairy proepisternum. [Old World and New World (introduced), now cosmopolitan] ...... **Musca domestica** ('complex')

[See text discussion for further detail.]

6   Specimens from Australia. ['Bush-fly']...... **Musca vetustissima**

[See annotation at couplet 8.]

—   Specimens from other areas 7

7   Specimens from Africa ...... **Musca sorbens** or **M. biseta**

[Experimental genetics shows these to be distinct species but there are no consistent morphological differences. In *sorbens* the male frons is usually broad (i.e. two or three times wider than the anterior ocellus) whereas the male head is typically almost holoptic in *biseta;* but both species show wide variation in frons width even at the same place.]

—   Specimens from other areas 8

8   Specimens from Mediterranean area, Arabian peninsula or southwestern Asia ... . **Musca sorbens**

—   Specimens from eastern Asia, Oriental region, Melanesia or Polynesia .... **Musca sorbens** or **M. vetustissima**

[Experimental genetics shows these to be distinct species but, as with *sorbens* and *biseta* in Africa, there are no consistent morphological differences. The male frons is similarly quite broad in typical *sorbens* but narrow-fronted male *sorbens* cannot be distinguished from the typically almost holoptic male *vetustissima*. Only *vetustissima*, however, is present in Australia.]

Faunal and taxonomic literature

Literature is scattered and there are few up-to-date comprehensive works. The following works assist in identification of adult flies belonging to families of some medical importance (except Glossinidae and Stomoxyinae treated in Chapters 9 and 10) but must be used with caution because now they are often partly outdated by less comprehensive but modern specialist literature: regional Diptera catalogues listed in Chapter 3 should be consulted for further leads to information.

Calliphoridae: Afrotropical (Zumpt, 1956a, Calliphorini and Chrysomyini, species; Zumpt, 1962, Madagascar); Nearctic (Hall, 1948, species; Shewell, 1987a, genera); Neotropical (Guimarães et al., 1983, myiasis-producing species); Oriental (Senior White et al., 1940, species); Palaearctic (Zumpt, 1956b, species); Papua New Guinea (Kurahashi, 1987, species).

Fanniidae: Afrotropical (Van Emden, 1941, species); Arabian peninsula (Pont, 1991, species, useful for Middle East); Australia (Pont, 1977, species); Nearctic (Chillcott, 1961, species); Palaearctic (Hennig, 1955–1964, species).

Muscidae: Afrotropical (see references in Pont, 1980); Arabian peninsula (Pont, 1991, species, useful for Middle East); Australasian (see references in Pont, 1989); Britain (Fonseca, 1968); Nearctic (Huckett and Vockeroth, 1987, genera); Oriental (Van Emden, 1965, species; Tumrasvin and Shinonaga, 1977, *Musca,* Thailand); Palaearctic (Hennig, 1955–1964, species).

Oestridae: Nearctic (Wood, 1987, genera); Neotropical (Guimarães et al., 1983, genera, some species); Old World (Zumpt, 1965, species).

Sarcophagidae: Afrotropical (Zumpt, 1972, species); Australia (Lopes, 1959, species); Madagascar (Reed, 1974, species); Nearctic (Shewell, 1987b, genera); Oriental (Senior White et al., 1940, species); Palaearctic (Verves, 1985, *Wohlfahrtia* species).

BIOLOGY OF SYNANTHROPIC FLIES

There is a very large literature on the biology and physiology of a few species (*Musca domestica, Calliphora vicina* and species of *Lucilia*) because they are easy to rear in the laboratory and have therefore become very useful 'model insects'.

Calyptrate flies are medically important because they suck blood (Glossinidae, Chapter 9; Stomoxyinae, Chapter 10), their larvae cause myiasis (Chapter 12) and the adults of some species are synanthropic; as such, the latter are abundant around human habitations. The biology of the common synanthropic species is discussed below.

The basic biology of the synanthropic species discussed in this chapter is fundamentally similar: eggs are laid directly on the medium in which the larvae will develop, although in some genera (e.g. some sarcophagids and calliphorids) the eggs hatch within the female and first-instar larvae are deposited; there are three larval stages and the mature larvae pupate, usually in the soil, retaining the tanned third larval skin as a protective puparium (Chapter 3). The time required for development is dependent on several factors including nutrition and moisture but the most important is usually temperature; e.g. at its optimum temperature of 25°C, *M. domestica* takes three to three and a half days from hatching of the eggs to pupation which can increase to more than 40 days at 15°C. The reproductive potential of many muscoid flies is legendary, the theoretical limits to increase being enormous. One calculation, using conservative rates of development and fecundity, estimated that a single female *Musca* laying eggs in the middle of April at the beginning of a north temperate summer would have 5.6 million offspring by the second week of September had they all lived. The principal constraints are larval breeding sites (although 1 kg of suitable medium produces 5000–10 000 adults of *M. domestica*) but other mortality factors include predators (especially ants in the tropics), pathogens (fungi of the Entomophthorales) and parasitoids. In tropical conditions *M. domestica* will have about 30 generations per year whereas there might be only ten generations in temperate areas.

To enable it to force its way out of the puparium and through thick layers of soil or other medium the newly emerging adult has an eversible sac – the ptilinum – on the front of its head (Chapter 3). Immediately after emergence the flies move upwards (negatively geotropic) and towards places away from the light (negatively phototropic) to allow the cuticle to harden, as in other insects after they have emerged. Knowledge of these preferred resting sites is important in control.

Mating takes place within 18–36 hours, the males being sexually mature before the females, and probably involves visual cues at long range and a series of pheromones immediately before copulation. At least two pheromones are used by *M. domestica*. One produced by the female attracts males and to a lesser extent females and is called muscalure; the other (unnamed) produced by males induces aggregation and receptivity of virgin females.

Most species locate the oviposition sites by attractants. These include carbon dioxide, ammonia and other volatile compounds emanating from decomposing materials (the various compounds being characteristic of different stages of decomposition).

Adults of most species will often feed on the same material as that in which larvae develop (decomposing organic material) but they are also attracted to flowers and other sweet-smelling places to feed on sugars. The method of food location by adult calyptrate flies differs – blow-flies use many olfactory cues over a consider-able distance, as evidenced by the large number of chemoreceptors on their antennae, whereas others such as *Musca domestica* use visual cues (e.g. attraction to dark spots) and search at random, changing the search pattern once in the vicinity of a potential food source. This is detected by humidity and smell. Liquid food is sucked up directly by capillary action through an elaborate system of pseudo-tracheae on the labella of the proboscis. Soluble foods are dissolved by saliva and the regurgitated contents of the crop before food is ingested.

The behaviour of various calyptrate flies (resting, dispersal, mating etc.) varies between species, but in general most are active only by day, preferring well-defined resting sites at night.

The following is a synopsis of the salient features of the biology of the principal genera which are most closely associated with human habitations.

Calliphoridae

This family contains a little over 1000 species worldwide. Most adults are brightly coloured, metallic blue or green, but some (such as the Congo floor-maggot, *Auchmeromyia senegalensis*) are yellowish or dull brown. The majority of species are oviparous but a few are viviparous. The medical interest of the family is principally as myiasis-producers (*Auchmeromyia, Cochliomyia, Chrysomya, Cordylobia*) (Fig. 11.2b) and to a lesser extent *Lucilia* and *Calliphora*). *Calliphora* and *Lucilia* are synanthropic flies very common around human habitations throughout the world. The biology of screw-worms is given in Spradbery (1991).

The two commonest species of *Calliphora* (bluebottles), *C. vicina* and *C. vomitoria*, are Holarctic in origin but have subsequently been transported around the world to other temperate regions (e.g. Australia and New Zealand). In areas of poor hygiene the females will lay batches of their yellow eggs on meat and under these circumstances the meat is termed 'fly blown'. The larvae are frequently carnivorous in their carrion habitat, feeding on the larvae of other species. *Calliphora uralensis*, a pest species from the southern USSR, rarely breeds in meat, preferring liquid excreta in latrines and cesspools. The adults of many species are frequently found in houses, feeding on food.

The adults of many species of *Chrysomya* (tropical blow-flies, Old World screw-worms) are brilliantly coloured and have a relatively broad thorax (Fig. 11.2a). Most species of the genus breed in carrion, where they are vigorously carnivorous, but some species such *C. putoria* (Afrotropical) and *C. megacephala* (Oriental and eastern Palaearctic) breed in wet faecal material and can even liquefy large masses of faeces; consequently they are commonly found breeding in latrines. The adults will feed on liquid faeces or liquefied carrion produced by the action of the larvae. Some species tenaciously walk over human skin. One of these, *C. bezziana*, is rather variable morphologically, but genetic study of this important myiasis-producing species over a range of sites from South Africa to New Guinea has provided no evidence for the existence of sibling species which could compromise use of the sterile male control technique (Strong and Mahon, 1991).

Lucilia species (greenbottles) are common in and around houses although some species (rarely) enter shaded areas to feed and oviposit. The eggs are laid on carrion, or in the case of the infamous *L. cuprina*, on the soiled fleeces of sheep from which they then invade living tissue to cause the condition known as sheep strike. The larvae rarely develop in faeces. Larval development is rapid and the mature larvae exhibit a distinct migration before pupating in the soil. The adults only occasionally land on man to feed on sores or secretions.

Fanniidae

This family, distributed mainly in the northern Holarctic and temperate Neotropical regions, contains about 265 species of which some 220 belong to *Fannia*. This is the only genus of medical importance; adults of a very few species have some hygienic importance and the larvae are occasionally involved in myiasis.

The commonest species is *F. canicularis*, which, with *M. domestica*, is one of the most abundant flies in and around human habitations. Males of *Fannia* tend to swarm outdoors and are not found much around human dwellings but the females of some species, especially the cosmopolitan *F. canicularis* (lesser house-fly) (Fig. 11.1b) and *F. scalaris* (latrine-fly), are often common inside houses, especially where there is poor sanitation; *F. scalaris* is often very abundant around latrine pits and over earth-closets and chemical closets. The females of some species are attracted to sweat and mucus, and it is in association with this habit that the North American species *F. benjamini* sometimes pesters humans by flying around the head; the flies are distinguished from other man-associated species of the genus by having yellow instead of black palps. *Fannia* breed in a wide range of decomposing organic materials, including excreta of all kinds, food, fruit and sometimes drinks such as milk. The eggs and larvae are well adapted for an aquatic life and therefore are found in quite liquid habitats, especially pools of semi-liquid faeces. It appears that they are unable to compete in drier faeces. They will even breed in urine-soaked clothing. A few species are attracted to carcasses, for instance the tropical *F. pusio* which will lay eggs on fresh meat. Attraction to human food, however, is unusual in any species, and this considerably limits the potential of *Fannia* species as transmitters of pathogens compared to house-flies (*Musca domestica*).

Sarcophagidae

This family contains about 2000 species and occurs worldwide. Some species are parasitic on arthropods or molluscs, a few on vertebrates, and females of almost all deposit young larvae instead of eggs (i.e. they are larviparous = viviparous). Two subfamilies are accepted in most classifications but only the Sarcophaginae contain species of medical importance. Larvae of the sarcophagine genera *Wohlfahrtia* and *Sarcophaga* are sometimes the cause of human myiasis (Chapter 12), but adults of these flies are harmless.

The genus *Sarcophaga* is not treated in the same way by all taxonomists, and this causes discrepancies in the literature which can be confusing. For long the name *Sarcophaga* was use for a large genus with many species of flesh-flies of the same general and easily recognized appearance – three broad black thoracic stripes and a grey and black chequered abdomen (Fig. 11.6a). Some specialists advocate increased fragmentation of this old and useful genus (easily understood by the generalist) with the unfortunate outcome that it now takes nearly 400 genera to classify the Sarcophagidae (Shewell, 1987b), females often cannot be identified to genus, and there are no longer any *Sarcophaga* species in the Afrotropical, Australasian and Nearctic faunas. On the restricted generic system the names for *Sarcophaga cruentata* and *S. crassipalpis*, species that sometimes cause human myiasis, become *Bercaea cruentata* and either *Parasarcophaga crassipalpis* or *Jantia crassipalpis*. We do not accept this profligate system here, and conform instead to the wide concept of *Sarcophaga* as recognized, for example, by Zumpt (1972), Dear (1980) and Pittaway (1991).

Muscidae

This is the second largest of the calyptrate families with over 3900 species worldwide (the Tachinidae having over 8000). Most species are inconspicuous but others make their presence felt as either agricultural pests (*Atherigona* – shoot-flies) or nuisance pests around humans and domestic animals. Some species bite and suck blood (*Stomoxys* and *Haematobia*); these belong to the subfamily Stomoxyinae and are discussed in Chapter 10. The biology of the family is comprehensively reviewed by Skidmore (1985).

Species of *Hydrotaea* are known as sweat-flies because of their annoying habit of landing and feeding on exudates of the eyes, nose and mouth. They sometimes even suck freely flowing blood, but do not bite. The most important species is *Hydrotaea irritans*. Larvae develop in faeces and carrion, and third-instar larvae are predatory on other larvae.

The genus *Musca* contains about 60 species but only a few are important synanthropic flies. The most common and widespread species is *M. domestica* (house-fly) (Fig. 11.1a) and the literature on this species is vast. West's (1951) monograph, although quite old now, is a useful source of basic information, and West and Peters (1973) include some 5720 references on this species alone up to 1972. A selected bibliography from 1972 to 1986 is given by Keiding (1986), who also gives a good summary of the biology.

Musca domestica has been studied in great detail and at least three 'subspecies' are recognized. *Musca domestica* occurs in all parts of the world but is least abundant in Africa. The 'subspecies' *M. d. curviforceps* and *M. d. calleva* are confined to Africa and are often found sympatrically; the former is endophilic and the latter exophilic. Detailed taxonomic work on the variation within *M. domestica* is described in Paterson (1964), Paterson and Norris (1970), Paterson (1974) and others.

The eggs of *M. domestica* are deposited on solid, moist and fermenting matter, but other species develop in wetter materials. Human faeces are only attractive in their drier solid state. Several species of *Musca* (including *M. sorbens*, *M. vetustissima*, *M. autumnalis*, *M. vitripennis*, *M. pattoni* and *M. fasciata*) will collect on

Figure 11.8 The calyptrate muscid fly *Musca sorbens* settled in typical abundance on an African of the Masai. The fact that such pestering flies are ignored instead of being continually brushed away enhances their potential as vectors of infection. (Photograph by Raymond Lewis and Denys Dawnay.)

suppurating sores or indeed any body secretion and therefore have the potential to transmit pathogens on such occasions. The 'bush-fly', *Musca vetustissima*, is a major pest over much of Australia and is closely related to *M. sorbens*, which together with *M. biseta* is a pest of equal magnitude and habits in Africa. Probably more importantly, some species commonly settle on humans and are a tremendous pest. People suffering huge numbers of flies often live in particularly insanitary conditions or closely with animals and become habituated to the flies crawling over them so that they no longer repel them (Fig. 11.8). In North America, *M. autumnalis* is known as the face-fly because it is such a pest around the heads of cattle. This Old World species was introduced into the Nearctic region as recently as 1952.

MEDICAL IMPORTANCE OF ADULT CALYPTRATE FLIES

The immature stages of many genera discussed in this chapter have a medical importance as causes of myiasis – the invasion of living mammalian tissues – and the larvae of some species (e.g. *Lucilia sericata*) have been used to clean wounds, topics discussed in Chapter 12. The remainder of this chapter therefore will be confined to the role of adults in the transmission of infection.

The propensity of the adults of some genera to feed on decaying materials (garbage, faeces, carrion, sewage) around houses and also on human food clearly gives them the *potential* for the transmission of pathogenic organisms. There are many genera of synanthropic flies for which few data exist beyond a rudimentary knowledge of their biology to incriminate them as medically or hygienically important. Further details of the medical importance of calyptrate flies can be found in Greenberg (1971, 1973) and Keiding (1986). The ubiquitous *Musca domestica* has been the subject of the most attempts to incriminate calyptrate flies in disease transmission. More than 100 different pathogens and parasites have been isolated from *M. domestica* (Greenberg, 1971) and it has been suggested that this fly transmits 65 of them. These infectious agents include: polio, hepatitis and trachoma viruses; cholera, *Salmonella*, *Shigella*, haemolytic streptococcus, pathogenic *Escherichia coli*, anthrax, diphtheria, tuberculosis, leprosy and yaws bacteria; *Entamoeba histolytica* protozoans and the helminths *Trichuris* and *Ancylostoma*. *Musca domestica* is not a biological vector of any of these, although it is a biological vector of some helminths (*Thelazia* species, eyeworms, see O'Hara and Kennedy, 1991) and its only possible significance could be as a mechanical vector of viruses and bacteria, and even of the eggs of helminths (Sulaiman et al., 1988).

There are three possible ways in which house-flies, and other calyptrate flies, can mechanically transmit pathogens.

(1) On the surface of the fly (legs, proboscis, mouthparts), as this is covered with many spines, hairs and microtrichia in which contaminated material can be trapped and transported. The survival of such trapped pathogens is short, however, because they become desiccated or are killed by the ultra-violet light in sunlight and grooming clears most pathogens within 24 hours. Relatively small numbers of pathogens can be transmitted to a human being (< 1 million) by the 'surface' route, and it is usually insufficiently contaminative to establish an infection. However, if food is contaminated the subsequent growth of the pathogens might reach the level of an infective dose.

(2) Regurgitation on food is a common prelude to feeding in *Musca* and similar flies. A small drop of the most recent meal is vomited onto the substrate and this could be an important route for infection by small pathogens; larger organisms, such as protozoal and helminths cysts, are usually filtered out by the network of pseudotracheae on the labella of the mouthparts.

(3) Ingestion and defaecation of pathogens is potentially one of the most important methods of transmission because the pathogen is protected while in the gut and is held for greater periods of time than in the previous two routes. However, there are some important limiting factors, including the ability of the pathogen to develop in the fly's gut. This survival is often dependent on the number of pathogens ingested; if there are less than 1000 *Salmonella* bacteria then none can be subsequently found in the fly's faeces. The natural gut flora out-competes many pathogens which would survive in sterile guts. Although the larvae of *Musca* develop in material teeming with micro-organisms, very few of these organisms are found in the gut of the adult, principally because much of the gut lining is shed or the conditions change during the pupal stage.

It is extremely important to distinguish between the isolation of a pathogen from an insect and its transmission (see Chapter 1). There have been numerous experiments in the laboratory on the transmission of pathogens by house-flies and there are some good data on the infective dose, survival rates, etc. but it is still not clear how these results relate to the field. Similarly, there have been many isolations of pathogens from wild-caught flies, although very few estimate the all-important pathogen load, the number of pathogens recovered, or the number of infective pathogens deposited by a naturally infected fly. Some of the most interesting data incriminating calyptrate flies in disease transmission come from field studies involving the control of flies. A classic example was a comparison between two towns in the southern USA one of which was sprayed with DDT and the other not; the treated town had a reduction in acute diarrhoeal disease of children caused by *Shigella* infection but infections due to *Salmonella* stayed the same (Lindsay et al., 1953). It is interesting to note in this context that as few as 100 *Shigella* bacteria are required to infect a human but many more are required to establish a *Salmonella* infection. Other types of study have measured changes in human disease before and after spraying insecticides or during the development of insecticide resistance by the house-flies. There are always considerable problems in interpreting these quasi-epidemiological studies; most are retrospective rather than prospective and it is not always clear whether the areas or towns being compared are truly matched. Clearly, there are many confounding factors in these types of studies and more work remains to be done to assess the potentially enormous public health significance of calyptrate flies.

Mention should be made here of the importance of calyptrate flies in forensic entomology, for dating the time of death of human cadavers in particular. This subject has been covered in detail by Smith (1986).

CONTROL

Whatever their true role in disease transmission the control of domestic calyptrate flies is very important to those subjected to the annoyance of these often tenacious pests. Where species are major veterinary pests (e.g. *Lucilia cuprina*, *Musca autumnalis*) specific methods of control have been developed.

The most important line of attack is on the breeding sites – mostly these will be around domestic dwellings and workplaces (e.g. abattoirs, slaughterhouses), particularly important being the sites where faeces and refuse are disposed. Control in the peridomestic environment alone will not be sufficient to eradicate the flies as there will be numerous breeding sites away from houses (e.g. animal dung or urine pools, rotting vegetation, carrion) and flies breeding in these sites will continually invade the controlled area. However, effective control can be achieved by diligence and widespread community cooperation. The disposal of refuse is a major problem facing many affected areas requiring substantial organization and infrastructure not usually available on the scale required for effective fly control.

Sources of human faeces as breeding sites can be controlled by appropriate sanitation such as the construction of fly-proof latrines. There has been considerable research on latrines for use in tropical countries; these need to satisfy the local population who use and maintain such facilities as well as the interests of those involved in fly control (Feacham et al., 1983). The treatment of faeces with various materials (e.g. oil, insecticides, chloride of lime) to make it less attractive to house-flies and the like is a relatively short-term solution – the treated material still needs to be disposed of and its subsequent use in agriculture can be limited.

Insecticides have been used extensively for fly control, and in many areas these remain the principal method of tackling the problem. However, the widespread use of insecticides has led to resistance against many classes of compound by some species (especially *Musca domestica*).

Several new approaches to fly control have been investigated, including biological control with predators (e.g. acarines), parasitoids of eggs and pupae, pheromone lures in conjunction with insecticides or chemosterilants, and microbial insecticides (*Bacillus thuringiensis*, serotype 1).

A succinct outline of control methods for *Musca* is given by Keiding (1986).

REFERENCES

Chillcott, J. G. 1961. A revision of the Nearctic species of Fanniinae (Diptera: Muscidae). *Canadian Entomologist* (Supplement) **14**: 1–295.

Dear, J. P. 1980. Family Sarcophagidae. Pp. 801–818 in Crosskey, R. W. (ed.), *Catalogue of the Diptera of the Afrotropical region*. 1437 pp. British Museum (Natural History), London.

Feacham, R. G., Bradley, D. J., Garelick, H. and **Duncan Mara, D.** (eds) 1983. *Sanitation and disease: health aspects of excreta and waste water management*. xxvii + 501 pp. John Wiley, Chichester.

Fonseca, E. C. M. d'A. 1968. Diptera Cyclorrhapha Calyptrata, Section (*b*) Muscidae. *Handbooks for the Identification of British Insects* **10** (4) (b): 1–119.

Greenberg, B. 1971. *Flies and disease*: vol. 1, *Ecology, classification and biotic associations*. viii + 856 pp. Princeton University Press, Princeton, New Jersey.

Greenberg, B. 1973. *Flies and disease*: vol. 2, *Biology and disease transmission*. x + 447 pp. Princeton University Press, Princeton, New Jersey.

Guimarães, J. H., Papavero, N. and **Pires do Prado, A.** 1983. As miiases na região neotropical (identificação, biologia, bibliografia). *Revista Brasileira de Zoologia* **1**: 239–416.

Hall, D. G. 1948. *The blowflies of North America*. 477 pp. Thomas Say Foundation, College Park, Maryland.

Hennig, W. 1955–1964. Muscidae. In Lindner, E. (ed.), *Die Fliegen der palaearktischen Region* **63b**: 1–1110. [In German: includes Fanniidae, as Fanniinae, on pp. 8–99.]

Huckett, H. C. and **Vockeroth, J. R.** 1987. Muscidae. Pp. 1115–1131 in McAlpine, J. F. (ed.), *Manual of Nearctic Diptera*: vol. 2, vi + pp. 675–1332. Research Branch, Agriculture Canada (Monograph No. 28).

Keiding, J. 1986. The house-fly – biology and control. Mimeographed document WHO/VBC/86.937, 63 pp. World Health Organization, Geneva.

Kurahashi, H. 1987. The blow flies of New Guinea, Bismarck Archipelago and Bougainville Island (Diptera: Calliphoridae). *Occasional Publications of the Entomological Society of Japan* **1**: 1–99.

Lindsay, D. R., Stewart, W. H. and **Watt, J.** 1953. Effect of fly control on diarrheal disease in an area of moderate morbidity. *Public Health Reports*, Washington, D.C. **68**: 361–367.

Lopes, H. de S. 1959. A revision of Australian Sarcophagidae (Diptera). *Studia Entomologica* (n. s.) **2**: 33–67.

McAlpine, J. F. 1981a. Morphology and terminology – adults. Pp. 9–63 in McAlpine, J. F., Peterson, B. V., Shewell, G. E., Teskey, H. J., Vockeroth, J. R. and Wood, D. M. (eds), *Manual of Nearctic Diptera*: vol. 1, vi + 674 pp. Research Branch, Agriculture Canada (Monograph No. 27).

McAlpine, J. F. 1981b. Key to families – adults. Pp. 89–124 in McAlpine, J. F., Peterson, B. V., Shewell, G. E., Teskey, H. J., Vockeroth, J. R. and Wood, D. M. (eds), *Manual of Nearctic Diptera*: vol. 1, vi + 674 pp. Research Branch, Agriculture Canada (Monograph No. 27).

McAlpine, J. F. (ed.) 1987. *Manual of Nearctic Diptera*: vol. 2, vi + pp. 675–1332. Research Branch, Agriculture Canada (Monograph No. 28).

McAlpine, J. F. 1989. Phylogeny and classification of the Muscomorpha. Pp. 1397–1518 in McAlpine, J. F. (ed.), *Manual of Nearctic Diptera*: vol. 3, vi + pp. 1333–1581. Research Branch, Agriculture Canada (Monograph No. 32).

O'Hara, J. E. and Kennedy, M. J. 1991. Development of the nematode eyeworm, *Thelazia skrjabini* (Nematoda: Thelazioidea), in experimentally infected face flies, *Musca autumnalis* (Diptera: Muscidae). *Journal of Parasitology* **77**: 417–425.

Paterson, H. E. 1964. Population genetic studies in areas of overlap of two subspecies of *Musca domestica* L. *Monographiae Biologicae* **14**: 244–254.

Paterson, H. E. 1974. The *Musca domestica* complex in Sri Lanka. *Journal of Entomology* (B) **43**: 247–259.

Paterson, H. E. and Norris, K. R. 1970. The *Musca sorbens* complex: the relative status of the Australian and two African populations. *Australian Journal of Zoology* **18**: 231–245.

Pittaway, A. R. 1991. *Arthropods of medical and veterinary importance: a checklist of preferred names and allied terms*. 178 pp. C. A. B. International, Wallingford.

Pont, A. C. 1977. A revision of Australian Fanniidae (Diptera: Calyptrata). *Australian Journal of Zoology* (Supplementary Series) **51**: 1–60.

Pont, A. C. 1980. Family Muscidae. Pp. 721–765 in Crosskey, R. W. (ed.), *Catalogue of the Diptera of the Afrotropical region*. 1437 pp. British Museum (Natural History), London.

Pont, A. C. 1989. Family Muscidae. Pp. 675–699 in Evenhuis, N. L. (ed.), *Catalog of the Diptera of the Australasian and Oceanian regions*. 1155 pp. Bishop Museum Press, Honolulu, and E. J. Brill, Leiden.

Pont, A. C. 1991. A review of the Fanniidae and Muscidae of the Arabian peninsula. *Fauna of Saudi Arabia* **12**: 312–365.

Reed, J. P. 1974. A revision of the Sarcophaginae of the Madagascan zoogeographical region, with a description of a new species (Diptera: Sarcophagidae). *Zeitschrift für angewandte Zoologie* **61**: 191–211.

Senior White, R., Aubertin, D. and Smart, J. 1940. *The fauna of British India, including the remainder of the Oriental region: Diptera* vol. 6, *Family Calliphoridae*. xiii + 288 pp. Taylor and Francis, London.

Shewell, G. E. 1987a. Calliphoridae. Pp. 1133–1145 in McAlpine, J. F. (ed.), *Manual of Nearctic Diptera*: vol. 2, vi + pp. 675–1332. Research Branch, Agriculture Canada (Monograph No. 28).

Shewell, G. E. 1987b. Sarcophagidae. Pp. 1159–1186 in McAlpine, J. F. (ed.), *Manual of Nearctic Diptera*: vol. 2, vi + pp. 675–1332. Research Branch, Agriculture Canada (Monograph No. 28).

Skidmore, P. 1985. *The biology of the Muscidae of the world*. [xii] + 550 pp. W. Junk Publishers, Dordrecht.

Smith, K. G. V. 1986. *A manual of forensic entomology*. 205 pp. British Museum (Natural History), London, and Cornell University Press, Ithaca, New York.

Spradbery, J. P. 1991. *A manual for the diagnosis of screw-worm fly*. i + 64 pp. CSIRO Division of Entomology, Canberra.

Strong, K. L. and Mahon, R. J. 1991. Genetic variation in the Old World screw-worm fly, *Chrysomya bezziana* (Diptera: Calliphoridae). *Bulletin of Entomological Research* **81**: 491–496.

Sulaiman, S., Sohadi, A. R., Yurms, H. and Iberahim, R. 1988. The role of some cyclorrhaphan flies as carriers of human helminths in Malaysia. *Medical and Veterinary Entomology* **5**: 72–80.

Tumrasvin, W. and Shinonaga, S. 1977. Studies on medically important flies in Thailand. III. Report of species biology to the genus *Musca* Linne, including the taxonomic key (Diptera: Muscidae). *Bulletin of Tokyo Medical and Dental University* **24**: 209–218.

Van Emden, F. I. 1941. Keys to the Muscidae of the Ethiopian region: Scatophaginae, Anthomyiinae, Lispinae, Fanniinae. *Bulletin of Entomological Research* **32**: 251–275.

Van Emden, F. I. 1965. *The fauna of India and adjacent countries: Diptera* vol. 7, *Muscidae*, Part 1. xiv + 647 pp. Government of India, Delhi.

Verves, Yu. G. 1985. Sarcophaginae [part]. Pp. 297–440 in Lindner, E. (ed.), *Die Fliegen der palaearktischen Region*, **64h** [part]. [In German: coverage of *Wohlfahrtia* on pp. 303–341.]

West, L. S. 1951. *The housefly: its natural history, medical importance and control.* xi + 584 pp. Comstock Publishing Company and Cornell University Press, Ithaca, New York.

West, L. S. and **Peters, O. B.** 1973. *An annotated bibliography of Musca domestica Linnaeus.* xiii + 743 pp. Dawsons, Folkestone and London, and Northern Michigan University, Marquette.

Wood, D. M. 1987. Oestridae. Pp. 1147–1158 in McAlpine, J. F. (ed.), *Manual of Nearctic Diptera*: vol. 2, vi + pp. 675–1332. Research Branch, Agriculture Canada (Monograph No. 28).

Zumpt, F. 1956a. Calliphoridae (Diptera: Cyclorrhapha). Part I: Calliphorini and Chrysomyiini. *Exploration du Parc National Albert: Mission G. F. de Witte (1933–1935)* **87**: 1–200.

Zumpt, F. 1956b. Calliphoridae. In Lindner, E. (ed.), *Die Fliegen der palaearktischen Region* **63i**: 1–140.

Zumpt, F. 1962. The Calliphoridae of the Madagascan region (Diptera). I. Calliphorinae. *Verhandlungen der Naturforschenden Gesellschaft in Basel* **73**: 41–100.

Zumpt, F. 1965. *Myiasis in man and animals in the Old World: a textbook for physicians, veterinarians and zoologists.* xv + 267 pp. Butterworths, London.

Zumpt, F. 1972. Calliphoridae (Diptera Cyclorrapha [sic]). Part IV. Sarcophaginae. *Exploration du Parc National des Virunga: Mission G. F. de Witte (1933–1935)* **101**: 1–264.

CHAPTER TWELVE

Diptera causing myiasis in man

Martin J. R. Hall and Kenneth G. V. Smith

Myiasis is the infestation of live human and vertebrate animals with dipterous larvae, which, at least for a certain period, feed on the host's dead or living tissue, liquid body-substances or ingested food (Zumpt, 1965). The different forms of myiasis have been classified in two ways. Firstly, in clinical terms, based upon the part of the host's body that is infested, and secondly, in parasitological terms, according to the type of host–parasite relationship (Patton, 1922a). The first classification can provide a convenient short-cut to identification of the fly species concerned for practical diagnosis, but the second gives a better understanding of the biology of the fly as a guide to treatment or prevention as well as providing information on the evolution of the habit. A clinical classification of myiasis based on the parts of the host affected, but which also takes account of the parasite–host relationship, is given in Table 12.1.

In a parasitological classification, there are two main groups of myiasis-producing species: obligatory parasites, which must develop on live hosts; and facultative parasites, which usually develop on decaying organic matter, such as carrion, faeces and rotting vegetation, but occasionally deposit their eggs or larvae on live hosts. The facultative species can be further split into primary, secondary and tertiary depending on whether they are able to initiate myiasis (primary) or only occur after other species have initiated it (secondary and tertiary) (Zumpt, 1965). A third group of species can cause accidental myiases when their eggs or larvae are ingested by the host. Zumpt (1965) termed these pseudomyiases.

Alternative or more refined classifications of myiasis, with detailed identification keys and further accounts of pathogenesis, are given by James (1947), Lane (1987), Leclercq (1969), Smith (1989) and Zumpt (1965). Guimarães et al. (1983) provide an extensive bibliography concerning the myiasis-producing flies of the Neotropics, with data on their identification, biology and control. Smith (1986b) provides keys to larvae specifically associated with forensic cases. Works useful for the identification of the larvae of Diptera in general are Ferrar (1987), Foote (1991), Smith (1989), Teskey (1981b) and, for the Far East, Ishijima (1967). Myiasis has had a much greater impact as an infestation of animals than of humans; there are no species of Diptera which are restricted to humans for their development. Other than *Dermatobia hominis*, the most important agents of myiasis able to complete their larval development in human tissues belong to the families Calliphoridae and Sarcophagidae. These species are most often found in cutaneous tissues or the head

Medical Insects and Arachnids Edited by Richard P. Lane and Roger W. Crosskey.
Published in 1993 by Chapman & Hall ISBN 0 412 40000 6

430 Diptera causing myiasis in man

Table 12.1 Classification of myiasis

The families and genera concerned are listed alphabetically and obligate and facultative relationships shown by the suffixes '(o)' and '(f)' respectively.

Group and subgroup	*Nature of infestation*	*Genera/families concerned*
Cutaneous myiasis		
Bloodsucking or sanguinivorous myiasis	Larvae attach to skin and suck blood or bite	*Auchmeromyia* (Calliphoridae) (o), Tabanidae (f), Therevidae (f)
Furuncular myiasis	Larvae penetrate skin and make boil-like swellings	*Cordylobia* (Calliphoridae) (o), *Dermatobia* (Oestridae) (o), *Wohlfahrtia* (Sarcophagidae) (o)
Creeping myiasis	Larvae tunnel in human epidermis, but do not complete development in man	Oestridae (Hypodermatinae and Gasterophilinae) (o)
Wound or traumatic myiasis	Larvae develop in wounds or lesions	Calliphoridae (o/f), Fanniidae (f), Muscidae (f), Phoridae (f), Sarcophagidae (o/f)
Body cavity myiasis		
Nasopharyngeal, auricular, lung and ophthalmomyiases	Eggs or larvae deposited in ear, eye, nose, sinuses and pharyngeal cavities	Calliphoridae (o/f), Muscidae (f), Oestridae (all four subfamilies) (o), Phoridae (f), Sarcophagidae (o/f)
Accidental myiasis		
Intestinal myiasis (enteric, rectal)	Larvae accidentally ingested or enter via rectum	Anisopodidae, Calliphoridae, Drosophilidae, Ephydridae, Fanniidae, Micropezidae, Muscidae, Phoridae, Piophilidae, Psychodidae, Sarcophagidae, Sepsidae, Stratiomyidae, Syrphidae, Therevidae, Tipulidae (all facultative)
Urogenital myiasis	Adults attracted to infected tissue or soiled clothing	Anisopodidae, Calliphoridae, Fanniidae, Muscidae, Sarcophagidae, Scenopinidae (all facultative)

cavities. However, many obligate myiasis species that are important pests of livestock, developing in the back (*Hypoderma*), head sinuses and nasal passages (*Oestrus*) or mouth and digestive tract (*Gasterophilus*), can also invade human tissues, but rarely develop beyond the first larval stage.

RECOGNITION AND ELEMENTS OF STRUCTURE

The pest stages of myiasis-causing species are the larvae; in contrast, the adults are generally harmless. This chapter will, therefore, concentrate on the larval stages, in particular on the mature third-stage larva. The larval stage is that most often encountered in cases of myiasis because the egg stage is of short duration or absent (e.g. Sarcophagidae and *Oestrus*), causes no damage to the host, and is therefore usually not detected. There are usually three larval stages in those families involved in myiasis (up to nine in Tabanidae) and, generally speaking, the first two stages are difficult to identify to species except by association with mature larvae or subsequently reared adults. Therefore, the keys in this chapter concentrate on third-stage larvae.

Larvae of the principal agents of myiasis of medical importance are of the typical calliphorid maggot form, with a pointed anterior and truncate posterior end. The main external features are shown in Fig. 12.1. The body of the larva comprises 12 segments: a small head segment, incompletely divided from a prothoracic segment, followed by a mesothoracic, a metathoracic and eight abdominal segments. There is no clear distinction between the thorax and the abdomen. The head segment is divided by a ventral furrow into left and right cephalic lobes, with the mouth opening at the base of the furrow. The mouthparts consist of a pair of mouth-hooks and associated sclerites, for the attachment of muscles, collectively known as the cephalopharyngeal skeleton (Fig. 12.1b,c). The head bears two pairs of peg-like sensory organs. Larvae do not have true segmental appendages and are thus technically legless (apodous). However, many have swellings to assist in locomotion, most noticeable in *Chrysomya albiceps*, *C. rufifacies*, *Fannia* and *Megaselia* (Figs 12.8c, 12.10 and 12.3e).

Respiration is achieved through spiracles which are simple openings connecting the outside air to the internal tracheal network. There is a pair of anterior spiracles on the prothoracic segment and a pair of caudal or posterior spiracles on the 12th, i.e. terminal, segment. Both provide useful taxonomic characters. The anterior spiracles of calliphorids and sarcophagids protrude through the body wall and divide into a fan-shaped series of finger-like lobes, each ending in a small aperture (Fig. 12.1a). The anterior spiracles are not visible in first-stage larvae.

The posterior spiracles usually comprise a pair of sclerotized plates, either set flat on the body cuticle of the last abdominal segment or raised on a process (Figs 12.1d,e and 12.2c). In the Sarcophagidae, they are set at the bottom of a deep cavity (Fig. 12.8b). The sides of this cavity can close to seal off the spiracles and prevent their contamination when the larva submerges in noxious media. The outer rim of the spiracular plate is more heavily sclerotized than the rest of the plate and is known as the peritreme. It can form a complete circle (ring) or be incomplete. On the portion of the plate towards the midline of second-stage and third-stage larvae is a structure called the button (Fig. 12.1e), which may or may not be clearly visible. This is the scar left from the spiracle of the previous stage after moulting (Fig. 12.1f). There are slits or pits on the surface of the spiracular plates for gaseous exchange. The posterior segment bears the anus surrounded by the anal plate, the cuticle of which is thinner than the rest of the body.

More detailed discussion of the morphology of larvae can be found in Ferrar (1987), Foote (1991) and Teskey (1981a).

REVIEW OF DIPTEROUS FAMILIES ASSOCIATED WITH HUMAN MYIASIS

The following review of the Diptera causing human myiasis is arranged by family, beginning with a key for family identification of larvae and a key to principal genera other than those of Oestridae. For the bots and warbles, the classification of

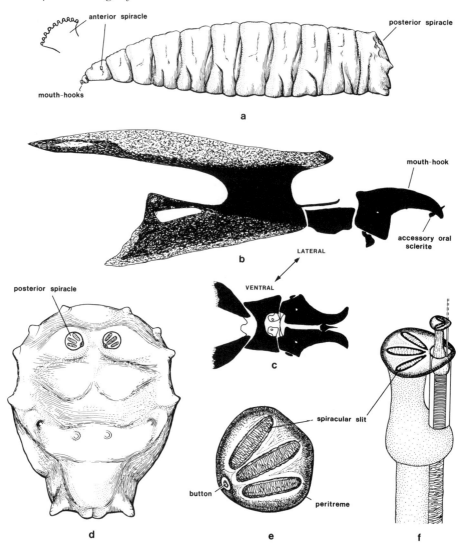

Figure 12.1 Elements of structure of a typical calyptrate fly maggot, illustrated from *Calliphora vicina* in the third (last) larval stage: (*a*) lateral view, showing the position of mouth-hooks and spiracles; (*b* and *c*) cephalopharyngeal skeleton, showing the mouth-hooks and accessory oral sclerite in lateral and ventral views; (*d*) end view, showing the posterior spiracles on the face of the last abdominal segment; (*e*) posterior spiracles, showing the features important in maggot identification; (*f*) schematic figure of the moulting process of the posterior spiracle during metamorphosis from the second-stage larva (two slits) to the third-stage larva (three slits).

Wood (1987) is used in which the single family Oestridae is recognized, containing four subfamilies previously ranked as families. Under each family heading information is given on biology and medical importance in addition to taxonomic information.

Family key for identification of dipterous larvae causing human myiasis

1 Larvae with an obvious head capsule (though this may be withdrawn into the thorax and become visible only under moderate pressure) 2

— Larvae without an obvious head capsule; mouth-hooks sometimes prominent but not enclosed in a definite head capsule.. 8
2 Head capsule distinct, larger than first thoracic segment, not retracted into thorax.. 3
— Head capsule small, partly retracted into thorax.. 7
3 Mandibles in vertical plane, parallel to each other, and often visible as a pair of hooks.. 4
— Mandibles in horizontal plane, opposed, so that their tips can be brought together... 6
4 Body surface smooth, white and elongate larva (giving a worm-like appearance) with a small, distinct head capsule, and behind it and visible through the cuticle, an internal rod; about 20 apparent segments due to subdivisions 5
— Body surface shagreened, i.e. roughened like shark-skin, body broad and flattened (Fig. 3.9c). [Occasional cause of enteric pseudomyiasis] .... STRATIOMYIDAE
5 Internal rod ending in a spatulate tip (Fig. 12.2a) [Normally in soil, especially sandy soil, and able to bite if handled] .... THEREVIDAE
— Internal rod pointed, not at all spatulate (Fig. 12.2b). [Occasionally involved in urogenital myiasis].. SCENOPINIDAE
6 Posterior spiracles surrounded by five anal lobes (Fig. 12.2e) [terminal segment ventrally with perianal shield consisting of thickened hypodermal cells and a sinuate lateral margin (Fig. 12.2d)]. [Occasionally involved in urogenital myiasis] .... ANISOPODIDAE
— Posterior spiracles without anal lobes and usually situated on a short siphon (Fig. 12.2c). [Occasionally involved in intestinal, urogenital and nasopharyngeal myiasis] .... PSYCHODIDAE
7 Mandibles oriented in horizontal plane, opposed so that their tips can be brought together. Posterior spiracles usually surrounded by a crown of finger-like processes (normally six) (Fig. 12.2f,g). [Occasionally involved in intestinal myiasis].. TIPULIDAE
— Mandibles oriented in vertical plane, parallel to each other. Posterior spiracles not surrounded by a crown of finger-like processes (Fig. 12.2h). [Occasional biters in paddyfields, see Chapter 8].. TABANIDAE
8 Posterior spiracles situated on a single breathing tube (sometimes bifurcated at end) 9
— Posterior spiracles on posterior face of terminal segment (sometimes each on a short stalk, but not on the same breathing tube) 10
9 Posterior spiracles close together or fused medially on a single tube which ranges in length from a short prominence (Fig. 12.3c) to a long retractile siphon (Fig. 12.3a). [Occasionally involved in intestinal myiasis] .... SYRPHIDAE
— Posterior spiracles widely separated at each end of a bifurcated tube (Fig. 12.3d). [Occasionally involved in intestinal myiasis]........ EPHYDRIDAE (*Teichomyza*)
10 Body with fleshy, tuberculate or spinous processes dorsally and laterally............ 11
— Body without such processes (though the integument itself sometimes with strong spines)... 13
11 Body more or less cylindrical and without long filiform processes but with shorter unbranched lateral and dorsal tubercles on segments 12

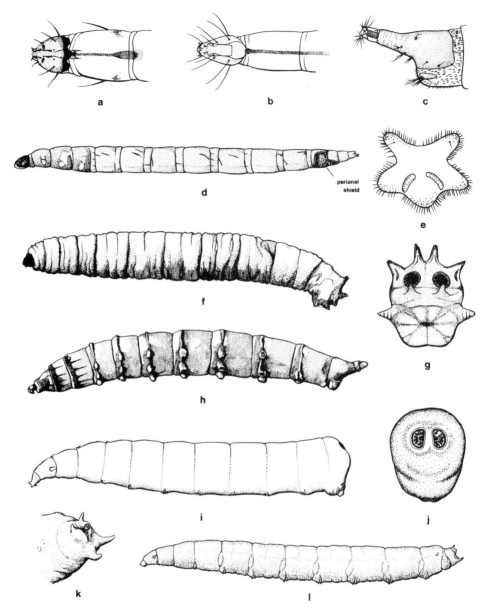

Figure 12.2 Dipterous larvae of some families of minor importance in myiasis: (*a*) Therevidae – head capsule of *Thereva* larva, dorsal view; (*b*) Scenopinidae – head capsule of *Scenopinus* larva, dorsal view; (*c*) Psychodidae – respiratory siphon of *Psychoda* larva, lateral view; (*d* and *e*) Anisopodidae – larva of *Sylvicola*, lateral view and end view of the last segment; (*f* and *g*) Tipulidae – larva of *Tipula*, lateral view and end view of the last segment; (*h*) Tabanidae – larva of *Tabanus*, lateral view; (*i* and *j*) Muscidae – larva of *Musca domestica*, lateral view and end view of the last segment; (*k* and *l*) Piophilidae – larva of a typical piophilid, showing the characteristic terminal segment in oblique posterior view and whole maggot in lateral view. All larvae in the last developmental stage.

— Body flattened and with filiform processes, these branched at least basally and appearing feathery on dorsal surface and sides of segments (Fig. 12.10). Posterior spiracles situated on stalks, each stalk with four lobes bearing three slits and button [Frequent in urogenital and intestinal myiasis] ....
.... FANNIIDAE*

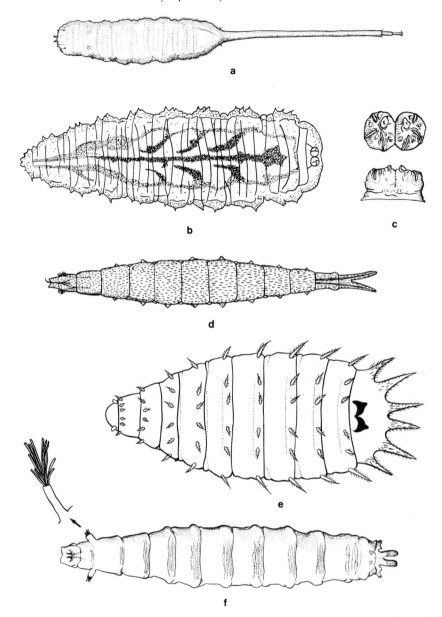

Figure 12.3 Dipterous larvae of some families of minor importance in myiasis: (*a*) Syrphidae – larva of *Eristalis*, dorsal view; (*b* and *c*) Syrphidae – larva of *Syrphus ribesii*, dorsal view of whole maggot and posterior spiracles in rear (upper) and dorsal (lower) view; (*d*) Ephydridae – larva of *Teichomyza fusca*, dorsal view; (*e*) Phoridae – typical larva of *Megaselia*, dorsal view; (*f*) Drosophilidae – typical larva of *Drosophila*, ventral view with enlarged detail of the anterior spiracles. All larvae in the last developmental stage.

12 Small larvae (up to 4 mm long) slightly flattened and with short processes on dorsal and lateral surfaces (Fig. 12.3*e*). Posterior spiracles on brown sclerotized tubercles, each with four slits arranged in facing pairs. [Occasionally in intestinal, urogenital and (recorded once) lung myiasis] .... PHORIDAE
— Larger larvae (up to 18 mm long), more nearly cylindrical and with large pointed fleshy processes laterally and dorsally (Fig. 12.8*c*). Posterior spiracles

in cleft on posterior face of terminal segment, consisting of flattened plates perforated by three slits. [Typically carrion feeders, but occurring in secondary myiasis in sheep and perhaps in wound myiasis in man]..................................
.... CALLIPHORIDAE (part)* (*Chrysomya albiceps* and *C. rufifacies*)
13 Posterior spiracles with a large number of small pores (Fig. 12.4*e*) or many short intertwined serpentine slits arranged in three groups on each spiracular plate (Fig. 12.4*f*) 14
— Posterior spiracles with up to three straight, curved or sinuous slits 16
14 Mouth-hooks well developed, strongly hooked (Fig. 12.4*d*) 15
— Mouth-hooks rudimentary. [Subcutaneous parasites of cattle and deer, occasionally causing creeping myiasis in man and malignant ophthalmomyiasis in first-stages].. OESTRIDAE (HYPODERMATINAE)
15 Body with weak spines restricted to limited areas. Posterior spiracles with many small pores (Fig. 12.4*e*). [Cause of nasal myiasis in sheep and occasionally benign ophthalmomyiasis in man] .... OESTRIDAE (OESTRINAE)
— Body with spines stronger and more evenly distributed. Posterior spiracles with many small serpentine slits (Fig. 12.4*f*). [Cause of dermal myiasis in rodents and man] .... OESTRIDAE (CUTEREBRINAE)
16 Posterior spiracles with three slits. [Third-stage larvae]...................................... 17
— Posterior spiracles with one or two slits. [First-stage or second-stage larvae exit here: they are not easily identified except by association with third-stage larva or adult or host] .... see under most likely family
17 Posterior spiracles with straight or arcuate slits 18
— Posterior spiracles with strongly sinuous slits (Fig. 12.11). [Involved in cutaneous, intestinal and urogenital myiasis] .... MUSCIDAE (part)*
18 Posterior spiracles not situated on fleshy processes, at most on inner, sloping surfaces of broad, fleshy, partly retractile mountings (Fig. 12.2*k*). Anterior spiracles terminating in short finger-like rays (not in *Dermatobia*) 19
— Posterior spiracles situated on short fleshy processes. Anterior spiracles prominent and terminating in long filamentous processes (Fig. 12.3*f*). [Occasional in intestinal myiasis] .... DROSOPHILIDAE
19 Anal segment without backwardly projecting processes, at most with a rosette of fleshy protuberances (e.g. Figs 12.1*d* and 12.8*d*) 20
— Anal segment with short fleshy, backwardly pointing processes (Fig. 12.2*k*,*l*). [Live larva can 'skip']. [Occasional in intestinal myiasis].......... PIOPHILIDAE
20 Posterior spiracles sunk in a deep cavity, which closes over and conceals them; slits more or less vertical and parallel 21
— Posterior spiracles visible, either exposed on surface, or set in a ring of tubercles; slits not vertical 23
21 Body with strong spines, grub-like (Figs 12.4*a*, 12.6*h*).. 22
— Body with short spinules and maggot-like (Fig. 12.8*a*) .... SARCOPHAGIDAE*
22 Posterior spiracles with slits bowed outwards at the middle (Fig.12.6*i*). Body ovate (Fig.12.6*h*) .... OESTRIDAE (GASTEROPHILINAE)
— Posterior spiracles with slits relatively straight. Body enlarged anteriorly and tapering posteriorly (Fig. 12.4*a*)......................OESTRIDAE (*Dermatobia hominis*)
23 Posterior spiracles with straight slits, all or part surrounded by a dark peritreme (Fig. 12.9*a*,*c*). [Occurring in dermal, subdermal, nasopharyngeal, intestinal and urogenital myiasis].. CALLIPHORIDAE (part)*

— Posterior spiracles with distinctly arcuate or distally bent slits; sometimes lacking a distinct, dark peritreme (Fig. 12.11). [Occasionally involved in intestinal myiasis] .... MUSCIDAE (part)*

Generic key to third-stage larvae of Calliphoridae, Fanniidae, Muscidae and Sarcophagidae causing myiasis in man

1 Body with obvious fleshy processes on dorsal and lateral surfaces (Fig. 12.8c) 2
— Body smooth or with short spines but without fleshy processes (Fig. 12.8a) 3
2 Posterior spiracles on short stalks, situated dorsally near anterior margin of terminal segment; stalks terminating in three lobes, each bearing a spiracular slit (Fig. 12.10) .... **Fannia**
— Posterior spiracles flush with posterior face of terminal segment (Fig. 12.8d) .... **Chrysomya** (part, 'hairy maggots')
3 Larva grub-like, i.e. fleshy and rounded at both ends 4
— Larva maggot-like, i.e. pointed at head end and truncate posteriorly 5
4 Cuticle without obvious spines. Anterior spiracles not projecting. Posterior spiracles very widely separated and with three short, straight, slits (Fig. 12.7a,b) .... **Auchmeromyia**
— Cuticle with obvious spines. Anterior spiracles in form of membranous stalks bearing finger-like processes. Posterior spiracles not widely separated and with serpentine slits (Fig. 12.7c,d,e) .... **Cordylobia**
5 Posterior spiracles with fully closed peritremal ring, sometimes more weakly near the button (Fig. 12.9c) 6
— Posterior spiracles with open peritremal ring (Fig. 12.9a) 8
6 Posterior spiracular slits straight and more or less parallel to each other (Fig. 12.9c) 7
— Posterior spiracular slits touching or very close together and strongly sinuous (Fig. 12.11a) .... **Musca domestica**
7 Cephalopharyngeal skeleton with pigmented accessory oral sclerite (Fig. 12.1b) . ... **Calliphora**
— Cephalopharyngeal skeleton without pigmented accessory oral sclerite .... **Lucilia**
8 Posterior spiracles not recessed in a cavity (Fig. 12.1d) 9
— Posterior spiracles recessed in a deep cavity and concealed when this is closed (Fig. 12.8b) .... **Sarcophaga** and **Wohlfahrtia**
9 Tracheal trunks from posterior spiracles without dark pigmentation 10
— Tracheal trunks from posterior spiracles with conspicuous dark pigmentation extending forwards as far as tenth or ninth segment (Fig. 12.9d: dissection may be necessary in a preserved specimen) .... **Cochliomyia hominivorax**
10 Posterior margin of segment 11 *with* dorsal spines 11
— Posterior margin of segment 11 *without* dorsal spines .... **Cochliomyia macellaria**
11 Posterior spiracles with distinct button (Fig. 12.9b). [Holarctic, almost entirely to north of Tropic of Cancer] 12

* For further identification see the following key.

— Posterior spiracles with indistinct button (Fig. 12.9a). [Tropics and subtropics of Old and New World] .... **Chrysomya** (part)
12 Posterior margin of segment 10 with dorsal spines. Length of larger tubercules on upper margin of posterior face of terminal segment distinctly greater than half width of a posterior spiracle (Fig. 12.9e) .... **Protophormia terraenovae**
— Posterior margin of segment 10 without dorsal spines. Length of larger tubercules on upper margin of posterior face of terminal segment less than half width of a posterior spiracle... **Phormia**

Oestridae

The Oestridae form a large and diverse family in which all species are obligate parasites of domestic animals or wildlife in their larval stages. Some cause serious pathological effects if they accidently invade human tissues. There are four subfamilies: Cuterebrinae, Oestrinae, Gasterophilinae and Hypodermatinae (Wood, 1987).

Cuterebrinae
Larvae of Cuterebrinae are mainly parasites of rodents and lagomorphs (rabbits, hares), occurring within subdermal cysts. Some species occur in monkeys, man and other mammals including domestic livestock. The subfamily is restricted to the New World and contains some 70 species in six genera. Studies of the biology of Cuterebrinae have been reviewed by Catts (1982).

Dermatobia hominis (tórsalo or human bot-fly) parasitizes man, cattle, dogs and a number of other wild and domestic mammals and birds. It occurs throughout Central and South America, where in many parts it is a very serious pest of cattle. There is a large literature on this species for which a partial bibliography is provided by Guimarães and Papavero (1966).

The female adopts a unique method of ensuring transportation of her eggs to the host. Packets of eggs, enclosed in 'cement', are deposited on day-flying mosquitoes such as *Psorophora*, on other flies (*Sarcophaga, Musca, Stomoxys*, see Fig. 12.12a) or even ticks (*Amblyomma*). Eggs can be laid directly onto insects in the vicinity of the host or on nearby vegetation if no suitable insects are available. This 'hitch-hiking' technique is known as phoresis. The warmth of the host induces the larva to hatch from the egg, after which, within five to ten minutes, it penetrates the host's skin. At the site of penetration a small nodule of host tissue develops around each larva, with a central breathing pore. Larvae have a distinctive shape, with attenuation of the posterior end (Fig. 12.4a), which renders them difficult to remove by manual pressure. This is most pronounced in the second stage (Fig. 12.13). The larva feeds for six to 12 weeks in man, then when mature emerges and drops to the ground to pupate (Lane et al., 1987).

The cutaneous swellings containing the larvae are itchy and can be periodically painful, but unless infected are usually of relatively short duration, though the exudations from the wound may acquire a fetid odour and be troublesome, soiling bedding and clothing. Infestation usually occurs on the head, though other parts of the body can be infested. Should larvae enter the eye a serious ophthalmomyiasis can result in the loss of an eye, or, rarely (in children) fatal brain damage.

Cases of *D. hominis* myiasis can be detected well beyond the natural range of the species in travellers who have recently visited endemic areas.

Cuterebra larvae parasitize rodents, rabbits, hares, monkeys and, occasionally, man (Baird et al., 1989). The immature stages and life history are not well known but it appears that eggs are probably laid on vegetation, sticks or other objects in the habitat occupied by potential hosts. Most infestations probably occur through the mucous membranes of the eye, nose, mouth or anal area, though one case of subcutaneous myiasis in the axilla of a boy has been reported.

Oestrinae
Oestrinae larvae develop in the head cavities (nasal or pharyngeal) of several species of mammals (Marsupialia, Proboscidea, Artiodactyla and Perissodactyla). The subfamily contains 34 species in nine genera. Members are found in all zoogeographical regions and there are some cosmopolitan species; most species occur in the Afrotropical and Palaearctic regions. Good general accounts of the family are given by Zumpt (1965, Old World), Grunin (1966, Palaearctic), Wood (1987, Nearctic) and Papavero (1977, World).

Females of *Oestrus ovis* deposit live first-stage larvae (Fig. 12.4c) into the nostrils of their hosts. The latter include several species of domestic animals, e.g. sheep, horse, donkey, goat, camel and 'game' animals. Rarely, larvae are deposited in the eye, mouth, nostrils or outer ear of man, especially people associated with sheep and goats. Typically the patient reports being struck in the eye by a flying insect or object. Up to 50 larvae have been removed from the conjunctival sac of a single patient. Fortunately (unlike *Hypoderma*, see below) larvae do not survive beyond the first-stage in man, but they can cause a more or less painful inflammation (usually diagnosed as acute catarrhal conjunctivitis) for as long as 10 days. Generally, the eyeball is not invaded and the course of the disease is benign, but respiratory distress is sometimes noticed (Mazzeo et al., 1987). Rarely, intra-ocular penetration can occur, causing more severe symptoms and effects (Rakusin, 1970). The infection is known as 'thimni' in Algeria and as 'tamné' in the Ahaggar Mountains of the central Sahara (Sergent, 1952). The larvae are not easy to see because of their small size and transparency but the blackish mouthparts are diagnostic (Fig. 12.4d). Comprehensive papers on oestrid ophthalmomyiasis in man are provided by Pampiglione (1958), Krümmel and Brauns (1956) and Zumpt (1963). Sometimes the larvae reach the nasal cavities causing swelling, pain and a frontal headache (e.g. Quesada et al., 1990). If they reach the throat, inflammation and swelling can make swallowing difficult but these symptoms soon pass.

In addition to *Oestrus ovis* other Oestrinae have been recorded attacking man, most commonly *Gedoelstia* species and *Rhinoestrus purpureus* (reported attacking man in the same manner as *O. ovis*), but some records have been due to misidentifications of *O. ovis* (James, 1947). If an oestrid other than *O. ovis* is suspected of attacking man the keys and descriptions in Zumpt (1965) or Wood (1987) should be used for identification.

Hypodermatinae
Larvae of Hypodermatinae (warble-flies) are obligate parasites of mammals (Rodentia, Lagomorpha and Artiodactyla). There are 32 species (in 11 genera), 29 of which occur in the Palaearctic, Afrotropical and Nearctic regions.

Females can each glue some 300–800 eggs to the hairs of the host, either singly (*Hypoderma bovis*) or in batches of up to 15 (Fig. 12.12b). The persistence of the flies can produce a severe behavioural response in the host, 'gadding', in which it

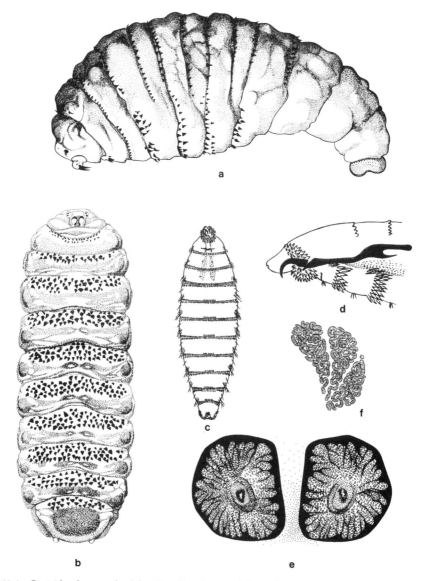

Figure 12.4 Oestridae larvae of subfamilies Oestrinae and Cuterebrinae: (*a*) *Dermatobia hominis*, third-stage larva in lateral view; (*b–e*) *Oestrus ovis*, third-stage larva in ventral view (*b*), first-stage larva in ventral view (*c*), mouthparts of first-stage larva in lateral view (*d*) and posterior spiracles of third-stage larva (*e*); (*f*) *Cuterebra emasculator*, posterior spiracle slits of third-stage larva (reproduced from Bennett, 1955).

gallops about in an effort to escape. The first-stage larvae penetrate the skin of the host and migrate through the body to particular sites under the skin where the second-stage and third-stage (Fig. 12.5*d*) larvae finally settle and form cysts (warbles).

Hypoderma bovis and *H. lineatum* generally parasitize bovines in the Northern Hemisphere but occasionally horse and man. *Hypoderma lineatum* has also been recorded from the yak (*Bos mutus*) in Central Asia and the bison (*Bison bison*) in North America.

More cases of human infestation by *H. lineatum* have been recorded than by *H. bovis*, and several reported cases of infestation of humans with larvae of *Hypoderma bovis* may in fact have been caused by confusion with *H. lineatum*: reliable differentiation of the two species is not easy, especially if the larvae have been damaged during extraction. Infestation of man probably occurs through handling cattle contaminated with recently emerged larvae as well as by direct oviposition. See for example James (1947), O'Rourke (1968) and Zumpt (1965).

Hypoderma lineatum and *H. bovis* have both been recorded as causing skin abscesses in man as in the normal hosts, but, in addition and more seriously, malignant ophthalmomyiasis can occur (e.g. Edwards et al., 1984) and even intracerebral myiasis (Kalelioğlu et al., 1989). Contamination could occur by wiping the hands across the eyes after handling cattle, or the flies might possibly deposit eggs on eyebrows or eyelashes. In addition to James (1947) and Zumpt (1965), cases of ophthalmomyiasis involving *Hypoderma* are reviewed by Krümmel and Brauns (1956).

The mouthparts of the first-stage larvae of *Hypoderma* are diagnostic and easily distinguishable from those of *Oestrus* by the toothed mandibles (Fig. 12.5a,b,c).

Other cases of hypodermatid myiasis include *H. diana* (deer bot-fly) (Fidler, 1987; James, 1947) and *Oedemagena tarandi* (reindeer warble-fly) which regularly causes ophthalmomyiasis in, for example, Norway (Kearney et al., 1991). *Oedemagena tarandi* is the only representative of its genus and its third-stage larva can be differentiated from those of *Hypoderma* by examination of the rows of spines at the anterior margins of the body segments: these are almost equally numerous on the dorsal and ventral surface in *Oedemagena*, but more numerous ventrally in *Hypoderma*.

The following key to *Hypoderma* larvae should be used cautiously for the early stages, and if possible Bishop et al. (1926), Grunin (1962) and Zumpt (1965) should be consulted. Other species occur in Cervidae, but are not well known and have not been recorded from man.

Key to larvae of *Hypoderma* species

1. Mouth-hooks well developed (Fig. 12.5a,b,c). Posterior spiracles each consisting of a simple circular opening. [Length 0.55–17 mm] .... First-stage larva 3
— Mouth-hooks rudimentary. Posterior spiracles with numerous openings 2
2. Posterior spiracles with 18–40 openings grouped in squarish or irregular masses and not surrounding a button. [Length 10–18 mm] .... Second-stage larva 5
— Posterior spiracles with numerous openings, kidney-shaped and partially surrounding the button (Fig. 12.5d,e). [Length 20–30 mm] .... Third-stage larva 7
3. Anterior end of mouth-hooks directed forward, blunt and posteriorly divided by a shallow incision into two blunt lobes (Fig. 12.5a) .... **H. diana**
— Anterior ends of mouth-hooks directed outwards and not divided into two lobes posteriorly 4
4. Mouth-hooks anteriorly divided into two blunt lobes, without a recurved pointed tooth (Fig. 12.5b) .... **H. bovis**
— Mouth-hooks anteriorly undivided, sharp, with a recurved pointed tooth some distance from the apex (Fig. 12.5c) .... **H. lineatum**
5. Body without patches of spinules on the dorsal surface 6
— Body with patches of spinules on the dorsal surface of some anterior segments... .... **H. diana**

442 *Diptera causing myiasis in man*

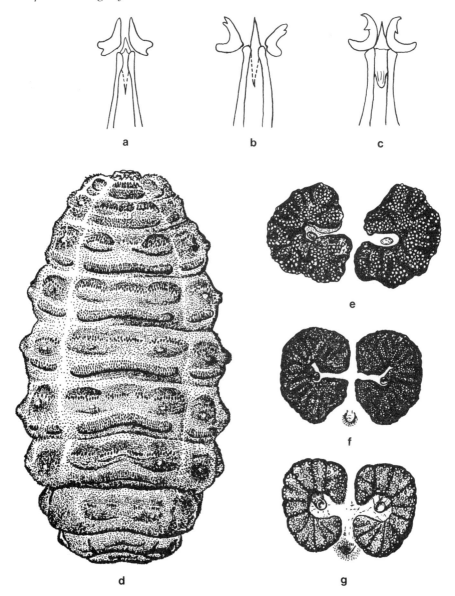

Figure 12.5 Oestridae larvae of subfamily Hypodermatinae (genus *Hypoderma*): (*a–c*) first-stage larval mouthparts in ventral view of *H. diana* (*a*), *H. bovis* (*b*) and *H. lineatum* (*c*); (*d*) third-stage larva of *H. bovis*, dorsal view; (*e–g*) posterior spiracles of third-stage larva of *H. diana* (*e*), *H. bovis* (*f*) and *H. lineatum* (*g*).

6   Posterior spiracles with stigmal plates orange or yellowish brown; stigmal rings or discs separated from each other, usually numbering 18–25 (range 12–40+) ....**H. lineatum**

—   Posterior spiracles with stigmal plates brown to black; stigmal rings or discs closely grouped together and fused, usually numbering 32–37 (range 19–50+) .... **H. bovis**

7   Posterior spiracular plate completely surrounded by tiny spines. [Hosts usually Bovidae] 8

— Posterior spiracular plate not surrounded by tiny spines or if some spines present then confined to small areas (Fig. 12.5e). [Hosts usually Cervidae] .... **H. diana**

8   Posterior spiracular plate with a narrow funnel-like channel (Fig. 12.5f) .... **H. bovis**

— Posterior spiracular plate with a much broader channel (Fig. 12.5g) .... **H. lineatum**

Gasterophilinae

Larvae of Gasterophilinae develop in the nasal sinuses and alimentary tracts of perissodactyl mammals (Equidae). There are some 18 species in five genera found mainly in the Palaearctic, Afrotropical and Oriental regions, though some *Gasterophilus* species have been accidentally introduced into other parts of the world.

Female flies deposit from 200 to 2500 eggs on skin and hairs of the host, or in the grass (Fig. 12.12c). Larvae enter the mouth either by their own movement or via the host's tongue during body licking or feeding. A good general account of their biology is provided by Zumpt (1965).

Occasionally man becomes infested with first-stage larvae of *G. haemorrhoidalis*, usually on the face and buttocks where they cause a 'creeping myiasis' in the skin (Grunin, 1955; James, 1947). Similar records to those of *G. haemorrhoidalis* involving *G. intestinalis* might in fact refer to other species of *Gasterophilus*. Flies have been observed ovipositing on humans, especially on the back of a hand holding a horse's mouth, and larvae of *G. pecorum* can be picked up by passing a moist hand over grass on which eggs had been laid. Larvae that penetrate the skin can cause unbearable itching and can burrow in the skin for at least 15 days. Experimental studies on first-stage larvae of *G. haemorrhoidalis*, *G. pecorum*, *G. nigricornis* and *G. inermis* have shown that larvae readily penetrate unbroken human skin in a few minutes but that those of *G. intestinalis* and *G. nasalis* do not do so, even after two hours. However, human cutaneous myiasis can be caused by *G. intestinalis* (Heath et al., 1968; Townsend et al., 1978). Clearly, critical identification is essential in reporting cases of myiasis involving these first-stage larvae.

The following key covers the larvae of those species likely to occur in man; other rarer and little-known species are treated by Zumpt (1965). Keys to the eggs of *Gasterophilus* species are given by Cogley (1991), James (1947) and Zumpt (1965).

Key to larvae of *Gasterophilus* species

1   Small (1 mm or less) larvae with posterior spiracles located terminally on a bifid protuberance (Fig. 12.6a,b,c) .... First-stage larva 2

— Larger (up to 20 mm) larvae with posterior spiracles opening on the terminal segment with two or three distinct slits .... Second-stage or third-stage larva 4

2   Body segments with anterior marginal spines in two rows and body with short inconspicuous hairs; second segment with hooks and spines equally prominent.. 3

— Body segments with anterior marginal spines in one closely set row and body with some long scattered hairs; second segment with very long ventral hooks (Fig. 12.6b) .... **G. nasalis**

3   Larva more elongated (Fig. 12.6c).. **G. intestinalis**
—   Larva less elongated (Fig. 12.6a)... **G. haemorrhoidalis**
4   Posterior spiracles with two slits... Second-stage larva 5
—   Posterior spiracles with three slits .... Third-stage larva 10
5   Body with zones of spines on most armed segments divided into two bands, each composed of two rows of spines and separated by a broad transverse bare space (Fig. 12.6d)... 6
—   Body with zones of spines not divided into two bands by a broad bare space (Fig. 12.6e).. 7
6   Posterior spiracles with slits in 10–13 transverse bands........................ **G. nasalis**
—   Posterior spiracles with slits arranged in four to six transverse bands .... **G. nigricornis**
7   Body slender, more cylindrical and with dorsal armature reaching tenth segment. Head segment with two isolated groups of denticles between antennal lobes and mouth-hooks... 8
—   Body relatively stout and conical, strongly broadened posteriorly (Fig. 12.6f) and with dorsal armature reaching eighth segment (rarely a few spines also on ninth segment). Head segment with denticles forming semicircular row between antennal lobes and mouth-hooks .... **G. pecorum**
8   Body without a spherical inflation on last two segments. Posterior spiracles with slits arranged in 16–20 transverse bands 9
—   Body with a spherical inflation on last two segments. Posterior spiracles with slits arranged in 12–14 transverse bands. .... **G. inermis**
9   Mouth-hooks uniformly curved dorsally (Fig. 12.6j). Body segments with spines of first row approximately twice as long as those of third row........................... **G. haemorrhoidalis**
—   Mouth-hooks not uniformly curved dorsally but with a shallow depression (Fig. 12.6h). Body segments with spines of first row approximately three times as long as those of third row.. **G. intestinalis**
10 Body segments with spines on ventral surface in one row................................... 11
—   Body segments with spines on ventral surface in two rows................................ 12
11 Body with first three segments more or less conical, each with sharp step-like constriction at hind margin; third segment always with a dorsal row of spines, and sometimes with variously developed ventral spines.................. **G. nasalis**
—   Body with first three segments more or less cylindrical, showing sharp constrictions posteriorly; third segment without spines dorsally or ventrally... **G. nigricornis**
12 Head segment with only lateral groups of denticles. Dorsal row of spines on the eighth segment not broadly interrupted medially; at least tenth segment with dorsal spines 13
—   Head segment with two lateral groups of denticles and one central group, the latter situated between the antennal lobes and the mouth-hooks (Fig. 12.6g). Dorsal rows of spines broadly interrupted medially on the seventh and eighth segments, tenth and eleventh segment without spines .... **G. pecorum**
13 Mouth-hooks uniformly curved dorsally (Fig. 12.6j). Body spines sharply pointed... 14
—   Mouth-hooks not uniformly curved dorsally, but with a shallow depression (Fig. 12.6h). Body spines with blunt tips .... **G. intestinalis**

Review of dipterous families associated with human myiasis 445

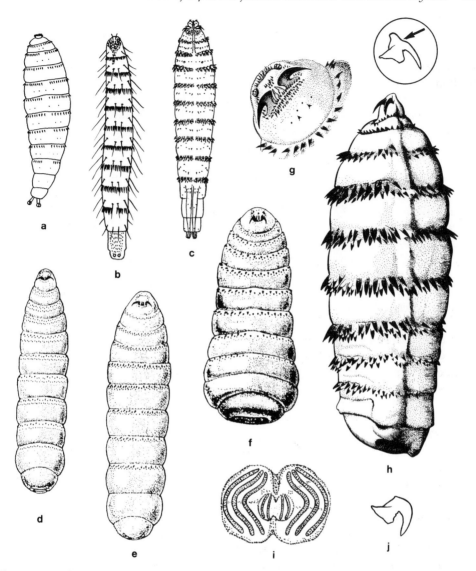

Figure 12.6 Oestridae larvae of subfamily Gasterophilinae: (a–c) first-stage larva of (a) *G. haemorrhoidalis*, (b) *G. nasalis*, (c) *G. intestinalis*; (d–f) second-stage larva of (d) *G. nasalis*, (e) *G. haemorrhoidalis*, (f) *G. pecorum*; (g) *G. pecorum* third-stage larval head in ventral view; (h) *G. intestinalis*, third-stage larva in ventrolateral view, with enlargement (circled) of the form of mouth-hook occurring in the second-stage and third-stage larvae (arrow indicating characteristic shallow depression of the outer margin); (i) *G. intestinalis*, posterior spiracles of third-stage larva; (j) *G. haemorrhoidalis*, mouth-hook of form found in second-stage and third-stage larvae (evenly curved on the outer margin).

14 Mouth-hooks strongly curved, their tips directed backwards and approaching base. Body segment 3 ventrally with two complete rows of spines; body segment 11 with one row of spines interrupted by a broad median gap .... **G. inermis**

— Mouth-hooks directed more laterally. Body segment 3 ventrally with one medially interrupted row of spines; body segment 11 with one row of a variable number of spines without medial interruption .... **G. haemorrhoidalis**

Calliphoridae

The Calliphoridae include some of the major species of obligate myiasis parasites that will develop to maturity in man, i.e. the Tumbu fly (*Cordylobia anthropophaga*), the New World screw-worm (*Cochliomyia hominivorax*) and the Old World screw-worm (*Chrysomya bezziana*). The role of screw-worm flies as agents of wound myiasis has been reviewed by Hall (1991). The Congo floor-maggot (*Auchmeromyia senegalensis*) is also in this family.

Cordylobia

The three species in this genus include *C. anthropophaga* (Fig. 12.7c), the Tumbu fly of continental Africa south of the Sahara. This species causes a boil-like (furuncular) type of myiasis like that produced by *Dermatobia hominis* in the Americas. It is one of the myiasis species most commonly found infesting man. Its life cycle has been well documented (Blacklock and Thompson, 1923). Females do not deposit their eggs directly onto a host. Instead the eggs are laid in batches of 200–300 on dry, shaded ground. Sandy situations are favoured, particularly where they have been contaminated with urine or faeces. Eggs are also laid on drying laundry hung out of direct sunlight – causing infestations when the clothes are next worn if these are not first ironed to kill the eggs. Larvae hatch after one to three days and can remain alive for 9–15 days without food, concealed just below the soil surface until activated by the body-heat or movement of a host. Once attached to a host they immediately burrow into the skin and remain at the entry site, enlarging within the furuncle over a period of about 8–15 days before emerging to fall to the ground and pupate. Symptoms of infestation often include an itching in the first two days, which then subsides and is replaced by more severe pain as the larva grows and becomes active in the boil-like lesion. Serous fluid exudes from the lesion which is often stained by blood and the larva's faeces. The posterior end of the larva is sometimes visible in the centre of the furuncle. Infestations can occur anywhere on the skin surface, arms, shoulders, torso, thighs, calves, axillae and scrotum. In general, only one or two larvae are found in an infestation. However, over 60 have been recorded, usually in individual lesions (Ockenhouse et al., 1990). Both man and animals can acquire a degree of immunity to *C. anthropophaga*.

Dogs are the domestic animals most frequently affected and are important reservoirs of infection. Many other domestic and wild animals can become infested, rats forming the main reservoirs of the flies in the field (Zumpt, 1965).

The Tumbu fly provides a good example of how, with rapid modern travel, medical staff encounter myiasis species far from their natural ranges. Tumbu fly myiasis has even been acquired by persons who have never visited Africa, but have presumably been in contact with larvae accidentally imported from there (Baily and Moody, 1985).

The related species, *C. rodhaini* (Lund's fly), closely resembles the Tumbu fly in appearance and life history, but infests man much less frequently (e.g. Kremer et al., 1970). Its range is the moister parts of tropical Africa, especially the areas of rain forest. The two species are easily distinguished by examination of the slits of the posterior spiracles in the third-stage larvae: these are only slightly sinuous in *C. anthropophaga* (Fig. 12.7d) but are markedly tortuous in *C. rodhaini* (Fig. 12.7e).

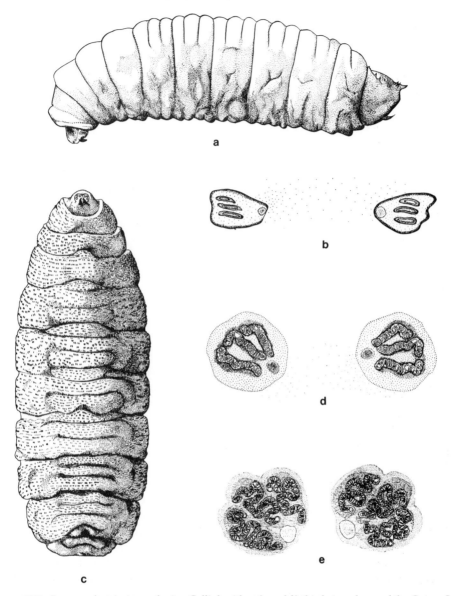

Figure 12.7 Larvae of myiasis-producing Calliphoridae: (*a* and *b*) third-stage larva of the Congo floor-maggot, *Auchmeromyia senegalensis* – lateral view (*a*) and posterior spiracles (*b*); (*c* and *d*) third-stage larva of the Tumbu fly, *Cordylobia anthropophaga*, ventral view (*c*) and posterior spiracles (*d*); (*e*) *Cordylobia rodhaini*, posterior spiracles of the third-stage larva.

Cochliomyia

Two of the four species in this genus are encountered in human myiasis: *Cochliomyia hominivorax*, the New World screw-worm fly, and *C. macellaria*, the secondary screw-worm fly. *Cochliomyia hominivorax* attacks man, cattle, horses, sheep, goats, pigs and dogs (James, 1947) and many species of wildlife (Baumgartner, 1988a).

The geographical distribution of *C. hominivorax* used to extend from the central and southern states of the USA, through Mexico, Central America, the Caribbean islands and northern countries of South America to Uruguay, northern Chile and northern Argentina (James, 1947) but following a major control programme in the USA and Mexico using the Sterile Insect Technique (SIT) the northern limit of *C. hominivorax* is now the border region of Mexico, Guatemala and Belize (Graham, 1985; Krafsur et al., 1987). In 1988, a population of *C. hominivorax* was discovered in Libya in North Africa and became the subject of a major SIT eradication programme; this was highly successful, and the last confirmed case of screw-worm myiasis was recorded in Libya in April 1991 (Lindquist et al., 1992).

Cochliomyia hominivorax is a true obligate parasite of mammals. Female screw-worms do not lay their eggs on carrion but at the edges of wounds on living mammals, or on mucous membranes associated with natural body openings such as the nostrils and sinuses, the eye orbits, mouth, ears and vagina. The species was first described from specimens taken from a fatal case of human nasal myiasis (*hominivorax* meaning man-eater) and many human cases are on record. They are still not uncommon in the endemic areas (Rawlins, 1988) and occur even in areas of introduction (Gabaj et al., 1989). The navels of newly born animals are a common site of infestation, as are the genital and perineal regions of their mothers (especially if traumatized during labour). Female flies lay an average of 200 eggs (range 10–490) in a flat, shingle-like, batch, all oriented in the same direction.

Within 24 hours of the eggs being laid, larvae emerge and immediately begin to feed on the underlying tissues, burrowing gregariously head downwards into the wound. As they feed the wound is enlarged and deepened, resulting in extensive tissue destruction. Infested wounds are attractive to other gravid females which lay further batches of eggs. The larvae reach maturity about five to seven days after hatching and leave the wound, falling to the ground into which they burrow and pupate.

The literature on the New World screw-worm is extensive and scattered, but it may be accessed rapidly by reference to the bibliography of Snow et al. (1981) and to an annual update (available from Dr. D. B. Taylor, USDA, Room 305, Plant Industry Building, East Campus, University of Nebraska, Lincoln, Nebraska 68583, USA).

Cochliomyia macellaria is found from Canada to Argentina and has often been implicated in human myiasis, but many of the records are based on misidentifications and refer to *C. hominivorax* (James, 1947). *Cochliomyia macellaria* develops primarily on carrion and, consequently, when larvae are involved in myiasis they are only secondary invaders which feed on the edge or surface of wounds. Immobile and debilitated persons are at most risk of acquiring an infestation of secondary screw-worm (Smith and Clevenger, 1986). The adults are common in slaughter-houses and outdoor markets. Females lay up to 1000 eggs in batches of 40–250; they often lay together, thereby producing masses of several thousand eggs. These hatch in about four hours and the larvae reach maturity in between 6 and 20 days.

Chrysomya

This Old World genus includes the Old World screw-worm fly, *Chrysomya bezziana*, an obligate parasite in wounds. Its life cycle and behaviour are remarkably like those of *Cochliomyia hominivorax*, the two screw-worm species appearing to occupy an identical parasitic niche in their respective ranges. Cases of human myiasis caused by *C. bezziana* are more common in India than in Africa (Patton, 1922a;

Zumpt, 1965). Adult female *C. bezziana* only oviposit on live mammals, depositing 150–500 eggs at wound sites or in body orifices (ears, nose, mouth and urogenital passages). Patton (1922b) describes human cases involving these orifices and involving wounds on the head, limbs and torso. The larvae hatch after 18–24 hours, moult once after 12–18 hours and a second time about 30 hours later. They feed for three to four days and then drop to the ground and pupate. The pupal stage lasts for seven to nine days in tropical conditions, but up to eight weeks in the subtropical winter months (Spradbery and Humphrey, 1988; Zumpt, 1965). The marked pathological reactions of cattle to infestation with *C. bezziana* are described in detail by Humphrey et al. (1980).

The Old World screw-worm occurs throughout much of Africa (from south of the Sahara to northern South Africa), the Indian subcontinent and South East Asia (from southern China through the Malay peninsula and the Indonesian and Philippine islands to New Guinea) (James, 1947; Sutherst et al., 1989; Zumpt, 1965). It has been introduced into several countries on the west coast of the Persian Gulf (Rajapaksa and Spradbery, 1989).

Chrysomya albiceps (Africa, southern Europe, Arabia, India and, recently, Central and South America) and *C. rufifacies* (Australasia, the Orient and, recently, Central and South America) are very similar facultative parasites which normally lay their eggs on carcasses. Their first-stage larvae feed on exudations of the decomposing flesh, but their second-stage and third-stage larvae are also predaceous, feeding on other blow-fly larvae. Both species are frequently involved in secondary myiasis in sheep, following an initial 'strike' by *Lucilia*. Neither species has yet been implicated in human myiasis, but *C. rufifacies* has been used in maggot therapy. The larvae are commonly known as 'hairy maggots' because of the fleshy projections on their bodies (Fig. 12.8c) which provide good characters for their separation: small spines are present on the stalks of at least some of the projections (especially most dorsal ones) of *C. rufifacies*, while the stalks of the projections of *C. albiceps* have no spines (Erzinçlioğlu, 1987). The larvae of *C. varipes* also have fleshy projections, but they have fewer than the hairy maggots, are smaller when they mature in similar conditions (11 mm length compared to 18 mm) and are known only from Australia (Kitching, 1976; Zumpt, 1965).

Chrysomya megacephala (India, Orient, Australasia and, recently, Africa and South and Central America) is commonly called the Oriental latrine-fly, because of its habit of breeding in faeces as well as on carrion and other decomposing organic matter. It can occur in large numbers around latrines and become a nuisance in slaughterhouses and open-air meat and fish markets. The larvae can cause a secondary myiasis of wounds of man and animals (Zumpt, 1965). The recent introduction of *Chrysomya* species into the Neotropics is discussed by Baumgartner (1988b).

Phormia and Protophormia

These closely related genera are, more or less, confined to areas north of the Tropic of Cancer. The important myiasis-causing species are *Phormia regina* and the more northern *Protophormia terraenovae*. They are very similar in appearance and habits, both usually breeding in carrion. Both cause wound myiasis and, in particular, *Protophormia terraenovae* can be a serious parasite of cattle, sheep and reindeer – although, unlike *Phormia regina*, it has not yet been recorded in humans (James, 1947; Smith, 1986b).

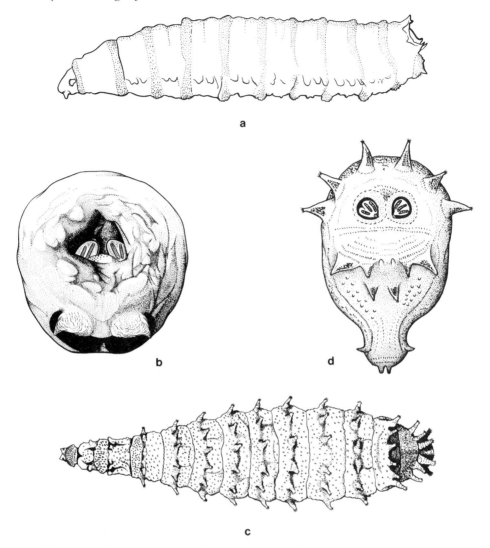

Figure 12.8 Larvae of myiasis-producing Sarcophagidae and Calliphoridae: (*a*) *Sarcophaga*, third-stage larva in lateral view; (*b*) *Sarcophaga*, rear view of the last segment of a second-stage larva showing the posterior spiracles deeply sunk in a cavity (atrium) on the posterior face; (*c* and *d*) *Chrysomya albiceps*, third-stage larva ('hairy maggot') in dorsal view (*c*) and in rear end view of the last segment to show the posterior spiracles flush with the body surface (*d*).

Lucilia

Members of this genus are responsible for the condition known as 'blow-fly strike' of sheep in several countries. In South Africa and Australia the species responsible is *Lucilia cuprina*, and in many temperate areas (including Europe and North America) the important species is *Lucilia sericata*.

The adult flies are metallic or coppery green and are, therefore, known collectively as 'greenbottles'. The life history of the two main species involved in myiasis is very similar. Females lay their eggs in wool, neglected and suppurating wounds, or on carcasses. Wool soiled with urine, faeces or blood is very attractive. The larvae hatch in a few days and feed on the tissues and their exudates. Development is much more rapid in the living warm-blooded animal than it is in carrion. In

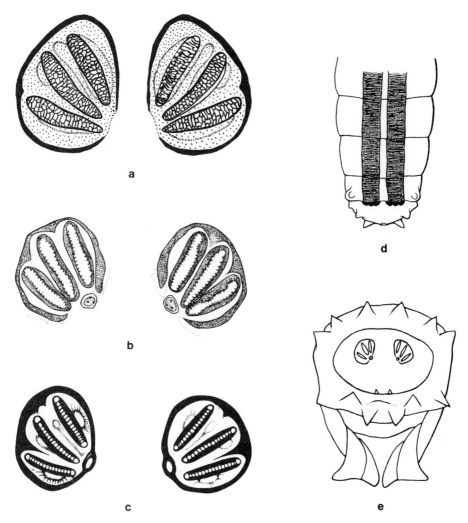

Figure 12.9 Identification features of third-stage larvae of some myiasis-producing Calliphoridae: (*a*) *Chrysomya bezziana*, posterior spiracles; (*b*) *Phormia regina*, posterior spiracles; (*c*) *Lucilia sericata*, posterior spiracles; (*d*) *Cochliomyia hominivorax*, schematic figure of pigmented dorsal tracheal trunks; (*e*) *Protophormia terraenovae*, tubercles on the posterior face of the last segment, end view.

addition to sheep, *Lucilia* can attack horses, cattle and man, the latter usually in cases that involve one or all of the following factors: personal neglect, old age, ill health and high fly density (Greenberg, 1984; Lukin, 1989).

Calliphora
There are numerous species in this cosmopolitan genus, known colloquially as 'bluebottles' because of the colour of the adult. The two most important species in myiasis are *Calliphora vicina* and *C. vomitoria* which have a similar biology. Females are attracted to any decaying material for oviposition; carrion is most suitable. Fresh meat intended for human consumption is also selected for egg laying. *Calliphora* species are usually only involved in myiasis as secondary invaders, but *C. vicina*, in particular, can be a primary invader. Zumpt (1965) reports several cases of human myiasis due to *Calliphora*.

Auchmeromyia

Bloodsucking larvae of the African species *Auchmeromyia senegalensis* (Fig. 12.7a), the Congo floor-maggot, are atypical myiasis species as they do not live on or in the host but suck the blood of sleeping humans (Noireau, 1992) and burrow-dwelling animals. This is termed sanguinivorous myiasis (Zumpt, 1965), but is more akin to the feeding of a mosquito or other bloodsucking fly than to the feeding of invasive myiasis larvae. The species is referred to here as *A. senegalensis* rather than *A. luteola*, in accordance with the correct nomenclature used by Pont (1980). Congo floor-maggots are thought to act as mechanical vectors of trypanosomes between warthogs (Geigy and Kauffman, 1977).

Sarcophagidae

The two important genera regarding myiasis in this family, the 'flesh flies', are *Wohlfahrtia* and *Sarcophaga*. Females are larviparous, depositing first-stage larvae instead of eggs. The two genera are virtually indistinguishable in the larval stage and, where possible, larvae should be reared through and identified from adults.

The most important agent of myiasis in the genus *Wohlfahrtia* is *W. magnifica*, an obligate parasite of warm-blooded vertebrates in southeastern Europe, southern and Asiatic USSR, the Middle East and North Africa. Some 120–170 larvae are deposited near to wounds or body openings of man and other animals, in particular sheep and camels (Higgins, 1985) but also goats, cattle, horses, donkeys, pigs, dogs and geese. The larvae feed and grow rapidly, maturing in five to seven days when they leave the wound for pupation (Kettle, 1990; Zumpt, 1965). The ear (e.g. Mellibovsky et al., 1987), eye (e.g. Baruch et al., 1982) and nose are most frequently infested in man and infestations can lead to deafness, blindness and even death (James, 1947; Zumpt, 1965). Oral myiasis can occur, especially in those with a habit of mouth-breathing (e.g. Konstantinidis and Zamanis, 1988).

Wohlfahrtia nuba also infests wounds of livestock in North Africa and the Middle East, but it probably feeds only on dead or diseased tissues rather than on living tissues (James, 1947). Therefore, it has been used in wound healing, even though it can cause myiasis in man and domestic animals (Zumpt, 1965). *Wohlfahrtia vigil* is a North American species whose larvae can penetrate unbroken skin provided it is thin and tender. Females are preferentially attracted to young animals to larviposit and human cases are almost always of babies under five months old (James, 1947). The larvae produce furuncles identical (Alexander, 1984) to those of *Dermatobia*, each lesion usually containing one larva but sometimes as many as five larvae. *Wohlfahrtia meigeni* is a Holarctic species with similar biology to that of *W. vigil*; it, too, can cause furuncular myiasis in babies (Haufe and Nelson, 1957, as *W. opaca*).

Flies in the genus *Sarcophaga* s.l. are very alike in all stages and extremely difficult to identify to species – with the consequence that many inaccurate accounts exist of species said to be associated with cases of myiasis. Many species breed in excrement, carrion and other decomposing organic matter and some are occasionally involved in myiasis, but little is known of their larval stages. *Sarcophaga cruentata* (syn. *haemorrhoidalis*) is the species most commonly associated with human myiasis. It is anthropophilic and has an almost worldwide distribution except for parts of Australasia. It can breed in decaying meat, rotten foodstuffs and similar organic matter, but faeces are its main larval habitat. Females will larviposit onto faeces as these are being deposited, thereby giving the false impression that the larvae were passed with the faeces. There are, however, many authentic cases of intestinal myiasis caused by *S. cruentata*, as larvae can be ingested on contaminated food and cause more or less severe symptoms of gastric disorder as they pass through the gut (Smith, 1986b; Zumpt, 1965). *Sarcophaga crassipalpis* (Holarctic, South America, South Africa, Australia) has been recorded infesting bed sores in an elderly hospital patient in Italy (Cutrupi et al., 1988).

Fanniidae

The characteristic larvae of the genus *Fannia* (Fig. 12.10) feed on a wide variety of substrates such as rotting plant material, leaf mould, fungi, decaying animal matter, foodstuffs and excrement and sometimes occur in the nests of mammals, birds and insects. Several species are involved in intestinal, urogenital, aural and dermal myiasis in man and other animals.

The body of the larva is flattened, broadest at the middle and narrowed anteriorly. Each segment, after the first, bears on each side a prominent lateral process, which is usually pectinate but can be almost simple (Lyneborg, 1970).

Fannia canicularis (the lesser house-fly) is a cosmopolitan species common in houses. The adults are strongly attracted to urine and are common in toilets, stables and pigsties. Larvae are common in human and animal excrement and especially common on poultry farms, where they breed in large accumulations of faeces and fowl carrion. Other media include foodstuffs such as fruits, smoked or salted meats, cheese and other milk products with which larvae can be accidentally ingested. This species often occurs in association with others in cases of urogenital, intestinal and wound myiasis, and its involvement in cases of ophthalmomyiasis has been reported (Tinne, 1970).

Fannia scalaris (the latrine-fly) is another cosmopolitan species, common in privies and cesspits, which is sometimes concerned in cases of urogenital, intestinal and aural myiasis. Larvae have been found in urine-soaked babies' nappies and cot-blankets.

Fannia incisurata (Holarctic, Mexico and Argentina) and *F. manicata* (Holarctic) have similar habits to *F. scalaris* and are sometimes a cause of intestinal myiasis; *F. incisurata* has also been reported in cases of aural myiasis.

Detailed data on *Fannia* and myiasis are provided by Zumpt (1965) and the forensic importance of the genus is considered by Smith (1986b). Lyneborg (1970) provides a key to the larvae of European species. Larvae of the species usually associated with myiasis can be identified by the following key.

Key to larvae of *Fannia* commonly involved in myiasis

1 Lateral processes pinnate almost to tips; dorsomedian processes less than half as long as lateral processes 2
— Lateral processes not pinnate but with small spinules at base (Fig. 12.10a). Dorsomedian processes similar to and almost as long as lateral processes .... **F. canicularis**
2 Dorsomedian processes represented by small button-like protuberances, not or hardly at all evident 3
— Dorsomedian processes well developed though small, spinulose (Fig. 12.10c) .... **F. scalaris**

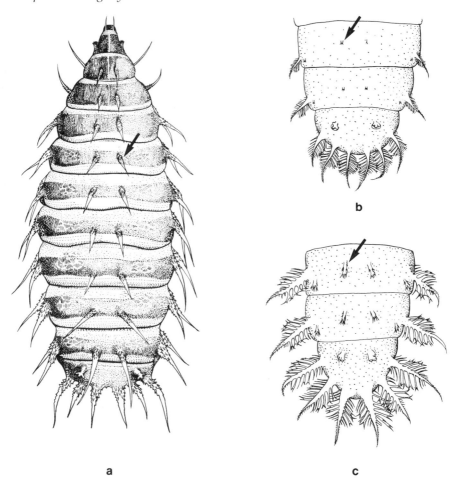

Figure 12.10 Third-stage larvae of *Fannia* (Fanniidae): (*a*) *F. canicularis*, whole larva in dorsal view showing the processes characteristic of the genus; (*b*) *F. manicata*, terminal segments in dorsal view; (*c*) *F. scalaris*, terminal segments in dorsal view. Arrows indicate the dorsomedian processes.

3   Lateral processes large and strongly developed, branches almost touching those of neighbouring processes .... **F. incisurata**
— Lateral processes relatively small, branches not nearly touching those of neighbouring process (Fig. 12.10*b*) .... **F. manicata**

Muscidae

The larvae of Muscidae develop in a wide range of habitats from decaying vegetable matter, wood, fungi and living plants, the nests and burrows of birds, mammals and insects to tree rot-holes, water margins and running water. Larvae of many species are carnivorous, at least in the later stages. Associated with this habit is a longer incubation period in the egg, which can hatch in the adult female of some species and be deposited as second-stage or even third-stage larvae. The biology of the immature stages of Muscidae is well covered by Skidmore (1985). The medical importance of adult Muscidae is treated in Chapter 11. The larvae of a few species cause myiasis in man.

Musca domestica (the house-fly) is a cosmopolitan, ubiquitous species the larva of which (Fig. 12.2*i*) can develop in horse, cow, human and poultry dung, or in material contaminated with excrement, decaying vegetable matter, garbage, decomposing foodstuffs, meat, carrion and even in urine. This is one of the few fly species of sufficient importance to have rated an individual monograph (West, 1951) and its own bibliography (West and Peters, 1973). Larvae have been reported in leg ulcers, nasal cavities of lepers, dermal wounds and intestinal and urogenital myiasis. Without doubt, maggots of *M. domestica* are swallowed commonly and normally passed again unnoticed. However, in a controlled experiment, nausea, vomiting, intestinal cramps and diarrhoea were observed in volunteers who swallowed live larvae in gelatine capsules (Kenney, 1945). Occasionally larvae have been reported in stools, especially of children, though rarely reported as accompanied by gastro-intestinal symptoms. Larvae have more frequently been reported in cases of myiasis where urine is involved, e.g. in neglected cots or nappies of infants and in the feet or footwear of incontinent geriatric patients.

Musca crassirostris larvae have been found in human intestines in India following Hindu purification ceremonies at which products of the cow, including fresh cow dung, are mixed and eaten. A similar case occurred in Italy where undigested grain from horse dung was eaten. The adult is bloodsucking (Chapter 10).

Larvae of *Stomoxys calcitrans*, the stable-fly (Fig. 10.1*b*), have been recorded in cases of intestinal and wound myiasis but Zumpt (1965) doubts that the species is of much importance as a cause of myiasis.

Muscina larvae have been reported in cases of myiasis but probably only one species has been correctly associated with man. Larvae of *Muscina stabulans* occur in rotting fungi, fruits, broken eggs, excrement and carrion, where they become carnivorous and even cannibalistic in the later stages. The species is involved in some cases of intestinal, urogenital, wound, cuticular, dermal, nasal, aural and ocular myiasis in animals, but is only occasionally involved in intestinal (probably pseudomyiasis) and rectal myiasis in man (Bernhard, 1987).

The cosmotropical *Synthesiomyia nudiseta* breeds in various decaying animal and vegetable materials such as human and animal cadavers, faeces and kitchen refuse. Larvae have been recorded in cases of secondary wound myiasis but are probably of little importance.

Larvae of *Hydrotaea* are carrion feeders, predaceous in later stages, and have an occasional forensic significance (Smith, 1986b).

The muscid genera referred to above may be identified by the following key.

Key to larvae of genera of Muscidae likely to be found in cases of myiasis in man

1 Posterior spiracles with strongly sinuous slits 2
— Posterior spiracles with straight or arcuate slits 4
2 Posterior spiracle slits each with three or more loops (Fig. 12.11*a*). Body with smoothly rounded posterior end (Fig. 12.2*i*) .... **Musca**
— Posterior spiracle slits each with two S-shaped loops. Body with setose ventral tubercules at posterior end 3
3 Body much deeper than broad at posterior end and with three prominent setulose tubercules. Posterior spiracle slits not surrounding button (Fig. 12.11*b*) .... **Synthesiomyia**

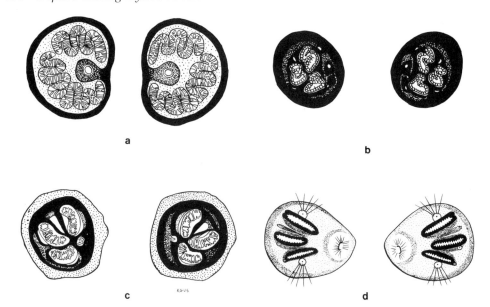

Figure 12.11 Posterior spiracles of third-stage larvae of myiasis-producing Muscidae: (*a*) *Musca domestica*; (*b*) *Synthesiomyia nudiseta*; (*c*) *Muscina stabulans*; (*d*) *Hydrotaea dentipes*.

— Body very little deeper than broad at posterior end and with moderately prominent setulose tubercules. Posterior spiracle slits surrounding button (Fig. 10.1*f*) .... **Stomoxys**

4 Posterior spiracle slits distinctly arcuate, widely separated and radiating from button (Fig. 12.11*c*). Spiracular plate heavily sclerotized, only slits and a small triangular scar translucent .... **Muscina**

— Posterior spiracle slits not arcuate, closer together and less strongly radiating from button (Fig. 12.11*d*). Spiracular plate lightly sclerotized .... **Hydrotaea**

Families of minor importance associated with myiasis

Phoridae (scuttle-flies)

Phoridae are small active flies with an almost worldwide distribution. They are easily recognized by their rather hump-backed appearance and running (scuttling) habit. Larvae (Fig. 12.3*e*) are found in accumulations of decaying organic matter, including faeces. They are dirty white, about 5 mm long, and the integument bears small fleshy tubercles. The posterior spiracles are situated on the top of two small processes towards the posterior end.

Phorid larvae can be ingested in food and then passed in the faeces. *Megaselia scalaris* and *M. rufipes* are the species usually recorded in cases of intestinal myiasis (Zumpt, 1965), but unless adults are reared from larvae specific identification is usually not possible, even by a specialist.

Acute urethral obstruction by larvae of *Megaselia scalaris* and a case where obstruction of the urinary flow by a urethral stone possibly contributed to urogenital myiasis by *M. scalaris* have been described (e.g. Meinhardt and Disney, 1989). *Megaselia spiracularis* has been incriminated in cases of intestinal myiasis in Japan (e.g. Kano et al., 1962) and in the first known case of lung myiasis in man

(Komori et al., 1978). Phorid larvae have also been recorded from a case of ophthalmomyiasis and cases of external, cutaneous myiasis.

Syrphidae (hover-flies, drone-flies)
Flies of this family are often brightly coloured, mimicking wasps and bees. Conspicuous as they are, the adults seldom come before the medical entomologist, but larvae are sometimes encountered in cases of intestinal myiasis. Larvae of *Eristalis*, 'rat-tailed maggots' (Fig. 12.3a), are frequently found in stagnant water that is heavily contaminated with organic matter. They also occur in gutters or drains where dead leaves and water accumulate, and in sewage. Sometimes, as they can occur in drinking water or in vegetable salads, they are ingested by humans. There are several reported cases in which larvae have been passed in the stools.

The main characteristic of the larva is the 'rat-tail', a long respiratory siphon bearing the posterior spiracles at the tip. This siphon is telescopic and enables the larva to breathe at various depths when submerged in liquid or semi-liquid media. *Eristalis* species occur throughout the world. The cosmopolitan drone-fly, *E. tenax*, is the most common species and has a larva of dirty-white coloration, measuring up to 2.5 cm in length, exclusive of the 'tail'. Larvae of *Helophilus* species have also been recorded in cases of intestinal myiasis. They closely resemble those of *Eristalis* but are distinguished by having undulating tracheal trunks instead of straight trunks as in *Eristalis*.

Larvae of the subfamily Syrphinae, particularly the genus *Syrphus* (e.g. Fig. 12.3b), frequently occur on vegetation, where they prey on aphids. When they infest salad ingredients they can be accidentally swallowed by man.

Psychodidae subfamily Psychodinae (moth-flies)
The Psychodinae form the principal subfamily of moth-flies. The subfamily Phlebotominae (sandflies) is treated in Chapter 4. Psychodine larvae (Fig. 17.2a) require mostly aquatic or subaquatic conditions and a few can cause myiasis. The tropicopolitan *Clogmia* (= *Telmatoscopus*) *albipunctatus* breeds in sewers, latrines, drain pipes and compost heaps and has been involved in cases of urogenital, nasopharyngeal and intestinal myiasis (e.g. Smith and Thomas, 1979). The cosmopolitan *Psychoda alternata*, the trickling filter-fly, is one of several species that occur in sewage bacteria beds. It has been recorded (as has *P. albipennis*) in cases of urogenital myiasis, in sputum and vomit. (See also Chapter 17.)

Other families
Larvae of *Tipula* (Tipulidae, crane-flies; Fig. 12.2f), known as leatherjackets, are normally found in soil, where they can do considerable damage to the roots of agricultural crops. They are dirty brownish grey in colour, and the black head capsule is capable of considerable retraction into the anterior part of the larva. The few recorded cases of intestinal myiasis caused by tipulid larvae are almost certainly the result of accidental ingestion of larvae with foodstuffs or with dirty drinking water.

Adult Anisopodidae (window-gnats) will oviposit on almost any moist surface and their larvae (Fig. 12.2d) have been recorded from a variety of habitats, including sewage works. *Sylvicola* (= *Anisopus*) *fenestralis* has been reported as causing intestinal and urogenital myiasis, the larvae being passed in urine and stools (Smith and Taylor, 1966).

The adults of Therevidae (stiletto-flies) are of no known medical importance but there are cases of larvae being coughed up or vomited, or biting and causing an allergic reaction (Smith, 1986a). Therevid larvae (Fig. 12.2a) resemble those of *Scenopinus* (below) and are voracious predators on the larvae of Coleoptera and earthworms in sandy to sandy-loam soils.

The family Stratiomyidae (soldier-flies) includes *Hermetia illucens* whose larvae (Fig. 3.7c) are occasionally involved in cases of enteric pseudomyiasis when they develop in overripe or decaying fruit and are accidentally ingested. The larva is a typical terrestrial stratiomyid, broad and rather flattened, and with a distinct narrow head; the body surface is clothed with short hairs and some transverse rows of bristles. Although primarily an American species, *Hermetia illucens* has been transported on ships to parts of the Australasian region, Europe, Africa and Asia.

Larvae of Scenopinidae (window-flies) (Fig. 12.2b) have been recorded in accidental myiasis in the maxillary sinus and in urogenital myiasis. They are yellowish white, slender, active and about 20 mm long when fully grown, and resemble therevid larvae (above).

Drosophilidae are commonly called fruit-flies but this name is more generally applied to members of the family Tephritidae and, for precision, it is preferable that the vernacular 'vinegar flies' is used for the Drosophilidae. The family contains many species, several of which have now become cosmopolitan (e.g. *Drosophila repleta*), largely through the agency of man. Drosophilidae breed in decaying and fermenting organic matter, including ripe and rotting fruit and fungi. The larvae (Fig. 12.3f) can be accidentally ingested, causing a transitory or false myiasis due to irritation of the bowel.

The larvae of Ephydridae (shore-flies) are mainly aquatic or semi-aquatic and their habitats include carrion and faeces. Larvae of *Teichomyza fusca* (Fig. 12.3d) occur in water containing a high concentration of organic matter and, when swallowed in foul water, can cause intestinal, urinary and rectal myiasis in man.

The family Piophilidae (Fig. 12.2k,l) includes the cosmopolitan *Piophila casei* whose larvae are called 'cheese-skippers' because of their skipping escape behaviour. This species is a common pest on cheese, ham and bacon and its larvae are occasionally ingested with cheese. They can survive the action of mammalian digestive enzymes and cause serious intestinal scarification before being voided alive.

The larvae of Sphaeroceridae (lesser dung-flies) are found in decaying matter of vegetable and animal origin, especially dung, including human faeces. The American species *Leptocera venalicia* has been known to cause human intestinal myiasis. Larvae of Sepsidae occur in excrement and rotting organic matter, including carrion and decaying food, and are occasionally involved in intestinal and urogenital myiasis. Larvae of Micropezidae are saprophagous in decaying vegetation or phytophagous on the root nodules of leguminous and other plants. *Calobata cibaria* has been recorded in a case of intestinal myiasis. Several genera of Anthomyiidae are pests of plants used as food, including roots that are eaten raw, and it is not surprising, therefore, that occasionally larvae are ingested and recorded in cases of intestinal myiasis. There are few genuine records due to the difficulty of identification. Larvae of the families of minor importance in myiasis that are not included in the present keys (Sphaeroceridae, Sepsidae, Micropezidae, Anthomyiidae) can be identified using Smith (1986b, 1989).

MAGGOT THERAPY

This chapter has so far considered fly larvae as causative agents of pathological conditions. However, certain larvae, mainly calliphorid, have been used to beneficial effect in the treatment of serious wound infections, particularly osteomyelitis (bone infection). The treatment, termed 'maggot therapy', has been reviewed recently by Sherman and Pechter (1988) and Leclercq (1990). Maggot therapy exploits the preference of certain myiasis species for diseased or necrotic tissues which are ingested by the larvae. At the same time, the larvae stimulate healing of the wound, possibly by a combination of: (a) the mechanical effects of their continuous movement (stimulating the production of both serous exudates, which flush bacteria from the wound, and healing granulation tissue, from viable tissues); (b) their feeding (including the enzymatic liquefaction of necrotic tissue as well as ingestion and digestion of bacteria); (c) their secretion of therapeutic agents; and (d) a generally therapeutic increase in wound alkalinity (Sherman and Pechter, 1988). Therapeutic products of maggots include allantoin, urea and ammonium bicarbonate (Bunkis et al., 1985) and bactericidal proteins secreted by *Proteus mirabilis*. Two of the latter, originally named mirabilicides (Greenberg, 1968), have been identified as phenylacetic acid and phenylacetaldehyde (Erdmann and Khalil, 1986).

The three species of blow-fly used most often in maggot therapy are *Lucilia sericata*, *L. illustris* and *Phormia regina*. The calliphorids, *Protophormia terraenovae*, *Calliphora vicina* and *Chrysomya rufifacies*, the muscid *Musca domestica* and the sarcophagid *Wohlfahrtia nuba* have also been used. In maggot production, eggs are first sterilized to prevent later contamination of the wound, and then placed onto a sterile food source. For treatment, 200–600 larvae are removed after two days and placed in a wound. When the full grown larvae cease feeding, after three to five days, they are replaced by another batch. This procedure is repeated for three weeks up to two months.

The height of popularity of maggot therapy was between 1920 and 1930, following extensive studies, such as that of Baer (1931), which were stimulated by the observation of the beneficial effects of even natural infestations of blow-flies on the wounds of soldiers in the First World War. The development of antibiotics prompted the decline of maggot therapy, yet it is still used in cases where antibiotic and surgical interventions have been unsuccessful (see Sherman and Pechter, 1988). It may well have a future in such cases, especially where resistance to antibiotics is encountered or in areas where there is limited access to such drugs or to expert surgery. Maggots are universally available and very cost effective (Bunkis et al., 1985).

CONTROL

Control of myiasis species can be considered at three levels: firstly, control or eradication of the fly population; secondly and if the former is not possible, avoidance of infestation; thirdly, treatment, if infestation has occurred because of lack of fly control and avoidance.

Adult fly control or eradication is beyond the scope of this chapter, but has been achieved with myiasis species, most notably with *Cochliomyia hominivorax* in large parts of the Americas (Krafsur et al., 1987) and with *Hypoderma bovis* and *H. lineatum* in Britain (Tarry, 1986), the latter two species now being detectable in Britain only

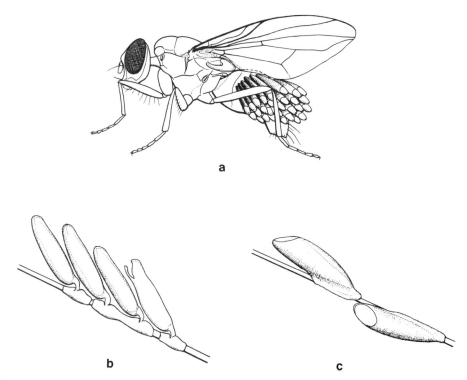

Figure 12.12 Eggs of some representative Oestridae: (*a*) *Dermatobia hominis* eggs (length about 1 mm) clustered on the abdomen of a calyptrate fly, legs and wing shown for one side only (schematized from Catts, 1982); (*b*) *Hypoderma lineatum* eggs glued to a hair of a host animal; (*c*) *Gasterophilus intestinalis* eggs glued to a hair of a host animal.

by serological analysis of the hosts (Sinclair et al., 1990; Tarry et al., 1992). The immune responses that permit such analysis and offer the prospect of vaccination to control myiases are reviewed by Baron and Colwell (1991).

Avoidance of infestation is a more attainable goal in most cases, through an understanding of the biology and behaviour of the flies, good standards of hygiene and common sense. James (1947) reports that it is dangerous to sleep outdoors between 10.00 and 16.00 hours in the summer months in areas infested by *Wohlfahrtia magnifica*. Similarly, in areas where the related *W. vigil* and *W. opaca* occur, babies should not be left to sleep outside. Ironing will kill eggs of *Cordylobia anthropophaga* laid on clothing and so prevent infestation by the larvae that would otherwise hatch. Dressing of wounds will prevent infestation by species that would otherwise deposit eggs or larvae into the wounds. Particular attention should be given to the wound dressings and plaster casts of the elderly or infirm since these can actually conceal infestations to which these persons are unable to respond (e.g. Abram and Froimson, 1987). Thorough cleansing of foodstuffs will help reduce the chance of accidental ingestion of fly larvae.

If the above measures fail and myiasis occurs and is diagnosed, then treatment basically consists of removal of the larvae and appropriate antibiotic follow-up to prevent secondary bacterial infections. Davis and Shuman (1982) and Lane (1987) provide concise information on treatment of myiases.

Larvae in furuncles can be stimulated to leave the lesion by occlusion of the opening with petroleum jelly, fats, water or similar substances, which prevent their respiration. This is very effective for removal of Tumbu fly larvae. As the larvae move towards the skin surface, they can be extracted by forceps or simple finger pressure. Larvae of *Dermatobia hominis* can be removed by pressure, with difficulty, or through a cruciform excision, but care should be taken to remove the whole larva, preferably alive, to avoid infection. A simple alternative to surgical proced-

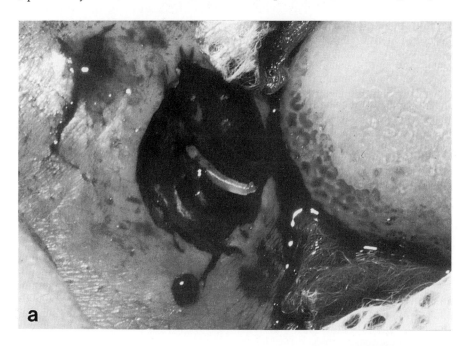

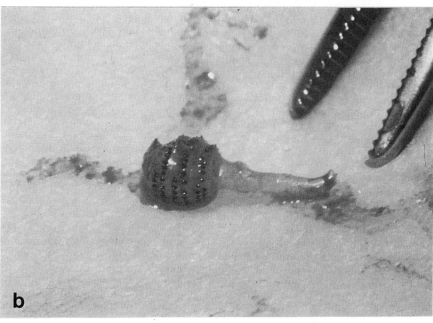

Figure 12.13 Second-instar larva of *Dermatobia hominis*; (*a*) posterior segments protruding from a myiasis wound in a human back during surgical removal; (*b*) the same larva (complete and undamaged) after removal.

ures for removal of *Dermatobia* larvae is the injection, by syringe, of about 2 ml of lidocaine hydrochloride underneath each furuncle (Nunzi et al., 1986). The pressure of the injection is sufficient to push the larvae out.

Larvae in wounds or body openings can be removed by applying a gauze pad soaked in chloroform or ether, or by irrigation of the area with 5–15% chloroform in light vegetable oil. In countries where the use of chloroform on humans is prohibited, an ethanol spray has been used with success (Arbit et al., 1986). Those maggots not purged can be removed by forceps, sometimes after surgery to expose deeply embedded specimens (Kersten et al., 1986).

Larvae in the eye orbit present particular problems. Those on the inner surfaces of the eyelids can be removed simply by forceps after eversion of the lids to expose them. Care should be taken when using an inspection light as larvae are generally photophobic and may be driven into less accessible sites. The most complex surgical procedures involve larvae that have invaded the eye. Syrdalen et al. (1982) review and summarize treatments for ophthalmomyiasis interna posterior. Dead larvae should not be removed if the eye is not reacting, but living larvae are best removed by pars plana vitrectomy, a technique used with success against *Hypoderma* and *Oedemagena* infestations.

COLLECTING, PRESERVING AND REARING MATERIAL

In addition to preparing some of the material from a case of myiasis for preservation, some of the eggs or larvae should be kept alive, where possible, and reared to the adult stage for a confirmatory diagnosis. This is especially important where only first-stage or second-stage larvae are collected, but it may be impossible with most of the obligate species. When culturing is attempted, containers that permit transfer of gases with the outside air should be used, e.g. the top being covered by gauze or tissue-paper secured by rubber bands. This will prevent asphyxiation and discourage the growth of mould. Care must be taken to keep the larvae and their food source moist to avoid desiccation. If mould becomes a problem, it can be countered by using a 0.24% formalin solution to moisten the medium, instead of plain water.

Larvae of many facultative species develop well on a diet of minced meat offered, ideally, in a small, open-topped container placed inside a larger, covered, container. When the larvae have finished feeding, they will crawl out of the smaller container and pupate in dry sand or sawdust placed on the floor of the larger container.

Pupae should be transferred to a cage with netted sides in which the adults can be examined as they emerge. A simple wire frame covered by a sleeve of muslin or mosquito-netting is suitable. Before killing the adults, they should be allowed to inflate their wings and fully harden their cuticle for one to two days. During this period their mature colouring will develop.

If adults are to be cultured and bred to produce a reference collection of eggs and larvae, they will require water and carbohydrate for their survival. Water can be supplied from a well-soaked wad of cotton wool in a small container, to prevent spillage. Honey, or a solution of sugar in water, will provide the necessary carbohydrate. A small amount of fresh meat should be placed in the cage daily for oviposition and to provide females with protein.

The importance of labelling cannot be over-emphasized. A specimen is of little scientific value if it is without a label giving details of where and when it was collected. This should state the origin of the myiasis case, not of the treating hospital. If larvae were recovered from a wound, then details should also be given of the location and nature of the wound.

Examination of larvae

Larvae are best examined in a fully extended condition. Live larvae placed directly into ethanol contract as they die, making it difficult to see the diagnostic features. To obtain relaxed, extended specimens there are two simple methods:

1. Live larvae can be dropped into hot water just below boiling point and then transferred to 80% ethanol for storage. If the water is actually boiling the larvae will tend to rupture.
2. Alternatively, larvae can be placed live into KAA, Kahle's fluid or acetic alcohol for up to 24 hours, to kill and fix them, then be rinsed and stored in 80% ethanol. These fixatives improve the preservation of internal structures and can actually be more convenient to use in the field away from means of heating water. Their formulae are:

KAA	Kerosene	1 part
	Ethanol (95%)	10 parts
	Glacial acetic acid	2 parts
Kahle's fluid	Formalin (35%)	6 parts
	Ethanol (95%)	15 parts
	Glacial acetic acid	2 parts
	Distilled water	30 parts
Acetic alcohol	Ethanol (90%)	3 parts
	Glacial acetic acid	1 part

For the preparation of material for microscopical examination of the mouthparts and spiracles, the material should first be 'cleared' by macerating the specimen in a 10% aqueous solution of potassium hydroxide (KOH) at room temperature for at least 15 minutes. Specimens that have been in alcohol for six months or more need a longer period in KOH, up to 12 hours at room temperature or a shorter time in warmed KOH. Small larvae should be put into the solution whole, with punctures to allow its penetration; larger larvae can be dissected and only the required parts macerated, taking care to keep all the parts of single specimens together. As the muscles soften they can be teased away with fine forceps or sharp needles. In order to avoid destroying the sclerites, care must be taken with dissecting instruments and with KOH.

When the muscle and fat-body have been cleared away, the specimens should be placed in glacial acetic acid for at least 15 minutes, to neutralize any residual KOH, and should then be rinsed well with 80% ethanol. Thorough rinsing with ethanol should be sufficient if acetic acid is not available. They are then ready for examination, mounting or storage.

There are many methods of mounting slides. One is to dehydrate the samples in absolute alcohol, transfer them to clove oil to clear, then mount in Canada balsam; alternatively, from absolute alcohol they can be mounted directly into Euparal.

Slides can only be viewed in one plane and this should be taken into account when mounting specimens. The cephalopharyngeal skeleton can be mounted as one preparation allowing a lateral view. Anterior spiracles can be mounted detached or with a portion of the surrounding cuticle. Posterior spiracles should be mounted as a pair to retain the relationship between them. The flat area of the posterior spiracular disc can be dissected off and mounted, taking care to note the orientation of the spiracles, their ventral and dorsal sides. With careful dissection, the posterior spiracles can be mounted together with the dorsal tracheal trunks that lead from them.

When an adult emerges from its puparium, dorsal and ventral flaps at the anterior end of the puparium open. These flaps should be retained with the rest of the puparium as they contain relics of the third-stage larva useful for identification. Inside the lower flap lie the sclerites of the third-stage cephalopharyngeal skeleton. The flaps might be covered by membranes, but can be cleared by KOH, as for larvae, and used to give an indication of the larval identity. The larval anterior spiracles can be found on the upper flap of the puparium, while the larval posterior spiracles can be studied at the posterior end of the puparium. Again, clearing with KOH can be helpful since these spiracles often become very heavily sclerotized in the puparium, making observation of the slits difficult.

Scanning electron microscopy can be a tremendous asset to the morphological study of fly larvae for identification and research purposes, the detailed images surpassing any produced by light microscopy (Spradbery, 1991). It has great potential for the study of groups of species that appear identical using other morphological techniques, e.g. *Sarcophaga* species (Aspoas, 1991). Careful preparation of the material is important to minimize physical distortion of the characters. Essentially, the material needs dehydration and then critical point drying before coating with gold/palladium (Grodowitz et al., 1982; Kitching, 1976; Ruiz-Martinez et al., 1989).

REFERENCES

Abram, L. J. and **Froimson**, A. I. 1987. Myiasis (maggot infection) as a complication of fracture management: case report and review of the literature. *Orthopedics* **10**: 625–627.

Alexander, J. O'D. 1984. *Arthropods and human skin.* x + 422 pp. Springer-Verlag, Berlin.

Arbit, E., **Varon**, R. E. and **Brem**, S. S. 1986. Myiatic scalp and skull infection with Diptera *Sarcophaga:* case report. *Neurosurgery* **18**: 361–362.

Aspoas, B. R. 1991. Comparative micromorphology of third instar larvae and the breeding biology of some Afrotropical *Sarcophaga* (Diptera: Sarcophagidae). *Medical and Veterinary Entomology* **5**: 437–445.

Baer, W. S. 1931. The treatment of chronic osteomyelitis with the maggot (larva of the blow fly). *Journal of Bone and Joint Surgery* **13**: 438–475.

Baily, G. G. and **Moody**, A. H. 1985. Cutaneous myiasis caused by larvae of *Cordylobia anthropophaga* acquired in Europe. *British Medical Journal* **290**: 1473–1474.

Baird, J. K., **Baird**, C. R. and **Sabrosky**, C. W. 1989. North American cuterebrid myiasis. Report of seventeen new infections of human beings and review of the disease. *Journal of the American Academy of Dermatology* **21**: 763–772.

Baron, R. W. and **Colwell**, D. D. 1991. Mammalian immune responses to myiasis. *Parasitology Today* **7**: 353–355.

Baruch, E., **Godel**, V., **Lazar**, M., **Gold**, D. and **Lengy**, J. 1982. Severe external ophthalmomyiasis due to larvae of *Wohlfahrtia* sp. *Israel Journal of Medical Sciences* **18**: 815–816.

Baumgartner, D. L. 1988a. Review of myiasis (Insecta: Diptera: Calliphoridae, Sarcophagidae) of Nearctic wildlife. *Wildlife Rehabilitation* **7**: 3–46.

Baumgartner, D. L. 1988b. Spread of introduced *Chrysomya* blowflies (Diptera: Calliphoridae) in the Neotropics with records new to Venezuela. *Biotropica* **20**: 167–168.

Bennett, G. F. 1955. Studies on *Cuterebra emasculator* Fitch 1856 (Diptera: Cuterebridae) and a discussion of the status of the genus *Cephenemyia* Ltr. 1818. *Canadian Journal of Zoology* **33**: 75–98.

Bernhard, K. V. 1987. Nachweise von Rektalmyiasis beim Menschen. *Angewandte Parasitologie* **28**: 59–61.

Bishopp, F. C., Laake, E. W., Brundrett, H. M. and **Wells, R. W.** 1926. The cattle grubs or ox warbles, their biologies and suggestions for control. *United States Department of Agriculture Bulletin* **1369**: 1–120.

Blacklock, [D.] B. and **Thompson, M. G.** 1923. A study of the tumbu-fly, *Cordylobia anthropophaga* Grünberg, in Sierra Leone. *Annals of Tropical Medicine and Parasitology* **17**: 443–502.

Bunkis, J., Gherini, S. and **Walton, R. L.** 1985. Maggot therapy revisited. *Western Journal of Medicine* **142**: 554–556.

Catts, E. P. 1982. Biology of New World bot flies: Cuterebridae. *Annual Review of Entomology* **27**: 313–338.

Cogley, T. P. 1991. Key to the eggs of the equid stomach bot flies *Gasterophilus* Leach 1817 (Diptera: Gasterophilidae) utilizing scanning electron microscopy. *Systematic Entomology* **16**: 125–133.

Cutrupi, V., Lovisi, A., Bernardi, A. and **Meggio, A.** 1988. Miasi, considerazioni su di un caso. *Rivista di Parassitologia* **3** [= 47] (1986): 185–188.

Davis, E. and **Shuman, C.** 1982. Cutaneous myiasis: devils in the flesh. *Hospital Practice* **1982** (December): 115–123.

Edwards, K. M., Meredith, T. A., Hagler, W. S. and **Healy, G. R.** 1984. Ophthalmomyiasis interna causing visual loss. *American Journal of Ophthalmology* **97**: 605–610.

Erdmann, G. R. and **Khalil, S. K. W.** 1986. Isolation and identification of two antibacterial agents produced by a strain of *Proteus mirabilis* isolated from larvae of the screwworm (*Cochliomyia hominivorax*) (Diptera: Calliphoridae). *Journal of Medical Entomology* **23**: 208–211.

Erzinçlioğlu, Y. Z. 1987. The larvae of some blowflies of medical and veterinary importance. *Medical and Veterinary Entomology* **1**: 121–125.

Ferrar, P. 1987. A guide to the breeding habits and immature stages of Diptera Cyclorrhapha. *Entomonograph* **8**: 1–907.

Fidler, A. H. 1987. Migrierende dermale Myiasis durch *Hypoderma diana*. *Mitteilungen der Österreichischen Gesellschaft für Tropenmedizin und Parasitologie* **9**: 111–119.

Foote, B. A. 1991. Order Diptera. Pp. 690–915 in Stehr, F. W. (ed.), *Immature insects*: vol. 2. xvi + 975 pp. Kendall/Hunt, Dubuque, Iowa.

Gabaj, M. M., Gusbi, A. M. and **Awan, M. A. Q.** 1989. First human infestations in Africa with larvae of American screw-worm, *Cochliomyia hominivorax* Coq. *Annals of Tropical Medicine and Parasitology* **83**: 553–554.

Geigy, R. and **Kauffman, M.** 1977. Experimental mechanical transmission of *Trypanosoma brucei* by *Auchmeromyia* larvae. *Protozoology* **3**: 103–107.

Graham, O. H. (ed.) 1985. Symposium on eradication of the screwworm from the United States and Mexico. *Miscellaneous Publications of the Entomological Society of America* **62**: 1–68.

Greenberg, B. 1968. Model for destruction of bacteria in the midgut of blow fly maggots. *Journal of Medical Entomology* **5**: 31–38.

Greenberg, B. 1984. Two cases of human myiasis caused by *Phaenicia sericata* (Diptera: Calliphoridae) in Chicago area hospitals. *Journal of Medical Entomology* **21**: 615.

Grodowitz, M. J., Krchma, J. and **Broce, A. B.** 1982. A method for preparing soft bodied larval Diptera for scanning electron microscopy. *Journal of the Kansas Entomological Society* **55**: 751–753.

Grunin, K. Ya. 1955. Horse botflies (Gastrophilidae). *Fauna of the USSR.* New Series No. 60, *Insects, Diptera* **17** (1). 95 [+1] pp. Izdatel'stvo Akademii Nauk SSSR, Moscow and Leningrad. [In Russian.]

Grunin, K. Ya. 1962. Warbleflies (Hypodermatidae). *Fauna of the USSR.* New Series No. 82, *Insects, Diptera* **19** (4). 237 [+1] pp. Izdatel' stvo Akademii Nauk SSSR, Moscow and Leningrad. [In Russian.]

Grunin, K. Ya. 1966. Oestridae. In Lindner, E. (ed.), *Die Fliegen der palaearktischen Region* **64a:** 1–96.

Guimarães, J. H. and **Papavero, N.** 1966. A tentative annotated bibliography of *Dermatobia hominis* (Linnaeus jr., 1781) (Diptera, Cuterebridae). *Archivos de Zoologia do Estado de São Paulo* **14:** 223–294.

Guimarães, J. H., Papavero, N. and **Pires do Prado, A.** 1983. As miíases na região neotropical (identificação, biologia, bibliografia). *Revista Brasileira de Zoologia* **1:** 239–416.

Hall, M. J. R. 1991. Screwworm flies as agents of wound myiasis. *World Animal Review* (Special Issue), **1991** (October): 8–17.

Haufe, W. O. and **Nelson, W. A.** 1957. Human furuncular myiasis caused by the flesh fly *Wohlfahrtia opaca* (Coq.) (Sarcophagidae: Diptera). *Canadian Entomologist* **89:** 325–327.

Heath, A. G. C., Elliott, D. C. and **Dreadon, R. G.** 1968. *Gasterophilus intestinalis,* the horse bot-fly as a cause of cutaneous myiasis in man. *New Zealand Medical Journal* **68:** 31–32.

Higgins, A. J. 1985. Common ectoparasites of the camel and their control. *British Veterinary Journal* **141:** 197–216.

Humphrey, J. D., Spradbery, J. P. and **Tozer, R. S.** 1980. *Chrysomya bezziana:* pathology of Old World screwworm fly infestations in cattle. *Experimental Parasitology* **49:** 381–397.

Ishijima, H. 1967. Revision of the third stage larvae of synanthropic flies of Japan (Diptera: Anthomyiidae, Muscidae, Calliphoridae and Sarcophagidae). *Japanese Journal of Sanitary Zoology* **18:** 47–100. [In English.]

James, M. T. 1947. The flies that cause myiasis in man. *United States Department of Agriculture Miscellaneous Publication* **631:** 1–175.

Kalelioğlu, M., Aktürk, G., Aktürk, F., Komsuoğlu, S. S., Kuzeyli, K., Tiğin, Y., Karaer, Z. and **Bingöl, R.** 1989. Intracerebral myiasis from *Hypoderma bovis* larva in a child. *Journal of Neurosurgery* **71:** 929–931.

Kano, R., Kaneko, K., Kawashima, K. and **So, N.** 1962. On some cases of myiasis. *Japanese Journal of Sanitary Zoology* **13:** 96–97. [In Japanese without English summary.]

Kearney, M. S., Nilssen, A. C., Lyslo, A., Syrdalen, P. and **Dannevig, L.** 1991. Ophthalmomyiasis caused by the reindeer warble fly larva. *Journal of Clinical Pathology* **44:** 276–284.

Kenney, M. 1945. Experimental intestinal myiasis in man. *Proceedings of the Society for Experimental Biology and Medicine* **60:** 235–237.

Kersten, R. C., Shoukrey, N. M. and **Tabbara, K. F.** 1986. Orbital myiasis. *Ophthalmology* **93:** 1228–1232.

Kettle, D. S. 1990. *Medical and veterinary entomology.* 658 pp. C.A.B. International, Wallingford.

Kitching, R. L. 1976. The immature stages of the Old-World screw-worm fly, *Chrysomya bezziana* Villeneuve, with comparative notes on other Australasian species of *Chrysomya* (Diptera, Calliphoridae). *Bulletin of Entomological Research* **66:** 195–203.

Komori, K., Hara, K., Smith, K. G. V., Oda, T. and **Karamine, D.** 1978. A case of lung myiasis caused by larvae of *Megaselia spiracularis* Schmitz (Diptera: Phoridae). *Transactions of the Royal Society of Tropical Medicine and Hygiene* **72:** 467–470. [Karamine as published on the title page of this work is an error of Katamine.]

Konstantinidis, A. B. and **Zamanis, D.** 1988. Gingival myiasis. *Journal of Oral Medicine* **42:** 243–245.

Krafsur, E. S., Whitten, C. J. and **Novy, J. E.** 1987. Screwworm eradication in North and Central America. *Parasitology Today* **3:** 131–137.

Kremer, M., Lenys, J., Basset, M., Rombourg, H. and **Molet, B.** 1970. Deux cas de myiase a *Cordylobia rodhaini* contractée au Cameroun et diagnostiquée an Alsace. *Bulletin de la Société de Pathologie exotique* **63**: 592–596.

Krümmel, H. and **Brauns, A.** 1956. Myiasis des Auges: medizinische und entomologische Grundlagen. *Zeitschrift für angewandte Zoologie* **43**: 129–190.

Lane, R. P. 1987. Flies causing myiasis. Pp. 1462–1468 in Manson-Bahr, P. E. C. and Bell, D. R. (eds), *Manson's tropical diseases*. Nineteenth edition. xviii + 1557 pp. Baillière Tindall, London.

Lane, R. P., Lovell, C. R., Griffiths, W. A. D. and **Sonnex, T. S.** 1987. Human cutaneous myiasis – a review and report of three cases due to *Dermatobia hominis*. *Clinical and Experimental Dermatology* **12**: 40–45.

Leclercq, M. 1969. *Entomological parasitology: the relations between entomology and the medical sciences*. xviii + 158 pp. Pergamon Press London.

Leclercq, M. 1990. Utilisation de larves de Diptères – Maggot Therapy – en médecine: historique et actualité. *Bulletin et Annales de la Société royale belge d'Entomologie* **126**: 41–50.

Lindquist, D. A., Abusowa, M. and **Hall, M. J. R.** 1992. The New World screwworm fly in Libya: a review of its introduction and eradication. *Medical and Veterinary Entomology* **6**: 2–8.

Lukin, L. G. 1989. Human cutaneous myiasis in Brisbane: a prospective study. *Medical Journal of Australia* **150**: 237–240.

Lyneborg, L. 1970. Taxonomy of European *Fannia* Larvae (Diptera, Fanniidae). *Stuttgarter Beiträge zur Naturkunde* **215**: 1–28.

Mazzeo, V., Ercolani, D., Trombetti, D., Todeschini, R. and **Gaiba, G.** 1987. External ophthalmomyiasis: report of four cases. *International Ophthalmology* **11**: 73–76.

Meinhardt, W. and **Disney, R. H. L.** 1989. Urogenital myiasis caused by scuttle fly larvae (Diptera: Phoridae). *British Journal of Urology* **64**: 547–548.

Mellibovsky, J., David, M., Gold, D. and **Lengy, J.** 1987. Myiasis of the head in a child. *Israel Journal of Medical Sciences* **23**: 298–299.

Noireau, F. 1992. Infestation by *Auchmeromyia senegalensis* as a consequence of the adoption of non-nomadic life by Pygmies in the Congo Republic. *Transactions of the Royal Society of Tropical Medicine and Hygiene* **86**: 329.

Nunzi, E., Rongioletti, F. and **Rebora, A.** 1986. Removal of *Dermatobia hominis* larvae. *Archives of Dermatology* **122**: 140.

Ockenhouse, C. F., Samlaska, C. P., Benson, P. M., Roberts, L. W., Eliasson, A., Malane, S. and **Menich, M. D.** 1990. Cutaneous myiasis caused by the African Tumbu fly (*Cordylobia anthropophaga*). *Archives of Dermatology* **126**: 199–202.

O'Rourke, F. J. 1968. Furuncular myiasis caused by warble fly (*Hypoderma*) larvae in patients from County Cork. *Journal of the Irish Medical Association* **61**: 19–20.

Pampiglione, S. 1958. Indagine epidemiologica sulla miasi congiuntivale umana da *Oestrus ovis* in Italia. Nota I: Inchiesta tra i medici italiani. *Nuovi Annali d'Igiene e Microbiologia* **9**: 242–263.

Papavero, N. 1977. The world Oestridae (Diptera), mammals and continental drift. *Series Entomologica* **14**: 1–240.

Patton, W. S. 1922a. Notes on the myiasis-producing Diptera of man and animals. *Bulletin of Entomological Research* **12**: 239–261.

Patton, W. S. 1922b. Some notes on Indian Calliphorinae. Part VII. Additional cases of myiasis caused by the larvae of *Chrysomyia bezziana* Vill., together with some notes on the Diptera which cause myiasis in man and animals. *Indian Journal of Medical Research* **9**: 654–682.

Pont, A. C. 1980. Family Calliphoridae. Pp. 779–800 in Crosskey, R. W. (ed.), *Catalogue of the Diptera of the Afrotropical region*. 1437 pp. British Museum (Natural History), London.

Quesada, P., Navarrete, M. L. and **Maeso, J.** 1990. Nasal myiasis due to *Oestrus ovis* larvae. *European Archives of Otorhinolaryngology* **247**: 131–132.

Rakusin, W. 1970. Ocular myiasis interna caused by the sheep nasal bot fly (*Oestrus ovis* L.). *South African Medical Journal* **44**: 1155–1157.

Rajapaksa, N. and Spradbery, J. P. 1989. Occurrence of the Old World screw-worm fly *Chrysomya bezziana* on livestock vessels and commercial aircraft. *Australian Veterinary Journal* **66**: 94–96.

Rawlins, S. C. 1988. Human myiasis in Jamaica. *Transactions of the Royal Society of Tropical Medicine and Hygiene* **82**: 771–772.

Robinson, W. H. 1971. Old and new biologies of *Megaselia* species (Diptera, Phoridae). *Studia Entomologica* **14**: 321–348.

Ruiz-Martinez, I., Soler-Cruz, M. D., Benitez-Rodriguez, R., Lopez, M. D. and Perez-Jimenez, J. M. 1989. Preparation of Dipteran larvae for scanning electron microscopy with special reference to myiasigen Dipteran species. *Scanning Microscopy* **3**: 387–390.

Sergent, E. 1952. La Thimni: myiase oculo-nasale de l'homme causée par l'oestre du mouton. *Archives de l'Institut Pasteur d'Algérie* **30**: 319–361.

Sherman, R. A. and Pechter, E. A. 1988. Maggot therapy: a review of the therapeutic applications of fly larvae in human medicine, especially for treating osteomyelitis. *Medical and Veterinary Entomology* **2**: 225–230.

Sinclair, I. J., Tarry, D. W. and Wassall, D. A. 1990. Reduction in UK warble infestation levels: serological survey, 1990. *Veterinary Record* **127**: 285.

Skidmore, P. 1985. *The biology of the Muscidae of the world.* [xii] + 550 pp. W. Junk Publishers, Dordrecht.

Smith, D. R. and Clevenger, R. R. 1986. Nosocomial nasal myiasis. *Archives of Pathology and Laboratory Medicine* **110**: 439–440.

Smith, K. G. V. 1986a. Larvae of Therevidae (Diptera) biting man. *Entomologist's Monthly Magazine* **122**: 115.

Smith, K. G. V. 1986b. *A manual of forensic entomology.* 205 pp. British Museum (Natural History), London, and Cornell University Press, Ithaca, New York.

Smith, K. G. V. 1989. An introduction to the immature stages of British flies: Diptera larvae, with notes on eggs, puparia and pupae. *Handbooks for the Identification of British Insects* **10** (14): 1–280.

Smith, K. G. V. and Taylor, E. 1966. *Anisopus* larvae (Diptera) in cases of intestinal and urinogenital myiasis. *Nature* **210**: 852.

Smith, K. G. V. and Thomas, V. 1979. Intestinal myiasis in man caused by larvae of *Clogmia* (= *Telmatoscopus*) *albipunctatus* Williston (Pschodidae, Diptera). *Transactions of the Royal Society of Tropical Medicine and Hygiene* **73**: 349.

Snow, J. W., Siebenaler, A. J. and Newell, F. G. 1981. Annotated bibliography of the screwworm, *Cochliomyia hominivorax* (Coquerel). *United States Department of Agriculture Science and Education Administration Agricultural Reviews and Manuals (Southern Series)* **14**: 1–32.

Spradbery, J. P. and Humphrey, J. D. 1988. The screw-worm fly: *Chrysomya bezziana. Proceedings of the 71st Annual Conference of the Association of Veterinary Inspectors of New South Wales*: 56–62.

Spradbery, J. P. 1991. *A manual for the diagnosis of screw-worm fly.* i + 64 pp. CSIRO Division of Entomology, Canberra.

Sutherst, R. W., Spradbery, J. P. and Maywald, G. F. 1989. The potential geographical distribution of the Old World screw-worm fly, *Chrysomya bezziana. Medical and Veterinary Entomology* **3**: 273–280.

Syrdalen, P., Nitter, T. and Mehl, R. 1982. Ophthalmomyiasis interna posterior: report of case by the reindeer warble fly larva and review of previous reported cases. *British Journal of Ophthalmology* **66**: 589–593.

Tarry, D. W. 1986. Progress in warble fly eradication. *Parasitology Today* **2**: 111–116.

Tarry, D. W., Sinclair, I. J. and Wassall, D. A. 1992. Progress in the British hypodermosis eradication programme : the role of serological surveillance. *Veterinary Record* **131**: 310–312.

Teskey, H. J. 1981a. Morphology and terminology – larvae. Pp. 65–88 in McAlpine, J. F., Peterson, B. V., Shewell, G. E., Teskey, H. J., Vockeroth, J. R. and Wood, D. M., *Manual of Nearctic Diptera*: vol. 1, vi + 674 pp. Research Branch, Agriculture Canada (Monograph No. 27).

Teskey, H. J. 1981b. Key to families – larvae. Pp. 125–147 in McAlpine, J. F., Peterson, B. V., Shewell, G. E., Teskey, H. J., Vockeroth, J. R. and Wood, D. M. (eds), *Manual of Nearctic Diptera*: vol. 1, vi + 674 pp. Research Branch, Agriculture Canada (Monograph No. 27).
Tinne, J. E. 1970. The invisible worm. *Lancet* **1970** (2): 1360.
Townsend, L. H., Hall, R. D. and **Turner, E. C.** 1978. Human oral myiasis caused by *Gasterophilus intestinalis* (Diptera: Gasterophilidae). *Proceedings of the Entomological Society of Washington* **80**: 129–130.
West, L. S. 1951. *The housefly: its natural history, medical importance, and control.* xi + 584 pp. Comstock Publishing and Cornell University Press, Ithaca, New York.
West, L. S. and **Peters, O. B.** 1973. *An annotated bibliography of Musca domestica Linnaeus.* xiii + 743 pp. Dawsons, Folkestone and London, and Northern Michigan University, Marquette.
Wood, D. M. 1987. Oestridae. Pp. 1147–1158 in McAlpine, J. F. (ed.), *Manual of Nearctic Diptera*: vol. 2, vi + pp. 675–1332. Research Branch, Agriculture Canada (Monograph No. 28).
Zumpt, F. 1963. Ophthalmomyiasis in man, with special reference to the situation in southern Africa. *South African Medical Journal* **37**: 425–428.
Zumpt, F. 1965. *Myiasis in man and animals in the Old World: a textbook for physicians, veterinarians and zoologists.* xv + 267 pp. Butterworths, London.

Part Two
OTHER INSECTS

CHAPTER THIRTEEN

Cockroaches (Blattaria)

N. R. H. Burgess

Some 4000 species of cockroaches have been described from most parts of the world, although the suborder Blattaria (which they form) is primarily tropical and subtropical. Most species live in forests where some are semi-aquatic, others burrow, bore into wood, or live in caves (cavernicolous); a few even live commensally with other insects, such as ants (Roth and Willis, 1957). They are predominantly nocturnal and are either omnivorous or vegetarian.

Cockroaches are among the most primitive of all winged insects and have changed little in appearance from their fossil remains of the carboniferous geological period some 250 million years ago. Despite superficial differences in appearance, there is a close affinity between cockroaches and praying mantises, and the two groups constitute the order Dictyoptera.

While most cockroaches have little contact with man, about a dozen species have become adapted to a peridomestic environment and have some degree of medical importance. In temperate regions they have colonized and flourished in man-made pseudo-tropical situations. The three most important species are *Blatta orientalis*, *Blattella germanica* and *Periplaneta americana*. Less abundant and widespread species are *Periplaneta australasiae* and *Supella longipalpa*.

The medical importance of cockroaches revolves around their dual infestation of houses, shops and other premises for storage, cooking and preparation of food on the one hand and sewers and rubbish dumps on the other. They have potential to transmit pathogens from one habitat to the other.

Because of their ubiquity, size and ease of laboratory colonization, cockroaches are used as 'models' for many fundamental studies on insects. Consequently, there is a considerable literature on their structure, biology and physiology: this is conveniently reviewed by Bell and Adiyodi (1981), Cornwell (1968, 1976) and Guthrie and Tindall (1968).

RECOGNITION AND ELEMENTS OF STRUCTURE

Typically, cockroaches are large, oval and dorsoventrally flattened insects easily recognized by their general appearance (Figs 13.1–13.4). The pest species range in length from 10 to 50 mm. The head bears a pair of compound eyes and ventral

Medical Insects and Arachnids Edited by Richard P. Lane and Roger W. Crosskey.
Published in 1993 by Chapman & Hall ISBN 0 412 40000 6

474 Cockroaches (Blattaria)

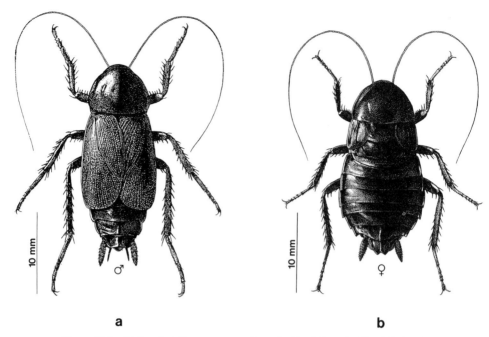

Figure 13.1 *Blatta orientalis*, common cockroach: (*a*) adult male; (*b*) adult female.

chewing mouthparts adapted for omnivorous feeding. Long filamentous antennae with numerous annulations distinguish cockroaches from beetles, insects with which they are sometimes confused by the layman. The thorax has three distinct segments, of which the first is evidenced dorsally by a large pronotum which is sometimes so well developed as to partially hide the head when seen from above. Each thoracic segment has a pair of strong legs covered in stout setae and terminating in paired claws; pad-like pulvilli are present on the last tarsal segment in some species. Two pairs of wings are present in the adults of most cockroaches, each with a network of veins – the latter giving rise to the name of the order Dictyoptera (Greek: *diktyon* = net, *pteron* = wing). The fore wings are thickened to a leathery texture and are closed scissors-like over the metathorax and abdomen, thus concealing the thin and often large, fan-like, hind wings. A few species, particularly *Blatta orientalis*, have greatly reduced wings (a condition known as brachyptery) and thus never fly. The abdomen is clearly segmented and its only appendages are paired cerci at the posterior end (visible projecting backwards in Figs 13.1*a,b* and 13.2*b*).

CLASSIFICATION AND IDENTIFICATION

Cockroaches and praying mantises were formerly grouped with phasmids (stick insects), grylloblattids, crickets, grasshoppers and mole-crickets in the order Orthoptera, but in modern times these two groups of insects have been considered an order of their own, the Dictyoptera. Nevertheless, all these groups are believed to have evolved from a common ancestor similar in appearance to present-day apterous Grylloblattodea (a tiny order of orthopteroid insects represented in the modern fauna by only 16 known species confined to the Holarctic realm).

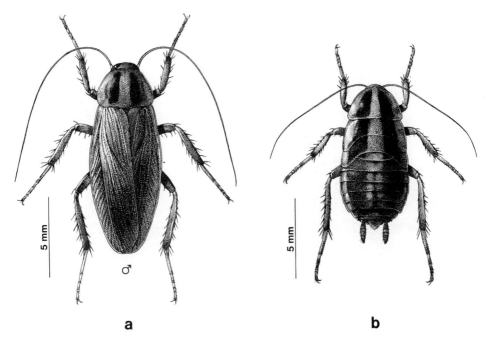

Figure 13.2 *Blattella germanica*, German cockroach: (a) adult male; (b) nymph.

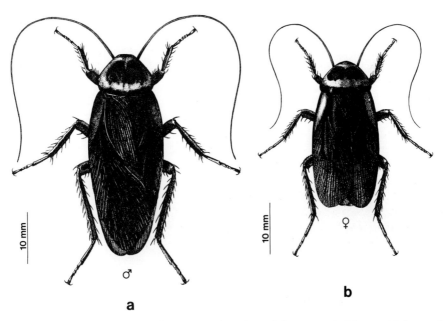

Figure 13.3 The two common peridomestic cockroaches of the genus *Periplaneta*: (a) *P. americana* (American cockroach), adult male; (b) *P. australasiae* (Australian cockroach), adult female.

The Dictyoptera are characterized morphologically by filiform antennae with many segments, biting mouthparts, five-segmented tarsi, fore wings thickened into 'tegmina' with a marginal costal vein, asymmetrical male genitalia, and reduced and concealed ovipositor; oothecae are secreted to enclose the eggs (a unique

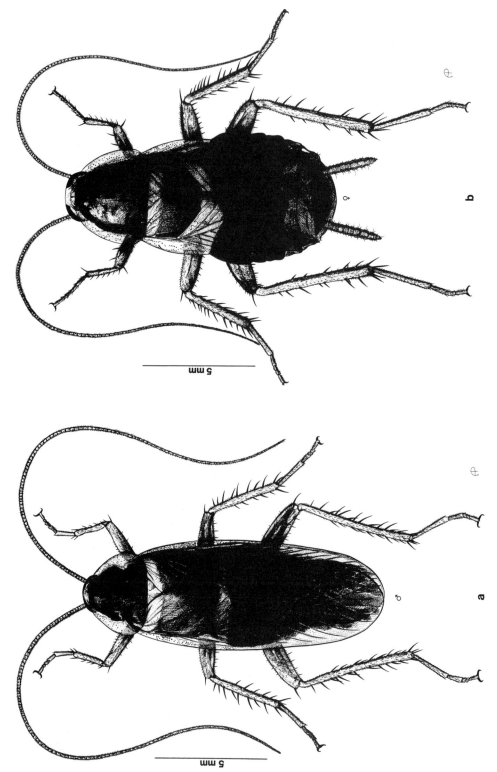

Figure 13.4 *Supella longipalpa*, brown-banded cockroach: (*a*) adult male; (*b*) adult female.

physiological attribute). The order is divided into two suborders: the Mantodea (mantises) with grasping raptorial fore legs, and the Blattaria (cockroaches) in which all the legs are alike (e.g. as in Fig. 13.3). Other differences are that in Mantodea the head is not covered by the pronotum and there are (usually) three ocelli, whereas in Blattaria the head is nearly or completely covered by the large, shield-like, pronotum and two ocelli are represented. Cockroach classification is by no means settled, not all authorities relying on the same criteria and some regarding the Blattaria as a full order and dividing it into more than 20 families. Most workers follow a more conservative classification in which four families are recognized: three of these, the Blaberidae, Blattellidae and Blattidae, contain species of some medical importance.

Identification key to peridomestic species of cockroach

The following key is only for recognition of adults of those species, mostly cosmopolitan, which are common in peridomestic environments. Immature specimens (nymphs) are similar to adults but completely without wings (e.g. Fig. 13.2b).

1 Fore wings well developed, reaching at least to end of abdomen (e.g. Fig. 13.3)... 2
— Fore wings absent or short, not reaching to end of abdomen (e.g. Fig. 13.4b) 5
2 Total length (front margin of head to end of fore wings) more than 18 mm 3
— Total length (front margin of head to end of fore wings) less than 17 mm 4
3 Fore wing with a pale yellow stripe along basal part of its anterior margin (Fig. 13.3b) .... **Periplaneta australasiae**
— Fore wing without a pale stripe along basal part of its anterior margin (Fig. 13.3a) .... **Periplaneta americana**
4 Pronotum with two broad longitudinal dark stripes (Fig. 13.2a). Fore wings uniformly coloured .... **Blattella germanica**
— Pronotum without dark stripes, brown with translucent lateral margins. Fore wings dark basally and pale apically (Fig. 13.4a) .... **Supella longipalpa** (male)
5 Legs straw-coloured. Pronotum with translucent lateral margins (Fig. 13.4b). Fore wings with pale bands (Fig. 13.4b). Body length less than 14 mm............... .... **Supella longipalpa** (female)
— Legs reddish brown to dark brown. Pronotum uniformly opaque (Fig. 13.1). Fore wings without pale bands. Body length more than 15 mm .... **Blatta orientalis**

Principal domestic species

Blatta orientalis (Blattidae) (Oriental or common cockroach)
Adults of this cockroach are about 20–25 mm in length and dark brown or black, the colour accounting for its familiar colloquial name of 'black beetle'. The wings are greatly reduced in the male (Fig. 13.1a) and vestigial in the female (Fig. 13.1b).

This cockroach is found in cooler conditions than *Blattella germanica* and is therefore more frequent than that species in cellars, basements and the cooler parts of kitchens and other catering facilities. Outdoors it inhabits yards, sewers and other drainage areas, and in temperate regions often infests rubbish tips. It moves more slowly than *B. germanica* and cannot climb efficiently because it has very small pulvilli. It is usually found on horizontal surfaces or very rough vertical surfaces such as brick or mud walls.

Blattella germanica (Blattellidae) (German cockroach)
Colloquial names for this cockroach include 'steamer', 'shiner' and 'croton bug'. The adult is 10–15 mm in length and has large fore wings which in the male cover the whole abdomen (Fig. 13.2a) but in the female leave the terminal segments exposed. The latter is broader than the male. Adults are pale brown to mid brown and have two broad dark longitudinal stripes on the pronotum: in nymphs, these stripes continue over the other two thoracic segments (Fig. 13.2b).

This species occurs in temperate regions in warm and moist conditions and, as well as sewers and drains, is therefore commonly present in kitchens and restaurants. It climbs efficiently and rests in crevices in vertical and horizontal surfaces. The dark streaks of its faeces are a clear indication of an infestation, especially where the faeces are spread into broad bands along its habitual pathways.

Periplaneta americana (Blattellidae) (American cockroach)
This is the largest of the domestic pest cockroaches, measuring 30–50 mm in length when mature (Fig. 13.3a). The wings of the male are slightly longer than those of the female, but they are well developed in both sexes and cover the abdomen. The general colour is reddish brown but there are prominent yellow patches on the sides of the pronotum. The sexes can be differentiated by the presence of a ventral keel at the tip of the abdomen in the female; males have a pair of ventral styles. The first five nymphal stages are a uniform pale brown, but in subsequent stages the yellow pronotal patches become more apparent.

This species is the most widespread and frequent cause of domestic cockroach infestation, particularly in the tropics and subtropics (where it is also common in rubbish tips and refuse areas). In temperate regions it is found in warm and humid conditions; it tends to infest food stores and catering premises indoors, drains and sewers outdoors. It moves rapidly, climbs efficiently, and is able to fly sluggishly in a warm atmosphere, often fluttering through open windows and around street lights at night.

Periplaneta australasiae (Blattellidae) (Australian cockroach)
This cockroach (Fig. 13.3b) is similar in appearance to *P. americana* but smaller (adult length 30–35 mm). It is reddish brown with yellowish pronotal patches extending onto the anterior margins of the fore wings. Both sexes are fully winged and the fore wings longer in the male than the female. The nymphs are conspicuously marked with yellow spots on the thorax and abdomen.

This species is second in importance to *P. americana* in terms of worldwide infestation, although it is not at all common in temperate regions. It prefers a warmer environment than *P. americana* and is typically found outdoors in the tropics, whence it spreads into human habitations. In temperate regions it is confined to heated greenhouses, where it can cause damage to commercial crops.

Occasional pest species

Blattella asahinai (Blattellidae)
Known as the 'Asian cockroach', this is essentially an outdoor species and is a common rural pest in many parts of South East Asia. Recently it has spread to the USA, probably having been introduced into Florida in 1984. *Blattella asahinai* readily flies and is attracted to light and pale-coloured walls, thus spreading into domestic situations. Its ability to disperse rapidly has contributed to this species becoming an important pest in subtropical environments.

Periplaneta fuliginosa (Blattellidae)
The adult of this species is 30–35 mm long, similar in appearance to other species of the genus but distinguished by its shiny dark brown colour. The wings are well developed in both sexes and cover the whole abdomen. It is a common pest in the southern United States.

Periplaneta brunnea (Blattellidae)
Although very similar in appearance to *P. americana*, this cockroach can be distinguished by its less conspicuous pale areas on the pronotum and the less pointed cerci of the adults. Its habits are very similar to those of *P. americana*.

Rhyparobia maderae (Blaberidae)
This pantropical species is known as the 'Madeira cockroach' (after its specific name) and is found in houses and outbuildings, open markets and (sometimes) storehouses. It can occur in huge numbers and be a serious local pest. Unlike some other primarily tropical pest species, it is not often found in temperate areas. For a long time the species was in the genus *Leucophaea* (known as *Leucophaea maderae*).

Supella longipalpa (= superlectilium) (Blattellidae)
Although its colour is variable, this cockroach is colloquially called the 'brown-banded cockroach' in reference to its typical colour pattern. It is a small species (adult length 10–15 mm) in which the wings fully cover the abdomen in the male (Fig. 13.4a) but not in the female (Fig. 13.4b). The species is a domestic pest in most parts of the tropics, but has spread to Australia and North America in the past 50 years. It is only occasionally found in European countries.

BIOLOGY

Cockroaches are exopterygote insects in which the life cycle is of the so-called indirect or incomplete type (Chapter 2). The female lays her eggs in a case (ootheca) containing from 12 to 50 eggs, depending on the species. The oothecae are very resistant to desiccation and in some cases impervious to many insecticides. Some species (e.g. *Blattella germanica*) carry the hardened ootheca for two to three weeks before depositing it. The first-instar nymphs which hatch from the oothecae are remarkably similar in appearance to the adult but without wings. There are 5–12 nymphal stages (instars), depending on species and, curiously, on sex; in *Blatta orientalis* the female passes through more instars, and takes longer to develop to the final moult, than the male.

Adult cockroaches live for a comparatively long period, up to two years or more where conditions are poor and food is not readily available. Without water, however, cockroaches die within a few weeks. The feeding habit is omnivorous, and the diet of cockroaches consists of almost any digestible material. Flight is dependent on temperature as well as food metabolism, so that whereas winged cockroaches often fly at night in warm tropical regions, flight is uncommon in cool temperate regions because the wing muscles cannot be sufficiently activated.

Different species vary in their reproductive capacity, and this might account for the observation that *Blatta orientalis* is gradually being replaced by *Blattella germanica* as the main pest species in many temperate parts of the world. Similarly, *Blattella asahinai* has successfully spread from South East Asia and established itself in the USA in the past decade, as a result of its rapid dispersal and high reproduction rate.

MEDICAL IMPORTANCE

Cockroaches are closely associated with man and are attracted by the moisture of various human foods and waste products (Laurence, 1983). Pest species are omnivorous, feeding on a wide range of organic material in their environment. Because of their abundance in sewers, latrines and drains, as well as in and around houses, they are potential carriers of faecal pathogens. A wide range of pathogens has been isolated from cockroaches (Roth and Willis, 1957, 1960, reviews; Burgess et al., 1973a). A summary is given in Table 13.1.

The actual medical significance of the carriage of such a wide range of pathogens by cockroaches depends in part on the behaviour of the insects themselves and in part on other means of transmission of these pathogens. The behaviour of cockroaches certainly lends itself to the spread of disease. If potentially pathogenic material is present in the environment (e.g. in latrines), cockroaches can take the organisms into their guts, or particles containing the pathogens can adhere to their feet, legs and body surface. If the cockroach then crawls over food utensils or food destined for human consumption, pathogens can be deposited while the cockroach moves about. While feeding, cockroaches often regurgitate fluid from the 'crop'; often this can contain organisms ingested in a previous meal. The habit of cockroaches of defaecating as they move about and feed also increases the chance of pathogens being deposited in areas where they might contaminate food or act as a source of infection to man.

Because of the nature of the organisms involved, it is extremely difficult to prove that cockroaches are the prime cause of any disease outbreak. Many of the pathogens isolated from cockroaches are spread by contaminative means between their human victims, especially those pathogens that cause food poisoning or other enteric conditions.

Some of the most important of the circumstantial evidence for the role of cockroaches in pathogen transmission is based on isolation of pathogens during outbreaks of disease. *Salmonella typhimurium* has been isolated from *Blattella germanica* infesting a children's ward in a Belgian hospital during an epidemic of gastroenteritis caused by this organism (Graffar and Mertens, 1950). *Salmonella*

Table 13.1 Pathogenic organisms found naturally infecting cockroaches

Group	Organism	Infection
Bacteria	*Escherichia coli*	Urogenital/intestinal infections
	Mycobacterium leprae	Leprosy
	Klebsiella pneumoniae	Pneumonia/URTI
	Proteus vulgaris	Gastroenteric tract infection
	Pseudomonas aeruginosa	Urinary and upper respiratory tract infections, wounds, burns
	Salmonella (including *S. typhi* and *S. typhimurium*)	Typhoid, other enteric fevers, gastroenteritis
	Serratia marcescens	Upper respiratory tract infection
	Shigella spp.	Dysentery, diarrhoea
	Staphylococcus aureus	Boils, abscesses
	Streptococcus faecalis	Faecal contamination
	Yersinia pestis	Plague
Fungus	*Aspergillus fumigatus*	Aspergillosis
Helminths	*Ancylostoma duodenale*	Hookworm infection
	Ascaris lumbricoides	Ascariasis (roundworm infection)
	Enterobius vermicularis	Pinworm infection
	Necator americanus	Hookworm infection
	Trichuris trichiura	Trichuriasis (whipworm infection)
Protozoan	*Entamoeba histolytica*	Dysentery
Viruses	Hepatitis virus	'Jaundice'
	Poliomyelitis virus	

bovis morbificans has been isolated in an Australian hospital in a similar outbreak (Mackerras and Mackerras, 1949). Other isolations have been of *Salmonella typhi* from cockroaches infesting patients' houses in Italy (Roth and Willis, 1957, p. 58), and of *Shigella dysenteriae* from *Blattella germanica* from the environs of patients in Northern Ireland (Burgess and Chetwyn, 1981). Laboratory studies have shown that the pathogenic serotype 0119 of *Escherichia coli* will be taken up by *Blatta orientalis* and will continue to be passed out in the faeces for 20 days (Burgess et al., 1973b).

Another facet of medical importance is that close association with cockroaches can cause allergic asthma in some people (Lan et al., 1988).

CONTROL

Cockroach control is based on three lines of attack: basic hygiene to reduce potential food sources, structural alterations to buildings to reduce cockroach access from sewers and drains and resting sites, and (most commonly) chemical control (Cornwell, 1976, review; Schal and Hamilton, 1990).

The construction of hospital buildings and their interiors, with numerous ducts, recesses and panelling, is a major factor contributing to the ease with which cockroach infestations are established and for the difficulties encountered in their control (Burgess, 1984).

REFERENCES

Bell, W. J. and Adiyodi, K. G. (eds) 1981. *The American cockroach.* xvi + 529 pp. Chapman & Hall, London and New York.

Burgess, N. R. H. 1984. Hospital design and cockroach control. *Transactions of the Royal Society of Tropical Medicine and Hygiene* **78**: 293–294.

Burgess, N. R. H. and Chetwyn, K. N. 1981. Association of cockroaches with an outbreak of dysentery. *Transactions of the Royal Society of Tropical Medicine and Hygiene* **75**: 332–333.

Burgess, N. R. H., McDermott, S. N. and Whiting, J. 1973a. Aerobic bacteria occurring in the hind-gut of the cockroach, *Blatta orientalis*. *Journal of Hygiene* **71**: 1–7.

Burgess, N. R. H., McDermott, S. N. and Whiting, J. 1973b. Laboratory transmission of Enterobacteriaceae by the oriental cockroach, *Blatta orientalis*. *Journal of Hygiene* **71**: 9–14.

Cornwell, P. B. 1968. *The cockroach*: vol. 1, *A laboratory insect and an industrial pest.* 391 pp. Hutchinson, London.

Cornwell, P. B. 1976. *The cockroach*: vol. 2, *Insecticides and cockroach control.* 557 pp. Associated Business Programmes, London.

Graffar, M. and Mertens, S. 1950. Le rôle des blattes dans la transmission des salmonelloses. *Annales de l'Institut Pasteur* **79**: 654–660.

Guthrie, D. M. and Tindall, A. R. 1968. *The biology of the cockroach.* 408 pp. Edward Arnold, London.

Lan, J. L., Lee, D. T., Wu, C. H., Chang, C. P. and Yeh, C. L. 1988. Cockroach hypersensitivity: preliminary study of allergic cockroach asthma in Taiwan. *Journal of Allergy and Clinical Immunology* **82**: 736–740.

Laurence, B. R. 1983. Flies, cockroaches and excreta. Pp. 495–500 in Feachem, R. G., Bradley, D. J., Garelick, H. and Duncan Mara, D. (eds), *Sanitation and disease: health aspects of excreta and wastewater management.* xxvii + 501 pp. John Wiley, Chichester.

Mackerras, I. M. and Mackerras, M. J. 1949. An epidemic of infantile gastro-enteritis in Queensland caused by *Salmonella bovis-morbificans* (Basenau). *Journal of Hygiene* **47**: 166–181.

Roth, L. M. and Willis, E. R. 1957. The medical and veterinary importance of cockroaches. *Smithsonian Miscellaneous Collections* **134** (10): 1–147.

Roth, L. M. and Willis, E. R. 1960. The biotic associations of cockroaches. *Smithsonian Miscellaneous Collections* **141** (whole volume): 1–470.

Schal, C. and Hamilton, R. L. 1990. Integrated suppression of synanthropic cockroaches. *Annual Review of Entomology* **35**: 521–551.

CHAPTER FOURTEEN

Bedbugs and kissing-bugs (bloodsucking Hemiptera)

C. J. Schofield and W. R. Dolling

The Hemiptera or 'true bugs' are widespread and numerous throughout tropical and temperate regions. Over 80 000 species are described, making this the largest of the exopterygote orders. The order has been traditionally divided into two suborders, Homoptera and Heteroptera, but a division into three or four suborders is now generally accepted. These names relate to the typical form of wings: the adults usually have the anterior pair of wings of harder consistency than the posterior pair (Hemiptera), either uniformly so (Homoptera), or with the apical portion more membranous than the remainder (Heteroptera). Many Heteroptera close their wings flat over the abdomen with the apical membranous areas overlapping (and so are sometimes referred to as 'cross-winged bugs') (Fig. 14.2), while the Homoptera often carry their wings raised like a roof over their backs. The Homoptera are all phytophagous, and include many groups of economically important crop pests such as aphids (greenfly or plant lice), psyllids (jumping plant lice), coccids (scale insects), cicadellids (leafhoppers), and aleurodids (whitefly). Many of these are vectors of plant viruses and mycoplasmas. Sometimes such pests occur in vast numbers and, when fragmented, may become inhalant allergens.

Hemiptera are characterized by piercing-and-sucking mouthparts. Some Hemiptera and most Heteroptera are adapted to feed on plant juices. Some Heteroptera are predators on insects and other invertebrates, piercing the invertebrate tissue to suck out the juices inside; some are adapted to feed on vertebrate blood. Some predatory bugs may occasionally take vertebrate blood, either indirectly via a bloodsucking prey – as in some Emesinae preying on blood-engorged mosquitoes (White et al., 1972) – or by biting vertebrates directly. Water bugs of the family Belostomatidae will occasionally bite (using their raptorial forelegs to hold the victim) and this gives them the common name of 'toebiters'. The bite is painful but otherwise considered harmless. Many other predatory bugs, especially of the families Reduviidae and Anthocoridae, will sometimes bite man – especially as a defensive measure – and can draw blood. Their bite is usually extremely painful and can provoke severe allergic reactions (see Ryckman (1979) and Ryckman and Bentley (1979) for annotated bibliography). Two heteropteran families, the Cimicidae and Polyctenidae (both closely related to the Anthocoridae),

Medical Insects and Arachnids Edited by Richard P. Lane and Roger W. Crosskey.
Published in 1993 by Chapman & Hall ISBN 0 412 40000 6

are entirely bloodsucking, while the family Reduviidae (assassin bugs) includes one subfamily, the Triatominae, which is also entirely bloodsucking.

The Cimicidae or bedbugs are small, virtually wingless insects living close to their vertebrate hosts; some are of economic importance as pests of fowl, while species that feed on humans, principally *Cimex lectularius* and *C. hemipterus*, are of some medical importance as nuisance insects that can contribute to chronic iron deficiency anaemia and may have a minor role in the transmission of some blood-borne viruses such as hepatitis-B. The closely related Polyctenidae are small ectoparasites of bats, unusual in being viviparous but with no known medical or economic importance. The Triatominae (sometimes known as 'kissing-bugs' or 'cone-nosed bugs') are renowned as vectors of *Trypanosoma cruzi*, causative agent of Chagas disease or South American trypanosomiasis (Brenner and Stoka, 1988a, b, c).

RECOGNITION AND ELEMENTS OF STRUCTURE

The Hemiptera are most easily recognized by their general body form and their piercing and sucking mouthparts, while closer examination of the mouthparts will indicate whether the bug is adapted to feed on plant juices, invertebrate fluids or vertebrate blood (Cobben, 1978). At rest, the proboscis is reflexed under the head, usually reaching or surpassing the first pair of legs. In most Hemiptera the proboscis is four-segmented and in many bugs which feed on plants extends the length of the body or even beyond. However, the proboscis is reduced to three segments in some homopteran families (e.g. Cicadidae, Psyllidae and Aleurodidae), in some predatory families (e.g. Anthocoridae, Nabidae and Reduviidae) and in all important bloodsucking groups (Cimicidae, Polyctenidae, Triatominae) (Fig.14.3a). Often the proboscis of predatory bugs is strongly sclerotized and sometimes curved, compared to bloodsucking bugs which have a finer straight proboscis.

The mouthparts are constructed on the same basic plan in all Hemiptera. The large, usually tubular and jointed labium forms a hollow sheath incorporating the much smaller elongate labrum. There are no labial palps, and the usual divisions of the labium in other insects are difficult to homologize with this highly modified structure, although it does bear a few sensory organs on its tip like those on the glossae and paraglossae of other insect mouthparts. The mandibles and maxillae are also highly modified. Each consists of a fine hairlike stylet and no trace of the usual subdivisions of the maxillae can be detected. Maxillary palps are lacking. The whole complex of stylets and sheath is termed the rostrum or proboscis, although these names are also applied to the only readily visible part, the labium.

When bloodsucking Hemiptera feed, the whole bundle of four stylets – normally enclosed within the labial sheath – can be protruded from the tip of the sheath and inserted into the host, but the labium never penetrates the epidermis. When the stylets have been guided into position some bugs partially or totally disengage them from the sheath to allow deep penetration of the host tissues. The outer mandibular stylets serve to pierce the host epidermis and to anchor the inner maxillary stylets in position. The two maxillary stylets are grooved and adhere together to form a tube along which saliva and food can pass, but there can be separate channels for the two fluids. The functional mouth can be a small gap between the stylets, at or near their apex in bloodsucking forms, or it may be a long slit, characteristic of predatory species (Barth, 1962; Bernard, 1974; Cobben, 1978; Lavoipierre et al., 1959). Hemipteran saliva is complex, containing enzymes such as hyaluronidase, phospholipase and various proteases in predaceous bugs and apyrase together with a complex mixture of vasodilators in bloodsucking species (Ribeiro, 1987).

In most Heteroptera the head projects forward (porrect). In addition to the compound eyes, two ocelli are present in adults, but these are absent in the immature stages. The antennae are usually four- or five-segmented, but in some reduviids such as *Microtomus* can appear to have ten or more segments because of secondary annulation. The thorax of adult bugs usually bears two pairs of wings, of which the fore (mesothoracic) pair usually overlies and protects the hind (metathoracic) pair when they are closed at rest. For this reason, the texture of the fore wings, either in the basal part or throughout, is usually tougher than that of the hind wings. Adults of the families Cimicidae and Polyctenidae are always brachypterous or micropterous, with the fore wings reduced to short undifferentiated flaps and the hind wings completely lacking. Some other Hemiptera have two or more forms of adults – some apterous, some macropterous and sometimes intermediate brachypterous forms.

Apart from the external genitalia there are no segmental abdominal appendages in bloodsucking Hemiptera. Cerci are completely absent.

Hemiptera are exopterygote or hemimetabolous insects, meaning that they undergo 'incomplete' or 'gradual' metamorphosis with the wings developing outside the body (as opposed to inside a pupal case). In families such as the Reduviidae and Cimicidae, the eggs have a conspicuous operculum or cap which is forced open when the first instar nymph emerges. Development then proceeds through several nymphal instars (usually five in Heteroptera) to the adult. All the nymphal stages are similar to the adults in morphology, behaviour, habitat and diet, except that they lack fully developed wings and have immature genitalia and gonads.

CLASSIFICATION AND IDENTIFICATION

Classification of the Hemiptera

The Hemiptera represent an ancient order dating from the Permian period (232–280 million years BP). Within the Permian, the Hemiptera radiated spectacularly, leaving a bewildering array of fossils – many just wing remains. But in addition to the divergence, convergent evolution has also been widespread, making interpretation difficult and conflicting (Wootton, 1981). This, together with the diversity of extant forms, has resulted in a number of conflicting arrangements for the higher classification of the order.

The early division of the order into two suborders, Homoptera and Heteroptera, with the latter divided into three series (Coleorrhyncha, Auchenorrhyncha and Sternorrhyncha) and the Heteroptera variously divided, has more or less survived to the present day. The classification of the Heteroptera has been reviewed in the light of cladistic methodology by Stys and Kerzhner (1975) and Cobben (1981). The current higher classification of the Hemiptera into four suborders can be summarized as follows:

Order: Hemiptera (true bugs)
Suborder 1: Coleorrhyncha – one family (mossbugs)
Suborder 2: Heteroptera – 75 families, mainly plant-sucking and predatory bugs. For general description and keys to families see China (1933), China and Miller (1959), Ghauri (1973), Schuh (1986), Southwood and Leston (1959).
Suborder 3: Auchenorrhyncha – 28 families, plant-sucking bugs including cicadas, lantern flies, leafhoppers etc.
Suborder 4: Sternorrhyncha – 41 families, plant-sucking bugs including whiteflies, aphids, scale insects and mealybugs. For general description and keys to families of Auchenorryncha and Sternorrhyncha see Evans (1946a, b, 1947, 1963) and Imms (1957).

Heteroptera alone have a gula; that is, the head capsule is closed ventrally by a bridge behind the rostrum so that the rostrum arises not at the hind margin of the head but some distance in front of it. In addition, the rostrum is often brought further forward by a considerable elongation of the head. In the other suborders the rostrum arises just in front of the prosternum (Coleorrhyncha and Auchenorrhyncha) or, apparently, between the anterior coxae (Sternorrhyncha).

The evolution of feeding habits within the Heteroptera has been subject to considerable discussion (see Cobben, 1978, 1979; Schuh, 1986; Sweet, 1979). Bloodsucking most probably derived from a predatory habit, with the intermediate stage perhaps being predation on nest-dwelling invertebrates. The Triatominae clearly represent a polyphyletic assemblage of bloodsucking Reduviidae (Schofield, 1988), while the Cimicidae and Polyctenidae may represent monophyletic or polyphletic assemblages of 'bloodsucking Anthocoridae' (cf. Schuh, 1986; Southwood and Leston, 1959). In both cases, the evolutionary trend seems likely to have gone from adventitious predatory bugs, to predators adapted to feed on guilds of invertebrates associated with nest-dwelling birds or mammals, and then to exploitation of the vertebrate host itself. In line with this interpretation (given more fully in Schofield, 1988), the three main obligate bloodsucking groups have arisen – the Cimicidae and Polyctenidae from the anthocorid line, and the Triatominae derived from various reduviid lines. For the medical entomologist, most of the complexities of hemipteran, particularly heteropteran, classification can be avoided by this simple evolutionary rationalization.

Family Cimicidae (bedbugs)

The Cimicidae form a well-defined family of bloodsucking bugs, closely related to the Polyctenidae and to the Anthocoridae. They are oval flattened insects without functional wings (Fig. 14.1a,e), although the fore wings remain represented in adults by two small pads on the dorsal surface of the thorax. Adults are about 5 mm long and 3 mm broad, while the nymphal stages range from tiny first instars barely 1 mm long, to the fifth instars just slightly smaller than the adults. Their colour ranges from yellowish to dark brown but often appears dark red when the bugs are recently fed. All five nymphal stages and both sexes of adults are obligate feeders on vertebrate blood.

There are 91 recognized species of Cimicidae, grouped in 23 genera forming six subfamilies (Ryckman et al., 1981; Usinger, 1966) (Table 14.1). Most are associated

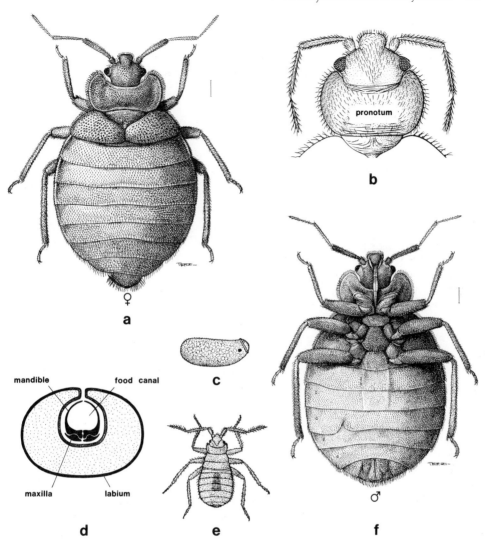

Figure 14.1 Bedbugs, genus *Cimex* (Cimicidae): (*a*) *C. lectularius*, common bedbug, female; (*b*) *C. hemipterus*, tropical bedbug, head and prothorax showing different shape of pronotum from that of the common bedbug; (*c*) egg of *C. lectularius*; (*d*) cross-section of *Cimex* rostrum (semi-diagrammatic); (*e*) newly hatched nymph of *C. lectularius*; (*f*) male of *C. lectularius*, undersurface.

with birds and/or bats, but two species – *Cimex lectularius* and *C. hemipterus* – are the familiar bedbugs commonly associated with man. Some bedbugs are economically important as pests of poultry, causing severe irritation, anaemia and weight loss, especially the common bedbug *C. lectularius* in North America, Europe and the USSR, *Haematosiphon inodorus* in Central America and *Ornithocoris toledoi* in Brazil. *Cimex columbarius* is also frequent in poultry houses, although it mainly infests pigeon cotes; *Oeciacus hirundinis* infests martin nests and sometimes bites man when it infests houses temporarily. *Leptocimex boueti* is mainly associated with bats in West Africa, but has also been reported biting man. Some cimicids associated with bats, including *C. pipistrelli* in Europe, are vectors of bat trypanosomes such as *Trypanosoma dionisii*, *T. incertum* and *T. vespertilionis* (Gardner and Molyneux, 1988a, b; Gardner et al. 1987; Van den Berghe et al., 1963).

Table 14.1 List of the subfamilies and genera of Cimicidae

Primicimicinae
   Bucimex (Chile: 1 species)
   Primicimex (USA, Mexico and Guatemala: 1 species)
Cimicinae
   Bertilia (Argentina and Chile: 1 species)
   Cimex (cosmopolitan: 21 species)
   Oeciacus (Nearctic region, western Palaearctic region: 2 species)
   Paracimex (Oriental region, southern Africa: 13 species)
   Propicimex (Argentina and Brazil: 2 species)
Cacodminae
   Aphrania (Afrotropical region, South East Asia: 5 species)
   Cacodmus (Africa, Middle East, Oriental region: 10 species)
   Crassicimex (tropical Africa, Madagascar, Cambodia: 3 species)
   Leptocimex (West Africa, Sudan, Middle East, Sri Lanka: 4 species)
   Loxaspis (tropical Africa and South East Asia: 6 species)
   Passicimex (tropical Africa: 1 species)
   Stricticimex (Afrotropical and Oriental regions, Egypt: 10 species)
Afrocimicinae
   Afrocimex (Equatorial Africa: 3 species)
Latrocimicinae
   Latrocimex (Brazil and Trinidad: 1 species)
Haematosiphoninae
   Caminicimex (Argentina and Uruguay: 1 species)
   Cimexopsis (mid-western and eastern USA: 1 species)
   Haematosiphon (southwestern USA, Mexico: 1 species)
   Hesperocimex (Nearctic region: 3 species)
   Ornithocoris (southeastern USA, South America: 2 species)
   Psitticimex (Argentina: 1 species)
   Synxenoderus (western USA: 1 species)

Bedbugs have long been recognized as pests and were mentioned, for example, in medieval European texts and in the classical Greek writings of Aristophanes, Aristotle and Dioscorides (Usinger, 1966). The common bedbug, *Cimex lectularius*, is a widespread household pest in most parts of the world. The tropical bedbug, *C. hemipterus* (= *C. rotundatus*) is also common in tropical parts of Asia, Africa and Central America; it differs from *C. lectularius* only in minor morphological detail (e.g. pronotal shape, Fig. 14.1*b*) and perhaps greater tolerance to higher temperatures (Newberry, 1990).

Both species lay their eggs in cracks and crevices of houses and outbuildings (behind torn wallpaper was a common site for *C. lectularius* in European houses). Each female bedbug lays about 200 eggs (Fig. 14.1*c*) which hatch after about ten days at 20°C. The nymphs (Fig. 14.1*e*) and adults usually feed at night when their hosts are sleeping, although they will feed during the day if conditions are favourable. Feeding behaviour, and hence development, is critically dependent on temperature and humidity. Bedbugs do not feed at temperatures less than 13°C, and will not feed if their tarsi encounter moisture (Bell and Schaefer, 1966). Bedbug eggs will not live longer than three months, and so do not survive the winter in unheated buildings in temperate regions. However, nymphs and adults can survive over a year without feeding (especially at temperatures below 18°C), reappearing from their crevices when hosts become available. Thus they can be easily transported inadvertently amongst old bedding material or hidden in cracks in old furniture, and this is probably the main mechanism for dissemination of household infestations. Heavily infested premises can generally be recognized by faecal spots on walls and furniture, and by the characteristic odour emitted from the adult metasternal scent glands. This odour is a complex mixture of mainly hexanol and octenol with several minor components, which appears to provide alerting and assembly messages to the bugs (Levinson and Bar Ilan, 1971; Levinson et al., 1974). It is reminiscent of the aroma of the spice coriander, which name derives from 'koris' the Greek word for 'bug'.

Experimental evidence shows that bedbugs can be infected with a range of human parasites and pathogens, including hepatitis-B, HIV and *Trypanosoma cruzi*. But although bedbugs have been suspected of complicity in the transmission of a wide range of human pathogens (see Ryckman et al., 1981 for bibliography), there is little evidence that bedbugs are significant as vectors of any human pathogen. Nevertheless, domestic infestations with *Cimex* can be very distressing, and the bites can be extremely irritating and prone to secondary infection after scratching. Allergic reactions to the bites are also common (Ryckman et al., 1981). Bedbugs will feed daily if given the opportunity, and, because of their numbers, can contribute to chronic iron-deficiency anaemia, especially in infants (Usinger, 1966; Venkatachalam and Belavady, 1962).

Occasionally, amongst vagrants in temperate climates who are unable to change their clothes very often, *C. lectularius* has been found behaving as an ectoparasite, breeding within the person's clothing and even occupying the space beneath overgrown toenails (J. W. Maunder, personal communication).

Family Polyctenidae (batbugs)

The family Polyctenidae comprises 32 species grouped in five genera (Maa, 1964; Ryckman and Sjogren, 1980). Of these, 16 species in four genera occur in the Old World, while the 16 species of *Hesperoctenes* occur in the New World. They are small ectoparasites of bats, superficially similar to nycteribiid flies, with no known medical importance. They lack eyes and ocelli and are always flightless, with the fore wings reduced to small flaps. The rostrum has three segments, and the tarsi – uniquely for Hemiptera – have four. Combs of stout regularly spaced setae (like the ctenidia of fleas) are present on some parts of the body. Fertilization is traumatic and haemocoelic, as in Cimicidae and many Anthocoridae. The bugs are parthenogenetic: there is a pseudoplacenta and the large nymphs are born at an advanced stage of development. Those of *Hesperoctenes* moult only three times before becoming adult.

Family Reduviidae and its subfamily Triatominae (kissing-bugs)

Most Reduviidae are predators on insects and other invertebrates. They are predominantly tropical, occupying a very wide range of terrestrial habitats and displaying a variety of hunting strategies and prey preferences. Over 6000 species are known (Maldonado Capriles, 1990), now grouped into 23 subfamilies (Table 14.2). Despite their great variety of body form reduviids can be distinguished from other superficially similar Hemiptera by the combination of a distinct neck, the laterally inserted four-segmented filiform antennae, a relatively short three-segmented

Table 14.2 Provisional arrangement of the subfamilies and tribes of the Reduviidae

(Phymatine assemblage)	(Emesine assemblage)
PHYMATINAE	EMESINAE
Phymatini	Emesini
Macrocephalini	Leistarchini
Carcinocorini	Ploiariolini
Themonocorini	Metapterini
ELASMODEMINAE	Deliastini
HOLOPTILINAE	Collartidini
Holoptilini	SAICINAE
Dasycnemini	Saicini
Aradellini	Visayanocorini
CENTROCNEMINAE	
	(Harpactorine assemblage)
(Reduviine assemblage)	BACTRODINAE
CETHERINAE	HARPACTORINAE
Cetherini	Harpactorini
Pseudocetherini	Apiomerini
Euphenini	Rhaphidosomini
REDUVIINAE	Tegeini (including Phonolibini)
STENOPODAINAE	Diaspidini
SPHAERIDOPINAE	Ectinoderini
CHRYXINAE	
VESCIINAE	(Ectrichodiine assemblage)
SALYAVATINAE	ECTRICHODIINAE
MANANGOCORINAE	
PIRATINAE	(Tribelocephaline assemblage)
PHYSODERINAE	TRIBELOCEPHALINAE
TRIATOMINAE	(Hammacerine assemblage)
Triatomini	HAMMACERINAE (= MICROTOMINAE)
Rhodniini	
Bolboderini	(Phimophorine assemblage)
Cavernicolini	PHIMOPHORINAE
Alberprosenini	Phimophorini
	Mendanocorini

proboscis that, at rest, does not extend beyond the prosternum, and the almost universal presence of a stridulatory groove on the prosternum. Adults of most Reduviidae also have two pairs of metathoracic scent glands, although one or the other pair (rarely both) can be reduced or absent. The dorsal Brindley's glands, which primarily secrete isobutyric acid in all species so far examined (Games et al., 1974; Schofield, 1979a), appear to be unique to the Reduviidae (Carayon, 1950). The ventral metathoracic (metasternal) glands are paired and open to the metacoxal cavities; this is in contrast to the single metathoracic gland apparatus of most other Heteroptera which opens to visible orifices on the metapleura or metasternum (Carayon, 1971; Schofield and Upton, 1978). The secretion of the metathoracic gland has been characterized only in the largest of the Triatominae – *Dipetalogaster maxima* – as 3-methyl-2-hexanone and is repellent to the bugs; it may act as an alarm pheromone (Rossiter and Staddon, 1983).

In many predatory reduviids, the fore legs are adapted to hold prey. Often the fore legs (and sometimes the mid legs) are strongly raptorial, equipped with spines, adhesive organs (the fossae spongiosa of early taxonomists) and/or glands secreting a glue-like substance (Readio, 1927).

Some predatory reduviids are frequently found in bird or mammal nests, preying on the abundant invertebrate fauna of the nest litter. Clearly it is advantageous for any predator to adapt to habitats where prey are abundant. Progressive adaptation to the more permanent of nests would involve some simplification of features more appropriate to wider-ranging species, and could lead to evolutionary adaptation to exploit the nest-building vertebrate itself as a food-source. As argued earlier (Schofield, 1988), this seems a plausible sequence of adaptation from which the bloodsucking habit could have arisen several times amongst the Reduviidae, to give rise to those reduviids now grouped within the subfamily Triatominae.

On morphological and developmental characteristics, the position of the Triatominae as a subfamily of the Reduviidae is very clear. Many of the morphological details used to distinguish Triatominae from other reduviids, e.g. the straight rostrum (Fig. 14.3a) addressed to the gula, have probably been derived in association with adaptations for feeding on vertebrate hosts (Cobben, 1978; Schofield, 1988). Originally, the haematophagous Reduviidae were listed within the subfamily Reduviinae (then known as Acanthaspidinae). In 1919, Jeannel formally differentiated the haematophagous genera as the tribe Triatomini, later elevated to subfamily level by Usinger (1939). Unfortunately, the name 'Triatomidae' entered the literature in 1926 but no characters were given then (or since) to justify family status. In spite of this, many parasitologists continue to use the term, referring to 'triatomid bugs'. The correct term is triatomine bugs, so long as these hemipterans are treated as a subfamily of Reduviidae.

There are now 118 species of Triatominae recognized on the basis of morphological characters (Table 14.3). They range from the tiny *Alberprosenia goyovargasi*, whose adults are just 5 mm long, to the giant *Dipetalogaster maxima* whose adults can reach 45 mm long.

Table 14.3 Checklist of the Triatominae and a synopsis of the geographical distribution of the species

Species in which natural infection with *Trypanosoma cruzi* has been reported are marked with an asterisk (*) and the following letters used for annotations: D, domestic populations reported; L, attraction to light reported from field observations; P, peridomestic populations reported; S, sylvatic populations known.

Tribe ALBERPROSENINI
Genus **Alberprosenia**
    A. goyovargasi – northeastern Venezuela [S, forest]
    A. malheiroi – Brazil (Pará) [S, forest]

Tribe BOLBODERINI
Genus **Belminus**
    B. costaricensis – Costa Rica, Mexico (Vera Cruz) [S]
    B. herreri – Brazil (Pará), Panama [S, forest]
    B. peruvianus – Peru (upper Marañón valley) [S]
    B. rugulosus – Colombia, Venezuela [S]
Genus **Bolbodera**
    B. scabrosa – Cuba [S]
Genus **Microtriatoma**
    *M. borbai** – Brazil (Paraná) [S, bromeliads]
    M. trinidadensis – Bolivia, Brazil (Mato Grosso, Pará), Colombia, Panama, Peru, Surinam, Trinidad, Venezuela [L,S]
Genus **Parabelminus**
    *P. carioca** – Brazil (Rio de Janeiro) [S, palms]
    P. yurupucu – Brazil (Bahia) [S, bromeliads]

Table 14.3 (Continued)

Tribe CAVERNICOLINI
Genus **Cavernicola**
    C. lenti – Brazil (Pará) [S, forest]
    C. pilosa – Brazil (Bahia, Espirito Santo, Matto Grosso, Pará), Colombia, Ecuador, Panama, Venezuela [S, bat caves]

Tribe RHODNIINI
Genus **Psammolestes**
    P. arthuri* – Colombia, Venezuela [S, birds' nests]
    P. coreodes – Argentina (Catamarca, Chaco, Córdoba, Corrientes, Entre Ríos, Formosa, Jujuy, La Rioja, Salta, Santa Fe, Santiago del Estero, Tucumán), Bolivia, Paraguay [S, birds' nests]
    P. tertius – Brazil (Bahia, Ceará, Goiás, Mato Grosso) [S, birds' nests]
Genus **Rhodnius**
    R. brethesi – Brazil (Amazonas, Pará), Colombia, Venezuela [S, palms]
    R. dalessandroi – Colombia [? S]
    R. domesticus* – Brazil (Bahia, Espirito Santo, Rio de Janeiro, Paraná, Santa Catarina, São Paulo) [S, bromeliads]
    R. ecuadoriensis* – Ecuador, northern Peru [D,P,S]
    R. nasutus* – Brazil (Ceará, Piauí, Rio Grande do Norte) [P,S]
    R. neglectus* – Brazil (Bahia, Distrito Federal, Goiás, Mato Grosso, Minas Gerais, Paraná, São Paulo) [D,P,S, palms]
    R. neivai – Colombia, Venezuela [S]
    R. pallescens – Colombia, Panama [D,P,S]
    R. paraensis* – Brazil (Pará) [S, forest]
    R. pictipes* – Bolivia, Brazil (Amazonas, Goiás, Mato Grosso, Pará), Colombia, Ecuador, French Guiana, Guyana, Peru, Surinam, Trinidad, Venezuela [L,S, palms]
    R. prolixus* – Colombia, Costa Rica, El Salvador, French Guiana, Guatemala, Guyana, Honduras, Mexico (Oaxaca, Chiapas), Nicaragua, Venezuela [L,P,D,S, palms]
    R. robustus* – Brazil (Amazonas, Pará), Colombia, Ecuador, French Guiana, Peru, Venezuela [S, palms and bromeliads]

Tribe TRIATOMINI
Genus **Dipetalogaster**
    D. maxima* – Mexico (Baja California) [S, rocks]
Genus **Eratyrus**
    E. cuspidatus* – Colombia, Ecuador, Guatemala, Panama, Venezuela [L,S]
    E. mucronatus* – Bolivia, Brazil (Amazonas, Mato Grosso, Pará), Colombia, French Guiana, Guyana, Peru, Surinam, Trinidad, Venezuela [L,S]
Genus **Linshcosteus**
    L. carnifex – India [S, only females known]
    L. chota – southern India [S, only males known]
    L. confumus – southern India (Bangalore, Mysore) [S, rocks]
    L. costalis – southern India (Bangalore, Mysore) [S, rocks]
    L. kali – southern India (Coimbatore, Madras) [S]
Genus **Panstrongylus**
    P. chinai* – Ecuador, Peru [D,L,S]
    P. diasi – Bolivia, Brazil (Bahia, Goiás, Minas Gerais, São Paulo) [S, dry conditions]
    P. geniculatus* – Argentina (Chaco, Corrientes, Formosa, Misiones, Santa Fe, Santiago del Estero), Bolivia, Brazil (Acre, Amapá, Amazonas, Bahia, Distrito Federal, Ceará, Espirito Santo, Goiás, Maranhão, Mato Grosso, Minas Gerais, Pará, Paraná, Rio de Janeiro, Rondônia, São Paulo), Colombia, Costa Rica, Ecuador, French Guiana, Guyana, Nicaragua, Panama, Paraguay, Peru, Surinam, Trinidad, Uruguay, Venezuela [L,P,S, humid conditions]
    P. guentheri* – Argentina (Buenos Aires, Catamarca, Chaco, Chubut, Córdoba, Corrientes, Entre Ríos, Jujuy, La Pampa, Mendoza, Neuquén, Río Negro, Salta, San Juan, San Luis, Santa Fe, Santiago del Estero, Tucumán), Bolivia, Paraguay [L,S]
    P. herreri* – Peru [D,P,S]

Table 14.3   *(Continued)*

*P. howardi** – Ecuador [D, rare]
*P. humeralis** – Panama [L,S]
P. lenti – Brazil (? Goiás) [only one specimen known]
*P. lignarius** – Brazil (Amazonas, Pará), Guyana, Surinam, Venezuela [S, forest]
*P. lutzi** – Brazil (Bahia, Ceará, Paraíba, Pernambuco, Piauí, Rio Grande do Norte) [D,S occasionally]
*P. megistus** – Argentina (Corrientes, Misiones), Brazil (Alagoas, Bahia, Ceará, Espirito Santo, Goiás, Maranhão, Mato Grosso, Minas Gerais, Pará, Paraíba, Paraná, Pernambuco, Piauí, Rio de Janeiro, Rio Grande do Norte, Rio Grande do Sul, Santa Catarina, São Paulo, Sergipe), Paraguay [D,L,P,S]
*P. rufotuberculatus** – Bolivia, Brazil (Amazonas, Mato Grosso, Pará), Colombia, Costa Rica, Ecuador, Mexico (Campeche), Panama, Peru, Venezuela [L,S]
P. tupynambai – Brazil (Rio Grande do Sul), Uruguay [S, under rocks]
   Genus **Paratriatoma**
P. hirsuta – Mexico (Baja California, Sonora), USA (Arizona, California, Nevada) [S, woodrat nests]
   Genus **Triatoma**
T. amicitiae – Sri Lanka [only one specimen known]
T. arthurneivai – Brazil (Bahia, Minas Gerais, Paraná, São Paulo) [S, rocks]
*T. barberi** – Mexico (Colima, Guerrero, Hidalgo, Jalisco, México DF, Michoacán, Morelos, Oaxaca, Puebla, Tlaxcala) [D,P,S]
T. bolivari – Mexico (Colima, Jalisco, Nayarit) [L, female unknown]
T. bouvieri – Nicobar Is, Philippines, Vietnam [S]
T. brailovskyi – Mexico (Colima, Jalisco, Nayarit) [L, female unknown]
*T. brasiliensis** – Brazil (Alagoas, Bahia, Ceará, Minas Gerais, Paraíba, Pernambuco, Piauí, Rio Grande do Norte, Sergipe [D,P,S, rock piles]
T. breyeri – Argentina (Catamarca, La Rioja) [L,P,S]
T. bruneri – Cuba [S]
*T. carrioni** – southern Ecuador, northern Peru [L,P,S]
T. cavernicola – Malaysia [S, bat caves]
*T. circummaculata** – Argentina (Buenos Aires), Brazil (Rio Grande do Sul), Uruguay [S, rocks]
*T. costalimai** – Brazil (Goiás) [L,S, rocks]
T. deanei – Brazil (Goiás) [S]
*T. delpontei** – Argentina (Catamarca, Chaco, Córdoba, Formosa, Jujuy, La Pampa, La Rioja, Salta, Santiago del Estero), Paraguay, Uruguay [P,S, birds' nests]
*T. dimidiata** – Belize, Colombia, Costa Rica, Ecuador, El Salvador, Guatemala, Honduras, Mexico (Campeche, Chiapas, Jalisco, Oaxaca, Puebla, Quintana Roo, San Luis Potosi, Tabasco, Vera Cruz, Yucatán), Nicaragua, northern Peru, Venezuela [D,L,P,S]
*T. dispar** – Colombia, Costa Rica, Ecuador, Panama [S, forest]
*T. eratyrusiformis** – Argentina (Catamarca, Chubut, Córdoba, La Pampa, La Rioja, Mendoza, Neuquén, Río Negro, San Juan, San Luis, Salta, Tucumán) [L,P,S]
T. flavida – Cuba [S, bat caves]
*T. gerstaeckeri** – Mexico (Chihuahua, Coahuila, Nuevo León, San Luis Potosi, Tamaulipas), USA (New Mexico, Texas) [P,S]
*T. guasayana** – Argentina (Buenos Aires, Catamarca, Chaco, Córdoba, Jujuy, La Pampa, La Rioja, Mendoza, Salta, San Juan, San Luis, Santa Fe, Santiago del Estero, Tucumán), Bolivia, Paraguay [P,S]
T. guazu – Paraguay [only one specimen known]
*T. hegneri** – Mexico (Quintana Roo) [S, rare]
T. incrassata – Mexico (Sonora), USA (Arizona) [S]
T. indictiva – Mexico (Chihuahua, Sinaloa), USA (Arizona, New Mexico, Texas) [S]
*T. infestans** – Argentina (Buenos Aires, Catamarca, Chaco, Chubut, Córdoba, Corrientes, Entre Ríos, Formosa, Jujuy, La Pampa, La Rioja, Mendoza, Misiones, Neuquén, Río Negro, Salta, San Juan, San Luis, Santa Fe, Santiago del Estero, Tucumán), Bolivia, Brazil (Alagoas, Ceará, Bahia, Goiás, Maranhão, Mato Grosso, Minas Gerais, Paraíba, Paraná, Pernambuco, Piauí, Rio de Janeiro, Rio Grande do Sul, Santa Catarina, São Paulo), Chile, Paraguay, southern Peru, Uruguay [D,P, except S in Cochabamba, Bolivia]

Table 14.3 *(Continued)*

T. lecticularia* – Mexico (Nuevo León), USA (Arizona, California, Florida, Georgia, Illinois, Kansas, Louisiana, Maryland, Missouri, New Mexico, North Carolina, Oklahoma, Pennsylvania, South Carolina, Tennessee, Texas) [S, occasionally D,P]

T. lenti* – Brazil (Bahia) [S, occasionally D,P]

T. leopoldi – Australia (northern Queensland), Indonesia (Sulawesi, West Irian), Papua New Guinea [L,S]

T. limai – Argentina (Córdoba) [S, rocks]

T. longipennis* – Mexico (Aguascalientes, Chihuahua, Colima, Jalisco, Nayarit, Sinaloa, Zacatecas) [S, occasionally P]

T. maculata* – Aruba, Bonaire, Brazil (Roraima), Colombia, Curacao, Guyana, Surinam, Venezuela [D,P,S]

T. matogrossensis* – Brazil (Mato Grosso) [P,S]

T. matsunoi – northern Peru [S, caves]

T. mazzottii* – Mexico (Durango, Guerrero, Michoacán, Nayarit, Oaxaca) [D,P,S]

T. melanocephala* – Brazil (Bahia, Paraíba, Pernambuco) [S, occasionally D]

T. mexicana – Mexico (Hidalgo, Querétaro) [L,S, only three specimens known]

T. migrans – India (Sikkim), Indonesia (Borneo, Java, Sumatra), Malaysia, Philippines, Sarawak, Thailand [S]

T. neotomae* – USA (Arizona, California, New Mexico, Texas) [S, woodrat nests]

T. nigromaculata* – Colombia, Venezuela [L,P,S]

T. nitida – Costa Rica, Guatemala, Honduras, Mexico (Yucatán) [S]

T. obscura – Jamaica [S]

T. oliveirai – Brazil (Rio Grande do Sul) [S, collected only twice]

T. pallidipennis* – Mexico (Colima, Guerrero, Jalisco, Michoacán, Morelos, Nayarit, Puebla) [D,P,S]

T. patagonica* – Argentina (Buenos Aires, Catamarca, Chubut, Córdoba, Corrientes, Entre Ríos, La Pampa, La Rioja, Mendoza, Neuquén, Río Negro, Salta, San Juan, San Luis, Santa Fe) [L,P,S]

T. peninsularis* – Mexico (Baja California) [L,S]

T. petrochii – Brazil (Bahia, Pernambuco, Rio Grande do Norte) [S, dry conditions]

T. phyllosoma* – Mexico (Oaxaca) [P,S]

T. picturata – Mexico (Colima, Jalisco, Nayarit, Oaxaca) [P,S]

T. platensis* – Argentina (Buenos Aires, Catamarca, Chaco, Córdoba, Corrientes, Entre Ríos, Formosa, Jujuy, La Pampa, La Rioja, Mendoza, Río Negro, San Juan, San Luis, Santa Fe, Santiago del Estero, Salta, Tucumán), southern Bolivia, Brazil (Rio Grande do Sul), southern Paraguay, Uruguay [S, birds' nests, occasionally P]

T. protracta* – Mexico (Baja California, Coahuila, Chihuahua, Durango, Nuevo León, San Luis Potosi, Sinaloa, Sonora, Tamaulipas, Zacatecas), USA (Arizona, California, Colorado, Nevada, New Mexico, Texas, Utah) [L,P,S especially woodrat nests]

T. pseudomaculata* – Brazil (Alagoas, Bahia, Distrito Federal, Ceará, Goiás, Minas Gerais, Paraíba, Pernambuco, Piauí, Rio Grande do Norte, Sergipe) [P,S, occasionally D]

T. pugasi – Indonesia (Java) [?S, rare]

T. recurva* – Mexico (Chihuahua, Nayarit, Sinaloa, Sonora), USA (Arizona) [S, woodrat nests]

T. rubida* – Mexico (Baja California, Nayarit, Sinaloa, Sonora, Vera Cruz), USA (Arizona, California, New Mexico, Texas) [S, woodrat nests]

T. rubrofasciata* – Andaman Is, Angola, Antigua, Argentina (Buenos Aires), Azores, Bahamas, Brazil (Alagoas, Bahia, Maranhão, Pará, Paraíba, Pernambuco, Rio de Janeiro, Rio Grande do Norte, São Paulo, Sergipe), Burma, Cambodia, Caroline Is, Central African Republic, China, Comoro Is, Cuba, Dominican Republic, French Guiana, Grenada, Guadelupe, Haiti, Hawaiian Is, Hong Kong, India (Assam), Indonesia (Borneo, Java, Sumatra), Jamaica, Japan (Okinawa), Madagascar, Malaysia, Martinique, Mauritius, Papua New Guinea, Philippines, Saudi Arabia, Seychelles, Sierra Leone, Singapore, South Africa, Sri Lanka, St Croix, St Vincent, Taiwan, Tanzania, Thailand, Trinidad, USA (Florida), Venezuela, Vietnam, Virgin Is [D,P,S, especially associated with rats]

Table 14.3 *(Continued)*

T. rubrovaria* – Argentina (Corrientes, Entre Ríos, Misiones), Brazil (Paraná, Rio Grande do Sul), Uruguay [S, rocks, occasionally P]
T. ryckmani – Guatemala, Honduras [S, rare]
T. sanguisuga* – USA (Alabama, Arizona, Arkansas, Florida, Georgia, Illinois, Indiana, Kansas, Kentucky, Louisiana, Maryland, Mississippi, Missouri, North Carolina, Ohio, Oklahoma, Pennsylvania, South Carolina, Tennessee, Texas, Virginia) [P,S]
T. sinaloensis* – Mexico (Sinaloa, Sonora) [S, woodrat nests]
T. sinica – China (Nanking) [?S]
T. sordida* – Argentina (Buenos Aires, Chaco, Córdoba, Corrientes, Formosa, Jujuy, La Rioja, Misiones, Salta, Santa Fe, Santiago del Estero, Tucumán), Bolivia, Brazil (Bahia, Goiás, Mato Grosso, Minas Gerais, Paraná, Pernambuco, Piauí, Rio Grande do Sul, Santa Catarina, São Paulo), Paraguay, Uruguay [D,P,S]
T. spinolai* – Chile [S rocks, P, occasionally D]
T. tibiamaculata* – Brazil (Bahia, Espirito Santo, Minas Gerais, Paraná, Rio de Janeiro, Santa Catarina, São Paulo, Sergipe) [S]
T. venosa* – Colombia [S, high altitudes about 1600–2200 m]
T. vitticeps* – Brazil (Bahia, Espirito Santo, Minas Gerais, Rio de Janeiro) [S, occasionally D]
T. williami* – Brazil (Goiás, Mato Grosso) [S, occasionally D]
T. wygodzinskyi – Brazil (Minas Gerais) [S]

The Triatominae form a relatively homogenous group, similar in form, biology and behaviour. Unlike many other groups of medically important arthropods, they present few taxonomic problems. All species are obligate bloodsuckers, and over half have been shown naturally or experimentally to be susceptible to infection with *Trypanosoma cruzi*, the causative agent of Chagas disease. Moreover, because of their similar biology, all species should be considered potential vectors of this parasite, although some are more efficient than others. Epidemiologically only about a dozen species have become sufficiently closely associated with man to represent a public health problem; of these, the most important vector species are *Triatoma infestans*, *Panstrongylus megistus*, *Rhodnius prolixus*, *T. brasiliensis* and *T. dimidiata* (Schofield, 1985b). The other species are mainly associated with nest-building birds and small mammals, and occasionally reptiles (for host lists see: Barretto, 1979; Ryckman, 1986c). It is clear, however, that many sylvatic species have the capacity to adapt to closer association with man, and many assume greater epidemiological significance as they invade and/or colonize rural houses from which the more domestic species have been eliminated by control campaigns.

The main anatomical features of adult triatominae are shown in Figs 14.2 and 14.3*a, e*. Adults and nymphs have a pair of compound eyes and antennae, and the position of insertion of the antennae relative to the eyes is used to distinguish important genera (Fig. 14.3*b–d*). Adult Triatominae also have a pair of light-sensitive ocelli just behind the eyes, but nymphs do not have ocelli. The background colour of Triatominae is usually black or brown with similarly coloured eyes, although white and red eye forms have been recorded for several species. Eye colour is genetically determined; red-eye is an autosomal recessive character.

Adults of many species have bands or patches of colour – red, brown, yellow, orange or pink – along the connexivum and similar markings on the legs and thorax. Both sexes have fully developed external genitalia, but those of the male are normally reflexed into a cuticular cavity and not usually visible. Thus, when viewed from above, the tip of the male abdomen usually appears smoothly rounded,

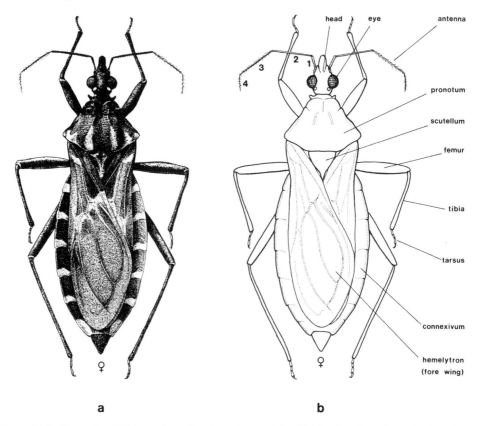

Figure 14.2 The reduviid kissing-bug *Panstrongylus megistus* (Triatominae), an important vector of Chagas disease: (*a*) general appearance of the adult female; (*b*) basic structure of the adult bug with the four antennal segments numbered. (Figure (*a*) reproduced from Lent and Wygodzinsky, 1979.)

compared to the pointed or lobed tip of the female abdomen (e.g. Fig. 14.2*a*). With the exception of the genitalia, both sexes have a similar appearance; female bugs are generally larger than males, although this is not always so.

Nymphs

Nymphs of Triatominae and most other reduviid subfamilies can be distinguished from those of other Hemiptera by their cone-shaped horizontal head and straight rostrum extending just to the first pair of legs. Amongst the Triatominae, adults and nymphs of all but *Cavernicola* and *Linshcosteus* have a prosternal stridulatory groove into which the tip of the proboscis reaches. The purpose of stridulation in Triatominae is unknown, but the sound production involves a wide spectrum of sonic and ultrasonic 'chirps' extending up to 100 kHz (Schofield, 1977). Triatomine nymphs differ from those of other reduviid subfamilies by their stout body and absence of either glandular hairs or abdominal scent glands. These glands are present on the urotergites of other reduviid nymphs except those of Saicinae and Emesinae, which have narrow elongate bodies. However, nymphs of Triatominae are much harder to distinguish from each other, especially since those collected from the wild may be 'camouflaged' by adhesion of fine material

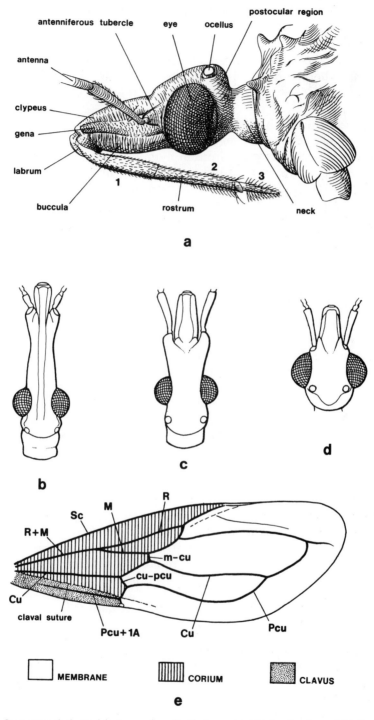

Figure 14.3 Some morphological features of medically important triatomine bugs: (*a*) left side view of the head of *Panstrongylus megistus* (segments of rostrum numbered); (*b–d*) comparison of head shape (dorsal view) in the three genera containing important vectors of Chagas disease, (*b*) *Rhodnius*, (*c*) *Triatoma* and (*d*) *Panstrongylus*; (*e*) typical wing structure of a reduviid bug. Note in (*b–d*) (not to scale) how the antennae are attached near the eyes on the short and robust head in *Panstrongylus*, far from the eyes and near the tip of the very long head in *Rhodnius*, and in an intermediate position in *Triatoma*.

from the substrate of their habitat. Nymphs of *T. costalimai*, for example, are black but when collected from typical habitats amongst limestone rocks appear pale grey because of limestone dust adhered into cuticular folds. Several authors have remarked on this so-called 'camouflage phenomenon' (e.g. Zeledón, 1981; Zeledón et al., 1973), whereby nymphs can be observed flicking dust over themselves; particles of up to 10 µm will then adhere in the cuticular architecture of the nymphs.

A further difficulty in the determination of triatomine nymphs is that many have not been formally described. Lent and Wygodzinsky (1979) provide keys to tribes and genera of Triatominae for fifth-instar nymphs (which will normally be applicable to third- and fourth instars) and for first-instar nymphs (which will generally key out second instars). In all cases, however, it is preferable to rear nymphs to adults in the laboratory (by feeding them on an appropriate laboratory host) in order to confirm their identification.

Classification and identification of Triatominae

The Triatominae can be uncontroversially grouped into five tribes and 14 genera (Table 14.3). Tribes and genera, and most species, can be easily distinguished on external morphological characters of adults; certain species can also be identified from the anatomy of the genitalia (Lent and Wygodzinsky, 1979). However, some species of *Rhodnius* are difficult to determine on these grounds. Barata (1981) provides a key to the eggs of *Rhodnius*, and biochemical analysis of adults and nymphs has now been proposed to clarify the distinctions between *R. prolixus*, *R. neglectus*, *R. robustus* and *R. pictipes* (Dujardin et al., 1988).

The typical chromosome complement of Triatominae is 2n = 20 + XY, although multiple X chromosomes seem quite common (Schofield, 1988; Usinger et al., 1966). There tends to be low chiasma frequency and the chromosomes are holocentric (i.e. with a diffuse centromere). Visualization of G-bands and C-bands is possible in chromosomes from embryonic and gonadic tissue (Maudlin, 1974; Perez et al., 1989) and may prove to have taxonomic value.

In their detailed monograph, Lent and Wygodzinsky (1979) list 112 species with keys (in English, Spanish and Portuguese) and descriptions of all but one, *Rhodnius dalessandroi* (which they imply may be of doubtful validity). To this list have been added *Alberprosenia malheiroi* (Serra et al., 1987), *Cavernicola lenti* (Barrett and Arias, 1985), *Triatoma bruneri* (Usinger, 1944: re-erected by Lent and Jurberg, 1981), *T. brailovskyi* (Martínez et al., 1984), *T. bolivari* (Carcavallo et al., 1987) and *T. matsunoi* (Fernandez-Loayza, 1989). *Triatoma gallardoi*, described from a single specimen in 1986, is now considered synonymous with *T. patagonica* (Carcavallo and Martínez, 1987).

Recent (but incomplete) checklists to the Triatominae, with synonymies, are given by Maldonado Capriles (1990) and by Ryckman (1986a) for South America, Ryckman (1984) for North and Central America and the West Indies, Ryckman and Casdin (1976) for western North America, Zarate and Zarate (1985) for Mexico, and by Ryckman and Archbold (1981) for Old World species. Etymology of many of the specific names is summarized by Ryckman (1986b), while the range of common names for Triatominae is given by Ryckman and Blankenship (1984). A pictorial key to the common Venezuelan species is given by Carcavallo and Tonn (1976).

Identification key to the genera of adult Triatominae

1. Ocelli inconspicuous, not raised, situated among coarse granules at general level of head surface.. 2
— Ocelli conspicuous, raised on distinct elevations on postocular region of head (Fig. 14.3a) 6
2. Head usually elongate and subconical, not strongly convex above when seen in side view; genae large and elongate, extending beyond end of clypeus by distance at least equal to clypeal width; antenniferous tubercles positioned far in front of eyes. Wing corium with veins distinct. Body surface rough (rugose and granulose)... 3
— Head ovoid and strongly convex dorsally when seen in side view; genae smaller, not extending beyond end of clypeus; antenniferous tubercles positioned near eyes. Wing corium with veins obsolete. Body surface smooth but hairy. [Panama and northern South America].................................. **Cavernicola**
3. Rostrum with first segment much shorter than second segment. Scutellum with its margins not produced near base to form more or less triangular lobes. Connexivum not ridged on outer edge.. 4
— Rostrum with first segment as long as or longer than second segment. Scutellum with each margin produced near base to form a large and more or less triangular lobe. Connexivum distinctly ridged along outer edge. [Central and northwestern South America].. **Belminus**
4. Scutellum triangular and with well-developed median posterior process (e.g. as in Fig. 14.2b)... 5
— Scutellum trapezoidal and without median posterior process (its hind margin straight). [Southeastern Brazil]... **Parabelminus**
5. Genae produced to a sharp point. Femora with spines on under surface. Tarsi with three segments. [Cuba] .... **Bolbodera**
— Genae not pointed, laterally compressed and with rounded apex. Femora without spines. Tarsi with two segments. [Costa Rica to northern half South America].. **Microtriatoma**
6. Head behind eyes with an area of many close-set swellings bearing hairs. Antennae attached near front of head (Fig. 14.3b)... 7
— Head behind eyes without such haired swellings. Antennae not attached near front end of head (Fig. 14.3c, d) 8
7. Head more or less triangular, somewhat flattened, clearly less than twice as long as its width across eyes; rostrum strongly flattened dorsoventrally and with cleft tip. Femora strongly widened and laterally compressed. [South America east of Andes].. **Psammolestes**
— Head more or less cylindrical, not flattened, at least twice as long as its width across eyes; rostrum almost cylindrical and with pointed tip. Femora of most species elongate and more or less cylindrical, never laterally compressed. [Widespread in Central and northern South America]...................... **Rhodnius**
8. Length more than 5 mm. Head longer than its width across eyes. Fore wings (hemelytra) without small branch vein connecting basal portion of vein R + M to vein Sc 9
— Length approximately 5 mm. Head very short and wide, not longer than its width across eyes. Fore wings (hemelytra) near base with small oblique branch vein connecting vein R + M to vein Sc. [Venezuela] ..... **Alberprosenia**

9 Head shape varied but usually more or less cylindrical (Fig. 14.3c) (if comparatively short then head and body with long and conspicuous hairs); antenniferous tubercles usually not close to eyes (if rather near, then head and body with long and conspicuous hairs) 10
— Head very short and wide (Figs 14.2a, 14.3d); antenniferous tubercles positioned very close to eyes (Fig. 14.3a). Head and body bare or with short setae pressed closely together. [Widespread in Neotropical region] .... **Panstrongylus**
10 Rostrum extending back as far as prosternum. Prosternum with median longitudinal groove (stridulatory sulcus) 11
— Rostrum not extending back beyond level of eyes. Prosternum without stridulatory sulcus. [India] .... **Linshcosteus**
11 Rostrum with first segment distinctly shorter than second, its apex not reaching beyond level of antenniferous tubercle. Scutellum with apical part not drawn out into sharply pointed spine of a length at least equal to body of scutellum.. 12
— Rostrum with first segment very long, almost as long as second and reaching point midway between antenniferous tubercle and eye. Scutellum with apical part formed into a sharply pointed spine equal in length to or longer than body of scutellum. [Central America and northern South America] .... **Eratyrus**
12 Head, body and appendages appearing bare or with short hairs, or only appendages with long hairs. Antenniferous tubercles positioned far in front of eyes, at or near middle of anteocular part of head. Fore femora of most species with two or more ventral denticles. Length 9.5–42 mm 13
— Head, body and appendages with abundant, long and curved, semi-erect hairs. Antenniferous tubercles positioned near to eyes. Fore femora without denticles. Length 12.5–14.5 mm. [Southwestern USA and northwestern Mexico]..... .... **Paratriatoma**
13 Very large bugs, length 33–45 mm. Femora without denticles. Connexivum with ventral plates imperceptible. [Mexico (Baja California)] .... **Dipetalogaster**
— Smaller bugs, length usually less than 30 mm (rarely 33 mm or more). Femora with or without denticles. Connexivum with ventral plates clearly defined (but sometimes very narrow). [Widespread in New World and Old World tropics and subtropics] .... **Triatoma**

Distribution of the Triatominae
Of the 118 recognized species of Triatominae (Table 14.3), 105 occur only in the New World, roughly from the Great Lakes in the USA to the south of Argentina (i.e. from latitude 42°N to 46°S) (Fig. 14.4). One genus, *Linshcosteus*, with five species, is confined to the Indian subcontinent; seven species of *Triatoma* are known only from South East Asia, and one species, *T. rubrofasciata*, is cosmopolitan in the tropics. Carcavallo (1988) lists the New World species by country distribution, ecological zone and habitat.

Given the abundance and diversity of the Triatominae in the New World, the origin of the Old World species has been the subject of considerable debate. The most recent explanation (Schofield, 1988) considers the Triatominae as a polyphyletic group, representing an assemblage of originally predatory reduviids in which the bloodsucking habit has probably arisen several times. Under this hypothesis,

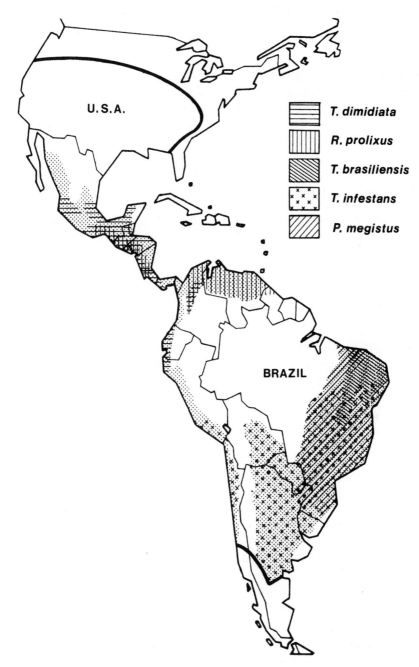

Figure 14.4 Map showing the approximate extent of human infection with *Trypanosoma cruzi* (stippled area), cause of Chagas disease, and of the distributions of the five most important vectors. Heavy lines in North and South America (at about 42°N and 46°S) mark the approximate northern and southern distribution limits of sylvatic Triatominae. *P.* = *Panstrongylus*; *R.* = *Rhodnius*; *T.* = *Triatoma*.

the Indian *Linshcosteus* is considered to have arisen independently from other Triatominae, while the other South East Asian species (*T. amicitiae*, *T. bouvieri*, *T. leopoldi*, *T. migrans*, *T. pugasi* and *T. sinica*) are considered to have developed from forms of *T. rubrofasciata* spread from northeastern Brazil.

In the New World, species of Triatominae are to be found in almost any habitat offering a degree of permanence, climatic shelter, and ready access to a blood source. The smaller genera of Triatominae tend to occupy more or less discrete geographic regions, as do the recognized species-groups within the genus *Triatoma* – supporting the idea of a polyphyletic origin for the subfamily. The most northerly species is *Triatoma sanguisuga*, widely distributed in the USA from Ohio and Pennsylvania to the Caribbean coast. It occurs in several sylvatic habitats such as hollow trees inhabited by raccoons and opossums, and in nests of woodrats and cotton rats. Peridomestic colonies of *T. sanguisuga* have been reported from stables, chicken coops and other animal enclosures, and the species has been reported infected with equine encephalitis virus (Kitselman and Grundmann, 1940). Adults of *T. sanguisuga* frequently enter houses and will readily bite man, but domestic colonies have not been reported.

To the south, *T. infestans* is recorded from houses in Melipilla in Chile (just south of Santiago), but occurs much further south in Argentina where it is presumably protected from cold climatic extremes by living exclusively inside houses.

BIOLOGY OF THE TRIATOMINAE

The Triatominae are defined on the basis of their obligate haematophagy. Their similarities of morphology, behaviour and reproductive strategy thus reflect the nature of their predatory ancestors coupled with behavioural and functional modifications demanded by adaptation to feed on vertebrate blood. The form of the proboscis, for example, is demanded by the need to penetrate the soft skin of vertebrates without causing undue disturbance to the host. Triatominae are relatively large insects that need to feed for up to 30 minutes to engorge fully, yet, unlike bloodsucking Diptera, they are unable to fly quickly from a disturbed host to avoid host defensive behaviour. Thus amongst Triatominae a painful bite can be regarded as a primitive character, while species more closely adapted to man, such as *T. infestans*, tend to have less perceptible bites. Species that feed mainly from diurnal hosts are nocturnal, feeding when the host is at rest, while species that feed from nocturnal hosts seem to have diurnal activity patterns (Constantinou, 1984a, b; Schofield, 1979b; Settembrini, 1984). There are several other examples of apparent convergence in the Triatominae, such as the relatively long and slender legs of many species inhabiting dry rocky areas, and the special (although structurally different) connexival modifications allowing great abdominal expansion in *T. spinolai* females and both sexes of *Dipetalogaster maxima* – both species that inhabit extremely arid regions.

Life cycle

The life cycle of Triatominae proceeds from eggs through five nymphal stages to the adult bugs. All stages of the nymphs and both sexes of adults are obligate blood-suckers, and usually occupy the same habitat. The life cycle of triatomine bugs is long compared to most other medically important insects. Even the smaller species such as *Rhodnius prolixus* take about three to four months to complete the egg-to-adult development cycle under good conditions. However, most species take 5–12 months to complete their development, depending on temperature and availability of blood-meals (Szumlewicz, 1976). Most species of Triatominae tolerate a range of atmospheric humidity between 30 and 80% – even species normally inhabiting more arid regions, such as *T. spinolai* from the Atacama region of Chile. However, some forest species such as *Panstrongylus geniculatus* are difficult to breed at humidities much less than 100% RH. Most species thrive at temperatures between 24 and 28°C; development is more rapid at higher temperatures but there is greater mortality and reproduction can be impaired (Okasha, 1970). Development is usually halted at temperatures below 16°C, while temperatures above 40°C are usually lethal.

Mating

Female bugs are ready to copulate two to three days after emergence from the fifth-instar nymph. Copulation is side-to-side, with the male slightly above the female, grasping the female genitalia with the pair of claspers on the genital capsule. Although females usually mate several times, a single mating can provide sufficient sperm which is then stored in the female's spermatheca. Sex pheromones are known only from *R. prolixus* and *T. mazzottii* (Baldwin et al., 1971; Ondarza et al., 1986; Rojas et al., 1990; Schofield and Moreman, 1976), where pheromones appear to mediate attraction of male bugs to copulating pairs. Laboratory crosses show that several species can produce interspecific hybrids (some of which are fertile) and wild hybrids have been reported between *T. infestans* and *T. platensis* (Usinger et al., 1966).

Adult bugs of some species can live for a year or more under good laboratory conditions, but various field studies suggest a mean adult lifespan of three to six months depending on species and conditions (cf. Gorla and Schofield, 1989). During this time, a fertilized female will lay 100–600 eggs. Egg-laying is dependent on blood intake, and usually begins a few days after each feed. Eggs are laid a few at a time, during the bug's circadian period of activity (Constantinou, 1984b). Bugs occupying arboreal habitats, such as *Psammolestes* and *Rhodnius*, tend to glue their eggs in batches to the substrate, while most other species lay their eggs loose, a few at a time. Eggs of *Psammolestes* are glued in clumps; this apparently offers some protection from microhymenopterous egg parasitoids since only the outer eggs of each clump become parasitized (Feliciangeli et al., 1978).

Eggs and nymphs

The eggs are oval, with an operculum through which the first-instar nymph will emerge. They are usually pearly-white, darkening to reddish pink as the embryo develops, although *Rhodnius* eggs are already pink when laid. The surface architecture of triatomine eggs is of taxonomic value (Cobben, 1968) and has been used to help distinguish between members of the *T. protracta* complex in North America (Ryckman, 1962), and members of the genus *Rhodnius* (Barata, 1981). The eggs hatch after about 10–30 days, depending on temperature.

Newly emerged nymphs are soft and pinkish, but the cuticle soon hardens and darkens, and within about two to three days the nymph is ready for its first blood-meal. At least one full blood-meal (or several smaller meals) is required for moulting to each successive nymphal stage, but nymphs can survive for several weeks without feeding if no host is available.

Successive nymphal stages differ from each other in minor morphological detail, but they can be readily distinguished by the size of the head capsule and thickness of the legs. These criteria can also distinguish the exuviae remaining after each moult. Overall body size is not a good criterion to distinguish nymphal stages because of the great size difference between fed and unfed bugs of the same stage. The wing pads are clearly visible on the fifth-stage nymphs, but can also be discerned on the fourth-stage nymphs. Future males and females can usually be determined by examination of the posterior sternites of fifth-, fourth- and sometimes third-stage nymphs (Gillett, 1935).

Blood-feeding

The five nymphal stages normally feed from the same variety of available hosts as the adult bugs. ATP and its analogues are the main stimulants of engorgement, detected in the ingested fluid by sensilla within the food canal (Friend and Smith, 1977, 1982). Some species appear to show particular host preferences from analysis of blood-meals. In many cases, however, this merely reflects the availability of hosts in the habitat occupied by the bugs. Most species of Triatominae taken from domestic or peridomestic habitats can be fed on a range of laboratory animals, including chicks, pigeons, mice, rats, guinea-pigs and rabbits. However, the more sylvatic species often have more precise host requirements (cf. Miles et al., 1981a).

If able to feed to repletion, triatomine nymphs may take up to nine times their own weight of blood, while adults take about two to four times their own weight. Thus fifth-stage nymphs usually take the largest quantity of blood, often in the range 300–1000 mg for species associated with man. The largest known species, *Dipetalogaster maxima*, can take 4 g or more. However, most species have considerable ability to withstand starvation, often surviving several months if no host is available.

Sometimes a single replete meal is sufficient to start the moult to the next stage. During or soon after feeding, the bugs begin to defaecate, voiding excess water from the blood-meal together with suspended spheres of uric acid (their nitrogenous waste product from metabolism) and haem from the undigested part of the previous blood-meal. Bug faeces on the walls of infested houses appear as very characteristic black and white streaks, due to the haem and uric acid respectively (Schofield et al., 1986).

The method by which triatomine bugs detect and orientate to their hosts is not entirely understood. Carbon dioxide appears to act as an alerting signal, but orientation and probing activity seems to be stimulated primarily by warm air (with or without CO_2) and/or radiant heat (for reviews, see Núñez, 1982; Schofield, 1979b). Freshly deposited faeces of triatomine nymphs appear to contain a volatile pheromone that attracts unfed nymphs, perhaps acting as an extra signal to indicate host availability (Schofield and Patterson, 1977; Ondarza et al., 1986).

Ingestion of blood is crucial to the development of Triatominae. When bug densities are high (relative to a fixed availability of hosts) the disturbance caused to the hosts can lead to bugs breaking off feeding before they are fully replete (Schofield, 1982). The resulting decline in their nutritional status leads to slower rates of nymphal development, reduced egg output by the female bugs, and an increased tendency for dispersive flight by adult males and females (Lehane and Schofield, 1982). Conversely, when bug densities are low, each bug obtains a full blood-meal and develops at its maximum rate. In this way, bug population densities are normally regulated in accordance with host availability (Schofield, 1985a). A recent review of many aspects of bloodfeeding behaviour has been provided by Núñez (1988).

Ecology

In sylvatic habitats such as opossum lodges or nests of furnariid birds, triatomine bug populations tend to be small, often fewer than ten bugs per nest. In larger habitats, such as the huge nests of colonial monk parrots *(Myopsitta monacha)* in Argentina (which can house up to 100 pairs of parrots) the bug populations can be much higher, while in rural houses in Latin America bug populations numbering several thousands have been recorded (e.g. Dias and Zeledón, 1955; Ponce and Ponce, 1990; Rabinovich et al., 1979).

Amongst the Triatominae, the presumed evolutionary trend has been for adaptation to more stable habitats offering an abundant and stable food source. In some cases, this has resulted in highly specific habitat and host associations. In terms of the epidemiology and control of Chagas disease, this tendency is of great importance, and can be conveniently described by five successive categories representing stages in relation to domestication, as follows (taken from Schofield, 1988):

1. Sylvatic species with highly specialized host and habitat associations. Examples include *Cavernicola pilosa* invariably found with bats, and the three species of *Psammolestes* which are found only in birds' nests of woven sticks (such as those of furnariid weaver birds). Such species are so specialized to particular sylvatic ecotopes that they are difficult to breed in the laboratory and seem unlikely to adapt to the domestic environment.
2. Sylvatic species with fairly specialized habitats, but whose adults are occasionally found in or around houses – often attracted to light. There are many examples, including members of the *T. protracta* complex, *T. rubrovaria*, *Eratyrus cuspidatus*, *P. geniculatus* and some species of *Rhodnius*.
3. Sylvatic species with more general ecotopes that occasionally colonize houses. Examples include *R. ecuadoriensis* and *R. neglectus*.
4. Species that retain sylvatic ecotopes but are also commonly associated with the domestic environment, and often colonize chicken houses and other animal enclosures. Perhaps the best examples are *T. sordida* and *T. pseudomaculata* in Brazil, *T. guasayana* in Argentina, *T. spinolai* in Chile, *T. maculata* in Venezuela, *P. herreri* in Peru, *R. pallescens* in Panama and *T. barberi* in Mexico. Several of these species are locally important as domestic vectors of *T. cruzi*.
5. Species highly adapted to the domestic environment. Foremost among these is *T. infestans*, restricted to the domestic and peridomestic environment throughout most of its geographical range. Other examples are *R. prolixus*, principal vector of *T. cruzi* in Venezuela and Colombia, *T. dimidiata* in parts of Central America, Colombia and Ecuador, and *T. brasiliensis* in the dry northeast of Brazil. *P. megistus*, once the most important domestic vector of *T. cruzi* in Brazil, has been replaced by *T. infestans* in many areas, but retains sylvatic ecotopes throughout its range and often colonizes houses in the more humid coastal parts of Brazil or where *T. infestans* has been eliminated by insecticide spraying.

MEDICAL IMPORTANCE

Reduviid bugs: Triatominae

Domestic infestations of triatomine bugs can be very stressful, especially where the householder is unable to afford remedial action. Domestic bug populations typically reach several hundred individuals of different stages, with each bug feeding every four to nine days. This translates to biting rates often in excess of 20 bites per person per night, representing a daily blood loss averaging 1–3 ml per person (Rabinovich et al., 1979; Schofield, 1981). The importance of the Triatominae rests with their capacity to transmit *Trypanosoma cruzi*, causative agent of Chagas disease or South American trypanosomiasis. Lewinsohn (1979) provides a good historical summary of the discovery of Chagas disease.

Trypanosoma cruzi is a common parasite of small mammals in the New World, and its distribution matches that of the New World species of Triatominae. Knowledge of its life cycle is reviewed by Zeledón (1988). The parasite has two reproductive phases, one in the cells of its mammalian host, and one in the posterior gut of its insect vector. However, the 'insect phase' of its reproductive cycle can also occur within the anal glands of opossums such as *Didelphis marsupialis*, a common reservoir host in whose nests triatomine bugs are frequently found (Deane et al., 1986). *Trypanosoma cruzi* does not infect birds (it is killed by a complement-mediated lysis in avian blood), and rarely causes severe pathological lesions in its wild mammal reservoir hosts.

Triatomine bugs become infected with *T. cruzi* by ingesting the parasites with their blood-meal. The parasites remain in the gut of infected bugs, where they develop and multiply; they do not successfully invade the bugs' haemocoel. *Trypanosoma rangeli*, a related parasite, is apparently non-pathogenic to man but often pathogenic to its bug vectors and develops in the haemocoel and salivary glands of some species of Triatominae, particularly *Rhodnius*. *Trypanosoma cruzi* is not transmitted by the bite of infected bugs, but is passed out with the bug's faeces. Thus the most efficient vector species are those that defaecate while feeding, so that the infective faeces are deposited on the skin or mucosa of the host. Scratching the site of the bite may help the parasites in the faeces to penetrate the abraded skin.

Although transmission by triatomine bugs is of greatest epidemiological significance, *T. cruzi* can also be transmitted by blood transfusion or organ transplant from infected donors, by eating infected material and sometimes across the placenta from an infected mother to the fetus (for review see Dias, 1979).

In man, infection with *T. cruzi* often leads to a small sore (chagoma) at the site of infection. If this site is around the eye, that eyelid can develop a marked swelling which is known as Romañas sign and is very characteristic of the acute stage of infection. Within a few days, fever and swollen lymph nodes can develop. This early acute phase of infection can be fatal (in up to 10–15% of cases), but more usually the patient survives to enter a symptomless phase which can last for many months or years. During this period, however, the parasites are invading most organs of the body, and chronic symptoms of the disease eventually develop in about 40–50% of those infected – often involving irreversible damage to heart (e.g. conduction blockage, arrhythmias, aneurysms, myocarditis) and intestine (e.g. mega-oesophagus, megacolon). In such cases the patient becomes progressively weaker, and can die from heart failure or other complications. At present, some

16–18 million people are estimated to be infected with *T. cruzi*, with 90–100 million at risk (World Health Organization, 1991).

From analysis of epidemiological history and trends it seems likely that in Brazil, at least, Chagas disease was a relatively recent phenomenon at the time of its discovery there in 1908. However, the disease and domestic populations of triatomine bug vectors may have been known to Andean cultures prior to European contact with Latin America. In Andean countries, triatomine bugs are commonly referred to by their Inca (i.e. quechua) names – 'vinchuca' (meaning that which lets itself fall, e.g. from the roof to a person sleeping below) or 'chirimacha' (meaning that which fears the cold). In contrast, there appears to be no word for the bugs in tupi-guarani, the language originally spoken by most tribes of the eastern part of the continent (in the modern guarani language, spoken mainly in Paraguay, the word for a triatomine bug is 'chinche guazu', from the Spanish 'chinche' for a bug, and 'guazu' the guarani term for big).

The spread of Chagas disease seems to be intimately linked with the changes in land use following European colonization of Latin America (Bucher and Schofield, 1984; Forattini, 1989; Schofield and Bucher, 1986; Schofield et al., 1982). As settlers moved into the interior of the continent they cleared land for crops and built simple houses using locally available materials – particularly walls made from adobe blocks or earth pressed to a wooden frame, with roofs of palm or grass thatch. Stored crops in these houses attract small mammals – rodents, opossums etc. – which are often infected with *T. cruzi*, while triatomine bugs could be brought in either by passive carriage of eggs and small nymphs within the fur of such mammals, or by adult bugs flying in at night. Eggs of *Rhodnius* are often adhered to palm leaves and so can be brought into houses roofed with palm thatch. The thatch roofs and the many cracks and crevices in mud walls of these houses provide excellent resting sites for the bugs, while the presence of man and domestic animals provides an abundant food source. Domestic bug populations can reach several hundred bugs of all stages, and infection rates with *T. cruzi* of over 50% are not unusual (Dias and Zeledón, 1955; Piesman et al., 1985; Rabinovich et al., 1979).

Bugs of other families

Many of the larger predaceous bugs will bite man when handled, as a means of defence. Large amounts of saliva can be injected, leading to intense pain at the time of biting and the development of necrotic lesions around the bites. Some smaller bugs, including plant-feeders, will probe exposed skin in certain circumstances, although the reasons for this are not clear. Some predaceous leaf-bugs (Miridae) and flower-bugs (Anthocoridae) will feed to repletion if allowed, becoming quite bloated with blood. Others seem merely to be sampling their substrate. More persistent biting by normally plant-feeding bugs in arid regions is sometimes reported – especially for species of lygaeid seed-bugs (e.g. *Leptodemus*) in Arabia and eastern Mediterranean countries. Usually these bugs appear unexpectedly in swarms and then move on after a few days, perhaps seeking moisture rather than food. Anthocorids living in warehouses, bakeries and similar buildings where there are infestations of stored-product pests for them to feed on, can reach high population densities at which their casual bites cause serious annoyance to people employed there. The culprits in such circumstances are frequently species of *Xylocoris* (Dolling, 1977; Ledger et al., 1982).

CONTROL

Because of its wide distribution, and the number of rural people at risk of infection, control of Chagas disease has become a public health priority in many Latin American countries. For practical purposes, chronic Chagas disease is incurable. Two drugs, nifurtimox and benznidazole, can be used for very early infections, but early diagnosis is difficult and adverse side-effects can occur. Treatment with either drug is lengthy and expensive. Moreover, because *T. cruzi* antigens can stimulate autoimmunity (immune attack on host tissues) the likelihood of developing a safe effective vaccine now seems very remote. Control of the disease therefore relies on interrupting transmission by eliminating the domestic triatomine bug vectors.

At a research level (and sometimes in field trials) most potential means of attacking triatomine bug populations have been considered. These have ranged from fumigants and other insecticides, housing modifications, traps (baited and unbaited), use of juvenile hormone mimics and insect growth regulators, repellents, chemosterilants, genetic manipulation and biological control agents (e.g. microhymenopteran egg parasitoids, nematodes such as *Neoaplectana* and fungi such as *Metarhizium*). For various reasons, mainly related to the population behaviour of the bugs, all these have failed except for certain insecticides and systems of house improvement (Schofield, 1985a, b).

Compounds from almost every class of organic insecticide have been tested against triatomine bugs. DDT was found to be ineffective, but in 1947, trials in Minas Gerais, Brazil, showed that the organochlorine BHC (HCH) was highly effective against domestic infestations of Triatominae (Dias and Pellegrino, 1948) and this became the main insecticide used in control campaigns, especially in Brazil and Argentina. Dieldrin, although more expensive and more toxic than BHC, was more widely used against *Rhodnius prolixus*. BHC, dieldrin and propoxur remained the most widely used insecticides against Triatominae until the advent of synthetic pyrethroids in the 1980s. Pyrethroids are highly effective against triatomine bugs, and have a much greater residual activity on mud walls compared to other classes of insecticides. Thus they can achieve satisfactory levels of control at much lower doses and with less frequent applications (Dias, 1987, 1988).

Chagas disease vector control campaigns typically involve three phases: preparatory, attack and vigilance. The preparatory phase involves sketch-mapping the areas to be treated and examining each house for signs of bug infestation. During the attack phase, all houses are treated in communities where infested houses have been reported – not just the houses in which bugs were found (this is because some houses may have had undetected infestations). After the attack phase is complete the houses are resurveyed for evidence of residual infestations and, if live bugs are found in more than 5% of houses, the attack phase repeated. During the vigilance phase the aim is to detect and selectively respray those houses where bugs are discovered. For this, mobilization of community support is now recognized to be of vital importance. A spray team is dispatched to respray infested houses. Often, neighbouring houses within a radius of 200 metres are also resprayed.

Insecticidal control campaigns against domestic Triatominae have been very successful in many regions of Latin America, especially in parts of Venezuela, Brazil, Argentina, Chile and Uruguay. However, when the continuity of such control campaigns is interrupted, there is always a risk that the bug populations may recover and spread back to their original levels. In some cases, it has been the original vector that has returned, but sometimes a different vector has replaced the species originally targeted for control. Examples include *T. maculata* replacing *R. prolixus* in parts of Venezuela, and *T. sordida* or *P. megistus* replacing *T. infestans* in parts of Brazil (Dias, 1987, 1988).

COLLECTING, PRESERVING AND REARING MATERIAL

Domestic and peridomestic Triatominae are normally collected using long blunt-ended forceps, and a torch to illuminate cracks where the bugs may be hiding. Irritant sprays (flushing-out agents) are sometimes used to encourage bugs to leave cracks to facilitate their capture; such sprays are generally prepared from dilute suspensions of various pyrethroid insecticides in water or kerosene (Pinchin et al., 1980). Domestic cimicids can be collected in much the same way. During collection and/or handling of wild-caught bugs, any contamination with fresh bug faeces or fluids from a damaged bug should be immediately swabbed with alcohol (or copious amounts of water if alcohol is not available).

Quantitative sampling of bugs in houses is difficult and subject to extreme bias in favour of the larger stages (Pinchin et al., 1981; Schofield, 1978). Baited traps have proved ineffective (e.g. Tonn et al., 1976) but artificial refuges such as the Gómez-Núñez box (Gómez-Núñez, 1965), the Cohen trap (H. L. Cohen, unpublished) or the Maria sensor (Wisnivesky-Colli et al., 1987) are sometimes used. These are cardboard boxes with holes, filled with folded paper, which are nailed to a house wall (usually in a bedroom) and examined after two to four weeks to see if any bugs have rested inside.

For sylvatic species, painstaking dissection of potential microhabitats (e.g. rockpiles, palm-tree crowns, bromeliads, bird nests, opossum lodges etc.) is often the only way to collect them. An alternative approach is to trap small mammals in the area (using cage-type live-traps), and then to attach a spool of cotton thread to the captured mammal, release it so it will pay out the thread behind it, and then track the mammal back to its nest which can then be examined for bugs (Miles, 1976; Miles et al., 1981a, b). Light-traps (visible, fluorescent, ultra-violet or infra-red) are useful for some sylvatic species (e.g. Ekkens, 1981; Martínez et al., 1984; Sjogren and Ryckman, 1966; Tonn et al., 1978). However, some flying bugs tend to drop short of the actual light, so that a trap should have a broad apron (e.g. large white sheet) extending along the ground in front of the light, with an observer ready to collect any bugs that land.

For identification, bugs can be killed with any standard killing agent (e.g. ether or chloroform) or heated (e.g. in tropical sunlight) to above 45°C for 10–20 minutes. Ideally, nymphs should be reared to adults in the laboratory to facilitate identification.

Laboratory colonies can be initiated from wild-caught bugs, but these must be handled carefully to avoid accidental transmission of *Trypanosoma cruzi*. Bugs should be placed in a strong wide-mouthed pot containing accordion-folded blotting paper and closed with two or more gauze tops. The pots should be stored over an oil bath in a locked room or incubator at around 25–28°C and 50–80% RH, either in darkness or a 12:12 hour light:dark cycle. Bugs should be maintained at fairly high densities in the pots. To facilitate exchange of essential gut symbionts between the bugs, the papers should be changed only if they become damp or heavily contaminated with faeces. The bugs are offered blood weekly or fortnightly from laboratory hosts or via a membrane feeder (Cedillos et al., 1982; Garcia et al., 1975; Langley and Pimley; 1978; Núñez and Segura, 1988).

The yield of colonies, especially of domestic species such as *T. infestans*, *R. prolixus* and *P. megistus*, depends primarily on the feeding opportunities given to the bugs: the better they are fed then the faster their development and the greater their reproductive output.

REFERENCES

Baldwin, W. F., Knight, A. G. and **Lynn, K. R.** 1971. A sex pheromone in the insect *Rhodnius prolixus* (Hemiptera: Reduviidae). *Canadian Entomologist* **103**: 18–22.

Barata, J. M. S. 1981. Aspectos morfológicos de ovos de Triatominae. II. – Caracteristicas macroscopicas e exocoriais de dez espécies do genero *Rhodnius* Stal, 1859 (Hemiptera–Reduviidae). *Revista de Saúde Pública* **15**: 490–542.

Barrett, T. V. and **Arias, J. R.** 1985. A new triatomine host of *Trypanosoma* from the central Amazon of Brazil: *Cavernicola lenti* n. sp. (Hemiptera, Reduviidae, Triatominae). *Memórias do Instituto Oswaldo Cruz* **80**: 91–96.

Barretto, M. P. 1979. Epidemiologia. Pp. 89–151 in Brener, Z. and Andrade, Z. (eds), *Trypanosoma cruzi e doença de Chagas*. 463 pp. Editora Guanabara Koogan, Rio de Janeiro.

Barth, R. 1962. Estudos anatómicos e histólogicos sôbre a subfamília Triatominae (Heteroptera, Reduviidae). XIX parte: estudo comparado das mandibulas de vários Triatominae e outros Reduviidae. *Memórias do Instituto Oswaldo Cruz* **60**: 91–101.

Bell, W. and **Schaefer, C. W.** 1966. Longevity and egg production of female bed-bugs, *Cimex lectularius*, fed various blood fractions and other substances. *Annals of the Entomological Society of America* **59**: 53–56.

Bernard, J. 1974. Mécanisme d'ouverture de la bouche chez l'hémiptère hématophage, *Triatoma infestans*. *Journal of Insect Physiology* **20**: 1–8.

Brenner, R. R. and **Stoka, A. de la M.** (eds) 1988a. *Chagas' disease vectors. I. Taxonomic, ecological and epidemiological aspects*. 155 pp. CRC Press Inc., Boca Raton, Florida.

Brenner, R. R. and **Stoka, A. de la M.** (eds) 1988b. *Chagas' disease vectors. II. Anatomic and physiological aspects*. 136 pp. CRC Press Inc., Boca Raton, Florida.

Brenner, R. R. and **Stoka, A. de la M.** (eds) 1988c. *Chagas' disease vectors. III. Biochemical aspects and control*. 156 pp. CRC Press Inc., Boca Raton, Florida.

Bucher, E. H. and **Schofield, C. J.** 1984. *Uso de la tierra y enfermedad de Chagas*. 14 pp. Centro de Aplicada, Cordoba, Argentina.

Carayon, J. 1950. Caractères anatomiques et position systématique des Hémiptères Nabidae. (Note préliminaire). *Bulletin du Museum national d'Histoire naturelle*, Paris (2) **22**: 95–101.

Carayon, J. 1971. Notes et documents sur l'appareil odorant metathoracique des Hémiptères. *Annales de la Société Entomologique de France* (n.s.) **7**: 737–770.

Carcavallo, R. U. 1988. The subfamily Triatominae (Hemiptera, Reduviidae): systematics and some ecological factors. Pp. 1–20 in Brenner and Stoka (1988a).

Carcavallo, R. U. and **Martínez, A.** 1987. Comentarios sobre *Triatoma gallardoi* Carpintero, 1986. *Chagas* **4**: 2.

Carcavallo, R. U. and **Tonn, R. J.** 1976. Clave gráfico de Reduviidae (Hemiptera) hematófagos de Venezuela. *Boletin de la Dirección de Malariologia y Saneamiento Ambiental* **16**: 244–265.

Carcavallo, R. U., Martínez, A. and **Palaez, D.** 1987. Una nueva especie de *Triatoma* Laporte de Mexico. *Chagas* **4**: 4–5.

Cedillos, R. A., Torrealba, J. W., Tonn, R. J., Mosca, W. and **Ortegón, A.** 1982. El xenodiagnostico artificial en la Enfermedad de Chagas. *Boletín de la Oficina Sanitaria Panamericana* **93**: 240–249.

China, W. E. 1933. A new family of the Hemiptera–Heteroptera with notes on the phylogeny of the suborder. *Annals and Magazine of Natural History* (10) **12**: 180–196.

China, W. E. and **Miller, N. C. E.** 1959. Check-list and keys to the families and subfamilies of the Hemiptera–Heteroptera. *Bulletin of the British Museum (Natural History)* (Entomology) **8**: 1–45.

Cobben, R. H. 1968. *Evolutionary trends in Heteroptera Part I. Eggs, architecture of the shell, gross embryology* [sic] *and eclosion*. 475 pp. Centre for Agricultural Publishing and Documentation, Wageningen.

Cobben, R. H. 1978. Evolutionary trends in Heteroptera Part II. Mouthpart-structures and feeding strategies. *Mededelingen Landbouwhogeschool Wageningen* **78** (5): 1–407.

Cobben, R. H. 1979. On the original feeding habits of the Hemiptera (Insecta): a reply to Merrill Sweet. *Annals of the Entomological Society of America* **72**: 711–715.

Cobben, R. H. 1981. The recognition of grades in Heteroptera and comments on R. Schuh's cladograms. *Systematic Zoology* **30**: 181–191.

Constantinou, C. 1984a. Photoreceptors involved in the entrainment of the circadian activity rhythm of the blood-sucking bug, *Rhodnius prolixus. Journal of Interdisciplinary Cycle Research* **15**: 195–202.

Constantinou, C. 1984b. Circadian rhythm of oviposition in the blood sucking bugs, *Triatoma phyllosoma, Triatoma infestans* and *Panstrongylus megistus* (Hemiptera: Heteroptera: Reduviidae). *Journal of Interdisciplinary Cycle Research* **15**: 203–212.

Deane, M. P., Lenzi, H. L. and **Jansen, A. M.** 1986. Double development cycle of *Trypanosoma cruzi* in the opossum. *Parasitology Today* **2**: 146–147.

Dias, E. and **Pellegrino, J.** 1948. Alguns ensaios com o "gammexane" no combate aos transmissores da doença de Chagas. *Brasil-Medico* **62**: 185–191.

Dias, E. and **Zeledón, R.** 1955. Infestação domiciliária em grau extremo por *Triatoma infestans. Memórias do Instituto Oswaldo Cruz* **53**: 473–486.

Dias, J. C. P. 1979. Mecanismos de transmissão. Pp. 152–174 in Brener, Z. and Andrade, Z. (eds), *Trypanosoma cruzi e doença de Chagas*. 463 pp. Editora Guanabara Koogan, Rio de Janeiro.

Dias, J. C. P. 1987. Control of Chagas disease in Brazil. *Parasitology Today* **3**: 336–341.

Dias, J. C. P. 1988. Controle de vetores da doença de Chagas no Brasil e riscos da reinvasão domiciliar por vetores secundários. *Memórias do Instituto Oswaldo Cruz* **83** (Supplement I, Special Issue): 387–391.

Dolling, W. R. 1977. *Dufouriellus ater* (Dufour) (Hemiptera: Anthocoridae) biting industrial workers in Britain. *Transactions of the Royal Society of Tropical Medicine and Hygiene* **71**: 355.

Dujardin, J. -P., Le Pont, F., Garcia-Zapata, M. T., Cardozo, L., Bermudez, H., Tibayrenc, M. and **Schofield, C. J.** 1988. *Rhodnius prolixus, R. neglectus, R. pictipes* and *Triatoma infestans*: an electrophoretic comparison. P. 83 in V Reunião sobre pesquisa aplicada em doença de Chagas, Araxá, Brazil.

Ekkens, D. B. 1981. Nocturnal flights of *Triatoma* (Hemiptera: Reduviidae) in Sabino Canyon, Arizona. I. Light collections. *Journal of Medical Entomology* **18**: 211–227.

Evans, J. W. 1946a. A natural classification of leaf-hoppers (Jassoidea, Hemiptera). Part 1. External morphology and systematic position. *Transactions of the Royal Entomological Society of London* **96**: 47–60.

Evans, J. W. 1946b. A natural classification of leaf-hoppers (Hemiptera, Jassoidea). Part 2: Aetalionidae, Hylicidae, Eurymelidae. *Transactions of the Royal Entomological Society of London* **97**: 39–54.

Evans, J. W. 1947. A natural classification of leaf-hoppers (Jassoidea, Homoptera). Part 3: Jassidae. *Transactions of the Royal Entomological Society of London* **98**: 105–271.

Evans, J. W. 1963. The phylogeny of the Homoptera. *Annual Review of Entomology* **8**: 77–94.

Feliciangeli, M. D., Fernández, E. and **Tonn, R. J.** 1978. A microhymenopteran parasite of eggs of *Psammolestes arthuri* (Hemiptera: Reduviidae) and observations of experimental parasitism of eggs of *Rhodnius prolixus* (Hemiptera: Reduviidae). *Journal of Medical Entomology* **14**: 593–594.

Fernández-Loayza, R. 1989. *Triatoma matsunoi:* nueva especie del norte peruano (Hemiptera, Reduviidae: Triatominae). *Revista Peruana de Entomologia* **31** (1988): 21–24.

Forattini, O. P. 1989. Chagas' disease and human behavior. Pp. 107–120 in Service, M. W. (ed.), *Demography and vector-borne diseases.* 402 pp. CRC Press Inc., Boca Raton, Florida.

Friend, W. G. and **Smith, J. J. B.** 1977. Factors affecting feeding by bloodsucking insects. *Annual Review of Entomology* **22**: 309–331.

Friend, W. G. and **Smith, J. J. B.** 1982. ATP analogues and other phosphate compounds as gorging stimulants for *Rhodnius prolixus. Journal of Insect Physiology* **28**: 371–376.

Games, D. E., Schofield, C. J. and **Staddon, B. W.** 1974. The secretion from Brindley's scent glands in Triatominae. *Annals of the Entomological Society of America* **67**: 820.

Garcia, E. de S., Macarini, J. D., Garcia, M. L. M. and **Ubatuba, F. B.** 1975. Alimentação de *Rhodnius prolixus* no laboratório. *Anais da Academia Brasileira de Ciências* **47,** 537–545.

Gardner, R. A. and **Molyneux, D. H.** 1988a. *Schizotrypanum* in British bats. *Parasitology* **97**: 43–50.

Gardner, R. A. and **Molyneux, D. H.** 1988b. *Trypanosoma (Megatrypanum) incertum* from *Pipistrellus pipistrellus:* development and transmission by cimicid bugs. *Parasitology* **96**: 433–447.

Gardner, R. A., Molyneux, D. H. and **Stebbings, R. E.** 1987. Studies on the prevalence of haematozoa of British bats. *Mammal Review* **17**: 75–80.

Ghauri, M. S. K. 1973. Hemiptera (bugs). Pp. 373–393 in Smith, K. G. V. (ed.), *Insects and other arthropods of medical importance.* xiv + 561 pp. British Museum (Natural History), London.

Gillett, J. D. 1935. The genital sterna of the immature stages of *Rhodnius prolixus* (Hemiptera). *Transactions of the Royal Entomological Society of London* **83**: 1–5.

Gómez-Núñez, J. C. 1965. Desarollo de un nuevo método para evaluar la infestación intradomiciliaria por *Rhodnius prolixus. Acta Cientifico Venezolano* **16**: 26–31.

Gorla, D. E. and **Schofield, C. J.** 1989. Population dynamics of *Triatoma infestans* under natural climatic conditions in the Argentine Chaco. *Medical and Veterinary Entomology* **3**: 179–194.

Imms, A. D. 1957. *A general textbook of entomology. Including the anatomy, physiology, development and classification of insects.* Ninth edition. x + 886 pp. Methuen, London, and Dutton, New York.

Kitselman, C. H. and **Grundmann, A. W.** 1940. Equine encephalomyelitis virus isolated from naturally infected *Triatoma sanguisuga* LeConte. *Kansas Agricultural Experiment Station Technical Bulletin* **50**: 1–15.

Langley, P. A. and **Pimley, R. W.** 1978. Rearing triatomine bugs in the absence of a live host and some effects of diet on reproduction in *Rhodnius prolixus* Stål (Hemiptera: Reduviidae). *Bulletin of Entomological Research* **68**: 243–250.

Lavoipierre, M. M. J., Dickerson, G. and **Gordon, R. M.** 1959. Studies on the methods of feeding of blood-sucking arthropods. I. – The manner in which triatomine bugs obtain their blood-meal, as observed in the tissues of the living rodent, with some remarks on the effects of the bite on human volunteers. *Annals of Tropical Medicine and Parasitology* **53**: 235–250.

Ledger, J. A., Rossiter, J. B. and **Oosthuizen, J. M. C.** 1982. Anthocorid bug bites in a Transvaal goldmine. *South African Medical Journal* **62**: 69–70.

Lehane, M. J. and **Schofield, C. J.** 1982. Flight initiation in *Triatoma infestans* (Klug) (Hemiptera: Reduviidae). *Bulletin of Entomological Research* **72**: 497–510.

Lent, H. and **Jurberg, J.** 1981. As espécies insulares de Cuba do gênero *Triatoma* Laporte (Hemiptera, Reduviidae). *Revista Brasileira da Biologia* **41**: 431–439.

Lent, H. and **Wygodzinsky, P.** 1979. Revision of the Triatominae (Hemiptera, Reduviidae), and their significance as vectors of Chagas' disease. *Bulletin of the American Museum of Natural History* **163**: 123–520.

Levinson, H. Z. and **Bar Ilan, A. R.** 1971. Assembling and alerting scents produced by the bedbug *Cimex lectularius* L. *Experientia* **27**: 102–103.

Levinson, H. Z., Levinson, A. R., Müller, B. and **Steinbrecht, R. A.** 1974. Structure of sensilla, olfactory perception, and behaviour of the bedbug, *Cimex lectularius,* in response to its alarm pheromone. *Journal of Insect Physiology* **20**: 1231–1248.

Lewinsohn, R. 1979. Carlos Chagas 1879–1934: the discovery of *Trypanosoma cruzi* and of American trypanosomiasis (foot-notes to the history of Chagas's disease). *Transactions of the Royal Society of Tropical Medicine and Hygiene* **73**: 513–523.

Maa, T. C. 1964. A review of the Old World Polyctenidae (Hemiptera: Cimicoidea). *Pacific Insects* **6**: 494–516.

Maldonado Capriles, J. 1990. Systematic catalogue of the Reduviidae of the world (Insecta: Heteroptera). *Caribbean Journal of Science* (Special Edition): x + 1–694.

Martínez, A., Carcavallo, R. U. and **Pelaez, D.** 1984. *Triatoma brailovskyi,* nueva espécie de Triatominae de Mexico. *Chagas* **1**: 39–42.

Maudlin, I. 1974. Giemsa banding of metaphase chromosomes in triatomine bugs. *Nature* **252**: 392–393.

Miles, M. A. 1976. A simple method of tracking mammals and locating triatomine vectors of *Trypanosoma cruzi* in Amazonian forest. *American Journal of Tropical Medicine and Hygiene* **25**: 671–674.

Miles, M. A., Souza, A. A. de and **Povoa, M.** 1981a. Chagas' disease in the Amazon basin. III. Ecotopes of ten triatomine bug species (Hemiptera: Reduviidae) from the vicinity of Belém, Pará State, Brazil. *Journal of Medical Entomology* **18**: 266–278.

Miles, M. A., Souza, A. A. de and **Póvoa, M.** 1981b. Mammal tracking and nest location in Brazilian forest with an improved spool-and-line device. *Journal of Zoology* **195**: 331–347.

Newberry, K. 1990. The tropical bedbug *Cimex hemipterus* near the southernmost extent of its range. *Transactions of the Royal Society of Tropical Medicine and Hygiene* **84**: 745–747.

Núñez, J. A. 1982. Food source orientation and activity in *Rhodnius prolixus* Stål (Hemiptera: Reduviidae). *Bulletin of Entomological Research* **72**: 253–262.

Núñez, J. A. 1988. Behavior of Triatominae bugs. Pp. 1–29 in Brenner and Stoka (1988b).

Núñez, J. A. and **Segura, E. L.** 1988. Rearing of Triatominae. Pp. 31–40 in Brenner and Stoka (1988b).

Okasha, A. Y. K. 1970. Effects of sub-lethal high temperature on an insect, *Rhodnius prolixus* (Stål). V. A possible mechanism of the inhibition of reproduction. *Journal of Experimental Biology* **53**: 37–45.

Ondarza, R. N., Gutiérrez-Martínez, A. and **Malo, E. A.** 1986. Evidence for the presence of sex and aggregation pheromones from *Triatoma mazzottii* (Hemiptera: Reduviidae). *Journal of Economic Entomology* **79**: 688–692.

Perez, R., Panzera, Y., Scafiezzo, S., Rydel, D., Mazzella, M. C., Panzera, F. and **Scvortzoff, E.** 1989. Cytogenetic analysis of six species of Triatominae (Hemiptera: Reduviidae). *Memórias do Instituto Oswaldo Cruz* **84** (Suppl. 2): 119.

Piesman, J., Mota, E., Sherlock, I. A. and **Todd, C. W.** 1985. *Trypanosoma cruzi:* association between seroreactivity of children and infection rates in domestic *Panstrongylus megistus* (Hemiptera: Reduviidae). *Journal of Medical Entomology* **22**: 130–133.

Pinchin, R., Fanara, D. M., Castleton, C. W. and **Oliveira Filho, A. M.** 1981. Comparison of techniques for detection of domestic infestations with *Triatoma infestans* in Brazil. *Transactions of the Royal Society of Tropical Medicine and Hygiene* **75**: 691–694.

Pinchin, R., Oliveira Filho, A. M. de and **Pereira, A. C. B.** 1980. The flushing-out activity of pyrethrum and synthetic pyrethroids on *Panstrongylus megistus,* a vector of Chagas's disease. *Transactions of the Royal Society of Tropical Medicine and Hygiene* **74**: 801–803.

Ponce, C. and **Ponce, E.** 1990. Indices de infeccion por *Trypanosoma cruzi* en triatominos de Honduras. *Boletín de la Dirección de Malariología y Saneamiento Ambiental* **33** (Supplemento no.1): 92.

Rabinovich, J. E., Leal, J. A. and **Feliciangeli de Piñero, D.** 1979. Domiciliary biting frequency and blood ingestion of the Chagas's disease vector *Rhodnius prolixus* Stahl [sic] (Hemiptera, Reduviidae), in Venezuela. *Transactions of the Royal Society of Tropical Medicine and Hygiene* **73**: 272–283.

Readio, P. A. 1927. Studies on the biology of the Reduviidae of America north of Mexico. *University of Kansas Science Bulletin* **17**: 5–291.

Ribeiro, J. M. C. 1987. Role of saliva in blood-feeding by arthropods. *Annual Review of Entomology* **32**: 463–478.

Rojas, J. C., Malo, E. A., Gutiérrez-Martínez, A. and Ondarza, R. N. 1990. Mating behavior of *Triatoma mazzottii* Usinger (Hemiptera: Reduviidae) under laboratory conditions. *Annals of the Entomological Society of America* **83**: 598–602.

Rossiter, M. and Staddon, B. W. 1983. 3-Methyl-2-hexanone from the triatomine bug *Dipetalogaster maximus* (Uhler) (Hemiptera; Reduviidae). *Experientia* **39**: 380–381.

Ryckman, R. E. 1962. Biosystematics and hosts of the *Triatoma protracta* complex in North America. (Hemiptera: Reduviidae) (Rodentia: Cricetidae). *University of California Publications in Entomology* **27**: 93–240.

Ryckman, R. E. 1979. Host reactions to bug bites (Hemiptera, Homoptera): a literature review and annotated bibliography. Part I. *California Vector Views* **26**: 1–24.

Ryckman, R. E. 1984. The Triatominae of North and Central America and the West Indies: a checklist with synonymy (Hemiptera: Reduviidae: Triatominae). *Bulletin of the Society of Vector Ecologists* **9**: 71–83.

Ryckman, R. E. 1986a. The Triatominae of South America: a checklist with synonymy (Hemiptera: Reduviidae: Triatominae). *Bulletin of the Society of Vector Ecologists* **11**: 199–208.

Ryckman, R. E. 1986b. Names of the Triatominae of North and Central America and the West Indies: their histories, derivations and etymology (Hemiptera: Reduviidae: Triatominae). *Bulletin of the Society of Vector Ecologists* **11**: 209–220.

Ryckman, R. E. 1986c. The vertebrate hosts of the Triatominae of North and Central America and the West Indies (Hemiptera: Reduviidae: Triatominae). *Bulletin of the Society of Vector Ecologists* **11**: 221–241.

Ryckman, R. E. and Archbold, E. F. 1981. The Triatominae and Triatominae-borne trypanosomes of Asia, Africa, Australia and the East Indies. *Bulletin of the Society of Vector Ecologists* **6**: 143–166.

Ryckman, R. E. and Bentley, D. G. 1979. Host reactions to bug bites (Hemiptera, Homoptera): a literature review and annotated bibliography. Part II. *California Vector Views* **26**: 25–49.

Ryckman, R. E. and Blankenship, C. M. 1984. The Triatominae and Triatominae-borne trypanosomes of North and Central America and the West Indies. A bibliography with index. *Bulletin of the Society of Vector Ecologists* **9**: 112–430.

Ryckman, R. E. and Casdin, M. A. 1976. The Triatominae of western North America, a checklist and bibliography. *California Vector Views* **23**: 35–52.

Ryckman, R. E. and Sjogren, R. D. 1980. A catalogue of the Polyctenidae. *Bulletin of the Society of Vector Ecologists* **5**: 1–22.

Ryckman, R. E., Bentley, D. G. and Archbold, E. F. 1981. The Cimicidae of the Americas and Oceanic Islands, a checklist and bibliography. *Bulletin of the Society of Vector Ecologists* **6**: 93–142.

Schofield, C. J. 1977. Sound production in some triatomine bugs. *Physiological Entomology* **2**: 43–52.

Schofield, C. J. 1978. A comparison of sampling techniques for domestic populations of Triatominae. *Transactions of the Royal Society of Tropical Medicine and Hygiene* **72**: 449–455.

Schofield, C. J. 1979a. Demonstration of isobutyric acid in some triatomine bugs. *Acta Tropica* **36**: 103–105.

Schofield, C. J. 1979b. The behaviour of Triatominae (Hemiptera: Reduviidae): a review. *Bulletin of Entomological Research* **69**: 363–379.

Schofield, C. J. 1981. Chagas disease, triatomine bugs, and blood-loss. *Lancet* **1981** (1): 1316.

Schofield, C. J. 1982. The role of blood intake in density regulation of populations of *Triatoma infestans* (Klug) (Hemiptera: Reduviidae). *Bulletin of Entomological Research* **72**: 617–629.

Schofield, C. J. 1985a. Population dynamics and control of *Triatoma infestans*. *Annales de la Société belge de Médecine tropicale* **65** (Suppl. 1): 149–164.

Schofield, C. J. 1985b. Control of Chagas' disease vectors. *British Medical Bulletin* **41**: 187–194.
Schofield, C. J. 1988. Biosystematics of the Triatominae. Pp. 285–312 in Service, M.W. (ed.), *Biosystematics of haematophagous insects*. xi + 363 pp. Clarendon Press, Oxford.
Schofield, C. J. and **Bucher, E. H.** 1986. Industrial contributions to desertification in South America. *Trends in Ecology and Evolution* **1**: 78–80.
Schofield, C. J. and **Moreman, K.** 1976. Apparent absence of a sex attractant in *Triatoma infestans* (Klug), a vector of Chagas' disease. *Transactions of the Royal Society of Tropical Medicine and Hygiene* **70**: 165–166.
Schofield, C. J. and **Patterson, J. W.** 1977. Assembly pheromone of *Triatoma infestans* and *Rhodnius prolixus* nymphs (Hemiptera: Reduviidae). *Journal of Medical Entomology* **13**: 727–734.
Schofield, C. J. and **Upton, C. P.** 1978. Brindley's scent-glands and the metasternal scent-glands of *Panstrongylus megistus* (Hemiptera, Reduviidae, Triatominae). *Revista Brasileira da Biologia* **38**: 665–678.
Schofield, C. J., Apt, W. and **Miles, M. A.** 1982. The ecology of Chagas disease in Chile. *Ecology of Disease* **1**: 117–129.
Schofield, C. J., Williams, N. G., Kirk, M. L., Garcia Zapata, M. T. and **Marsden, P. D.** 1986. A key for identifying faecal smears to detect domestic infestations of triatomine bugs. *Revista da Sociedade Brasileira de Medicina Tropicale* **19**: 5–8.
Schuh, R. T. 1986. The influence of cladistics on Heteropteran classification. *Annual Review of Entomology* **31**: 67–93.
Serra, R. G., Atzingen, N. C. B. and **Serra, O. P.** 1987. Nueva especie del genero *Alberprosenia* Martinez y Carcavallo, 1977, del estado de Pará, Brasil (Hemiptera, Triatominae). *Chagas* **4** (1): 3.
Settembrini, B. P. 1984. Circadian rhythms of locomotor activity in *Triatoma infestans* (Hemiptera: Reduviidae). *Journal of Medical Entomology* **21**: 204–212.
Sjogren, R. D. and **Ryckman, R. E.** 1966. Epizootiology of *Trypanosoma cruzi* in south-western North America. Part VIII: nocturnal flights of *Triatoma protracta* (Uhler) as indicated by collections at black light traps (Hemiptera: Reduviidae: Triatominae). *Journal of Medical Entomology* **3**: 81–92.
Southwood, T. R. E. and **Leston, D.** 1959. *Land and water bugs of the British Isles*. xi + 436 pp. Frederick Warne, London and New York.
Stys, P and **Kerzhner, I.** 1975. The rank and nomenclature of higher taxa in recent Heteroptera. *Acta Entomologica Bohemoslovaca* **72**: 65–79.
Sweet, M. H. 1979. On the original feeding habits of the Hemiptera (Insecta). *Annals of the Entomological Society of America* **72**: 575–579.
Szumlewicz, A. P. 1976. Laboratory colonies of Triatominae, biology and population dynamics. *Pan American Health Organization Scientific Publication* **318**: 63–82.
Tonn, R. J., Espinola, H., Mora, E. and **Jimenez, J. E.** 1978. Trampa de luz negra como método de captura nocturna de triatominos en Venezuela. *Boletín de La Dirección de Malariologia y Saneamiento Ambiental* **18**: 25–30.
Tonn, R. J., Otero, M. A. and **Jimenez, J.** 1976. Comparación del método hora hombre con la trampa Gómez-Núñez en la busqueda de *Rhodnius prolixus*. *Boletín de la Dirección de Malariologia y Saneamiento Ambiental* **16**: 269–275.
Usinger, R. L. 1939. Descriptions of new Triatominae with a key to genera (Hemiptera, Reduviidae). *University of California Publications in Entomology* **7**: 33–56.
Usinger, R. L. 1944. The Triatominae of North and Central America and the West Indies and their public health significance. *Public Health Bulletin* **288**: 1–83.
Usinger, R. L. 1966. *Monograph of Cimicidae (Hemiptera–Heteroptera)*. xi + 585 pp. Entomological Society of America, College Park, Maryland.
Usinger, R. L., Wygodzinsky, P. and **Ryckman, R. E.** 1966. The biosystematics of Triatominae. *Annual Review of Entomology* **11**: 309–330.
Van den Berghe, L., Chardome, M. and **Peel, E.** 1963. An African bat trypanosome in *Stricticimex brevispinosus* Usinger, 1959. *Journal of Protozoology* **10**: 135–138.

Venkatachalam, P. S. and **Belavady, B.** 1962. Loss of haemoglobin iron due to excessive biting by bed bugs. A possible aetiological factor in the iron deficiency anaemia of infants and children. *Transactions of the Royal Society of Tropical Medicine and Hygiene* **56**: 218–221.

White, G. B., Boreham, P. F. L. and **Dolling, W. R.** 1972. Synanthropic emesine bugs (Reduviidae, Emesinae) as predators of endophilic mosquitoes. *Transactions of the Royal Society of Tropical Medicine and Hygiene* **66**: 535–536.

Wisnivesky-Colli, C., Paulone, I., Perez, A., Chuit, R., Gualtieri, J., Solarz, N., Smith, A. and **Segura, E. L.** 1987. A new tool for continuous detection of triatomine bugs, vectors of Chagas' disease, in rural households. *Medicina*, Buenos Aires **47**: 45–50.

Wootton, R. J. 1981. Palaeozoic insects. *Annual Review of Entomology* **26**: 319–344.

World Health Organization 1991. Control of Chagas disease. Report of a WHO Expert Committee. *WHO Technical Report Series* **811**: vi + 1–95.

Zarate, L. G. and **Zarate, R. J.** 1985. A checklist of the Triatominae (Hemiptera: Reduviidae) of Mexico. *International Journal of Entomology* **27**: 102–127.

Zeledón, R. 1981. *El Triatoma dimidiata y su relacion con la enfermedad de Chagas*. 146 pp. Editorial Universidad Estatal a Distancia, San Jose, Costa Rica.

Zeledón, R. 1988. Life cycle of *Trypanosoma cruzi* in the insect vector. Pp. 59–75 in Brenner and Stoka (1988b).

Zeledón, R., Valerio, C. E. and **Valerio, J. E.** 1973. The camouflage phenomenon in several species of Triatominae (Hemiptera: Reduviidae). *Journal of Medical Entomology* **10**: 209–211.

CHAPTER FIFTEEN

Lice (Anoplura)

Joanna Ibarra

Lice are parasitic insects which spend their entire life cycle on the host, never voluntarily leaving it except to transfer to a new host; both sexes of the adult and all nymphal stages suck blood. They are highly host specific, so if an animal louse transfers to a human it will be unable to establish an infestation. It is something of a shock to the Westerner in the last decade of the twentieth century to discover that wherever there is a human population, irrespective of the level of cultural and social development, human lice are present.

Humans are parasitized by three species of lice, all members of the Anoplura (sucking lice). There are two genera, *Pediculus*, containing *P. capitis* (the head louse) and *P. humanus* (the body or clothing louse), and *Pthirus* with a single species *P. pubis*. The name of the latter genus has often been spelt *Phthirus*, but the original spelling *Pthirus* has been ruled correct (International Commission on Zoological Nomenclature, 1987).

Each species of human louse specializes in infesting particular parts of the body. The head louse (*Pediculus capitis*) occurs in head hair and cements its eggs on head hair shafts whereas the closely related clothing louse (*Pediculus humanus*) lays its eggs on clothing fibres and visits the body to feed. *Pthirus pubis* is found on any coarse body hair, including the eyelashes, and occasionally on head hair. The latter, as its specific name suggests, is closely associated with genital hair, and is popularly known as the crab-louse. The three human lice are distributed worldwide. *Pediculus humanus* is limited to sectors of the population unable to change their clothes, usually because they possess only one set.

Infestation with any of the three lice is described as pediculosis, but strictly speaking infestation with crab-lice is better termed phthiriasis (spelt with a 'phth'). It is common in medical and paramedical literature to use the term 'lousy' for a person with lice. In legal terminology, lice are 'vermin of the person' and people with lice are described as 'verminous'.

Although neglected infestation with lice causes clinical symptoms, the medical importance of lice is principally in the life-threatening diseases *P. humanus* transmits: classical epidemic typhus, louse-borne relapsing fever and trench fever. Although many epidemics have historically been widespread and devastating (Busvine, 1976), epidemics are today confined to impoverished and disaster-struck regions. They occur in all continents other than Australia and Antarctica but are particularly significant in Africa (Gratz, 1985a). Lice are obligate parasites which

Medical Insects and Arachnids Edited by Richard P. Lane and Roger W. Crosskey.
Published in 1993 by Chapman & Hall ISBN 0 412 40000 6

are almost always transmitted by inter-host contact, so it is most important to consider the social organization of a human population at risk during control campaigns. Much of the modern concern with lice has focused on these insects as a community health problem. Consequently, much of the literature is semi-popular, rather than the more formal intensive scientific research found in other groups of medically important arthropods.

RECOGNITION AND ELEMENTS OF STRUCTURE

Lice are wingless and dorsoventrally flattened insects which possess numerous adaptations to an ectoparasitic life style (e.g. well-developed claws). Species found on humans are translucent and vary in colour from pale beige to dark grey according to the hair and skin colour on which they hatch. The insects darken considerably after feeding and the passage of the blood-meal is clearly visible through the cuticle.

The three species of lice found on humans all belong to the Anoplura (sucking lice). There is, however, considerable disagreement over the status of the Anoplura. They have long been treated as a suborder of the Phthiraptera, together with the Mallophaga (chewing lice) and Rhynchophthirina, but more recently some authors (e.g. Ferris, 1951; Kim and Ludwig, 1978) treat the Anoplura as a separate order independent of the order Mallophaga and its suborder Rhynchophthirina. A cladistic analysis by Lyal (1985) supports the view that the Anoplura are best treated as a suborder in a classification recognizing four suborders (Amblycera, Ischnocera, Rhynchophthirina and Anoplura).

The Anoplura (= Siphunculata) are small insects ranging in length from 0.5 to 8 mm. The head is narrow, the antennae have five segments, and the mouthparts are highly modified, comprising fine stylets concealed in a ventral pouch. The three thoracic segments are fused together. In contrast, the rest of the Phthiraptera have conspicuous mandibulate mouthparts on a broad head, and the thorax consists of at least two separate segments. There are about 490 species of Anoplura. Kim and Ludwig (1978) group these into 42 genera and 15 families and give keys to families and diagnoses of the genera. Note, however, that some specialists consider the Kim and Ludwig classification to contain an unnecessarily large number of families each with very few species.

Until recently, the two genera of medical importance, *Pthirus* and *Pediculus*, were placed in the same family, the Pediculidae, but Kim and Ludwig – from their phylogenetic analysis using morphological combined with host data – conclude that these genera have evolved on two distinct lineages which warrant their independent family status. Thus the Pediculidae contains *Pediculus* with only two species (*P. capitis* and *P. humanus*) which are found on Cebidae (New World monkeys), Pongidae (gibbons and great apes) and Hominidae. The Pthiridae contains *Pthirus* with a single species found on *Homo* (Hominidae) and *Gorilla* (Pongidae). The coevolution of lice and their mammal hosts has been discussed by Kim (1988).

Pediculus (head and clothing lice)

Pediculus species (Fig. 15.1a) are slender insects, longer than they are wide, which grow to a maximum length of about 3 mm. The male is slightly smaller than the female and can be distinguished by the presence of dark transverse bands on the

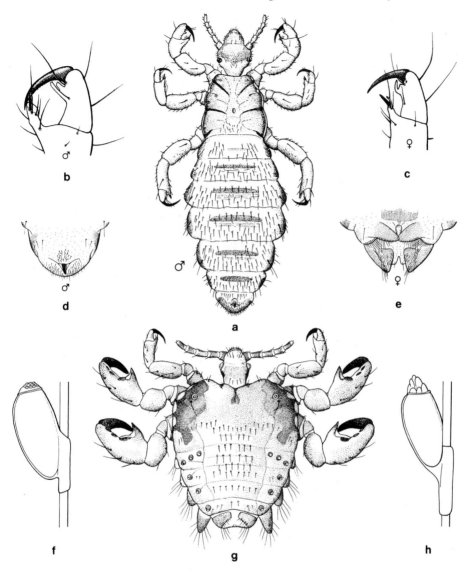

Figure 15.1 The lice parasites of humans: (*a–e*) *Pediculus humanus*, (*a*) dorsal view of male, (*b*) tibial claw of male, (*c*) tibial claw of female, (*d*) terminalia of male, (*e*) terminalia of female; (*f*) *Pediculus capitis*, egg on hair shaft; (*g–h*) *Pthirus pubis*, (*g*) dorsal view, (*h*) egg on hair-shaft. (Figures (*a–c*) redrawn from Keilin and Nuttall, 1930; (*d, e*) from Ferris, 1951; (*f, h*) from Hase, 1931.)

dorsum of the abdomen. The posterior end of the male abdomen is rounded, the ventral surface curving upwards so as to bring the anus and sexual orifice to the upper surface. In the female the abdomen terminates in two large posterior lobes, which give it a bilobed appearance; in ventral view the paired gonopods are visible as well as a shield-shaped pigmented area (Fig. 15.1*d,e*).

The thoracic segments are fused together, and each of the short legs ends in a disproportionately large claw, adapted in size and shape for clinging onto hair or fibres (Fig. 15.1*a*). There are considerable sexual differences in the legs, the fore leg of the male being much broader and bearing a larger tibial 'thumb' and tarsal claw (Fig. 15.1*b,c*). The shape of the claw within a population of lice has considerable

520 Lice (Anoplura)

epidemiological significance. For example, in the USA where there is a substantial black American minority, the prevalence of head infestation is 35 times higher amongst the numerically dominant Caucasians than in the black American population (Juranek, 1985). A major contributory factor is thought to be the adaptation of the claw grasp to the predominant hair type in the host population; Caucasian hair is round in cross-section whereas Afro-Caribbean hair is oval.

The specific status of the *Pediculus* lice on man has been a matter of discussion and controversy for a long time. The most recent evidence for treating the body and head louse as two distinct species comes from an analysis of natural double-infestations on humans in Ethiopia (Busvine, 1978). This showed no overlap in morphometric characters, particularly the length of the tibia of the middle leg, between lice from the head and body of the humans affected (Fig. 15.2). There were no intermediate specimens in the double-infestations although intermediates could be produced by experimental cross-mating in the laboratory. When offered a choice of hair or cloth fibres for oviposition, *P. capitis* chooses hair and *P. humanus* chooses cloth (Buxton, 1948). The two species have a different ecology and medical importance. Only *P. humanus* (clothing or body louse) is known to transmit pathogens to man. A thorough understanding of the genetic relationship of these two lice is clearly significant in studies of insecticide efficacy and resistance.

Pthirus (crab lice)

Pthirus pubis is much rounder than *Pediculus*, being as long as it is wide. It resembles a minute crab (Fig. 15.1g), hence its colloquial name. The claws are larger than in

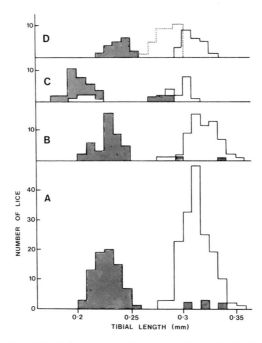

Figure 15.2 Differentiation of *Pediculus humanus* from *P. capitis* on the basis of tibial lengths of lice collected from patients in Ethiopia. Shaded areas are based on specimens from the head, unshaded areas on specimens from clothing: (A) female lice from Addis Ababa; (B) male lice from Addis Ababa; (C) lice from Anuak and Nuer people; (D) lice from laboratory strains, with an F1 cross indicated by the dotted line. In (C) and (D) the sexes are combined. (Redrawn from Busvine, 1978.)

Pediculus and, together with the wider leg span, are adaptations for clinging to sparsely distributed coarse body hairs (pubic and axillary hair, and eyelashes).

The eggs of human lice are pearly white, readily reflecting the colour of their surroundings. They are about 1 mm long and differ in shape between *Pediculus* and *Pthirus* (Fig. 15.1*f,h*). At the distal end of the egg is a perforated operculum providing for exchange of air and moisture to the embryo. Eggs are glued firmly onto a hair or cloth fibre and are laid with the operculum pointing towards the tip of the shaft or thread. The glue of *P. capitis* is so tenacious that it cannot be dissolved without destroying the hair. The integument of the insect is tough and the cuticle of the egg more so. The egg turns glistening white as it fills with air when the first-stage nymph begins to hatch. The eggs are commonly called 'nits', but there is some confusion over the term; in some publications it refers both to developing eggs and empty eggshells, but in others is used only for the empty eggshells. Popularly, louse infestation is often referred to simply as 'nits'. The presence of eggs is important in the diagnosis of an infestation.

Detailed descriptions of the morphology of sucking lice are given in Buxton (1948), Ferris (1951) and Kim and Ludwig (1978).

The zoogeography of lice has been extensively studied in relation to the evolution and radiation of their hosts. Because, like humans, chimpanzees (*Pan troglodytes*) have a *Pediculus* and gorillas (*Gorilla gorilla*) a pubic louse (*Pthirus*), it seems that these primate lice antedate the divergence of the ape and anthropoid stocks (Traub, 1980).

BIOLOGY

The incrimination of *Pediculus* as the vector of typhus in the first decade of this century and the increased danger of typhus epidemics during both world wars gave a considerable impetus to the study of louse biology. Much of this work focused on *P. humanus*, not only because it was an important vector, but also because a large number of the lice in a clothing louse infestation can be collected easily from the host's garments, whereas head and crab lice must be collected from a person. By the mid-1940s mass rearing methods were developed in which *P. humanus* were adapted to feed on rabbits (Culpepper, 1944). The limitations of observations made on rabbit-adapted lice, far removed from their natural habitat, must be considered when interpreting laboratory studies. The masterly review of human louse biology by Buxton (1948) remains an essential reference. More recent studies have been summarized in Pan American Health Organization (1973).

Oviposition begins about 24 hours after mating. The female uses the two gonopods and grasping tubercles at the end of the abdomen to position an egg on a hair or cloth fibre. A single egg with a rapidly drying cement is expelled and some females die by accidentally gluing themselves down with the eggs or by blocking the opening of the oviduct – so causing subsequent swelling and internal trauma. *Pediculus capitis* tends to lay eggs singly on a hair (Fig. 15.3), usually on hairs around the nape of the neck and behind the ears, whereas *P. humanus* eggs are laid in clusters on the seams of garments, waistbands or even in animal-hair body-ornaments (e.g. Sholdt et al., 1979). *Pthirus pubis* lays several eggs on a single hair.

There are contradictory reports of the number of eggs laid after mating. Studies in the USA during the 1970s established a mean of 56 fertile eggs laid after a single insemination, but Maunder (1983) nevertheless continued to maintain that a female

Figure 15.3 Scanning electron microscope photograph of the egg and nymph of *Pediculus capitis*. (Reproduced courtesy of M. Smith.)

must mate between ovipositing each egg to ensure its fertility. A female *P. capitis* lays about six eggs in 24 hours, and a total of about 55 eggs in her average nine-day adult life (Buxton, 1948). *Pediculus humanus* seems to lay about double this number of eggs. The scanty data on *Pthirus pubis* suggest that its female lays three eggs in 24 hours.

Incubation of the eggs is brought about by body heat and takes from seven to ten days. Removal of infested clothing slows the process, and eggs of *P. humanus* are unlikely to hatch at all if the same infested clothes are worn for only three hours each day. The first-instar nymph uses minute cutting hooks to perforate the operculum. The first-instar nymph must feed on blood shortly after hatching and thereafter approximately every five hours. Failure to obtain blood causes lice to die quickly, and *Pediculus capitis* will die of starvation after 55 hours and *P. humanus* after 85 hours at 23°C (Busvine, 1985); death of both species occurs even sooner at temperatures higher than this.

Despite their semi-liquid diet of blood, lice are continually under threat of dehydration. The spiracles situated on the abdomen have a closing mechanism and are important in reducing water loss from the respiratory system (Hase, 1931). The faeces are excreted as a fine black dust both during and after feeding.

Pediculus capitis takes seven to ten days to become adult, moulting three times in the process; *P. humanus* takes a similar time for development but longer if the infested garments are only worn for a few hours each day.

Accurate information on louse biology is important in epidemiological studies of *P. capitis* infestations and the design of appropriate control measures. For example, to incubate successfully, an egg requires a temperature between 24°C and 37°C with an optimum between 29°C and 32°C. From these data it has been concluded that most eggs are affixed to hair shafts near or touching the scalp. On the basis that the hair grows about 0.4 mm per day, Mellanby (1942a) suggested that it is possible to calculate when an egg was laid by its distance from the scalp. Clinical trials of insecticides have been conducted with this assumption built into the protocol (e.g. Donaldson and Logie, 1986) and the value of health education about lice measured using the ratio between extinct and active cases calculated on the basis of the position of the eggs (Maunder, 1988). However, field observations show that eggs are laid where long hair is near or touching the scalp, so some viable eggs could be along the shaft at a considerable distance from the scalp. When the host is perspiring a louse moves away down the hair to avoid the humidity. Therefore, suggestions that eggs found more than one centimetre from the scalp should be regarded as non-viable are clearly incorrect. There is clearly still a great need for accurate scientific study of the biology of this familiar insect.

MEDICAL IMPORTANCE

An essential reference, although somewhat dated, is the review of the medical importance and control of lice and louse-borne diseases issued by the Pan American Health Organization (1973) which includes many small papers each accompanied by a discussion. The most thorough and extensive epidemiological study of both head and body lice from any one country is that of Sholdt et al. (1979) for Ethiopia, which also includes a good bibliography.

Biting pests

In many parts of the world infestations with lice are a public health problem. This is not by any means restricted to developing countries and the head louse, *P. capitis*, is perceived as a growing problem as much in developed and industrialized countries. In the USA, the 1985 data of the Centers for Disease Control showed that there were more cases of head louse infestation (pediculosis capitis) amongst school-aged children than there were of all other communicable diseases combined, except the common cold. In Britain, conservative estimates suggest that at least one primary school child in ten catches head lice in a year, and, more seriously, nearly a tenth of sufferers catch them several times in succession. In two schools in southern England, Ibarra (1989) found an incidence of head lice of 63% in one school and 21% in the other by a questionnaire method. Most cases are treated as soon as they are discovered, so it is not the natural symptoms of infestation that are considered to be the major clinical problem: there is now concern over the possible consequences to individuals of repeated pesticide treatment of patients and their contacts (Altschuler and Kenney, 1986).

Social distress of families forever infested with lice is an additional strain in the western world where pediculosis patients are the victims of lice and of the censorious attitude to the lousy. Catching lice in these circumstances is associated in the public mind with self-inflicted neglect, perpetuating the common myth that *P. capitis* infestations grow out of dirt and implying that cleanliness is a protection from lice. In the absence of well-designed epidemiological studies, there is considerable debate over the current prevalence of infestation with *P. capitis*, especially in developing countries where national statistics are not available. Mellanby (1941, 1942a, b) conducted some key epidemiological studies in England which showed that a policy of treating infested school children without ensuring that other family members are freed from lice is a self-defeating control strategy. In one survey in the USA, 59% of infested children had at least one other infested household member (Juranek, 1977). Most such studies show that females are more likely to catch head lice after puberty than males.

Although it has been recorded that the *P. humanus* burden of an individual can exceed several thousand hatched insects (Sholdt et al. (1979) recorded 21 500 on one person), and of *P. capitis* several hundred, the norm is ten or less (Buxton, 1948; Mellanby, 1942a). Contrary to popular belief, itching of the bites is not the first sign of infestation; irritation caused by louse faeces often occurs earlier. It can take time before the body reacts to louse saliva (Buxton, 1948) so as much as three months often elapses before there is any reaction to the bites.

In developed countries, *P. humanus* is usually only found on homeless vagrants where heavy infestations ('vagabond's disease') present as darkened, hardened areas of the skin (morbus errora) due to continuous biting and scratching, often following the seam lines of garments worn close to the skin.

Pthirus pubis is more sedentary than *P. capitis* or *P. humanus* (Burgess et al., 1983) and a tendency to bite repeatedly in the same places can cause the formation of bluish spots (maculae caeruleae) (Alexander, 1984). Black spots of louse faeces in underwear are a sign of *P. pubis* infestation. The few studies that have been made in developed countries suggest that the prevalence of *P. pubis* infestations is increasing (e.g. Fisher and Morton, 1970). These data are collected at sexually transmitted disease (STD) clinics so it is not surprising that nearly half the patients recorded also have another STD. It seems that within the home it is common for *P. pubis* to be transferred to the eyelashes of young children from, for example, the beards or hairy arms of older members of the family. Faeces in the eyelashes can be a cause of blepharitis. All three species of lice infesting humans can cause intense itching (pruritis) and be a cause of uncontrolled scratching which produces excoriations and secondary infection. Differential diagnosis to distinguish louse infestation from scabies is important in the choice of appropriate treatment regime.

Role as disease vectors

Typhus and louse-borne relapsing fever are still a problem in Africa, South America and Central Asia (Gratz, 1985b).

Typhus and other rickettsial infections

Typhus is caused by *Rickettsia prowazeki* and humans are the sole reservoir of the disease. *Pediculus humanus* acquires the pathogen from the blood of an infected human during the first ten days of illness. The *Rickettsia* multiplies in the lumen and epithelial cells of the louse midgut. These cells eventually rupture, releasing enormous numbers of pathogens to be passed in the louse faeces. The louse dies some seven to ten days after infection. *Rickettsia* in the faeces remains infective to humans for up to three months, invading through scratches and abrasions, and by inhalation. Thus typhus can be caught by a person without an infestation of lice. However, a recovered patient can develop recrudescent typhus (Brill Zinsser disease, bereavement fever) and become infectious to lice once again.

The onset of the disease in humans is associated with chills and a cough, which can be accompanied by a sparse itching rash beginning on the soles of the feet and spreading upwards towards the trunk. Fever rises rapidly with a profound headache. Broad spectrum antibiotics, such as tetracycline, bring a quick recovery.

Another *Rickettsia*, *R. quintana,* causes five-day fever or trench fever, named after the devastating impact it had on soldiers in the First World War. The mode of transmission is similar to that of typhus and the disease is characterized by two successive acute attacks of fever.

Murine typhus, usually much milder than classical typhus and producing only sporadic cases, is a zoonosis caused by *R. mooseri*. The reservoirs of infection are rats and other rodents. The main vectors are fleas, but *P. humanus* can transmit the pathogen from human to human through its infected faeces. The flea is seemingly unaffected by *R. mooseri* whereas *R. prowazeki* is pathogenic to lice – a fact which suggests that classical typhus evolved from the older murine typhus (Buxton, 1948).

Louse-borne relapsing fever
The relapsing fevers of humans are caused by spirochaetes of the genus *Borrelia*. Unlike tick-borne relapsing fever caused by *B. duttoni,* louse-borne relapsing fever is not a zoonosis and the only reservoirs are humans. In ticks the pathogen *B. duttoni* is trans-ovarially transmitted but in lice the causative agent of louse-borne relapsing fever, *B. recurrentis,* is not hereditary. Instead, *P. humanus* becomes infected by biting a human with the disease. About four days after digestion of the infective meal the spirochaetes begin to circulate in the haemolymph, where they slowly multiply. Transmission to humans depends on crushing an infected insect, either on the skin or when 'cracking' lice with the teeth or fingers. It is unusual for a person free from lice to acquire a *B. recurrentis* infection.

Pediculus can carry the bacteria responsible for impetigo in their gut and can transmit it mechanically.

CONTROL

Pediculus capitis (head lice)

One of the main obstacles to be overcome in head louse control is difficulty in diagnosing light infestations. The use of a fine-toothed comb on wet or oiled hair to dislodge a louse greatly increases the chance of louse detection. For this purpose, plastic detector-combs are more comfortable than a metal nit-comb (which has much finer teeth) and equally effective. Lice detached from the hair are much easier to see than lice on the head.

Nymphs and adults can be washed out and combed from the hair, but if they are very numerous an insecticide should also be used. Vigilance should be kept for hatching eggs if these are not removed with a metal nit-comb or 'cracked' between the finger nails, especially as the major impediment to head louse control lies in the egg stage. To demonstrate that lice are not necessarily damaged by ordinary combing mention should be made of the Australian aborigines who are expert in controlled 'louse farming'. Their aim is to ensure that everyone has a few lice since the lice are held in high esteem. Regular hairwashing keeps down the irritation caused by the faeces, and in order not to lose lice in the rinsing water, the aborigines comb them off first and put them back on afterwards (Trigger, 1981).

Malathion and carbaryl are routinely used in Britain and several other countries where head lice are now resistant to both DDT and lindane (Maunder, 1971). In addition, pyrethroids have recently become available for controlling infestations. Lotion formulations applied to dry hair undiluted and left on for some hours are much more effective in penetrating the eggs than shampoo formulations which are diluted with water and rinsed off after only a few minutes. Caution must be used in interpreting the results of *in vitro* tests and *in vivo* nurse-made visual inspections following treatment as proof of efficacy of an insecticide because the former are performed on rabbit-adapted *P. humanus* colonies and the latter produce a high rate of false negatives (Mellanby, 1941). Studies in Britain have shown that application for 12 hours of a malathion lotion is effectively ovicidal, and clinical trials in Central America confirm this (Meinking et al., 1986). Resistance to malathion is an emerging problem in louse control in Britain.

The high number of patients who have to be treated repeatedly can be caused by reinfestation after successful treatment via undiagnosed contacts, unsuccessful treatment (usually because of incomplete egg-kill by the product used) or by unsuccessful treatment as a result of insecticide resistance.

Treatment of a whole community as a strategy for preventing infestations before they occur is now being advocated in the USA but this is not advisable (Edman and Clark, 1990). Targeting health education to the community at risk and concentrating a control campaign into a short period of maximum effort is the most economically effective approach (Combescot, 1990; Ibarra, 1991).

Pediculus humanus (clothing lice)

Hatched lice usually remain on infested clothing when it is removed, and any lice left clinging to the body can be brushed or washed off. Lice and eggs will die if the clothing remains unworn for 17 days, preferably isolated in a polythene bag. Boiling will destroy lice and eggs but the same result can be achieved by putting the dry infested clothing into a tumble-drier at a temperature in excess of 60°C for 15 minutes. However, these methods have little practical use in the conditions where *P. humanus* most often thrives – i.e. conditions of extreme deprivation and where fuel might not be available for boiling clothes. When epidemic typhus threatens, insecticides are necessary. DDT made a dramatic debut by halting the Second World War typhus outbreak in Europe, but resistance to this insecticide limits its use today. In a comprehensive account, Gratz (1985b) stresses the need for assessment of insecticide resistance prior to control campaigns. Clearly, accurate identification is necessary to ensure that the appropriate species is being tested.

Pthirus pubis (crab lice)

Water-based insecticide lotions used against *P. capitis* are effective against crablice. All the hairy parts of the body, including the head hair, should be treated. Assessment of the susceptibility of *P. pubis* to insecticides is long overdue. Treatment of crab lice on the eyelashes should be made by gently smearing a small amount of petroleum jelly along the closed lashes twice a day for ten days. This will kill the nymphs as they hatch. No attempt should be made to remove the eggs as 'nits' will disappear fairly quickly as eyelashes fall and regrow.

REFERENCES

Alexander, J. O'D. 1984. *Arthropods and human skin.* 422 pp. Springer-Verlag, Berlin.
Altschuler, D. Z. and **Kenney, L. R.** 1986. Pediculicide performance, profit and the public health. *Archives of Dermatology* **122**: 259–261.
Burgess, I., Maunder, J. W. and **Myint, T. T.** 1983. Maintenance of the crab louse, *Pthirus pubis,* in the laboratory and behavioural studies using volunteers. *Community Medicine* **5**: 238–241.
Busvine, J. R. 1976. *Insects, hygiene and history.* 262 pp. Athlone Press, London.
Busvine, J. R. 1978. Evidence from double infestations for the specific status of human head and body lice (Anoplura). *Systematic Entomology* **3**: 1–8.
Busvine, J. R. 1985. Pediculosis: biology of the parasites. Pp. 163–174 in Orkin, M. and Maibach, H. I. (eds), *Cutaneous infestations and insect bites.* xv + 321 pp. Marcel Dekker, New York.
Buxton, P. A. 1948. *The louse. An account of the lice which infest man, their medical importance and control.* Second edition. viii + 164 pp. Edward Arnold, London.
Combescot, C. 1990. Epidemiologie actuelle de la pédiculose a *Pediculus capitis. Bulletin de l'Académie nationale de Médecine* **174**: 231–237.
Culpepper, G. H. 1944. The rearing and maintenance of a laboratory colony of the body louse. *American Journal of Tropical Medicine* **24**: 327–329.
Donaldson, R. J. and **Logie, S.** 1986. Comparative trial of shampoos for treatment of head infestation. *Journal of the Royal Society of Health* **106**: 39–40.
Edman, J. D. and **Clark, J. M.** 1990. The prophylactic use of pediculicides: a formula for resistance development. *Progress* **6** (1): 5.
Ferris, G. F. 1951. The sucking lice. *Memoirs of the Pacific Coast Entomological Society* **1**: ix + 1–320.
Fisher, I. and **Morton, R. S.** 1970. *Phthirus pubis* infestation. *British Journal of Venereal Diseases* **46**: 326–329.
Gratz, N. G. 1985a. Epidemiology of louse infestations. Pp. 187–198 in Orkin, M. and Maibach, H. I. (eds), *Cutaneous infestations and insect bites.* xv + 321 pp. Marcel Dekker, New York.
Gratz, N. G. 1985b. Treatment resistance in louse control. Pp. 219–230 in Orkin, M. and Maibach, H. I. (eds), *Cutaneous infestations and insect bites.* xv + 321 pp. Marcel Dekker, New York.
Hase, A. 1931. Siphunculata; Anoplura; Aptera. Läuse. *Biologie des Tiere Deutschlands* **30**: 1–58.
Ibarra, J. 1989. Head lice in schools. *Health at School* **4**: 147–151.
Ibarra, J. 1991. Bug busting: CHC's programme of action against head lice. *Shared Wisdom: Journal of Community Hygiene Concern* **2**: 16–19.
International Commission on Zoological Nomenclature 1987. *Official lists and indexes of names and works in zoology.* 366 pp. International Trust for Zoological Nomenclature, London.
Juranek, D. D. 1977. Epidemiology of lice. *Journal of School Health,* June: 346–364.
Juranek, D. D. 1985. *Pediculus capitis* in school children. Pp. 199–211 in Orkin, M. and Maibach, H. I. (eds), *Cutaneous infestations and insect bites.* xv + 321 pp. Marcel Dekker, New York.
Keilin, D. and **Nuttall, G. H. F.** 1930. Iconographic studies of *Pediculus humanus. Parasitology* **22**: 1–10 [with 8 unpaginated plates].
Kim, K. C. 1988. Evolutionary parallelism in Anoplura and eutherian mammals. Pp. 91–114 in Service, M. W. (ed.), *Biosystematics of haematophagous insects.* xi + 363 pp. Clarendon Press, Oxford.

Kim, K. C. and Ludwig, H. W. 1978. The family classification of the Anoplura. *Systematic Entomology* **3**: 249–284.

Lyal, C. H. C. 1985. Phylogeny and classification of the Psocodea, with particular reference to the lice (Psocodea: Phthiraptera). *Systematic Entomology* **10**: 145–165.

Maunder, J. W. 1971. Resistance to organochlorine insecticides in head lice and trials using alternative compounds. *Medical Officer* **125**: 27–29.

Maunder, J. W. 1983. The appreciation of lice. *Proceedings of the Royal Institution of Great Britain* **55**: 1–31.

Maunder, J. W. 1988. Updated community approach to head lice. *Journal of the Royal Society of Health* **108**: 201–202.

Meinking, T. L., Taplon, D., Kalter, D. C. and Eberle, M. W. 1986. Comparative efficacy of treatments for pediculosis capitis infestations. *Archives of Dermatology* **122**: 267–271.

Mellanby, K. 1941. The incidence of head lice in England. *Medical Officer* **65**: 39–43.

Mellanby, K. 1942a. Natural populations of the head-louse (*Pediculus humanus capitis*: Anoplura) on infected children in England. *Parasitology* **34**: 180–184.

Mellanby, K. 1942b. Relation between size of family and incidence of head lice. *Public Health* **56**: 31–32.

Pan American Health Organization 1973. Proceedings of the International Symposium on the control of lice and louse-borne diseases. *Pan American Health Organization Scientific Publication* **263**: x + 1–311.

Sholdt, L. L., Holloway, M. L. and Fronk, W. D. 1979. *The epidemiology of human pediculosis in Ethiopia*. xii + 150 pp. Navy Disease Vector Ecology and Control Center, Jacksonville, Florida.

Traub, R. 1980. The zoogeography and evolution of some fleas, lice and mammals. Pp. 93–172 in Traub, R. and Starcke, H. (eds), *Fleas*. x + 420 pp. Balkema, Rotterdam. [Published 'Proceedings of the International Conference on Fleas, Ashton Wold/Peterborough/UK, 21–25 June 1977'.]

Trigger, D. S. 1981. Blackfellows, whitefellows and head lice. *Australian Institute of Aboriginal Studies Newsletter*, March: 63–72.

CHAPTER SIXTEEN

Fleas (Siphonaptera)

Robert E. Lewis

Fleas are small, wingless insects with a holometabolous (complete) metamorphosis. They are laterally compressed, with numerous bristles and appear shiny, varying from yellowish brown to almost black. Adult fleas are obligate bloodsucking parasites of warm-blooded vertebrates and 94% of the known species occur on mammals, the remaining 6% on birds. The larvae are elongate, lack eyes and legs, and most are sparsely covered with long setae. They are not normally parasitic but feed on organic material in the nest or lair of the host, including pellets of blood produced by feeding adults. Pupation takes place in a silken cocoon spun by the third-instar larva.

There are approximately 2500 described species and subspecies in this order, and these are grouped into 239 genera in 15 families. It is thought that there are another 500 species and subspecies awaiting description. The group is worldwide, and has representatives in such inhospitable locations as the Arctic and Antarctica; it occurs from sea level to high altitudes wherever satisfactory hosts are present.

The number of species of fleas that are of medical importance is a matter of debate. Only a few species are directly associated with disease transmission to humans and these are normally limited to those having direct access to both human and wild animal populations. Much of the literature dealing with disease transmission by fleas is based on experimental laboratory infection and transmission studies which probably do not reflect the situation in nature. In addition, the dynamics of host-to-host transmission in nature may be quite different from those of host-to-human transmission. Thus a large population of poor vector species could maintain the zoonosis in a wild host population without posing a threat to an adjacent human population. Conversely, human intrusion into an area where the wild host population is under the stress of an enzootic vastly increases the chances of acquiring the disease, even if the fleas are poor vectors to man.

RECOGNITION AND ELEMENTS OF STRUCTURE

The terminology used to describe flea morphology is complicated and although it is relatively consistent within the flea literature, through the domination of early workers such as Hopkins, Jordan and Rothschild, it is not easily understood by the non-specialist. The principal features used in identification of the medically important adult fleas are given below and illustrated in Fig.16.1.

Medical Insects and Arachnids Edited by Richard P. Lane and Roger W. Crosskey.
Published in 1993 by Chapman & Hall ISBN 0 412 40000 6

The internal anatomy and histology of fleas have been described in a remarkable work by Rothschild et al. (1986) and the morphology, biology and identification of larvae have been comprehensively reviewed by Elbel (1991).

As in all other adult insects the body is divided into head, thorax and abdomen. Unlike most other insects, however, the body is laterally compressed and fleas are viewed from the side for purposes of identification.

The shape of the head varies considerably and is useful for distinguishing some species (e.g. Fig. 16.10). Approximately midway along the gently curving anterior margin of the frons is a small bump, the frontal tubercle, which is variable in shape and structure (Fig. 16.12). It is absent or deciduous in a number of species. The gena occupies the ventral portion of the anterior part of the head, extends backwards (genal lobe, Fig. 16.5a) towards the antennal fossa, and varies in shape between genera (Fig. 16.13). A conspicuous comb of spines is often present on the gena, the genal ctenidium. The comb is either along the ventral edge of the gena, when it is termed the horizontal ctenidium (Fig. 16.10), or along its posterior margin, when it is termed an oblique or vertical ctenidium (Fig. 16.15). The spines of the ctenidia are heavily sclerotized outgrowths of the cuticle. They do not arise from a distinct pore (alveolus) and are therefore not technically setae.

The piercing–sucking mouthparts, often termed the fascicle, are ventrally placed and consist of three stylets: a pair of fine maxillary laciniae (bearing coarse teeth in some species) and the single (i.e. unpaired) epipharynx. These are held in place by the labial palps, which are usually five-segmented.

Eyes are variable in adults and range from well-developed and heavily pigmented to vestigial or absent, particularly in those species that are nest fleas. Close to the eye is a row of setae called the ocular row. There are sometimes additional rows of setae on the preantennal portion of the head. In some families the tentorium can be seen extending slightly in front of the eye. The antennae lie in deep grooves (antennal fossae) each side of the head and have three segments. The third antennal segment, termed the antennal clavus, is usually club-like and subdivided into nine segments (Fig. 16.14e); in rhadinopsyllids and some pulicids, however, two or more of the flagellar segments have become fused, thereby reducing the apparent number of segments.

The three thoracic segments are clearly represented by three dorsal sclerites (pronotum, mesonotum and metanotum) and complementary ventral plates. Several of the pleural plates (e.g. mesepisternum, metepisternum, lateral metanotal area) vary in shape and are thus taxonomically important (Fig. 16.2d). A pronotal ctenidium is present in some genera (Fig. 16.7a), even in the absence of a genal ctenidium, but never the reverse. In most species the metanotum does not bear a ctenidium, but where a comb is present it usually consists of only a few marginal spinelets. Combs are never present on the middle thoracic segment, the mesothorax. An internal rod, the pleural rod (Fig. 16.1), divides the mesopleuron into an anterior mesepisternum and a posterior mesepimeron. Its presence is characteristic of some genera, e.g. *Xenopsylla*.

The legs bear numerous taxonomic features, and consist of a large basal coxa, a small trochanter, the femur, tibia and tarsus. The tarsus is divided into five segments termed tarsomeres. The apical pretarsus consists of paired claws or ungues and an unguitractoral plate (Fig. 16.8). The apical claw-bearing segment is sometimes called the distitarsomere.

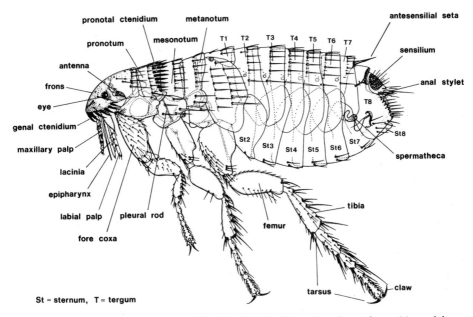

Figure 16.1 Morphology of a generalized flea (female). The illustration shows the positions of the genal and pronotal combs (ctenidia), and the pleural rod, important in taxonomy because of their presence or absence in different species.

The abdominal segments are clearly divided into dorsal tergites (T1–T10) and ventral sternites (St2–St9) (Fig. 16.1). Roman numerals are usually used to denote the segments in the literature (e.g. Hopkins and Rothschild, 1953, 1956, 1962, 1966, 1971). The tergites are not usually modified but in some genera they bear internal thickenings (incrassations). The posterior margin of T7 in both sexes bears a variable number of well-developed antesensilial setae, except in the males of *Rhadinopsylla* species. In males, T8 can vary in size from species to species, but in females it is always large and covers most or all of the terminal modified segments (Fig. 16.4). The so-called modified abdominal segments in adult fleas include both the tergites and sternites beyond segment 7, and their modifications must be considered separately for the sexes.

In females T8 is usually well developed and covers most if not all of the terminal portion of the abdomen. In some cases the configuration and chaetotaxy of its caudal margin can be useful in taxonomic discrimination. Sternum 8 is reduced in most cases to a slightly sclerotized lobe which projects posteriorly between the large lateral lobes of T8. Sternum 9 is even more reduced and membranous and the membrane between the sternites on either side of the body bears the external opening to the vagina and internal sex organs. Specialists do not agree in their interpretation of the segments which make up the composite sclerite bearing the sensilium, the anal lobes and the anal stylets in female fleas (Fig. 16.1). Very likely, T9, T10 and St10 are all involved. The sensilium consists of a flat or convex plate bearing spicules and a variable number of dome-shaped structures from which trichobothria project (Fig. 16.11*b*). It is obviously sensory, but its exact role has yet to be determined. Behind the sensilium is a pair of anal stylets which usually bear a single long apical seta and a number of shorter subapical bristles. The plate is

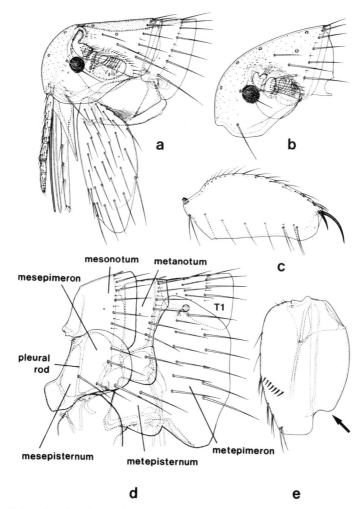

Figure 16.2 Oriental rat flea, *Xenopsylla cheopis*: (*a*) male head, prothorax and fore coxa; (*b*) female head; (*c*) female hind femur, outer side; (*d*) thoracic sclerites and their terminology; (*e*) hind coxa (point of abrupt contraction of outer margin arrowed).

terminated by a cone-shaped projection bearing the dorsal and ventral anal lobes and the anus. Internally, the genital system of the female includes the vagina, bursa copulatrix, spermathecal duct and one or two spermathecae (Fig. 16.8*a*). The latter are usually divided into a heavily sclerotized bulga and a less sclerotized finger-like projection, the hilla (Fig. 16.6*a*). There are also a number of glands associated with these structures that are not visible in specimens that have been cleared and mounted.

The modifications of the terminal abdominal segments of the male are much more complicated. The male intromittent organ is termed the aedeagus (Fig. 16.3*d*). It is an extremely complex structure of obscure derivation and is seldom used in identification. However, associated structures derived from the terminal tergites and sternites are the main structures employed in taxonomic discrimination. As in the female, St8 can be reduced to the extent that it encloses the remaining genital structures and it may bear modifications that are useful in identification, such as

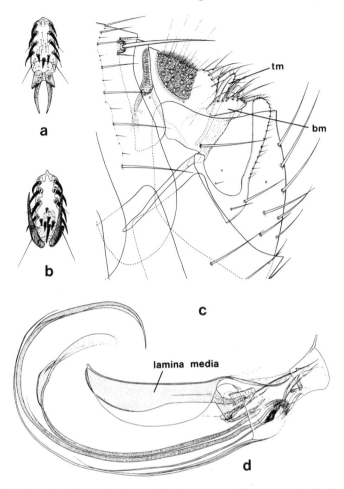

Figure 16.3 Oriental rat flea, *Xenopsylla cheopis*: (*a*) plantar setae of last segment of male fore tarsus; (*c*) male terminalia; (*d*) aedeagus. *Ctenocephalides felis damarensis*: (*b*) plantar setae of last segment of fore tarsus (male).

spicules and characteristic chaetotaxy. Sternum 8 is missing or vestigial in some genera, but when present it usually takes the form of a posteriorly projecting lobe with characteristic setation, and often other specializations. Sternum 9 is a paired structure, fused ventrally and with proximal and distal arms (Fig. 16.5*b*, 16.17). The configuration and chaetotaxy of the latter have taxonomic importance. The entire region is dominated by a large, arching, structure collectively referred to as the clasper, which is thought to be derived from T9. Anteriorly, the sclerite consists of a dorsal apodeme and pair of ventral apodemes which collectively make up the manubrium of the clasper. Posteriorly, a pair of lobes (basimeres) project that are usually referred to as the fixed processes and that usually bear an articulated movable process, the telomere (Fig. 16.7*d*). In the Pulicoidea each fixed process usually bears two telomeres; a sensilium is present in the male, as are dorsal and ventral anal lobes, but anal stylets are absent.

Fleas are well adapted morphologically for an ectoparasitic life on their hosts. Very often these adaptations are host specific and thus exhibit a considerable

534 *Fleas (Siphonaptera)*

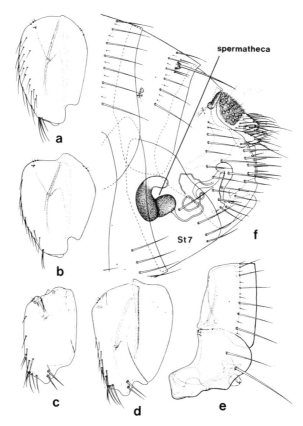

Figure 16.4 (*a* and *b*) inner side of hind coxa of male *Citellophilus simplex* s.str. and *Nosopsyllus fasciatus*, respectively; (*c* and *d*) outer side of mid coxa of female *Ctenocephalides felis* s.str. and *Nosopsyllus fasciatus*, respectively; (*e*) *Pulex irritans*, male mesothorax; (*f*) *Xenopsylla cheopis*, female terminalia.

degree of 'co-ordinated evolution' with the hosts (Holland, 1964; Traub, 1980a). However, Marshall (1980) argues that the function of such combs is not to prevent dislodgement from the host but to protect highly mobile joints. The adaptive radiation to hosts, especially mammals, is comparable to that of other ectoparasites (lice, and Streblidae, Hippoboscidae and Oestridae in the Diptera) (Price, 1980).

CLASSIFICATION AND IDENTIFICATION

The higher classification of the Siphonaptera has remained fairly constant for the past 50 years or more, 15 or 16 families being generally recognized depending how genera are assigned. Karl Jordan developed a classification into 17 families and this was formally set down by Hopkins and Rothschild (1953). This number has since been reduced to 16, however, by inclusion of the Amphipsyllidae as a subfamily of the Leptopsyllidae. Smit (1982) proposed a somewhat different arrangement resulting in only 15 families, but it is too soon to know how fully this arrangement will be accepted by specialists. Most of the recent monographs have followed the modified Jordan classification (Hopkins and Rothschild, 1953). The two systems are listed below for comparison, and a compromise between the two adopted for this chapter.

Classification and identification 535

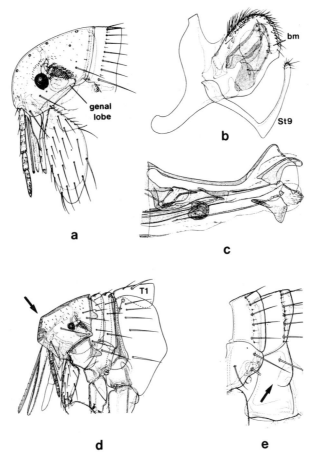

Figure 16.5 *Pulex irritans*: (*a*) male head, prothorax and fore coxa; (*b*) paramere and sternum 9; (*c*) aedeagus. *Echidnophaga gallinacea*; (*d*) female head and thorax (arrow marking angulation of the frons). *Synosternus pallidus*; (*e*) female mesothorax and metathorax (arrow marking incomplete differentiation of metanotum and metepisternum). bm = basimere.

Jordan and others	Smit (1982)
Tungidae	Tungidae
Pulicidae	Pulicidae
Vermipsyllidae	Vermipsyllidae
Rhopalopsyllidae	Rhopalopsyllidae
Malacopsyllidae	Malacopsyllidae
Hystricopsyllidae	Hystricopsyllidae
Macropsyllidae	Ctenophthalmidae
Stephanocircidae	Stephanocircidae
Pygiopsyllidae	Pygiopsyllidae
Hypsophthalmidae	Chimaeropsyllidae
Coptopsyllidae	Coptopsyllidae
Ancistropsyllidae	Ancistropsyllidae
Xiphiopsyllidae	Xiphiopsyllidae
Ischnopsyllidae	Ischnopsyllidae
Ceratophyllidae	Ceratophyllidae
Leptopsyllidae	
Amphipsyllidae	

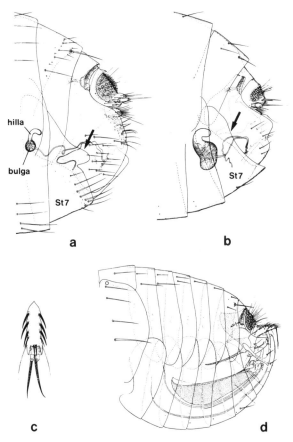

Figure 16.6 *Pulex irritans*: (a) female terminalia (shape of sternum 7 hind margin arrowed). *Echidnophaga gallinacea*: (b) female terminalia (shape of sternum 7 hind margin arrowed); (c) last hind tarsal segment showing plantar setae (female); (d) male abdomen.

Although Jordan did not use superfamilies, Smit (1982) assembled his 15 families into five superfamilies, namely Vermipsylloidea including only the Vermipsyllidae, the Pulicoidea including Tungidae and Pulicidae, the Malacopsylloidea including the Rhopalopsyllidae and Malacopsyllidae, the Hystricopsylloidea including the six families Hystricopsyllidae to Coptopsyllidae in Smit's list above, and the Ceratophylloidea including the remaining four families (Ancistropsyllidae to Ceratophyllidae).

All families except Ceratophyllidae have been covered in the series of British Museum (Nature History) flea catalogues, volumes I–V by Hopkins and Rothschild (1953, 1956, 1962, 1966, 1971), volume VI by Mardon (1981) and volume VII by Smit (1987). Although not treated in the same format, the Ceratophyllidae have been covered by Traub et al. (1983) and Lewis (1990). Invalid genus-group and species-group names, in the technical meaning of the *International Code of Zoological Nomenclature*, have been dealt with by Lewis and Lewis (1989).

Identification

Taxonomic characters used for identification are found in practice only in the hard-parts morphology, and as yet fleas have been little studied from the viewpoint of

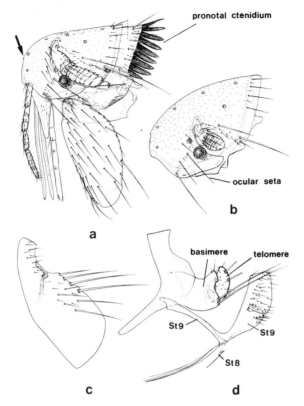

Figure 16.7 Northern rat flea, *Nosopsyllus fasciatus*: (*a*) male head, prothorax and fore coxa (arrow marking frontal tubercle); (*b*) female head; (*c*) main tergum 8; (*d*) paramere and sterna 8 and 9.

cytological or biochemical taxonomy. The features concerned most in identification keys are those listed above as elements of structure.

It is mainly the fleas of terrestrial mammals, representing nearly 90% of the known species, that are of medical importance, and such species figure largely in studies on plague transmission, for example. Bat fleas and bird fleas are therefore not considered here. However, the European chicken flea, *Ceratophyllus gallinae*, can occasionally be a nuisance.

Several species of fleas parasitic on synanthropic hosts have become virtually cosmopolitan and, being common and secondarily associated with man, are of prime medical importance. Some of these are the most widely used species for experimental purposes. A separate and fully illustrated identification key for these common fleas is therefore provided here. It should be noted, however, that like most other fleas these common species have related forms which resemble them closely – and that it is not difficult to give a flea the wrong name! An example of the confusion created by misidentifications is illustrated by the studies of bubonic plague in Sri Lanka by L. F. Hirst, who, with the assistance of N. C. Rothschild, demonstrated that two closely related species of *Xenopsylla* were involved in the reservoir of this disease on the Indian subcontinent; relevant works are summarized in Hirst (1953). Another example concerned two closely related species of *Pulex*. Until quite recently most specimens of this genus from North America were identified as the human flea, *P. irritans*, because differences between this species and the related *P. simulans* had not been recognized.

538 Fleas (Siphonaptera)

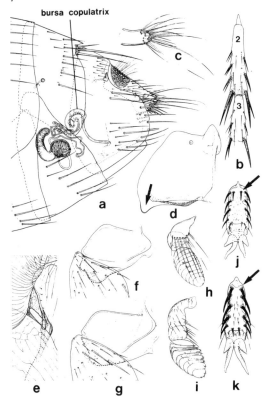

Figure 16.8 (a) *Nosopsyllus fasciatus*, female terminalia; (b) *Nosopsyllus fasciatus*, male second and third hind tarsal segments; (c) *Amphalius runatus* s.str. female anal stylet; (d) *Delostichus talis*, female metasternum (anteroventral projection arrowed); (e) rod-like structure between metepimeron and basal sternum occurring in *Stivalius* (Key B couplet 5 and Key F couplet 1); (f, g) prosternosome and fore coxal base in male *Rhopalopsyllus lugubris* s.str. and *Polygenis litargus*, respectively; (h, i) antenna of female *Delostichus talis* and male *Polygenis litargus*, respectively; (j, k) plantar setae of last hind tarsal segment of *Amphipsylla rossica* and *Orchopeas howardi* s.str., respectively (positions of first pair of plantar setae arrowed).

Flea identification is not helped by the fact that certain structures are liable to show a fair amount of individual variation of which the true nature can only be understood after studying long series of specimens. Moreover, structurally abnormal specimens can be encountered. Females of many species cannot be accurately identified in the absence of associated males.

It is not feasible to construct a single, simple, concise and foolproof key to the numerous species of fleas that are proven or potential vectors of disease organisms and the keys provided here (except for Key A) serve only to identify the genus of such fleas. They include about a third of the genera of terrestrial mammal fleas. In addition to Key A for identification of species commonly associated with man, five separate geographically restricted keys (Keys B–F) are given to simplify identification: these cover Eurasia, Africa, North America, Central and South America, and Australasia respectively. Even so, keys to selected genera are never fully satisfactory and must be used with care. It would be wise to send specimens to a specialist for confirmation of doubtful identifications.

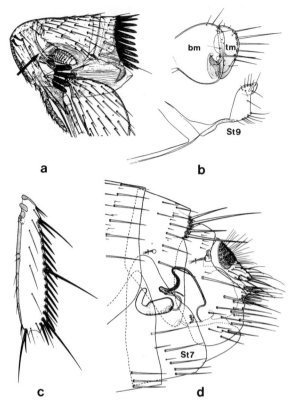

Figure 16.9 *Leptopsylla segnis*: (*a*) male head, prothorax and fore coxal base; (*b*) paramere and sternum 9; (*c*) female hind tibia; (*d*) female terminalia. bm = basimere, tm = telomere.

Key A: Identification key to cosmopolitan species commonly associated with man

The following key includes seven species widely associated with humans, their pets and their domiciles. These are species most likely to be encountered by public health workers and others who are not specialists in Siphonaptera. They are 'metropolitan' fleas in the sense that they are the only species likely to be found in major urban settings.

1 Ctenidia absent 2
— Ctenidium present at least on pronotum (as in Figs 16.1, 16.7*a*, 16.9*a*) 4
2 Pleural rod absent (e.g. as in Fig. 16.4*e*) 3
— Pleural rod present (Fig. 16.2*d*). [Antesensilial seta inserted distinctly before hind margin of T7 (Figs 16.3*c*, 16.4*f*. Male paramere as in Fig. 16.3*c*, basimere rather broad, with hind margin straight or slightly concave and bearing several slender setae. St9 straight and widened towards its apex (Fig. 16.3*c*). Aedeagus as in Fig. 16.3*d*, its lamina media rather broad and slightly concave on dorsal margin. Female St7 as in Fig. 16.4*f*. Spermatheca with bulga somewhat longer than broad, not wider than base of hilla, ventral margins of bulga and hilla more or less in line (Fig. 16.4*f*)] .... **Xenopsylla cheopis**

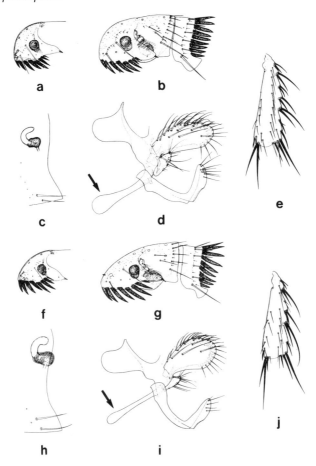

Figure 16.10 Morphological differences between the dog flea, *Ctenocephalides canis* (*a–e*) and cat flea, *Ctenocephalides felis* s.str. (*f–j*). Front of male head (*a* and *f*); female head and pronotum (*b* and *g*); spermatheca (*c* and *h*); paramere and sternum 9 (manubrium shape arrowed) (*d* and *i*); hind tibia (*e* and *j*).

3 Frons angulate (Fig. 16.5*d*). Head behind antenna with two setae (and in female usually with small lobe). [Genal lobe pointed backwards (Fig. 16.5*d*). Laciniae very broad and coarsely serrated. Thoracic length on dorsal margin less than length of T1 (Fig. 16.5*d*). Hind tarsus with plantar setae of last segment (distitarsomere) as in Fig. 16.6*c* and claws without large basal projection. Male genitalia as in Fig. 16.6*d*. Female terminal abdominal segments and genitalia as in Fig. 16.6*b*] .... **Echidnophaga gallinacea**

— Frons smoothly rounded (Fig. 16.5*a*). Head behind antenna with only one strong seta. [Ocular seta placed below eye, latter conspicuous. Genal margin with one small pseudo-spinelet (rarely two and sometimes none). Male paramere with broad basimere covering doubled telomere (Fig. 16.5*b*). Aedeagus as in Fig. 16.5*c*. Female St7 with posterior outline as in Fig. 16.6*a* (arrowed) (usually hardly visible in mounted specimens). Spermatheca with globular bulga and curved hilla (Fig. 16.6*a*)].. **Pulex irritans**

4 Genal ctenidium present (Figs 16.9*a*, 16.10*b,g*) 5

— Genal ctenidium absent (Fig. 16.7*a*). [Head without setae in front of row of three below eye, but with one or a few small setae above this row (Fig. 16.7*a*). Hind

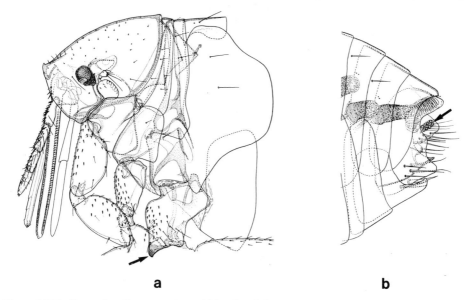

Figure 16.11 Jigger flea, *Tunga penetrans*: (*a*) head and thorax, tergum 1 and sternum 2 (arrow marking characteristic tooth-like projection of hind coxa); (*b*) terminalia of unexpanded female (arrow marking the trichobothria).

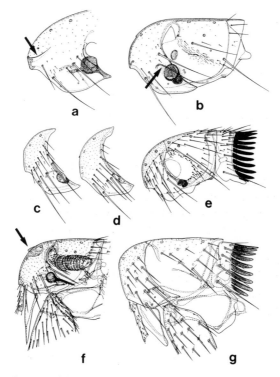

Figure 16.12 Head and pronotal features: (*a*) *Coptopsylla lamellifera dubinini*, front of head showing transverse slit (arrowed); (*b*) *Ophthalmopsylla volgensis palestinica*, head of female (tentorial rod arrowed); (*c*) *Frontopsylla wagneri* s.str., front of male head; (*d*) *Foxella ignota* s.str., front of male head; (*e*) *Afristivalius torvus*, head and pronotum of female; (*f*) *Listropsylla fouriei*, head of male (frontal tubercle arrowed); (*g*) *Stenistomera macrodactyla*, head and prothorax of male.

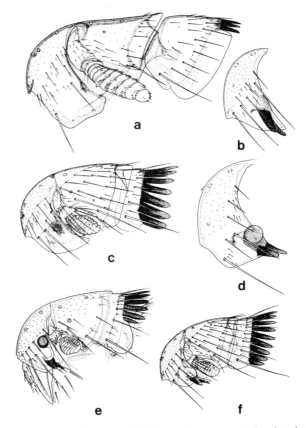

Figure 16.13 Head and pronotal features: (*a*) *Xiphiopsylla daemonicola*, head and pronotum of male; (*b*) *Meringis shannoni*, front of male head, (*c*) *Catallagia dacenkoi*, head and pronotum of female; (*d*) *Mesopsylla tuschkan andruschkoi*, front of male head; (*e*) *Chiastopsylla quadrisetis*, head and pronotum of female; (*f*) *Neopsylla setosa spinea*, head and pronotum of female.

tarsus of male with apical setae of second segment (tarsomere) not reaching back as far as end of third segment (Fig. 16.8*b*). Male abdomen with vestigial St8 (Fig. 16.7*d*), T8 strongly rounded from behind last dorsal marginal seta (Fig. 16.7*c*), and St9 and paramere as in Fig. 16.7*d*. Female terminalia with posterior margin of St7 slanting, spermatheca with hilla not markedly narrowed apically, and bursa copulatrix with forwardly curved duct and strongly coiled apical part (Fig. 16.8*a*)].......................... **Nosopsyllus fasciatus**

5 Genal ctenidium formed of eight or nine spines, oriented horizontally (Fig. 16.10*a,f*) 6

— Genal ctenidium formed of four spines, oriented vertically (Fig. 16.9*a*). [Head with two of setae near frontal angle spiniform (Fig. 16.9*a*). Tibiae with a row of spiniform setae along posterodorsal margin forming a false comb (Fig. 16.9*c*). Male paramere and St9 as in Fig. 16.9*b*. Female St7 and spermatheca as in Fig. 16.9*d*] .... **Leptopsylla segnis**

6 Head strongly convex anteriorly in both sexes and not noticeably elongate (Fig. 16.10*a,b*). Hind tibia with eight seta-bearing notches along dorsal margin (Fig. 16.10*e*). Male with paramere as in Fig. 16.10*d*, manubrium strongly dilated towards its apex. Spermatheca with apical part of hilla quite long (Fig. 16.10*c*) .... **Ctenocephalides canis**

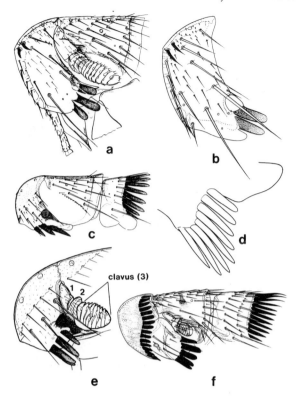

Figure 16.14 Head and pronotal features: (*a*) *Leptopsylla algira* s.str., male head; (*b*) *Peromyscopsylla scotti*, front of male head; (*c*) *Ctenophthalmus calceatus cabirus*, head and pronotum of male; (*d*) *Hystricopsylla occidentalis* s.str., genal ctenidium; (*e*) *Neotyphloceras crassispina* s.str., head of female; (*f*) *Sphinctopsylla inca*, head and pronotum of female.

— Head not strongly convex anteriorly and distinctly elongate (especially in female) (Fig. 16.10*f,g*). Hind tibia with six seta-bearing notches along dorsal margin (Fig. 16.10*j*). Male with paramere as in Fig. 16.10*i*, manubrium only slightly dilated towards its apex. Spermatheca with apical part of hilla short (Fig. 16.10*h*) .... **Ctenocephalides felis felis**

Key B: Identification key to medically important genera in Eurasia

1 Ctenidia absent 2
— Ctenidium present at least on pronotum (as in Fig. 16.7*a*) 4
2 Frons without a transverse slit (e.g. as in Fig. 16.12*b*) 3
— Frons with a transverse slit on its lower part (Fig. 16.12*a*, arrowed) ....
    .... **Coptopsylla**
3 Frons smoothly rounded in profile (Fig. 16.2*a,b*). Pleural rod present (Fig. 16.2*d*).
    .... **Xenopsylla**
— Frons angulate in profile (Fig. 16.5*d*, arrowed). Pleural rod absent ....
    .... **Echidnophaga**
4 Genal ctenidium absent 5
— Genal ctenidium present 16
5 Rod-like structure absent between metepimeron of thorax and basal sternum of abdomen 6

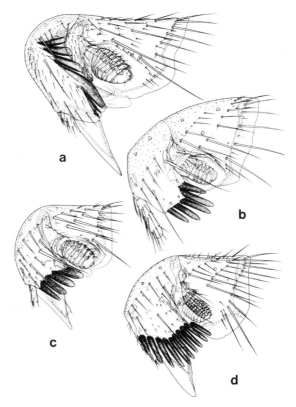

Figure 16.15 Head and pronotal features: (*a*) *Dinopsyllus wansoni*, head of male; (*b*) *Rhadinopsylla li ventricosa*, head of female; (*c*) *Adoratopsylla intermedia* s.str., head of male; (*d*) *Stenoponia sidimi*, head of male.

— Rod-like structure present between metepimeron of thorax and basal sternum of abdomen (Fig. 16.8*e*) .... **Stivalius**

6 Tentorial rod visible in front of eye (Fig. 16.12*b*, arrowed). Ocular row with upper seta inserted beside or near antennal fossa (Fig. 16.12*b*) 7

— Tentorial rod absent. Ocular row with uppermost seta inserted in front of eye (Fig. 16.7*a,b*).. 10

7 Last segment of hind tarsus (distitarsomere) with first pair of plantar setae forward of and nearly in line with second pair (Fig. 16.8*k*, arrowed).............. 8

— Last segment of hind tarsus with first pair of plantar setae placed between members of second pair, so making transverse row of four setae (Fig. 16.8*j*, arrowed) .... **Amphipsylla**

8 Male paramere with telomere usually more or less triangular (Fig. 16.16*a,d*). Spermatheca with bulga either not globular or not sharply demarcated from hilla 9

— Male paramere with telomere elongate and not very broad (Fig. 16.17*d*). Spermatheca with globular bulga sharply demarcated from hilla (Fig. 16.20*a*). .... **Paradoxopsyllus**

9 Eye with dark posteroventral part relatively large (Fig. 16.12*b*). Male with St8 much reduced and modified (Fig. 16.16*a*) and basimere with one acetabular seta (same figure, arrowed). Spermatheca with bulga differentiated from hilla by clear interruption of internal striae (Fig. 16.19*a*).............. **Ophthalmopsylla**

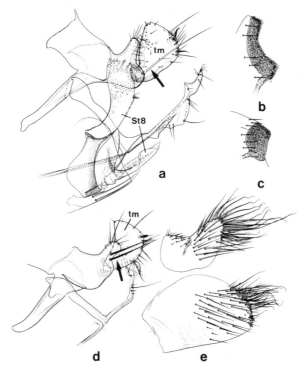

Figure 16.16 (*a*) *Ophthalmopsylla volgensis palestinica*, paramere and sterna 8 and 9; (*b, c*) *Megabothris rectangulatus*, spiracular fossa of tergum 8, in male and female respectively; (*d, e*) *Frontopsylla wagneri* s.str., paramere with sternum 9, and segment 8, respectively. tm = telomere. Arrows indicate acetabular setae (Key B couplet 9).

— Eye with dark posteroventral part smaller (Fig. 16.12*c*). Male with St8 unmodified (Fig. 16.16*e*) and basimere with two acetabular setae (same figure, arrowed). Spermatheca with bulga and hilla virtually undifferentiated (Fig. 16.19*f*) .... **Frontopsylla**

10 Male telomere without posteroventral elongation. Anal stylet of female with at most three or four lateral setae in addition to well-differentiated apical seta 11

— Male telomere with conspicuous posteroventral elongation (Fig. 16.17*e*). Female anal stylet with numerous setae (Fig. 16.8*c*). Female bursa copulatrix very large, wide and curved.. **Amphalius**

11 Mid and hind coxae on inner side with vestiture of thin setae laterally on basal third (Fig. 16.4*a*)... 12

— Mid and hind coxae on inner side without such setae on basal third (Fig. 16.4*b*).. 13

12 Labial palp reaching beyond apex of fore coxa. Male T8 without area spiculosa (Fig. 16.18*b*). Female with three or four antesensilial setae.................................. .... **Oropsylla (Oropsylla)**

— Labial palp usually not reaching beyond apex of fore coxa. Male T8 with dorsal area spiculosa (Fig. 16.18*a*). Female with two antesensilial setae .... **Citellophilus**

13 Male St8 fully developed and usually setose (e.g. Fig. 16.18*e, f*). Female bursa copulatrix not coiled 14

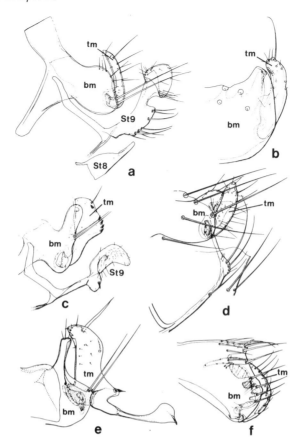

Figure 16.17 Terminalia features of male: (a) *Oropsylla montana*, paramere with sterna 8 and 9; (b) *Craneopsylla minerva* s.str. apex of paramere; (c) *Orchopeas sexdentatus pennsylvanicus*, paramere with sternum 9; (d) *Paradoxopsyllus teretifrons*, paramere with sterna 8 and 9; (e) *Amphalius runatus* s.str., apex of paramere; (f) *Sphinctopsylla ares*, apex of paramere. bm = basimere, tm = telomere.

— Male St8 vestigial (Fig. 16.7d) and usually without setae. Female bursa copulatrix with a forwardly curved duct and strongly coiled apical part (Fig. 16.8a) .... **Nosopsyllus**

14 Abdominal T8 with spiracular fossa not greatly enlarged 15

— Abdominal T8 with spiracular fossa much enlarged, especially in female (Fig. 16.16b,c) .... **Megabothris**

15 Male St8 long and narrow and with long membranous process (Fig. 16.18f). Spermatheca with bulga longer than hilla and latter with large apical papilla (Fig. 16.20f) .... **Amalaraeus**

— Male St8 with triangular base and elongate apical part with or without setae (Fig. 16.18e). Spermatheca with bulga pear-shaped or cylindrical and hilla without apical papilla (Fig. 16.20b) .... **Ceratophyllus (Monopsyllus)**

16 Terga 2–7 each with more than one row of setae 17

— Terga 2–7 each with one row of setae (Fig. 16.1). [Genal ctenidium formed of eight or nine spines (in common species) and orientated horizontally (Fig. 16.10a,f)] .... **Ctenocephalides**

17 Genal ctenidium formed of two spines 18

— Genal ctenidium formed of more than two spines 19

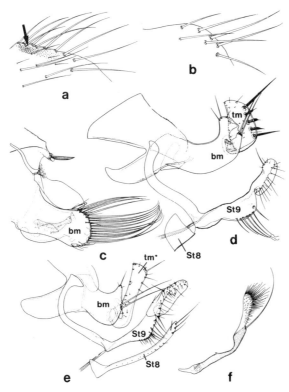

Figure 16.18 Terminalia features of male: (*a*) *Citellophilus tesquorum sungaris*, dorsum of T8 showing area spiculosa (arrowed); (*b*) *Oropsylla silantiewi*, dorsum of T8 without area spiculosa; (*c*) *Plocopsylla wolffsohni*, apex of paramere showing hair-fringed basimere; (*d*) *Malaraeus telchinus*, paramere with sterna 8 and 9; (*e*) *Ceratophyllus anisus*, paramere with sterna 8 and 9; (*f*) *Amalaraeus penicilliger mustelae*, sternum 8. bm = basimere, tm = telomere.

18 Genal spines not or only slightly overlapping (Fig. 16.13*d*). [Tibiae without false comb of spiniform setae] .... **Mesopsylla**
— Genal spines positioned with base of anterior spine overlapping that of posterior spine (Fig. 16.13*f*) .... **Neopsylla**
19 Tergum 1 without a ctenidium. Genal ctenidium formed of three to eight spines 20
— Tergum 1 with a well-developed ctenidium. Genal ctenidium formed of 9–15 spines (Fig. 16.15*d*) .... **Stenoponia**
20 Genal ctenidium formed of three spines and oriented horizontally, spines usually sharp-pointed (Fig. 16.14*c*) .... **Ctenophthalmus**
— Genal ctenidium formed of four to eight spines and at least hindmost part oriented vertically, spines with bluntly rounded tips (Fig. 16.15*b*) .... **Rhadinopsylla**

Key C: Identification key to medically important genera in the Afrotropical region

1 Pronotal ctenidium absent 2
— Pronotal ctenidium present (Fig. 16.7*a*) 5
2 Hind coxa with a row or patch of spiniform setae on inner side, its anteroventral corner rounded and its outline shape sharply narrowed below middle of

548 *Fleas (Siphonaptera)*

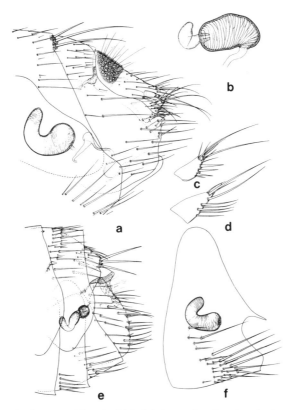

Figure 16.19 Terminalia features of female: (*a*) *Ophthalmopsylla volgensis palestinica*, complete terminalia; (*b*) *Plocopsylla ulysses*, spermatheca; (*c, d*) anal sternum of *Orchopeas leucopus* and *Opisodasys keeni*, respectively; (*e*) *Sphinctopsylla ares*, complete terminalia; (*f*) *Frontopsylla wagneri* s.str., spermatheca and sternum 7.

    posterior margin (Fig. 16.2*e*, arrowed). Sensilium with more than eight sensory pits each side. [Pronotum dorsally shorter than mesonotum. Genal lobe without setae behind eye (Fig. 16.5*a*)] 3
— Hind coxa without spiniform setae, its anteroventral corner projecting downwards as a triangular tooth (Fig. 16.11*a*, arrowed) and its outline shape without such narrowing. Sensilium with eight sensory pits each side (Fig. 16.11*b*, arrowed). [Jiggers or sand-fleas].. **Tunga**
3 Thorax with ventral edge of metanotum not completely differentiated from metepisternum (Fig. 16.5*e*, arrowed) 4
— Thorax with ventral edge of metanotum completely differentiated from metepisternum (Fig. 16.2*d*) .... **Xenopsylla**
4 [Specimens from Africa or Middle East] .... **Synosternus**
— [Specimens from Madagascar] .... **Synopsyllus**
5 Genal ctenidium absent 6
— Genal ctenidium present 7
6 Frontal tubercle very large and triangular, situated in a deep cavity (Fig. 16.12*f*, arrowed). Head with a long slender spine behind eye, latter well developed (Fig. 16.12*f*). Abdomen with marginal spinelets on anterior terga ....
.... **Listropsylla**

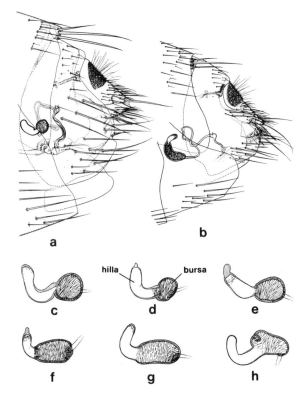

Figure 16.20 Female terminalia and spermatheca: (*a*) *Paradoxopsyllus teretifrons*, complete terminalia; (*b*) *Ceratophyllus anisus*, complete terminalia; (*c–h*) spermathecae of *Oropsylla montanus* (*c*), *Thrassis acamantis* s.str. (*d*), *Oropsylla silantiewi* (*e*), *Amalaraeus penicilliger mustelae* (*f*), *Malaraeus telchinus* (*g*) and *Craneopsylla minerva* s.str. (*h*).

— Frontal tubercle small. Head without such spine behind eye, latter inconspicuous. Abdomen without marginal spinelets on anterior terga .... **Xiphiopsylla**
7 Terga each with more than one row of setae 8
— Terga each with one row of setae (Fig. 16.1). [Genal ctenidium formed of eight or nine spines (in common species) and oriented horizontally (Fig. 16.10*a,f*)] .... **Ctenocephalides**
8 Genal ctenidium formed of two or three spines and positioned below level of antenna (e.g. as in Fig. 16.14*a*) 9
— Genal ctenidium formed of five spines and positioned mainly in front of antenna and obliquely parallel to slope of frontal margin (Fig. 16.15*a*) .... **Dinopsyllus**
9 Genal ctenidium formed of three spines 10
— Genal ctenidium formed of two partly overlapping spines (Fig. 16.13*e*) (these sometimes reduced to one or absent) .... **Chiastopsylla**
10 Genal ctenidium oriented vertically and with blunt-tipped spines (Fig. 16.14*a*). Frons with two or three short spiniform setae near frontal angle (Fig. 16.14*a*) .. .. **Leptopsylla**
— Genal ctenidium oriented horizontally and usually with pointed spines (Fig. 16.14*c*). Frons without spiniform setae .... **Ctenophthalmus**

Key D: Identification key to medically important genera in North America

1. Ctenidia absent (except vestigial genal comb present in *Pulex*).......... 2
— Ctenidium present at least on pronotum 4
2. Frons with frontal tubercle. Mid coxa with outer internal ridge (Fig. 16.4*d*) 3
— Frons without frontal tubercle (Fig. 16.5*a*). Mid coxa without such ridge (Fig. 16.4*c*) .... **Pulex**
3. Terga 2–7 with one row of setae.......... **Anomiopsyllus**
— Terga 2–7 with additional setae anterior to main row.......... **Polygenis**
4. Genal ctenidium absent.......... 5
— Genal ctenidium present 16
5. Terga 2–7 with additional setae anterior to main row 6
— Terga 2–7 with one row of setae .... **Hoplopsyllus**
6. Head with tentorial rod visible in front of eye (Figs 16.12*g*, 16.13*c*). [Mesonotum without marginal spinelets].......... 7
— Head without tentorial rod in front of eye.......... 8
7. Head with two rows of setae on its preantennal part (Fig. 16.13*c*); frons profile rounded. Hind coxa broad .... **Catallagia**
— Head with more than two rows of setae on its preantennal part (Fig. 16.12*g*); frons profile more angular. Hind coxa very narrow .... **Stenistomera**
8. Eye large and dark 9
— Eye vestigial and pale (Fig. 16.12*d*) .... **Foxella**
9. Basal abdominal sternum with not more than two lateral setae.......... 10
— Basal abdominal sternum with a patch of lateral setae on its upper half .... **Oropsylla (Opisocrostis)**
10. Mid and hind coxae with thin setae laterally on basal third of inner side (as in Fig. 16.4*a*).......... 11
— Mid and hind coxae without such setae (as in Fig. 16.4*b*).......... 13
11. Male St8 setose and not reduced; telomere variously shaped but not long, slender and curved forwards. Spermatheca not or only indistinctly constricted between bulga and hilla. Anal stylet with two or more lateral setae.......... 12
— Male St8 without setae and much reduced; telomere long and slender, curved forwards (Fig. 16.17*a*). Spermatheca sharply constricted between bulga and hilla (Fig. 16.20*c*). Anal stylet with one long lateral seta .... **Oropsylla (Diamanus)**
12. Male St8 rather short and fairly broad. Female with two or three antesensilial setae. Spermatheca with bulga shorter than broad and hilla usually longer than bulga, apical papilla of hilla often small or absent (Fig. 16.20*d*) .... **Thrassis**
— Male St8 narrow. Female with three or four antesensilial setae. Spermatheca with bulga at least as long as wide and hilla at most only slightly longer than bulga, hilla with large apical papilla (Fig. 16.20*e*) .... **Oropsylla (Oropsylla)**
13. Hind tarsus with first pair of plantar setae (Fig. 16.8*k*, arrowed) on last segment nearer midline than other pairs.......... 14
— Hind tarsus with first pair of plantar setae on last segment not noticeably nearer midline than other pairs.......... 15
14. Male St8 without setae; telomere with a row of four or five short black spiniform setae close together along hind margin (Fig. 16.17*c*). Anal sternum of female with lower margin convex near middle (Fig. 16.19*c*).......... **Orchopeas**

— Male St8 with at least one seta; telomere with longer spiniform setae. Anal sternum of female with lower margin nearly straight (Fig. 16.19d).................. **Opisodasys**
15 Male St8 narrow and often reduced, triangular at base and apical part with or without a membranous process. Spermatheca with bulga elongate, pyriform or cylindrical and much longer than hilla (Fig. 16.20b) ....
.... **Ceratophyllus (Monopsyllus)**
— Male St8 vestigial (Fig. 16.18d). Spermatheca with bulga oval and not or hardly at all longer than hilla (Fig. 16.20g).. **Malaraeus**
16 Genal ctenidium formed of two spines 17
— Genal ctenidium formed of five to nine spines (Fig. 16.14d). [Pronotal ctenidium formed of numerous spines] .... **Hystricopsylla**
17 Genal ctenidium with non-overlapped spines lying parallel to each other and both fully visible (Fig. 16.14b). Frontal profile angulate and frons with two or three spiniform setae near point of angulation (Fig. 16.14b)..................................
.... **Peromyscopsylla**
— Genal ctenidium with one spine overlapping other spine and largely concealing it from view (Fig. 16.13b). Frontal profile smoothly rounded and frons without such spiniform setae (Fig. 16.13b)... **Meringis**

Key E: Identification key to medically important genera in Central and South America

1 Pronotal ctenidium absent 2
— Pronotal ctenidium present... 7
2 Hind coxa without small spiniform setae on inner side (Fig. 16.11a). Sensilium with eight sensory pits each side (Fig. 16.11b) 3
— Hind coxa and sensilium without this combination of characters 4
3 Hind coxa with anteroventral corner projecting downwards as a triangular tooth (Fig. 16.11a, arrowed). [Jiggers or sand-fleas].. **Tunga**
— Hind coxa without such apical tooth .... **Hectopsylla**
4 Antennal clavus with asymmetrical segments (Fig. 16.8i). Metasternum without anteroventral projection or almost so... 5
— Antennal clavus with symmetrical segments (Fig. 16.8h). Metasternum with well-developed anteroventral projection (Fig. 16.8d, arrowed). [Labial palp with five segments] .... **Delostichus**
5 Prosternosome not projected strongly downwards between fore coxae (Fig. 16.8g)... 6
— Prosternosome projected strongly downwards between fore coxae (Fig. 16.8f).....
.... **Rhopalopsyllus**
6 Labial palp not reaching apex of fore coxa .... **Polygenis**
— Labial palp reaching beyond apex of fore coxa .... **Tiamastus**
7 Genal ctenidium absent.. 8
— Genal ctenidium present 9
8 Terga 2–7 with one row of setae .... **Hoplopsyllus**
— Terga 2–7 with additional setae anterior to main row .... **Plusaetis**
9 Terga 2–7 with additional setae anterior to main row 10
— Terga 2–7 with one row of setae .... **Cediopsylla**
10 Head with a long vertically oriented anterior ctenidium in addition to genal ctenidium (Fig. 16.14f) 11

552 *Fleas (Siphonaptera)*

— Head with only a genal ctenidium (Figs 16.14e, 16.15c) 13
11 Male basimere without fringe of setae. Spermatheca with hilla not projecting into bulga 12
— Male basimere with apical fringe of long setae (Fig. 16.18c). Spermatheca with subcylindrical bulga and hilla usually projecting well into bulga (Fig. 16.19b) . ... **Plocopsylla**
12 Male telomere with blunt apex (Fig. 16.17b). Spermatheca with elongate bulga provided with a tuberculoid incrassation (Fig. 16.20h) .... **Craneopsylla**
— Male telomere with tapering apex and (usually) semilunar shape (Fig. 16.17f). Spermatheca with globular bulga lacking any trace of tuberculoid incrassation (Fig. 16.19e) .... **Sphinctopsylla**
13 Genal ctenidium formed of four fully visible and uniformly spaced spines (Fig. 16.15c)... **Adoratopsylla**
— Genal ctenidium formed of four spines of which lowermost is almost entirely covered and concealed by neighbouring spine (Fig. 16.14e)................................ .... **Neotyphloceras**

Key F: Identification key to medically important Australasian genera

1 Pronotal ctenidium absent. [Frons with smoothly rounded profile (Fig. 16.2a,b)]. .... **Xenopsylla**
— Pronotal ctenidium present. [Eye situated above (not in front of) base of fore coxa (Fig. 16.12e). Rod-like structure present between metepimeron of thorax and basal sternum of abdomen (Fig. 16.8e)]...................................... **Stivalius s.l.**

Geographical distribution

While the order Siphonaptera is worldwide, some of the families are quite restricted in their occurrence and only a few could be said to be cosmopolitan. The 15 families are listed below in alphabetical order, with information on geographical distribution and host preferences. The geographical distribution and phylogeny of fleas support the concepts of austral faunal relationships, transatlantic connections and other aspects of the theory of continental drift. The siphonapteran relationships among the austral continents are at the level of subfamily or family, not the genus. A masterly review of the geography of fleas and their use in deducing mammal phylogeny and geographic radiation is given by Traub (1980b).

Ancistropsyllidae: The three species are restricted to the western and northwestern Oriental region (Indian and Indo-Chinese subregions). Two are known only from the type specimens, one from a civet cat from Cambodia, the other from Thailand from a muntjak (deer). The third, from Nepal and India, parasitizes various species of artiodactylans (Cervidae and Bovidae). (See Hopkins and Rothschild, 1971.)

Ceratophyllidae: This large family includes 22.5% of all flea species in 43 currently recognized genera. It is typically Holarctic and relatively few genera and species occur in the southern hemisphere. The only Antarctic flea belongs to this family. Seventy-seven of the 515 species and subspecies are parasites of birds, the remainder being associated with small rodents. Two or three species are essentially cosmopolitan due to human transport. (See Traub et al., 1983; Lewis, 1990.)

Chimaeropsyllidae: The eight genera contain 28 taxa distributed in Africa south of the Sahara. The species are associated with small rodents, insectivores and macroscelid. (See Hopkins and Rothschild, 1956.)

Coptopsyllidae: The single genus *(Coptopsylla)* contains 25 named taxa distributed through the more arid areas of the Palaearctic region from northeast Africa to Mongolia. Its members are all parasites of gerbilline rodents. (See Hopkins and Rothschild, 1956.)

Ctenophthalmidae: This large family currently contains 38 genera and 751 species and subspecies. Although primarily Holarctic, representatives of the family occur on all continental landmasses except Antarctica; only a few occur in Australia and South America. Rodents and insectivores are the preferred hosts. (See Hopkins and Rothschild, 1962, 1966.)

Hystrichopsyllidae: There are eight genera collectively containing 54 named species and subspecies. Four of these are uniquely Australasian, one Neotropical, another western Palaearctic and the remaining two Holarctic. The Australian forms parasitize marsupials and rodents; those on other continents are found on rodents and insectivores. (See Hopkins and Rothschild, 1962, 1966.)

Ischnopsyllidae: The 20 genera in this family contain 129 species and subspecies. All are exclusively parasites of bats, and representatives occur on all of the continental landmasses except Antarctica. (See Hopkins and Rothschild, 1962.)

Leptopsyllidae: This diverse family includes 29 genera and 331 named taxa. Two of the six genera belonging to the Leptopsyllinae occur in the southeastern Oriental region (Indo-Malayan subregion), two in Madagascar and two are Holarctic, with elements extending into Africa; one species is cosmopolitan. The remaining species belong to the Amphipsyllinae and are essentially Holarctic. Most leptopsyllids are parasites of rodents, but a few parasitize birds. (See Hopkins and Rothschild, 1971.)

Malacopsyllidae: There are two genera, each with one species. They are restricted to the Neotropical, where they parasitize edentates (armadillos). (See Smit, 1987.)

Pulicidae: The 22 genera and 180 described taxa are parasites of a broad range of hosts including insectivores, carnivores, hyraxes and rodents, and sometimes occur as 'accidentals' on a multitude of mammals and birds. At least five species are cosmopolitan, having been transported worldwide by man. Pulicids occur on all of the continental landmasses except Antarctica. (See Hopkins and Rothschild, 1953.)

Pygiopsyllidae: Except for one genus and four species that occur in the Neotropical region, the 37 genera and 183 species and subspecies are distributed from Africa eastwards through India and southeastern Asia to Australia and New Zealand. They are parasites of a broad range of hosts, including monotremes, marsupials and rodents. Three genera include parasites of birds, one of sea birds on sub-Antarctic islands. (See Mardon, 1981.)

Rhopalopsyllidae: Primarily restricted to the Neotropical and parts of the southern Nearctic regions, the 139 taxa are grouped into ten genera. Most are parasites of marsupials and rodents. The species of one genus are parasites of sea birds nesting on the coasts of southern South America, southern Africa, Australia, New Zealand and the circum-Antarctic islands. (See Smit, 1987.)

Stephanocircidae: The nine genera and 51 species are known as 'helmeted' fleas and show a distinct Gondwanian distribution. The Stephanocircinae contains two genera with seven species that parasitize marsupials in Australasia. The other seven genera are Neotropical and include parasites of marsupials and rodents. (See Hopkins and Rothschild, 1956.)

Tungidae: Sometimes treated as a subfamily of the Pulicidae, three of the four genera are nearly confined to the New World tropics but have representatives in China and Japan. The remaining genus contains two species that parasitize warthogs and pangolins in the Afrotropical region. The New World species parasitize a broad range of mammalian and avian hosts and one species of *Tunga* is now well established in tropical Africa, having presumably been transported there on livestock. (See Hopkins and Rothschild, 1953.)

Vermipsyllidae: This is a Holarctic family containing three genera and 38 species and subspecies. Two of the genera are restricted to Central Asia and western China and contain species parasitic on ungulates. The third genus is Holarctic and all but one of its 25 taxa parasitize carnivores. (See Hopkins and Rothschild, 1956.)

Xiphiopsyllidae: The eight species of *Xiphiopsylla* (the only genus) are confined to higher elevations of East Africa, where they are parasites of rodents. (See Hopkins and Rothschild, 1956.)

Faunal and taxonomic literature

The flea fauna of the world has been treated by Hopkins and Rothschild (1953, 1956, 1962, 1966, 1971), Mardon (1981), Smit (1987), Lewis (1972, 1973b, 1974a–d, 1990) and Lewis and Lewis (1985). However, there are a number of monographic treatments of the fauna of specific countries or zoogeographical areas that contain information beyond that in the references cited above. Following is a list of some of these by region and, where practical, by subregion. Some areas of the world are poorly known, or at least poorly monographed, and the publications cited here are not equal in their coverage, but all contain keys, drawings and other useful information about the fauna of the area. Additional information can be found in Smit (1978).

Palaearctic: There are a few general papers on 'subregions': western Europe (Beaucournu and Launay, 1990); European USSR (Skalon, 1988); Central Asia and Kazakhstan (Ioff et al., 1965); Asia Minor (Peus, 1977); far eastern USSR (Ioff and Skalon, 1954). Other specialized literature is as follows: Afghanistan (Lewis, 1973a); British Isles (Smit, 1957); Caucasus (Tiflov et al., 1977); China (Liu, 1986); Czechoslovakia (Rosický, 1957); Egypt (Lewis, 1967); France (Beaucournu and Launay, 1990); Germany (Jancke, 1938; Peus, 1972); Iraq (Hubbard, 1960); Israel (Theodor and Costa, 1967); Italy (Berlinguer, 1964); Japan (Sakaguti, 1962; Sakaguti and Jameson, 1962); Korea (Tipton et al., 1972); Middle East (Lewis and Lewis, 1990); Mongolia (Goncharov et al., 1989); Morocco (Hastriter and Tipton, 1975); Poland (Skuratowicz, 1967); Saudi Arabia (Lewis, 1982); Sweden (Brinck-Lindroth and Smit, 1971); Ukraine (Yurkina, 1961).

Afrotropical: General Africa review (Haeselbarth, 1966); southern Africa (de Meillon et al., 1961; Marcus, 1961); Malagasy subregion (Lumaret, 1962); Angola (Ribeiro, 1974).

Oriental: Indian subcontinent (Iyengar, 1973); South East Asia (Traub, 1972).

Australasian: Australia (Traub and Dunnet, 1973; Dunnet and Mardon, 1974); New Guinea (Holland, 1969); New Zealand (Pilgrim, 1980; Smit, 1979).

Nearctic: Alaska, Canada, Greenland (Holland, 1949, 1985); Mexico (Morales, 1983; Traub, 1950); USA (eastern USA, Fox, 1968; western USA, Hubbard, 1947; Pacific northwest, Lewis et al., 1988); North America (Johnson, 1961).

Neotropical: Central America (Traub, 1950); South America (Johnson, 1957); Colombia (Méndez, 1977); Mexico (Morales, 1983); Panama (Tipton and Méndez, 1966); Venezuela (Tipton and Machado-Allison, 1972).

BIOLOGY

The developmental cycle (Fig. 16.21) from egg through larval and pupal stages to adult is not known for most fleas and the following account is a generalization based on those species for which this information is available. The life cycle in the temperate regions is probably an annual one, but even this is not known for certain, although species that have been colonized, such as *Xenopsylla cheopis*, breed constantly under optimum laboratory conditions (Smit, 1977). The life cycle of most fleas takes place in the dwelling place of the host. The reproduction of the rabbit flea *Spilopsylla cuniculi* is closely tuned to the reproduction of the host using the host's hormones as cues (Rothschild and Ford, 1964). A female can produce several hundred eggs or more; a cat flea, for instance, can under optimal conditions lay about 25 eggs a day for at least three or four weeks or even longer, totalling some 700–900 eggs during its lifetime. Most eggs are laid during the female's sojourn on the host or while in the latter's nest or lair. The majority will therefore land in the right place for the development of the larvae. The eggs usually hatch after about five days and the larvae feed on organic debris with their chewing mouthparts. Blood is an additional requirement for the larvae of some species. This is supplied by adult fleas which, during feeding, eject faeces consisting of the remnants of digested blood of a previous meal, followed by droplets of virtually undigested blood. Some larvae even prod adult fleas to produce faecal blood, which they then suck up. Moreover, larvae can be predators and scavengers and they can attack, kill and devour other small, weak arthropods present in nest material or even feed on dead adult fleas. Larvae are sometimes found on host animals, e.g. in the fur of dirty dogs and cats, on human beings of unclean habits and on nestling birds. In southeast Australia and Tasmania the larvae of *Uropsylla tasmanica* actually live in burrows which they excavate in the skin of their marsupial hosts. After two or three weeks, during which they moult twice, larvae are fully grown and spin cocoons of silk produced by the salivary glands. Dust and other fine particles adhere to the freshly spun silk of the cocoon. Two or three days later the cocooned larva, or prepupa, sheds its cuticle and transforms into a pupa.

The duration of the pupal stage depends on the ambient temperature but is usually one or two weeks. The adult flea, after emerging from the pupal cuticle,

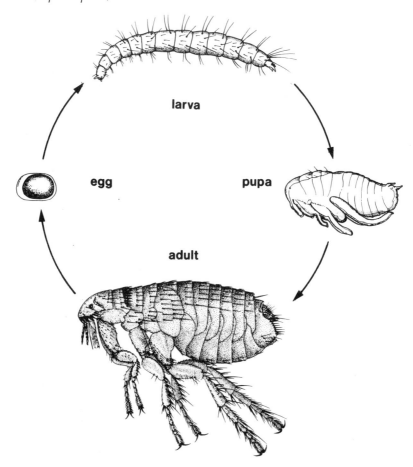

Figure 16.21 Life cycle of a typical flea.

requires a stimulus (usually vibration) to induce it to leave the cocoon. In the absence of such a stimulus it can remain alive, but inactive, within the cocoon for long periods. This peculiarity explains why the first person or animal to enter a dwelling or nest that had been uninhabited for a long time can suddenly be attacked by innumerable fleas. Fleas can fast for long periods and this enables them to spend a considerable time searching for a host after their original host has died or has vacated its nests. The adults of a number of species can copulate immediately after emergence from the cocoon and egg production begins after one or more days. The females require nourishment before the first batch of eggs is laid. In some species there is often no apparent linkage between the breeding season of the host and that of the flea but in others, e.g. rabbit flea *Spilopsylla cuniculi*, the pattern of behaviour and development is completely interwoven with that of the host species.

Most fleas are not strictly host specific but are nest specific, since they are to a large extent more dependent on factors determined by the environment governing their development than on the type of animal from which the adults obtain their food. Many fleas feed, or at least try to do so, on any available animal, though this is usually the one that built and occupies the nest. While feeding on a typical host is useful for the individual, keeping it alive and thereby prolonging its chances of coming into contact with a primary host, this might not be advantageous for the population as a whole since fertility can be seriously impaired by feeding to repletion on 'foreign' blood.

Whether a certain mammal or bird is a suitable host for fleas depends on the composition (food and shelter for larvae) and microclimate (temperature and humidity) of its nest or lair. It follows that those mammals and birds which do not build or occupy nests have living-places unsuitable for the development of fleas. Those that do not return regularly to some sort of home cannot have fleas of their own unless these fleas are strongly modified and adapted for a close association with the body of such hosts.

MEDICAL IMPORTANCE

Synopsis of medically important genera

The following list includes all those genera that have at least some medical importance and have been included in the keys. For each genus there is a statement of the family to which it belongs, the principal reference, the number of included taxa, the geographical distribution and the vector status. Quite a number of flea species are treated as two or more subspecies (frequently unjustifiably); the current number of subspecies (abbreviated ssp.) is shown in addition to the number of species.

There is some disagreement over family limits in the mammalian orders, but the comprehensive review of mammalian classification in Honacki et al. (1982) is the system mainly followed here.

Adoratopsylla (Ctenophthalmidae) [Hopkins and Rothschild, 1966]: This is a South American genus with two subgenera, *Adoratopsylla* s.str. with three species (five ssp.) and *Tritopsylla* with two species (three ssp.). The hosts are marsupials of the family Didelphidae (American opossums). *Adoratopsylla intermedia copha* is a vector of plague in Ecuador.

Amalaraeus (Ceratophyllidae) [Traub et al., 1983]: This is a temperate and cold-zone genus of the Palaearctic and Nearctic regions. There are seven species (14 ssp.) and the hosts are arvicoline murid rodents. *Amalaraeus penicilliger* ssp. is a suspected vector of haemorrhagic nephroso-nephritis and lymphocytic choriomeningitis in the Asiatic USSR.

Amphalius (Ceratophyllidae) [Traub et al., 1983]: The four species (ten ssp.) of this genus are distributed through the eastern Palaearctic and northwestern Nearctic regions, where they feed on pikas, lagomorphs of the genus *Ochotona* (Ochotonidae). *Amphalius runatus* is a vector of erysipeloid in Asiatic USSR.

Amphipsylla (Leptopsyllidae) [Hopkins and Rothschild, 1971]: Though found mainly in the colder parts of the Palaearctic region, this genus (total 37 species, 25 ssp.) includes three taxa that are represented in the Nearctic region. The hosts are arvicoline (= microtine) murid rodents (voles and lemmings). *Amphipsylla rossica* is a plague vector in European USSR, *A. primaris mitis* a plague vector in Asiatic USSR and Mongolia (where it also transmits erysipeloid).

Anomiopsyllus (Ctenophthalmidae) [Hopkins and Rothschild, 1962]: Members of this Nearctic genus (12 species, five ssp.) are parasites of woodrats (*Neotoma*) (Muridae:

Hesperomyinae). *Anomiopsyllus nudatus hiemalis* is a vector of plague in Texas, as is *A. nudatus* in the western USA.

Catallagia (Ctenophthalmidae) [Hopkins and Rothschild, 1962]: This genus (25 species, three ssp.) is mainly Nearctic but a few species occur in the northern and eastern Palaearctic region. The hosts are cricetine and arvicoline murid rodents. *Catallagia decipiens* is a vector of plague in the western USA.

Cediopsylla (Pulicidae) [Hopkins and Rothschild, 1953]: The four species (two ssp.) of this genus are Nearctic and Neotropical. The hosts are Leporidae, cottontail rabbits of the genus *Sylvilagus*. *Cediopsylla spillmani* is a plague vector in Peru.

Ceratophyllus (Ceratophyllidae) [Traub et al., 1983]: This mainly Holarctic genus has six subgenera and 62 species (25 ssp.). The hosts are mainly squirrels and other rodents. Only fleas of the subgenus *Monopsyllus* appear to be associated with disease transmission, although *Ceratophyllus (C.) gallinae* and *C. (Emmareus) garei* have been reported as carriers of Omsk haemorrhagic fever virus. *Ceratophyllus (Monopsyllus) anisus* is thought to be a vector of pseudotuberculosis, erysipeloid and listeriosis, and *C. (M.) indages* a vector of tick-borne encephalitis in Asiatic USSR.

Chiastopsylla (Chimaeropsyllidae) [Hopkins and Rothschild, 1956]: This is a South African genus with 14 species (three ssp.). The hosts are rodents, principally African swamp rats (Otomyinae). The genus includes the plague vector *Chiastopsylla rossi*.

Citellophilus (Ceratophyllidae) [Traub et al., 1983]: The members of this Palaearctic genus (13 species, 12 ssp.) feed mainly on ground-dwelling sciurid rodents (*Spermophilus*) but sometimes on marmots (*Marmota*). In Asiatic USSR, *Citellophilus lebedewi* is a vector of plague, and *C. tesquorum* a vector of plague and erysipeloid.

Coptopsylla (Coptopsyllidae) [Hopkins and Rothschild, 1956]: This genus of 19 species (eight ssp.) is distributed from the Mediterranean area to Central Asia. The hosts are gerbilline rodents. *Coptopsylla bairamaliensis* and *C. lamellifera* are plague vectors in Central Asia.

Craneopsylla (Stephanocircidae) [Hopkins and Rothschild, 1956]: This genus contains only *Craneopsylla minerva*. The hosts are rodents and marsupials, and the species is a plague vector in Argentina.

Ctenocephalides (Pulicidae) [Hopkins and Rothschild, 1953]: The 11 species (three ssp.) of this genus are found mainly in Africa and Eurasia, but two are cosmopolitan. The hosts are principally carnivores, though some African species occur on hares, hyraxes and ground-squirrels. The fleas also occur on goats in the Mediterranean area and the Oriental region. The cosmopolitan *Ctenocephalides felis felis* and *C. canis* are plague vectors, as are *C. felis strongylus* in East Africa and *C. orientis* in the Oriental region. *Ctenocephalides canis* and *C. felis felis* in Europe and eastern USA, and *C. felis strongylus* in East Africa, are intermediate hosts for the cestode tapeworm *Dipylidium caninum*.

Ctenophthalmus (Ctenophthalmidae) [Hopkins and Rothschild, 1966]: This genus (155 species, 157 ssp.) is concentrated in the Palaearctic and Afrotropical regions but has a few representatives in North and Central America and in the Oriental region. The hosts are mainly murine rodents (rats) but sometimes insectivores or other small mammals. The following are vectors of plague in the areas shown: *Ctenophthalmus breviatus* (European USSR), *C. calceatus cabirus* (East Africa), *C.*

dolichus (Central Asia), *C. phyris* (Central Africa), *C. pollex* and *C. secundus* (European USSR). Pathogenic agents isolated from these fleas are those of tularaemia and erysipeloid from *C. teres* and *C. wladimiri* in the European USSR, pseudotuberculosis and tick-borne encephalitis from *C. congenerioides congenerioides* in far eastern USSR, listeriosis from *C. orientalis* in European USSR, haemorrhagic nephroso-nephritis from *C. assimilis assimilis* in Asiatic USSR and *C. agyrtes* ssp., *C. orientalis* and *C. solutus solutus* in the European USSR, and lymphocytic choriomeningitis from *C. assimilis assimilis* in Asiatic USSR.

Delostichus (Rhopalopsyllidae) [Smit, 1987]: The seven species in this genus are Neotropical. The hosts are rodents. *Delostichus talis* is a vector of plague in Argentina.

Dinopsyllus (Ctenophthalmidae) [Hopkins and Rothschild, 1966]: This genus has two subgenera, both Afrotropical, *Dinopsyllus* s.str. with 26 species and *Cryptoctenopsyllus* with one species. The hosts are rodents. *Dinopsyllus ellobius* is a plague vector in South Africa and *D. lypusus* a plague vector in Central and East Africa.

Echidnophaga (Pulicidae) [Hopkins and Rothschild, 1953]: The 20 species (two ssp.) of this genus are distributed through Eurasia (mainly warmer areas), Africa and Australia. They include one cosmopolitan species. The hosts are usually rodents, marsupials, carnivores and warthogs, but one cosmopolitan species is found on birds (especially chickens) as well as rats, carnivores and larger insectivores. Plague vectors include *E. gallinacea* (USA) and *E. oschanini* (Asiatic USSR and Mongolia); the former also transmits murine typhus in the southern USA. *Echidnophaga larina* is an intermediate host for *Dipylidium caninum* in East Africa.

Euhoplopsyllus (Pulicidae) [Hopkins and Rothschild, 1953]: This genus (three species, six ssp.) is distributed from Afghanistan through eastern Asia to Greenland. The hosts are Leporidae (hares and rabbits). *Euhoplopsyllus glacialis affinis* is a plague vector in southwestern USA, *E. andensis* and *E. manconis* are plague vectors in Peru.

Foxella (Ceratophyllidae) [Traub et al., 1983]: This is a Nearctic genus with two subgenera, *Foxella* s.str. with one species (11 ssp.) and *Afoxella* with three species. The hosts are rodents of the Geomyidae (pocket gophers). *Foxella ignota* ssp. is a plague vector in the southwestern USA.

Frontopsylla (Leptopsyllidae) [Hopkins and Rothschild, 1971]: This is a Eurasian genus with four subgenera of which only *Frontopsylla* s.str. (21 species, 27 ssp.) has any medical importance. *Frontopsylla semura* is a vector of plague in the European USSR, and *F. wagneri* and *F. luculenta* are suspected vectors of erysipeloid in the Asiatic USSR; the latter is also suspected of the transmission of listeriosis and salmonellosis in the Asiatic USSR.

Hectopsylla (Pulicidae) [Hopkins and Rothschild, 1953]: The 11 species of this genus are all Neotropical. Except for two species on birds, they occur on rodents. *Hectopsylla suarezi* is a plague vector in Ecuador and an undetermined species in Peru.

Hoplopsyllus (Pulicidae) [Hopkins and Rothschild, 1953]: There are two species in this Nearctic genus, of which *H. anomalus* is a plague vector in the western USA. The hosts are Leporidae and ground-squirrels (*Spermophilus*).

Hystricopsylla (Hystricopsyllidae) [Hopkins and Rothschild, 1962]: This is a Palaearctic and Nearctic genus with 18 species (18 ssp.). The hosts are mainly rodents. *Hystricopsylla occidentalis linsdalei* and *H. dippiei* ssp. transmit plague in the western USA.

Leptopsylla (Leptopsyllidae) [Hopkins and Rothschild, 1971]: The two subgenera in this genus, *Leptopsylla* s.str. (nine species, 16 ssp.) and *Pectinoctenus* (six species, two ssp.), are Palaearctic, Nearctic and Afrotropical. The hosts are murine and arvicoline rodents. *Leptopsylla aethiopica* transmits plague in Africa; *L. segnis* is believed to transmit murine typhus, tickborne encephalitis and erysipeloid in Asiatic USSR, southeastern USA and China, and Poland, respectively, and *L. pavlovskii* to transmit salmonellosis in Asiatic USSR.

Listropsylla (Ctenophthalmidae) [Hopkins and Rothschild, 1966]: One of the nine species (four ssp.) of this African genus, *Listropsylla dorippae,* transmits plague in southern Africa. The hosts are rodents.

Malaraeus (Ceratophyllidae) [Traub et al., 1983]: This is a Nearctic genus of three species. They feed on murine rodents (rats). *Malaraeus sinomus* and *M. telchinus* are plague vectors in the western USA.

Megabothris (Ceratophyllidae) [Traub et al., 1983]: This genus, distributed in the Palaearctic and Nearctic regions, has four subgenera, *Megabothris* s.str. (eight species, three ssp.), *Amegabothris* (five species), *Gebiella* (three species, two ssp.) and *Kueichenlipsylla* (two species). The hosts are arvicoline, cricetine and murine rodents. *Megabothris (A.) abantis* and *M. (A.) clantoni* transmit plague in the western USA. The causative agent of tularaemia has been isolated from *M. (G.) rectangulatus* in northern Europe, the pseudotuberculosis bacillus from *M. (M.) calcarifer* in the Russian Far East, lymphocytic choriomenigitis from *M. (G.) rectangulatus* in Asiatic USSR, and haemorrhagic nephroso-nephritis from *M. (G.) turbidus* in the European USSR.

Meringis (Ctenophthalmidae) [Hopkins and Rothschild, 1962]: This is a Nearctic genus of 17 species. The hosts are heteromyid rodents (pocket mice and kangaroo-rats). *Meringis shannoni* is a vector of plague in the western USA.

Mesopsylla (Leptopsyllidae) [Hopkins and Rothschild, 1971]: The seven species (13 ssp.) of this genus are distributed through the Mediterranean area and Central Asia. The hosts are rodents of the family Dipodidae (jerboas). *Mesopsylla apscheronica* and *M. tuschkan* transmit plague in Central Asia.

Neopsylla (Ctenophthalmidae) [Hopkins and Rothschild, 1962]: This is a mainly Palaearctic and northern Nearctic genus of 47 species (28 ssp.). The hosts are rodents, especially murids and sciurids. *Neopsylla bidentatiformis* is a vector of plague is Asiatic USSR and Mongolia, *N. inopina* in the western USA, *N. mana* in Mongolia and *N. setosa* in European USSR. Isolations of pathogens from fleas of this genus include pseudotuberculosis bacilli from *N. bidentatiformis* in Asiatic USSR, erysipeloid from *N. bidentatiformis* and *N. pleskei* in Asiatic USSR, *Salmonella* from *N. bidentatiformis* and *N. pleskei orientalis* in Asiatic USSR, *Staphylococcus aureus* from *N. pleskei orientalis* in Asiatic USSR, and Q fever rickettsiae from *N. pleskei* in Asiatic USSR.

Neotyphloceras (Ctenophthalmidae) [Hopkins and Rothschild, 1966]: This is a Neotropical genus containing two species (three ssp.). These occur on rodents and marsupials. *Neotyphloceras rosenbergi* is a vector of plague in Ecuador.

Nosopsyllus (Ceratophyllidae) [Traub et al., 1983]: This is a mainly Palaearctic genus but has a few species in Africa, one that is cosmopolitan and one that is almost cosmopolitan. There are four subgenera, *Nosopsyllus* s.str. (30 species, 16 ssp.), *Gerbillophilus* (20 species, 17 ssp.), *Nosinius* (one species) and *Penicus* (one species). The hosts are rodents, including squirrels. Several species are vectors of plague, namely *N. aralis* ssp. (Asiatic USSR), *N. consimilis* (European USSR), *N. fasciatus* (USA), *N. laeviceps* (European USSR), *N. mokrzeckyi* (European USSR), *N. nilgiriensis* (India), *N. tersus* (Asiatic USSR) and *N. turkmenicus* (Asiatic USSR). Pseudotuberculosis bacilli have been isolated from *N. consimilis* (European USSR) and erysipeloid from *N. fasciatus* (USSR). *Nosopsyllus fasciatus* is known to be an intermediate host of *Hymenolepis diminuta* in England, Argentina and Australia.

Ophthalmopsylla (Leptopsyllidae) [Hopkins and Rothschild, 1971]: This Palaearctic genus has three subgenera, of which only *Ophthalmopsylla* s.str. (five species, 16 ssp.) is of medical importance. The hosts are rodents, especially Dipodidae (jerboas). *Ophthalmopsylla volgensis* is a plague vector in the European and Asiatic USSR. *Listeria* has been isolated from *O. volgensis* in the European USSR and *Salmonella* from *O. kukuschkini* in the Asiatic USSR.

Opisodasys (Ceratophyllidae) [Traub et al., 1983]: Only one species of this Nearctic genus (total eight species, two ssp.), *O. nesiotus*, is of medical importance as a vector of plague in the western USA. The hosts are mainly arboreal sciurids.

Orchopeas (Ceratophyllidae) [Traub et al., 1983]: The ten species (13 ssp.) of this genus occur on arboreal sciurids and cricetine rodents in the Nearctic. *Orchopeas leucopus* and *O. neotomae* are plague vectors in the southwestern USA and *O. sexdentatus* in western USA.

Oropsylla (Ceratophyllidae) [Traub et al., 1983]: This is a Holarctic genus with four subgenera of which only *Oropsylla* s.str. (eight species), *Oropsylla (Diamanus)* (one species) and *Oropsylla (Opisocrostis)* (four species, two ssp.) are of medical importance. The hosts are ground-dwelling sciurid rodents (*Spermophilus* and *Marmota*). The following are vectors of plague: *O. (O.) idahoensis*, *O. (O.) rupestris* and *O. (D.) montana* (western USA), *O. (O.) ilovaiskii* (European USSR), *O. (O.) silantiewi* (Asiatic USSR and Mongolia) and *O. (Op.) brunneri*, *O. (Op.) hirsuta*, *O. (Op.) labis*, *O. (Op.) tuberculata* (western USA). Erysipeloid bacilli have been isolated from *O. silantiewi* in Asiatic USSR and Mongolia.

Paradoxopsyllus (Leptopsyllidae) [Hopkins and Rothschild, 1971]: The 40 species (two ssp.) of this genus occur on rodents in Asia. Plague vectors are *Paradoxopsyllus curvispinus* in Japan and Asiatic USSR, *P. dashidorzhii* in Mongolia and *P. teretifrons* in Asiatic USSR.

Peromyscopsylla (Leptopsyllidae) [Hopkins and Rothschild, 1971]: This is a Holarctic genus of 19 species (18 ssp.) The fleas occur on murine and cricetine rodents. *Peromyscopsylla hesperomys adelpha* is a vector of plague in the southwestern USA.

Pleochaetis (Ceratophyllidae) [Traub et al., 1983]: This genus contains three species from the western Nearctic region. The hosts are cricetine rodents. *Pleochaetis exilis* is a plague vector in the southern USA.

Plocopsylla (Stephanocircidae) [Schramm and Lewis, 1988]: This is a genus of 28 species distributed throughout the Andes of South America. The hosts are rodents and marsupials. *Plocopsylla hector* is a vector of plague in Ecuador.

Plusaetis (Ceratophyllidae) [Traub et al., 1983]: The 12 species (four ssp.) of this genus occur in the Nearctic and Neotropical regions. They feed on rodents. *Plusaetis dolens quitanus* and *P. equatoris* are vectors of plague in Peru.

Polygenis (Rhopalopsyllidae) [Smit, 1987]: The distribution of this genus (nine species, 17 ssp.) extends from the southern Nearctic region through the Neotropical region. It is classified into four subgenera. The hosts are mainly rodents, but the fleas occur also on marsupials. Some members of *Polygenis* s.str. are vectors of plague, namely *P. brachinus* (Peru), *P. gwyni* (southeastern USA), *P. litargus* (Ecuador and Peru) and *P. platensis cisandinus* (Argentina).

Pulex (Pulicidae) [Hopkins and Rothschild, 1953; Hopla, 1980a]: This genus, mainly distributed through the Nearctic and Neotropical regions, has two subgenera, *Pulex* s.str. (three species) and *Juxtapulex* (three species). *Pulex irritans*, a species of the nominate subgenus, is cosmopolitan (Fig. 16.22b). Its hosts are various large and coarse-coated mammals such as pigs, canids, mustelids, deer, tapirs and peccaries, but the species also occurs on humans. The species is a vector of plague in the USA, but possibly not as important as *P. simulans* in maintaining the disease in the sylvatic cycle. The same flea has been associated with erysipeloid in Asiatic USSR and Mongolia, and is an intermediate host of *Dipylidium caninum* in Italy and Switzerland.

Rhadinopsylla (Ctenophthalmidae) [Hopkins and Rothschild, 1962]: This is a Holarctic genus with six subgenera, *Rhadinopsylla* s.str. (nine species), *Ralipsylla* (two species, four ssp.), *Micropsylla* (two species, two ssp.), *Micropsylloides* (one species), *Sinorhadinopsylla* (one species) and *Actenophthalmus* (46 species, six ssp.). The hosts are mainly rodents. *Rhadinopsylla (Rh.) cedestis* and *R. (Rh.) ucrainica* are vectors of plague in the European USSR, and *R. (Ra.) li ventricosa* a plague vector in Asiatic USSR.

Rhopalopsyllus (Rhopalopsyllidae) [Smit, 1987]: The seven species (nine ssp.) of this genus are found in the Neotropical and southern Nearctic regions and feed on rodents. An as yet unidentified species is a vector of plague in Argentina.

Sphinctopsylla (Stephanocircidae) [Hopkins and Rothschild, 1956]: This Neotropical genus includes six species. They feed on rodents. *Sphinctopsylla mars* is a vector of plague in Peru.

Stenistomera (Ctenophthalmidae) [Hopkins and Rothschild, 1962]: This genus contains three species, of which *Stenistomera alpina* and *S. macrodactyla* are vectors of plague in the western USA.

Stenoponia (Ctenophthalmidae) [Hopkins and Rothschild, 1962]: This is a Holarctic genus of 16 species (14 ssp.). The hosts are rodents. Three species are vectors of plague, *Stenoponia conspecta* and *S. vlasovi* in Asiatic USSR, and *S. tripectinata tripectinata* in Asia Minor and European USSR.

Stivalius (Pygiopsyllidae) [Mardon, 1981]: The six species of *Stivalius* are distributed through the Oriental region. They occur on rodents. *Stivalius ahalae* is a plague vector in South East Asia and *S. cognatus* in Java.

Synopsyllus (Pulicidae) [Hopkins and Rothschild, 1953]: The five species of this genus are confined to Madagascar. They occur on rodents and insectivores. *Synopsyllus fonquernii* is a vector of plague.

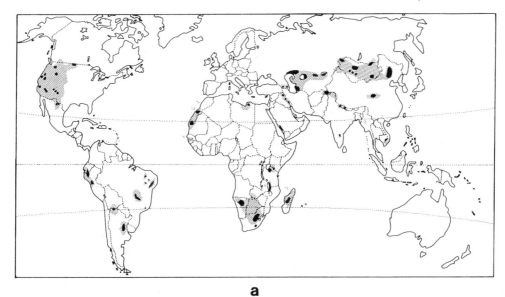
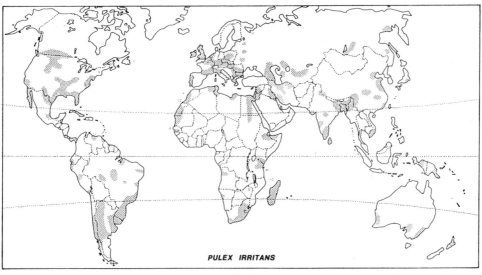

Figure 16.22 Geographical distribution of: (*a*) plague; (*b*) the human flea *Pulex irritans*. In the upper figure solid black represents areas of frequent transmission and stipple represents areas of infrequent or suspected transmission. (Maps reproduced from World Health Organization, 1989.)

Synosternus (Pulicidae) [Hopkins and Rothschild, 1953]: This genus contains seven species (two ssp.) distributed throughout Africa and southwestern Asia. They are found on rodents and insectivores (Erinaceidae). *Synosternus pallidus* is a plague vector in West Africa.

Thrassis (Ceratophyllidae) [Traub et al., 1983]: This western Nearctic genus of 11 species (24 ssp.) is associated mainly with ground-dwelling sciurid rodents of the genera *Spermophilus* (ground-squirrels) and *Marmota* (marmots). *Thrassis acamantis*,

T. arizonensis, T. bacchi, T. fotus, T. francisis, T. pandorae, T. petiolatus and *T. stanfordi* are vectors of plague in the western USA (Stark, 1970).

Tiamastus (Rhopalopsyllidae) [Smit, 1987]: The seven species of this Neotropical genus are mostly found on rodents. *Tiamastus cavicola* transmits plague in Ecuador and Peru.

Tunga (Tungidae) [Hopkins and Rothschild, 1953]: The species of this genus are found in the Americas, tropical Africa and eastern Asia. They are placed in two subgenera, *Tunga* s.str. with seven species and *Brevidigita* with two species. The hosts are edentates, man, domestic livestock and rodents. *Tunga (T.) penetrans* causes tungiasis of humans in Central and South America and tropical Africa.

Xenopsylla (Pulicidae) [Hopkins and Rothschild, 1953]: This genus (77 species, 14 ssp.) is distributed throughout the warmer parts of the Old World; *X. cheopis* is cosmopolitan (Fig. 16.23a). The hosts are mainly murine rodents (rats). The following species transmit plague: *X. astia* (Oriental region, Fig. 16.23b), *X. brasiliensis* (Central Africa, Fig. 16.23b), *X. buxtoni* (Asia Minor), *X. cheopis* (cosmopolitan, between latitudes 45°N and 45°S, Fig. 16.23a), *X. conformis*, *X. gerbilli*, *X. hirtipes*, *X. nuttalli* and *X. skrjabini* (Asiatic USSR), *X. eridos*, *X. hirsuta*, *X. philoxera*, *X. phyllomae*, *X. pirlei* and *X. versuta* (South Africa), *X. nubica* (northern Africa) and *X. vexabilis* (Australasia, Hawaii). *Xenopsylla cheopis* is reportedly a vector of erysipeloid in the USSR and of murine typhus in the southern USA and European and Asiatic USSR. It is possibly an intermediate host of *Hymenolepis diminuta* in Australia, Argentina and Mexico.

Xiphiopsylla (Xiphiopsyllidae) [Hopkins and Rothschild, 1956]: This East African genus includes eight species (two ssp.) with murine rodent hosts. *Xiphiopsylla lippa* is a vector of plague in Central Africa.

Fleas and health

Fleas can affect human health in several ways : as biting pests, burrowing 'jiggers', parasite intermediate hosts and vectors of pathogens.

Flea bites

When a flea inserts its mouthparts into the skin of the host, saliva is injected and blood is pumped up by dilatation of part of the oesophagus. The saliva prevents coagulation of blood and is the cause of the skin reaction, manifested as erythema and oedema, to the act of bloodsucking. A flea puncture in human skin can cause intense itching for one or more days and is characterized by a tiny dark spot (purpura pulicosa), which can be visible for days, surrounded by a patch of swollen and reddish skin (roseola pulicosa). The first flea puncture causes no observable skin reactions in a host but induces hypersensitivity. When the host is repeatedly bitten by fleas over a long period the skin reactions are at first of a delayed type. Then for some time there will be an immediate reaction followed by delayed reaction. This disappears eventually, leaving an immediate reaction which itself will finally also no longer be apparent. A state of non-reactivity (immunity) has then been reached. The host's reaction to feeding involves an antigen–antibody complex and appears to influence the development and survival of microorganisms that can be injected into the skin by the flea.

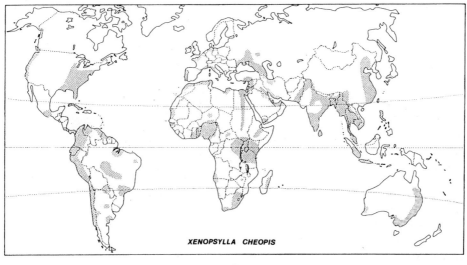
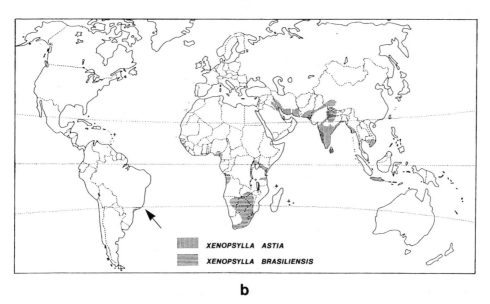

Figure 16.23 Geographical distribution of three important vectors of plague: (*a*) *Xenopsylla cheopis*; (*b*) *Xenopsylla astia* and *Xenopsylla brasiliensis*. Arrow indicates the restricted area of *X. brasiliensis* in South America. (Maps reproduced from World Health Organization, 1989.)

Jiggers
The so-called jiggers, chigoes or sand-fleas constitute excellent examples of the extreme evolutionary modification a flea can undergo, structurally and behaviourally. Nine species of such jiggers are known but only one of them, *Tunga penetrans* (Fig. 16.24*a*) occurring in Central and South America and in tropical Africa, has any medical importance. The larvae of these extremely small fleas develop in dry sandy soil at places frequented by the hosts. They pass through only two instars and development from egg to adult takes about three weeks under favourable conditions. The tiny, freshly emerged adults are very agile and actively search for a

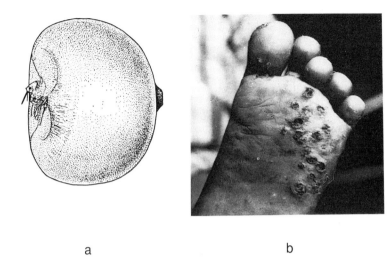

Figure 16.24 The jigger flea, *Tunga penetrans*: (*a*) gravid female removed from host skin; (*b*) a case of human tungiasis in the foot of a West African child.

host. The females usually attach themselves to the feet of fairly large mammals, man (Fig. 16.24*b*) and pigs being particularly suitable hosts. The soft skin between the toes or under the toe-nails is especially favoured. Heavy infestations can occur on the hands and arms, particularly around the elbow, and in the genital region. The female buries herself in the skin of the host with the aid of the strong and serrated stylets of the mouthparts. The tip of her abdomen remains exposed and these abdominal segments contain the large spiracles through which the flea breathes, as well as the anus and the genital opening. The abdomen gradually distends during feeding and ovulation, ultimately attaining the size and shape of a small pea, 1000 times her entire original body size. The process of expansion, which takes eight to ten days, is not primarily a result of the maturation of eggs, since this does not begin until the final stage of expansion. A total of several thousand eggs is produced, and each of them is ejected so that it has a good chance of falling on the ground away from the host. When the female dies she remains embedded in the skin, where her presence often causes an inflammation which eventually results in the dead flea being expelled by ulceration. Neglected or improperly treated lesions often become secondarily infected with other organisms, sometimes causing loss of digits, septicaemia or tetanus. Infestation by this flea is commonly referred to as tungiasis.

Fleas as intermediate hosts
Several parasitic worms are associated with fleas. One of the commonest tapeworms of dogs and cats is the double-pored tapeworm, *Dipylidium caninum*, which is also an occasional human parasite. Cat fleas, dog fleas and human fleas are important intermediate hosts. Worm eggs are discharged in the faeces of the cat or dog, and flea larvae, usually the third instar, swallow them along with the organic matter upon which larvae feed. The worm larvae hatch in the mid-gut of the flea larva and penetrate the gut wall into the haemocoel. During the pupal and adult stages of the flea, the parasites increase in size and become encapsulated infective larvae (cysticercoids). A mammal becomes infected by swallowing infected fleas when, for example, licking its fur. Human cases usually involve children or infants

that become infested by ingesting adult fleas that are infected. The rat tapeworm, *Hymenolepis diminuta*, with rat fleas as intermediate hosts, is also an occasional human parasite.

Fleas as vectors
Many species of flea are potential vectors of microorganisms present in the blood of their hosts, and several have actually been incriminated. However, the grounds on which fleas are incriminated as vectors are not always clear. If man enters the mammal/bird–pathogen–flea cycle (zoonosis), he can become infected and if susceptible show clinical symptoms. The role played by fleas as vectors of diseases transmitted from man-to-man (anthroponoses) is not clear. Inter-human transmission of bubonic plague during the Middle Ages, involving the human flea and possibly the dog flea and cat flea, might have been more commonplace than is usually acknowledged in the literature. Human-to-human transmission by *Pulex irritans* has been reported from Kurdistan.

Fleas transmit protozoans, bacteria, rickettsiae and viruses between mammals but the main diseases transmitted to man are bacterial or rickettsial. Various pathogens have been isolated from different species of fleas but their vectorial role is of little or no known significance.

Microorganisms can be transmitted from host to host by contaminated mouthparts, saliva, regurgitations from the alimentary canal or via infected faeces. The relative importance of mechanical transmission by contaminated mouthparts is unclear, but at least one virus is known to be transmitted in this way. Regurgitation by fleas whose proventriculus is blocked by plague bacilli is the common mode of plague transmission to humans. Once established in the human population, much of the human-to-human transmission probably involves respiratory aerosols from infected persons; hence the ensuing disease is known as pneumonic plague. The rickettsiae causing murine typhus are usually transmitted via infected faeces. Finally, ingestion of infected fleas during grooming or by accident can also result in infection.

In general fleas do not demonstrate the highly evolved relationships with pathogens and parasites so commonly seen with mosquitoes and other biting flies, or with ticks, and mites. No pathogens develop anywhere in the flea except in the digestive tract, and none except worms evidently invades the haemocoel, salivary glands or gonads.

The single known exception to this is the trans-ovarial and trans-stadial passage of rickettsiae causing murine typhus (Azad, 1990); with this exception, trans-ovarial and trans-stadial transmission of microorganisms is evidently not possible – although trans-stadial passage is known for some helminths. This has been interpreted by some as evidence of a fairly recent origin for the order Siphonaptera.

Fleas are involved in the transmission of the following diseases:

Plague (Yersinia pestis): This is primarily an infection of rodents (Pollitzer, 1954), and occurs in two recognizable forms. The urban form of the disease usually develops into an acute and rapidly fatal septicaemia. When a rodent dies of the disease, its infected fleas, usually species of *Xenopsylla* (distribution Fig. 16.23*a,b*), have to attack other available animals and man can then be incidentally infected. Normally, rodent fleas are reluctant to feed on man. The bubonic form of plague is transmitted either by regurgitation of bacilli contained in the digestive tract of fleas, the proventriculus of which usually becomes temporarily blocked by bacteria some time after infection, or by direct contamination from the mouthparts. The incidence of human plague cases has decreased markedly since the last great Indian epidemic of 1898–1918 when over 10 million people died. However, there is a vast reservoir of the plague bacillus (Fig. 16.22a) in the sylvatic form of the disease in many parts of the world where the bacillus is an integral part of the ecosystem (Bibikova and Zhovtyi, 1980, USSR; Cavanaugh and Williams, 1980) and plague outbreaks still occur now and again. The disease is almost always curable if treated in time, but outbreaks might prove less easily preventable in future as rats and their fleas become more resistant to pesticides (Pollitzer, 1966). Temporary personal protection can be conferred by vaccination.

Tularaemia (Francisella tularensis): The pathogen has been isolated from fleas in Scandinavia, the USSR and USA, although the usual vectors of this plague-like infection are ticks and some biting flies (Tabanidae). Fleas can acquire and retain the pathogen but they do not, experimentally at least, transmit it by bite to a clean host (Hopla, 1980a). This example indicates the extreme caution that must be applied to interpreting data and reports of isolations of pathogens from fleas and other insects. Uncritical compilations of these reports abound in the literature.

Pseudotuberculosis (Yersinia pseudotuberculosis): The pathogen of pseudotuberculosis has been isolated from rat fleas in the far eastern USSR.

Erysipeloid (Erisipelothrix rhusiopathiae): This infection is usually transmitted to man through skin abrasions but the causative agent has been isolated from 12 species of fleas in Asiatic USSR and four species in the European part of the USSR.

Fleas have also been suggested as vectors of widespread animal and human diseases such as listeriosis (*Listeria monocytogenes*), brucellosis (*Brucella melitensis*) and salmonellosis (*Salmonella enteritidis* and *S. typhimurium*) simply because bacteria have been isolated from them. Fleas are confirmed vectors of the rickettsial infection murine typhus (*Rickettsia typhi*), a worldwide infection of murine rodents which in man is clinically milder than the louse-borne epidemic typhus. Fleas are the principal vectors from rodent to rodent and rodent to man. A most critical and comprehensive review is given by Traub et al. (1978). Other, tickborne rickettsiae (*Rickettsia conori* causing boutonneuse fever and *Coxiella burneti* causing Q fever) have been isolated from fleas. The virus of myxomatosis (*Fibromavirus myxomatosis*), not a disease of humans, is transmitted mechanically via contaminated mouthparts by a number of bloodsucking insects, especially mosquitoes. In Britain the disease almost eliminated the wild rabbit (*Oryctolagus cuniculus*) with the rabbit flea, *Spilopsylla cuniculi,* as the principal vector. This flea was introduced into Australia to control the European rabbit as a vector of myxomatosis along with native mosquitoes. The experiment met with mixed results (Shepherd, 1980). The virus does not reproduce in the flea (Bibikova, 1977). Of the other viruses isolated from fleas, only in Omsk haemorrhagic fever virus transmission are fleas suspected of playing a significant role (Traub et al., 1983).

CONTROL

Methods of flea control are dependent upon the context of the problem (Gratz, 1980, review). The most commonly encountered problem involves the infestation of homes and other buildings by fleas brought in on pets or livestock. Treating infested pets with insecticides and common sanitation measures such as vacuuming and making sure that the pet bedding is clean are usually sufficient to prevent an indoor infestation from developing.

Over the past decade a number of insect growth regulators (IGRs) have been developed, such as methoprene (isopropyl (E, E)-11-methoxy-3, 7, 11-trimethyl-2, 4-dodecadienoate) which are effective in controlling household infestations of fleas. Advantages include a 75-day residual activity indoors, the colourless and odourless nature of the chemical, and its high LD_{50} (acute oral dose 34 600 mg/kg for rats); the last means that it is essentially nontoxic to warm-blooded vertebrates. When combined with a low dilution of an adulticide such preparations are extremely effective.

Flea control might also be required in foci of sylvatic plague. This disease is endemic in many parts of the world and becomes a public health problem where human populations impinge upon the reservoir rodent populations. Under these conditions two options are available: control of the fleas directly with insecticides, or control of the rodent hosts. In restricted locations, such as city parks, college campuses and military installations, rodent control is often the most effective and least expensive alternative. In larger and less circumscribed foci, treatment techniques vary with the habitat. In areas with little vegetation, where entrances to rodent burrows are visible, the burrows can be treated directly with an appropriate insecticide. Where vegetation and other ground cover conceals the burrow entrances, bait-boxes provide adequate control. These consist of devices containing grain or other baits that have been treated with an insecticide. In the process of feeding or garnering the bait, the rodent picks up the insecticide on its fur and this ultimately proves fatal to its fleas.

Since all species of fleas are not equally effective as vectors of disease, control programmes are usually preceded by surveys to determine the species involved and the need for treatment. While plague has been reported from a number of mammalian species, it is usually concentrated in the terrestrial sciurid population, specifically ground-squirrels, prairie dogs and marmots, and particular attention should be paid to these hosts during surveys.

COLLECTING, PRESERVING AND REARING MATERIAL

As with other groups of ectoparasitic arthropods, host animals are the primary source of specimens. Hosts can be captured alive and anesthetized briefly while they are brushed for their parasites. Such humane procedures are not always an option to the collector in the field, and more frequently hosts are shot or trapped, depending on their size. For small rodents and insectivores the standard break-back mousetrap called the 'Museum Special' is better than any other type of trap since 100 or more of them can be set by a single collector, they are compact and they kill instantly. Traps should be checked approximately two to three hours after sunset and again at 23.00 h or so.

When collecting hosts, care must be taken to place each host animal in a separate container to prevent parasite transfer from one host to another. Small plastic bags can be used, but cloth bags are preferable since they soak up body fluids and can be washed when necessary. Bags can be placed in a large can or other closed container and sprinkled with chloroform to immobilize the ectoparasites.

Each host is removed from its bag while the bag is examined for parasites that have left the body of the host. The host body is then brushed, combed or tapped against the surface until ectoparasites cease to appear. Ears, nostrils, the mouth and other body openings should be examined for fleas that may have entered in an attempt to escape the fumigant. Exposed parts of the host such as the ears, nose, feet and tail should be examined for fleas that may be attached by their mouthparts or embedded in the host's tissues.

Fleas can be stored temporarily or permanently in 75–80% ethyl alcohol. Some workers suggest that for permanent storage the specimens should be kept dry in vials and held in place with non-absorbent cotton.

Fleas of larger hosts with a heavy pelage are more easily collected if not immobilized; instead they can be obtained simply by blowing on the host and aspirating the fleas that come out of the fur. This technique is particularly effective after the host's body has cooled.

Nesting material taken from the lair, or from guano piles beneath bat roosts, can yield large numbers of adult fleas if kept in paper bags or other containers in the laboratory. Samples should be kept at room temperature and sprinkled with water periodically to keep them moist but not wet. They should be checked for adults every few days until adults cease to emerge.

For most fleas, accurate identification requires that they be cleared and mounted on microscope slides. All specimens from a single host should be kept together in the same container (vial, watchglass) during processing through the following reagents:

1. 10% KOH solution at room temperature until the body contents dissolve and the specimens become somewhat transparent (usually 1–2 days). Specimens should never be boiled in this solution to speed up the process.
2. 5–10% solution of glacial acetic acid for 30–60 minutes to neutralize the KOH. (If permanent mounts are not needed, cleared specimens can be mounted in water under a coverglass for identification and then preserved in alcohol.)
3. Two changes of absolute alcohol, followed by 50% xylene in alcohol for 60 minutes and finally two changes of pure xylene for 60 minutes each. Specimens are then mounted in Canada balsam. Before mounting it might be necessary to tease the appendages apart and away from the body.

Locating specimens on a slide under the microscope is time-consuming and an effort should be made to be consistent in positioning the specimens so that they are all in essentially the same position. This is easily accomplished by using a template outlining the slide and with an X in the desired position.

Rearing

Most colonies in existence today are of species of economic importance, either as vectors of disease or as parasites of domestic animals. A detailed list of references to rearing methods is given in Smit (1977) and Marshall (1981). Recently some success has been achieved in feeding adult cat fleas through a membrane.

REFERENCES

Azad, A. F. 1990. Epidemiology of murine typhus. *Annual Review of Entomology* **35**: 553–569.
Beaucournu, J.-C. and **Launay, H.** 1990. Les puces (Siphonaptera) de France et du Bassin méditerranéen occidental. *Faune de France et Régions Limitrophes* **76**: 1–548.

Berlinguer, G. 1964. *Aphaniptera d'Italia. Studio monografico.* xv + 318 pp. "Il Pensiero Scientifico" Editore, Rome. [In Italian.]

Bibikova, V. A. 1977. Contemporary views on the interrelationships between fleas and the pathogens of human and animal diseases. *Annual Review of Entomology* **22**: 23–32.

Bibikova, V. A. and **Zhovtyi, I. F.** 1980. Review of certain studies of fleas in the USSR, 1967–1976. Pp. 257–272 in Traub and Starcke (1980).

Brinck-Lindroth, G. and **Smit, F. G. A. M.** 1971. The Kemner collection of Siphonaptera in the Entomological Museum, Lund with a check-list of the fleas of Sweden. *Entomologica Scandinavica* **2**: 269–286.

Cavanaugh, D. C. and **Williams, J. E.** 1980. Plague: some ecological interrelationships. Pp. 245–256 in Traub and Starcke (1980).

de Meillon, B., Davis, D. H. S. and **Hardy, F.** 1961. *Plague in southern Africa*: vol. 1, *The Siphonaptera (excluding Ischnopsyllidae).* viii + 280 pp. Government Printer, Pretoria.

Dunnet, G. M. and **Mardon, D. K.** 1974. A monograph of Australian fleas (Siphonaptera). *Australian Journal of Zoology* (Supplementary Series) **30**: 1–273.

Elbel, R. E. 1991. Order Siphonaptera. Pp. 674–689 in Stehr, F. W. (ed.), *Immature insects*: vol. 2, xvi + 975 pp. Kendall/Hunt, Dubuque, Iowa.

Fox, I. 1968. *Fleas of eastern United States.* vii + 191 pp. Hafner Publishing Co., New York and London. [Facsimile edition of 1940 work published by Iowa State College Press, Ames, Iowa.]

Goncharov, A. I., Romashcheva, T. P., Kotti, B. I., Bavaasan, A. and **Zhigmid, S.** 1989. *Keys to the fleas of the Mongolian National Republic.* 415 pp. Ulan-Bator. [In Russian.]

Gratz, N. G. 1980. Problems and developments in the control of flea vectors of disease. Pp. 217–240 in Traub and Starcke (1980).

Haeselbarth, E. 1966. Order Siphonaptera. Pp. 117–212 in Zumpt F. (ed.), The arthropod parasites of vertebrates in Africa south of the Sahara (Ethiopian Region). 3. Insecta excl. Phthiraptera. *Publications of the South African Institute for Medical Research* **13** (52): 1–283.

Hastriter, M. W. and **Tipton, V. J.** 1975. Fleas (Siphonaptera) associated with small mammals of Morocco. *Journal of the Egyptian Public Health Association* **50**: 79–169.

Hirst, L. F. 1953. *The conquest of plague. A study of the evolution of epidemiology.* xvi + 478 pp. Clarendon Press, Oxford.

Holland, G. P. 1949. The Siphonaptera of Canada. *Dominion of Canada Department of Agriculture Technical Bulletin* **70**: 1–306.

Holland, G. P. 1964. Evolution, classification, and host relationships of Siphonaptera. *Annual Review of Entomology* **9**: 123–146.

Holland, G. P. 1969. Contribution towards a monograph of the fleas of New Guinea. *Memoirs of the Entomological Society of Canada* **61**: 1–77.

Holland, G. P. 1985. The fleas of Canada, Alaska and Greenland (Siphonaptera). *Memoirs of the Entomological Society of Canada* **130**: 1–630.

Honacki, J. H., Kinman, K. E. and **Koeppl, J. W.** (eds) 1982. *Mammal species of the world: a taxonomic and geographic reference.* ix + 694 pp. Allen Press and Association of Systematics Collections, Lawrence, Kansas.

Hopkins, G. H. E. and **Rothschild, M.** 1953. *An illustrated catalogue of the Rothschild collection of fleas (Siphonaptera) in the British Museum (Natural History). With keys and short descriptions for the identification of families, genera, species and subspecies. I. Tungidae and Pulicidae.* xv + 361 pp. British Museum (Natural History), London.

Hopkins, G. H. E. and **Rothschild, M.** 1956. *An illustrated catalogue of the Rothschild collection of fleas (Siphonaptera) in the British Museum (Natural History). With keys and short descriptions for the identification of families, genera, species and subspecies of the order. II. Coptopsyllidae, Vermipsyllidae, Stephanocircidae, Macropsyllidae, Ischnopsyllidae, Chimaeropsyllidae and Xiphiopsyllidae.* xi + 445 pp. British Museum (Natural History), London.

Hopkins, G. H. E. and **Rothschild, M.** 1962. *An illustrated catalogue of the Rothschild collection of fleas (Siphonaptera) in the British Museum (Natural History). With keys and short descriptions for the identification of families, genera, species and subspecies of the order. III. Hystricopsyllidae*

(Anomiopsyllinae, Hystricopsyllinae, Neopsyllinae, Rhadinopsyllinae and Stenoponiinae). ix + 560 pp. British Museum (Natural History), London.

Hopkins, G. H. E. and Rothschild, M. 1966. *An illustrated catalogue of the Rothschild collection of fleas (Siphonaptera) in the British Museum (Natural History). With keys and short descriptions for the identification of families, genera, species and subspecies of the order. IV. Hystricopsyllidae (Ctenophthalminae, Dinopsyllinae, Doratopsyllinae and Listropsyllinae)*. viii + 549 pp. British Museum (Natural History), London.

Hopkins, G. H. E. and Rothschild, M. 1971. *An illustrated catalogue of the Rothschild collection of fleas (Siphonaptera) in the British Museum (Natural History). With keys and short descriptions for the identification of families, genera, species and subspecies of the order. V. Leptopsyllidae and Ancistropsyllidae*. viii + 530 pp. British Museum (Natural History), London.

Hopla, C. E. 1980a. A study of the host associations and zoogeography of *Pulex*. Pp. 185–207 in Traub and Starcke (1980).

Hopla, C. E. 1980b. Fleas as vectors of tularemia in Alaska. Pp. 287–300 in Traub and Starcke (1980).

Hubbard, C. A. 1947. *Fleas of western North America. Their relation to the public health*. ix + 533 pp. Iowa State College Press, Ames, Iowa.

Hubbard, C. A. 1960. Fleas and plague in Iraq and the Arab world. Part 2. *Iraq Natural History Museum Publications* **19**: 1–143.

Ioff, I. G. and Skalon, O. I. 1954. *Keys to the fleas of eastern Siberia, the Far East and adjacent regions*. 275 pp. Medgiz, Moscow. [In Russian.]

Ioff, I. G. and Tiflov, V. E. 1954. *Keys to the Aphaniptera (Suctoria-Aphaniptera) of southeastern USSR*. 201 pp. Antiplague Institute, Stavropol'. [In Russian.]

Ioff, I. G., Mikulin, M. A. and Skalon, O. I. 1965. *Keys to the fleas of central Asia and Kazakhstan*. 370 pp. 'Meditsina', Moscow. [In Russian.]

Iyengar, R. 1973. The Siphonaptera of the Indian subregion. *Oriental Insects* (Supplement) **3**: 1–102.

Jancke, O. 1938. Flohe oder Aphaniptera (Suctoria): Lause oder Anoplura (Siphunculata). *Tierwelt Deutschlands* **35**: 1–78.

Johnson, P. T. 1957. A classification of the Siphonaptera of South America. With descriptions of new species. *Memoirs of the Entomological Society of Washington* **5**: 1–299.

Johnson, P. T. 1961. A revision of the species of *Monopsyllus* Kolenati in North America (Siphonaptera, Ceratophyllidae). *United States Department of Agriculture Technical Bulletin* **1227**: 1–69.

Lewis, R. E. 1967. The fleas (Siphonaptera) of Egypt. An illustrated and annotated key. *Journal of Parasitology* **53**: 863–885.

Lewis, R. E. 1972. Notes on the geographical distribution and host preferences in the order Siphonaptera. Part 1. Pulicidae. *Journal of Medical Entomology* **9**: 511–520.

Lewis, R. E. 1973a. Siphonaptera collected during the 1965 Street Expedition to Afghanistan. *Fieldiana: Zoology* **64**: xi + 1–161.

Lewis, R. E. 1973b. Notes on the geographical distribution and host preferences in the order Siphonaptera. Part 2. Rhopalopsyllidae, Malacopsyllidae and Vermipsyllidae. *Journal of Medical Entomology* **10**: 255–260.

Lewis, R. E. 1974a. Notes on the geographical distribution and host preferences in the order Siphonaptera. Part 3. Hystricopsyllidae. *Journal of Medical Entomology* **11**: 147–167.

Lewis, R. E. 1974b. Notes on the geographical distribution and host preferences in the order Siphonaptera. Part 4. Coptopsyllidae, Pygiopsyllidae, Stephanocircidae and Xiphiopsyllidae. *Journal of Medical Entomology* **11**: 403–413.

Lewis, R. E. 1974c. Notes on the geographical distribution and host preferences in the order Siphonaptera. Part 5. Ancistropsyllidae, Chimaeropsyllidae, Ischnopsyllidae, Leptopsyllidae and Macropsyllidae. *Journal of Medical Entomology* **11**: 525–540.

Lewis, R. E. 1974d. Notes on the geographical distribution and host preferences in the order Siphonaptera. Part 6. Ceratophyllidae. *Journal of Medical Entomology* **11**: 658–676.

Lewis, R. E. 1982. Insects of Saudi Arabia. Siphonaptera. A review of the Siphonaptera of the Arabian peninsula. *Fauna of Saudi Arabia* **4**: 450–464.

Lewis, R. E. 1990. The Ceratophyllidae: currently accepted valid taxa (Insecta: Siphonaptera). *Theses Zoologicae* **13**, 267 pp., Koeltz Scientific Books, Koenigstein.

Lewis, R. E. and **Lewis, J. H.** 1985. Notes on the geographical distribution and host preferences in the order Siphonaptera. Part 7. New taxa described between 1972 and 1983, with a supraspecific classification of the order. *Journal of Medical Entomology* **22**: 134–152.

Lewis, R. E. and **Lewis, J. H.** 1989. A catalogue of invalid and questionable genus-group and species-group names in the Siphonaptera (Insecta). *Theses Zoologicae* **11**, 263 pp. Koeltz Scientific Books, Koenigstein.

Lewis, R. E. and **Lewis, J. H.** 1990. An annotated checklist of the fleas (Siphonaptera) of the Middle East. *Fauna of Saudi Arabia* **11**: 251–276.

Lewis, R. E., Lewis, J. H. and **Maser, C.** 1988. *The fleas of the Pacific Northwest*. 296 pp. Oregon State University Press, Corvallis, Oregon.

Liu, Z. (ed.) 1986. *Fauna Sinica: Insecta: Siphonaptera.* 1334 pp. Science Press, Beijing [= Peking]. [In Chinese.]

Lumaret, R. 1962. Insectes Siphonaptères. *Faune de Madagascar* **15**: 1–109.

Marcus, T. 1961. The bat fleas of southern Africa (Siphonapt.: Ischnopsyllidae). *Journal of the Entomological Society of Southern Africa* **24**: 173–211.

Mardon, D. K. 1981. *An illustrated catalogue of the Rothschild collection of fleas (Siphonaptera) in the British Museum (Natural History). With keys and short descriptions for the identification of families, genera, species and subspecies of the order. VI. Pygiopsyllidae.* 298 pp. British Museum (Natural History), London.

Marshall, A. G. 1980. The function of combs in ectoparasitic insects. Pp. 79–87 in Traub and Starcke (1980).

Marshall, A. G. 1981. *The ecology of ectoparasitic insects.* xvi + 459 pp. Academic Press, London.

Méndez, E. 1977. Mammalian–Siphonapteran associations, the environment, and biogeography of mammals of southwestern Colombia. *Quaestiones Entomologicae* **13**: 91–182.

Morales, J. C. M. 1983. *Estado actual del conocimiento de los Siphonaptera de Mexico.* 122 pp. Universidad Nacional Autonoma de Mexico, Mexico City.

Peus, F. 1972. Zur Kenntnis der Flohe Deutschlands (Schluss) (Insecta, Siphonaptera). IV. Faunistik und Ökologie der Saugertierflohe. *Zoologische Jahrbucher* (Systematik) **99**: 408–504.

Peus, F. 1977. Flohe aus Anatolien und anderen Ländern des Nahes Ostens. *Abhandlungen der Zoologisch-Botanischen Gesellschaft in Wien* **20**: (1976) 1–111.

Pilgrim, R. L. C. 1980. The New Zealand flea fauna. Pp. 173–184 in Traub and Starcke (1980).

Pollitzer, R. 1954. *Plague.* 698 pp. World Health Organization, Geneva [No. 22 in WHO Monograph Series].

Pollitzer, R. 1966. *Plague and plague control in the Soviet Union. History and bibliography through 1964.* xiii + 478 pp. Fordham University, Bronx, New York.

Price, P. W. 1980. The extent of adaptive radiation in fleas. Pp. 69–78 in Traub and Starcke (1980).

Ribeiro, H. 1974. Sifonápteros de Angola (Insecta, Siphonaptera). Estudo sistemático e dados bioecológicos interessando à epidemiologia do peste. *Anais de Instituto de Higiene e Medicina Tropical* **2**: 3–199.

Rosický, B. 1957. Blechy-Aphaniptera. *Fauna ČSR* **10**: 1–439. [In Czech.]

Rothschild, M. and **Ford, B.** 1964. Breeding of the rabbit flea (*Spilopsyllus cuniculi* Dale) controlled by the reproductive hormones of the host. *Nature* **201**: 103–104.

Rothschild, M., Schlein, Y. and **Ito, S.** 1986. *A colour atlas of insect tissues via the flea.* 184 pp. Wolfe Publishing, London.

Sakaguti, K. 1962. *A monograph of the Siphonaptera of Japan.* viii + 255 pp. Nippon Printing and Publishing Co., Osaka.

Sakaguti, K. and **Jameson, E. W.** 1962. The Siphonaptera of Japan. *Pacific Insects Monograph* **3**: 1–169.

Schramm, B. A. and **Lewis, R. E.** 1988. A taxonomic revision of the flea genus *Plocopsylla* Jordan, 1931 (Siphonaptera: Stephanocircidae). *Theses Zoologicae* **9**, 157 pp. Koeltz Scientific Books, Koenigstein.

Shepherd, R. 1980. The European rabbit flea *Spilopsyllus cuniculi* (Dale) in Australia – its use as a vector of myxomatosis. Pp. 301–307 in Traub and Starcke (1980).

Skalon, O. I. 1988. Order Siphonaptera (Aphaniptera, Suctoria). Pp. 1311-1385 in Bei-Bienko, G. Ya. (ed.), *Keys to the insects of the European part of the USSR*: vol. 5 (Diptera and Siphonaptera) (2): xxi + 1505 pp. Smithsonian Institution Libraries and National Science Foundation, Washington, D.C. [English translation of work first published in Russian by 'Nauka', Leningrad, 1969.]

Skuratowicz, W. 1967. Pchly-Siphonaptera (Aphaniptera). *Klucze do Oznaczania Owadów Polski* [Keys for the identification of Polish Insects] **29**: 1–141. [In Polish.]

Smit, F. G. A. M. 1957. Siphonaptera. *Handbooks for the Identification of British Insects* **1** (16): 1–94.

Smit, F. G. A. M. 1977. Rearing – references to descriptions of methods of rearing fleas. *Flea News* **12**: [2–4, unnumbered].

Smit, F. G. A. M. 1978. Siphonaptera: bibliography of faunistic key-works, catalogues and lists. *Flea News* **13**: 2–9 (numbered pagination but actually pages 7–15 in newsletter issue).

Smit, F. G. A. M. 1979. The fleas of New Zealand (Siphonaptera). *Journal of the Royal Society of New Zealand* **9**: 143–232.

Smit, F. G. A. M. 1982. Siphonaptera. Pp. 557–563 in Parker, S. P. (ed.), *Synopsis and classification of living organisms*: Vol. 2. 1232 pp. McGraw-Hill, New York.

Smit, F. G. A. M. 1987. An illustrated catalogue of the Rothschild collection of fleas (Siphonaptera) in the British Museum (Natural History). With keys and short descriptions for the identification of families, genera, species and subspecies of the order VII. Malacopsylloidea (Malacopsyllidae and Rhopalopsyllidae). 380 pp. Oxford University Press and British Museum (Natural History), Oxford and London.

Stark, H. E. 1970. A revision of the flea genus *Thrassis* Jordan 1933 (Siphonaptera: Ceratophyllidae). With observations on ecology and relationship to plague. *University of California Publications in Entomology* **53**: 1–184.

Theodor, O. and **Costa, M.** 1967. *A survey of the parasites of wild mammals and birds in Israel. Part 1. Ectoparasites.* 117 pp. Israel Academy of Sciences and Humanities, Jerusalem.

Tiflov, V. E., Skalon, O. I. and **Rostigaev, B. A.** 1977. *Keys to the fleas of the Caucasus.* 278 pp. Antiplague Institute, Stavropol'. [In Russian.]

Tipton, V. J. and **Machado-Allison, C. E.** 1972. Fleas of Venezuela. *Brigham Young University Science Bulletin* (Biological Series) **17** (6): 1–115.

Tipton, V. J. and **Méndez, E.** 1966. The fleas (Siphonaptera) of Panama. Pp. 289–385 in Wenzel, R. L. and Tipton, V. J. (eds). *Ectoparasites of Panama.* xii + 861 pp. Field Museum of Natural History, Chicago.

Tipton, V. J., Southwick, J. W., Ah, H. -s. and **Yu, H. -s.** 1972. Fleas of Korea. *Korean Journal of Parasitology* **10**: 52–63. [In English.]

Traub, R. 1950. Siphonaptera from Central America and Mexico. A morphological study of the aedeagus with descriptions of new genera and species. *Fieldiana: Zoology Memoirs* **1**: 1–127.

Traub, R. 1972. The Gunong Benom Expedition 1967. 11. Notes on zoogeography, convergent evolution and taxonomy of fleas (Siphonaptera), based on collections from Gunong Benom and elsewhere in South-East Asia. 1. New taxa (Pygiopsyllidae, Pygiopsyllinae). *Bulletin of the British Museum (Natural History)* (Zoology) **23**: 201–305.

Traub, R. 1980a. Some adaptive modifications in fleas. Pp. 33–67 in Traub and Starcke (1980).

Traub, R. 1980b. The zoogeography and evolution of some fleas, lice and mammals. Pp. 93–172, in Traub and Starcke (1980).

Traub, R. and **Dunnet, D. M.** 1973. Revision of the siphonapteran genus *Stephanocircus* Skuse, 1893 (Stephanocircidae). *Australian Journal of Zoology* (Supplement) **20**: 41–128.

Traub, R., Rothschild, M. and **Haddow, J. F.** 1983. *The Rothschild collection of fleas. The Ceratophyllidae: key to the genera and host relationships. With notes on their evolution, zoogeography and medical importance.* xv + 288 pp. [Privately published, distributed by Academic Press, London.]

Traub, R. and **Starcke, H.** (eds) 1980. *Fleas.* x + 420 pp. Balkema, Rotterdam. [Published 'Proceedings of the International Conference on Fleas, Ashton Wold/Peterborough/UK, 21–25 June 1977'.]

Traub, R., Wisseman, C. L. and **Farhang-Azad, A.** 1978. The ecology of murine typhus – a critical review. *Tropical Diseases Bulletin* **75**: 237–317.

World Health Organization 1989. *Geographical distribution of arthropod-borne diseases and their principal vectors.* WHO/VBC/89.967, 134 pp. World Health Organization, Geneva.

Yurkina, V. I. 1961. Fleas. *Fauna of Ukraine* **17** (4): 1–152. Ukrainian Academy of Science, Kiev. [In Ukrainian.]

CHAPTER SEVENTEEN

Insects of minor medical importance

Kenneth G. V. Smith

Insects have a medical significance in addition to direct transmission of pathogenic organisms or as biting pests. There are many non-biting insects which are the source of potent allergens (reviews in Bellas, 1982; Levine and Lockley, 1981; Wirtz, 1984) and of venoms and toxins (Frazier, 1969; Habermann, 1972; Minton, 1974; Tu, 1977, 1984) or which invade the human body. Flies of many families have been reported to bite man, although often in such cases it seems they are only obtaining moisture from the skin, e.g. Tipulidae (McCrae, 1967); some, however, actually bite and take blood. In a less direct, but nevertheless clinically significant role, insects are the cause of entomophobias or delusory parasitosis (Mumford, 1982; Smith, 1987).

The medical importance of the insect groups considered here varies considerably from one part of the world to another.

DIPTERA

Chironomidae (non-biting midges)
These midges bear a superficial resemblance to mosquitoes but can be distinguished by their wing venation and lack of biting mouthparts (Fig. 17.1a). Chironomidae are among the most abundant Diptera in aquatic environments, and their emergence *en masse* from bodies of standing fresh water such as dam impoundments and waste stabilization lagoons (Grodhaus, 1967, keys) can give rise to hypersensitivity in human populations inhabiting surrounding areas (Bellas, 1982). This problem is well documented for parts of the Nile valley in northern and central Sudan, where *Cladotanytarsus lewisi* (local name 'nimitti') is responsible for allergic rhinitis and bronchial asthma (Cranston et al., 1981). The low molecular weight haemoglobins present in the larvae (Baur et al., 1982) are potent allergens to those occupationally handling larvae. These compounds probably become airborne in the shed meconium of emerging adults during the chironomid season in Sudan and elsewhere. The problem of sensitization to Chironomidae may be more widespread than is currently thought.

Larvae (Fig. 17.1b) can also be found in drinking water systems; here they are usually of no medical importance, but their sudden appearance can indicate malfunction and contamination of a water supply.

Medical Insects and Arachnids Edited by Richard P. Lane and Roger W. Crosskey.
Published in 1993 by Chapman & Hall ISBN 0 412 40000 6

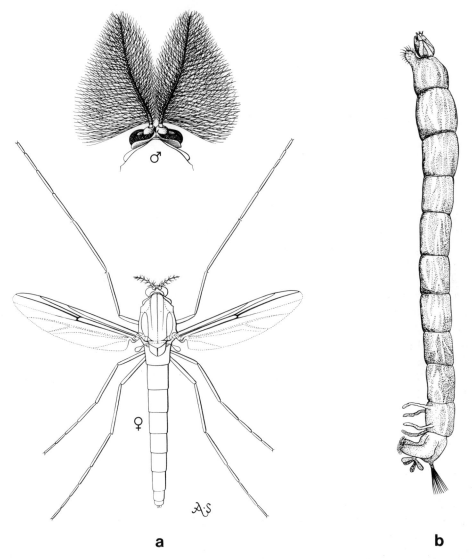

Figure 17.1 Non-biting midges, family Chironomidae: (*a*) adult female of *Chironomus*, with head of male (above) at larger scale to show characteristic plumed antennae of this sex; (*b*) larva of a typical chironomid (*Chironomus dorsalis*), left side view.

Chaoboridae (phantom midges)
Swarms of *Chaoborus edulis* over Lake Victoria and Lake Nyasa in East Africa can be so dense that adults appear as a black cloud which moves with the wind. Besides becoming a nuisance, such swarms can be alarming and even dangerous to people in their path. Fishermen in small boats can even be suffocated before they can escape from the flies. In the past, bodies of these midges were compressed into 'Kungu cake' and eaten by the local population (Bodenheimer, 1951).

Psychodidae (moth-flies)
The characteristically hairy wings closed over the body like a tent (Fig. 17.2*c*) serve to distinguish the non-biting moth-flies from the bloodsucking sandflies

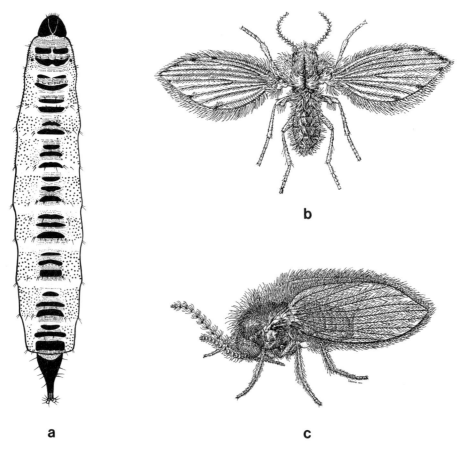

Figure 17.2 Moth-flies, family Psychodidae: (*a*) larva of a *Psychoda* species; (*b*) *Psychoda* with outstretched wings (drawn from *P. alternata*); (*c*) *Psychoda* at rest, showing characteristic oblique tent-like wing posture.

(Phlebotominae) and other nematocerous Diptera. Several species of the genus *Psychoda* (e.g. *P. alternata*, Fig. 17.2*b*) are associated with sewage works, where they breed in the trickle-filter beds, their larvae feeding on the bacteria there. The larvae (Fig. 17.2*a*) prevent the beds from becoming choked with bacteria and are thus beneficial. The adults however can occur in such vast numbers that they become a nuisance to sewage work staff and people in the neighbourhood, e.g. in the Transvaal (Ordman, 1946).

Anisopodidae (window-gnats)

Anisopodidae have no medical importance as adults but they are occasionally reported as a nuisance in the vicinity of sewage works, where they breed in the percolating filters, and can cause myiasis as larvae. They can be confused with mosquitoes but the lack of scales, the short proboscis and the wing venation distinguish them.

Phoridae (scuttle-flies)

Adult phorids are small, reddish brown to black flies with a distinctive wing venation (Fig. 17.3*c*), and those of *Megaselia halterata*, the mushroom-fly, can annoy

Insects of minor medical importance 579

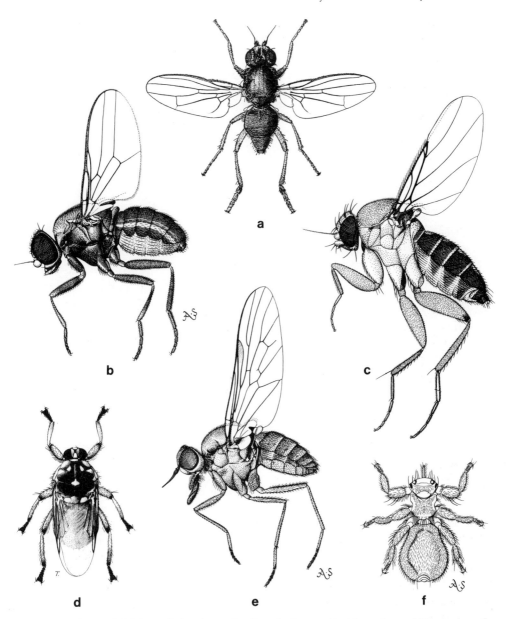

Figure 17.3 Some adult Diptera belonging to families of minor medical importance: (*a*) lesser dung-fly, *Copromyza equina* (Sphaeroceridae); (*b*) eye-fly, *Siphunculina funicola* (Chloropidae); (*c*) scuttle-fly, *Megaselia scalaris* (Phoridae); (*d*) louse-fly, *Hippobosca rufipes* (Hippoboscidae); (*e*) snipe-fly, *Spaniopsis longicornis* (Rhagionidae); (*f*) sheep-ked, *Melophagus ovinus* (Hippoboscidae).

pickers and cause allergic reactions such as bronchial asthma (Kern, 1938; Truitt, 1951) as well as causing a nuisance by entering houses in the vicinity of mushroom farms.

Rhagionidae and Athericidae (snipe-flies)
The family Rhagionidae contains several species recorded as biting man. Species of *Symphoromyia* are vicious, if occasional, biters in the Palaearctic and Nearctic

regions. They alight quite silently and inflict a painful bite (Turner, 1979: review) and a severe allergic reaction may even occur. Rhagionids have been suspected as mechanical vectors of tularaemia (Mills, 1943) and anaplasmosis (Hoy and Anderson, 1978). Shemanchuk and Weintraub (1961) point out that intermittent probing of these flies when taking a blood-meal increases the probability of their acting as mechanical vectors.

Three genera of Athericidae contain species which suck blood: *Atherix* in the Holarctic realm (e.g. Webb, 1977, Nearctic species), *Suragina* in the Far East and Central America, and *Spaniopsis* (Fig. 17.3e) in Australia. Nagatomi (1962) has discussed the biology and provided a general bibliography: it lists his several works on snipe-flies in Japan (where *Atherix* has been recorded biting man).

Ephydridae (shore-flies)

Adults can be recognized by the unusually broad mouth-opening, strongly convex face and wing venation in which the anal vein, anal cell and second basal cell are lacking. Most species breed in fresh or brackish water or wet grassland. Some are troublesome as sweat-feeders or eye-flies (e.g. *Chlorichaeta tuberculosa*) (Bohart and Gressitt, 1951). In the USA, some species that breed in salt waters (e.g. *Ephydra cinerea* and *Hydropyrus hians*) occur in vast numbers and are sometimes a sufficient nuisance to affect the tourist trade.

Chloropidae (eye-flies, frit-flies)

This is a large family of small flies (length often about 2 mm) that are usually associated as immature stages with plant material. Chloropids can be recognized by their large and (usually) distinct frontal (= ocellar) triangle on the head, scarcity of strong head bristles and reduced wing venation. *Hippelates* is the most important genus from the medical point of view but is atypical in that it has numerous head bristles; however, it has unusually strong and curved tibial spurs and these provide a characteristic identifying feature.

The adults of some species of *Hippelates* (New World) and *Siphunculina* (Old World) are attracted to human body secretions and will feed on sweat, open sores and infected and weeping eyes. The flies scrape the surface with the spiny tips of their mouthparts and often provoke such irritation as to increase the quantity of discharging fluid – to which still more eye-flies are attracted. Some species have been incriminated as mechanical vectors of yaws and conjunctivitis (Kumm and Turner, 1936) and of streptococcal skin infections in Panama and Trinidad (Bassett, 1970; Taplin et al., 1965, p. 549). Apart from their role as vectors they can be extremely irritating when flying very close to the eyes. In the southeastern USA *H. pusio* is commonly associated with man, and *H. collusor* in the southwest USA. In the Oriental region, *Siphunculina funicola* (Fig. 17.3b) is the most frequent species around man. The larvae feed on grass-stems and roots, or live amongst decaying vegetation in the soil.

Sphaeroceridae (lesser dung-flies)

Sphaeroceridae are small, dark flies (Fig. 17.3a) with a characteristic wing venation, which breed in decaying organic matter of vegetable and animal origin, including both animal and human excrement. Adult *Leptocera caenosa* are frequently found in houses in temperate regions and can usually be traced to small accumulations of sewage arising from faulty or damaged drainage systems (Fredeen and Taylor,

1964). *Salmonella* have been isolated from Sphaeroceridae collected in open markets in warmer climates, e.g. Mexico (Greenberg, in Harwood and James, 1979).

Hippoboscidae (keds, louse-flies, flat-flies)
These are ectoparasites of birds and mammals. While man is never a preferred host, there are many records of these flies biting man. *Crataerina pallida, Hippobosca equina, H. camelina, H. variegata, H. rufipes* (Fig. 17.3d), *Melophagus ovinus* (Fig. 17.3f), *Lipoptena cervi* and *Pseudolynchia canariensis* are species occasionally reported biting man. Bites are most frequent when humans are associated with horses, cattle, sheep, deer and even pigeons. The deer-ked (*Lipoptena cervi*) can be a serious nuisance in Finland (Hackman, 1979) and Russia (Ivanov, 1975), where its bites can cause dermatitis.

Nycteribiidae and Streblidae (bat-flies)
These are ectoparasites of bats. Since some bats transmit rabies (Manson-Bahr and Bell, 1987), the flies could possibly be involved in the maintenance of the virus in bat populations (Harwood and James, 1979).

HYMENOPTERA

Many members of this order can inflict a severe sting, or, as in some ants, a painful bite. The stinging apparatus is the modified female ovipositor which is used to inject venom into the victim. Wasps and ants retract their stings after use, and can thus sting repeatedly, but honeybees are rarely able to do so as the sting has barbs which hold it so firmly that the bee's abdomen ruptures when it tries to pull the sting out. The bee's poison gland, which is attached to the sting, will continue injecting venom after separation. Some parasitic Hymenoptera have very long ovipositors used for depositing eggs in larval hosts living in rather inaccessible situations; while they cannot normally penetrate human skin, some Ichneumonidae may attempt to use the ovipositor as a sting.

Fatalities from the stings of Hymenoptera are rare, for instance in England and Wales fewer than five cases a year are reported (Edwards, 1980). However, in the USA, bee and wasp stings cause over half the deaths from venomous animals; snakes account for only 30% (Parrish, 1963). A good introduction to the biology and classification of the Hymenoptera is provided by Gauld and Bolton (1988). Hymenopteran venoms are discussed by Blum (1981), Bücherl and Buckley (1972), Minton (1974), Tu (1977, 1984), and, specifically for Hymenoptera, by Piek (1986) and Habermann (1972).

Apidae and Vespidae (bees and wasps)
The effects of bee and wasp stings are of little significance to most victims and usually only produce local swellings and itching or mild pain which is easily soothed. Medical assistance is normally only necessary for stings in the mouth, especially the throat, or multiple stinging where cases of anaphylactic shock can be fatal. Massive anaphylaxis causes muscular paralysis and can be treated by parenteral adrenaline. Some individuals, such as beekeepers, become desensitized to bee stings but others can react more violently until they become so hypersensitive that their lives are in serious danger. A review of the world literature is given by Barr (1971) and an account of bee sting allergy by Frankland (1976).

Bee and wasp venoms are complicated mixtures composed of proteins such as the enzymes phospholipase A and hyaluronidase, with haemolytic and spreader effects, acetyl choline and histamine. The histamine is responsible for the immediate effects of smarting and swelling at the site of the sting, and in sensitized subjects, for the release of tissue histamine throughout the body causing burning and other unpleasant sensations far removed from the site of the sting. The venoms of wasps (*Vespula, Dolichovespula*) and hornets (*Vespa*) differ from bee venom in containing a higher proportion of histamine and also serotonin (5-hydroxytryptamine).

The introduction of the African honeybee (*Apis mellifera adansonii*) into South America to boost the honey production of the long established European bees has caused much concern through its increased attacks on man and domestic animals. Honeybees are not native to the Americas, and most wild and 'domesticated' honeybees in Brazil were a mixture of two races introduced there in 1839, viz. the black European bee (*A. m. mellifera*) and the Italian bee (*A. m. ligustica*). *Apis m. adansonii*, though normally aggressive in its native Africa, is manageable with due care and respect, but since its introduction into Brazil in 1956 it has become a serious problem. 'Africanized honeybees' soon spread to other parts of South America and have been moving northwards at a rate of about 400 km per year; they have now reached Mexico and the southern USA. When disturbed, hundreds of bees will attack anyone or anything within 100 metres of the nest and will reputedly pursue victims for over a kilometre. The venom is no more toxic than that of *A. m. ligustica*. Over 350 humans and an unknown number of domestic animals have died in Latin America as a result of the stings. Michener (1975) and Taylor (1977) give good accounts of the early development of the 'killer bees' problem and a symposium produced in 1988 includes 38 contributions on 'killer bees' (Needham et al., 1988).

The honey produced by hive bees and wild honeybees is sometimes toxic if particular poisonous plants are foraged.

Some stingless social bees of the tropics (e.g. *Melipona*) can be very aggressive and their mass biting (and in some Neotropical species, squirting of a caustic fluid) exceedingly unpleasant (Wille, 1983, biology). The stingless 'sweat-bees' of the genus *Trigona* are strongly attracted to human perspiration and are very annoying in the African savanna regions because of the large numbers which hover around the body and alight on the skin, especially of the face and hands. Occasionally, because of their numbers and persistence, they have been mistaken for blackflies of the *Simulium damnosum* complex, to which there is a slight superfical resemblance.

The stings of paper-wasps of the genus *Polistes* cause many fatalities in the New World. The wasps build their nests in trees, shrubs, under the eaves and in similar sites around buildings in rural areas. In some parts of the USA and Mexico these wasps can be more troublesome than *Vespula* or *Vespa*, especially in agricultural areas where the harvesting of fruit crops can be interrupted (Harwood and James, 1979). The biology and control of social wasps is dealt with by Edwards (1980), and both social and solitary wasps by Spradbery (1973).

Formicidae (ants)
Ants can bite, sting and squirt formic acid. Usually the effects of ant stings are mild but ants, like wasps, are capable of multiple stinging and this can induce anaphylactic shock. Some species will clear a house of bedbugs and other pests, while the jaws of some species (e.g. *Myrmecia gulosa*) have been used to close

wounds in lieu of stiches in primitive surgery.

Most ant venoms are proteins but not in the fire-ant *Solenopsis richteri* s.l. of the southeastern USA, the venom of which has necrotic effects similar to the bites of *Loxosceles* spiders. When a fire-ant colony is disturbed, ants erupt and a person can receive as many as 5000 stings in a matter of seconds; death can quickly result (Gurney, 1975). Other fire-ants and harvester-ants are troublesome in the USA. In Australia, bulldog-ants (*Myrmecia*) are a nuisance, and throughout the tropics and warmer temperate regions many species of the subfamily Myrmecinae commonly sting man.

Under certain conditions, protein-feeding ants such as the Pharaoh ant (*Monomorium pharaoensis*) could become mechanical vectors when they feed on soiled linen and discarded bandages in hospitals. These ants have been found to harbour *Salmonella*, *Pseudomonas*, *Staphylococcus*, *Streptococcus* and *Clostridium* pathogens (Beatson, 1972). *Monomorium* quickly colonizes suitable areas in buildings to become a considerable nuisance (Wilson, 1971, pp. 40–41).

COLEOPTERA

Canthariasis and Scarabiasis

Beetles are second to Diptera in importance as invaders of the human body, dead or alive. As carrion fauna they can be of forensic interest in establishing post-mortem interval (time of death), a subject fully covered in Smith (1986). Invasion of the living body (so-called pseudoparasitism) by beetles or their larvae is called canthariasis or scarabiasis and many case histories have been reported in the literature (Theodorides, 1950, for review). Most invasions appear to be accidental, yet the scarabaeid dung-beetles *Onthophagus bifasciatus* (Fig. 17.4c), *O. unifasciatus*, *Caccobius mutans* and *C. vulcanus* (Fig. 17.4b) have been reported many times from the bowels of children, especially in Sri Lanka – where such invasions are considered 'not uncommon' (Edirisinghe, 1988). Fletcher (1924) suggested that these beetles enter the anus when the child is asleep, causing diarrhoea and griping, but there is considerable disagreement over the portal of entry since dirt-eating (geophagy) is common in malnourished children with scarabiasis. Other cases of intestinal infestation involve species normally found in foodstuffs, such as the larvae of Dermestidae (larder-beetles) or Tenebrionidae (meal-worms) (Ebeling, 1975; Pérez-Íñigo, 1974). In these cases it is probable that eggs or small larvae were ingested in raw or imperfectly cooked food; they do little harm since they do not usually survive mastication and the normal digestive processes.

A few cases of urinogenital infestation by Coleoptera have been reported, e.g. a clerid beetle involved in Sudan and a larva of the ptinid *Niptus* in Finland (Leclercq, 1969). Occasional cases of nasal canthariasis have also been reported, including the result of using an aspirator or 'pooter' (Hurd, 1954). Harwood and James (1979) suggest that adult and larval beetles may be found alive in the nasal sinuses, having been directly inhaled or having crawled there following the consumption of contaminated food. Larvae of the carpet beetle *Anthrenus scrophulariae* have been known to crawl into the ears of sleeping persons, a trait also described for adults of the scarabaeids *Cyclocephala borealis* and *Autosericea castanea* which affected 176 boy scouts camping in Pennsylvania, USA (Maddock and Fehn, 1958).

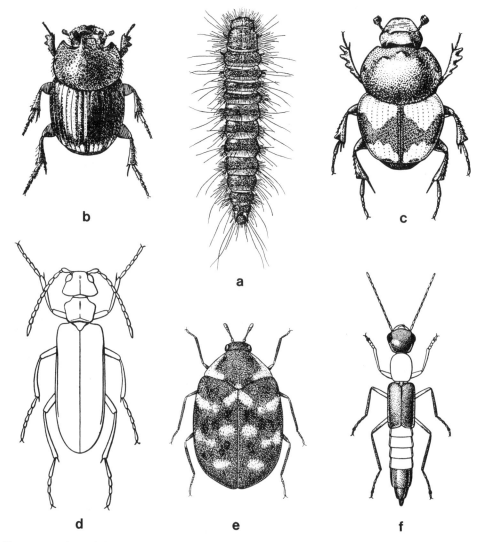

Figure 17.4 Some Coleoptera belonging to families of minor medical importance: (*a*) larva of the larder-beetle, *Dermestes lardarius* (Dermestidae); (*b, c*) two dung-beetles concerned in human scarabiasis, respectively *Caccobius vulcanus* and *Onthophagus bifasciatus* (Scarabaeidae); (*d*) so-called 'Spanish fly', *Lytta vesicatoria* (Meloidae); (*e*) carpet-beetle of genus *Anthrenus* (Dermestidae, *A. verbasci* illustrated); (*f*) blister-beetle of genus *Paederus* (Staphylinidae, *P. sabaeus* illustrated).

Conjunctivitis

Certain minute Staphylinidae (*Anotylus*, *Atheta*, *Oxytelus* and *Paederus*) sometimes fly in large numbers and accidentally enter the eye, causing a severe burning sensation or even temporary blindness ('Nairobi eye'). The burning is thought to be the result of a defensive secretion released by the beetle (Frank and Kanamitsu, 1987).

Dermatitis and blisters

The larvae of some Dermestidae (Fig. 17.4*a*) have urticating hairs which penetrate the skin, so liberating histamine and causing dermatitis. If inhaled, the hairs can cause difficulty in breathing and if ingested give rise to inflammation of the mouth and palate (erucic stomatitis). Dockers manually unloading infected cargoes have been severely affected.

Some species of the family Meloidae (blister-beetles) are an important source of medicinal cantharidin. The best known is the 'Spanish fly' (*Lytta vesicatoria*) (Fig. 17.4*d*), but others such as *Epicauta hirticornis* and *Mylabris cichorii* of India are also involved. The body fluids of these beetles contain vesicating substances which cause blisters if rubbed on the skin; the African *Mylabris alterna* has even been known to cause deaths in Zimbabwe when misused by witchdoctors in aphrodisiacal preparations (Hall, 1985). In the family Oedemeridae, *Ananta bicolor* (= *Sessinia collaris*) and *Ananta decolor* (known as coconut beetles in some Polynesian islands) can cause severe blistering (Herms, 1925). Some Staphylinidae including species of *Paederus* (Fig. 17.4*f*) cause urticaria and blistering of the skin (Frank and Kanamitsu, 1987; Mhalu and Mandara, 1981).

Body fluids from the larvae of *Diamphidea nigroornata* (Chrysomelidae) are used by South African bushmen as a lethal arrow poison, death resulting from general paralysis. Extracts of some carabid larvae are also used as arrow poisons by the Kalahari bushmen (Freyvogel, 1972; Jolivet, 1967).

Biting beetles
Occasionally, certain beetles occur in such numbers or aggregations as to become a severe nuisance, e.g. the well-known ladybirds (Coccinellidae). Hodek (1973) records biting by aggregations of *Semiadalia undecimnotata* on mountain tops. Herman and Cukerman (1982) give details of two patients whose symptoms included purplish pigmentation of the skin and blistering of the feet after contact with coccinellids. However, ladybird bites are usually trivial and these reactions might be due to sensitivity to chemical exudations of ladybirds by 'reflex bleeding' from the leg joints. These beetles are toxic and distasteful to many animals because of the alkaloids, histamines and quinolenes they contain.

Transmission of pathogens
More than 40 species of vertebrate-infesting nematodes have been recorded to use beetles, mainly Scarabaeidae and Tenebrionidae, as intermediate hosts (Hall, 1929). The tapeworm *Hymenolepis diminuta* is common in rats and is occasionally found in man. Its intermediate hosts are stored-product beetles, including *Tenebrio molitor* (see also Lepidoptera and Dermaptera below). Another occasional parasite of man is *Gongylonema pulchrum*, which develops in various scarabaeid beetles of the genus *Aphodius*. Parasitic worms of the phylum Acanthocephala require a vertebrate definitive host and, respectively, an insect or crustacean as developmental host according to whether they have a terrestrial or aquatic life cycle (DeGiusti, 1971). The acanthocephalid worm *Macracanthorhynchus hirudinaceus* develops in the larvae of cockchafers, occasionally in pigs and rarely in man, and has tenebrionid beetles (*Blaps mucronata*) as an intermediate host.

Some coprophagous and necrophagous beetles such as *Dermestes, Attagenus* and *Anthrenus* (Fig. 17.4*e*) are believed to spread virulent spores of anthrax (*Bacillus anthracis*).

LEPIDOPTERA

Sometimes lepidopterous larvae are accidentally ingested in food such as vegetables or stored food products, but there is little documentation concerning any medical significance this might have. Clearly, if swallowed larvae are poisonous (e.g. Zygaenidae) or covered in urticating hairs, then some of the reactions discussed below might follow. The rat tapeworm, *Hymenolepis diminuta,* which occasionally affects man, has the meal-moth *Pyralis farinalis* as an intermediate host.

Urtication by caterpillars and moths
The caterpillars of certain moths, especially Arctiidae, Bombycidae, Eucliidae, Lasiocampidae, Limacodidae, Lithosiidae, Lymantriidae, Megalopygidae, Noctuidae, Notodontidae, Saturniidae, Sphingidae and Thaumetopoeidae, and certain butterflies (Morphoidea and Nymphalidae) possess urticating or 'nettling' hairs which can cause dermatitis. General reviews of the subject are given by Bücherl and Buckley (1972), Lamas and Peres (1987), Quiroz (1978), Rothschild et al. (1970), Southcott (1978, Australasia) and Wirtz (1984).

The urticating hairs are usually connected to glands which release poison into the wound made by the hairs. The poison liberates histamine (Kawamoto, 1978; de Jong and Bleumink, 1977a, b; de Jong et al., 1975, 1976; Valette and Huidobro, 1954, 1956; Valle et al., 1954). The intensity of the irritation varies with the species of caterpillar and the sensitivity of the victim, but usually the symptoms are transitory. The irritation is more severe when the hairs reach a mucous membrane or the eye, where they can give rise to nodular conjunctivitis. If inhaled, detached caterpillar hairs can cause dyspnoea (laboured breathing) and if ingested can cause mouth irritation. Urticating hairs can also become attached to the cocoon when the larva pupates, and later to the adult moth. The hairs of some species retain their urticating properties long after being shed by the caterpillars.

Secretions can be ejected by some caterpillars which, when they are handled, cause transitory symptoms such as irritation or inflammation of the skin.

The most important urticating moth is the Venezuelan saturniid *Lonomia achelous,* the larvae of which (Fig. 17.5a) inject a powerful anticoagulant if handled or even brushed against; serious bleeding can result (Arocha-Piñango and Laryisse, 1969; Marsh and Arocha-Piñango, 1971). Rubber-tappers in northern Brazil have an occupational, long-term, disability of the fingers after contact with caterpillars of *Premolis semirufa* (Arctiidae) causing lesions in the peripheral connective tissue (Dias and Azevedo, 1973). Other well-known urticating larvae include *Megalopyge* species (flannel-moths, Megalopygidae) in tropical South America and the southern USA, *Euproctis* species (brown-tail moth, Lymantriidae) in Europe, USA, Asia and the Far East, *Thaumetopoea* species (processionary moths, Thaumetopoeidae) in Europe and the Mediterranean area, and *Ochrogaster* (bag-shelter moth, Notodontidae) in New Guinea and northern Australia. Larvae of the following genera are troublesome in the eastern USA: *Automeris, Dirphia* and *Hemileuca* (Saturniidae), and *Sibine* and *Euclea* (Limacodidae) (Harwood and James, 1979).

Urticaria from adult moths seems to be important only in South America where moths of the genus *Hylesia* (Saturniidae) occur (Gusmão et al., 1961). In this genus the barbed or spiny urticating setae are confined to the female moth (Harwood and James, 1979). Allergic reactions to airborne emanations of Lepidoptera have been investigated in Japan by Kino and Oshima (1978) and reviewed for North America by Wirtz (1984).

Eye-moths
Moths of the families Pyralidae, Geometridae and Noctuidae include species that frequent the eyes of various mammals, where they feed on lachrymal secretions

Insects of minor medical importance 587

Figure 17.5 Two Lepidoptera of medical importance: (*a*) caterpillar of the South American saturniid moth *Lonomia achelous*, an insect able if touched to induce serious bleeding in humans after injecting a powerful anticoagulant; (*b*) the South East Asian noctuid moth *Calyptra eustrigata* in the act of re-ingesting, while the proboscis remains in the wound, human blood regurgitated a few seconds earlier (photograph by H. Bänziger).

(Fig. 17.6). Of 20 species recorded on mammals in Thailand, six affected humans (Bänziger and Büttiker, 1969). Slight pain is caused by the feeding habit and the eye can become inflamed. Mechanical transmission of pathogens is a possibility. Bänziger (1971) traces the hypothetical development of this habit.

Bloodsucking moths
Based on examination of the stomach contents of eye-frequenting noctuid moths in South East Asia, Büttiker (1959) first reported that *Loboscraspis griseifusa* and *Arcyophora sylvatica* occasionally ingest blood. There are two groups of these moths: those which pierce the skin and suck blood, and those which scrape the skin and

Figure 17.6 The Oriental pyralid eye-moth *Filodes fulvodorsalis* imbibing lachrymal secretions from a human eye (photograph by H. Bänziger).

suck blood. The noctuid *Calyptra eustrigata* (Fig. 17.5b) belongs to the former group and has a strong proboscis which enables it to pierce the skin of a number of mammalian hosts in the wild, and of man under laboratory conditions (Bänziger, 1986). Other species in the genus are well-known in southern Asia and Africa as fruit-piercing moths. So far, bloodsucking has been observed only in male moths (Bänziger, 1983, and references).

THYSANOPTERA (THRIPS)

Thrips are minute, narrow insects with elongate, fringed wings and rasping–sucking mouthparts. They are mostly sap-sucking insects on plants but some are capable of penetrating the skin and sucking blood, e.g. *Karnyothrips flavipes,* a predator of scale insects in the Mediterranean subregion. They can cause itching and rashes through skin pricking, or inflammation in the eyes, ears or throat. This usually happens when their foodplants dry up under adverse climatic conditions. Mass flights of thrips can also occur in hot sultry weather, and this has earned them the name of 'thunder flies'. The species usually involved are widely distributed and include some feeding on cereals and grasses (notably *Chirothrips aculeatus* and *Limothrips cerealium*), the pear thrips (*Taeniothrips inconsequens*), and the onion thrips (*Thrips tabaci*). Other species recorded as biting are *Limothrips denticornis* (Germany), *Gynaikothrips ficorum* (Algeria), a *Heliothrips* species (Sudan), and *Thrips imaginis* (Australia). Lewis (1973) gives a good account of the Thysanoptera, and Southcott (1986) reviews biting by thrips, especially in the Australasian region.

INCIDENTAL INSECTS

Zygentoma (Thysanura) bristletails are wingless primitive insects which run about, usually at night, but scurry away from the light. The silverfish (*Lepisma saccharina*) and the firebrat (*Thermobia domestica*) are common domestic insects feeding on farinaceous debris. They are usually harmless but there is an unusual case in the Persian Gulf of a thysanuran invading the ear, causing tinnitus and earache.

Psocoptera (booklice) occur in vegetable and organic debris including foodstuffs and books. Allergic reactions such as rhinitis and bronchial asthma have been reported (Rijckeart et al., 1981).

Mass emergence of some aquatic insects such as Ephemeroptera (mayflies) and Trichoptera (caddis flies) (e.g. in the region of the Great Lakes of North America) results in a large number of exuviae. These cast skins are fragmented and, when windborne, can become inhalant allergens. The emerging adults can also become a nuisance or hazard if present in sufficient numbers (Corbet, 1966).

Some earwigs (Dermaptera) are reported to have drawn blood with their 'pincers' (Bishopp, 1961; Southcott, 1986). *Anisolabis* carries the cestode *Hymenolepis diminuta*.

INSECTS AND HYGIENE

A full survey of this aspect of entomology is provided by Busvine (1980) and Ebeling (1975). A useful survey of insects in food and as food, and their significance for human health, is given by Gorham (1979). Keys to insects infesting stored food are provided by Mound (1989).

Insects as a potential source of food are currently receiving some attention. A classic survey is given by Bodenheimer (1951) and recent information can be gleaned from *The Food Insects Newsletter* issued by the Department of Entomology, University of Wisconsin.

Swarms of insects indoors can be a nuisance or interfere with equipment. These are usually Diptera of the families Anisopodidae, Chloropidae, Sphaeroceridae, Phoridae, Calliphoridae (*Pollenia,* cluster fly) and some Muscidae. Oldroyd (1964) discusses the different ways insects enter houses.

Entomophobia, or the fear of insects, has been surveyed by Olkowski and Olkowski (1976) and Mumford (1982), but delusory parasitosis (sometimes confused with entomophobia) has a large and disparate literature. True delusory parasitosis is an imaginary infestation in which no actual insects are present. Patients sometimes present with rashes and sores, but these have usually been inflicted by scratching or by substances used to 'treat' the skin. Care has to be taken that an actual infestation, perhaps of mites, is not involved. Some useful conclusions and literature surveys of this condition are given by Lyell (1983), Mester (1975, 1977), Paulson and Petrus (1969) and Smith (1987).

REFERENCES

Arocha-Piñango, C. L. and **Layrisse, M.** 1969. Fibrinolysis produced by contact with a caterpillar. *Lancet* **1969** (1): 810–812.

Bänziger, H. 1971. Bloodsucking moths of Malaya. *Fauna – the Zoological Magazine,* Rancho Mirage **1**: 5–16.

Bänziger, H. 1983. A taxonomic revision of the fruit-piercing and blood-sucking moth genus *Calyptra* Ochsenheimer (= *Calpe* Treitschke) (Lepidoptera: Noctuidae). *Entomologica Scandinavica* **14**: 467–491.

Bänziger, H. 1986. Skin-piercing blood-sucking moths IV: biological studies on adults of 4 *Calyptra* species and 2 subspecies (Lep., Noctuidae). *Mitteilungen der Schweizerischen Entomologischen Gesellschaft* **59**: 111–138.

Bänziger H. and **Büttiker, W.** 1969. Records of eye-frequenting Lepidoptera from man. *Journal of Medical Entomology* **6**: 53–58.

Barr, S. E. 1971. Allergy to Hymenoptera stings. Review of the world literature: 1953–1970. *Annals of Allergy* **29**: 49–66.

Bassett, D. C. J. 1970. *Hippelates* flies and streptococcal skin infection in Trinidad. *Transactions of the Royal Society of Tropical Medicine and Hygiene* **69**: 138–147.

Baur, X., Dewair, M., Fruhmann, G., Aschauer, H., Pfletschinger, J. and **Braunitzer, G.** 1982. Hypersensitivity to chironomids (non-biting midges): localisation of the antigenic determinants within certain polypeptide sequences of haemoglobins (erythrocruorins) of *Chironomus thummi thummi* (Diptera). *Journal of Allergy and Clinical Immunology* **69**: 66–76.

Beatson, S. H. 1972. Pharaoh's ants as pathogen vectors in hospitals. *Lancet* **1972** (1): 425–427.

Bellas, T. E. 1982. *Insects as a cause of inhalant allergies. A bibliography.* Second edition. CSIRO Australia Division of Entomology Report No. 25. 60 pp. CSIRO, Canberra.

Bishopp, F. C. 1961. Injury to man by earwigs (Dermaptera). *Proceedings of the Entomological Society of Washington* **63**: 114.

Blum, M. S. 1981. *Chemical defenses of arthropods.* xii + 562 pp. Academic Press, New York.

Bodenheimer, F. S. 1951. *Insects as human food. A chapter of the ecology of man.* 352 pp. Junk, The Hague.

Bohart, G. E. and **Gressitt, J. L.** 1951. Filth-inhabiting flies of Guam. *Bernice P. Bishop Museum Bulletin* **204**: vii + 1–152.

Bücherl, W. and **Buckley, E. E.** (eds) 1972. *Venomous animals and their venoms: vol. 3, Venomous invertebrates.* xxii + 537 pp. Academic Press, New York and London.

Busvine, J. R. 1980. *Insects and hygiene. The biology and control of insect pests of medical and domestic importance.* Third edition. vii + 568 pp. Chapman & Hall, London and New York.

Büttiker, W. 1959. Blood-feeding habits of adult Noctuidae (Lepidoptera) in Cambodia. *Nature* **184**: 1167.

Corbet, P. S. 1966. A quantitative method of assessing the nuisance caused by non-biting aquatic insects. *Canadian Entomologist* **93**: 683–687.

Cranston, P. S., El Rab, M. O. G. and **Kay, A. B.** 1981. Chironomid midges as a cause of allergy in the Sudan. *Transactions of the Royal Society of Tropical Medicine and Hygiene* **75**: 1–4.

DeGiusti, D. L. 1971. Acanthocephala. Pp. 140–157 in Davis, J. W. and Anderson, R. C. (eds), *Parasitic diseases of wild mammals.* x + 364 pp. Iowa State University Press, Ames, Iowa.

de Jong, M. C. J. M. and **Bleumink, E.** 1977a. Investigative studies of the dermatitis caused by the larva of the brown-tail moth, *Euproctis chrysorrhoea* L. (Lepidoptera, Lymantriidae) III. Chemical analysis of skin reactive substances. *Archives for Dermatological Research* **259**: 247–262.

de Jong, M. C. J. M. and **Bleumink, E.** 1977b. Investigative studies of the dermatitis caused by the larva of the brown-tail moth, *Euproctis chrysorrhoea* L. (Lepidoptera, Lymantriidae) IV. Further characterization of skin reactive substances. *Archives for Dermatological Research* **259**: 263–281.

de Jong, M. C. J. M., Bleumink, E. and **Nater, J. P.** 1975. Investigative studies of the dermatitis caused by the larva of the brown-tail moth (*Euproctis chrysorrhoea* Linn.) I. Clinical and experimental findings. *Archives for Dermatological Research* **253**: 287–300.

de Jong, M. C. J. M., Hoedemaecker, P. J., Jongebloed, W. L. and **Nater, J. P.** 1976. Investigative studies of the dermatitis caused by the larva of the brown-tail moth (*Euproctis* *chrysorrhoea* Linn.) II. Histopathology of skin lesions and scanning electron microscopy of their causative setae. *Archives for Dermatological Research* **255**: 177–191.

Dias, L. B. and **Azevedo, M. C. de** 1973. Pararama, doença causada por larvas de lepidóptero: aspectos experimentais. *Boletin de la Oficina Sanitaria Panamericana* **75**: 197–203.

Ebeling, W. 1975. *Urban entomology*. viii + 695 pp. Division of Life Sciences, University of California.

Edirisinghe, J. S. 1988. Scarabiasis. *Tropical Doctor* **18**: 47–48.

Edwards, R. 1980. *Social wasps. Their biology and control*. 398 pp. Rentokil, East Grinstead.

Fletcher, T. B. 1924. Intestinal Coleoptera. *Indian Medical Gazette* **59**: 296–297.

Frank, J. H. and **Kanamitsu, K.** 1987. *Paederus*, sensu lato (Coleoptera: Staphylinidae): natural history and medical importance. *Journal of Medical Entomology* **24**: 155–191.

Frankland, A. W. 1976. Bee sting allergy. *Bee World* **57**: 145–150.

Frazier, C. A. 1969. *Insect allergy: allergic and toxic reactions to insects and other arthropods*. xiv + 493 pp. Warren H. Green, St Louis, Missouri.

Fredeen, F. J. H. and **Taylor, M. E.** 1964. Borborids (Diptera: Sphaeroceridae) infesting sewage disposal tanks, with notes on the life cycle, behaviour and control of *Leptocera (Leptocera) caenosa* (Rondani). *Canadian Entomologist* **96**: 801–808.

Freyvogel, T. A. 1972. Poisonous and venomous animals in East Africa. *Acta Tropica* **29**: 401–451.

Gauld, I. and **Bolton, B.** (eds) 1988. *The Hymenoptera*. xi + 332 pp. British Museum (Natural History), London and Oxford University Press, Oxford.

Gorham, J. R. 1979. The significance for human health of insects in food. *Annual Review of Entomology* **24**: 209–224.

Grodhaus, G. 1967. Identification of chironomid midges commonly associated with waste stabilization lagoons in California. *California Vector Views* **14**: 1–12.

Gurney, A. B. 1975. Some stinging ants. *Insect World Digest* **1975** (September/October): 19–25.

Gusmão, H. H., Forattini, O. P. and **Rotberg, A.** 1961. Dermatite provocado por Lepidopteros do genero *Hylesia*. *Revista do Instituto de Medicina de São Paulo* **3**: 114–120.

Habermann, E. 1972. Bee and wasp venoms. The biochemistry and pharmacology of their peptides and enzymes are reviewed. *Science* **177**: 314–322.

Hackman, W. 1979. Älglusflugans, *Lipoptena cervi*, invandringshistoria i Finland. *Entomologisk Tidskrift* **100**: 208–210.

Hall, M. C. 1929. Arthropods as intermediate hosts of helminths. *Smithsonian Miscellaneous Collections* **81** (15): 1–77.

Hall, M. J. R. 1985. The blister beetle: a pest of man, his animals and crops. *Zimbabwe Science News* **19**: 11–15.

Harwood, R. F. and **James, M. T.** 1979. *Entomology in human and animal health*. Seventh edition. vi + 548 pp. Macmillan, New York, Collier Macmillan, Toronto and Baillière Tindall, London.

Herman, J. and **Cukerman, R.** 1982. Contact dermatitis due to ladybirds. *Practitioner* **226**: 311.

Herms, W. B. 1925. Entomological observations on Fanning and Washington Islands, together with general biological notes. *Pan-Pacific Entomologist* **2**: 49–54.

Hodek, I. 1973. *Biology of Coccinellidae*. 260 pp. Junk, The Hague, and Academia Publishing House of the Czechoslovak Academy of Sciences, Prague.

Hoy, J. B. and **Anderson, J. R.** 1978. Behavior and reproductive physiology of blood-sucking snipe flies (Diptera: Rhagionidae: *Symphoromyia*) attacking deer in northern California. *Hilgardia* **46**: 113–168.

Hurd, P. D. 1954. "Myiasis" resulting from the use of the aspirator method in the collection of insects. *Science* **119**: 814–815.

Ivanov, V. I. 1975. Anthropophily in *Lipoptena cervi* L. (Diptera: Hippoboscidae). *Meditsinskaya Parazitologiya i Parazitarnye Bolezni* **44**: 491–495. [In Russian.]

Jolivet, P. 1967. Les Alticides vénéneux de l'Afrique du Sud. *Entomologiste* **23**: 100–111.

Kawamoto, F. 1978. Studies on the venomous spicules and spines of moth caterpillars II. Pharmacological and biochemical properties of the spicule venom of the Oriental tussock moth caterpillar, *Euproctis subflava*. *Japanese Journal of Sanitary Zoology* **29**: 175–183. [In Japanese with English summary.]

Kern, R. 1938. Asthma due to sensitization to a mushroom fly (*Aphiochaeta agarici*). *Journal of Allergy and Clinical Immunology* **61**: 10–16.

Kino, T. and **Oshima, S.** 1978. Allergy to insects in Japan. I. The reaginic sensitivity to moth and butterfly in patients with bronchial asthma. *Journal of Allergy and Clinical Immunology* **61**: 10–16.

Kumm, H. W. and **Turner, T. B.** 1936. The transmission of yaws from man to rabbits by an insect vector, *Hippelates pallipes* Loew. *American Journal of Tropical Medicine* **16**: 245–267.

Lamas, G. and **Peres, J. E.** 1987. Lepidopteros de importancia medica. *Diagnostico* **20**: 121–125.

Leclercq, M. 1969. *Entomological parasitology: the relations between entomology and medical sciences.* xviii + 158 pp. Pergamon Press, London.

Levine, M. I. and **Lockley, R. F.** 1981. *Monograph on insect allergy.* 84 pp. Typecraft, Pittsburgh.

Lewis, T. 1973. *Thrips: their biology, ecology and economic importance.* xv + 349 pp. Academic Press, London and New York.

Lyell, A. 1983. Delusions of parasitosis. *British Journal of Dermatology* **108**: 485–499.

Maddock, D. R. and **Fehn, C. F.** 1958. Human ear invasions by adult scarabaeid beetles. *Journal of Economic Entomology* **51**: 546–547.

Manson-Bahr, P. E. C. and **Bell, D. R.** (eds) 1987. *Manson's tropical diseases.* Nineteenth edition. xvii + 1557 pp. Baillière Tindall, London.

Marsh, N. A. and **Arocha-Piñango, C. L.** 1971. Observations on a saturniid moth caterpillar causing severe bleeding in man. *Proceedings of the Royal Entomological Society of London* (Journal of Meetings, C) **36**: 9–10.

McCrae, A. W. R. 1967. Unique record of crane flies biting man. *Uganda Journal* **31**: 128.

Mester, H. 1975. Induzierter 'Dermatozoenwahn'. *Psychiatria Clinica* **8**: 339–348.

Mester, H. 1977. Das Syndrom des Wahnhaften Ungezieferbefalls. *Angewandte Parasitologie* **18**: 70–84.

Mhalu, F. S. and **Mandara, M. P.** 1981. Control of an outbreak of rove beetle dermatitis in an isolated camp in a game reserve. *Annals of Tropical Medicine and Parasitology* **75**: 231–234.

Michener, C. D. 1975. The Brazilian bee problem. *Annual Review of Entomology* **20**: 399–416.

Mills, H. B. 1943. An outbreak of the snipe fly *Symphoromyia hirta*. *Journal of Economic Entomology* **36**: 806.

Minton, S. A. 1974. *Venom diseases.* 235 pp. C. C. Thomas, Springfield.

Mound, L. (ed.) 1989. *Common insect pests of stored products. A guide to their identification.* ix + 68 pp. British Museum (Natural History), London.

Mumford, J. 1982. Entomophobia: the fear of arthropods. *Antenna* **6**: 156–157.

Nagatomi, A. 1962. Studies in the aquatic snipe flies of Japan Part V. Biological notes (Diptera, Rhagionidae). *Mushi* **36**: 103–149.

Needham, G. R., Page, R. E., Delfinado-Baker, M. and **Bowman, C. E.** (eds) 1988. *Africanized honey bees and bee mites.* xviii + 572 pp. Ellis Horwood, Chichester.

Oldroyd, H. 1964. *The natural history of flies.* xiv + 324 pp. Weidenfeld & Nicolson, London.

Olkowski, H. and **Olkowski, W.** 1976. Entomophobia in the urban ecosystem, some observations and suggestions. *Bulletin of the Entomological Society of America* **22**: 313–317.

Ordman, D. 1946. Sewage filter flies (*Psychoda*) as a cause of bronchial asthma. *South African Medical Journal* **20**: 32–35.

Parrish, H. M. 1963. Analysis of 460 fatalities from venomous animals in the United States. *American Journal of Medical Science* **245**: 129–141.

Paulson, M. J. and **Petrus, E. P.** 1969. Delusions of parasitosis: a psychological study. *Psychosomatics* **10**: 111–120.

Pérez-Íñigo, C. 1974. Dípteros y coleopteros pseudoparásitos del intestino humano. *Graellsia* **27** (1971): 161–176.

Piek, T. (ed.) 1986. *Venoms of the Hymenoptera. Biochemical, pharmacological and behavioural aspects.* xi + 570 pp. Academic Press, London.

Quiroz, A. D. 1978. Venoms of Lepidoptera. Pp. 555–611 in Bettini, S. (ed.), *Arthropod venoms.* xxxiii + 977 pp. Springer-Verlag, Berlin.

Rijckaert, G., Thiel, C. and **Fuchs, E.** 1981. Silberfischen und Staulausen als Allergene. *Allergologie* **4**: 80–86.

Rothschild, M., Reichstein, T., von Euw, J., Aplin, R. and **Harman, R. R. M.** 1970. Toxic Lepidoptera. *Toxicon* **8**: 293–299.

Shemanchuk, J. A. and **Weintraub, J.** 1961. Observations on the biting and swarming of snipe flies (Diptera: *Symphoromyia*) in the foothills of southern Alberta. *Mosquito News* **21**: 238–243.

Smith, K. G. V. 1986. *A manual of forensic entomology.* 205 pp. British Museum (Natural History), London and Cornell University Press, Ithaca, New York.

Smith, K. G. V. 1987. Arthropod dermatoses, stings, bites, allergies and neuroses. Pp. 926–930 in Manson-Bahr, P. E. C. and Bell, D. R. (eds), *Manson's tropical diseases.* Nineteenth edition. viii + 1557 pp. Baillière Tindall, London.

Southcott, R. V. 1978. Lepidopterism in the Australian region. *Records of the Adelaide Children's Hospital* **2**: 87–173.

Southcott, R. V. 1986. Medical ill-effects of Australian primitive winged and wingless insects. *Records of the Adelaide Children's Hospital* **3**: 277–356.

Spradbery, J. P. 1973. *Wasps. An account of the biology and natural history of social and solitary wasps with particular reference to those of the British Isles.* xvi + 408 pp. Sidgwick & Jackson, London.

Taplin, D., Zaias, N. and **Rebell, G.** 1965. Environmental influences on the microbiology of the skin. *Archives of Environmental Health* **11**: 546–550.

Taylor, O. R. 1977. The past and possible future spread of Africanized honeybees in the Americas. *Bee World* **58**: 19–30.

Theodorides, J. 1950. The parasitological, medical and veterinary importance of Coleoptera. *Acta Tropica* **7**: 48–60.

Truitt, G. W. 1951. The mushroom fly as a cause of bronchial asthma. *Annals of Allergy* **9**: 513–516.

Tu, A. T. 1977. *Venoms: chemistry and molecular biology.* x + 560 pp. John Wiley, New York.

Tu, A. T. (ed.) 1984. *Handbook of natural toxins*: vol 2, *Insect poisons, allergens and other invertebrate venoms.* xv + 732 pp. Marcel Dekker, New York.

Turner, W. J. 1979. A case of severe human allergic reaction to bites of *Symphoromyia* (Diptera: Rhagionidae). *Journal of Medical Entomology* **15**: 138–139.

Valette, G. and **Huidobro, H.** 1954. Pouvoir histamino-liberateur du venin de la Chenille processionnaire du Pin (*Thaumetopoea pityocampa* Schiff.). *Comptes Rendus des Séances de la Société de Biologie et de ses Filiales* **148**: 1605–1607.

Valette, G. and **Huidobro, H.** 1956. Pouvoir histaminoliberateur du venin de la Chenille processionnaire du Pin (*Thaumetopoea pitiocampa* [sic] Schiff.). *Comptes Rendus des Séances de la Société de Biologie et de ses Filiales* **150**: 658–661.

Valle, J. R. Picarelli, Z. P. and **Prado, J. L.** 1954. Histamine content and pharmacological properties of crude extracts from setae of urticating caterpillars. *Archives Internationales de Pharmacodynamie et de Therapie* **98**: 324–334.

Webb, D. W. 1977. The Nearctic Athericidae (Insecta: Diptera). *Journal of the Kansas Entomological Society* **50**: 473–495.

Wille, A. 1983. Biology of the stingless bees. *Annual Review of Entomology* **28**: 41–64.

Wilson, E. O. 1971. *The insect societies.* x + 548 pp. Belknap Press of Harvard University Press, Cambridge, Massachusetts.

Wirtz, R. A. 1984. Allergic and toxic reactions to non-stinging arthropods. *Annual Review of Entomology* **29**: 47–69.

Part Three
ARACHNIDS

CHAPTER EIGHTEEN

Ticks and mites (Acari)

M. R. G. Varma

The class Arachnida contains arthropods which possess neither antennae nor mandibles. It includes such diverse forms as spiders, scorpions, ticks and mites and is found in many temperate and tropical regions. Of the eleven subclasses of the Arachnida, nine are completely predaceous and have mouthparts adapted for a predatory existence. The subclasses Opiliones and Acari, however, are exceptions to the rule of total predation in arachnids.

Although the study of ticks and mites as a distinct discipline started in eighteenth- and nineteenth-century Europe, the ancient Egyptians were aware of ticks and their medical importance. Tick fever is referred to in a papyrus scroll dated 1550 BC. The Greeks, too, knew of ticks and mites and the latin term 'acarus', meaning a kind of mite, is derived from the classical Greek 'akari' and the late Greek 'akares', meaning small, short, tiny. But it was not until 1735 that Linnaeus used the generic name *Acarus*, although the terms 'Akari' and 'mites' originated much earlier.

The subclass Acari to which the ticks and mites belong, is a heterogeneous group, exhibiting a diversity of habits; some members are exclusively plant feeders, others are parasitic on vertebrates and invertebrates. The life cycle consists of four stages, the egg and three mobile stages, the hexapod larva, octopod nymph and the adult. The earliest fossil mite, *Protacarus crani*, dates from the mid-Palaeozoic era. No fossil ticks have been discovered yet. About 30 000 species of ticks and mites have been described, belonging to 2000 genera, but about half a million species of Acari are believed to exist.

A major account of mammalian diseases associated with arachnids is given by Nutting (1984).

RECOGNITION, STRUCTURE AND CLASSIFICATION OF THE ACARI

The subclass Acari and the subclass Araneae (spiders) can be easily distinguished from other arachnids by the absence of somatic segmentation. In spiders the head and thorax are fused to form the cephalothorax which carries the mouthparts and the legs. The abdomen is connected to the cephalothorax by a narrow pedicel. In the Acari there is no such division, although the body can have sutures and annulations. Good general accounts of the Acari are given by Baker and Wharton (1952), Krantz (1978), McDaniel (1979) and Woolley (1988) with special emphasis on mites.

Medical Insects and Arachnids Edited by Richard P. Lane and Roger W. Crosskey. Published in 1993 by Chapman & Hall ISBN 0 412 40000 6

598 *Ticks and mites (Acari)*

The typical acarine body (Fig. 18.1) consists of an anterior gnathosoma and a posterior idiosoma. The region of the body carrying the legs is the podosoma and the region behind the legs, the opisthosoma. There are four pairs of legs in nymphs and adults; the larvae have only three pairs. The legs are arranged in two sets, the anterior two pairs attached to the propodosoma and the posterior two pairs to the

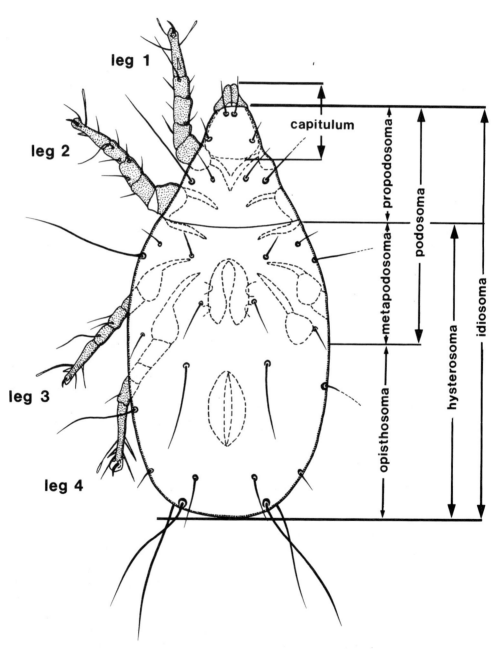

Figure 18.1 Divisions of the body of an acarine arthropod. Capitulum = gnathosoma.

metapodosoma. The gnathosoma and the propodosoma are known collectively as the proterosoma; the metapodosoma together with the opisthosomal region forms the hysterosoma.

The gnathosoma is more commonly referred to, particularly in ticks, as the capitulum or false head (Fig. 18.3b). Although less complicated than the feeding apparatus of many insects and composed of fewer components, the gnathosoma is a highly specialized organ. It is essentially a food-carrying tube connected to the oesophagus. The roof of the tube is the epistome and the lateral walls are made up of the coxae of the paired palpi. The floor of the tube forms the subcapitulum or basis capituli, of which the anterior extension is formed by the unpaired median hypostome: in ticks this structure has rows of backwardly directed teeth on the ventral side. Lying above the buccal cavity are paired chelicerae, often retractile appendages adapted for piercing or cutting, which are two-segmented in ticks and three-segmented in mites. The palps are jointed sensory appendages bearing thigmotactic and chemosensory hairs which help in locating food.

The acarine idiosoma performs the same functions as the abdomen, thorax and parts of the head of insects. It is unsegmented, but can have grooves or sutures, and is either soft or partially covered dorsally and ventrally with sclerotized shields which protect the animal from desiccation and predators. Externally, the idiosoma carries structures which have locomotory, respiratory, sensory and secretory functions. The locomotory structures are the legs. These are attached to the ventral side of the idiosoma and have seven segments: the coxa (epimere), trochanter, femur, genu, tibia, tarsus and pretarsus (the last carrying paired claws and/or a median empodium). The coxae are sometimes fused to the body wall. Larvae possess only the first three pairs of legs, of which the posterior pair appears in the first nymphal instar. Although primarily ambulatory, legs (particularly the anterior pair) can be used as sensory structures for food location, for grasping mates during mating and for capturing prey.

The stigmata (= spiracles) are respiratory structures found on the idiosoma. They are taxonomically important as their presence or absence and their position provide major diagnostic characters for separating the suborders of the Acarina. The stigmata are the external openings of a complex tracheal system. In many small mites, the astigmatid mites (scabies mites, mange mites, house-dust mites), and in larval ticks, the stigmata are absent and respiration takes place through the integument. In some other astigmatid mites, genital papillae associated with genital tracheae are respiratory in function. Paired openings between coxae 1 and 2 in trombiculid mite larvae, known as claparede organs or urstigmata, serve a similar function. Nymphs and adults of trombiculid mites and hair-follicle mites have a single pair of stigmata behind or between the cheliceral bases. The gamasid mites also have a single pair of spiracles but placed mediolaterally between the coxae of legs 2 and 4; ticks possess well-developed spiracles mediolaterally behind leg 4.

The acarine idiosoma bears a variety of cuticular setae which are either mechanoreceptors or chemoreceptors; these are often arranged in distinct patterns on the body and are taxonomically useful. Simple eyes (ocelli) or other photosensitive organs are present in many acarines. On the ventral side of some mites and argasid ticks, between or near the coxae of the first two pairs of legs, are simple pores which are the external openings of an elaborate internal system of canals forming the coxal glands. The coxal glands have an osmoregulatory function and maintain the water balance and ionic concentration of the haemolymph.

Acarine taxonomy has not attained the sophistication of insect taxonomy. The classification of Krantz (1978) is the most satisfactory to date. It recognizes two orders, the Acariformes and Parasitiformes. The former contains three suborders, the Acaridida, Actinedida and Oribatida, and the latter contains four suborders, the Mesostigmata (= Gamasida), Ixodida, Holothyrida and Opilioacarida. Of these suborders, only the Acaridida, Actinedida, Mesostigmata and Ixodida contain species of medical importance. The major diagnostic character in acarine taxonomy is the presence or absence of stigmata and their position and structure when present, though the structures are notoriously difficult to see. To the medical and veterinary entomologist, the obligatorily bloodsucking ticks are the most important group and have often been treated separately from the mites, but, the Ixodida (ticks) are only one of the many branches of the acarine tree.

The four suborders containing ticks and mites important in health can be distinguished by the following key.

Identification key to the suborders of Acari

1 Hypostome of capitulum without barbs. Stigmata present or absent, when present not opening on stigmatal plates; if stigmata lateral to coxae 2 and 3 then with peritremes (Fig. 18.11b). Tarsi of first pair of legs without sensory pit 2
— Hypostome of capitulum with backwardly directed barbs (Fig. 18.4). Stigmatal plates present behind coxae of fourth pair of legs or laterally above coxae of legs 2 and 3; stigmata without peritremes. Tarsi of first pair of legs with sensory pit.. IXODIDA (= METASTIGMATA)
2 Idiosoma without conspicuous shields. Legs with coxae fused to body wall. Palps without apotele 3
— Idiosoma with sclerotized areas forming distinct shields (evident by pale brown to mid-brown colour). Legs with free coxae articulated to idiosoma. Palps with apotele (Fig. 18.12). [Stigmata with peritremes] ....
.... MESOSTIGMATA (= GAMASIDA)
3 Palps small, inconspicuous, pressed against sides of hypostome. Chelicerae usually chelate. Legs usually with three claws (i.e. median claw present in addition to outer pair) and with complex pulvillus (varying from pad-like to trumpet-like). Body never worm-like. Stigmata absent..
.... ACARIDIDA (= ASTIGMATA)
— Palps usually well developed. Chelicerae usually adapted for piercing, sometimes chelate (pincers-like). Legs with one or two claws, without complex pulvillus. Body sometimes worm-like (as in *Demodex*, Fig. 18.14b). Stigmata present or absent, when present positioned between bases of chelicerae or on upper surface of propodosoma .... ACTINEDIDA (= PROSTIGMATA)

REVIEW OF THE SUBORDER IXODIDA (TICKS)

The suborder Ixodida, containing about 800 species, is a small group compared to the other groups of acarines but is extremely important because of the economic impact of these ectoparasites on livestock and man, particularly when they act as vectors of various pathogenic organisms. The suborder has three families, the

Ixodidae (hard ticks), Argasidae (soft ticks) and the Nuttalliellidae. The last family contains only *Nuttalliella namaqua*, a species with unknown hosts (Hoogstraal and Aeschlimann, 1982) which occurs in swallows' nests among rocks in southern Africa and Tanzania. Only the female of *N. namaqua* is known and its position in the Ixodida cannot be clarified until the male and immatures are discovered.

Ticks probably evolved as parasites of reptiles in the warm and humid climate of the Palaeozoic and Mesozoic eras, feeding on the plentiful supply of large and glabrous reptiles living gregariously. Two main 'lines' or families evolved, the argasid line (family Argasidae) and the ixodid line (family Ixodidae). With the appearance of birds and mammals during the Tertiary sub-era, there was a change from reptilian hosts, although many primitive ticks have retained a predilection for reptiles. The genus *Hyalomma*, and subsequently *Dermacentor*, *Rhipicephalus*, *Boophilus* and other related ticks that parasitize mammals also appeared during Tertiary times. Tick and mammal coevolution is discussed by Hoogstraal and Kim (1985).

One consequence of coevolution of ticks with mammalian hosts has been a reduction in tick size, but some large argasids are parasites of large hosts such as warthogs and hyenas. Smaller argasids are represented by parasites of birds and small mammals. Ixodid adaptation to mammals has consisted of a reduction in body size and a shortening of the palps. Primitive ixodids, such as *Ixodes*, *Aponomma* and *Amblyomma*, have retained their large size and/or elongate palps.

The Argasidae include three subfamilies of medical importance, all worldwide in distribution: the Argasinae, with the single genus *Argas* (56 species), the Ornithodorinae containing over 100 species of *Ornithodoros*, and the Otobinae with two species of *Otobius*. In the Ixodidae, the primitive subfamily Ixodinae is the largest and is widely distributed; it contains only *Ixodes* (217 species). The Amblyomminae (126 species), comprising the genera *Aponomma* and *Amblyomma*, are tropical and subtropical; they have retained the primitive structural features and most are parasites of reptiles. The Haemaphysalinae includes the single genus *Haemaphysalis* with 155 species, a few of which have retained primitive characters such as elongate palps. They are found throughout much of the world, but only five species occur in the New World. The Hyalomminae, with the single genus *Hyalomma*, has 30 species, all confined to the Palaearctic, Afrotropical and Oriental regions. Although still retaining the elongate palps of early ixodids, very few *Hyalomma* are parasitic on reptiles and many have adapted to the arid and semi-arid environments of savannas and steppes. The last subfamily of the Ixodidae, the Rhipicephalinae, has eight genera and 114 species and is the most recently evolved. Most of its members are tropical and confined to the Afrotropical and Oriental regions; however, several *Dermacentor* species are Holarctic and a few occur in the Neotropical region. Rhipicephaline ticks feed exclusively on mammals and many infest large wandering mammals. Two genera, *Boophilus* and *Margaropus*, utilize a single mammal species for all their parasitic stages.

Recognition and elements of structure

Ticks are by far the biggest acarines and their size serves to distinguish them from the mites. Many mites are microscopic; some of the larger ones reach the size of small unfed ixodid nymphs, but the toothed hypostome, the sensory organs (Haller's organs) on the tarsi of the first pair of legs, the absence of claws on the palps are characters unique to ticks. The one-host cattle ticks of the genus *Boophilus* are among the smallest ticks, the unfed adults often reaching 2 or 3 mm long. The other extreme is represented by the genus *Amblyomma* containing the largest ticks; the unfed females grow to as much as 8 mm long. There is also an increase in size with the blood-meal, more noticeable with the ixodid ticks which imbibe considerable quantities of blood; females of some of the larger *Amblyomma* can increase from just under 1 cm to over 2.5 cm in length, and from about 0.04 g to over 4.0 g in weight after a blood-meal. Male ticks are usually smaller than females, and male ixodids (which imbibe very little blood) show little increase in size when they feed. The morphology, physiology and behaviour of ticks is given in Obenchain and Galun (1982) and Sauer and Hair (1986).

The morphological characters important for identification of argasid and ixodid ticks are illustrated in Figs 18.2 and 18.3.

The surface cuticle of ticks is important in their identification and its features consequently have a specialized terminology. Mammillae are minute, regular, usually hemispherical elevations on the integument of argasids such as *Ornithodoros* (Fig. 18.2) and are quite distinct from granulations on the integument of *Otobius*, which are much less regular. Punctations are circular depressions which dot the integument and frequently bear hairs. Their development varies, and the scutum is described as coarsely or finely punctate according to whether they are marked or slight. The scutum of both sexes of ixodid ticks of the genera *Dermacentor* (Fig. 18.6a), *Amblyomma* (Fig. 18.6b), and some species of *Rhipicephalus,* have 'enamelled' pigmented areas which form symmetrical patterns and can be white, grey or orange. Such ticks are referred to as ornate.

There are several 'grooves', linear depressions or furrows, paired or unpaired, found mainly on the ventral side of both argasids and ixodids. Such grooves are referred to by their position, i.e. anal groove, marginal groove, posteromedian groove. Their depth is greatly influenced by the degree of engorgement and they become scarcely discernible in fully fed ticks. In contrast, 'folds' are ridges on the integument of argasid ticks in which the prominence depends on the degree of engorgement.

Uniform, rectangular regions on the posterior margin of the body of ixodid ticks are termed festoons (Fig. 18.3d) and are separated by grooves. They are distinct in unfed specimens, but difficult to see in distended females. In argasids, small areas of the integument prominent on the dorsum but sometimes absent ventrally mark the attachment points of dorsoventral muscles; they form discs (Fig. 18.3e) and are usually arranged in a symmetrical fashion.

The most conspicuous part of the head is the capitulum (gnathosoma, false head) (Fig. 18.4). It projects forwards in ixodid ticks and is clearly visible from above (Fig. 18.3a–c), but is ventral and therefore not visible from above in argasids (except in larvae) (Figs 18.2 and 18.3e). In ixodids, the terms longirostrate and brevirostrate are sometimes used to refer to species with long and short capitula. The basal portion of the capitulum is referred to as the basis capituli and is articulated with the body. The shape of the basis capituli is of considerable diagnostic value (Fig. 18.5), especially the protruding posterolateral angles (cornua) of the dorsal side and the corresponding ventral outgrowths (auriculae).

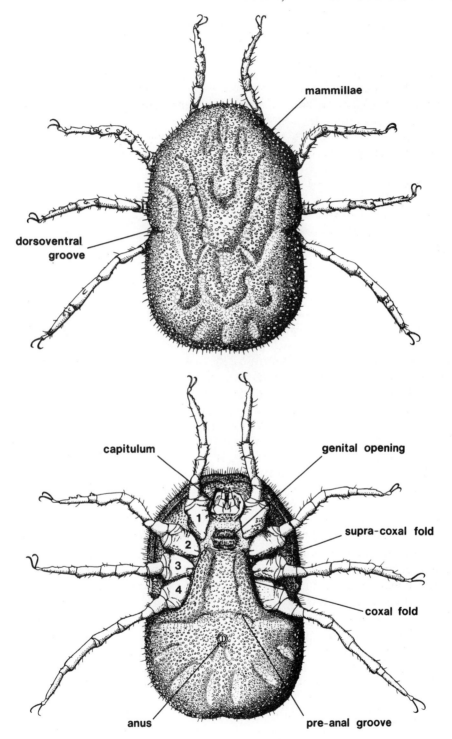

Figure 18.2 Basic morphology of an argasid (soft) tick. Dorsal view (upper figure) and ventral view (lower figure) of the eyeless tampan, *Ornithodoros moubata*, vector of Central African relapsing fever.

604 *Ticks and mites (Acari)*

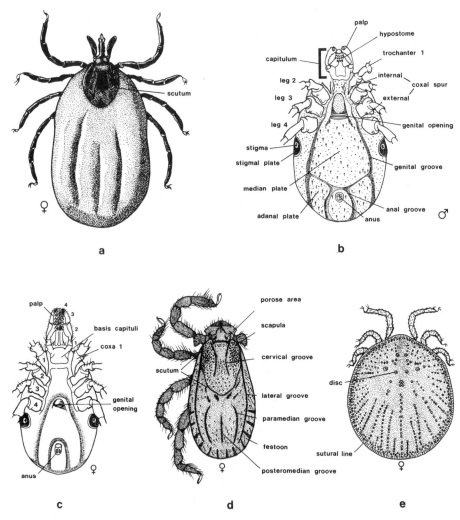

Figure 18.3 External features of ticks used in identification: (*a*) engorged female of *Ixodes ricinus* (sheep tick, family Ixodidae), dorsal view to show the scutum; (*b*) male of *Ixodes ricinus*, ventral view (legs incomplete); (*c*) female of *Ixodes ricinus*, ventral view (legs incomplete); (*d*) female of *Haemaphysalis leachi* (yellow dog tick), dorsal view (legs shown on one side only); (*e*) female of *Argas vespertilionis* (Argasidae), a soft tick.

The hypostome is the median ventral part of the capitulum, immovably attached to the basis capituli and bearing denticles (recurved teeth) on the ventral side for anchoring the mouthparts in the host's skin. The dentition of the hypostome is indicated by figures; thus 3/3 means three longitudinal files of denticles on each side of the midline.

Either side of the capitulum are two dorsal chelicerae, used for piercing the host. The palps (palpi, pedipalps) are paired appendages attached to the anterior part of the basis capituli. When opposed, they protect the upper surface of the hypostome and chelicerae. They consist of four segments, numbered 1 to 4 from the base; only 2 and 3 are prominent and clearly visible in the ixodids. The depression or cavity on the ventral side of argasids in which the capitulum is accommodated is the camerostome.

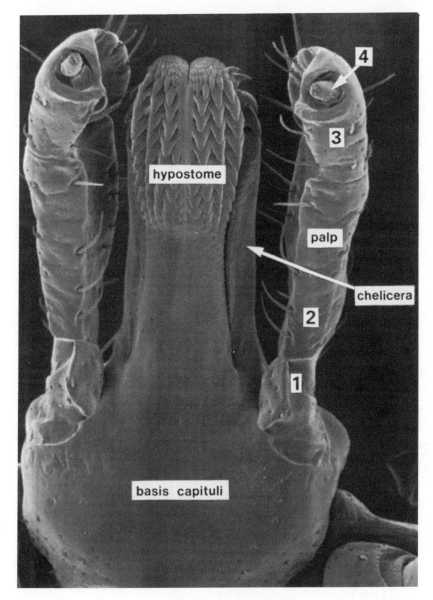

Figure 18.4 Capitulum of a typical ixodid tick (*Amblyomma*) as seen by scanning electron microscope.

Two more or less depressed areas, the porose areas, often lacking distinct boundaries, are present on the dorsum of the basis capituli of female ixodids. The space between them is referred to as the 'interval'. The pores on the surfaces are the external openings of numerous ducts of glands which produce secretions for protecting the eggs.

The scutum is a sclerotized plate on the dorsum of ixodid ticks. The size is stated by giving the length first (including the scapula, the anterior angles of the scutum projecting forward on each side of the basis capituli) and then the greatest width. The pseudoscutum is the portion of the male scutum corresponding in shape and position to the female scutum. It is not a definite structure, but is sometimes

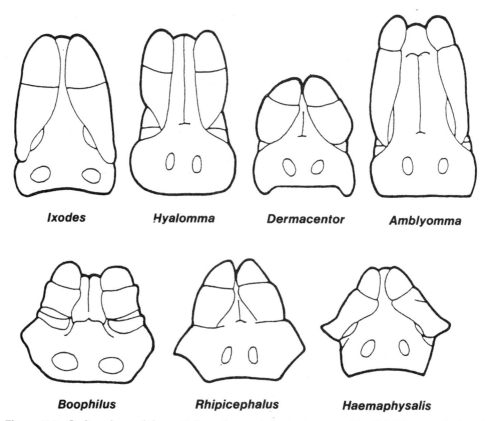

Figure 18.5 Outline shape of the capitulum of seven important genera of ixodid ticks, seen in dorsal view.

demarcated by a ridge. Near the middle of the scutum in the male and posterior to the scutum in female ixodid ticks are paired circular areas, the foveae. These are the external openings of the dermal glands.

Male ixodids have grooves running along the sides of the scutum, termed the lateral grooves. In female ixodids, the corresponding grooves start at the posterolateral borders of the scutum and are called marginal grooves. Cervical grooves run backwards from the scapulae (Fig. 18.3d) in both male and female ixodids and form the anterior edges of cervical pits.

The basal segment of the leg, the coxa, is described as trenchant when it has a sharp posterior margin. Coxal spurs, spur-like or spine-like projections from the posterior surfaces or posterior margins of the coxae, are sometimes present; when there are two such spurs, they are termed internal and external coxal spurs (Fig. 18.3b) and the coxa as a whole is described as bifid.

Large sclerotized 'plates' are present on the ventral surface of *Ixodes* males (Fig. 18.3b), but do not rise above surface level and are thus distinct from the 'shields' on the ventral surface of males of other genera. Shields are substantial chitinous structures associated with the anus in males of *Rhipicephalus*, *Hyalomma* and *Dermacentor*. Depending on their position relative to the anus, they are referred to as adanal shields, accessory shields and subanal shields.

The spiracles (stigmata) are paired structures leading into a tracheal respiratory system. Each is often surrounded by a stigmal (or spiracular) plate (Fig. 18.3b) which is circular, oval or comma-shaped and enclosed in a broad sclerotized frame. The plate sometimes has circles of fine pores.

The sutural line (Fig. 18.3e) is a definite line around the periphery of *Argas* species demarcating the dorsal from the ventral surface. It is clearly visible even in engorged specimens.

Three genital structures are taxonomically important: the genital opening (vulva), situated anteriorly on the ventral side (immediately behind the capitulum); the genital apron, a lightly sclerotized flap originating in front of, and covering, the genital opening; and the genital grooves, which start at the genital orifice and extend backwards to the anal groove.

The anus is the external opening of the alimentary tract, located posteroventrally, and consists of two reversible flaps enclosed in a ring or frame. The anal groove surrounds the anus, either anteriorly (as in *Ixodes* ([Prostriata]) or posteriorly (as in other ixodid genera [Metastriata]).

Larval ticks are hexapod and lack spiracles. Nymphs and adults are octopod and have spiracles. A small dorsal shield or scutum restricted to the anterior part of the idiosoma is present in larval, nymphal and adult female ixodids, but is obscured and difficult to see in a fully engorged female because of the curvature of the body. In the male ixodid, the scutum covers the entire dorsal side and thus there is sexual dimorphism in ixodids. Argasids lack a scutum in all stages. Sex differentiation is difficult, except in *Ornithodoros*, in which the sexes can be distinguished by the shape of the genital opening (curved in males and straight in females). The immature stages of ticks are very similar morphologically to the adults and a late nymphal stage of an argasid can be difficult to distinguish from a small adult; examination of the ventral side and absence of the genital opening will confirm the immature status of a tick. In ixodids, the porose areas on the dorsal side of the capitulum are present in females but absent in immatures and males.

Identification

Although morphological characters are more than adequate for separating the families, genera and most species of ticks, the existence of species complexes, size variations and intermediate forms between closely related species can sometimes make specific identification difficult. Scanning electron microscopy (SEM) of larvae, nymphs and of the female genital area has helped in morphological studies of closely related species (Pegram and Walker, 1988). Other methods used in tick taxonomy when morphological identification is difficult are cytotaxonomy (Oliver, 1972) and isoenzymes (Hunt and Hilburn, 1985) and cuticular hydrocarbon analysis (Hunt, 1986).

For many parts of the world, identification of immature stages of ticks at species level is difficult or impossible. Since all stages are parasitic and all can acquire and transmit pathogens, specific identification of all stages and the association of immatures with adults are important. In practice, this is best done by rearing progeny from single identified females. Alternatively, immature ticks collected from natural habitats or from animals can be reared individually to adults, which can then be identified with certainty to species.

Identification key to families and genera of ticks

The following key omits the Nuttalliellidae, a family with only one known species, *Nuttalliella namaqua* of the Afrotropical region (Krantz, 1978).

1 Capitulum (mouthparts) ventral and invisible when specimen seen from above (except in larva) (Fig. 18.2). Scutum absent (Figs 18.2 and 18.3e), dorsum of body with leathery integument. Stigmatal plates small, situated anteriorly to coxae of fourth pair of legs. Eyes, if present, in lateral folds. ['Soft ticks'] .... ARGASIDAE 2
— Capitulum (mouthparts) projecting anteriorly and visible when specimen seen from above (Figs 18.3a and 18.6). Scutum present, covering dorsum completely (male) (Fig. 18.6b) or anteriorly (female) (Fig. 18.3a). Stigmatal plates large, situated posteriorly to coxae of fourth pair of legs (Fig. 18.3b). Eyes, if present, exposed near lateral margins of scutum. ['Hard ticks'].......................... IXODIDAE 4
2 Body periphery undifferentiated, without a definite suture distinguishing dorsal from ventral surface (Fig. 18.2).. 3
— Body periphery flattened and usually structurally different from dorsum, with a definite suture distinguishing dorsal from ventral surface (Fig. 18.3e) .... **Argas**
3 Adult: integument granular; hypostome vestigial. Nymph: integument beset with spines; hypostome well developed .... **Otobius**
— Adult and nymph: integument leathery and mammillated (Fig. 18.2a) or tuberculate, without spines; hypostome well developed .... **Ornithodoros**
4 Anal groove extending anteriorly around anus, if only weakly (Fig. 18.3b) 5
— Anal groove lying entirely posterior to anus, either well defined or indistinct. [Palps usually inornate] 6
5 Anal groove less distinct anteriorly than posteriorly around anus. [Palps ornate] .... **Cosmiomma**
— Anal groove equally distinct anteriorly and posteriorly around anus (Fig. 18.3b). .... **Ixodes**
6 Eyes absent... 7
— Eyes present 8
7 Palps long and narrow, at least three times as long as their width, segment 2 without outer angulation at base.. **Aponomma**
— Palps short and broad, about twice as long as wide, segment 2 with obvious outer angulation at base (Fig. 18.5)... **Haemaphysalis**
8 Palps wider than long or at most only slightly longer than their width (Fig. 18.5) 9
— Palps much longer than wide (Fig. 18.5)... 13
9 Basis capituli usually hexagonal dorsally. [Medium-sized or small ticks, usually without colour pattern] 10
— Basis capituli rectangular dorsally (Fig. 18.5). [Large ticks with definite colour pattern (Figs 18.6a and 18.7b)]... **Dermacentor**
10 Festoons absent. Stigmatal plates (Fig. 18.3b) round or oval. Anal groove faint or obsolete. [Males very small] 11
— Festoons present (Fig. 18.3d). Stigmatal plates with tail-like protrusion. Anal groove distinct. [Males of medium size]... 12
11 Palps without ridges. Male with massive leg segments..................... **Margaropus**
— Palps with dorsal and lateral ridges. Male with normal legs .... **Boophilus**

12 Basis capituli with very strong and pointed lateral angles. Male without ventral plates. Male with coxae of fourth legs much enlarged .... **Rhipicentor**
— Basis capituli without such pronounced lateral angles (Fig. 18.5). Male with ventral plates. Male with coxae of fourth legs normal .... **Rhipicephalus**
13 Palps with second segment less than twice as long as third segment (Fig. 18.5). Scutum without pattern. [Male with adanal and subanal plates]........................ .... **Hyalomma**
— Palps with second segment more than twice as long as third segment (Fig. 18.5). Scutum with pattern (most species) (Fig. 18.6b). [Male without ventral plates] ....**Amblyomma**

Faunal and taxonomic literature (ticks)

General: Bibliography (Hoogstraal, 1970–1982, list of world literature to beginning of 1980s); catalogue (Doss et al., 1974, world genera and species); distribution (Doss et al., 1978, world species). Entrée to the world taxonomic literature by group and area is provided by Sims (1980). Estrada-Peña (1989) has begun an intended catalogue series under the title Index-Catalog of the Ticks (Acarina: Ixodoidea) with a volume on *Haemaphysalis*.

Palaearctic: Arabia (Hoogstraal and Kaiser, 1959; Hoogstraal et al., 1981); Balkans (Oswald, 1939); Britain (Arthur, 1963); China (Cheng and Pang, 1982); Egypt (Hoogstraal and Kaiser, 1958); Italy (Starkoff, 1958); Libya (Hoogstraal and Kaiser, 1960); USSR (Anastos, 1957; Pomerantzev, 1950).

Afrotropical: Ivory Coast (Aeschlimann, 1967); Kenya (Walker, 1974); Madagascar (Hoogstraal, 1953; Uilenberg et al., 1980); Mozambique (Theiler, 1943); Somalia (Pegram, 1976); South Africa (Bedford, 1934); Sudan (Hoogstraal, 1956); Tanzania (Yeoman and Walker, 1967); Uganda (Smith, 1969a,b; Matthysse and Colbo, 1987); Yemen (Hoogstraal and Kaiser, 1959); Zaire (Nuttall, 1916); Zambia (MacLeod, 1970); *Ixodes* (Arthur, 1965). *Rhipicephalus* (Pegram and Walker, 1988).

Oriental: China (Cheng and Pang, 1982); India (Sharif, 1928; Trapido et al., 1964, *Haemaphysalis*); Indonesia (Anastos, 1950; Kohls, 1957b, Borneo); Malaysia (Audy et al., 1960; Kohls, 1957b); Philippines (Kohls, 1950); Vietnam (Toumanoff, 1944).

Australasian: Australia (Roberts, 1970); Micronesia (Kohls, 1957a); New Zealand (Dumbleton, 1953).

Nearctic: North America (Cooley and Kohls, 1944, Argasidae; Cooley and Kohls, 1945, *Ixodes*).

Neotropical: Belize (Varma, 1973); Central America (Cooley and Kohls, 1944, Argasidae); Cuba (Cooley and Kohls, 1944); Panama (Fairchild et al., 1966).

Medically important genera and species

Argas
This genus is found in arid regions with long dry seasons. Most species are nocturnal and associated with birds. Larval *Argas,* unlike nymphs and adults, remain attached to their hosts and feed for several days. *Argas persicus* is a parasite of poultry and wild birds; it originated in the Palaearctic region but has been introduced with chickens into most parts of the world, except possibly South

America. The tick breeds and shelters in cracks and crevices in poultry houses and has considerable economic importance as a vector of *Borrelia anserina* and *Aegyptianella pullorum* among poultry; heavy tick infestations can kill the birds through exsanguination. It is known to attack man, causing painful bites.

Argas arboreus, a species at one time confused with *A. persicus*, is widespread in the Afrotropical region and infests herons and other medium-sized wading birds. The ticks hide under the bark of trees in heron rookeries and climb up to the birds' nests to feed. The species is an arbovirus vector, transmitting Quaranfil virus in Egypt, Nigeria and South Africa.

Argas reflexus is known as the pigeon tick because of its close association with pigeons, although other avian and mammalian hosts (including man) are frequently attacked especially in the vicinity of unoccupied pigeon cotes. It is abundant in the Middle or Near East, from where it has spread into Europe, southwestern Russia and most of Asia; there are a few records from the New World. In Egypt, Quaranfil virus has been isolated from the subspecies *A. reflexus hermanni*.

Otobius

This genus contains two North American species, *O. megnini* parasitic on cattle and *O. lagophilus* parasitic on rabbits. The integument is spiny in the nymphs and granulated in the adults. There is no lateral sutural line and no clear lateral margin demarcating the dorsal and ventral surfaces of the body (a distinction from *Argas*). *Otobius megnini*, the spinose ear tick, originated in the New World, from where it has been introduced in Africa and India. It is restricted to arid, hot areas and is a serious pest of cattle and horses, but other livestock and man are also parasitized. The larvae enter the ear cavities of animals, where they feed and moult into spinose nymphs. The nymphs attach and feed for several months before dropping off the host and creeping into cracks and crevices in stables and animal shelters where they moult into the fiddle-shaped non-feeding adults. There are several records of nymphal infestation in the ears of man.

Ornithodoros

Like *Argas*, most *Ornithodoros* ticks inhabit restricted habitats such as dens, caves, burrows and nests and are therefore not usually available to domestic livestock; a few important species, however, are parasitic on man and livestock and are found in stables, or places where the host animals rest. The genus can be distinguished from the other two genera of argasids by the mammillated integument and the absence of a distinct lateral margin to the body. These ticks are reservoirs and vectors of human relapsing fever in several parts of the world. They are also associated with the *Iridovirus* causing African Swine fever. Several related arboviruses, some of them causing human disease, have been isolated from *Ornithodoros* parasitizing seabirds. *Ornithodoros savignyi* is the eyed tampan or sand tampan. It is one of the few argasids with eyes and is the most widely distributed species of *Ornithodoros*, ranging from Sri Lanka and India westwards to Arabia, North, East and southern Africa to Namibia. It is an outdoor tick of arid and semi-arid environments, found near the soil surface where cattle, camels and other livestock rest in the shade of trees and stone fences. It will attack man readily and can transmit relapsing fever spirochaetes experimentally, but is almost certainly not involved in transmission of the disease in nature. *Ornithodoros moubata*, the 'eyeless tampan'

(Fig. 18.2), is very similar in appearance to *O. savignyi,* and in those parts of Africa where the distribution of the two species overlaps, *O. moubata* can be distinguished by the absence of laterally placed eyes. It is widely distributed in East, Central and southern Africa, where it is the vector of relapsing fever (*Borrelia duttoni*). A nocturnal feeder, the species is found in huts, including the occupants' possessions, and outdoors in warthog burrows. The tick is common on travel routes, particularly in rest-houses, and can be carried long distances in mats and bed-rolls. What was considered one species is now believed to be a species complex, of which *O. moubata* s.str. with two subspecies is the most important; *O. moubata moubata* feeds predominantly on man and chickens, *O. moubata porcinus* on warthogs, antbears and porcupines. Both subspecies have domestic and wild populations and it is the domestic populations which are important in disease transmission. In Africa, the tick transmits African Swine fever among pigs. *Ornithodoros erraticus* is confined to northwest Africa, Spain and Portugal, and feeds on rodents. However, large populations can be found in pigsties, where the ticks feed on pigs and transmit African Swine fever virus. The species is also an important vector of relapsing fever (*Borrelia hispanica*). *Ornithodoros tholozani* (= *O. papillipes*) is found in the arid zone from Libya eastwards to the western provinces of China, where it transmits relapsing fever (*Borrelia persica*). It feeds mainly on domestic animals but also bites humans and birds. On islands in the Arabian Gulf it is associated with marine birds but will eagerly attack humans; irritant bullae develop at the sites of the bite with associated intense pruritis, headache and fever (possibly caused by Zirqua virus).

The distribution of *O. talaje* extends from South and Central America northwards to the southern states of the USA. The usual hosts are wild rodents and livestock, but the species feeds readily on man. It inflicts very painful bites and is a vector of relapsing fever in Guatemala, Panama and Colombia. Another Central and South American species, *Ornithodoros rudis* (syn. *O. venezuelensis*), lives in the walls of huts, feeds avidly on humans and is the most important vector of relapsing fever (caused by *Borrelia venezuelensis*) in Colombia, Ecuador, Panama and Venezuela. Three species of *Ornithodoros,* viz. *O. hermsi* of the Rocky Mountain and Pacific coast states of the USA, *O. parkeri* of western USA and *O. turicata* of western USA and northern Mexico, primarily attack rodents but will also feed on humans; they are vectors, respectively, of *Borrelia hermsi, B. parkeri* and *B. turicatae* in man.

Ornithodoros coriaceus ('tlaaja', 'pajaroello') is a large species that ranges from California to Mexico and is notorious for attacking humans, cattle and deer.

Similarly, *O. rostratus* ('quanco'), a South American tick, is an avid biter of man, domestic animals and peccaries. It does not transmit relapsing fever, but its bites are painful and can lead to pruritis and inflammation.

Ixodes
This genus differs from other ixodids by the position of the anal groove, which surrounds the anus anteriorly. The rest of the ixodids have the anal groove surrounding the anus posteriorly. The mouthparts are long and the males have several shields or plates which almost cover the entire ventral surface. Those *Ixodes* species which are polyphagous and attack a wide variety of hosts have the greatest economic importance as disease vectors, e.g. the *Ixodes ricinus/persulcatus* group.

Ixodes ricinus, the European sheep tick or castor bean tick (Fig. 18.3*a,b*), is found in rough pasture and woodland throughout its range from Ireland and Britain to western and central Europe and European USSR, and from Algeria and Morocco eastwards to the Caspian Sea and northern Iran. The life cycle takes two to four years to complete. The immature stages are found on rodents and birds, but the adults attack larger mammals such as sheep, cattle and man. This tick is the vector of the arbovirus of Louping ill (primarily of sheep and cattle, but affecting man also) in Britain and Ireland, of tick-borne encephalitis virus (TBE) in Europe, and of *Borrelia burgdorferi* which causes Lyme disease (Lane et al., 1991). *Ixodes persulcatus*, the taiga tick, is the major vector of Russian Spring Summer encephalitis virus (RSSE) and of Lyme disease in the USSR. The female of *I. persulcatus* is morphologically very similar to *I. ricinus* but has a straight or wavy, instead of arched, genital opening. *Ixodes persulcatus* is widespread in Russia, from the Baltic Sea to the Russian Far East, and in Japan. The western distribution overlaps that of *I. ricinus*, but the species is more cold-hardy than *I. ricinus;* the life cycle and feeding habits of the two species are generally similar.

Ixodes dammini belongs to the *I. ricinus* group and is widely distributed in northeastern USA and adjacent Canada. The immatures bite rodents, particularly the white-footed mouse (*Peromyscus leucopus*), and man. The white-tailed deer (*Odocoileus virginianus*) is the main host of the adults and the range of the tick appears to be extending, probably because of proliferation of the deer (Lane et al., 1991). It is the vector of Lyme disease in the eastern USA (in western USA the vector is the related *I. pacificus*), the most important vector-borne disease in North America, and of human babesiosis caused by *Babesia microti*. *Ixodes holocyclus* in the coastal areas of Queensland and New South Wales in Australia, and *I. rubicundus*, the Karoo tick in the humid hill and mountain areas of South Africa, cause tick paralysis in animals and man. *Ixodes holocyclus* is also the vector of Queensland tick typhus (*Rickettsia australis*).

Haemaphysalis
These ticks inhabit humid, well-vegetated, biotopes and parasitize wild and domestic animals. They are small acarines with short mouthparts and often have the second palpal segment expanded laterally (salient) beyond the rectangular basis capituli. Species that infest livestock are confined to Eurasia and tropical Africa, where the immatures feed on rodents and other small mammals, and on birds. One species, the winter-active *H. inermis* of northern Iran, southwestern USSR and central and southern Europe, is exceptional among the ixodids in that its larval stage completes its feeding in only a few hours.

Haemaphysalis spinigera is the most important vector of the arbovirus causing Kyasanur Forest disease in monkeys and man in southern India, but the tick itself has a wider distribution in India and Sri Lanka. It flourishes in areas where forests have been cleared and cattle, the major and amplifying hosts of the adults, have been introduced. The immatures feed on small forest rodents, monkeys and man. Man is most at risk from bites of these ticks, and of infection, during the dry premonsoon period when villagers enter the forest to gather firewood. *Haemaphysalis leachi* (Fig. 18.3d) is a widely distributed tick that infests carnivores, particularly dogs. In South Africa, urban cases of boutonneuse fever or South African tick typhus caused by *Rickettsia conori*, are associated with contamination of skin or eyes with infected ticks crushed between the fingers while being removed from dogs.

Rhipicephalus
All species of this genus occur in Eurasia and Africa and are reddish or blackish brown ticks with hexagonal basis capituli and short mouthparts. The fore coxa is bifid in both sexes, and in the males paired plates are found on each side of the anus. The species are three-host ticks, but a few such as the red-legged tick *R. evertsi evertsi* have a two-host life cycle.

Rhipicephalus appendiculatus, the brown ear tick of Africa, is found in wooded and shrubby grassland from southern Sudan to South Africa but is absent from West Africa. It is an important parasite of domestic livestock as well as large antelopes, and has been studied extensively. Climatic databases and satellite-derived vegetation cover data have been used to predict its distribution in Africa (Perry et al., 1990). Although a three-host tick, considerable numbers of immatures and adults infest the same hosts. The ears of animals are the favourite sites of adults, but 'overflow' populations of adults and immatures can be found elsewhere. The species transmits East Coast fever (*Theileria parva*) among cattle and the virus of Nairobi sheep disease among sheep in East Africa. It is an important vector of African tick typhus in the South African veldt and is an avid man-biter.

From its homeland in Africa, *R. sanguineus* (kennel tick or brown dog tick) (Fig. 18.7a) has been transported to much of the world and survives in heated buildings in urban communities, even in cold regions such as Canada and Scandinavia, feeding on domestic dogs kept in the house. In the Mediterranean basin it is the chief vector of boutonneuse fever (*Rickettsia conori*).

Dermacentor
Ticks of this genus are medium-sized to large acarines with rectangular basis capituli, short mouthparts and usually an ornate pattern. The fore coxa is bifid in both sexes and coxa 4 is greatly enlarged in males. The latter sex has no ventral plates. Most species are three-host ticks, but *D. nitens,* the tropical horse tick of America, and *D. albipictus,* the winter tick or moose tick of Canada and parts of the USA, utilize only one host during the life cycle. *Dermacentor andersoni* (Fig. 18.6a), the Rocky Mountain wood tick, is confined to North America, where it extends from western Nebraska to the eastern slopes of the Cascades and from northern Arizona and New Mexico to British Columbia and Manitoba. It is the chief vector of Rocky Mountain spotted fever (*Rickettsia rickettsi*) and Colorado tick fever in these areas. Practically every rodent is parasitized by immature stages, but adults are found on wild and domestic herbivores and eagerly bite man. *Dermacentor variabilis,* the American dog tick, is distributed east of the Rocky Mountains and in California, Mexico and Canada; it is particularly abundant along the eastern coast of the USA. The hosts of immatures are rodents; adults parasitize wild and domestic carnivores, including dogs, and also man. The species is vector of Rocky Mountain spotted fever in the eastern USA, where dogs infested with the tick bring the infection close to homes – with the consequence that (unlike in western USA) women and children suffer from the disease.

Dermacentor marginatus, D. silvarum and *D. nuttalli* are some of the more important Eurasian species and the chief vectors of Siberian tick typhus (*Rickettsia sibirica*). Lowland forest and shrubby areas from Kazakhstan to central Europe are the typical habitats of *D. marginatus; D. silvarum* ranges from the eastern limits of

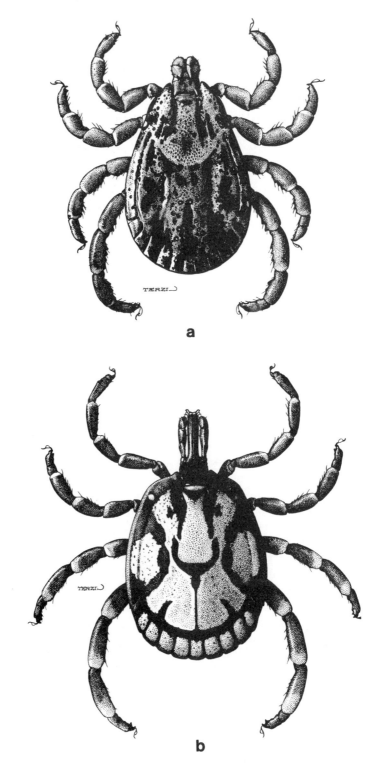

Figure 18.6 Adults in dorsal view of two species of ixodid ticks involved in transmission of human disease: (*a*) Rocky Mountain wood tick, *Dermacentor andersoni*; (*b*) bont tick, *Amblyomma hebraeum*.

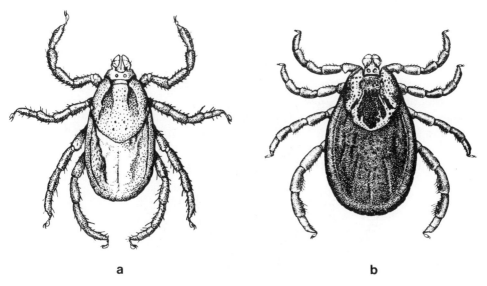

Figure 18.7 Adults in dorsal view of two species of ixodid ticks involved in transmission of human disease: (*a*) brown dog tick, *Rhipicephalus sanguineus*; (*b*) *Dermacentor reticulatus*.

D. marginatus to western Siberia, and *D. nuttalli* from central Siberia to Mongolia and China southwards into central Asia and Tibet.

Dermacentor reticulatus (= *D. pictus*) (Fig. 18.7*b*) has the same distribution as *D. marginatus* and is found in mixed and deciduous forests. It has been introduced into southern England.

Amblyomma
This genus contains large and beautifully ornamented ticks in which the males lack ventral plates. The long mouthparts inflict painful bites on man and deep penetrating wounds on animals which often become secondarily infected. Attached ticks feeding on man are difficult to remove because of the long mouthparts, but gentle traction and a twisting movement make removal easier. Many of the economically important species infest livestock, and *A. variegatum*, the tropical bont tick ('bont' refers to the banding pattern on the legs), is an important vector of heartwater fever (*Cowdria ruminantium*) among cattle in Africa.

Amblyomma hebraeum (Fig. 18.6*b*), the South African bont tick, ranges from South Africa northwards into Zimbabwe and Mozambique. The immatures feed on both small and large mammals and the adults on large, wild and domestic animals. It is the vector of tick typhus (*R. conori*) in the South African veldt where larvae and nymphs swarm towards, and feed avidly on, man.

Amblyomma americanum, the Lone Star tick, is widely distributed from central and eastern USA to South America. All stages attack man and in south-central and southeastern USA the species transmits Rocky Mountain spotted fever.

The Cayenne tick, *A. cajennense*, distributed from South America and the Caribbean to southern Texas, is the vector of Rocky Mountain spotted fever to man in Central America, Colombia and Brazil. Immatures attack man readily, producing intensely itching granulomatous lesions which can take several months to heal.

Hyalomma

The medium-sized ticks of this genus have a long capitulum and subtriangular/rectangular basis capituli. The males have ventral plates on each side of the anus. Most species are adapted to arid or semi-arid regions. Populations within a species can show considerable variation in size and morphological characters which make accurate identification difficult.

Hyalomma marginatum marginatum is widely distributed in the USSR and southern Europe and is the vector of the virus that causes Crimean–Congo haemorrhagic fever in these regions. Birds are important hosts of the immature stages, which are carried by migratory birds between various parts of Europe and Africa.

BIOLOGY OF TICKS

General accounts of tick biology can be found in Wilde (1978) and Sonenshine (1991).

Life cycles

All ticks are obligate temporary ectoparasites of vertebrates and require a bloodmeal for their development. The life cycles are complex and involve regular alternation of blood-feeding (parasitic) and free-living stages with a change of hosts. This gorging–fasting pattern makes ticks 'two different organisms' (Needham and Teel, 1991). The on-host existence, which is less than 10% of a tick's life span, is adapted for blood-feeding and the off-host part is concerned with life-extending strategies which conserve water and energy resources and increase the chances of a blood-feed. One or more hosts and one or more years can be necessary to complete the life cycle. Typically, the tick life cycle consists of one inactive (egg) and three mobile blood-feeding stages (larva, nymph, adult). There are four types of life cycle which differ according to the number of host changes and moults.

The multi-host developmental cycle is characteristic of argasids which inhabit restricted habitats and which feed on the same animal several times or on several animals (of the same or different species) during their life cycle. The single larval stage feeds once (except some larval *Ornithodoros* which do not feed) before moulting into first-stage nymphs. There are two to seven nymphal instars each of which feeds once before moulting away from the host to the next stage. Adults mate away from the host and feed several times but there are no further moults. The female lays a small batch of eggs (400–500) after each feed. The restricted and sheltered habitats of these ticks mean that host-finding hazards are reduced. Argasid habitats are sometimes closely associated with man and domestic animals, e.g. chicken coops and pigeon lofts, pigsties, or with native hosts, but can be remote from human habitations in birds' nests, in sea-bird colonies on the ground, in loose soil, under semi-desert trees or tree bark, in animal burrows and caves.

Ixodid life cycles (Fig. 18.8) can be one-host, two-host or three-host depending on the number of animals the ticks parasitize during their development. Most ixodids are field-inhabiting (i.e. non-domestic) ticks and their macrobiotopes are forests and pastures, where there is a plentiful supply of host animals, or microbiotopes in ground cover vegetation where there are suitable microclimatic conditions for

Biology of ticks 617

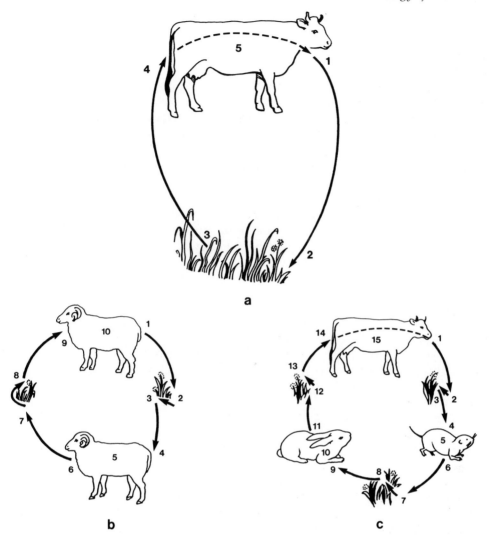

Figure 18.8 Life cycles of ticks: (*a*) one-host cycle, in which fed female detaches (**1**), oviposits and dies (**2**), active larva moves to vegetation tip (**3**) and climbs on host (**4**), tick then develops from larva to nymph and adult with all three stages feeding on the same host (**5**); (*b*) two-host cycle, sequence as in one-host cycle up to the nymphal stage (**5**), then fed nymph detaches from host (**6**), develops to adult in vegetation (**7**), adult moves to vegetation tip (**8**), attaches to second host (**9**) and feeds (**10**); (*c*) three-host cycle, sequence as in one-host cycle up to larva feeding on host (**5**), then larva detaches (**6**) and moults to nymph in vegetation (**7**), active nymph moves to vegetation tip (**8**) and climbs on second host (**9**), feeds on second host (**10**), detaches (**11**), moults into adult in vegetation (**12**), adult moves to vegetation tip (**13**) and climbs onto third host (**14**) and feeds (**15**).

survival and development during the off-host phase. A few ixodids are creatures of restricted habitats and are found in the nests and burrows of their hosts.

Host-finding by field-inhabiting ixodids is a risky business. Unlike insects, ticks move only very short distances and (usually) do not actively seek a host. 'Questing' ticks can detect the proximity of a host, but physical contact is necessary before the ticks will transfer to the hosts. Many questing ticks perish before they can even find a host. Furthermore they are subject to predation and extreme climatic conditions while on the ground. Ixodid ticks have compensated for these risks by reducing the number of feeds and/or the number of hosts and by producing a huge number of offspring.

Engorged female ixodids, after dropping off host animals onto ground vegetation, start ovipositing in about a week. Oviposition can be delayed by several weeks until environmental conditions are suitable for survival of eggs, which are extremely susceptible to desiccation. The sticky mass of several thousand eggs is laid over a period of several days, after which the spent female dies. The larvae hatch from the eggs in 10–20 days, but the eggs of some species diapause and larvae take considerably longer, two months in some cases, to hatch. This is a biological adaptation to ensure that hatching occurs only when climatic conditions are favourable for larval survival and host-finding. The larvae remain inactive in the ground vegetation for several days while the cuticle hardens and then ascend vegetation to gather in clusters questing for hosts, usually small mammals and ground-inhabiting birds. On making contact with a host, the larvae quickly disperse over the body and attach to 'groom-free' sites. Blood-feeding takes about four to six days to complete before engorged larvae drop to the ground and moult into nymphs. The nymphal premoult period usually lasts for two to three weeks. Once moulting starts, it is usually complete in a few days. Like the freshly hatched larvae, freshly hatched nymphs spend several days on the ground, hardening the cuticle and digesting and excreting food reserves from the previous blood-meal. Questing nymphs board an animal host as soon as they make contact, irrespective of whether or not it belongs to the same species as the larval host, and then seek suitable places at which to attach and feed. Feeding lasts for about a week. Nymphs are also found on smaller mammals and birds. However, the immatures of a few ixodid species are better adapted for feeding either on small or large vertebrates; many species of medical and veterinary importance are of this type. Drop-off and premoult development of fed nymphs follows the larval pattern, but the periods are slightly longer. After a period of inactivity freshly moulted adults quest and find a host, usually a large mammal. Most adult ixodids quest at the tips of vegetation and are easy to collect but some, such as species of *Amblyomma*, do not have this habit and cannot be collected by conventional dragging or flagging. Except in *Ixodes*, mating of ixodid ticks occurs on the host. (*Haemaphysalis longicornis* and a few other species though are parthenogenetic.) Females engorge fully only after mating. Males remain attached for several weeks or months, sometimes mating with several females; the turnover of females is therefore a more accurate indicator of the seasonal dynamics of tick populations than simple tick counts. The fed females drop to the ground and oviposit before dying. This is a basic three-host ixodid life cycle and involves three different hosts of the same or different species (Fig. 18.8c), or the same individual three times if it remains in the vicinity of the questing larvae, nymphs and adults. About 600 of the 650 or so ixodid species have a three-host life cycle.

The evolution of the two-host and one-host life cycles is basically associated with adaptation to adverse environmental conditions or to the habits of host animals. Some *Hyalomma* and *Rhipicephalus* species found in steppes or savannas where there are long dry seasons and little rainfall have a two-host life cycle (Fig. 18.8b). The fed larvae do not drop to the ground, but remain attached to the host. After moulting, the nymphs escape from the larval exuviae and move a short distance before attaching to the same animal. The engorged nymphs drop to the ground and moult into adults which then seek a second, and usually larger, host on which to feed. The fed females drop off and oviposit before dying.

In the one-host life cycle (Fig. 18.8a), shown by *Boophilus, Margaropus* and the tropical horse tick *Anocentor nitens*, all stages parasitize large wandering foraging animals such as cattle, deer and horses. Host-finding in such situations is a chancy business and the ticks have overcome this by completing their development on the same individual that they boarded as larvae. Larval feeding and moulting, as well as nymphal feeding and moulting, occur on the host. The mated and engorged female drops to the ground and lays eggs.

In humid tropical climates the life cycle is uninterrupted but shows a distinct seasonality of stages. The tropical one-host cattle ticks with their short developmental cycles undergo two or three generations per year, but most tropical three-host ixodids have a univoltine cycle with one generation per year.

Nest-burrow argasids live in relatively stable microhabitats and can feed and reproduce continuously throughout the year. But in this group, as in nest-burrow ixodids in cooler climates, development can be so seasonally adapted that a generation sometimes takes a year or more. Two-year and three-year developmental cycles are characteristics of ticks of colder climates (e.g. *Ixodes persulcatus* and *I. ricinus*) and infestation, moulting, oviposition and egg development are limited to the warm season.

An important epidemiological factor in the life history of ticks is their exceptional longevity and remarkable power to survive starvation. This is particularly true of *Ornithodoros* in which some species can survive for up to 11 years without a bloodmeal. Many ixodids can also live for long periods, up to year or more.

Tick pheromones

Considerable progress has been made during the past 20 years in analysing tick pheromones and studying their effects on tick behaviour (Sonenshine, 1986: review). Ticks produce three types of pheromones: assembly pheromones, non-specific and of low volatility which induce assembly or cluster formation; aggregation–attachment pheromones, species-specific volatile compounds which regulate cluster formation of the same species on feeding sites; and sex pheromones, non-specific and species-specific pheromones which attract males to sexually active females for copulation.

Guanine excreted by ticks has been identified as the assembly pheromone of many argasids and ixodids. It is deposited as the ticks move around in their environment and arrests the movement of other individuals, so leading to assembly or clustering.

Aggregation–attachment pheromones are volatiles produced only by feeding males of a few species of *Amblyomma*, e.g. *A. variegatum* and *A. hebraeum*. They attract males and females of the same species. This chemical attracting improves mating success, since females will attach and feed only when they sense the pheromone and conspecific clusters are formed at the start of feeding. The aggregation–attachment pheromone is probably a mixture of compounds in which *o*-nitrophenol is responsible for searching and aggregating behaviour and methyl salicylate and pelargonic acid elicit grasping and attaching responses. The source of the compounds is not known, but the most likely sites are the dorsal foveal glands.

Sex pheromones regulate mating behaviour in a hierarchy of events. Species integrity is maintained and wasteful interspecific matings are avoided because conspecific mates recognize and respond only to signals from other partners, rejecting all other signals. This is particularly important when individuals of more than one species are attached at the same sites on the same animal or when species have sympatric distributions. Sex pheromones are produced only by feeding females. In soft ticks, a non-specific pheromone of unidentified biochemistry appears to be present in the coxal fluid of females; males will attempt to copulate with females of other species coated with coxal fluid of their conspecific mates and in these circumstances species integrity is maintained by ecological isolating mechanisms. Female sex pheromones in ixodid ticks apparently control the sequential events of the elaborate courtship ritual which culminates in conspecific copulation and deposition of the spermatophore in the female genital opening. The volatile sex pheromone, 2, 6-dichlorophenol, produced by the foveal glands of the female, controls the earlier events, i.e. excitation of sexually active feeding males, detachment and orientation of the males to the pheromone-producing females, and mounting of the females. In *Dermacentor* species, the final events of courtship, i.e. gonopore location, recognition of the conspecific female and spermatophore transfer are controlled by a second sex ('aphrodisiac') pheromone, the genital sex pheromone in the female vulva which the male detects by probing the vulva with his cheliceral digits. The pheromone is probably produced by the greatly expanded lobular accessory glands of the feeding females.

Mating behaviour

Mating of argasids and the Ixodinae takes place away from the host, but on the host in the rest of the Ixodidae. The sexually mature male, stimulated by the sex pheromone produced by the sexually mature female, climbs on her back and moves around the posterior end of her body until the under surfaces of the sexes are in a facing position and his capitulum is near her genital aperture. They entwine their legs; the female takes no active part in mating and remains motionless till copulation is complete. The male oscillates his mouthparts in the female genital opening several times to distend it prior to sperm transfer. Males and females can remain *in copula* for some hours before sperm transfer, and in *Ixodes ricinus* the copulatory act can last up to a week. The male ejects a mass of sperm into the centre of a balloon-shaped spermatophore. Since ticks have no external genitalia, the male grasps the spermatophore with his chelicerae and implants it in the female gonopore. The mass of sperm is then pushed into the female genital tract.

In argasids, blood-meal and subsequent production of coxal fluid by the female seems to be necessary for attracting the males and for successful copulation. Argasid males copulate several times, sometimes daily over several days or even twice during the same day. Successful inseminations increase with the number of copulations and blood-meals. In female argasids, which feed repeatedly, each act of copulation may precede a blood-meal. Copulation also takes place after feeding and even during oviposition. In the genus *Ixodes*, unfed males can fertilize females, but in other ixodids it is necessary for males to attach and feed for several days for mating to be successful. Ixodid males can also mate repeatedly with females of different 'batches' which board a host and attach. Multiple matings of female ixodids are rare, since the female feeds only once, copulates, oviposits and then dies.

Host specificity

In the context of ticks, host specificity can be defined as an association between a tick species and a vertebrate species or as an association which is essential for the continual survival of populations of that particular tick species (Hoogstraal and Aeschlimann, 1982). Strict host specificity is found in most tick species, but can be obscured when physiologically acceptable feral or domestic animals intrude into areas of the primary host–parasite association. Only about 10% of tick species feed on domestic animals, other populations of the same species being found on wild and domestic animals sharing the same biotopes. For example, the brown ear tick *Rhipicephalus appendiculatus* occurs on wild buffalo and cattle in East Africa. There are several degrees of specificity, ranging from strict specificity, where all stages parasitize only one species or group of vertebrates, to moderate specificity or no specificity at all. Six argasid species are strictly host specific to reptiles and parasitize tortoises and lizards. Among the ixodids, almost all *Aponomma* species and about a third of *Amblyomma* species specifically parasitize reptiles. Birds are the primary hosts of most species of *Argas*, about 10% of *Ornithodoros* species, several *Ixodes* species, and of immatures of some ixodids. Terrestrial mammals are parasitized by many ticks and most genera contain one or more strictly host-specific species, for example the ticks that parasitize Australian monotremes and marsupials and Madagascan insectivores. More than 50 species in the genera *Argas*, *Ornithodoros* and *Ixodes* are totally host specific on bats. Rodents are primary hosts to many *Ornithodoros* species and to the immatures of about half the ixodid species. Adults of certain *Haemaphysalis* and *Rhipicephalus* are primarily parasites of carnivores. The Artiodactyla are parasitized by at least 190 ixodid species (not *Aponomma* and *Rhipicentor*) and six argasid species (one of *Otobius* and five of *Ornithodoros*). Only a very few ixodids are strictly primate host specific, e.g. *Ixodes schillingsi* on African monkeys and *Haemaphysalis lemuris* on Malagasy lemurs, although many ticks, particularly the immatures, will feed on monkeys and man and are important in disease transmission. Finally there are some ixodids (*Amblyomma americanum*, *A. cajennense*, *Ixodes persulcatus*, *I. ricinus* and *I. holocyclus*) which will feed on any available hosts including man ('non-particular specificity'). Species are important in the transmission of zoonotic infections to man.

Host-seeking and feeding

Feeding is an important phase in the life cycle of ticks during which pathogens are acquired and transmitted. Interference with feeding is therefore a potential strategy for controlling ticks and tick-borne infections. Feeding is usually taken to mean ingestion, but it is really a sequence of smoothly integrated behavioural patterns which start with hunger and end with satiation. Ultrastructural and electrophysiological studies have identified many of the receptors involved in feeding. They are located on the tarsi of the first pair of legs (Haller's organ), palps, chelicerae and scutum. In an excellent paper, Waladde and Rice (1982) defined the sequence of events concerned in feeding and the following is a summary of their observations. The events start with appetence and host engagement during which the 'hunter' ticks actively hunt or seek a host and the 'ambusher' ticks gather at vantage points ready to engage a host when it comes within reach. Several species are hunters and extreme examples are arid-region ticks such as some *Hyalomma* species and *Ornithodoros savignyi* which run to their hosts. Carbon dioxide is an important stimulus eliciting hunting behaviour, a fact which has been utilized for collecting ticks over distances of 20 m or more using a CO_2-baited trap. A typical example of an ambusher tick is *Boophilus microplus* which climbs up vegetation to a height appropriate for contact with passing cattle. Host stimuli such as odour, CO_2 or vibrations elicit the 'questing' behaviour where the first pair of legs is waved in the direction of the stimulus. But the ticks do not leave their position and contact has to be made for them to board the hosts. Vibrations can be specific stimuli, and any with a frequency close to that produced by a host animal is likely to result in appetence behaviour; for example, larvae of the cattle tick *B. microplus* will respond to 80–800 Hz vibrations as these are in the range of sounds produced by cattle grazing in green grass. Other stimuli eliciting appetence behaviour and host engagement are visual images, contact, shadowing and radiant heat. Once the host is engaged, the next steps are exploration, penetration and attachment. Exploration involves moving over the host to find an attachment site. The stimuli responsible for this are not clearly understood. They could be non-specific, ticks attaching to any part of the body of any animal. But in the case of ticks which are host specific or which have predilection sites such as the ears or the perineal region, there may be specific unknown stimuli. After the exploratory phase ends, penetration starts, with the cheliceral digits being introduced into the skin (attachment) and brought into contact with host body fluids – which are then 'tasted' (ingestion). A favourable input will lead to completion of penetration and the start of engorgement; an unfavourable input results in withdrawal of the mouthparts and resumed exploration. The engorgement process is different in argasids and ixodids. Argasids, with exception of *Argas* larvae, feed rapidly and it is difficult to know when engorgement has begun, but slow-feeding ixodid ticks excrete granules of digested blood and this indicates the start of engorgement.

Females of most ixodids will not engorge fully unless they have mated. The last two events in the feeding sequence are detachment, withdrawal and disengagement of the mouthparts and leaving the host. The stimuli for this are unknown, although the stretching of the cuticle during the last stage of engorgement, and activation of stretch receptors in the dorsoventral muscles, could trigger responses from the synganglion. Leaving the host in the case of ixodids has a regular pattern, the 'drop-off rhythm', and appears to be governed by the activity cycle of the host animals.

The mid-gut of ticks, unlike that of insects, consists of several radiating blind sacs or diverticulae which can be clearly seen in starved ticks. This structural peculiarity is necessary for accommodating the large quantity of ingested blood. Argasids feed within an hour, but during this short time they take enough blood to increase twelve-fold in weight. Each nymphal instar feeds once but males and females take multiple feeds. Concentration of the blood-meal is effected by filtering off excess water through specialized organs (the paired coxal glands) and two drops of coxal fluid are produced at the end of engorgement, either on the host just before detachment or away from the host after detachment.

Ixodids are slow feeders which take several days to engage. Secure attachment during this prolonged period is facilitated by secretion from the salivary glands of a cement substance which hardens around the buried mouthparts, or by possession of long mouthparts which penetrate deep into the skin. Engorged ixodid females can weigh up to 200 times as much when fed as when unfed, but the actual weight of blood ingested could be much higher than this weight difference implies – because concentration of the blood takes place during feeding by return of excess water to the host through special salivary gland cells and by excretion of digested blood. The considerable body expansion during the final rapid feeding phase is made possible by growth of the cuticle. Males of some ixodid species do not feed, but males which mate on the host need a small blood-meal to mature the sperm.

Oviposition

Argasid females undergo several gonotrophic cycles and oviposition follows each of the several blood-meals. Ixodid females, on the other hand, have only one gonotrophic cycle following the single blood-meal. Female ticks have to be inseminated and fully engorged before oviposition takes place. If the blood-meal is below a certain threshold, oviposition will fail to take place and such ticks if they become detached will start host-seeking behaviour. Engorged females are positively geotropic and will seek shelter in the substrate. The pre-oviposition period usually lasts a few days, but ticks of temperate regions often enter a reproductive diapause in which oviposition is delayed through several weeks of inclement weather. Oviposition lasts from a few days to a few weeks according mainly to the temperature. The oviposition process is complex. The capitulum of engorged ixodid females is bent to a vertical position in relation to the rest of the body and the anterolateral part lowered so as to position the genital aperture below the capitulum. This reorientation does not take place in argasid females. The hypostome is pressed firmly against the body wall. A structure called Gene's organ is now everted and inflated between the basis capituli and the mouthparts. The egg is deposited between the two lateral horns of the Gene's organ and coated with a waxy waterproofing secretion prior to elevation of the capitulum and transport of the egg to the anterodorsal part of the female's body. Ixodid females additionally protect their eggs by coating them with an anti-autoxidant from the glands of the porose areas, the openings of which are situated on the dorsum of the basis capituli. Each egg is voided individually in a laying process lasting 3–12 minutes. Ixodid eggs usually stick together and very often the egg-mass sticks to the female's dorsum and mouthparts. If left undisturbed, the female remains in contact with the eggs and does not move away.

MEDICAL IMPORTANCE OF TICKS

Ticks are primarily parasites of wild animals and only about 10% of the species feed on domestic animals. But these relatively few species have prospered on domestic livestock and are responsible for considerable economic loss to farmers in developed countries as well as to rural herdsmen in developing countries. They cause damage and loss of blood in animals and transmit diseases. During the second half of the last century vast areas of grassland in the Americas, Africa and Australia were opened up for introduction of European cattle breeds and increased meat production. The rapid increase in cattle populations which followed resulted in spectacular and colossal losses from tick-borne infections. Such losses are rare these days but ticks and tick-borne diseases still constitute an important constraint on successful livestock farming in many areas of the world. Native cattle in developing countries, however, have developed some immunity through years of contact with ticks and disease agents.

Many ticks will feed opportunistically on humans, some more avidly than others. *Ornithodoros moubata* of East, Central and South Africa, is perhaps the only species which can be considered as predominantly a man-biter and is found in huts. It will feed on chickens as well as people. Tick species which are indiscriminate in their choice of hosts are the most important as vectors of pathogens to man.

There are several factors which make ticks efficient vectors of pathogens. Rapid feeding by argasids and firm host-attachment by slow-feeding ixodids prevents dislodgement and removal by the host. The slow feeding by ixodids gives them ample time to ingest large numbers of pathogens from an infected host and to transmit the infection to a new host; it also allows the dispersal of infected ticks to new areas while attached to the hosts. Similar opportunities for acquiring and transmitting pathogens by argasids are provided by the multiple feeds by nymphs, and by both females and males. Blood-feeding at least once by each stage gives more opportunities to acquire and transmit a variety of pathogens, while a wide host range makes a blood-meal more certain. Many ixodids have a very high reproductive potential and can live for long periods; the ability to starve in argasids even surpasses that of ixodids and they can go without a blood-meal for years. This is of great survival value and also ensures survival of pathogens in infected ticks for long periods. Trans-stadial and trans-ovarial transmission of many pathogens in ticks makes them true reservoirs of infection and some tick-borne pathogens, such as spirochaetes in argasids, can be maintained in nature in the absence of vertebrate hosts by trans-stadial and trans-ovarial passage through several generations.

Tick-borne infections in nature are focal in their occurrence. A hut, log cabin or cave harbouring infected *Ornithodoros* ticks can provide foci of relapsing fever. In the more general biotopes in which ixodids are found, foci of infection are maintained by wild animals and their tick ectoparasites. Large concentrations of infected ticks can be found where the host animals rest. In these maintenance foci, which are often cryptic or hidden, pathogen, vertebrate host and tick vector have reached a balanced relationship in which man plays no part. Ecological changes such as clearance of forests, cultivation of crops and formation of fringe habitats, and intrusion of man into these areas, result in tick–man contact and outbreaks of disease. Man is a tangential and accidental host for pathogen and tick and a dead-end host so far as further transmission is concerned.

Another factor in the epidemiology of tick-borne diseases is change in human behaviour. General affluence since the Second World War has produced societies with time and money for leisure activities such as camping, walking in forests or mushroom-picking; there is also an increasing demand, particularly among the young, for adventure holidays in the bush. Such activities bring people into contact with foci of infection. The apparent spread of tick-borne diseases, such as tick-borne encephalitis in Europe and the appearance of 'new' diseases such as Kyasanur Forest disease in southern India and Lyme disease in the USA and Europe, are due to a combination of ecological changes and human behavioural factors. Increased awareness among clinicians has led to better diagnostic techniques.

Transmission of disease agents by ticks

Ticks transmit viral, rickettsial, bacterial and protozoal diseases; these are all zoonoses affecting wild or domestic animals. Infection of ticks with pathogens does not appear to harm the ticks. Development of viral, rickettsial and bacterial agents in ticks is propagative (multiplication only).

Following the infective feed, there is disseminated infection involving most of the tissues of the tick, including the synganglion. Transmission is inoculative, by salivary secretion as with tick-borne arboviruses, tick-borne rickettsiae and Lyme disease. Such disseminated infection does not occur in the case of insects such as fleas and lice infected with rickettsiae, where the organisms are confined to the gut and excreted with faeces while the insects feed. Transmission of relapsing fever spirochaetes by soft ticks is inoculative through saliva as well as contaminative, by the secreted infective coxal fluid contaminating the bite wound. Transmission of relapsing fever spirochaetes by human lice, on the other hand, is always contaminative, brought about by crushing the body of infected lice on the host and release of spirochaetes from the haemocoel. Development of Protozoa in ticks is cyclopropagative and transmission is inoculative through saliva. With some tick-borne pathogens, for example Lyme disease, it can take a day or two of attachment before infected ticks can transmit, and prompt removal of attached ticks before this time will lessen the chances of infection.

Ticks as a cause of disease

Attached feeding ticks are often not noticed, but after the ticks have detached there can be severe local reactions at the sites of bites; in extreme cases these can lead to granulomatous lesions. Lesions are often found where clothing is tight, for example around the waist, and can be accompanied by an intense itching; they often take several months to heal.

Erythema chronicum migrans (ECM)
This is a skin lesion which follows the bites of ticks infected with *Borrelia burgdorferi* and usually precedes Lyme disease. Typically, the lesion consists of an annular erythema with a central clearing surrounded by a red migrating border and usually reaches a diameter of about 15 cm.

Tick paralysis
This is a rapidly fatal but easily cured disease of man and animals. It is mainly associated with ixodid ticks (43 species in ten genera worldwide) and is caused by injection of a neurotoxin in the saliva of female ticks when they bite. Some *Argas* species also produce a paralysing toxin. The toxin, which might be different in different species, disrupts nerve synapses in the spinal cord and blocks the neuromuscular junctions. The tick involved in human tick paralysis is *Dermacentor andersoni* in western North America, *D. variabilis* in eastern North America, *Amblyomma americanum* in southern USA, *Ixodes holocyclus* in Australia and *Ixodes rubicundus* in South Africa. In five to seven days after attachment the patient develops fatigue, numbness of the legs and muscular pain. A flaccid paralysis rapidly extends from the lower to the upper extremities. If the attached tick is not found and removed, the condition worsens, expressing itself as difficulty in

swallowing and lingual and facial paralysis. Convulsions follow and death results from respiratory failure. The disease is usually seen in children, but occurs also in adults. The attached tick should be found and removed carefully. Pulling it off forcibly leaves the mouthparts embedded in the skin and so allows the symptoms to continue. Removal of *D. andersoni* from the patient will lead to rapid improvement, but unfortunately this is not so with paralysis induced by *I. holocyclus*, from which the patient can eventually die.

Ticks as vectors of viruses

More than 100 arboviruses are known to be associated with 116 tick species; 32 of these species are argasids and 84 ixodids (Hoogstraal, 1980). In many cases, the association is based solely on isolation of virus, and the role of many of the viruses in the causation of human and animal diseases in unknown. Hoogstraal (1966, 1973) has provided excellent reviews of ticks and tick-borne viruses. Among the more important human diseases caused by tick-borne viruses are Colorado tick fever (Reoviridae, genus *Orbivirus*), tick-borne encephalitides (Flaviviridae, *Flavivirus*), Kyasanur Forest disease (also *Flavivirus*) and Crimean–Congo haemorrhagic fever (Bunyaviridae, genus *Nairovirus*). All are zoonoses. Transmission to humans is by the bite of infected ticks and trans-ovarial transmission occurs in some of the ticks involved.

Colorado tick fever (CTF)
This is a zoonosis existing as foci in the Rocky Mountain states and South Dakota of the USA and in western Canada. The main vector is *Dermacentor andersoni*, in which the prevalence of infection can be 10% or higher, although the virus has been isolated from other species. The reservoirs are various small mammals such as ground-squirrels (*Spermophilus*) and chipmunks (*Tamias*). Transmission to man is by the bite of adult ticks. There is no trans-ovarial passage of the virus in the tick.

Tick-borne encephalitides
The encephalitides transmitted by ticks are in two forms. The far eastern form, called Russian Spring Summer encephalitis (RSSE), is associated with taiga forest in eastern USSR and northeastern China and is transmitted by *Ixodes persulcatus*; the western form, known as Central European or western tick-borne encephalitis (TBE), occurs in Europe and western USSR in association with coniferous and temperate deciduous forests and is transmitted by *Ixodes ricinus*. Some other tick species are involved in transmission to a minor extent. In the western USSR, where the two main vectors coexist, both forms of encephalitis can occur. Many human cases are contracted during holidays and weekend trips made by city people to the forests, and the reported 'spread' of TBE in Europe is almost certainly due to increased tick–human contact during leisure activities. Transmission is by tick bite, although humans can become infected with TBE by drinking milk or eating cheese made from the milk of infected goats. The viruses are maintained in enzootic foci by trans-ovarial passage in ticks.

Kyasanur Forest disease (KFD)
This disease was discovered in 1957 in southern India following investigations into the death of monkeys in Kyasanur Forest and of illness and death in the human

population in villages adjacent to the forest. Human cases were associated with wood-gathering and cattle-grazing in the forest. The major vector is *Haemaphysalis spinigera*, a species widely distributed in the forests of southern and central India and Sri Lanka. The immature stages of this tick usually bite birds, small forest mammals and monkeys, but man is also bitten. Cattle are important hosts of the adult ticks. Transmission of the virus is by tick bite. There is apparently no natural trans-ovarial transmission in *H. spinigera*. *Haemaphysalis turturis* is another species which maintains the enzootic cycle and there have been isolations from still other forest ticks. Since 1957, outbreaks have occurred in other areas, all in southern India and all in forests, in association with deforestation and the intrusion of man into already existing foci of infection.

Crimean–Congo haemorrhagic fever (CHF)
Hoogstraal (1979) has reviewed the epidemiology of this disease. The first cases were recognized among Soviet military personnel in the Crimea in 1944–45; since then numerous outbreaks have occurred in southern USSR, Bulgaria and Pakistan. The discovery of an identical virus causing human and animal disease in the Congo (now Zaire) led to the disease being renamed Crimean–Congo haemorrhagic fever. The disease is now known to be enzootic in steppe, savanna, semi-desert and foothill biotopes in eastern and central Europe, most of European and Asian USSR, parts of the Oriental region and in Africa from Egypt to South Africa and from Nigeria to Kenya. At least 27 tick taxa are associated with maintenance of the virus. The infection is maintained among cattle by species of *Boophilus* and among small mammals (such as hares and hedgehogs) and larger artiodactyls by *Hyalomma* species. Other ticks such as *Amblyomma variegatum* and *Rhipicephalus* species probably serve as link vectors between wild and domestic animals. Human epidemics are associated with aggressively host-seeking ticks of the *Hyalomma marginatum* complex and with *Hyalomma anatolicum anatolicum*, ticks which attain great numbers during certain periods. Infection is usually by tick bite or by crushing infected ticks on the skin. Trans-ovarial transmission occurs mainly in *Hyalomma* ticks. Unusually severe weather in winter and spring reduces the tick populations, with the result that virus circulation reverts from epizootic to enzootic intensity.

Ticks as vectors of rickettsiae

Tick-borne rickettsioses of man have been reviewed by Hoogstraal (1967) and acarine-borne rickettsioses of the Old World by Řeháček and Tarasevich (1988). The major rickettsial diseases that ticks transmit to man are Rocky Mountain spotted fever and various forms of tick typhus. Of lesser significance is an infection known as Q fever.

Rocky Mountain spotted fever (RMSF)
Rocky Mountain spotted fever, also known as Tobia fever and São Paulo fever, was described originally from the Rocky Mountains of the USA but is widely distributed in the Nearctic and Neotropical regions. The disease is now most prevalent in eastern USA, where – because of encroachment of residential areas into abandoned fields and woodlands with much tick and rodent activity – it has become suburban/rural and highly prevalent in women and children. Fortunately, antibiotics have now reduced considerably the formerly high mortality rate.

Transmission is by tick bite and there is trans-stadial and trans-ovarial passage of rickettsiae in the ticks. The vector in western USA is the Rocky Mountain wood tick, *Dermacentor andersoni,* and in eastern USA the American dog tick, *D. variabilis.* The immatures of both these species feed on small mammals, but adults of *D. andersoni* parasitize large wild and domestic herbivores and adults of *D. variabilis* attack wild and domestic carnivores; however, adults of both species feed secondarily on other mammals, including man. The vector in the Neotropical region is the Cayenne tick, *Amblyomma cajennense.*

Apart from the ticks just mentioned, the rickettsiae of Rocky Mountain spotted fever have been isolated from the cosmopolitan dog tick, *Rhipicephalus sanguineus,* and this tick might be involved in the maintenance of natural foci.

Heavy rickettsial infections can kill the female ticks or reduce their egg output.

Boutonneuse fever, Siberian tick typhus and Queensland tick typhus
Boutonneuse fever, caused by *Rickettsia conori,* is so-called because it produces button-like lesions at the tick attachment sites. It is widespread in Africa, European and North African areas near the Mediterranean, in Israel and most countries of South East Asia. It is known by various local names, including Marseilles fever, Crimean tick typhus, South African tick typhus, Kenya tick typhus and Indian tick typhus. The main vectors are *Amblyomma hebraeum* (in the South African veldt) and *Rhipicephalus sanguineus,* but *Hyalomma leachi* and *Rhipicephalus appendiculatus* can also be involved. In France the disease is so closely associated with rabbits that the prevalence dropped dramatically following decimation of the rabbit population by myxomatosis (Le Gac, 1966). Transmission is by tick bite or by contact of the hands with skin and eyes after accidental crushing of ticks removed from dogs; urban cases are associated with dogs and dog ticks. Trans-ovarial passage of rickettsiae in the ticks occurs.

Siberian tick typhus, caused by *Rickettsia sibirica,* ranges from the far eastern Pacific seaboard of the USSR westwards as far as Armenia. The most important vectors are *Dermacentor marginatus, D. silvarum, Hyalomma concinna* and *Rhipicephalus sanguineus.* The last-named is the vector in the western parts of the disease distribution. Transmission is by tick bite and there is trans-stadial and trans-ovarial passage of the rickettsiae in the ticks.

Queensland tick typhus, caused by *Rickettsia australis,* is related to mite-borne rickettsial pox and is restricted to coastal and densely forested areas of northeastern Australia. The vector is *Ixodes holocyclus,* an unusually indiscriminate feeder. Transmission is by bite of the tick and can occur within a few hours of attachment. Both trans-stadial and trans-ovarial passage of rickettsiae occur. The Flinders Island spotted fever recently reported from Tasmania (Stewart, 1991) is probably different from Queensland tick typhus.

Q fever
The causative organism of Q fever, *Coxiella burneti,* is enzootic in cattle, sheep and goats in most parts of the world and is transmitted among domestic animals by ixodid ticks. The rickettsiae occur in large numbers in milk, urine, faeces and fetal fluids of infected animals. Humans become infected occasionally by contamination or inhalation and suffer a moderately severe pneumonia which is rarely fatal.

Tick-borne borrelioses

Two human diseases caused by borreliae are transmitted by ticks: relapsing fever with *Ornithodoros* vectors and Lyme disease with *Ixodes* vectors.

Tick-borne (endemic) relapsing fever

This disease is endemic across southern Asia from Israel to western China, in Iberia and North Africa, parts of western and equatorial Africa, widely in eastern and southern Africa, in central and western USA, Mexico and Central America, and in northwestern South America and northern Argentina (Fig. 18.9). It is caused by several species of *Borrelia* and transmitted by several species of *Ornithodoros* ticks. The distributions of the principal vectors are shown in Fig. 18.9. Several hundred cases occur each year. The foci of infection are restricted, e.g. to situations such as huts, caves and log cabins. Infection can occur in very young children. There is little seasonal fluctuation in the incidence of the disease because of the relatively stable environments in which the ticks live. Trans-stadial and trans-ovarial infection rates are high and the ticks are the true reservoirs of infection. Transmission is by tick bite (saliva) and also sometimes through contamination of the bite wound with infective coxal fluid produced by feeding ticks just before they detach; vector ticks which do not produce coxal fluid while on the host (e.g. some of the New World *Ornithodoros*) transmit solely by bite.

Lyme disease

This disease is caused by *Borrelia burgdorferi*. It is widespread in the USA, from where reports of several hundred cases every year make Lyme disease the most important tick-borne infection of man in North America. The disease is now known also in the Old World, and has been recognized in the British Isles and continental Europe, USSR, northeastern China, Japan and South Africa. The known vectors are *Ixodes dammini* in northeastern USA, *I. pacificus* in California, *I. ricinus* in Britain, Europe and parts of USSR, and *I. persulcatus* in eastern USSR, China and Japan. All these species feed indiscriminately on a variety of small animals when in the immature stages and on large mammals (such as deer and livestock) when adult. *Ixodes holocyclus* is a suspected vector in Australia, from where a few cases of Lyme disease have been reported, but this needs confirmation. Transmission of Lyme disease is by tick bite, but one to three days of attachment can elapse before the tick is able to transmit the *Borrelia* pathogen. Trans-ovarial transmission in *Ixodes* host ticks has been demonstrated but the rate is very low. The relationship between the vectors and the pathogens in North America and Europe is reviewed by Lane et al. (1991).

Ticks as vectors of human babesiosis

Babeşiosis in domestic livestock is an important veterinary problem, but human cases of the disease are rare. From Europe and USSR only about half a dozen cases have been reported; all have been in splenectomized individuals and fatal. The *Babesia* species involved is *B. divergens* and the vector *Ixodes ricinus*. In contrast to

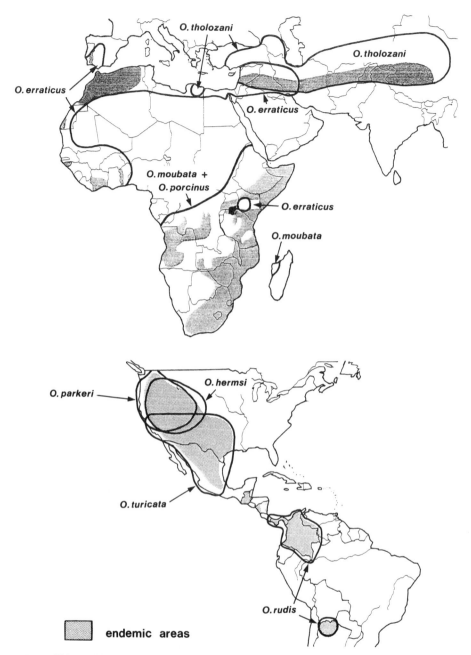

Figure 18.9 Old World and New World distribution of endemic tick-borne relapsing fever and its principal *Ornithodoros* vectors.

European babesiosis, American human babesiosis (caused by the rodent piroplasm *Babesia microti*) is usually a benign infection; it occurs in spleen-intact and asplenic individuals. The focus of the American form of the disease lies in the northeastern United States, where the vector is *Ixodes dammini*. Transmission is by tick bite.

Trans-stadial and trans-ovarial passage occurs in *I. ricinus* infected with *B. divergens*, but there is apparently trans-stadial passage of *B. microti* in *I. dammini* only from larvae to nymphs (neither trans-stadial passage from nymphs to adults nor trans-ovarial passage has been reported).

INTRODUCTION TO MITES: RECOGNITION AND STRUCTURE

The mites hugely outnumber the ticks, the latter accounting for only about 800 species among the 30 000 described species of Acari. They display tremendous diversity of form, behaviour and habitat selection, and can be ground-dwelling, aerial or aquatic, free-living predators, feeders on plants, plant derivatives or other organic matter, and sometimes parasites of invertebrates or vertebrates. More than one type of habit can be represented in a single family. Some predaceous mites are beneficial to man and have been used for the attempted biological control of houseflies in manure and for controlling mites of horticultural importance. Stored grain and other stored products are often infested by various kinds of mites which, although not parasitic, can cause dermatitis in man if they or their dead bodies or faeces are contacted; other clinical effects can occur if these mites are swallowed or inhaled.

Species which are parasitic on vertebrates, including man, and which have potential medical importance, form only a small fraction of the total number of mite species. The habit of vertebrate parasitism probably evolved through ancestral mites making contact with long-stay nest-dwelling (nidicolous) hosts and evolving as nest parasites, until constant host parasitism developed – as is apparently happening among the bird mites of the genus *Dermanyssus* (Krantz, 1978). Many mite families have members which are medically important, feeding on human blood, lymph, sebaceous secretions of digested human host tissue which they ingest by puncturing the skin or by imbibition of fluids from surface lesions. There are some mites which are endoparasitic and found in nasal passages, tracheae or lungs of various mammals, birds and reptiles. None of these mites has been recorded from humans.

The life cycle of mites is very similar to that of ticks. The egg is succeeded by a hexapod larva, octopod nymphal stages and adult. Female mites produce a small number of relatively large eggs, from which the larvae emerge. There can be one, two or three nymphal stages, called respectively the protonymph, deutonymph and tritonymph (e.g. Fig. 18.17). At least one of the developmental stages is usually inactive and develops without feeding. In some astigmatid mites (those without respiratory stigmata), the second nymphal stage (deutonymph) is highly resistant to environmental conditions and is known as a hypopus.

Generally speaking, mites of medical importance, except the hair follicle mites (*Demodex* species) and the scabies mite (*Sarcoptes scabiei*), are not host-species specific. In contrast, the hair follicle mites are highly host specific. Scabies mites that attack animals can also attack humans, but are unable to establish permanent infestations. Good general accounts of mites are given by Baker and Wharton (1952), Baker et al. (1956), Krantz (1978), Strandtmann and Wharton (1958) and Woolley (1988).

Many mites are microscopic or just visible to the naked eye. Their intake of blood or tissue juices is modest and they do not show the dramatic increase in size after feeding that characterizes the ticks. The lack of a prominent, well-developed, toothed hypostome distinguishes mites from ticks. The body has the same general structure, however, as in ticks and shows similar division into anterior capitulum (= gnathosoma) and posterior idiosoma (Fig. 18.1). The capitulum carries the palps and chelicerae. The last segment of the palps usually carries a claw-like structure, the apotele, which is absent in ticks. The chelicerae can be stylet-like and used for piercing, or claw-like and adapted for tearing. The tritosternum found mid-ventrally near the front end is not a part of the capitulum, but in some predatory mites it directs overflow of prey fluids towards the mouth region. The idiosoma ('body') is unsegmented, but various sutures and grooves can be present on its surface, as in the scabies mite (Fig. 18.13) and the hair follicle mite (Fig. 18.14b). The idiosoma is sometimes soft and unsclerotized, but many mites have one or two sclerotized dorsal shields; the anterior one of these can be much reduced and present only as an elongate sclerite (the crista metopica of the family Glycyphagidae). Ventrally, the idiosoma is sometimes soft, but usually the anal and genital regions have some type of surrounding sclerite – the anal shield and genitoventral (= epigynal) shield respectively. As in ticks, telescoping of the terminal tergites has resulted in shift of the anus away from the hind end of the body to a ventral position near the genital region. The tips of the legs (tarsi) are provided with a membranous structure, the empodium, situated between the claws. The stigmata of mesostigmatic (gamasid) mites are usually associated with elongate sclerotized processes, the peritremes (Fig. 18.11b).

CLASSIFICATION AND IDENTIFICATION OF MITES

Mites that have medical importance belong to three suborders of Acari, the Mesostigmata (which together with suborder Ixodida or ticks form the order Parasitiformes) and the Acaridida and Actinedida, which together constitute the order Acariformes. Mites of the suborder Mesostigmata can be readily distinguished from ticks by having a single pair or stigmata between coxae 2 and 4 (usually associated with elongate peritremes), by possession of a claw on the palpal tarsus, by the hypostome serving only as part of the floor of the capitulum, and by the absence of a sensory pit on the tarsus of leg 1. Higher classification of mites is based primarily on the presence or absence of stigmata and their position on the body.

Identification

Identification depends on morphological characters for which examination by compound microscope is routinely necessary. Scanning electron microscopy (SEM) has been used to observe external structure (Woolley, 1970) and appears to have considerable potential in mite studies. Chromosomes and enzymes have been investigated by Wang (1988) as taxonomic aids for trombiculid mites, but are not yet practical tools in mite identification.

In general, mite identification is extremely difficult and largely a matter for the specialist. Identification keys are inevitably complicated and often not very helpful for recognizing the relatively few species of mites of medical and public health importance. The following key is provided simply as a guide for the medical entomologist and must be used with care. (Note that the genitoventral shield mentioned in the key and labelled on Fig. 18.10d is often simply called the genital plate.)

Identification key to principal mites of medical importance

1 Stigmata absent posterior to second pair of legs (coxae 2) 2
— Stigmata present as one lateral pair between bases of legs 2 and 4 3
2 Legs without claws. Palps with two segments. [Stigmata absent] 12
— Some or all legs with claws. Palps with more than two segments. [Stigmata present or absent, when present opening anteriorly] 14
3 Chelicerae long and whip-like, tapering apically, chelae at tips absent or very small.. 4
— Chelicerae not long and whip-like, shorter and stronger and chelae at tips blade-like (with or without teeth) 5
4 Dorsal surface of body with two shields, a small posterior shield present in addition to main shield. Anal shield egg-shaped and with central anal opening (Fig. 18.10d) .... **Liponyssoides sanguineus** (Dermanyssidae)
— Dorsal surface of body with one shield. Anal shield not egg-shaped and with anal opening at posterior end (Fig. 18.10b) ....
.... **Dermanyssus gallinae** (Dermanyssidae)
5 Dorsal shield not nearly covering dorsal body surface (Figs 18.11a,d and 18.12c). Genitoventral shield narrowed posteriorly (Fig. 18.11b,c). Chelicerae with toothless chelae... 6
— Dorsal shield virtually covering dorsal body surface. Genitoventral shield not narrowed posteriorly. Chelicerae usually with toothed chelae 8
6 Dorsal shield broad, its setae short 7
— Dorsal shield narrow and tapering posteriorly (Fig. 18.11d), its setae long ....
.... **Ornithonyssus bacoti** (Macronyssidae)
7 Sternal shield with two pairs of setae (Fig. 18.11b) ....
.... **Ornithonyssus sylviarum** (Macronyssidae)
— Sternal shield with three pairs of setae ....
.... **Ornithonyssus bursa** (Macronyssidae)
8 Genitoventral shield widened posteriorly (Fig. 18.12a,b), with more than one pair of setae 9
— Genitoventral shield not widened posteriorly, with one pair of setae ....
.... **Hirstionyssus isabellinus** (Laelapidae)
9 Body densely covered with setae 10
— Body with few setae (these arranged in transverse rows) 11
10 Genitoventral shield with pear-shaped outline (Fig. 18.12a) ....
.... **Haemogamasus pontiger** (Laelapidae)
— Genitoventral shield with larger subcircular outline (Fig. 18.12b) ....
.... **Eulaelaps stabularis** (Laelapidae)
11 Genitoventral shield with concave posterior margin, surrounding anterior part of anal shield................................... **Laelaps echidninus** (Laelapidae)
— Genitoventral shield with convex posterior margin, not surrounding anterior part of anal shield .... **Laelaps nuttalli** (Laelapidae)
12 Legs of normal length and all similar in structure... 13
— Legs short and stubby, anterior two pairs ending in discs, posterior two pairs ending in long bristles (Fig. 18.13)................... **Sarcoptes scabiei** (Sarcoptidae)

13 Body with two pairs of long posterior marginal setae (Fig. 18.14*a*) ....
   .... **Dermatophagoides** species (Pyroglyphidae)
— Body with four pairs of long posterior marginal setae ....
   .... **Tyrophagus** species (Acaridae)
14 Body not unusually elongate, with setae 15
— Body unusually elongate and with worm-like annulations, without setae (Fig. 18.14*b*) .... **Demodex** species (Demodicidae)
15 Capitulum and palps conspicuous. Body with feathery setae (Fig. 18.16). [Parasitic in larval stage (three pairs of legs)] ....
   .... species of TROMBICULIDAE
— Capitulum and palps inconspicuous. Body with simple, non-feathery, setae (Fig. 18.15*d,g*). [Not parasitic on vertebrates: females usually with gravid body, Fig. 18.15*e*)] .... **Pyemotes tritici** (Pyemotidae)

Faunal and taxonomic literature

Literature on mites is very scattered and there are very few works covering regions or countries even for particular families or genera. Sims (1980, Acari on pp. 44–65) is an important starting point for faunal works on all mites, and Krantz (1978) reviews the classification with illustrated family definitions and keys on a worldwide basis. For mites of medical importance, the following will be of use: Azad (1986, synopsis of mites of public health importance); Fain (1968, Sarcoptidae); Fain et al. (1990, keys, distribution of non-vector species); Hughes (1976, stored food and household mites worldwide, keys, illustrations); Vercammen-Grandjean (1968, chiggers, Far East); Vercammen-Grandjean and Langston (1976, Trombiculidae worldwide): Wang (1988, *Leptotrombidium* vectors in China); Wang and Yu (1992, *Leptotrombidium*, key to Chinese species); Wen (1984, Trombiculidae of China); Yunker (1973, synopsis of common world species and useful bibliography).

BIOLOGY AND MEDICAL IMPORTANCE OF MITES

The biology of mites is so varied that it is impossible to give a general account. Instead, the following synopsis deals with mites which are either parasitic on vertebrates or which have public health importance even though they are not parasitic. The biology and medical importance are covered together in this section, taking the suborders and families concerned in turn.

Suborder Mesostigmata (= Gamasida)

Mites in this suborder are relatively large, from 0.2 to 2.0 mm in length, and can easily be seen with the naked eye. Many are predators and some have been used in the biocontrol of red spider-mites of glasshouse plants, and of bush-flies (*Musca vetustissima*) in Australia. Those which parasitize man belong to three families, the Dermanyssidae, Macronyssidae and Laelapidae. They are haematophagous and in nature feed on birds and rodents, but many are not host specific in certain circumstances – attacking unusual hosts such as man, for example, when birds leave an infested nest or rodents are destroyed. Damage results when large numbers feed on the host. The adults can survive for several months without a blood-meal. Only *Liponyssoides sanguineus* is a confirmed vector of pathogens.

Dermanyssidae

Dermanyssid mites are parasites of birds and mammals. The adults are about one millimetre long. They become bright red after a blood-meal, the colour becoming darker as the blood is digested. The chelicerae are long and stylet-like (Fig. 18.10a,b).

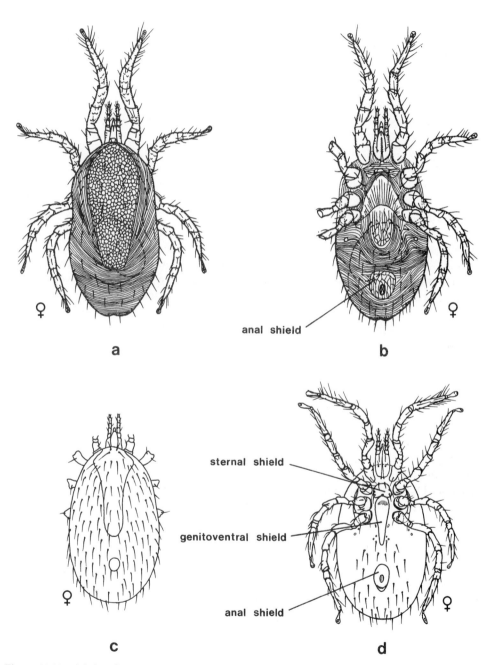

Figure 18.10 Adults of two species of parasitic mites of the family Dermanyssidae: (*a, b*) *Dermanyssus gallinae* (red poultry mite), female in dorsal view (*a*) and ventral view (*b*); (*c, d*) *Liponyssoides sanguineus* (house-mouse mite), female in dorsal view (*c*) and ventral view (*d*). Some legs are incomplete in (*b*) and only the leg bases are shown in (*c*).

Dermanyssus gallinae (Fig. 18.10*a,b*) is a cosmopolitan species known as the red poultry mite or chicken mite; it attacks chickens and turkeys, and wild birds such as pigeons, sparrows and starlings. When the normal hosts are not available, the mites will attack mammals (including man) and their bites can be painful and irritating. The unfed mites are greyish and about 0.7 mm in length, but after a blood-meal (usually taken during the night) they enlarge to a length of about one millimetre and become bright red. The species can be distinguished from the closely related *Liponyssoides sanguineus* (Fig. 18.10*c,d*) by having a single dorsal shield. The eggs are deposited in cracks or under debris in poultry houses or birds' nests. The entire life cycle from egg to adult can take as little as one week. Thousands of mites can be present in infested premises, and in poultry houses lead to reduced egg laying and even to hens abandoning their eggs. Very heavy infestations lead to exsanguination and death of the birds. Infestations in houses, hospitals and offices, sometimes reported, usually come from birds' nests in attics and under eaves; removal of nests and insecticidal treatment of walls of invaded rooms will normally eradicate the mites. *Dermanyssus gallinae* is putatively a vector of the St Louis encephalitis arbovirus, but this needs confirmation.

Liponyssoides sanguineus (Fig. 18.10*c,d*) is commonly known as the house-mouse mite and is generally similar to the poultry mite. It was originally described from Egypt as a parasite of small rodents and is widespread. Although the preferred host is the house mouse (*Mus musculus*), rats and other rodents are also frequently attacked. The life cycle lasts 18–23 days. The mite is a nest-dweller and occurs on the host only to feed. The species is a vector of *Rickettsia akari* causing rickettsial pox in man. This disease was first recognized and associated with house mice and *L. sanguineus* in the USA during the 1940s, but cases (in some outbreaks as many as a thousand) have been reported from the Ukraine, South Korea, South Africa and equatorial Africa.

Macronyssidae

This family contains bloodsucking mites of birds, mammals and reptiles. *Ornithonyssus*, the genus of greatest interest, has a single dorsal shield which is either broad or narrow but always has a posteriorly tapering shape. The related genus *Ophionyssus* contains *O. natricis*, an ectoparasite of snakes, and differs from *Ornithonyssus* in having two dorsal shields.

Ornithonyssus bacoti (Fig. 18.11*c,d*) is commonly known as the tropical rat mite although it was originally discovered on the brown rat (*Rattus norvegicus*) in Egypt. It occurs in tropical and temperate regions, mainly in seaports, infesting mice and rats. Adults are about one millimetre long and have long setae on their narrow dorsal shield. The mite bites man readily in rat-infested buildings, particularly granaries and groceries, causing painful irritant pustules up to 18 mm in diameter. Heavily infested rodents can die from exsanguination. The species is the vector in rodents of the filarial worm *Litomosoides carinii*. Although various pathogens have been isolated from, or experimentally transmitted by, this mite, it is not considered an important vector.

Ornithonyssus bursa (Fig. 18.12*c*) is known as the tropical fowl mite and occurs in tropical and subtropical regions, where it attacks poultry but also pigeons, sparrows and mynah birds; like *Dermanyssus gallinae*, it is a serious poultry pest. The species readily bites man, causing temporary irritation, but such attacks usually occur only when the bird hosts desert their nests in buildings. The mites, however, cannot survive for long periods away from their hosts.

Biology and medical importance of mites 637

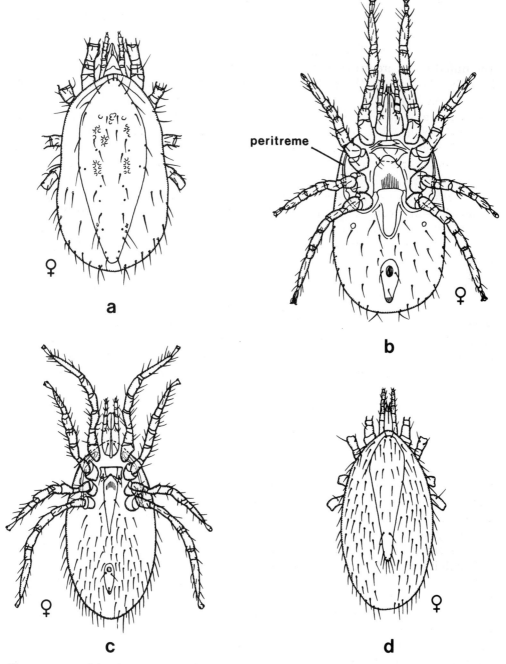

Figure 18.11 Adults of two species of parasitic mites of the family Macronyssidae: (*a*, *b*) *Ornithonyssus sylviarum* (northern fowl mite), female in dorsal view (*a*) and ventral view (*b*); (*c*, *d*) *Ornithonyssus bacoti* (tropical rat mite), female in ventral view (*c*) and dorsal view (*d*). Only the leg bases are shown in (*a*) and (*d*).

Ornithonyssus sylviarum (Fig. 18.11*a,b*) is a serious pest of poultry and wild birds in the northern temperate regions of Europe and North America and is hence called the northern fowl mite. It is also common, however, in southern Australia and has been recovered from wild birds in South Africa. The sternal shield has only two

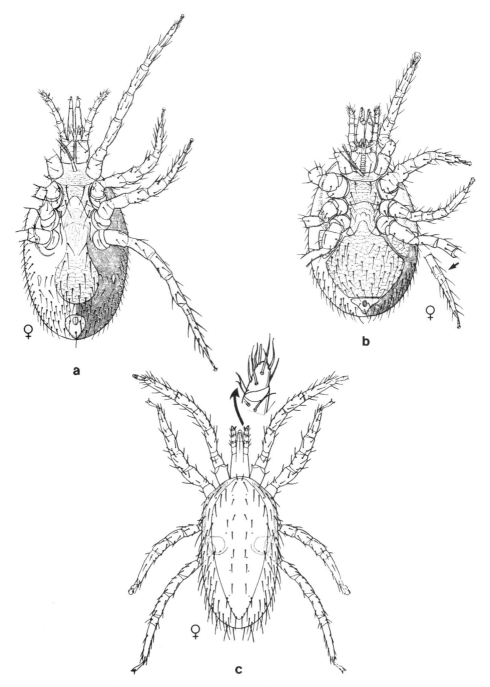

Figure 18.12 Adults of three species of parasitic mites: (*a*) *Haemogamasus pontiger* (Laelapidae), female in ventral view; (*b*) *Eulaelaps stabularis* (Laelapidae), female in ventral view; (*c*) *Ornithonyssus bursa* (tropical fowl mite, Macronyssidae), female in dorsal view, with enlargement of the tip of the palp showing the apotele (arrowed). (Figures (*a*) and (*b*) from Baker et al., 1956.)

pairs of setae and this distinguishes *O. sylviarum* from *O. bursa*, which has three pairs. The entire life cycle can take only one week to complete and is spent on the hosts. Mite populations build up rapidly on the birds and the adult mites make heavily infested birds appear grey or black; such birds lose weight and eventually die from loss of blood. In the absence of the preferred hosts, the mites will attack rodents and humans.

Laelapidae
This is a large family that includes both free-living and parasitic forms. All the parasitic species are haematophagous. The adults have a single dorsal shield. They are found on the bodies or in the burrows of rodent hosts. Five species – *Hirstionyssus isabellinus, Haemogamasus pontiger, Eulaelaps stabularis, Laelaps echidninus* and *Laelaps nuttalli* – are found worldwide and are important parasites of commensal and wild rodents. *Haemogamasus pontiger* (Fig. 18.12a) and *Eulaelaps stabularis* (Fig. 18.12b) have been reported to bite man in the absence of their normal hosts, causing irritation and dermatitis. During the Second World War, *H. pontiger* was suspected in England of causing extensive dermatitis among soldiers who slept on straw-filled mattresses.

Suborder Acaridida

The members of this suborder are small (0.2–1.2 mm in length), thin-skinned mites without any obvious shields. The coxae are sunk into the ventral body wall and are referred to as epimeres. The palps have only two segments and the chelicerae are usually shaped like pincers. Stigmata and tracheae are absent, hence the old name 'astigmatid' mites. Of the three families of medical importance, only the Sarcoptidae are parasitic.

Sarcoptidae
These mites are skin parasites of mammals and birds and spend their entire life cycle in burrows in the skin. They are globose mites with finely striated cuticle and cutting chelicerae. They cause the condition called scabies (or mange) in mammals and 'scaly leg' in birds. The most important species is the scabies mite or itch mite, *Sarcoptes scabiei*. Another species, *Notoedres cati*, causes mange in cats and also a transient dermatitis in man.

Scabies and *Sarcoptes scabiei* are comprehensively reviewed by Arlian (1989). The mite causes scabies in man and mange in other mammals that it infests; these include primates, horses, tapirs, wild and domestic bovid ruminants, pigs, camels, carnivores (including dogs and lions), rabbits and guinea pigs. Despite this diversity of potential host, a detailed study of scabies mites from man and various other animals has failed to reveal any species-level characters which differentiate mites from different hosts; *Sarcoptes scabiei* is therefore now considered to be one species. Populations found on different host species differ physiologically more than morphologically and are referred to as forms (that on man, for instance, is *S. scabiei* form *hominis*). Those from one host species do not establish themselves on another. Humans can become infested from horses or dogs (animal scabies) but such infestations are mild and disappear spontaneously.

Scabies mites are very small, males and females being about 0.2 mm and 0.4 mm in length respectively, and possess striated cuticle which bears various specialized dorsal scales and bristles (Fig. 18.13). The legs are arranged in two groups; the anterior pairs in both sexes end in stalked pulvilli which, because of their disc-shaped appearance, are called suckers. The pulvilli help the mite to grip the host's skin and so aid movement. In the female, the two posterior pairs of legs end in long

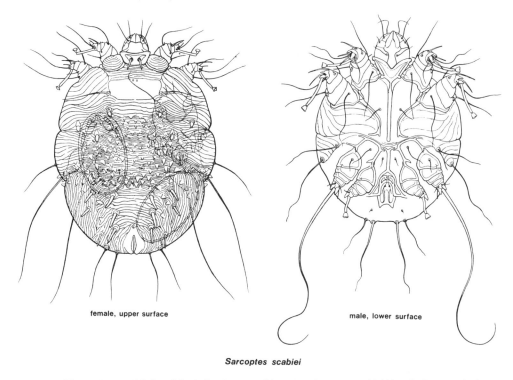

Figure 18.13 Adults of the itch mite or scabies mite, *Sarcoptes scabiei* (family Sarcoptidae).

bristles, but in the male only the third pair ends in bristles (the fourth resembling the first two pairs in having stalked pulvilli).

The newly inseminated female scabies mite moves rapidly about on the skin surface, progressing up to 2.5 cm per minute, and selects a suitable site for making a burrow. The burrow, usually about a centimetre long, is confined within the horny layer of the skin and is driven parallel to the skin surface. The mite uses its chelicerae and the sharp edges of the first two pairs of legs to cut into the skin. Both sexes burrow but only the female makes permanent winding burrows, and the burrow of the male is no more than a millimetre long. The female lays a few relatively large eggs while 'excavating' the burrow and is found among faeces at the burrow end. Burrowing occurs at a rate of 0.5–5.0 mm per day. The hexapod larvae that hatch from the eggs move to the skin surface to find shelter, and probably food, in the host's hair follicles. The two nymphal stages that precede the adult are also found in hair follicles. Mating probably occurs on the host, and the entire life cycle takes 10–14 days.

Human scabies is widespread in the tropics but is not confined there. Increases in the incidence of the disease appear to occur in 15–20 year cycles, and this cycling is probably due to fluctuating levels of immunity in the population. Persons of all ages are affected. In developing countries prevalence levels are highest in poor communities and in children: for example, in urban Dhaka (Dacca), Bangladesh, 77% of children under five years old have scabies (Stanton et al., 1987). In developed countries, all age groups are affected and infestation in children can reach 5%. Apparent recent resurgence of scabies in Western countries (suggested by better diagnosis than formerly) may be due to greater population movement and altered lifestyle among young adults leading to frequent and intimate personal contact. Scabies is transmitted from person to person only by close personal and prolonged contact, such as holding hands. Transmission also occurs when people are sleeping together and is very common in families, dormitories, mental institutions and nurseries. The mites perish rapidly away from the human body.

Most mite burrows occur in the interdigital and elbow skin, but skin of the scrotum, penis, breasts, knees and buttocks is also frequently infested; the face and scalp are rarely affected, except in children. A period of 3–4 weeks often elapses before itching, which is an immunologically mediated response to the presence of mites and mite faeces, develops. A few mites – a typical scabies case might have fewer than 20 – are enough to produce intense itching (particularly at night). Relief by scratching can result in rupture of the mite burrows and removal or death of the mites. The burrows often become secondarily infected with bacteria. In infested persons, an extensive rash characterized by erythema and follicular papules develops that can cover areas where there are no mites. In immunocompromised individuals, who do not respond to the presence of mites by itching and scratching, the mites can build up very large populations – more than a thousand mites per person. Such a condition, which produces a scaly crusted skin, is popularly known as 'crusted scabies' or 'Norwegian scabies' and is highly contagious.

Pyroglyphidae
This family contains five genera, of which one, *Dermatophagoides*, has considerable medical importance. The pyroglyphids occur on the skin surface of mammals and birds and in their nests. They are scavengers which usually feed on skin detritus, but they are also found in food products and are a common constituent of dust from granaries. *Dermatophagoides* mites have been known to burrow into the scalp of humans and produce intensely irritating small red papules (Traver, 1951). In the 1960s they were first recognized as an important cause of house dust allergy.

Dermatophagoides pteronyssinus (Fig. 18.14a) is the European house dust mite and *D. farinae* the American house dust mite. Together with *Euroglyphus maynei*, these are the mites most frequently encountered in house dust. *Dermatophagoides pteronyssinus* is the most common species in houses and occurs in every inhabited continent of the world. The mites, minute and difficult to see with the naked eye, complete their life cycle in about three weeks if conditions are optimum for them. They like dampness, and are common in beds, blankets, pillows and floor rugs and carpets, where they feed on shed human skin; thousands can be found in almost any sample of house dust. The mites are more abundant in beds than anywhere else in the house, and freely become airborne when beds are made or pillows 'fluffed up'. Consequently, they are easily inhaled, and inhalation of house mites is a cause of allergic reactions in the respiratory tract which result in asthma and rhinitis. Allergens are contained not only in the bodies of living and dead mites, but also in mite body fragments and faeces. The problem of house dust mites and the allergies they cause is reviewed by Wharton (1976) and Fain et al. (1990).

Acaridae
Species of this family, and the related family Glycyphagidae, have several long setae on the posterior part of the body. They infest various stored food products, feeding on organic material such as grain, copra, dried fruits, vanilla pods, dried

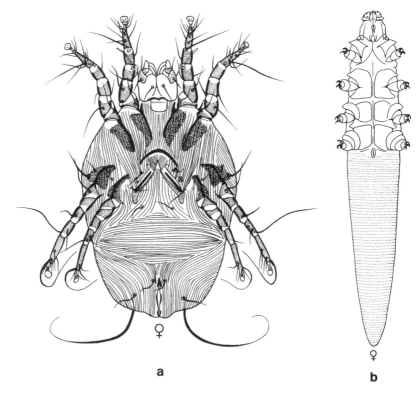

Figure 18.14 Adults of two species of parasitic mites: (*a*) *Dermatophagoides pteronyssinus* (European house dust mite, Pyroglyphidae), female in ventral view; (*b*) follicle mite of the genus *Demodex*, female in ventral view. *Demodex* figure slightly schematic and based on Krantz (1978).

eggs and cheese (Hughes, 1976). Handling these products can cause a contact dermatitis, well known as the occupational hazard called 'grocer's itch'. More specifically, *Tyrophagus putrescentiae* (Fig. 18.15*a,b*) causes copra itch, *Carpoglyphus lactis* causes dried fruit dermatitis, and *Glycyphagus domesticus* causes the typical grocer's itch. *Glycyphagus destructor* (Fig. 18.15*c*) is an inhabitant of hay and can cause rhinitis, itching eyes and wheezing in persons handling hay. Acarid and glycyphagid mites if swallowed with food cause intestinal upsets, or can enter the urinary system or respiratory tract where they are liable to cause irritation or respiratory symptoms – conditions known, respectively, as intestinal acariasis, urinary acariasis and pulmonary acariasis. The mites can be found in faeces, urine and sputum, but there is no evidence that they breed in these bodily products.

Suborder Actinedida

This is a complex and heterogeneous group containing terrestrial species that are plant-feeding, predatory or parasitic, and also aquatic species. The larvae of some free-living species are parasites of vertebrates. The body of these mites is weakly sclerotized and ranges in length from 0.1 to 10 mm. Stigmata are present and situated either forwards on the capitulum or just behind it on the anterior part of the idiosoma (hence the name Prostigmata sometimes used for these mites). The chelicerae can be blade-like, as in *Trombicula*, or stylet-like as in *Pyemotes*. Three of

Biology and medical importance of mites 643

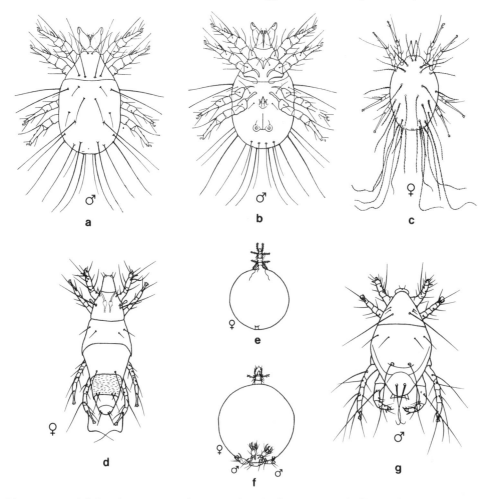

Figure 18.15 Adults of some mites of occasional medical importance: (*a*, *b*) *Tyrophagus putrescentiae* (Acaridae), male in dorsal view (*a*) and ventral view (*b*); (*c*) *Glycyphagus destructor* (Glycyphagidae), female; (*d–g*) *Pyemotes tritici* (hay-itch mite, Pyemotidae), female (*d*), female in gravid condition (*e*), gravid female with males clustered around her genital opening (*f*), male (*g*). The acarid and glycyphagid mites shown are a cause of dermatitis in humans ('grocer's itch').

the 50 or so families in the suborder have some medical importance, namely the Demodicidae, Trombiculidae and Pyemotidae, though mites of some other families – such as *Chyletiella parasitivorax* (Chyletidae) which normally attacks rabbits and cats – occasionally attack man. The family Tetranychidae, including plant-feeding spider-mites of considerable horticultural importance, belongs to the suborder.

Demodicidae
Members of the genus *Demodex* are minute (about 0.1–0.4 mm long) and have an annulated worm-like body (Fig. 18.14*b*). They live head downwards in the hair follicles and sebaceous glands of man and of a wide range of wild and domestic animals. *Demodex* species on different hosts form a group of sibling species which are highly host specific, though more than one species can occur in different tissues

of the same host. Two species infest man. *Demodex folliculorum* lives in the hair follicles, where it spends its entire life cycle, but the stubbier-bodied *D. brevis* inhabits the sebaceous glands. Infestations occur mainly in the facial area, eyelids and around the nose. The pathological significance is hard to assess, especially as the mites are found both in healthy and diseased persons. Most infestations are benign, but in extreme cases there can be dry erythema with follicular scaling; when this occurs in the region of the eyelids it can cause blepharitis. Granulomatous acne could result from facial infestations.

Trombiculidae
Larvae of trombiculid mites are known variously as chiggers, harvest bugs and scrub itch mites and are parasites of mammals (including man) and birds. They are circular or oval (Fig. 18.16), creamy white to bright red, and very tiny – measuring only about 0.25 mm long; as individuals, their small size makes them difficult to see, but larvae feeding in clusters are easily seen. The principal characteristics are a prominent capitulum with toothed chelicerae, conspicuous palps with apposable claws, and a sclerotized scutum bearing (usually) seven setae. The nymphs and adults are free-living predators which live in soil and feed on soil-inhabiting arthropods and their eggs. They have a figure-of-eight shape (Fig. 18.17), measure one millimetre in length, and have a dense covering of red hairs which gives them a velvety appearance – hence their common name of velvet mites. Most of the currently recognized species have been described from the larval stage, so it is necessary to rear nymphs and adults in the laboratory in order to associate them correctly with larvae of known identity. The shape of the scutum, and the shape and disposition of setae on its surface, provide important diagnostic characters for genera and species. Over 1200 species have now been described. About 20 of these are known to attack man, either causing dermatitis or transmitting disease.

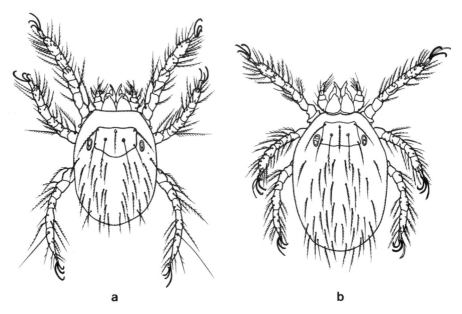

Figure 18.16 Parasitic larval stage of two species of mites of the family Trombiculidae: (*a*) *Neotrombicula autumnalis* (harvest mite); (*b*) *Leptotrombidium akamushi*, vector of scrub typhus in Japan.

The normal hosts of medically important trombiculid larvae are rodents and birds, but the mites will readily attack man. Females lay eggs in damp but well-drained soil, such as river banks, scrub jungle, grassy fields, or neglected gardens with rank growth of grass and weeds. The eggs hatch into six-legged larvae which cluster at the tip of grass and fallen leaves and are stimulated to climb onto passing animal hosts when these brush against such vegetation. Once aboard, the larvae select suitable sites for attachment and feeding, on rodents often clustering inside the ears and on birds around the eyes. On humans they mostly attach where clothing is tight, as around the waist or genitals. The larvae feed on host tissues, sucking up liquefied tissue digested by their saliva. Blood is not ingested in this process, but a few blood cells can sometimes be found in the mite's stomach.

Larvae remain on the skin surface, but the formation of a feeding tube (stylostome) whose hardened walls lead into the skin from the point of attachment enables them to feed continuously. Feeding takes several days, depending on the species, after which the fully fed larvae drop off the host to continue their complex life cycle (Fig. 18.17). The larva develops into an inactive protonymph, which in turn develops into an active predatory deutonymph. A further inactive stage, the

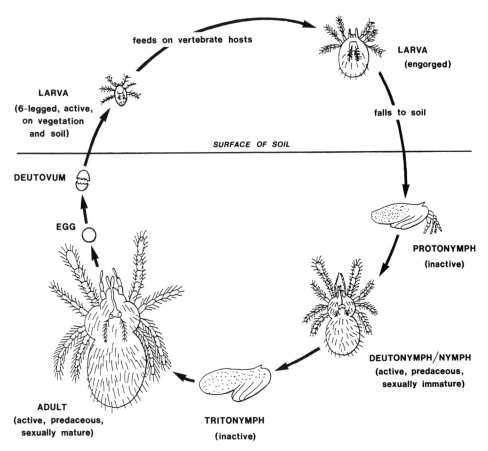

Figure 18.17 The life cycle of a trombiculid mite such as that of the *Leptotrombidium* vectors of scrub typhus.

tritonymph, is followed by the predatory adult. The adults are active, and females become inseminated when they happen to walk over the stalked spermatophores which the males deposit on the substrate.

In the tropics, the life cycle is completed in about 40 days and there is more than one generation per year, but in temperate regions there is only one annual generation and larvae are seasonal, appearing in summer and early autumn.

Dermatitis Worldwide, several species of chiggers, though not transmitting any disease, bite man and often cause an intense itchy dermatitis leading to pustules and wheals a few hours after exposure. The condition is known as trombidiosis or scrub itch. *Neotrombicula autumnalis* (Fig. 18.16a) is the 'harvest mite' of Europe, so-called because the parasitic larvae are active at harvest time in summer and early autumn. It is also known as 'lepte autumnal' and 'aoutat'. The normal hosts are small mammals and birds, but humans are bitten when walking or working in infested fields. In the New World, *Eutrombicula alfreduggesi,* the American chigger mite (also called 'thalzahuatal' and 'bicho colorado'), and *E. batatas*, commonly attack man although their usual hosts are also small mammals and birds.

Disease transmission Trombiculid mites transmit scrub typhus (Tsutsugamushi disease), a zoonotic infection caused by *Rickettsia tsutsugamushi* (= *R. orientalis*). The disease was first described from Japan, but is present in parts of Pakistan and India, China, Korea, much of mainland South East Asia, Indonesia, Philippines, coastal Queensland, New Guinea and some other Melanesian islands, and in certain islands of the Indian Ocean (Fig. 18.18). According to Osuga et al. (1991) it also occurs in Africa. The ecology of the disease is reviewed in Traub and Wisseman (1974).

The disease is found associated with a wide range of biotopes, including flooded river banks (in Japan), scrub, dense but disturbed forests in South East Asia (the result of slash-and-burn cultivation), semi-deserts (in Pakistan) and alpine reaches (in the Himalayas). A more appropriate name for it than 'scrub typhus' is therefore 'chigger-borne rickettsiosis'. The vector mite and the disease most typically occur in 'fringe habitats', i.e. at the meeting point of two major vegetational zones such as forest and scrub.

Larvae of the vector species infest rodents and insectivores and the distribution of the mites is dependent on the home ranges of the hosts – which do not usually overlap. Mite colonies therefore tend to be isolated from each other and to occur as 'mite islands'. Humans get infected when they intrude into an infected mite island. Transmission of the rickettsiae is by bite of infected mites, but only larval mites can acquire and transmit infection since (as in trombiculids generally) only the larval stage of vector species is parasitic. Trans-stadial and trans-ovarial passage in the mites of the rickettsiae that cause the disease is obligatory for maintenance of infection, and the mites are considered to be the main reservoir. The bites produce no irritation, but a lesion, the 'eschar', usually develops at the site of each bite made by an infected chigger.

The genus *Leptotrombidium* contains all the important vectors. *Leptotrombidium deliense* is the principal vector over most of the distribution of the disease (Fig. 18.18) and *L. akamushi* the chief vector in Japan. Other important vectors are *L. fletcheri* in Malaysia, New Guinea and the Philippines, *L. arenicola* on sandy beaches in Malaysia, *L. pallidum* and *L. scutellaris* in limited areas of Japan, and *L. pavlovskyi* in the Far East of the USSR.

Biology and medical importance of mites 647

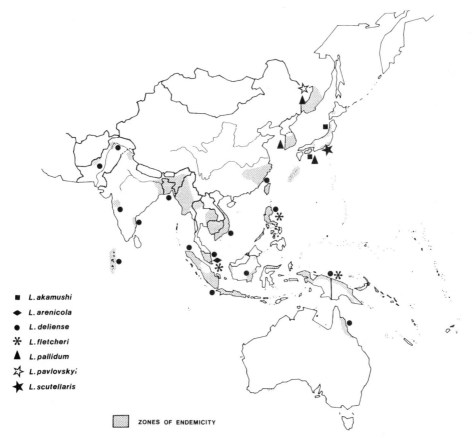

■ L. akamushi
◆ L. arenicola
● L. deliense
✳ L. fletcheri
▲ L. pallidum
☆ L. pavlovskyi
★ L. scutellaris

▨ ZONES OF ENDEMICITY

Figure 18.18 Distribution of species of *Leptotrombidium* which act as vectors of scrub typhus (Tsutsugamushi disease, chigger-borne rickettsiosis) in areas where the disease is endemic.

Pyemotidae
Mites in this family are lightly sclerotized and have reduced palps and stylet-like chelicerae. They are not normally parasitic on man, but *Pyemotes tritici* (erstwhile *P. ventricosus*) (Fig. 18.15d–g), the grain mite or hay itch mite, readily bites agricultural workers and other people who handle infested grain and straw or other infested stored products – though this widespread mite is usually a parasite of the larvae of grain moths, grain beetles, cow pea weevils and rice weevils. The mites do not burrow into the skin, but their bites can cause a severe itchy skin eruption characterized by the development of vesicles which, if scratched, burst and readily become infected by bacteria. The mites have an interesting life cycle on their insect hosts. The fertilized female attaches to an insect larva and swells up enormously as the eggs develop inside her (Fig. 18.15e). The eggs hatch into larvae and the young mites then complete their development inside the female's body, only adult males and females emerging from her genital orifice. The emerged males remain parasitic on their mother, clinging round her genital opening (Fig. 18.15f) for some time and mating with the females as these emerge. The fertilized female then moves away to find another insect larva.

CONTROL OF TICKS AND MITES

Control of ticks

Chemical methods have been the mainstay of tick control for several years and have been directed at tick parasites of livestock. Every group of chemical insecticides, from arsenicals to synthetic pyrethroids, has been used in attempts to control ticks, but insecticidal resistance has always subsequently developed to one or more of them. The number of resistant tick species rose from two in 1947 to at least 15 by 1985. Insecticidal treatment of livestock usually involves dipping the animals in insecticide-filled dip tanks or spraying their entire body surface, but more novel methods include use of insecticide-impregnated ear-tags and pour-on insecticide formulations. Treatment of tick-infested areas by chemicals is neither cost-effective nor environmentally desirable for control of ixodid ticks whose habitats cover extensive ranges, but is often effective against argasid ticks living in restricted habitats (such as *Ornithodoros* ticks in huts). These ticks can also be eradicated without chemical insecticides by filling in cracks and crevices in walls where they hide.

The constraints of chemical control, the most important of which is resistance, have focused attention on alternative methods such as the breeding of livestock that are naturally resistant to tick infestation and on the use of tick vaccines. A genetically engineered vaccine against the one-host cattle tick *Boophilus microplus* has been developed (Rand et al., 1989) and laboratory experiments suggest the feasibility of a vaccine against the three-host African species *Rhipicephalus appendiculatus* (Varma et al., 1990). A holistic approach, using a combination of methods such as conventional insecticide application, ear-tags, exclusion of wild animal hosts from paddocks, selective breeding of animals for tick resistance and induction of host resistance by vaccines, will ultimately be the most successful control strategy.

So far as human infestation is concerned, the best way to prevent attack by ticks is avoidance of tick-infested areas. The use of insecticide-impregnated clothing or repellent creams on the legs will prevent ticks from climbing onto the body and attaching to the skin. Tucking trouser-bottoms into long socks made of closely woven material also helps to prevent ticks from contacting the skin.

A general account of tick control strategies is given in Griffiths and McCosker (1990).

Control of mite infestations

Mite infestations are of very different kinds according to the diverse biology of the different species involved. Space allows only for brief comments here about scabies mite, chiggers, house dusts mites and red poultry mite.

To diagnose scabies it is necessary to inspect the skin for burrows and confirm the presence of mites by extracting them and examining them under a microscope. This can be done by carefully opening the burrows and removing the mites from the burrow ends with a pin or mounted needle (to which they easily stick). Dark faeces in the burrows make them easy to see in light-skinned persons. Treatment consists in covering the entire body from the neck downwards with an insecticidal lotion containing 0.5% malathion or 25% benzyl benzoate; the latter, however, is not well tolerated by children. The lotion should be kept on for a day. If itching persists, a second treatment should be given a week later. Crotamiton (Eurax, containing 10%

N-ethyl-0-crotontoluide) is also very effective. Sulphur ointment is slow but effective, though it can produce dermatitis and being greasy is unpleasant to use. Tetraethylthiuram monosulphide (Tetmosol) is effective as a lotion but of little value in soaps.

For safety from chiggers, personal protection by impregnation of socks and trouser-legs is recommended, using a repellent such as benzyl benzoate, dimethyl phthalate (DMP) or diethyltoluamide (DEET).

House dust mites can be killed by insecticidal treatment of beds and sofas, but this should be followed by thorough vacuum-cleaning to remove the dead mites and their products. Use of carpets made from synthetic fibres instead of wool can be advantageous, as there is some evidence that the static charge which the fibres develop holds particles of dust and so decreases the risk of mite inhalation (Price et al., 1990). Covering bedding fabric with a polyurethane coating ('Ventflex') makes it impermeable to mite antigens and reduces the amount available for direct inhalation from bedding (Owen et al., 1990).

Infestations of red poultry mite can normally be eradicated by insecticidal treatment of the walls of invaded rooms and by removal of birds' nests (the usual source of the mites) (see earlier).

COLLECTING, PRESERVING AND REARING MATERIAL OF TICKS AND MITES

Collection of ticks

Argasidae
Argasid ticks, except for the larvae of *Argas* (which remain attached to hosts for several days), are found in restricted habitats of their hosts and rarely move very far. *Ornithodoros* ticks can be found in loose dry soil of huts, animal burrows and animal resting places (e.g. shade under trees or cracks and crevices in mud-walled animal shelters) and can be collected by passing soil through a metal sieve; those that parasitize sea birds can be found by turning over rocks on which the birds roost. Nest material can be dried for collecting other bird-associated argasids. Ticks can be collected from cracks and crevices and from wooden structures, after first narcotizing them with tobacco smoke blown into such hiding places. Those that harbour under tree bark, such as *Argas arboreus*, can be collected after removing bark and examining the underside, and sometimes from exposed tree trunks. Some species are attracted by CO_2, and dry ice (solid CO_2) can be used successfully in the collection of burrow-dwelling ticks. Collections are quantified and compared on the percentage of infested premises and numbers of ticks collected from each, either by unit volume of soil or by 'man-hour' of collection. (There is little seasonal fluctuation in numbers of argasids since their microhabitats are relatively stable.)

Ixodidae
Collecting methods for unfed stages of ixodid ticks and for the parasitic feeding stages are different. Unfed stages occur in the biotopes where host animals forage or rest, e.g. pastures or forest, or in host burrows or nests in the case of nidicolous species.

Unfed questing ticks are collected by dragging a one-metre-square piece of lint or blanket attached to a string over vegetation. Where this is not practicable, on uneven ground and over shrubs, the blanket is attached to a pole and waved like a flag to make contact with the vegetation. However, 'dragging' or 'flagging' should not be done continuously over long distances or for a long time as this allows collected ticks to be rubbed off, back into the vegetation. Blanket-dragging is quantified by pulling the blanket over a measured area or swathe at a uniform pace, the unit drag. Collections are expressed as the number of ticks collected per man-hour or unit time of actual dragging or flagging.

Another way of collecting unfed questing ticks is to use CO_2. Many species are attracted by carbon dioxide and can be collected in a CO_2-baited trap, a conical device containing subliming dry ice; ticks attracted to the trap become stuck to adhesive tape attached to the trap base. The main advantage of this method is that it is independent of temperature or sunlight and captures ticks at times of the year when drag collecting is ineffective. The effective sampling area of the trap is 25 m^2 and more ticks can be collected by CO_2 trapping than by dragging or flagging (Kinzer et al., 1990).

The parasitic stages of ixodids, which remain attached to animals and feed for several days, can be collected directly from the host. However, such ectoparasite collections can represent only a small fraction of the ixodid population in any area; foraging animals move over large areas and act as 'biodrags' or 'bioflags', collecting and concentrating ticks on their bodies that originate from large areas difficult or time-consuming to sample. Individual ticks can be collected directly from a host animal by placing the tip of curved forceps under the anterior end of the tick immediately above the animal's skin and pulling gently. Alternate rotation of the tick's body clockwise and anticlockwise minimizes the risk of the mouthparts being left in the skin. Special care is needed, however, with immature and partially fed ticks as these are harder to remove than adults and engorged ticks.

If time is limited and only a rough idea of the tick fauna of an area is required, examination of livestock in villages, markets and abattoirs is helpful. However, domestic animals carry only about 10% of all tick species. Wild animals support a diversity of ticks and small mammals and birds are often parasitized by immature stages of species whose adults occur on larger wild and domestic animals. Hunters can be recruited to preserve ticks removed from shot or trapped wild animals. Immature stages from small shot, netted or trapped mammals or birds can be collected by individually wrapping the animals in lint (fluffy side inwards), placing them in plastic bags and leaving them in a refrigerator or in the cold outdoors overnight; the ticks then detach, leave the cooling bodies of the animals and get entangled in the fluff.

Living trapped animals should be lightly anaethetized with ether and examined for attached ticks. On small mammals these are usually found around the eyes, on the snout and on both surfaces of the ear pinnae; on birds they are found mostly around the eyes and the ear openings, at the base of the beak, and under the wings. As unfed or slightly fed larvae and nymphs can be difficult to see it is easiest to obtain these if they are allowed to finish feeding and drop from the host. The host animal can either be put into a cage (with food) set over a tray of water, or into a cylindrical cage which is closed at each end with wire-mesh and enclosed in a stout cloth bag secured with a draw-string; fed ticks will then drop into the water or into the bag. Engorged larvae and nymphs should be kept alive so they can develop to nymphs and adults; the moulted exuviae should be preserved for specific identification and for correct association of immature stages with adults.

Ectoparasite collections from small animals are quantified and compared by infestation rate (number of animals with ticks/total number of animals examined) and number per infested animal, but this is often not possible with large animals because of the difficulty of collecting all the ticks, particularly when they are larvae or are distributed all over the host's body. Many species, however, have favourite sites of attachment (predilection sites): for example, adults of the tropical horse tick (*Anocentor nitens*) and the brown ear tick (*Rhipicephalus appendiculatus*) feed inside the ear pinnae and along their fringes. Collections from such defined areas are referred to as the standard count, which for most species is the number on the head and neck. Other counts are total counts, i.e. the number of ticks from the entire body surface, or half-body counts for which either the left or right side of the body is examined and all attached ticks collected. Large animals such as cattle usually need to be immobilized ('thrown') or confined in a crush-pen before the ticks can be collected from them and a count made.

Ectoparasite collections should record the locality, nearest map coordinates, date of collection, number and species of host animal, host weight and sex, method of host collection (animal shot, netted, trapped or from market, abattoir, village), tick species, tick sex and life stage, and the state of tick engorgement. The fate of collected ticks should also be noted, e.g. held for moulting/egg laying, preserved, or processed for pathogen isolation. (Ixodids particularly require high humidity for survival. Ticks collected from vegetation or animals, if to be kept alive, should be placed in tubes that contain some damp plaster-of-Paris or a few blades of fresh grass.)

Preserving and examining ticks
Forcible removal of ticks often results in damaged or lost capitula, making identification even to genus difficult. Flat, unfed ticks collected by dragging are the easiest to identify. With engorgement, it becomes progressively more difficult to see the capitulum and scutum clearly. Ticks collected from animals are often encrusted with dried blood or exudates or have host tissue attached to their mouthparts. Live ticks with attached host tissue will clean themselves in an hour or so if kept in a tube. Ticks encrusted with blood or exudates can be cleaned in 70% ethanol with a soft artist's brush on which the bristles have been cut short to uniform length. All stages can be preserved in 70% ethanol, but for 'ornate' ticks a preservative such as Pampel's fluid (2 ml of glacial acetic acid, 6 ml of 40% formalin, 30 ml of distilled water, 15 ml of 95% ethanol) will prevent the scutal pattern from fading. Live ticks can be placed directly in preservative fluids without needing to be killed first in hot water. For identification, adults should be examined dry under a dissecting microscope, preserved specimens being allowed to dry so they are free of any surface sheen produced by the liquid preservative. Smaller larvae and nymphs are best examined under a compound microscope after they have been slide-mounted in a gum chloral mixture such as Hoyer's solution (recipe below).

Rearing ticks
Most ticks of medical importance are not host specific and all stages can be maintained in the laboratory on rabbits and guinea pigs. Species routinely maintained on these hosts include *Amblyomma hebraeum* and *A. variegatum*, *Dermacentor reticulatus*,

Haemaphysalis spinigera, *Ixodes ricinus* and *Rhipicephalus sanguineus* and *R. appendiculatus*. The same animal should not be used for repeated infestations as ticks induce protective immunity in host animals and this can impair the feeding performance of subsequent batches of ticks. The methods for rearing the rapid-feeding argasids and slow-feeding ixodids are slightly different, but for both groups it is advisable to confine the ticks in feeding bags or capsules, either for the entire feeding period (ixodids and slow-feeding *Argas* larvae) or until the ticks have attached (rapid-feeding argasids). Keeping caged animals on a moated table with oil in the moat will prevent the escape of stray loose ticks. The bird-feeding *Ornithodoros* and *Argas* species can be fed on pigeons, using feeding capsules fixed to the axillary region of the birds (Kaiser, 1966). Mammal-biting *Ornithodoros* can be fed on the ears of an immobilized rabbit after placing the animal in a restraining cage. Feeding-bags made of heavy duty drill, or aluminium feeding capsules for ixodids, are fixed to the animals' backs (the ticks have predilection sites on their natural hosts but will feed readily on the backs of laboratory rabbits and guinea pigs). It is not necessary to shave the feeding area on the host animal, although shaving makes it easier to fix the feeding-bags securely with adhesive tape.

Unfed ticks of all stages, and engorged ticks, can be kept in glass or clear plastic specimen tubes closed with fine nylon netting. Egg-laying females are best kept in Petri dishes, from which eggs can be harvested and tubed when oviposition has ceased. As many ticks require high humidity for survival, the tubes should be placed in desiccator cabinets with salt solutions to provide the required humidity. Alternatively, ticks can be kept in tubes closed at one end with damp plaster-of-Paris (six parts) and activated charcoal (one part), plugged with cotton wool in nylon gauze, and stood upright in a tray of wetted sand to maintain a high humidity. To obviate the risk of ticks escaping, rearing tubes should be handled on moated table tops or (if these are not available) in white enamel trays containing water. Wetted camel-hair brushes are the best implements for picking up larvae and nymphs, forceps for picking up adults.

Collecting and preserving mites

Parasitic mites either attach themselves to a host for several days (chiggers) or only for short periods – in which case they can be found moving around in the host's fur or feathers. Many parasitic mites breed and remain in the burrows or nests of the hosts, boarding them only to feed. Collection methods therefore depend greatly on the mites' habits. Nest material or burrow litter is processed by using a modified Tullgren apparatus based on the Berlese funnel (see Azad, 1986). This consists of an ordinary metal funnel 30–50 cm in diameter with a sieve or screen inside and an incandescent bulb suspended above. Samples (no more than 12 litres in volume) are placed on the screen and held there for at least four days. Heat from the bulb slowly dries out the material and this forces the mites to move down the funnel into a collecting jar screwed to the bottom of the apparatus and containing 70% alcohol. Care is needed, as too much heat from the bulb kills the mites and too little prolongs the collection period. A frosted 75-watt bulb is satisfactory. Where there is no electricity the apparatus can be used outside in a place where heat from the sun can dry out the samples.

Collecting methods for obtaining parasitic mites attached to, and feeding on, small wild mammal and bird hosts are essentially similar to those described above for ticks. Living animals must first be collected by trapping, shooting or netting. When trapping, the use of 'live' traps is advisable, otherwise the mites will leave the dead, cooling body of the host and be lost. Mites from freshly dead animals can be collected by wrapping the individual animal in lint (fluffy side inwards) in securely closed plastic bags and leaving them overnight in the cold. The following morning the lint is opened and the detached mites caught on it removed with the moistened tip of a fine camel-hair brush. They can then be preserved in 70% ethanol. Trapped live animals should be anaesthetized and shaken over a white enamel tray, the fur being combed to dislodge any mites moving about on the body.

Trombiculid mite larvae remain attached to their hosts for several days, and these are best collected by placing the live infested animal in a wire-mesh cage over a tray of water – after first lightly anaesthetizing the animal and examining it for attached mites. The engorged mites which drop into the water can be picked up with a brush and preserved or used for rearing. To collect unfed trombiculid mites in an infested area a useful method is to place discs of black laminate or porcelain on the ground. Chiggers crawl onto these in a few minutes and can be picked up with a brush. Nymphs and adults, on the other hand, can be collected by floatation: samples of surface soil are mixed with water and the mites simply float to the surface.

For collecting house dust mites, dust samples are removed from mattresses and settees with a small portable vacuum-cleaner to which is attached a small polythene sampling tube (measuring about 7.5 × 2.5 cm). On completion of sampling, the tubes are tightly closed. In the laboratory the dust is shaken out into a Petri dish and examined under a dissecting microscope at ×40 magnification. After removing the live mites, for preservation or use in rearing, the dead mites are collected by covering the dust sample in the Petri dish with 90% lactic acid and placing the dish at 50°C for one to two days. The sample is then mixed with distilled water and centrifuged for five minutes at 2000 rpm. The supernatant fluid is removed and examined for dead mites.

Mites can be preserved for short periods (up to three months) in 70% alcohol, but after this time they tend to harden. It is therefore advisable to collect them directly into, or to transfer them into, another preservative, such as Oudeman's fluid; this consists of glycerine (five parts), 70% alcohol (87 parts) and glacial acetic acid (eight parts). Soft-bodied or weakly sclerotized mites can be placed directly on a microscope slide in a drop of mountant such as Hoyer's medium (made from 50 ml distilled water, 30 g crystalline gum arabic, 200 g chloral hydrate and 20 ml glycerine). Sclerotized mites are first cleared by placing them for a few hours to several days, depending on the degree of sclerotization, in lactophenol (50 parts lactic acid, 25 parts phenol crystals, 25 parts distilled water). After washing in several changes of water to remove all traces of lactophenol, the specimens are orientated in a drop of mounting medium on a slide and covered with a coverglass. Mounted specimens are then 'baked' in a 50°C incubator for several days and the coverglass afterwards ringed with clear nail varnish before microscopical examination and study. Only one specimen should be mounted on each slide.

Rearing mites
Almost any container can be used for rearing trombiculid mites, but it is necessary to maintain a high humidity as damp moist conditions are required for completion of the life cycle. Small porous earthenware pots about 8 cm wide and 6 cm deep are cheap and very satisfactory. The pots can be kept moist by placing them in a tray of

wet sand. The bottom of the container needs to be covered with suitable substrate such as regular or medium-sized vermiculite to a depth of about 0.5 cm (Farrell and Wharton, 1949). This forms a porous and moist bed through which all stages of the mite can move easily. The pots should be covered with clear glass plates. When nymphs are seen to be active they can be provided with food put into the pots to feed both nymphs and adults. Mosquito eggs make good food, and disaggregated egg rafts of *Culex* mosquitoes have been used successfully in rearing *Leptotrombidium deliense*. Unfed larvae ascend the side of the pot and congregate round its rim under the glass covering plate, whence they can be removed into water using the moistened tip of a fine brush. Once in water they will form a tight cluster which can then be picked up with a brush and placed directly on the ears of a mouse confined in a tubular metal wire-mesh tube suspended over water. After about six hours the mouse is removed from the tube and placed (with food and drinking water) in a metal wire-mesh cage standing in a tray of water. The water in the tray is examined and changed twice a day, any floating engorged mites being removed to start a new breeding-pot.

Culture methods have been successful for the house dust mite *Dermatophagoides farinae* in which mites are kept in cells between glass plates and fed with flakes of fish food.

REFERENCES

Aeschlimann, A. 1967. Biologie et écologie des tiques (Ixodoidea) de Côte d'Ivoire. *Acta Tropica* **24**: 281–405.

Anastos, G. 1950. The scutellate ticks, or Ixodidae, of Indonesia. *Entomologica Americana* **30**: 1–144.

Anastos, G. 1957. *The ticks or Ixodides of the U.S.S.R.: a review of the literature.* vi + 397 pp. Department of Health, Education and Welfare, Washington, D.C. [= Public Health Service Publication No. 548.]

Arlian, L. G. 1989. Biology, host relations, and epidemiology of *Sarcoptes scabiei*. *Annual Review of Entomology* **34**: 139–161.

Arthur, D. R. 1963. *British ticks.* ix + 213 pp. Butterworths, London.

Arthur, D. R. 1965. *Ticks of the genus Ixodes in Africa.* viii + 348 pp. Athlone Press, London.

Audy, J. R., **Nadchatram, M.** and **Lim, B.-l.** 1960. Malaysian parasites XLIX. Host distribution of Malayan ticks (Ixodoidea). [In Malaysian parasites XXV–XLIX.] *Studies from the Institute for Medical Research*, Kuala Lumpur **29**: 225–246.

Azad, A. F. 1986. Mites of public health importance and their control. Mimeographed document WHO/VBC/86.931, 52 pp. World Health Organization, Geneva.

Baker, E. W. and **Wharton, G. W.** 1952. *An introduction to acarology.* xiii + 465 pp. Macmillan, New York.

Baker, E. W., **Evans, T. M.**, **Gould, D. J.**, **Hull, W. B.** and **Keegan, H. L.** 1956. *A manual of parasitic mites of medical or economic importance.* 170 pp. National Pest Control Association, New York.

Bedford, G. A. H. 1934. South African ticks. Part I. *Onderstepoort Journal of Veterinary Science and Animal Industry* **2**: 49–99.

Cheng, G. and **Pang, D.** 1982. Identification of Acari in China. Pp. 725–838 in Lu, B. S. (ed.), *Identification handbook for medically important animals in China.* 956 pp. People's Health Publication Company, Beijing. [In Chinese.]

Cooley, R. A. and **Kohls, G. M.** 1944. The Argasidae of North America, Central America and Cuba. *American Midland Naturalist Monograph* **1**: 1–152.

Cooley, R. A. and Kohls, G. M. 1945. The genus *Ixodes* in North America. *National Institutes of Health Bulletin* **184**: iii + 1–246. United States Government Printing Office, Washington D.C.

Doss, M. A., Farr, M. M., Roach, K. F. and Anastos, G. 1974. Ticks and tickborne diseases I. Genera and species of ticks. *Index-Catalogue of Medical and Veterinary Zoology (Special Publication)*: vol. 3, Part 1 Genera A–G, 429 pp.; Part 2 Genera H–N, 593 pp.; Part 3, Genera O–X, 329 pp. Agricultural Research Service, United States Department of Agriculture, Washington D.C.

Doss, M. A., Farr, M. M., Roach, K. F. and Anastos, G. 1978. Ticks and tickborne diseases IV. Geographical distribution of ticks. *Index-catalogue of Medical and Veterinary Zoology (Special Publication)*: vol. 3, 648 pp. Science and Education Administration, United States Department of Agriculture, Washington D.C.

Dumbleton, L. J. 1953. *The ticks (Ixodoidea) of the New Zealand sub-region*. 28 pp. [+ 7 unnumbered pp. figs]. Wellington. [= Cape Expedition Series Bulletin 14.]

Estrada-Peña, A. 1989. Indice-catalogo de las garrapatas (Acarina: Ixodoidea) en el mundo [also title in English]: vol. i: genero *Haemaphysalis*. iv + 932 pp. Secretariado de Publicaciones de la Universidad de Zaragoza, Zaragoza.

Fain, A. 1968. Études de variabilité de *Sarcoptes scabiei* et une révision des Sarcoptidae. *Acta Zoologica et Pathologica Antverpiensia* **47**: 1–196.

Fain, A., Guérin, B. and Hart, B. J. 1990. *Mites and allergic disease*. 190 pp. Allerbio, Varennes en Argonne, France.

Fairchild, G. B., Kohls, G. M. and Tipton, V. J. 1966. The ticks of Panama (Acarina: Ixodoidea). Pp. 167–219 in Wenzel, R. L. and Tipton, V. J. (eds), *Ectoparasites of Panama*. xii + 861 pp. Field Museum of Natural History, Chicago.

Farrell, C. E. and Wharton, G. W. 1949. A culture medium for chiggers (Trombiculidae). *Journal of Parasitology* **35**: 435.

Griffiths, R. B. and McCosker, P. J. (eds) 1990. Proceedings of the FAO Expert Consultation of revision of strategies for the control of ticks and tick–borne diseases (Rome, 25–29 September 1989). *Parassitologia* **32**: 1–209.

Hoogstraal, H. 1953. Ticks (Ixodoidea) of the Malagasy faunal region. Their origins and host-relationships; with descriptions of five new *Haemaphysalis* species. *Bulletin of the Museum of Comparative Zoology* (Harvard) **111**: 37–113.

Hoogstraal, H. 1956. *African Ixodoidea. I. Ticks of the Sudan.* ii + 1101 pp. Bureau of Medicine and Surgery, Department of the Navy [Washington, D.C].

Hoogstraal, H. 1966. Ticks in relation to human diseases caused by viruses. *Annual Review of Entomology* **11**: 261–308.

Hoogstraal, H. 1967. Ticks in relation to human diseases caused by *Rickettsia* species. *Annual Review of Entomology* **12**: 377–420.

Hoogstraal, H. 1970–1982. *Bibliography of ticks and tickborne diseases:* vol. 1, 499 pp. (1970); vol. 2, 495 pp. (1970); vol. 3, 435 pp. (1971); vol. 4, 354 pp. (1972); vol. 5 (Part I), 492 pp. (1974); vol. 5 (Part II), 455 pp. (1978); vol. 6, 407 pp. (1981); vol. 7, 219 pp. (1982).

Hoogstraal, H. 1973. Viruses and ticks. Pp. 349–390 in Gibbs, A. J. (ed.), *Viruses and invertebrates*. xiii + 673 pp. North Holland Publishing Company, Amsterdam. [= North Holland Research Monographs, Frontiers of Biology, 31.]

Hoogstraal, H. 1979. The epidemiology of tick-borne Crimean–Congo hemorrhagic fever in Asia, Europe, and Africa. *Journal of Medical Entomology* **15**: 307–417.

Hoogstraal, H. 1980. Established and emerging concepts regarding tick-associated viruses, and unanswered questions. Pp. 49–63 in Vesenjak-Hirjan, J., Porterfield, J. S. and Arslanagic, E. (eds), Arboviruses in the Mediterranean countries. *Zentralblatt für Bakteriologie, Mikrobiologie und Hygiene* Abteilung I (Supplement) **9**: xv + 1–332. Gustav Fischer, Stuttgart. [Report of Sixth Symposium of the Federation of European Microbiological Societies.]

Hoogstraal, H. and Aeschlimann, A. 1982. Tick–host specificity. *Mitteilungen der Schweizerischen Entomologischen Gesellschaft* **55**: 5–32.

Hoogstraal, H. and **Kaiser, M. N.** 1958. The ticks (Ixodoidea) of Egypt. A brief review and keys. *Journal of the Egyptian Public Health Association* **33**: 51–85.

Hoogstraal, H. and **Kaiser, M. N.** 1959. Ticks (Ixodoidea) of Arabia. With special reference to the Yemen. *Fieldiana: Zoology* **39**: 297–322.

Hoogstraal, H. and **Kaiser, M. N.** 1960. Observations on ticks (Ixodoidea) of Libya. *Annals of the Entomological Society of America* **53**: 445–457.

Hoogstraal, H. and **Kim, K. C.** 1985. Tick and mammal coevolution, with emphasis on *Haemaphysalis*. Pp. 505–568 in Kim, K. C. (ed.), *Coevolution of parasitic arthropods and mammals*. xiv + 800 pp. John Wiley, New York.

Hoogstraal, H., Wassef, H. Y. and **Büttiker, W.** 1981. Ticks (Acarina) of Saudi Arabia: Fam. Argasidae, Ixodidae. *Fauna of Saudi Arabia* **3**: 25–110.

Hughes, A. M. 1976. *The mites of stored food and houses*. Second edition. 400 pp. Her Majesty's Stationery Office, London. [= Ministry of Agriculture, Fisheries and Food Technical Bulletin 9.]

Hunt, L. M. 1986. Differentiation between three species of *Amblyomma* ticks (Acari: Ixodidae) by analysis of cuticular hydrocarbons. *Annals of Tropical Medicine and Parasitology* **80**: 245–249.

Hunt, L. M. and **Hilburn, L. R.** 1985. Biochemical differentiation between species of ticks (Acari: Ixodidae). *Annals of Tropical Medicine and Parasitology* **79**: 525–532.

Kaiser, M. K. 1966. The subgenus *Percicargus* (Ixodoidea, Argasidae, *Argas*). 3. The life cycle of *A. (P.) arboreus*, and a standardized rearing method for argasid ticks. *Annals of the Entomological Society of America* **59**: 496–502.

Kinzer, D. R., Presley, S. M. and **Hair, J. A.** 1990. Comparative efficiency of flagging and carbon dioxide-baited sticky traps for collecting the lone star tick *Amblyomma americanum* (Acarina: Ixodidae). *Journal of Medical Entomology* **27**: 750–775.

Kohls, G. M. 1950. Ticks (Ixodoidea) of the Philippines. *National Institutes of Health Bulletin* **192**: 1–28. United States Government Printing Office, Washington, D.C.

Kohls, G. M. 1957a. Insects of Micronesia: Acarina: Ixodoidea. *Insects of Micronesia* **3**: 84–104.

Kohls, G. M. 1957b. Malaysian parasites XVIII. Ticks (Ixodoidea) of Borneo and Malaya. [In Malaysian parasites XVI–XXXIV.] *Studies from the Institute for Medical Research*, Kuala Lumpur **28**: 65–94.

Krantz, G. W. 1978. *A manual of acarology*. Second edition. vii + 509 pp. Oregon State University, Corvallis.

Lane, R. S., Piesman, J. and **Burgdorfer, W.** 1991. Lyme borreliosis: relation of its causative agent to its vectors and hosts in North America and Europe. *Annual Review of Entomology* **36**: 587–609.

Le Gac, P. 1966. Répercussion de la myxomatose sur la fièvre exanthématique boutonneuse méditerranéenne. *Bulletin of the World Health Organization* **35**: 142–147.

MacLeod, J. 1970. Tick infestation patterns in the southern province of Zambia. *Bulletin of Entomological Research* **60**: 253–274.

Matthyesse, J. G. and **Colbo, M. H.** 1987. *The ixodid ticks of Uganda: together with species pertinent to Uganda because of their present known distribution*. viii + 426 pp. Entomological Society of America, Maryland, USA.

McDaniel, B. 1979. *How to know the mites and ticks*. viii + 335 pp. Wm C. Brown, Dubuque, Iowa.

Needham, G. R. and **Teel, P. D.** 1991. Off-host physiological ecology of ixodid ticks. *Annual Review of Entomology* **36**: 659–681.

Nuttall, G. H. F. 1916. Ticks of the Belgian Congo and the diseases they convey. *Bulletin of Entomological Research* **6**: 313–352.

Nutting, W. B. (ed.) 1984. *Mammalian diseases and arachnids:* vol. 1, *Pathogen biology and clinical management*, [xv] + 277 pp.; vol. 2, *Medico-veterinary, laboratory, and wildlife diseases, and control*, [xiv] + 280 pp. CRC Press Inc., Boca Raton, Florida.

Obenchain, F. D. and **Galun, R.** 1982. *Physiology of ticks*. xii + 509 pp. Pergamon Press, Oxford.

Oliver, J. H. 1972. Cytogenetics of ticks (Acari: Ixodoidea), 6. Chromosomes of *Dermacentor* species in the United States. *Journal of Medical Entomology* **9**: 177–182.

Osuga, T., Kimura, M., Goto, H., Shimada, K. and **Suto, T.** 1991. A case of tsutsugamushi disease probably contracted in Africa. *European Journal of Clinical Microbiology and Infectious Diseases* **10**: 95–96.

Oswald, B. 1939. On Yugoslavian (Balkan) ticks (Ixodoidea). *Parasitology* **31**: 271–280.

Owen, S., Morganstern, M., Hepworth, J. and **Woodcock, A.** 1990. Control of house dust mite antigen in bedding. *Lancet* **1990** (335): 396–397.

Pegram, R. G. 1976. Ticks of the northern regions of the Somali Democratic Republic. *Bulletin of Entomological Research* **66**: 345–363.

Pegram, R. G. and **Walker, J. B.** 1988. Clarification of the biosystematics and vector status of some African *Rhipicephalus* species (Acarina: Ixodidae). Pp. 61–76 in Service, M. W. (ed.), *Biosystematics of haematophagous insects*. xi + 363 pp. Clarendon Press, Oxford.

Perry, B. D., Lessard, P., Norval, R. A. I., Kundert, K. and **Kruska, R.** 1990. Climate, vegetation and the distribution of *Rhipicephalus appendiculatus* in Africa. *Parasitology Today* **6**: 100–104.

Pomerantzev, B. I. 1950. Ixodid ticks (Ixodidae). *Fauna of the USSR*. New Series. No. 41, Arachnids **4** (2), 223 pp. Akademii Nauk SSSR, Moscow and Leningrad. [In Russian: English translation, 199 pp., American Institute of Biological Sciences, Washington, D.C.]

Price, J. A., Pollock, I., Little, S. A., Longbottom, J. L. and **Warner, J. O.** 1990. Measurement of airborne mite antigen in homes of asthmatic children. *Lancet* **1990** (336): 895–897.

Rand, K. N., Moore, T., Sriskantha, A., Spring, K., Tellam, R., Willadsen, P. and **Cobon, G. S.** 1989. Cloning and expression of a protective antigen from the cattle tick *Boophilus microplus*. *Proceedings of the National Academy of Sciences of the United States of America* **86**: 9657–9661.

Řeháček, J. and **Tarasevich, I. V.** 1988. *Acari-borne rickettsiae and rickettsioses in Eurasia*. 343 pp. Veda Publishing House of the Slovak Academy of Sciences, Bratislava.

Roberts, F. H. S. 1970. *Australian ticks*. 267 pp. Commonwealth Scientific and Industrial Research Organization, Australia.

Sauer, J. R. and **Hair, J. A.** (eds) 1986. *Morphology, physiology, and behavioral biology of ticks*. 510 pp. Ellis Horwood, Chichester, and John Wiley, New York.

Sharif, M. 1928. A revision of the Indian Ixodidae with special reference to the collection in the Indian Museum. *Records of the Indian Museum* **30**: 217–344.

Sims, R. W. (ed.) 1980. *Animal identification: a reference guide*: vol. 2, Land and freshwater animals (*not* insects). x + 120 pp. British Museum (Natural History), London, and John Wiley, Chichester.

Smith, M. W. 1969a. Variations in tick species and populations in the Bugisu District of Uganda. Part I. The tick survey. *Bulletin of Epizootic Diseases of Africa* **17**: 55–75.

Smith, M. W. 1969b. Variations in tick species and populations in the Bugisu District of Uganda. Part II. The effects of altitude, climate, vegetation and husbandry on tick species and populations. *Bulletin of Epizootic Diseases of Africa* **17**: 77–106.

Sonenshine, D. E. 1986. Tick pheromones. *Current Topics in Vector Research* **2** (1984): 225–263.

Sonenshine, D. E. 1991. *Biology of ticks*. Vol. 1. xix + 447 pp. Oxford University Press, New York and Oxford.

Stanton, B., Khanam, S., Nazrul, H., Nurani, S. and **Khair, T.** 1987. Scabies in urban Bangladesh. *Journal of Tropical Medicine and Hygiene* **90**: 219–226.

Starkoff, O. 1958. *Ixodoidea d'Italia: studio monografico*. 384 pp. [+ 1 unnumbered p. addenda]. 'Il Pensiero Scientifico' Editore, Rome.

Stewart, R. S. 1991. Flinders Island spotted fever: a newly recognized endemic focus of tick typhus in Bass Strait. Part 1. Clinical and epidemiological features. *Medical Journal of Australia* **154**: 94–99.

Strandtmann, R. W. and **Wharton, G. W.** 1958. A manual of mesostigmatid mites parasitic on vertebrates. *University of Maryland Institute of Acarology Contribution* **4**: xi + 1–330 [+ 1–69 figures].

Theiler, G. 1943. *Notes on the ticks off domestic stock from Portuguese East Africa.* 55 pp. [+ 14 unnumbered pp. of figs]. Imprensa Nacional de Moçambique, Lourenço Marques [= Maputo].

Toumanoff, C. 1944. *Les tiques (Ixodoidea) de l'Indochine: recherches faunistiques avec indications sur les Ixodides des pays voisins: notions générales sur la biologie et les moyens combaître ces Acarines.* ii + 220 pp. Société des Imprimeries et Librairies Indochinoises, Saigon.

Trapido, H., Varma, M. G. R., Rajagopalan, P. K., Singh, K. R. P. and **Rebello, M. J.** 1964. A guide to the identification of all stages of *Haemaphysalis* ticks of South India. *Bulletin of Entomological Research* **55**: 249–270.

Traub, R. and **Wisseman, C. L.** 1974. The ecology of chigger-borne rickettsiosis (scrub typhus). *Journal of Medical Entomology* **11**: 237–303.

Traver, J. R. 1951. Unusual scalp dermatitis in humans caused by the mite, *Dermatophagoides* (Acarina, Epidermoptidae). *Proceedings of the Entomological Society of Washington* **53**: 1–25.

Uilenberg, G., Hoogstraal, H. and **Klein, J.-M.** 1980. Les tiques (Ixodoidea) de Madagascar et leur role vecteur. *Archives de l'Institut Pasteur de Madagascar* (Numéro Special) (1979): 1–153.

Varma, M. G. R. 1973. Ticks (Ixodidae) of British Honduras. *Transactions of the Royal Society of Tropical Medicine and Hygiene* **67**: 92–102.

Varma, M. G. R., Heller-Haupt, A., Trinder, P. K. E. and **Langi, A. O.** 1990. Immunization of guinea-pigs against *Rhipicephalus appendiculatus* adult ticks using homogenates from unfed immature ticks. *Immunology* **71**: 133–138.

Vercammen-Grandjean, P. H. 1968. *The chigger mites of the Far East (Acarina: Trombiculidae and Leeuwenhoekiidae): an illustrated key and a synopsis; some new tribes, genera and subgenera.* 135 pp. U.S. Army Medical Research and Development Command, Washington, D.C.

Vercammen-Grandjean, P. H. and **Langston, R.** 1976. *The chigger mites of the world:* vol. 3, *Leptotrombidium* complex: Section A, *Leptotrombidium* s.str., pp. 1–612 [+ 2 unnumbered pp.]; Section B, *Trombiculindus*, *Hypotrombidium* and *Ericotrombidium*, plus heterogenera, pp. 613–1061; Section C, Iconography, [4 unnumbered pp.] + 298 pp. George Williams Hooper Foundation, University of California, San Francisco.

Waladde, S. M. and **Rice, M. J.** 1982. The sensory basis of tick feeding behaviour. *Current Themes in Tropical Science* **1**: 71–118.

Walker, J. B. 1974. *The ixodid ticks of Kenya: a review of present knowledge of their hosts and distribution.* xi + 220 pp. Commonwealth Institute of Entomology, London.

Wang, D.-q. 1988. Biosystematic problems in relation to the vectors of scrub typhus in China. Pp. 177–191 in Service, M. W. (ed.), *Biosystematics of haematophagous insects.* xi + 363 pp. Clarendon Press, Oxford.

Wang, D.-q. and **Yu, Z.-z.** 1992. Chigger mites of the genus *Leptotrombidium*: key to species and their distribution in China. *Medical and Veterinary Entomology* **6**: 389–395.

Wen, T. (ed.) 1984. *Sand mites of China (Acariformes: Trombiculidae and Leeuwenhoekiidae).* 2 + ix + 370 pp. Xue Lin Publishing House, China.

Wharton, G. W. 1976. House dust mites. *Journal of Medical Entomology* **12**: 577–621.

Wilde, J. K. H. (ed.) 1978. *Tick-borne diseases and their vectors.* xix + 573 pp. Centre for Tropical Veterinary Medicine, University of Edinburgh.

Woolley, T. A. 1970. Some observations on external anatomy of oribatid mites by the scanning electron microscope. *Bioscience* **20**: 1253–1257.

Woolley, T. A. 1988. *Acarology: mites and human welfare.* xix + 484 pp. John Wiley, New York.

Yeoman, G. H. and **Walker, J. B.** 1967. *The ixodid ticks of Tanzania; a study of the zoogeography of the Ixodidae of an East African country.* xii + 215 pp. Commonwealth Institute of Entomology, London.

Yunker, C. 1973. Mites. Pp. 425–492 in Flynn, R. J. (ed.), *Diseases of laboratory animals.* xvi + 884 pp. Iowa State University Press, Ames, Iowa.

CHAPTER NINETEEN

Spiders and scorpions (Araneae and Scorpiones)

J. L. Cloudsley-Thompson

Spiders (Araneae) and scorpions (Scorpiones) are two orders of the class Arachnida, subphylum Chelicerata. In addition to ticks and mites (Acari), the Arachnida also include whip-scorpions (Thelyphonida), tailless whip-scorpions (Amblypygi), wind-scorpions or camel-spiders (Solifugae), false-scorpions (Pseudoscorpiones), harvestmen (Opiliones) and some other less important orders. None of these minor orders has any medical significance. Larger Solifugae are alleged to give very painful bites, but this is a most unusual occurrence and these arachnids are not venomous; *Mastigoproctus giganteus*, the vinegaroon of southern USA and Mexico and the largest of the whip-scorpions, ejects a defensive spray from a movable knob at the base of its flagellum that can blister sensitive human skin and stain the fingers but in other respects the species is harmless.

Most spiders give poisonous bites but few have jaws sufficiently powerful to penetrate human skin; and even among these the venoms are, with few exceptions, harmless. In the case of the larger tarantulas (Mygalomorphae) the urticating hairs that clothe the body can cause more pain than the bite. The stings of scorpions, on the other hand, are often very painful and those of several species can even endanger human life.

Spiders are widely distributed throughout the terrestrial world. Over 30 000 species have been described, but the total probably exceeds 100 000. Spiders are grouped into approximately 60 families whose interrelations are still a matter of controversy – over 18 different systems of classification have been proposed during the present century. All spiders, except for members of the family Uloboridae, possess a pair of poison glands; but only 20–30 species are dangerously poisonous to humans. The most important of these are, undoubtedly, the North American black widow *Latrodectus mactans* (Theridiidae) and congeneric species in other parts of the world.

Medical Insects and Arachnids Edited by Richard P. Lane and Roger W. Crosskey.
Published in 1993 by Chapman & Hall ISBN 0 412 40000 6

Another much-feared species is the American brown recluse spider *Loxosceles reclusa* (Sicariidae), while the Australian funnel-web *Atrax robustus* (Dipluridae) is also dangerous. The South American *Phoneutria fera* (Ctenidae) is not only extremely venomous but, in contrast to most other dangerously poisonous spiders, is very aggressive (Bücherl, 1971a).

According to the recent work of Sissom (1990) about 1400 species and subspecies of scorpions, the order Scorpiones, have been described. They are mostly found in arid tropical regions, where several are dangerously poisonous and can cause death to human beings. Junqua and Vachon (1968) compiled a list of 79 species reputed to be of medical importance in that their stings often result in serious illness or death, but Polis (1990) states that fewer than 25 are lethal. All these deadly species belong to the family Buthidae. The toxicity of several dangerous species is known to vary in different parts of their range. Economic conditions and the availability of public health services can affect the relative medical importance of different scorpion species (Keegan, 1980).

RECOGNITION AND ELEMENTS OF STRUCTURE

Except in mites, the arachnid body is divided into two regions or tagmata, the prosoma (cephalothorax) which bears the mouthparts and limbs, and the opisthosoma (abdomen) (Fig. 19.1) which contains the caeca and gonads. The prosoma is nearly always unsegmented, as in spiders and scorpions: the opisthosoma is segmented in scorpions (Fig. 19.2) but only in more primitive spiders, the Liphistiomorphae. Disappearance of opisthosomal segmentation in Araneae is one of many characters which indicate that spiders are more highly evolved than scorpions. Indeed, the body form of members of the order Scorpiones has barely changed for over 400 million years, since the Silurian geological period (Sissom, 1990), and in their segmentation scorpions resemble the order Eurypterida (= Gigantostraca) which evolved from marine Cambrian ancestors and became extinct in Permian times (Cloudsley-Thompson, 1988).

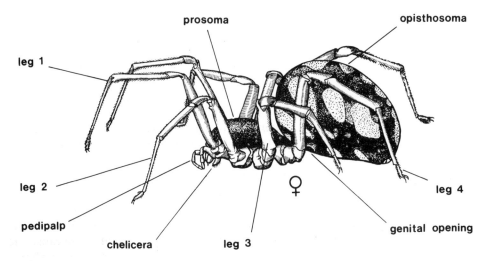

Figure 19.1 Basic morphology of a typical spider. The specimen shown is a female of *Latrodectus tredecimguttatus*.

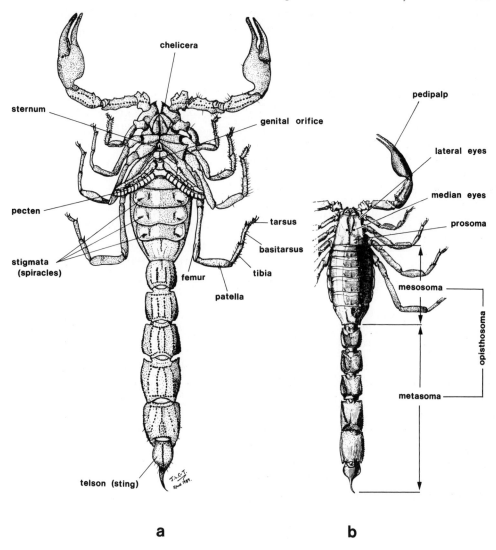

Figure 19.2 Basic morphology of a typical scorpion: (*a*) ventral view; (*b*) dorsal view. The species illustrated is *Androctonus australis* (Buthidae). Terminology for the leg segments is that of Hjelle (1990).

The two tagmata of arachnids are joined between the sixth and seventh segments: in spiders, the seventh is narrowed to a stalk-like pedicel. The abdomen of scorpions is divided into a wide anterior mesosoma (preabdomen) of eight segments, and a long posterior metasoma (postabdomen) or cauda (tail) of five segments with a sting (telson) at its end (Fig. 19.2). The prosoma has a cover or carapace, and consists of a presegmental acron and six segments, indicated by their appendages. These appendages appear initially as embryonic appendage buds: the first pair consists of the chelicerae or jaws, which arise behind the stomodaeum. Because the pharynx is directed posteriorly, however, the chelicerae lie anterior to, or above, the mouth. They are made up of no more than three articles or segments and, in most orders, take the form of claw-like pinchers. In spiders and the tailless whip-scorpions or whip-spiders (Amblypygi = Phrynichida), however, they are subchelate – that is, the chelate finger (the distal segment) has been lost and the chelicerae are piercing structures. In spiders and Solifugae, where the chelicerae are prehensile organs, they are used to hold and kill the prey, to squeeze out nutrient juices, and as defensive weapons. In scorpions, on the other hand, the chelicerae are small and the prey is gripped by the claws of the pedipalps which form the second pair of prosomal appendages. The pedipalps of spiders are tactile organs and, in the male (Fig. 19.4a), are adapted for reproduction. The poison glands of spiders open at the tips of the distal segments of the chelicerae.

Behind the pedipalps are four pairs of walking legs. (The number is reduced to three in larval mites and Ricinulei.) In spiders, the first or second pair of legs can be especially long and used either as feelers or to grasp the prey.

The sting of scorpions is curved and pointed. Its base is enlarged and contains a pair of poison glands which open near the tip. On the ventral side of the mesosoma, immediately behind the genital opercula, are the pectines – comb-like sense organs which are not found in any animals other than scorpions (Fig. 19.2). Each pecten articulates with a chitinous plate representing the second sternite, and is provided with complex musculature. The lamellae and teeth (Fig. 19.3a,b) vary in number from three to more than 40, depending upon the sex and species of the scorpion, and are richly supplied with nerves (Cloudsley-Thompson, 1968). In most spiders some posterior abdominal sternites have been lost, resulting in the location of the spinnerets at the tip of the abdomen.

Arachnid sense organs consist of innervated sensory hairs or setae, trichobothria and slit sense organs. Trichobothria are long, delicate, movable hairs inserted in the centres of circular membranes. In scorpions they are innervated by the dendrite of a single nerve and move in one plane, and in spiders, by the dendrites of three nerves and move in various directions. Slit sense organs are narrow crevices covered by thin membrane. All these three types of tactile sensillum respond also to vibrations and have proprioceptive functions (that is to say, respond to internal stimuli). The eyes of arachnids are simple ocelli. Scorpions have one pair of median eyes and two to five pairs of lateral eyes (Fig. 19.2) on the edge of the carapace; in spiders there are usually eight eyes, sometimes six, rarely four or two.

Lung-books, air-filled respiratory cavities opening on the undersurface of the body, occur both in spiders and in scorpions. Scorpions and mygalomorph spiders have four pairs, but most other spiders have two. In scorpions, the lung-books open to the exterior by means of stigmata (Fig. 19.2): in spiders the slit-shaped entrances to the lung-books can be seen at the anterior end of the ventral side of the opisthosoma. The lung slits lie behind the lung-books, which appear whitish and opaque (Foelix, 1982). In addition to lung-books, spiders also possess tubular tracheae. These lie in the third abdominal segment and open to the exterior by one or two small stigmata or spiracles. Most spiders have a single stigma in front of the spinnerets. Tracheae are extremely variable and are most elaborate in small spiders which are especially susceptible to desiccation.

In some species of scorpions the tail is thicker in males than in females but, in others, it is difficult, and sometimes impossible, to distinguish the sexes without recourse to dissection. The sexes of adult spiders, on the other hand, can easily be recognized. In male spiders, the tarsi of the pedipalps are greatly enlarged (Fig. 19.4a) and modified for reproduction. They are used to transfer sperm to the females. It is usually possible to see, with the naked eye, whether or not a spider is

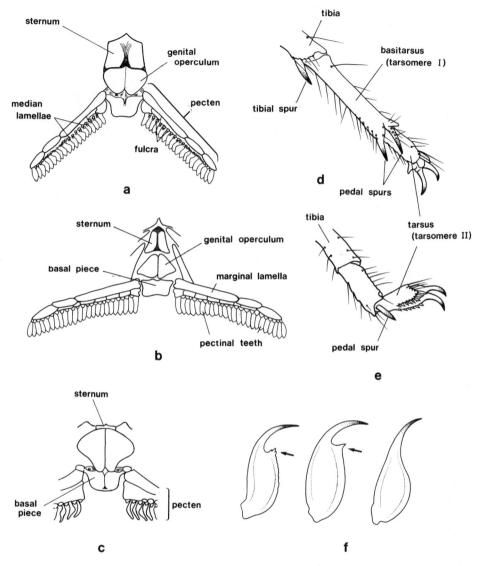

Figure 19.3 Important morphological features of scorpions used as key characters for family identification: (*a–c*) sternal region of (*a*) Vaejovidae (*Vaejovis spinigerus* illustrated), (*b*) Buthidae (*Centruroides exilicauda* illustrated) and (*c*) Bothriuridae (*Bothriurus* sp. illustrated); (*d, e*) terminal segments of last walking leg (leg 4) in (*d*) *Androctonus australis* (Buthidae) and (*e*) *Scorpio maurus* (Scorpionidae); (*f*) side views of the telson (sting) showing condition with (left and centre, arrowed) and without (right) a tubercle or spine ('subaculear protuberance').

an adult male. The external genitalia of the females (Fig. 19.4*b*) are known as epigynes. They receive the palps of the males during mating and are situated centrally at the anterior end of the ventral side of the abdomen, behind the lungbooks. The palps and epigynes of adult spiders are brown and sclerotized: their shapes are the most important of all the diagnostic characters used for the determination of species. When epigynes are absent (all Mygalomorphae and some Araneomorphae), female spiders can seldom be identified with certainty.

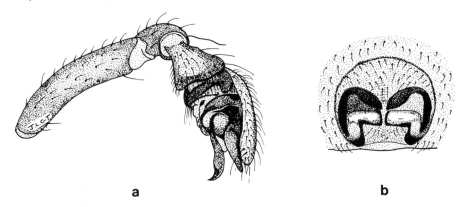

Figure 19.4 Two features of a typical spider associated with mating: (*a*) pedipalp of male; (*b*) epigyne of female. The species illustrated is *Steatoda* (= *Lithyphantes*) *albomaculata*.

In spiders of families such as the Filistatidae, Eresidae, Acanthoctenidae, Uloboridae, Psechridae, Amaurobiidae, Dictynidae etc., a sieve-like plate, known as the cribellum, can be seen in front of the spinnerets when the spider is lying on its back. This produces fine, fluorescent, bluish hackled threads of silk which are combed onto the web by a row of fine, hooked spines, the calamistrum, on the metatarsus of the fourth pair of legs. These families, however, have no medical significance.

Ancestral spiders had four pairs of spinnerets but, among extant suborders, this ancestral trait is retained only in some Liphistiomorphae. In Mygalomorphae there are two or three pairs of spinnerets, and sometimes even only one pair; while Araneomorphae usually have three pairs which develop embryonically from the extremities of body segments 10 and 11 (Foelix, 1982). The silk glands terminate in little spigots on the surface of each spinneret. The spinnerets are equipped with well-developed musculature and are extremely mobile, their functioning being enhanced by movements of the abdomen.

The chelicerae of spiders consist of two parts: a stout base and a movable articulated fang. The fang normally rests in a groove of the basal segment, like the blade of a pen-knife. When a spider strikes, however, the fangs are erected and penetrate the body of the prey or aggressor, while poison is injected through a minute opening at the tip. In the more primitive suborders of spiders (Liphistiomorphae and Mygalomorphae) the chelicerae strike forwards and downwards (Fig. 19.5). This paraxial articulation is known as orthognath. In the suborder Araneomorphae, which includes more highly evolved families, the chelicerae work against one another, like forceps or pliers. Diaxial articulation of this kind is known as labidognath. Both sides of the cheliceral groove are often armed with cuticular teeth which serve to buttress the movable fang and also macerate the body of the prey. Not only are the chelicerae used for subduing prey and in defence, but they are employed in different families for digging burrows (Ctenizidae), carrying egg-cocoons (Pisauridae), transporting prey (Araneidae), interlocking during mating (Dictynidae and Tetragnathidae), and for stridulating (Linyphiidae).

On the ventral side of the prosoma of spiders lies the sternum. Although this is derived from four fused sternites, it is undivided. Just anterior to the sternum is a small median plate, the labium, whose shape is of taxonomic importance. The coxae

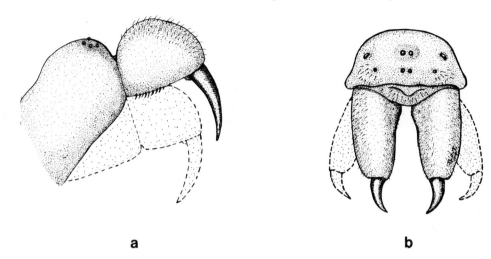

Figure 19.5 Mode of articulation of spider chelicerae: (*a*) orthognath, in which the prosoma is raised with fangs extended and prey impaled with a downward motion; (*b*) labidognath, in which the chelicerae pivot sideways towards one another.

of the pedipalps are adapted to form chewing mouthparts or maxillae. In Liphistiomorphae and Mygalomorphae they are only slightly modified; but, in Araneomorphae, they are broadened laterally and, in most families, the anterior rims are clearly serrated. The inner sides of the maxillae are fringed with a dense tuft of hair or scopula which acts as a filter when liquefied food is sucked into the mouth.

Four pairs of walking legs are articulated between the carapace and sternum. Each leg consists of seven segments, namely coxa, trochanter, femur, patella, tibia, metatarsus and tarsus; the last carries two or three claws. The front two pairs of legs are frequently relatively long and the first pair, in particular, is often used to probe the environment. The tarsal claws are usually serrated: when there are three of them, in web-building spiders, the small median claws are used to grasp the threads of silk. Many hunting spiders possess dense tufts of hairs, or scopulae, under their claws. These scopulae enable their owners to climb smooth vertical surfaces because each hair has 500–1000 'end feet' which enhance physical adhesion, especially by capillarity (Foelix, 1982).

Scorpions possess seven pairs of appendages: the chelicerae, pedipalps, pectines and four pairs of legs (Fig. 19.2). The chelicerae consist of three segments: the basal portion, movable finger and fixed finger. The number, relative size and arrangement of teeth on these fingers are used in classification. Each pedipalp consists of six segments: coxa, trochanter, femur, patella, tibia and tarsus (of which the proximal portion is known as the 'hand'). The size, location and number of teeth, as well as various keels, tubercles and lobes on the hand are of taxonomic importance (Keegan, 1980). As in spiders, the legs consist of seven segments. The terminology of the leg segments has been inconsistent and Hjelle (1990) has tabulated seven systems. Hjelle's system is used here, as shown in Fig. 19.2*a*.

In addition to those already mentioned, morphological characters of taxonomic importance in scorpions include the shape and proportions of the sternum (Fig. 19.3*a–c*), the presence or absence of an accessory spine on the telson (Fig. 19.3*f*), the presence or absence of tibial spurs (Fig. 19.3d,e), and the number of lateral eyes. Specific determination frequently depends upon the distribution of trichobothria on, and the spinations of, the pedipalps, the number of pectinal teeth, the granulations of the cephalothorax and the colour of the body.

Comprehensive accounts of the anatomy and morphology of spiders and scorpions are given by Bücherl (1971a,b), Hjelle (1990), Kaestner (1968), Millot (1949) and Millot and Vachon (1949).

CLASSIFICATION AND IDENTIFICATION

The systematics and identification both of spiders and of scorpions is by no means simple. Certain species are easily recognizable but much taxonomic research is still urgently needed, especially on the mygalomorph spiders and scorpions.

Spiders

Spiders are grouped into three suborders according to whether the abdomen is clearly segmented, the way in which the chelicerae are articulated, the presence or absence of an epigyne, and of a cribellum and calamistrum (Kaestner, 1968). Some arachnologists do not consider the cribellate families to form a homogeneous group; others regard them as a separate evolutionary branch showing parallel evolution with non-cribellate spider families.

It must be appreciated that it is only possible here to provide an extremely superficial, over-simplified and very incomplete account – mainly in the form of the annotated key given below – to show where medically important species fit in the classifications and how they can be partly identified. For further details the reader is referred to the chapter on poisonous spiders by Bücherl (1971a), a work specially useful for its inclusion of keys and family diagnoses. Smith (1987) lists 650 tarantulas and gives 500 descriptions, while Schmidt (1986) provides keys and descriptions for the Mygalomorphae. No single book can possibly attempt to encompass the taxonomy of all the Araneomorphae, however. Further information has to be obtained from monographs and books devoted to the spider faunas of different countries. A useful account of the principal families of spiders, with references, can be found in Levi (1982).

Identification key to suborders and families of spiders containing medically important species (based partly on Bücherl, 1971a)

1 Abdomen without visible segmentation. Chelicerae orthognath (paraxial) or labidognath (diaxial).. 2
— Abdomen clearly segmented dorsally. Chelicerae orthognath.................................
.... Suborder LIPHISTIOMORPHAE (= MESOTHELAE)
   [One family, Liphistiidae, harmless to humans.]
2 Chelicerae orthognath (Fig. 19.5a). Lung-books in two pairs. Female without epigyne.................... Suborder MYGALOMORPHAE (= ORTHOGNATHA) 3
— Chelicerae labidognath, working like forceps against each other (Fig. 19.5b). One pair of lung-books. Female with epigyne ....
.... Suborder ARANEOMORPHAE (= LABIDOGNATHA) 5

3 Claw tufts present. Tarsi with two claws. Posterior spinnerets much shorter than the abdomen 4
— Claw tufts absent. Tarsi with three claws. Posterior spinnerets about as long as the abdomen.. DIPLURIDAE 3a
    3a Upper claws pectinated in a double row. [South America] .... genus **Trechona**
    — Upper claws pectinated in a single row. [Australia] .... genus **Atrax**
    [Important species are *A. formidabilis* and *A. robustus* (Fig. 19.7b).]
4 Terminal segment of posterior spinnerets as long as or longer than the preceding segment..THERAPHOSIDAE
    [Hairy spiders most of which are considered harmless despite large size. Includes tarantulas and bird-eating spiders such as genera *Aphonopelma, Avicularia, Dugesiella, Eurypelma, Grammostola, Lasiodora, Sericopelma* and *Theraphosa*.]
— Terminal segment of posterior spinnerets distinctly shorter than the preceding segment... BARYCHELIDAE
    [*Harpactirella* of South Africa contains 11 venomous species including *H. karroica* and *H. treleaveni*.]
5 Eyes eight, unicolorous or of different colours. Chelicerae free 6
— Eyes six, in three pairs and pearly white. Chelicerae fused at base, each with a lamella on its margin facing the fang .... LOXOSCELIDAE (= SICARIIDAE)
    [Brown spiders. *Loxosceles* includes over 50 very venomous species in Eurasia, Africa and America. Most serious is *L. reclusa* of USA.]
6 Tarsi with two claws (sometimes obscured by hair tufts) 7
— Tarsi with three claws (not hidden) 9
7 Eyes in three rows 8
— Eyes in two rows of four .... CLUBIONIDAE
    [Worldwide, including many species of small ground-hunting spiders. *Chiracanthium punctorium* (Europe, Asia), *C. diversum* and *C. brevicalcaratum* (Australia, Fiji, Hawaii) and *C. inclusum* (North America and West Indies) dangerous.]
8 Eye formula 2–4–2: median pair of eyes in second row larger than eyes of first row, eyes of third row largest... CTENIDAE
    [Wandering spiders. *Phoneutria* includes about a dozen species (mostly Brazil) of which *P. fera, P. nigriventer* and *P. rufibarbis* best known.]
— Eye formula 4–2–2: median pair of eyes of first row very large, eyes of second row inconspicuous... SALTICIDAE
    [Includes over 400 genera of jumping spiders abundant in tropics. A few *Dendryphantes* species, especially *D. noxiosus* of Bolivia, harmful to humans.]
9 Eye formula 4–4. Third claw pectinated. [Web-building spiders]...................... 10
— Eye formula 4–2–2. Chelicerae with distinct boss or condyle. Tarsi with third claw smooth or bearing a single tooth... LYCOSIDAE
    [Includes over 2000 species of wolf-spiders of which a dozen or so are harmful species of *Lycosa* (= *Tarentula*).]
10 Abdomen globular and with scanty hair. Fourth tarsus with a comb of curved and serrated black bristles... THERIDIIDAE
    [Includes about 1500 species of comb-footed spiders in many genera. Dangerous species are blackish with red, yellow or grey markings. *Steatoda* (= *Lithyphantes*) and *Latrodectus* are venomous to man. Medically most important are South American *S. ancoratus* and *S. andinus* (latter is 'cirari' of Bolivia, Paraguay and Chile), *L. mactans* (Fig. 19.7a) of North and Central America, *L. curacaviensis* from southern Canada to Patagonia, *L. hasselti* of South East Asia and Australasia, and *L. tredecimguttatus* of the Mediterranean subregion.]

— Abdomen oblong with normal pilosity. Fourth tarsus without a comb ....
.... ARANEIDAE (= ARGIOPIDAE)
[Includes several thousand species of orb-web spiders. A few harmful to humans, most importantly *Glyptocranium* (= *Mastophora*) *gasteracanthoides* of South America.]

Scorpions

Monographic treatments of the order Scorpiones have been given by Kraepelin (1899), Werner (1934–35) and Kaestner (1940) but are now considerably outdated. An outline account of the principal families, but without references, has recently been given by Francke (1982). Very recently, Sissom (1990) has reclassified the order into nine families instead of the six which have for long been recognized by most authorities. However, the additional families in Sissom's treatment result only from more restricted interpretation of family limits, and here it is considered preferable to adopt the familiar and well-tried six-family system – especially as the three other families in the Sissom system have no medical importance.

Scorpion families are classified mainly on the shape of the sternum, presence and position of spurs on the terminal leg segments, the presence or absence of a tooth or spine on the lower surface of the telson (known in specialist literature as the subaculear tubercle or protuberance), and the number of pairs of lateral eyes. Modern classifications lay particular stress on the number and disposition of the trichobothria (mentioned earlier), following a major study of the taxonomic value of these sense organs by Vachon (1974). In families other than Buthidae the femur of the pedipalp has fewer than ten trichobothria of which only one is on the inner surface, whereas in the Buthidae (the family containing the scorpions most dangerous to man) the pedipalp femur bears ten or more trichobothria of which four or five are on the inner surface (Sissom, 1990).

The following key has been adapted from those of Bücherl (1971b), Sheals (1973) and Keegan (1980) with characters verified against those listed in the diagnostic family characterizations given by Sissom (1990). (See the latter work for illustrated detail of morphological characters, especially trichobothrial patterns.)

Identification key to the families of scorpions

The following key aims to help in the identification to family of any scorpion thought to be of medical concern. It does not reliably cover some unimportant minor genera which do not fully conform to the main characters of the families in which specialists place them.

1   Sternum subtriangular or pentagonal, usually longer than wide and quite conspicuous (Fig. 19.3*a,b*). Telson with or without tooth (subaculear protuberance) (Fig. 19.3*f*). Tibiae with or without spur 2
—   Sternum forming a very short transverse plate, much wider than its length and very inconspicuous (Fig. 19.3*c*). Telson tooth absent. Tibiae without spur. [Australia, southern Africa, South America] .... BOTHRIURIDAE
2   Sternum approximately pentagonal, its sides subparallel (Fig. 19.3*a*). Tibiae without spur. Legs with one or two pedal spurs.* Telson with or without tooth. Pedipalp femur bearing fewer than ten trichobothria (one on inner surface) 3
—   Sternum subtriangular (Fig. 19.3*b*). Tibial spur present at least on third and/or fourth (last) pairs of walking legs (Fig. 19.3*d*). Legs with two pedal spurs

(Fig. 19.3*d*). Telson usually with tooth-like tubercle (Fig. 19.3*f*, arrowed). Pedipalp femur with ten or more trichobothria (four or five on inner surface). [Tropical and warm temperate parts of Old and New World, including Mediterranean subregion] .... BUTHIDAE

3   Legs with one pedal spur (Fig. 19.3*e*). Telson with or without tooth (subaculear protuberance)... 4

—   Legs with two pedal spurs (as in Buthidae, Fig. 19.3*d*). Telson without tooth (subaculear protuberance) 5

4   Telson with tooth-like tubercle or spine (subaculear protuberance) (Fig. 19.3*f*, arrowed). [Southwestern United States, Mexico and Central America, northern South America, Caribbean islands, Middle East] .... DIPLOCENTRIDAE

—   Telson without subaculear protuberance, concave outline of its tip smoothly curved (Fig. 19.3*f*, right). [Africa and Middle East, Oriental region, Australia, Meso-America]... SCORPIONIDAE
    [As used here this family includes the Ischnuridae as recognized by Sissom (1990). Genera are Old World except *Opisthacanthus* found partly in Central and South America.]

5   Usually two pairs of lateral eyes. [Southern Europe, North Africa, Middle East, southwestern United States and Mexico to South America]....... CHACTIDAE
    [As used here this family includes the Oriental genus *Chaerilus* which Sissom (1990) treats as the monogeneric family Chaerilidae.]

—   Usually three pairs of lateral eyes. [Southwestern Canada to northern South America, southwest Asia through India to Malaya and Borneo]......................... .... VAEJOVIDAE
    [Closely related families Chactidae and Vaejovidae are only dubiously distinct. They have traditionally been distinguished by the number of pairs of lateral eyes, but this character is not wholly reliable and sometimes even variable at specific level. See Sissom (1990, p. 103). The Iuridae as recognized by Sissom (1990) is part of the same familial complex distinguished by having a dark tooth on the inner margin of the movable finger of the chelicera which is absent in other Chactidae/Vaejovidae.]

* Pedal spurs are positioned between tarsomeres 1 and 2, i.e. by the end of the basitarsus (tarsomere 1) (Fig. 19.3*d,e*).

Faunal and taxonomic literature

So few spiders are dangerously poisonous that it would be pointless to list books and monographs devoted to the arachnid fauna of regions in which these occur. I have surveyed the arachnid fauna of the Sahara (Cloudsley-Thompson, 1984), while Vachon (1952) has provided a superb monograph of the known Saharan scorpions, with keys and detailed descriptions. Lamoral and Reynders (1975) have published an annotated, alphabetical list of all scorpion taxa described or recorded from the Afrotropical faunal region, namely sub-Saharan Africa and adjacent islands. The scorpions of Israel and Sinai are described by Levy and Amitai (1980), those of Arabia by Vachon (1979), while an account of the Caucasian species can be found in Byalynitskii-Birulya (1917, 1964). Between them, these monographs provide keys to and descriptions of nearly all the Old World scorpions of known medical importance, since, apart from *Androctonus australis*, which does not cause many deaths there, the only really dangerous species in India and Pakistan is *Buthotus tamulus*. In any case, Pocock's (1900) book encompasses the scorpion fauna of

Pakistan, India, Sri Lanka and Burma. None of the Australian Buthidae is considered to be of medical significance. Lawrence (1955) has provided a monograph on the scorpions of southern Africa.

Neotropical scorpions have received extensive treatment from Mello-Leitão (1945); Francke (1978) has published a revision of the Diplocentridae and (1985) a conspectus of all the genus-group names used for scorpions, whether valid or not, with complete bibliographic citations.

BIOLOGY

The biology of Arachnida is a field of outstanding interest. Knowledge of the biology of spiders is summarized by Foelix (1982), while earlier accounts have been given by Bristowe (1971), Cloudsley-Thompson (1968), Gertsch (1979), Kaestner (1968), Main (1976), Millot (1949) and Savory (1928). Scorpions have attracted somewhat less attention than spiders but they, too, have recently received comprehensive treatment (Polis, 1990). Earlier works include those of Bücherl (1971b), Byalynitskii-Birulya (1917, 1964), Cloudsley-Thompson (1968), Kaestner (1968) and Vachon (1952).

Life history

Both spiders and scorpions are dioecious, and the sexes are separate, except in rare instances when the females are parthenogenetic and males may be absent. In spiders, the females are nearly always larger than the males and sexual dimorphism is especially obvious in tropical orb-weavers. The sexes are very similar among scorpions. Because of their small size, male spiders require fewer moults, and mature earlier than females.

Before oviposition takes place, female spiders spin silken cocoons in which the eggs are deposited. In the simplest cases, only a few silk threads are wrapped around the eggs; but typical cocoons consist of a basal pad and a cover plate which enclose and protect the egg mass within. Web-spinners, such as *Latrodectus*, usually attach their egg sacs to the undersides of leaves, or place them in a retreat in the web. Funnel-web spiders (*Atrax* and *Trechona*) enlarge the lower parts of their funnels to take the eggs, while the larger Theraphosidae and Lycosidae deposit them in the burrows. Smaller lycosids carry egg cocoons attached to their spinnerets. After hatching, young wolf-spiders climb onto their mothers' backs where they remain until after the first moult.

Scorpions, too, are sub-social in the sense that the young ride on their mother's back until after the first moult – during which time they subsist on the egg yolk remaining in their alimentary canals. From the second instar onwards, both spiders and scorpions live solitary lives and tend to avoid one another. Scorpions and larger ground-living spiders tend to be nocturnal, spending the day in burrows or sheltered retreats under rocks, stones and fallen logs from which they emerge at night and wander around in search of prey.

Mating

Arachnids evolved from marine ancestors which probably transmitted sperm enclosed in spermatophores. Scorpions deposit a stalked spermatophore during a courtship dance in which the male grasps the claws of the female with his own chelate pedipalps. From time to time he vibrates his pedipalps and body, a movement that has been described as 'juddering'. The two scorpions move back and forward in a 'promenade-à-deux', during which the pectines of the male are spread out and swept over the substrate until they detect a stone or rock to which a spermatophore can be attached. The male pulls his mate over this so that the paired sperm containers are situated exactly in front of her genital opening. She opens the cover of the spermatophore by a sudden movement of her body and the sperm is injected into her genital atrium. The male immediately releases the female and, in some species, she subsequently consumes the empty spermatophore (Cloudsley-Thompson, 1988; Polis and Sissom, 1990; Schaller, 1979).

Indirect sperm transfer in spiders has evolved from indirect transfer of spermatophores via the substrate. The male ancestral spiders probably deposited spermatophores on the ground and transferred them to the female by means of claws on the pedipalps. From this simple beginning has been derived the complex palpal insemination found in spiders today. Before mating takes place, the male spider spins a tiny triangular web with silk from his posterior spinnerets on which a drop of seminal fluid is deposited. This is drawn into the pedipalps and inserted into the genital orifice of the female after courtship. Jumping-spiders (Salticidae) and wolf-spiders (Lycosidae), which hunt their prey by sight, indulge in visual courtship displays. In short-sighted hunting spiders, such as *Dysdera* (Dysderidae) and Clubionidae, which detect their prey by the sense of touch, courtship is primarily tactile; while, in tube-dwelling and web-building spiders, which are stimulated by vibrations of the silk caused by movements of the prey, the courtship of the male consists in tapping a distinct code on the web of the female, as a result of which he is not himself mistaken for prey. It is believed that courtship displays have evolved from chemotactic searching motions of the male for the female (see reviews in Bristowe, 1971; Cloudsley-Thompson, 1988).

Predators and prey

Despite stings and venomous bites, scorpions and large spiders are vulnerable to predatory birds and mammals, as well as to members of their own kind and to insect parasitoids. The urticating hairs of theraphosid spiders (Cooke et al., 1972) are undoubtedly more effective in deterring predatory attack than are the chelicerae. These hairs can cause great discomfort to human beings and, if the face is touched after stroking one of the creatures, sight may even be endangered. When irritated, Theraphosidae will comb out their opisthosomal hairs in clouds with their back legs, and may even develop bald patches on their abdomens in consequence. It is probable that avoidance or predation is an important factor responsible for their primarily nocturnal behaviour. Desert forms tend to emerge from their burrows at night, under the stimulus of circadian biological clocks, while forest species are more often 'sit-and-wait' predators which are relatively inactive and far less rhythmic in their behaviour. Many day-active spiders achieve concealment from their enemies by crypsis, while some tropical orb-web builders place narrow ribbons of silk or stabilimenta across their webs. These help to disguise the spider and may also make the web conspicuous so that it is not destroyed inadvertently by passing birds.

672 *Spiders and scorpions (Araneae and Scorpiones)*

Except for certain mites, most arachnids are predatory, feeding principally upon insects and other arthropods. Scorpions are able to detect the movements of potential prey with the aid of trichobothria on the legs which are sensitive to ground vibrations, while the pedipalpal trichobothria respond to air currents. When the prey has been caught, it is held firmly with the claws and stung if it struggles; the chelicerae then tear it apart, one alternately holding on while the other pinches and pulls.

MEDICAL IMPORTANCE

Distribution of spiders and scorpions of medical importance

The four most dangerous genera of spiders, in respect to the numbers of deaths and the sickness they cause, are *Atrax* (Dipluridae), *Harpactirella* (Barychelidae), *Loxosceles* (Sicariidae) and *Latrodectus* (Theridiidae). Their distribution throughout the world, already mentioned, is summarized in Fig. 19.6b. They are virtually restricted to tropical and subtropical regions or to regions with a Mediterranean climate, along with the dangerous species of *Lycosa* (Lycosidae) and *Phoneutria* (Ctenidae) as follows (Bücherl, 1971a):

Atrax	eastern Australia, New Zealand (Fig. 19.7b)
Harpactirella	South Africa
Latrodectus mactans	from USA to Argentina and Hawaii (Fig. 19.7a)
L. tredecimguttatus	Mediterranean subregion, Middle East, southern Russia, Arabia and Ethiopia
L. cinctus	southern and eastern Africa
L. menavodi	Madagascar
L. hasselti	South East Asia, Australia and New Zealand
L. pallidus	New Zealand, eastern Mediterranean subregion, Middle East and USSR
L. curacaviensis	from southern Canada to Patagonia
Loxosceles	southwestern North America, Central and South America
Lycosa	Central and South America
Phoneutria	Brazil (Rio Grande do Sul)

Dangerous genera of scorpions have a similar distribution (Fig. 19.6a). Their ranges can be summarized as follows (Bücherl, 1971b):

Androctonus	from Morocco and Senegal eastward to India (Fig. 19.8b)
Buthus	Mediterranean subregion, Middle East and eastern Africa
Centruroides	southern states of USA, Mexico and Central America (Fig. 19.8d)
Hottentotta	northern and northeastern Africa, Middle East
Leiurus	eastern Africa, Middle East (Fig. 19.8c)
Mesobuthus	southern and central Asia
Parabuthus	Sudan to South Africa
Tityus	South America and Trinidad (Fig. 19.8a)

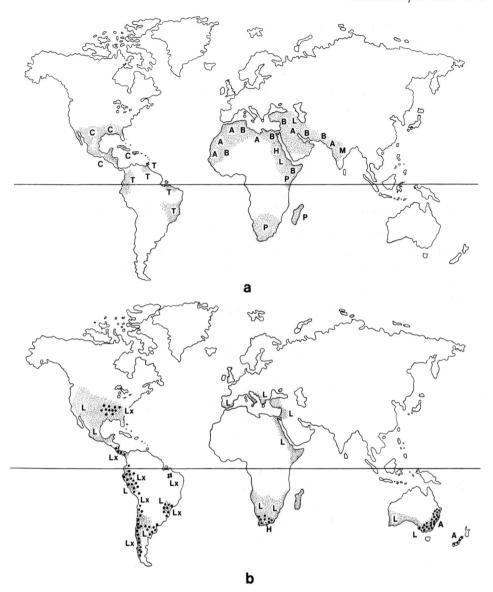

Figure 19.6 World distribution of important venomous scorpions and spiders: (*a*) scorpions of the genera *Androctonus* (A), *Buthus* (B), *Centruroides* (C), *Hottentotta* (H), *Leiurus* (L), *Mesobuthus* (M), *Parabuthus* (P) and *Tityus* (T); (*b*) spiders of the genera *Atrax* (A), *Harpactirella* (H), *Latrodectus* (L) and *Loxosceles* (Lx). Larger dots on the spider map (*b*) show more limited ranges of genera which are overlapped by wider range of *Latrodectus*.

Poisoning by spider bites

Native British spiders have seldom been responsible for biting people and, when they have done so, the poison has had almost no effect, e.g. that of the purse-web spider *Atypus affinis* (Mygalomorphae), wolf-spiders (Lycosidae), or garden spiders *Araneus diadematus*. A ground-hunting spider of the genus *Dysdera* is alleged on one

674   *Spiders and scorpions (Araneae and Scorpiones)*

Figure 19.7   Two of the spiders most dangerous to man: (*a*) *Latrodectus mactans*, the black widow spider of North America; (*b*) *Atrax robustus*, the Sydney funnel-web spider of Australia.

occasion to have given a painful bite to the ankle of a building worker, and the mouse-coloured house spider *Scotophaeus blackwalli* (Gnaphosidae) has been suspected of causing a bite to a woman's face that eventually left a small permanent scar (Duffey and Green, 1975).

In July 1974, reports were received that workmen on maintenance work at a sewage farm near Birmingham were being bitten by tiny spiders emerging from the filter-beds, which dropped down their necks or crawled up their sleeves. The bites caused irritations and swelling. The species responsible was found to be *Leptorhoptrum robustum* (Linyphiidae), a spider of fens, marshes, moorlands and mountains which can reach huge densities in the extremely artificial environment of a sewage filter-bed (Duffey and Green, 1975).

Figure 19.8 Four of the scorpions most dangerous to man: (*a*) *Tityus trinitatis* of Trinidad; (*b*) *Androctonus crassicauda* of the Middle East; (*c*) *Leiurus quinquestriatus* of North Africa and the Middle East; (*d*) *Centruroides suffusus* of Mexico.

Apart from the dangerous *Latrodectus tredecimguttatus*, discussed below, the only other European spiders whose bites are of any medical significance are *Chiracanthium punctorium* (Clubionidae) and *Lycosa tarentula* (Lycosidae). The venom of *Steatoda payculliana* (Theridiidae) has been shown experimentally to be harmful to guinea pigs, but this timid little spider has never been known to bite a human being. *Chiracanthium punctorium* lives in southern and mid-Europe. The consequence of the bite is severe local pain, swelling and redness with a small area of necrosis at the site of the bite. Regional lymph nodes can become enlarged and tender, and the patient affected with moderate temperature, chest discomfort and general malaise. Some patients have been bitten while working in the fields, others indoors when dressing. The famous *L. tarentula* may also cause necrosis, but bites are very rare because the spider spends its life in a burrow only venturing out occasionally to hunt its prey (Maretić, 1975).

The venom of *Phoneutria* species is probably the most active of any spider venom as far as most mammals are concerned. Its action is essentially neurotoxic, while that of *Lycosa* is cytotoxic with local necrotic effects only. In relation to the frequency of species, the bites of *Loxosceles* and *Latrodectus* are the most important. *Loxosceles* venom is cytotoxic and haemolytic (Schenberg and Pereira Lima, 1971), that of *Latrodectus* is neurotoxic. Only female black widow spiders secrete enough venom to make them dangerous. The literature on latrodectism is worldwide and relates to hundreds of cases (Bücherl, 1971a; Maretić, 1971, 1975; Minton, 1968; Sheals, 1973). Thorp and Woodson (1945) reported 55 deaths from bites in the United States between 1726 and 1943, and hundreds of others must have occurred in various countries of the world. At the same time, it must be remembered that spiders are often blamed for bites and stings the true causes of which may well be other venomous arthropods.

Principal scorpions of medical importance

The principal genera and species of scorpions of medical importance are cited below. As in the case of spiders, whenever one species is harmful to human beings, congeneric species are usually dangerous also: but only those convincingly implicated are mentioned here. None of the Chactidae or Bothriuridae is of medical importance. Of the Diplocentridae, only *Nebo hierichonticus* is even of minor significance while, with the possible exception of *Hemiscorpion lepturus*, the family Scorpionidae contains no dangerous species. Vaejovidae of the genera *Vaejovis* and *Hadrurus* can deliver painful stings, but the effects are nearly always local. The only really dangerous scorpions belong to the family Buthidae. Genera of this family are represented in both the Old and the New World, and the most dangerous of these to human beings are *Androctonus*, *Buthus*, *Centruroides*, *Hottentotta* and *Mesobuthus*, *Leiurus*, *Parabuthus* and *Tityus*. In general, scorpions with massive claws (Scorpionidae, Chactidae, Diplocentridae, Bothriuridae) are relatively harmless, while more dangerous species (Buthidae, Vaejovidae) have slender chelae. Colour is irrelevant as far as the toxicity of the venom is concerned.

Family Buthidae

Genus Androctonus Fat-tailed scorpions of this genus are the most dangerous of any in the Palaearctic Eremic zone. Throughout North Africa and probably much of the Middle East, *A. australis* (Fig. 19.2) is responsible for many more deaths than any other species of the genus. The colour varies from yellow to brown in different regions, and sometimes the claws and terminal tail segments are darker than the rest of the body. *Androctonus crassicauda* (Fig. 19.8b) is found from Turkey to the Arabian peninsula, its colour varying from dark brown to black. Its venom is much less toxic to human beings than is that of *A. australis*. Congeneric species of potential medical importance include *A. bicolor*, a darkly coloured species common in Egypt, Israel and Jordan, the black *A. aeneas* of Morocco, Algeria and Tunisia, *A. amoreuxi* found in the Middle East, *A. hoggarensis* of Algeria and northern Nigeria, and *A. mauretanicus* of Morocco.

Genus Buthus *Buthus occitanus*, including its subspecies, is widely distributed in southern Europe, the Middle East and North Africa. It is the only species in the genus of significant medical importance, but its toxicity differs markedly in different regions. In France, for example, it does not cause severe symptoms whereas, in Algeria, the venom produces intense pain and shock from which children occasionally die.

Genera Hottentotta and Mesobuthus The species concerned here were until recently placed in *Buthotus*, but synonymy of this generic name with *Hottentotta* (Francke, 1985) means that the latter name should now be used (Polis, 1990).

About 20 species of *Hottentotta* occur throughout Africa and the Middle East. The sub-Saharan *H. minax* is found under the bark of acacia trees in Sudan, beneath stones in Eritrea, and infests houses in Chad. Its sting is said to be painful. *Mesobuthus*, also with about 20 species, occurs through Asia from Iran to Korea but only the Indian *M. tamulus* has significant medical importance. However, *M. martensi* is of medical interest in Korea and Manchuria, its venom having a rather toxic effect.

Genus Leiurus *Leiurus quinquestriatus* (Fig. 19.8c) is one of the most common species of scorpion in the Sahel region of Africa. In Sudan it is known as the 'Omdurman scorpion', but it is actually found from Turkey, throughout the Middle East into Egypt, Libya, Algeria and the Sudan. Although it is smaller (up to 9 cm in length) and less heavily built than *A. australis* (10 cm or more), and produces considerably less venom, the toxicity of its poison is much greater, and *L. quinquestriatus* constitutes a health problem wherever it is found.

Genus Parabuthus Approximately 30 species of this genus are known, and all but a few are found only in South Africa. Several of them, such as *P. liosoma*, *P. granulatus*, *P. pallidus* and *P. fulvipes* are very venomous. Although their stings usually result only in severe local pain, systemic effects and deaths have occasionally been reported, especially among children.

Genus Centruroides This genus vies with *Androctonus* as containing the world's most dangerous scorpions. Species occur in Mexico, USA, the West Indies, Central and South America. *Centruroides exilicauda* (syn. *sculpturatus*) is the only really dangerous species of scorpion in the United States. In Mexico, *C. elegans*, *C. limpidus*, *C. infamatus* and *C. suffusus* are also dangerous. *Centruroides suffusus*, the 'Alacran de Durango', is especially notorious as a cause of human death. It varies from yellow to brown, has two dark longitudinal stripes on the tergites of the preabdomen, and attains a length of 9 cm (Fig. 19.8d).

Genus Tityus Species of this Caribbean and South American genus are similar in size and behaviour to species of *Centruroides*. Of a total of more than 100 species, the following are regarded as being dangerously venomous: *T. serrulatus* (Brazil), *T. bahiensis* (Brazil and Argentina), *T. trinitatis* (Trinidad and Venezuela), *T. trivittatus* (Brazil, Paraguay, Uruguay and Argentina), and *T. cambridgei* (Guyana). The prosoma and all except the last segment of the preabdomen of *T. serrulatus* vary from yellow to blackish brown, with dark markings on each, while the postabdomen and appendages are mostly pale yellow. The medical importance of this species results from the toxicity of its venom and its habits, which are similar to those of *C. exilicauda*. It is especially common in urban environments where it is a frequent cause of scorpion sting. Adults can attain a length of 7 cm, and the species is unique in being parthenogenetic, males never having been found. Second in medical importance is *T. bahiensis* which also infests houses. The cephalothorax and preabdomen are uniformly brown and the postabdomen and appendages reddish brown. *Tityus trinitatis* (Fig. 19.8a) is also extremely venomous but does not come into houses.

Poisoning by scorpion stings

Scorpion venoms are of two kinds: those which produce local and usually transitory symptoms seldom lasting more than a day or so, and those producing systemic symptoms. The poison of even closely related species may differ greatly in toxicity while, at the same time, the symptoms following stings by dangerous scorpions throughout the world are remarkably similar. Pain and swelling are frequent at the site of the sting, especially with species such as *Hadrurus hirsutus* in the southwestern United States, *Buthus occitanus* in North Africa and *Isometres maculatus* in Australia. Violent local reaction, however, is unrelated to generalized symptoms, which include restlessness and excitement, drooling of saliva, fever, convulsions and respiratory distress. Stings by *Tityus serrulatus* often cause muscular palsy that can result in permanent weakness of a limb (Diniz, 1971; Minton, 1968). Recent work on scorpion toxins is reviewed by Simard and Watt (1990).

Mortality from scorpion stings is much higher among children than adults. For instance, the death rate in Trinidad is about 25% in children under five years of age, but only 0.25% among adults. In Brazil, deaths due to *Tityus serrulatus* are about 0.8–1.4% of adults stung, 3–5% among school children and 15–20% among very young children. About 50% of children admitted into Omdurman General Hospital suffering from the stings of *Leiurus quinquestriatus* succumb, but they would probably not be brought there unless they were in a bad state. It is generally agreed that administration of antivenims is the most effective treatment. Most stings require little treatment: a cube of ice over the site of the sting may reduce pain and cut down the circulation. Atropine is recommended as a parasympatholytic drug and, when convulsions occur, sodium phenobarbitol given intravenously can be helpful. Oxygen and artificial respiration may also be required (Balozet, 1971; Bücherl, 1971b; Keegan, 1980; Stahnke, 1966).

CONTROL

It is simpler to avoid being bitten by spiders or stung by scorpions than it is to control the numbers of these animals. For instance, it has been estimated that 90% of cases of latrodectism in Texas occur in men bitten while using outdoor privies. No doubt if webs were removed from beneath the seats, the spiders would not be stimulated to attack the delicate portion of the human anatomy causing them to vibrate. In Europe, most bites from *Latrodectus tredecimguttatus* occur during the harvest when the spiders are carelessly pressed by the hands or arms of harvesters, or climb into their clothes or shoes (Maretić, 1971, 1975).

In Australia, the victims of *Atrax robustus* are mostly small children bitten whilst crawling beneath the raised floors of houses (Main, 1976). Although insecticides have been used to control these spiders, it would be simpler and better for the environment merely to wire off such dangerous places. Again, most people suffering attack from *Loxosceles* are bitten while dressing. They could avoid this by shaking their clothes before putting them on. Similarly, most scorpion stings occur when people get dressed or put on their shoes without checking to see if there is a scorpion present.

Scorpions can be prevented from entering houses by raising the floor level at least 20 cm above the ground, by surrounding the house with horizontal glazed ceramic tiles (which scorpions cannot climb) and by not allowing plants to trail against the walls of the house. Potential scorpion refuges, such as logs and piles of bricks, should be removed from the immediate vicinity of human dwellings (Keegan, 1980). A little care can obviate considerable pain and discomfort.

COLLECTING AND PRESERVING MATERIAL

Spiders can be collected with a robust insect sweep-net, by beating vegetation onto thick polythene sheeting, by sieving, pitfall traps and Berlese funnels. They can be captured with a glass tube, or sucked up with a pooter. For small spiders, a very serviceable pooter can be made from a 50 ml pipette with the ends of the tubular sections cut off and plastic tubes placed over them. Because spiders are so aggressive, they should be blown out of the pooter into individual glass tubes with polythene caps immediately after they have been caught.

Scorpions are to be found beneath rocks, logs and refuse of various kinds. When the entrance holes of their burrows have been located, they can be dug out. It is easiest to lift them by the tail, using long forceps, and place them individually in jars or boxes of suitable size.

Spiders and scorpions, like other arachnids, are best killed and preserved in 70% or 80% ethyl alcohol. The only disadvantage of this medium is that it is volatile and evaporates rapidly: if specimens are allowed to dry out, they become distorted and cannot be identified. Some workers add 5% glycerine as an insurance against complete dehydration, but this makes the specimens sticky, and disturbances of the refractive index occur when they are transferred to glycerine-free alcohol for examination. Moreover, if they do dry out completely, mould grows on the glycerine. Contraction of the legs, when living animals are placed in alcohol, can be prevented by killing them first with ethyl acetate vapour. Isopropyl alcohol diluted with distilled water to 50% is an acceptable alternative to ethyl alcohol and, in recent years, propylene phenoxytol has been used successfully as a preservative for arachnids. Although it has a bactericidal and fungicidal action, it is not a fixative and does not resist autolysis of the cells. It should be diluted to 1% with hot water and stirred vigorously. A better post-fixative preservative is obtained by mixing propylene glycol (10 parts), propylene phenoxytol (1 part) and distilled water (89 parts). The propylene glycol and phenoxytol must be mixed together before water is added.

In past times it was customary to store small specimens in jars of 70% or 80% alcohol, in separate tubes sealed with bungs of cotton wool. Today, it is more usual to store them separately in polythene or glass tubes with polythene closures. It is usually necessary to store larger specimens individually in polythene or glass tubes or bottles with polythene or other suitable caps. Corks are *not* recommended as they discolour the alcohol and do not prevent drying out. Rubber stoppers lower the pH of the preservative and often develop a poor seal. Labels with full data should always be placed *inside* the tubes with the specimens to which they refer. Preservation of scorpions is discussed by Sissom et al. (1990).

REFERENCES

Balozet, L. 1971. Scorpionism in the Old World. Pp. 349–371 in Bücherl, W. and Buckley, E. E. (eds), *Venomous animals and their venoms*: vol. 3, *Venomous invertebrates*. xxii + 537 pp. Academic Press, New York.

Bristowe, W. S. 1971. *The world of spiders*. Revised edition. xvi + 304 pp. Collins, London.

Bücherl, W. 1971a. Spiders. Pp. 197–277 in Bücherl and Buckley (1971).

Bücherl, W. 1971b. Classification, biology, and venom extraction of scorpions. Pp. 317–347 in Bücherl and Buckley (1971).

Bücherl, W. and **Buckley, E. E.** (eds) 1971. *Venomous animals and their venoms*: vol. 3, *Venomous invertebrates*. xx + 537 pp. Academic Press, New York and London.

Byalynitskii-Birulya, A. A. 1917. Arachnoidea arthrogastra caucasica. Part I. Scorpiones. *Zapiski Kavkazskaya Muzeya (Mémoires du Musée de Caucase)* (A) **5**: iii + 1–253 (index viii). [In Russian: for English translation see next reference.]

Byalynitskii-Birulya, A. A. 1964. *Arthrogastric arachnids of Caucasia. Part I. Scorpions.* 170 pp. Israel Programme for Scientific Translations, Jerusalem. [English translation of Byalynitskii-Birulya, 1917, see above.]

Cloudsley-Thompson, J. L. 1968. *Spiders, scorpions, centipedes and mites*. Second edition. xv + 278 pp. Pergamon Press, Oxford.

Cloudsley-Thompson, J. L. 1984. Arachnids. Pp. 175–204 in Cloudsley-Thompson, J. L. (ed.), *Sahara desert* (Key Environments Series No. 2). x + 348 pp. Pergamon Press, Oxford.

Cloudsley-Thompson, J. L. 1988. *Evolution and adaptation of terrestrial arthropods*. x + 141 pp. Springer-Verlag, Berlin.

Cooke, J. A. L., Roth, V. D. and **Miller, F. H.** 1972. The urticating hairs of theraphosid spiders. *American Museum Novitates* **2498**: 1–43.

Diniz, C. R. 1971. Chemical and pharmacological properties in *Tityus* venoms. Pp. 311–315 in Bücherl and Buckley (1971).

Duffey, E. and **Green, M. B.** 1975. A linphyiid spider biting workers on a sewage-treatment plant. *Bulletin of the British Arachnological Society* **3**: 130–131.

Foelix, R. F. 1982. *Biology of spiders*. vi + 306 pp. Harvard University Press, Cambridge, Mass. and London.

Francke, O. F. 1978. Systematic revision of diplocentrid scorpions (Diplocentridae) from circum-Caribbean lands. *Special Publications. The Museum of Texas Tech University* **14**: 1–92.

Francke, O. F. 1982. Scorpiones. Pp. 73–75 in Parker, S. P (ed.), *Synopsis and classification of living organisms*: vol. 2. 1232 pp. McGraw-Hill, New York.

Francke, O. F. 1985. Conspectus genericus scorpionorum 1758–1982 (Arachnida: Scorpiones). *Occasional Papers. The Museum of Texas Tech University* **98**: 1–32.

Gertsch, W. J. 1979. *American spiders*. Second edition. xiii + 274 pp. Van Nostrand Reinhold, New York.

Hjelle, J. T. 1990. Anatomy and morphology. Pp. 9–63 in Polis (1990).

Junqua, C. and **Vachon, M.** 1968. Les arachnides vénineux et leurs venins. État actuel des recherches. *Mémoires. Academie royale des Sciences d'Outre-Mer* (Classe des Sciences naturelles et médicales) (n.s.) **17** (5): 1–136.

Kaestner, A. 1940. 1. Ordnung des Arachnida: Scorpiones. Pp. 117–240 in Kükenthal, W. and Krumbach, T. (eds), *Handbuch der Zoologie* **3** (2) (1, Chelicerata) (1931–1940): xxx + 1–240.

Kaestner, A. 1968. *Invertebrate zoology*: vol. 2, *Arthropod relatives, Chelicerata, Myriapoda*. vii + 472 pp. Interscience Publications, New York.

Keegan, H. L. 1980. *Scorpions of medical importance*. ix + 140 pp. University Press of Mississippi, Jackson, Mississippi.

Kraepelin, K. 1899. Arachnoidea. Scorpiones und Pedipalpi. *Das Tierreich*, vol. 8. xviii + 265 pp. Friedlander, Berlin.

Lamoral, B. H. and **Reynders, S. C.** 1975. A catalogue of the scorpions described from the Ethiopian faunal region up to December 1973. *Annals of the Natal Museum* **22**: 489–576.

Lawrence, R. F. 1955. Solifugae, scorpions and Pedipalpi. *South African Animal Life* **1**: 152–262.

Levi, H. W. 1982. Araneae. Pp. 77–95 in Parker, S. P. (ed.), *Synopsis and classification of living organisms*: vol. 2. 1232 pp. McGraw-Hill, New York.

Levy, G. and **Amitai, P.** 1980. *Fauna Palaestina. Arachnida I. Scorpiones.* 130 pp. Israel Academy of Sciences and Humanities, Jerusalem.

Main, B. Y. 1976. *Spiders.* 296 pp. Collins, Sydney.

Maretić, Z. 1971. Latrodectism in Mediterranean countries, including South Russia, Israel, and North Africa. Pp. 299–309 in Bücherl and Buckley (1971).

Maretić, Z. 1975. European araneism. *Bulletin of the British Arachnological Society* **3**: 126–130.

Mello-Leitão, C. de 1945. Escorpiões Sul-Americanos. *Arquivos do Museu nacional*, Rio de Janeiro **40**: 1–468.

Millot, J. 1949. Ordre des Araneides (Araneae). Pp. 589–743 in Grassé, P. P. (ed.), *Traité de Zoologie*: vol. 6, *Onychophores – Tardigrades – Arthropodes – Trilobitomorphes – Chélicérates*. 979 pp. Masson, Paris.

Millot, J. and **Vachon, M.** 1949. Ordre des Scorpions. Scorpionides Latreille, 1810. Scorpiones Hemprich et Ehrenberg, 1810. Scorpiides C. L. Koch, 1837. Pp. 386–436 in Grassé, P. P. (ed.), *Traité de Zoologie*: vol. 6, *Onychophores – Tardigrades – Arthropodes – Trilobitomorphes – Chélicérates*. 979 pp. Masson, Paris.

Minton, S. A. 1968. Venoms of desert animals. Pp. 487–516 in Brown, G. W. (ed.), *Desert biology*: vol. 1, xvii + 635 pp. Academic Press, New York and London.

Pocock, R. I. 1900. Arachnida. In Blanford, W. T. (ed.), *The fauna of British India, including Ceylon and Burma*. xii + 279 pp. Taylor & Francis, London.

Polis, G. A. (ed.) 1990. *The biology of scorpions.* xxiii + 587 pp. Stanford University Press, Stanford, California.

Polis, G. A. and **Sissom, W. D.** 1990. Life history. Pp. 161–223 in Polis (1990).

Savory, T. H. 1928. *The biology of spiders.* xx + 376 pp. Sidgwick & Jackson, London.

Schaller, F. 1979. Significance of sperm transfer and formation of spermatophores in arthropod phylogeny. Pp. 587–608 in Gupta, A. P. (ed.), *Arthropod phylogeny*. xx + 762 pp. Van Nostrand Reinhold, New York.

Schenberg, S. and **Pereira Lima, F. A.** 1971. *Phoneutria nigriventer* venom-pharmacology and biochemistry of its components. Pp. 279–297 in Bücherl and Buckley (1971).

Schmidt, G. 1986. *Vogelspinnen. Lebensweise, Bestimmungsschlussel, Haltung und Zucht.* 87 pp. Albrecht Philler Verlag, Minden.

Sheals, J. G. 1973. Arachnida (scorpions, spiders, ticks, etc.). Pp. 417–472 in Smith, K. G. V. (ed.), *Insects and other arthropods of medical importance.* xiv + 561 pp. British Museum (Natural History), London.

Simard, J. M. and **Watt, D. D.** 1990. Venoms and toxins. Pp. 414–444 in Polis (1990).

Sissom, W. D. 1990. Systematics, biogeography, and paleontology. Pp. 64–160 in Polis (1990).

Sissom, W. D., Polis, G. A. and **Watt, D. D.** 1990. Field and laboratory methods. Pp. 445–461 in Polis (1990).

Smith, A. M. 1987. *The tarantula. Classification and identification guide.* 178 pp. Fitzgerald Publishing, London.

Stahnke, H. L. 1966. *The treatment of venomous bites and stings.* Revised edition. v + 117 pp. Arizona State University, Tempe, Arizona.

Thorp, R. W. and **Woodson, W. D.** 1945. *Black widow. America's most poisonous spider.* xi + 222 pp. University of North Carolina Press, Chapel Hill, North Carolina.

Vachon, M. 1952. *Études sur les scorpions.* 482 pp. Institut Pasteur d'Algérie, Algiers.

Vachon, M. 1974. Etudes des caractères utilisés pour classer les familles et les genres des Scorpions (Arachnides). 1. La trichobothriotaxie en Arachnologie. Sigles trichobothriaux et types de trichobothriotaxie chez les Scorpions. *Bulletin du Museum national d' Histoire naturelle*, Paris (3) (Zoologie) **104** (1973): 857–958. [This work is usually misdated 1973. It was not published until 1974, as is made evident from the printing completion date of 31 January 1974 stated on p. 958.]

Vachon, M. 1979. Arachnids of Saudi Arabia: Scorpiones. *Fauna of Saudi Arabia* **1**: 30–66.

Werner, F. 1934–1935. Scorpiones, Pedipalpi. In Bronns, H. G. (ed.), *Klassen und Ordnungen des Tierreichs*: vol. 5 (4) (8). 490 pp. Akademische Verlagsgesellschaft, Leipzig.

Scientific names index

All scientific names of organisms mentioned in text, figures and tables are listed in this index. To help the user who might be unfamiliar with any name, and to provide quick access to where the organism concerned is classified, the name of a higher group (usually order or class) has been added in parenthesis. To make the index as comprehensive as possible, the semi-vernacular uses of scientific names (with lower case initial letters) have been captured as equivalent to formal Latin category names: for example, muscomorph is indexed as Muscomorpha, triatomines as Triatominae, gerbilline as Gerbillinae, argasids as Argasidae. English or other vernacular names for organisms can be found in the Subject index.

Abonnencius (Diptera) 91
Acacia (Leguminosae) 97
Acalyptratae (Diptera) 60, 65, 67, 69, 70
Acanthaspidinae (Hemiptera) 491
Acanthocephala 585
Acanthoctenidae (Araneae) 664
Acari (Arachnida) 38, 597–658, 659
Acaridae (Acari) 634, 641–2, 643
Acaridida (Acari) 600, 632, 639–42
Acariformes (Acari) 600, 632
Acarina (Acari) 599, 609
Acartophthalmidae (Diptera) 66
Acarus (Acari) 597
Achatina (Mollusca: Gastropoda) 154, 194
Acroceridae (Diptera) 65
Actenophthalmus (Siphonaptera) 562
Actinedida (Acari) 600, 632, 642–7
Adersia (Diptera) 321, 324
Adlerius (Diptera) 83, 86, 91, 94, 96, 99, 107, 108
Adoratopsylla (Siphonaptera) 552, 557
  A. intermedia 544
  A. intermedia copha 557

Aedeomyia (Diptera) 125, 130, 133, 134, 136, 140, 154, 157, 168, 171, 172, 191
  A. catasticta 133, 134, 147
Aedeomyiini (Diptera) 140
Aedes (Diptera) 6, 16, 55, 120, 127, 131, 134, 136, 137, 138, 139, 140, 143, 144, 145, 148–52, 153, 154, 158, 163, 171, 172, 173, 175, 176, 178, 179, 180, 181, 184, 185, 186, 187, 188, 189, 191, 192, 193, 194, 200, 205, 206, 207, 210, 211, 212, 217, 220, 252, 301
  A. aegypti 7, 121, 123, 139, 148, 149, 166, 174, 175, 176, 177, 178, 183, 185, 189, 190, 194, 195, 206, 207, 208, 210, 211, 212, 213, 217, 221
  A. aegypti aegypti 14, 149
  A. aegypti formosus 14, 149
  A. aegypti 'queenslandensis' 149
  A. africanus 149, 178, 179, 181, 193, 207
  A. albopictus 149, 182, 190, 194, 206, 208, 210
  A. atlanticus 151
  A. atropalpus 151, 176, 182, 194
  A. australis 18
  A. bromeliae 149, 194, 207
  A. campestris 186

A. canadensis 186
A. candidoscutellum 193
A. cantans 176, 185, 186
A. caspius 151, 161, 181, 191
A. cataphylla 178
A. chemulpoensis 193
A. communis 16
A. cooki 166, 167, 203
A. cumminsii 178
A. detritus 151, 161, 181, 186, 191
A. diantaeus 151
A. dobodurus 190
A. domesticus 178
A. dorsalis 186
A. dupreei 136, 184
A. excrucians 121, 167
A. fijiensis 203
A. flavescens 189
A. fluviatilis 194
A. fulvus pallens 178
A. furcifer 151, 207
A. futunae 166, 203
A. geniculatus 151, 193
A. harinasutai 203
A. hebrideus 166
A. hendersoni 151, 161
A. hexodontus 151, 184
A. impiger 176, 196
A. infirmatus 178
A. ingrami, 178
A. irritans 194
A. katherinensis 166
A. kesseli 166, 203
A. kiangsiensis 203

A. *kochi* 151, 203
A. *luteocephalus* 186, 193, 207
A. *malayensis* 167
A. *mariae* 181, 194
A. *melanimon* 191
A. *natronius* 182, 189, 192
A. *neoafricanus* 207
A. *nigripes* 176, 196
A. *nigromaculis* 191
A. *niveus* 151, 203
A. *niveus* group 205, 208
A. *oceanicus* 203
A. *opok* 207
A. *palmarum* 190, 194
A. *pembaensis* 181, 194
A. *pernotatus* 166
A. *pionips* 151
A. *poicilius* 151, 203, 205
A. *polynesiensis* 150, 166, 167, 176, 182, 203, 205
A. *pseudoafricanus* 193
A. *pseudoscutellaris* 150, 166, 167, 203, 205, 206
A. *punctodes* 151, 178
A. *punctor* 167
A. *rotumae* 166
A. *rusticus* 136
A. *samoanus* 203
A. *scapularis* 202
A. *scapularis* group 173
A. *scutellaris* 150, 166, 167
A. *scutellaris* complex 165, 194
A. *scutellaris* group 166–7, 176, 208
A. *seoulensis* 193
A. *sierrensis* 121, 151, 187, 193
A. *simpsoni* 167, 207
A. *sollicitans* 139, 151, 177, 181, 184, 192
A. *stimulans* 184
A. *tabu* 166, 167
A. *taeniorhynchus* 139, 151, 181, 184, 192, 202
A. *taylori* 151, 207
A. *theobaldi* 184
A. *thibaulti* 193
A. *togoi* 151, 186, 194, 203, 205
A. *tongae* 166, 203
A. *triseriatus* 151, 161, 174, 182, 186, 187, 193, 196, 206
A. *tutuilae* 203
A. *upolensis* 166, 203
A. *varipalpus* 174
A. *vexans* 150, 166, 184, 186
A. *vigilax* 151, 184, 192, 203, 205

A. *vigilax vansomerenae* 188
A. *vittatus* 151, 187, 194
A. *wauensis* 152
A. *woodi* 137
Aedimorphus (Diptera) 150, 171
Aedini (Diptera) 140, 186
Aegophagamyia (Diptera) 321
Aegyptianella pullorum (Rickettsiales) 610
Aepyceros melampus (Mammalia: Artiodactyla) 370
Afoxella (Siphonaptera) 559
Afristivalius torvus (Siphonaptera) 541
Afrocimex (Hemiptera) 488
Afrocimicinae (Hemiptera) 488
Agromyzidae (Diptera) 66
Alanstonea (Diptera) 171
Alberprosenia (Hemiptera) 491, 499
A. *goyovargasi* 491
A. *malheiroi* 491, 498
Alberprosenini (Hemiptera) 490, 491
Alcelaphus buselaphus (Mammalia: Artiodactyla) 370
Aleurodidae (Hemiptera) 483, 484
Alouatta (Mammalia: Primates) 207
Alphavirus (virus) 208
Amalaraeus (Siphonaptera) 546, 557
A. *penicilliger* 557
A. *penicilliger mustelae* 547, 549
Amaurobiidae (Araneae) 664
Amblycera (Phthiraptera) 518
Amblyomma (Acari) 438, 601, 602, 605, 606, 609, 615, 618, 619, 621
A. *americanum* 615, 621, 625
A. *cajennense* 615, 621, 628
A. *hebraeum* 614, 615, 619, 628, 651
A. *variegatum* 206, 615, 619, 627, 651
Amblyomminae (Acari) 601
Amblyospora (Protozoa: Microspora) 195
Amblypygi (Arachnida) 38, 659, 662
Amegabothris (Siphonaptera) 560
Amphalius (Siphonaptera) 545, 557

A. *runatus* 538, 546, 557
Amphipsylla (Siphonaptera) 544, 557
A. *primaris mitis* 557
A. *rossica* 538, 557
Amphipsyllidae (Siphonaptera) 534, 535
Amphipsyllinae (Siphonaptera) 553
Ananta (Coleoptera)
A. *bicolor* 585
A. *decolor* 585
Anaphlebotomus (Diptera) 91, 94, 95, 99
Anasolen (Diptera) 250
Ancala (Diptera) 313, 322
A. *africana* 313
Ancistropsyllidae (Siphonaptera) 535, 536, 552
Ancylostoma (Nematoda: Strongyloidea) 424
A. *duodenale* 481
Androctonus (Scorpiones) 672, 673, 676, 677
A. *aeneas* 676
A. *amoreuxi* 676
A. *australis* 661, 663, 669, 676, 677
A. *bicolor* 676
A. *crassicauda* 675, 676
A. *hoggarensis* 676
A. *mauretanicus* 676
Anisolabis (Dermaptera) 589
Anisopodidae (Diptera) 64, 430, 433, 434, 457, 578, 589
Anisopus (Diptera) 457
Anocentor nitens (Acari) 619, 651
Anomiopsyllus (Siphonaptera) 550, 557
A. *nudatus* 558
A. *nudatus hiemalis* 558
Anopheles (Diptera) 120, 124, 127, 129, 130, 131, 132, 134, 136, 137, 138, 139, 140, 142, 143, 145, 146, 158, 161, 162, 163, 168, 172, 173, 175, 176, 177, 178, 180, 181, 183, 184, 188, 189, 191, 192, 193, 194, 197, 198–9, 200, 201, 202, 205, 208, 209, 210, 211, 212, 214, 216, 220, 221
A. *aconitus* 145, 191, 198, 202
A. *albimanus* 145, 179, 188, 198, 200, 202, 212, 213, 216, 221
A. *albitarsis* 145, 165, 198

Scientific names index 685

A. annularis 198
A. anthropophagus 199, 202, 203
A. aquasalis 145, 181, 191, 198, 202
A. arabiensis 145, 163, 164, 183, 191, 192, 198, 202
A. argyritarsis 145, 198
A. atroparvus 164, 180, 183, 197, 198
A. aztecus 164, 198
A. balabacensis 198, 199, 202
A. bambusicolus 145
A. bancroftii 199, 203
A. barberi 143, 187, 193
A. barbirostris 129, 200, 202, 203, 204, 205
A. barianensis 193
A. bellator 145, 193, 198, 202
A. beklemishevi 162, 164
A. braziliensis 198
A. bwambae 163, 164
A. campestris 198, 203
A. caroni 180
A. cinereus 142
A. claviger 143, 187, 191, 194, 198
A. crucians 121, 142, 173
A. cruzii 145, 198
A. culicifacies 142, 145, 181, 184, 191, 194, 197, 198
A. culicifacies complex 141, 165
A. darlingi 145, 165, 191, 198, 202
A. dirus 141, 145, 180, 198, 202, 209
A. dirus complex 165
A. donaldi 198, 202, 203
A. earlei 164, 183
A. engarensis 162
A. farauti 145, 165, 199, 203, 216
A. flavirostris 198, 202
A. fluviatilis 145, 198
A. franciscanus 183
A. freeborni 145, 164, 183, 184, 191, 198
A. funestus 6, 145, 192, 198, 200, 202, 208
A. gambiae 13, 141, 145, 162, 164, 179, 188, 191, 192, 197, 198, 200, 202, 209, 213, 216
A. gambiae complex 141, 163–4, 172, 175, 200, 208
A. hackeri 193

A. hamoni 180
A. hancocki 202
A. hermsi 164
A. hilli 199
A. hispaniola 198
A. hyrcanus 145
A. hyrcanus group 162
A. implexus 179
A. jeyporiensis 198, 199, 202
A. karwari 199
A. koliensis 165, 199, 203
A. kweiyangensis 202, 203
A. labranchiae 164, 198
A. letifer 198, 202
A. leucosphyrus 145, 198, 202
A. leucosphyrus complex 165
A. leucosphyrus group 173
A. ludlowae 198
A. maculatus 145, 192, 198, 202, 209
A. maculatus complex 165
A. maculipennis 141, 145, 164, 183
A. maculipennis complex 162, 164, 172
A. mangyanus 198
A. marshalli complex 165
A. martinius 164
A. melanoon 164
A. melas 163, 164, 181, 191, 197, 198, 202
A. merus 13, 163, 164, 181, 191, 198, 202
A. messeae 164, 198
A. minimus 145, 189, 192, 198, 199, 202, 209, 216
A. moucheti 198
A. multicolor 192, 198
A. neivai 198
A. nigerrimus 198, 202, 203
A. nili 145, 192, 198, 202
A. nuneztovari 6, 145, 165, 198, 216
A. occidentalis 164
A. omorii 193
A. pattoni 198, 199
A. pauliani 202
A. pharoensis 184, 198
A. philippinensis 188, 198, 199, 202, 216
A. plumbeus 142, 187, 193
A. pseudopunctipennis 145, 192, 198
A. pulcherrimus 184, 187, 198
A. punctimacula 198
A. punctipennis 183
A. punctulatus 145, 165, 192, 199, 203, 216
A. punctulatus complex 165

A. quadriannulatus 163, 164
A. quadrimaculatus 138, 141, 198
A. quadrimaculatus complex 165
A. sacharovi 164, 198
A. sergentii 145, 198
A. sicaulti 164
A. sinensis 142, 145, 191, 198, 199, 203, 210, 212
A. smithi 192
A. stephensi 145, 163, 179, 181, 184, 191, 194, 198, 212, 221
A. subalpinus 164
A. subpictus 191, 199, 202
A. sundaicus 181, 191, 198, 199
A. superpictus 145, 192, 198
A. tessellatus 198, 202
A. triannulatus 198
A. umbrosus 188
A. vagus 192, 202
A. vanhoofi 192
A. varuna 198
A. walkeri 184
A. wellcomei 202
A. whartoni 198, 202
Anophelinae (Diptera) 58, 121, 132, 134, 136, 139, 140, 142–5, 146, 162, 172, 173
Anoplura (Phthiraptera), 517–28
Anotylus (Coleoptera) 584
Anthocoridae (Hemiptera) 483, 484, 486, 489, 507
Anthomyiidae (Diptera) 66, 195, 411, 458
Anthomyzidae (Diptera) 66
Anthrenus (Coleoptera) 584, 585
  A. scrophulariae 583
  A. verbasci 584
Aphodius (Coleoptera) 585
Aphonopelma (Araneae) 667
Aphrania (Hemiptera) 488
Apicomplexa (Protozoa) 241
Apidae (Hymenoptera) 581
Apioceridae (Diptera) 65
Apiomerini (Hemiptera) 490
Apis (Hymenoptera)
  A. mellifera adansonii 582
  A. mellifera ligustica 582
  A. mellifera mellifera 582
Aponomma (Acari) 601, 608, 621
Apterygota (Insecta) 40
Arachnida 32, 33, 36–8, 595–680

Aradellini (Hemiptera) 490
Araneae (Arachnida) 38, 597, 659–82
Araneidae (Araneae) 664, 668
Araneomorphae (Araneae) 663, 664, 665, 666
Araneus diadematus (Araneae) 673
Araucnephia (Diptera) 250
Araucnephioides (Diptera) 250
Archaeognatha (Insecta) 39, 40
Arctiidae (Lepidoptera) 586
Arcyophora sylvatica (Lepidoptera) 587
Argas (Acari) 601, 607, 608, 609, 610, 621, 622, 625, 649, 652
  A. arboreus 610, 649
  A. persicus 609, 610
  A. reflexus 610
  A. reflexus hermanni 610
  A. vespertilionis 604
Argasidae (Acari) 599, 601, 602, 603, 604, 607, 608, 609, 610, 616, 619, 620, 621, 622, 623, 624, 648, 649
Argiopidae (Araneae) 668
Armigeres (Diptera) 135, 140, 147, 152, 153–4, 161, 171, 175, 181, 185, 188, 193, 194
  A. angustus 181
  A. annulitarsis 181
  A. dolichocephalus 154, 181, 190, 193
  A. flavus 181
  A. kuchingensis 191
  A. subalbatus 154, 189, 191
Armillifer (Pentastomida) 36
  A. armillatus 36
  A. moniliformis 36, 37
Armilliferidae (Pentastomida) 36
Arrenuridae (Acari) 180
Arthropoda 4, 30–47
Artiodactyla (Mammalia) 439, 552, 621, 627
Arvicanthis (Mammalia: Rodentia) 109
Arvicolinae (Mammalia: Rodentia) 557, 558, 560
Ascaris lumbricoides (Nematoda: Ascaridoidea), 481
Aschiza (Diptera) 59, 65, 67, 69
Ascodipteron (Diptera) 74
Asilidae (Diptera) 58, 62, 65, 73, 74, 324
Asiloidea (Diptera) 65, 67
Asilomorpha (Diptera) 65, 67

Aspergillus fumigatus (Fungi) 481
Asteiidae (Diptera) 66
Astigmata (Acari) 600
Ateles (Mammalia: Primates) 207
Athericidae (Diptera) 64, 69, 579, 580
Atherigona (Diptera) 422
Atherix (Diptera) 580
Atheta (Coleoptera) 584
Atrax (Araneae) 667, 670, 672, 673
  A. formidabilis 667
  A. robustus 660, 667, 674, 678
Atrichopogon (Diptera) 296
Attagenus (Coleoptera) 585
Atylotus (Diptera) 322
Atypus affinis (Araneae) 673
Auchenorrhyncha (Hemiptera) 485, 486
Auchmeromyia (Diptera) 411, 412, 420, 430, 437, 452
  A. luteola (= *senegalensis*) 11, 452
  A. senegalensis 11, 412, 420, 446, 447, 452
Aulacigastridae (Diptera) 66
Austenina (Diptera) 340
Australimyzidae (Diptera) 66
Austroconopinae (Diptera) 295, 296, 297
Austroconops (Diptera) 288, 291, 295, 297
  A. mcmillani 297
Austromansonia (Diptera) 155
Austrophlebotomus (Diptera) 90, 91, 93, 99
Austrosimulium (Diptera) 243, 250, 251, 254, 259, 268
  A. australense 268
  A. pestilens 261, 264, 266, 268
  A. ungulatum 263, 268
Automeris (Lepidoptera) 586
Autosericea castanea (Coleoptera) 583
Avaritia (Diptera) 301, 304
Avicennia (Avicennaceae, mangrove) 191, 193
Avicularia (Araneae) 667
Axymyiidae (Diptera) 64
Axymyiomorpha (Diptera) 64
Ayurakitia (Diptera) 171

Babesia (Protozoa: Apicomplexa) 629
  B. divergens 629, 631
  B. microti 612, 630, 631
Bacillus (Bacteria)

  B. anthracis 325, 399, 585
  B. sphaericus 212, 213
  B. thuringiensis 211, 213, 278, 279, 302, 426
Bactrodinae (Hemiptera) 490
Bartonella bacilliformis (Bacteria) 78, 104
Barychelidae (Araneae) 667, 672
Belminus (Hemiptera) 491, 499
  B. costaricensis 491
  B. herreri 491
  B. peruvianus 491
  B. rugulosus 491
Belostomatidae (Hemiptera) 483
Beltranmyia (Diptera) 293
Bembecidae (Hymenoptera) 324
Bercaea cruentata (Diptera) 422
Bertilia (Hemiptera) 488
Bibio (Diptera) 60
Bibionidae (Diptera) 60, 64
Bibionoidea (Diptera) 64
Bibionomorpha (Diptera) 64
Bironella (Diptera) 140, 141, 142, 144
Bison bison (Mammalia: Artiodactyla) 440
Blaberidae (Dictyoptera) 477, 479
Blaps mucronata (Coleoptera) 585
Blatta orientalis (Dictyoptera) 473, 474, 477, 479, 480, 481
Blattaria (Dictyoptera) 473–82
Blattella (Dicytoptera)
  B. asahinai 479, 480
  B. germanica 473, 475, 477, 478, 480
Blattellidae (Dicytoptera) 477, 478, 479
Blattidae (Dictyoptera) 477
Blephariceridae (Diptera) 64, 73
Blephariceroidea (Diptera) 64
Blephariceromorpha (Diptera) 64
Bolbodera (Hemiptera) 491, 499
  B. scabrosa 491
Bolboderini (Hemiptera) 490, 491
Bombycidae (Lepidoptera) 586
Bombyliidae (Diptera) 65, 73, 324
Bombylioidea (Diptera) 65
Boophilus (Acari) 601, 602, 606, 608, 619, 627
  B. microplus 622, 648

Boophthora (Diptera) 12, 250
  B. erythrocephala (= *Simulium erythrocephalum*) 12
Borrelia (Spirochaeta) 20, 525, 629
  B. anserina 610
  B. burgdorferi 325, 612, 625, 629
  B. duttoni 525, 611
  B. hermsi 611
  B. hispanica 611
  B. parkeri 611
  B. persica 611
  B. recurrentis 399, 525
  B. turicatae 611
  B. venezuelensis 611
Bos mutus (Mammalia: Artiodactyla) 440
Bothriuridae (Scorpiones) 663, 668, 676
Bothriurus (Scorpiones) 663
Bouvieromyiini (Diptera) 316, 321
Bovidae (Mammalia: Artiodactyla) 369, 370, 371, 552, 639
Brachycera (Diptera) 55, 58, 59, 61, 62, 63, 64, 67, 68–70
Brachyconops (Diptera) 295
Braulidae (Diptera) 66
Braunsiomyia (Diptera) 321
Brevidigita (Siphonaptera) 564
Brucella melitensis (Bacteria) 568
Bruchomyiinae (Diptera) 79, 89, 92
Brugia (Nematoda: Filarioidea) 205
  B. malayi 200, 201, 203–4, 205
  B. timori 200, 201, 204, 205
Brumptomyia (Diptera) 89, 90, 91, 93
Bucimex (Hemiptera) 488
Bunyaviridae (virus) 626
Buthidae (Scorpiones) 660, 661, 663, 668, 669, 670, 676
Buthotus (Scorpiones) 677
  B. tamulus 669
Buthus (Scorpiones) 672, 673, 676
  B. occitanus 676, 678
Byssodon (Diptera) 250

Caccobius (Coleoptera)
  C. mutans 583
  C. vulcanus 583, 584
Cacodminae (Hemiptera) 488
Cacodmus (Hemiptera) 488

Calliphora (Diptera) 406, 407, 409, 411, 413, 419, 420, 437, 451
  C. erythrocephala (= *vicina*) 413
  C. uralensis 420
  C. vicina 56, 62, 413, 419, 420, 432, 451, 459
  C. vomitoria 413, 420, 451
Calliphoridae (Diptera) 25, 66, 70, 71, 72, 405, 406, 409, 410, 412, 413, 418, 420–1, 429, 430, 431, 436, 437, 446–52, 459, 589
Calliphorini (Diptera) 418
Calobata cibaria (Diptera) 458
Calyptra eustrigata (Lepidoptera) 587, 588
Calyptratae (Diptera) 54, 56, 62, 66, 67, 69, 70, 390, 403–28, 432, 460
Cambarus (Crustacea) 34
Camillidae (Diptera) 66
Caminicimex (Hemiptera) 488
Cancraedes (Diptera) 194
Canidae (Mammalia: Carnivora) 105, 562
Capensomyia (Diptera) 91, 96, 99
Carabidae (Coleoptera) 585
Carcinocorini (Hemiptera) 490
Carnidae (Diptera) 66
Carnoidea (Diptera) 66
Carpoglyphus lactis (Acari) 642
Catachlorops (Diptera) 322
Catallagia (Siphonaptera) 550, 558
  C. dacenkoi 542
  C. decipiens 558
Cavernicola (Hemiptera) 492, 496, 499
  C. lenti 492, 498
  C. pilosa 492, 505
Cavernicolini (Hemiptera) 490, 492
Cebidae (Mammalia: Primates) 207, 518
Cecidomyiidae (Diptera) 64, 72
Cediopsylla (Siphonaptera) 551, 558
  C. spillmanni 558
Cellia (Diptera) 145, 202
Celyphidae (Diptera) 66
Centrocneminae (Hemiptera) 490
Centruroides (Scorpiones) 672, 673, 676, 677
  C. elegans 677

  C. exilicauda 663, 677
  C. infamatus 677
  C. limpidus 667
  C. sculpturatus (= *exilicauda*) 677
  C. suffusus 675, 677
Ceratophyllidae (Siphonaptera) 535, 536, 552, 557, 558, 560, 561, 562, 563
Ceratophylloidea (Siphonaptera) 536
Ceratophyllus (Siphonaptera) 546, 551, 558
  C. anisus 547, 549, 558
  C. gallinae 537, 558
  C. garei 558
  C. indages 558
Ceratopogonidae (Diptera) 55, 57, 58, 64, 71, 74, 75, 100, 288—309
Ceratopogoninae (Diptera) 295, 296, 297
Ceratopogonini (Diptera) 295
Cercopithecidae (Mammalia: Primates) 207
Cervidae (Mammalia: Artiodactyla) 443, 552
Cetherinae (Hemiptera) 490
Cetherini (Hemiptera) 490
Chactidae (Scorpiones) 669, 676
Chaerilidae (Scorpiones) 669
Chaerilus (Scorpiones) 669
Chaetocruiomyia (Diptera) 192
Chagasia (Diptera) 136, 137, 140, 141, 142, 144
  C. bathana 144, 162
Chaoboridae (Diptera) 60, 64, 120, 139, 195, 577
Chaoborus (Diptera) 60
  C. edulis 577
Cheiridium (Arachnida) 184
Chelicerata (Arthropoda) 32, 33, 36, 38, 659
Chiastopsylla (Siphonaptera) 549, 558
  C. quadrisetis 542
  C. rossi 558
Chilopoda (Myriapoda) 33, 35, 37
Chimaeropsyllidae (Siphonaptera) 535, 553, 558
Chinius (Diptera) 85, 90, 91, 93
  C. julianensis 84

Chiracanthium (Araneae)
  C. brevicalcaratum 667
  C. diversum 667
  C. inclusum 667
  C. punctorium 667, 675
Chironomidae (Diptera) 61, 64, 68, 73, 74, 120, 288, 576, 577
Chironomoidea (Diptera) 4, 64, 68
Chironomus (Diptera) 577
  C. dorsalis 577
Chirothrips aculeatus (Thysanoptera) 588
Chlorichaeta tuberculosa (Diptera) 580
Chloropidae (Diptera) 66, 69, 579, 580, 589
Chlorotabanus (Diptera) 322, 323, 324
Chrysomelidae (Coleoptera) 585
Chrysomya (Diptera) 403, 407, 412, 413, 420, 437, 438, 448–9
  C. albiceps 414, 431, 436, 449, 450
  C. bezziana 6, 414, 421, 446, 448, 449, 451
  C. megacephala 405, 414, 421, 449
  C. putoria 421
  C. rufifacies 414, 431, 436, 449, 459
  C. varipes 449
Chrysomyini (Diptera) 418
Chrysopinae, *see* Chrysopsinae
Chrysops (Diptera) 55, 56, 310, 313, 314, 315, 316, 317–21, 321, 322, 323, 325, 326, 327
  C. caecutiens 317
  C. centurionis 317, 318, 320, 326
  C. dimidiatus 317, 318, 320, 325, 326
  C. distinctipennis 315, 318, 319, 325
  C. fixissimus 311
  C. griseicollis 315, 318, 319
  C. langi 318, 320, 326
  C. longicornis 315, 317, 318, 319, 325
  C. silaceus 315, 317, 318, 320, 324, 325, 326
  C. streptobalius 318, 319
  C. zahrai 318, 320
Chrysopsinae (Diptera) 311, 314, 316, 321, 322, 323

Chrysopsini (Diptera) 316, 321
Chryxinae (Hemiptera) 490
Chyletidae (Acari) 643
Chyletiella parasitivorax (Acari) 643
Chyromyidae (Diptera) 66
Cicadellidae (Hemiptera) 483
Cicadidae (Hemiptera) 484
Cimex (Hemiptera) 488, 489
  C. columbarius 487
  C. hemipterus 484, 487, 488
  C. lectularius 484, 487, 488, 489
  C. pipistrelli 487
  C. rotundatus 488
Cimexopsis (Hemiptera) 488
Cimicidae (Hemiptera) 483, 484, 485, 486–9, 509
Cimicinae (Hemiptera) 488
Citellophilus (Siphonaptera) 545, 558
  C. lebedewi 558
  C. simplex 534
  C. tesquorum 558
  C. tesquorum sungaris 547
Cladotanytarsus lewisi (Diptera) 576
Clarias fuscus (Pisces) 212
Cleridae (Coleoptera) 583
Clogmia albipunctatus (Diptera) 457
Clostridium (Bacteria) 583
Clubionidae (Araneae) 667, 671, 675
Clusiidae (Diptera) 66
Cnephia (Diptera) 250, 251
  C. dacotensis 263
  C. pecuarum 268
Cnesia (Diptera) 250
Cnesiamima (Diptera) 250
Coccinellidae (Coleoptera) 585
Cochliomyia (Diptera) 407, 412, 413, 420, 447–8
  C. hominivorax 414, 437, 446, 447, 448, 451, 459
  C. macellaria 414, 437, 447, 448
Coelomomyces (Fungi) 195
Coleoptera (Insecta) 39, 40, 458, 583–5
Coleorrhyncha (Hemiptera) 485, 486
Collartidini (Hemiptera) 490
Collembola (Insecta) 33, 39, 40
Colocasia (Araceae, coco-yam) 154, 193
Connochaetes gnou (Mammalia: Artiodactyla) 370
Conopidae (Diptera) 65

Conopoidea (Diptera) 65
Copromyza equina (Diptera) 579
Coptopsylla (Siphonaptera) 543, 553, 558
  C. bairamaliensis 558
  C. lamellifera 558
  C. lamellifera dubinini 541
Coptopsyllidae (Siphonaptera) 535, 536, 553, 558
Coquillettidia (Diptera) 129, 134, 135, 137, 139, 140, 141, 145, 146, 154–5, 156, 169, 173, 174, 175, 176, 178, 181, 187, 188, 191, 196, 202, 209, 211, 217, 218, 301
  C. crassipes 204
  C. fuscopennata 174, 179
  C. metallica 178
  C. perturbans 121, 155
  C. richiardii 155, 187
  C. venezuelensis 155
Cordylobia (Diptera) 407, 411, 412, 420, 430, 437, 446
  C. anthropophaga 405, 413, 446, 447, 460
  C. rodhaini 412, 446
  C. ruandae 413
Corethrellidae (Diptera) 64
Coromyia (Diptera) 91
Cosmiomma (Acari) 608
Cowdria ruminantium (Rickettsiales) 615
Coxiella burneti (Rickettsiales) 568, 628
Craneopsylla (Siphonaptera) 552, 558
  C. minerva 546, 549, 558
Crassicimex (Hemiptera) 488
Crataerina pallida (Diptera) 581
Crematogaster (Hymenoptera) 178
Cricetinae (Mammalia: Rodentia) 558, 560, 561
Crozetia (Diptera) 247, 250, 263
Crustacea (Arthropoda) 31, 32, 33, 34–5
Cryptochetidae (Diptera) 66
Cryptoctenopsyllus (Siphonaptera) 559
Ctenidae (Araneae) 660, 667, 672
Ctenizidae (Araneae) 664
Ctenocephalides (Siphonaptera) 546, 549, 558
  C. canis 540, 542, 558
  C. felis 534
  C. felis damarensis 533

C. felis felis 534, 540, 543, 558
C. felis strongylus 558
Ctenophthalmidae (Siphonaptera) 535, 553, 557, 558, 559, 560, 562
Ctenophthalmus (Siphonaptera) 547, 549, 558
  C. agyrtes 559
  C. assimilis 559
  C. breviatus 558
  C. calceatus cabirus 543, 558
  C. congeneroides 559
  C. dolichus 559
  C. orientalis 559
  C. phyris 559
  C. pollex 559
  C. secundus 559
  C. solutus 559
  C. teres 559
  C. wladimiri 559
Culex (Diptera) 16, 120, 123, 126, 127, 128, 131, 136, 137, 138, 140, 143, 145, 146, 148, 150, 157, 158–9, 160, 163, 165, 169, 172, 173, 175, 176, 178, 181, 184, 185, 191, 193, 194, 202, 205, 209, 211, 217, 220, 654
  C. abominator 158
  C. aikenii 185
  C. alogistus 158, 185
  C. annulirostris 159, 203
  C. antennatus 202
  C. australicus 166
  C. autogenicus 166
  C. bitaeniorhynchus 188, 203
  C. brevipalpis 193
  C. cinereus 191
  C. demeilloni 6
  C. fatigans (= *quinquefasciatus*) 158
  C. gaudeator 158, 185
  C. gelidus 158, 195
  C. globocoxitus 166
  C. kyotoensis 193
  C. molestus 166, 192, 202
  C. nebulosus 182
  C. nigripalpus 158, 178
  C. pallens 166, 182, 203
  C. peus, see *C. stigmatosoma*
  C. pilosus 158, 185
  C. pipiens 121, 141, 158, 159, 165, 166, 177, 179, 182, 183, 185, 192, 194, 202, 203, 213
  C. pipiens complex 159, 165–6, 167
  C. quinquefasciatus 121, 158, 159, 165, 166, 175, 177, 179, 182, 189, 191, 192, 194, 195, 200, 202, 203, 204, 210, 213, 217, 219, 221, 301
  C. restuans 158
  C. sinensis 188
  C. sitiens 167, 192, 203
  C. starckeae 192
  C. stigmatosoma 182, 191
  C. taeniopus 158, 185
  C. tarsalis 158, 175, 178, 182, 183, 184, 191
  C. tenaguis 189
  C. thalassius 192
  C. theileri 159
  C. tigripes 131
  C. tritaeniorhynchus 158, 175, 180, 191, 209, 212
  C. univittatus 159
  C. vishnui 158, 167, 179
Culicella (Diptera) 159, 184, 185
Culicidae (Diptera) 14, 15, 55, 57, 58, 61, 64, 68, 74, 75, 79, 90, 100, 120–240
Culicinae (Diptera) 121, 123, 132, 134, 135, 136, 139, 140, 141, 142, 143, 144, 145–61, 162, 172, 188
Culicini (Diptera) 139, 140, 146, 148–52, 167
Culiciomyia (Diptera) 172
Culicoidea (Diptera) 64, 68
Culicoides (Diptera) 24, 57, 78, 252, 267, 288, 289, 290, 291, 292, 293, 294, 295, 297, 298, 299, 300, 301, 302, 303, 304
  C. actoni 301
  C. adersi 300
  C. arakawae 293, 299, 304
  C. barbosai 301
  C. brevitarsis 293, 299, 301, 302
  C. dycei 302
  C. fulvithorax 292
  C. fulvus 301
  C. furens 293, 300, 304
  C. grahamii 6, 292, 300
  C. imicola 292, 300, 301
  C. imicola group 302
  C. impunctatus 298
  C. inornatipennis 292, 300
  C. insignis 293, 300, 301
  C. insinuatus 301
  C. marksi 302
  C. melleus 298
  C. milnei 292
  C. milnei group 300, 301
  C. mississippiensis 299
  C. nubeculosus 298, 304
  C. obsoletus 300, 301
  C. oxystoma 293, 299, 301, 302
  C. paraensis 293, 301
  C. phlebotomus 293, 300
  C. pulicaris 292, 300
  C. puncticollis 292
  C. riethi 304
  C. robertsi 300
  C. schultzei 299, 304
  C. schultzei group 299, 301, 302
  C. stellifer 300
  C. variipennis 293, 298, 299, 301, 302, 304
  C. venustus 300
  C. wadai 301
Culicoidini (Diptera) 295, 297
Culicomorpha (Diptera) 4, 55, 64, 67, 68
Culiseta (Diptera) 137, 140, 145, 157, 159–60, 169, 172, 175, 176, 181, 184, 185, 194
  C. alaskaensis 159
  C. annulata 160, 177, 183, 185, 187
  C. dyari 160
  C. fraseri 160
  C. impatiens 183
  C. incidens 159, 191
  C. inornata 160, 174, 183, 188
  C. longiareolata 160
  C. melanura 160
  C. morsitans 159, 176, 177
  C. silvestris 159
Culisetini (Diptera) 140
Curtonotidae (Diptera) 66
Cuterebra (Diptera) 439
  C. emasculator 440
Cuterebrinae (Diptera) 410, 436, 438–9, 440
Cyclocephala borealis (Coleoptera) 583
Cyclops (Crustacea) 34, 35
  C. quadricornis 35
Cyclorrhapha (Diptera) 63, 64, 65, 66, 67
Cydistomyia (Diptera) 322
Cynolebias (Pisces) 212
Cypselosomatidae (Diptera) 65

Dacnoforcipomyia (Diptera) 295, 296
Dampfomyia (Diptera) 91
Dasybasis (Diptera) 322, 323
Dasychela (Diptera) 322

Dasycnemini (Hemiptera) 490
Dasyhelea (Diptera) 296
Dasyheleinae (Diptera) 295, 296
Decapoda (Crustacea) 34
Deinocerites (Diptera) 123, 125, 130, 131, 136, 140, 148, 157, 160–1, 168, 173, 176, 178, 194
   D. cancer 160, 174, 175, 178
   D. dyari 178
   D. pseudes 161
Deliastini (Hemiptera) 490
Delostichus (Siphonaptera) 551, 559
   D. talis 538, 559
Demeillonius (Diptera) 91, 95
Demodex (Acari) 38, 631, 634, 642, 643–4
   D. brevis 644
   D. folliculorum 644
Demodicidae (Acari) 634, 643–4
Dendryphantes (Araneae) 667
   D. noxiosus 667
Dermacentor (Acari) 601, 602, 606, 608, 613, 620
   D. albipictus 613
   D. andersoni 613, 614, 625, 626, 628
   D. marginatus 613, 615, 628
   D. nitens 613
   D. nuttalli 613, 615
   D. pictus (= *reticulatus*) 615
   D. reticulatus 615, 651
   D. silvarum 613, 628
   D. variabilis 613, 625, 628
Dermanyssidae (Acari) 633, 634, 635–6
Dermanyssus (Acari) 631
   D. gallinae 633, 635, 636
Dermaptera (Insecta) 40, 589
Dermatobia (Diptera) 430, 436, 462
   D. hominis 429, 436, 438, 460, 461
Dermatophagoides (Acari) 634, 641
   D. farinae 641, 654
   D. pteronyssinus 641, 642
Dermestes (Coleoptera) 585
   D. lardarius 584
Dermestidae (Coleoptera) 583, 584
Deuterophlebiidae (Diptera) 64
Diachlorini (Diptera) 316, 322
Diachlorus (Diptera) 322, 323
   D. ferrugatus 327

Diamanus (Siphonaptera) 550, 561
Diamphidea nigroornata (Coleoptera) 585
Diaspidini (Hemiptera) 490
Diastatidae (Diptera) 66
Diceromyia (Diptera) 151, 171
Dichelacera (Diptera) 322, 323
Dicladocera (Diptera) 322
Dictynidae (Araneae) 664
Dictyoptera (Insecta) 39, 40, 473, 474, 475
Didelphidae (Mammalia: Marsupialia) 557
Didelphis marsupialis (Mammalia: Marsupialia), 506
Dinopsyllus (Siphonaptera) 549, 559
   D. ellobius 559
   D. lypusus 559
   D. wansoni 544
Diopsidae (Diptera) 65
Diopsoidea (Diptera) 65
Dipetalogaster (Hemiptera) 492
   D. maxima 490, 491, 492, 502, 504
Diphyllobothrium (Cestoda) 35
Diplocentridae (Scorpiones) 669, 670, 676
Diplopoda (Myriapoda) 32, 33, 35, 37
Diplura (Insecta) 33, 40
Dipluridae (Araneae) 660, 667, 672
Dipodidae (Mammalia: Rodentia) 560, 561
Diptera (Insecta) 4, 14, 15, 17, 39, 40, 42, 43, 46, 49–469, 576–81, 589
Dipylidium caninum (Cestoda) 20, 558, 559, 562, 566
Dirofilaria (Nematoda: Filarioidea) 241
   D. immitis 151, 205
Dirphia (Lepidoptera) 586
Dixa (Diptera) 60, 61
Dixidae (Diptera) 60, 64, 68, 120, 139, 144
Dolichopodidae (Diptera) 65, 195
Dolichovespula (Hymenoptera) 582
Dorylinae (Hymenoptera) 397
Dracaena (Agavaceae) 193
Dracunculus (Nematoda: Dracunculoidea) 34
   D. medinensis 34
Drosophila (Diptera) 435

   D. repleta 458
Drosophilidae (Diptera) 66, 69, 430, 435, 436, 458
Dryomyzidae (Diptera) 66
Dugesiella (Araneae) 667
Duttonella (Protozoa: Sarcomastigophora) 373
Dysdera (Araneae) 671, 673
Dysderidae (Araneae) 671

Echidnophaga (Siphonaptera) 543, 559
   E. gallinacea 535, 536, 540, 559
   E. larina 559
   E. oschanini 559
Ectemnia (Diptera) 246, 249, 250
Ectinoderini (Hemiptera) 490
Ectrichodiinae (Hemiptera) 490
Edentata (Mammalia) 105, 553, 564
Edwardsellum (Diptera) 4, 5, 7, 250
Eichhornia (Diptera) 188
Elasmodeminae (Hemiptera) 490
Embioptera (Insecta) 40
Emesinae (Hemiptera) 483, 490, 496
Emesini (Hemiptera) 490
Emmareus (Siphonaptera) 558
Empididae (Diptera) 65, 74
Empidoidea (Diptera) 65
Endopterygota (Insecta) (= Holometabola) 4, 39, 40
Entamoeba histolytica (Protozoa) 424, 481
Enterobius vermicularis (Nematoda: Oxyuroidea) 481
Entomophthorales (Fungi) 419
Ephemeroptera (Insecta) 40, 43, 589
Ephydra cinerea (Diptera) 580
Ephydridae (Diptera) 66, 69, 73, 195, 430, 433, 435, 458, 580
Ephydroidea (Diptera) 66
Epicauta hirticornis (Coleoptera) 585
Epistylis (Protozoa: Ciliophora) 195
Equidae (Mammalia: Perissodactyla) 443
Equus burchellii (*quagga*) (Mammalia: Equidae) 370
Eratyrus (Hemiptera) 492, 500

E. cuspidatus 492, 505
E. mucronatus 492
Eresidae (Araneae) 664
Eretmapodites (Diptera) 130, 134, 140, 147, 152, 154, 169, 171, 176, 177, 181, 184, 185, 188, 193, 194
E. chrysogaster 154, 178
E. grahami 154, 190, 194
E. subsimplicipes 154, 188, 189, 194
Erinaceidae (Mammalia: Insectivora) 563
Eristalis (Diptera) 435, 457
E. tenax 457
Erysipelothrix rhusiopathiae (Bacteria) 568
Escherichia coli (Bacteria) 424, 481
Esenbeckia (Diptera) 321, 323
Euclea (Lepidoptera) 586
Eucliidae (Lepidoptera) 586
Euhoplopsyllus (Siphonaptera) 559
  E. andensis 559
  E. glacialis affinis 559
  E. manconis 559
Eulaelaps stabularis (Acari) 633, 638, 639
Eumelanomyia (Diptera) 158
Euphenini (Hemiptera) 490
Euphlebotomus (Diptera) 91, 94, 95, 97, 107, 108
Euphorbia (Euphorbiaceae) 189
Euproctis (Lepidoptera) 586
Euroglyphus maynei (Acari) 641
Eurychoromyiidae (Diptera) 66
Eurypelma (Araneae) 667
Eurypterida (Arthropoda) 660
Eusimuliini 5
Eusimulium (Diptera) 250
Eutrombicula (Acari)
  E. alfreduggesi 646
  E. batatas 646
Evandromyia (Diptera) 91
Exopterygota (Insecta) (= Hemimetabola) 39, 40, 479, 483, 485

Fannia (Diptera) 61, 404, 407, 409, 416, 417, 421, 431, 437, 453, 454
  F. benjamini 416, 421
  F. canicularis 403, 404, 416, 417, 421, 453, 454
  F. incisurata 453, 454
  F. manicata 453, 454
  F. pusio 421

F. scalaris 403, 416, 417, 421, 453, 454
Fanniidae (Diptera) 70, 404, 409, 411, 416, 418, 421, 430, 434, 437, 453–4
Fergusoninidae (Diptera) 66
Fibromavirus myxomatosis (virus) 568
Ficalbia (Diptera) 125, 140, 154, 156, 172, 178, 191
  F. minima 156
Ficalbiini (Diptera) 140
Fidena (Diptera) 321
Filistatidae (Araneae) 664
Filodes fulvodorsalis (Lepidoptera) 588
Finlaya (Diptera) 131, 151, 173, 193, 194, 202
Flaviviridae (virus) 626
Flavivirus (virus) 626
Forcipomyia (Diptera) 288, 291, 292, 294, 295, 296, 298, 300, 301
  F. anabaenae 295, 296
  F. lefanui 291, 292
  F. taiwana 304
Forcipomyiinae (Diptera) 291, 295, 296
Formicidae (Hymenoptera) 582
Foxella (Siphonaptera) 550, 559
  F. ignota 541, 559
Francisella tularensis (Bacteria) 325, 568
Frontopsylla (Siphonaptera) 545, 559
  F. luculenta 559
  F. semura 559
  F. wagneri 541, 545, 548, 559
Furnariidae (Aves) 505

Galago senegalensis (Mammalia: Primates) 207
Galindomyia (Diptera) 140, 148, 157, 161
  G. leei 160, 161
Gamasida (Acari) 599, 600, 632, 634
Gambusia (Pisces) 212
  G. affinis 212
  G. affinis affinis 212
  G. affinis holbrooki 212
Gasterophilidae (Diptera) 340
Gasterophilinae (Diptera) 410, 430, 436, 438, 443–5
Gasterophiloidea (Diptera) 340
Gasterophilus (Diptera) 430, 443
  G. haemorrhoidalis 443, 444, 445

G. inermis 443, 444, 445
G. intestinalis 410, 443, 444, 445, 460
G. nasalis 443, 444, 445
G. nigricornis 443, 444
G. pecorum 443, 444, 445
Gebiella (Siphonaptera) 560
Gedoelstia (Diptera) 439
Geometridae (Lepidoptera) 586
Geomyidae (Mammalia: Rodentia) 559
Geophilimorpha (Myriapoda) 36
Geoskusea (Diptera) 194
Gerbillinae (Mammalia: Rodentia) 79, 553, 558
Gerbillophilus (Siphonaptera) 561
Gigantodax (Diptera) 242, 250, 259
Gigantostraca (Arthropoda) 660
Glossina (Diptera) 16, 55, 57, 333–82, 407, 408, 409
  G. austeni 341, 342, 344, 345, 346, 347, 348, 350, 351, 355, 356, 360, 361, 382
  G. austeni austeni 346
  G. austeni mossurizensis 346
  G. brevipalpis 340, 342, 343, 344, 345, 346, 347, 349, 352, 353, 356, 357, 359, 360, 368
  G. caliginea 343, 344, 345, 351, 354, 355, 359, 376
  G. frezili 343, 344, 345, 347, 349, 352, 353, 357, 358
  G. fusca 343, 349, 350, 358, 360, 361
  G. fusca congolensis 343, 344, 345, 346, 353, 354, 357, 369
  G. fusca fusca 343, 344, 345, 353, 354, 357, 374
  G. fusca group 337, 338, 340, 341, 342, 343, 344, 345, 346, 347, 348, 352, 353, 357, 358, 362, 363, 366, 367, 368, 369, 370, 382
  G. fuscipes 342, 343, 351, 354, 360, 361, 376
  G. fuscipes fuscipes 344, 345, 346, 347, 351, 354, 355, 359, 362, 368, 371, 377, 378
  G. fuscipes martinii 344, 345, 346, 351, 354, 355, 359
  G. fuscipes quanzensis 344, 345, 346, 351, 354, 355, 359
  G. fuscipleuris 342, 343, 344,

G. fuscipleuris (cont'd) 345, 350, 352, 353, 357, 358, 360, 361
G. haningtoni 343, 344, 345, 350, 353, 357, 358
G. longipalpis 344, 345, 346, 347, 349, 351, 355, 356, 360, 361
G. longipennis 335, 342, 343, 344, 345, 347, 348, 352, 353, 357, 358, 359, 360, 362, 369
G. medicorum 343, 344, 345, 347, 349, 352, 353, 357, 358, 374
G. morsitans 339, 346, 349, 351, 355, 356, 360, 361, 365, 368, 370
G. morsitans centralis 344, 345, 346, 347, 370, 371, 377
G. morsitans group 341, 342, 343, 344, 345, 347, 348, 352, 356, 357, 363, 364, 368, 369, 370, 372, 377, 378, 380, 381, 382
G. morsitans morsitans 344, 345, 346, 347, 366, 368, 369, 370, 372, 377
G. morsitans submorsitans 344, 345, 347, 362, 366, 372, 373, 375
G. nashi 343, 344, 345, 350, 352, 353, 358
G. nigrofusca 343, 347, 350, 352, 353, 357, 358
G. nigrofusca hopkinsi 344, 345, 350
G. nigrofusca nigrofusca 343, 344, 345, 349, 350
G. pallicera 346, 350, 359
G. pallicera newsteadi 344, 345, 346, 351, 354, 355, 359
G. pallicera pallicera 336, 344, 345, 346, 349, 350, 354, 355, 359
G. pallidipes 344, 345, 346, 347, 349, 351, 354, 356, 360, 361, 366, 367, 368, 369, 370, 371, 377
G. palpalis 342, 343, 349, 351, 354, 360, 361, 369, 376
G. palpalis gambiensis 344, 345, 347, 354, 355, 359
G. palpalis group 340, 341, 342, 343, 344, 345, 346, 347, 348, 352, 355, 357, 362, 364, 366, 367, 368, 369, 370, 372, 376, 378, 380, 381, 382

G. palpalis palpalis 343, 344, 345, 346, 347, 354, 355, 359, 369, 370, 371, 372, 374, 375
G. schwetzi 343, 344, 345, 349, 352, 353, 357, 358, 360, 361
G. severini 343, 344, 345, 350, 352, 353, 357, 358
G. swynnertoni 339, 344, 345, 346, 347, 351, 356, 360, 361, 366, 368, 377
G. tabaniformis 343, 344, 345, 348, 349, 353, 357, 358, 360, 361, 369
G. tachinoides 342, 344, 345, 346, 347, 349, 350, 354, 355, 359, 360, 362, 366, 368, 369, 371, 376
G. vanhoofi 343, 344, 345, 350, 352, 353, 358, 359
Glossinidae (Diptera) 55, 57, 61, 66, 68, 70, 71, 74, 75, 333–88, 403, 408, 418, 419
Glossininae (Diptera) 340
Glycyphagidae (Acari) 632, 641, 642
Glycyphagus (Acari)
  G. destructor 642, 643
  G. domesticus 642, 643
Glyptocranium gasteracanthoides (Araneae) 668
Gnaphosidae (Araneae) 674
Gnathostoma (Nematoda: Spiruroidea) 35
Gomphostilbia (Diptera) 250
Gongylonema pulchrum (Nematoda: Spiruroidea) 585
Goniops (Diptera) 325
Gonyaulax (Protozoa: Sarcomastigophora) 34
Gorilla gorilla (Mammalia: Primates) 518, 521
Grabhamia (Diptera) 152
Grammostola (Araneae) 667
Grassomyia (Diptera) 82, 83, 86, 91, 95
Greniera (Diptera) 250
Grylloblattodea (Insecta) 40, 474
Gymnopaidinae (Diptera) 249
Gymnopais (Diptera) 242, 247, 250, 262, 264
Gynaikothrips ficorum (Thysanoptera) 588

Habronema microstoma (Nematoda: Spiruroidea) 399
Hadrurus (Scorpiones) 676
  H. hirsutus 678
Haemagogus (Diptera) 41, 125, 137, 140, 150, 152–3, 170, 173, 175, 181, 186, 193, 206, 207, 220
  H. albomaculatus 207
  H. anastasionis 189
  H. capricornii 152, 207
  H. equinus 152, 185, 186, 189, 207
  H. janthinomys 152, 207
  H. leucocelaenus 152, 207
  H. spegazzinii 152, 207
Haemaphysalinae (Acari) 601
Haemaphysalis (Acari) 601, 606, 608, 609, 612, 621
  H. inermis 612
  H. leachi 604, 612
  H. lemuris 621
  H. longicornis 618
  H. spinigera 612, 627, 652
  H. turturis 627
Haematobia (Diptera) 55, 73, 389, 390, 391, 392, 393, 396–7, 398, 407, 422
  H. exigua 390, 393, 395, 396
  H. irritans 390, 393, 394, 395, 396, 399
  H. minuta 394
Haematobosca (Diptera) 391
Haematopota (Diptera) 56, 310, 314, 317, 321, 322, 323
  H. maculosifacies 312
Haematopotini (Diptera) 71, 312, 316, 322
Haematosiphon (Hemiptera) 488
  H. inodorus 487
Haematosiphoninae (Hemiptera) 488
Haemogamasus pontiger (Acari) 633, 638, 639
Haemoproteus (Protozoa: Apicomplexa) 300
Hammacerinae (Hemiptera) 490
Harpactirella (Araneae) 667, 672, 673
  H. karroica 667
  H. treleaveni 667
Harpactorinae (Hemiptera) 490
Harpactorini (Hemiptera) 490
Hearlea (Diptera) 250

Hebridosimulium (Diptera) 250
Hectopsylla (Siphonaptera) 551, 559
  H. suarezi 559
Heizmannia (Diptera) 123, 125, 140, 152, 153, 169, 170, 172, 181, 185, 186, 193
Helcocertomyia (Diptera) 91, 98, 109
Heleomyzidae (Diptera) 66
Heliconia (Heliconiaceae) 161, 193, 195
Helidomermis (Nematoda: Mermithoidea) 298
Heliothrips (Thysanoptera) 588
Hellichiella (Diptera) 250
Helophilus (Diptera) 457
Helosciomyzidae (Diptera) 66
Hemicnetha (Diptera) 250
Hemileuca (Lepidoptera) 586
Hemimerus (Dermaptera) 44
Hemimetabola (Insecta) (= Exopterygota) 39, 40, 485
Hemiptera (Insecta) 39, 40, 195, 483–516
Hemiscorpion lepturus (Scorpiones) 676
Hepatocystis (Protozoa: Apicomplexa) 300
  H. kochi 300
Hermetia (Diptera) 59
  H. illucens 458
Hertigia (Diptera) 89, 91, 93
Hesperocimex (Hemiptera) 488
Hesperoctenes (Hemiptera) 489
Hesperomyinae (Mammalia: Rodentia) 558
Heteromyidae (Mammalia: Rodentia) 560
Heteromyiini (Diptera) 295
Heteroptera (Hemiptera) 483, 485, 486, 490
Hexapoda (Arthropoda) 32, 33, 38
Hilarimorphidae (Diptera) 65
Himalayum (Diptera) 250
Hippelates (Diptera) 580
  H. collusor 580
  H. pusio 580
Hippobosca (Diptera) 44, 409
  H. camelina 581
  H. equina 581
  H. rufipes 579, 581
  H. variegata 581
Hippoboscidae (Diptera) 66, 70, 71, 74, 340, 408, 409, 534, 579, 581
Hippoboscoidea (Diptera) 66, 69, 70, 72, 340

Hippocentrum (Diptera) 317, 320, 321, 322
Hirstionyssus isabellinus (Acari) 633, 639
Hodgesia (Diptera) 123, 130, 137, 140, 154, 157, 191
Hodgesiini (Diptera) 140
Holoconops (Diptera) 295
Holometabola (Insecta) (= Endopterygota) 4, 39, 40, 529
Holoptilinae (Hemiptera) 490
Holoptilini (Hemiptera) 490
Holothyrida (Acari) 600
Hominidae (Mammalia: Primates) 518
Homo (Mammalia: Hominidae) 518
Homoptera (Hemiptera) 483, 485
Hoplopsyllus (Siphonaptera) 550, 551, 559
  H. anomalus 559
Hottentotta (Scorpiones) 672, 673, 676, 677
  H. minax 677
Howardina (Diptera) 173
Huaedes (Diptera) 152
Hyalomma (Acari) 601, 606, 609, 616, 618, 622, 627
  H. anatolicum 627
  H. concinna 628
  H. leachi 628
  H. marginatum 616
  H. marginatum complex 627
Hyalomminae (Acari) 601
Hybomitra (Diptera) 322
Hydrocharis (Hydrocharitaceae, frog's bit) 138
Hydropyrus hians (Diptera) 580
Hydrotaea (Diptera) 422, 455, 456
  H. dentipes 456
  H. irritans 422
Hylesia (Lepidoptera) 586
Hymenolepis diminuta (Cestoda) 561, 567, 585, 586, 589
Hymenoptera (Insecta) 39, 40, 324, 508, 581–3
Hypoderma (Diptera) 430, 439, 441, 442, 462
  H. bovis 439, 440, 441, 442, 443, 459
  H. diana 441, 442, 443
  H. lineatum 440, 441, 442, 443, 459, 460
Hypodermatinae (Diptera) 410, 430, 436, 438, 439–43

Hypsophthalmidae (Siphonaptera) 535
Hystricopsylla (Siphonaptera) 551, 560
  H. dippiei 560
  H. occidentalis 543
  H. occidentalis linsdalei 560
Hystricopsyllidae (Siphonaptera) 535, 536, 553, 560
Hystricopsylloidea (Siphonaptera) 536

Ichneumonidae (Hymenoptera) 581
Icosiella neglecta (Nematoda: Filarioidea) 300
Idiophlebotomus (Diptera) 90, 91, 93
Ingramia (Diptera) 130, 134, 135, 146, 156, 193
Insecta (Arthropoda) 4, 32, 33, 38–46
Insectivora (Mammalia) 553, 558, 562, 563, 569, 621
Inseliellum (Diptera) 250
Iridovirus (virus) 610
Ironomyiidae (Diptera) 65
Ischnocera (Phthiraptera) 518
Ischnopsyllidae (Siphonaptera) 535, 553
Ischnuridae (Scorpiones) 669
Isometres maculatus (Scorpiones) 678
Isopoda (Crustacea) 73
Isoptera (Insecta) 39, 40
Iulus (Myriapoda) 35
Iuridae (Scorpiones) 669
Ixodes (Acari) 601, 606, 607, 608, 609, 611, 618, 620, 621, 629
  I. dammini 612, 629, 630
  I. holocyclus 612, 621, 625, 626, 628, 629
  I. pacificus 612, 629
  I. persulcatus 612, 619, 621, 626, 629
  I. ricinus 604, 611, 612, 619, 620, 621, 626, 629, 631, 652
  I. ricinus group 612
  I. ricinus/persulcatus group 611
  I. rubicundus 612, 625
  I. schillingsi 621
Ixodida (Acari) 600–16, 632
Ixodidae (Acari) 601, 602, 604, 605, 606, 607, 608, 615, 616, 617, 618, 619, 620,

Ixodidae (Acari) (cont'd) 621, 622, 623, 624, 625, 628, 648, 649–51
Ixodinae (Acari) 601, 620
Ixodoidea (Acari) 609

Janthinosoma (Diptera) 152
Jantia crassipalpis (Diptera) 12, 422
Johnbelkinia (Diptera) 131, 140, 147, 148, 170
Juxtapulex (Siphonaptera) 562

Karnyothrips flavipes (Thysanoptera) 588
Kasaulius (Diptera) 91, 94, 95
Kerteszia (Diptera) 145, 173, 193, 202
Klebsiella pneumoniae (Bacteria) 481
Kobus (Mammalia: Artiodactyla) 370
Kueichenlipsylla (Siphonaptera) 560

Labidognatha (Araneae) 665, 666
Laelapidae (Acari) 633, 634, 638, 639
Laelaps (Acari)
   L. echidninus 633, 639
   L. nuttalli 633, 639
Lagenidium (Fungi) 195
Lagomorpha (Mammalia) 438, 439, 557
Larroussius (Diptera) 84, 86, 87, 91, 94, 95, 97, 99, 107, 108, 113
Lasiocampidae (Lepidoptera) 586
Lasiodora (Araneae) 667
Lasiohelea (Diptera) 288, 290, 291, 292, 294, 295, 296, 298, 299, 300, 301, 304
Latrocimex (Hemiptera) 488
Latrocimicinae (Hemiptera) 488
Latrodectus (Araneae) 667, 670, 672, 673, 676
   L. cinctus 672
   L. curacaviensis 667, 672
   L. hasselti 667, 672
   L. mactans 659, 667, 672, 674
   L. menavodi 672
   L. pallidus 672
   L. tredecimguttatus 660, 667, 672, 675, 678
Lauxaniidae (Diptera) 66
Lauxanioidea (Diptera) 66

Leicesteria (Diptera) 154
Leishmania (Protozoa: Sarcomastigophora) 6, 21, 78, 97, 99, 102, 103, 104, 105, 106, 107, 108, 398
   L. aethiopica 97, 105, 108
   L. amazonensis 103, 109
   L. braziliensis 91, 105, 106, 109
   L. chagasi 105, 109
   L. donovani 97, 105, 108
   L. garnhami 109
   L. guyanensis 104, 105, 106, 109
   L. infantum 97, 104, 105, 108, 109
   L. major 97, 104, 105, 106, 107, 108, 110
   L. mexicana 105, 109
   L. panamensis 103, 105, 106, 109
   L. peruviana 98, 109
   L. pifanoi 109
   L. tropica 97, 104, 105, 107, 108, 399
   L. venezuelensis 109
Leistarchini (Hemiptera) 490
Leiurus (Scorpiones) 672, 673, 676, 677
   L. quinquestriatus 675, 677, 678
Lepidoptera (Insecta) 39, 40, 585–8
Lepiselaga (Diptera) 322, 323
   L. crassipes 327
Lepisma saccharina (Insecta: Zygentoma) 589
Leporidae (Mammalia: Lagomorpha) 558, 559
Leptocera (Diptera)
   L. caenosa 580
   L. venalicia 458
Leptocimex (Hemiptera) 488
   L. boueti 487
Leptoconopinae (Diptera) 291, 295, 296
Leptoconops (Diptera) 55, 57, 288, 291, 294, 295, 296, 297, 298, 299, 300, 302, 303
   L. bequaerti 301
   L. kerteszi 304
   L. rhodesiensis 291
   L. spinosifrons 302
Leptodemus (Hemiptera) 507
Leptopsylla (Siphonaptera) 549, 560
   L. aethiopica 560
   L. algira 543
   L. pavlovskii 560
   L. segnis 539, 542, 560

Leptopsyllidae (Siphonaptera) 534, 535, 553, 557, 559, 560, 561
Leptopsyllinae (Siphonaptera) 553
Leptorhoptrum robustum (Araneae) 674
Leptotrombidium (Acari) 634, 645, 646, 647
   L. akamushi 644, 646, 647
   L. arenicola 646, 647
   L. deliense 646, 647, 654
   L. fletcheri 646, 647
   L. pallidum 646, 647
   L. pavlovskyi 646, 647
   L. scutellaris 646, 647
Leucocytozoon (Protozoa: Apicomplexa) 241, 300
   L. caulleryi 300
   L. simondi 241
   L. smithi 241
Leucophaea maderae (Dictyoptera) 479
Levitinia (Diptera) 250
Levua (Diptera) 194
Lewisellum (Diptera) 250
Lilaea (Diptera) 321
Limacodidae (Lepidoptera) 586
Limatus (Diptera) 125, 140, 146, 147, 148, 170
   L. flavisetosus 178
Limnesiidae (Acari) 180
Limothrips (Thysanoptera)
   L. cerealium 588
   L. denticornis 588
Linguatula serrata (Pentastomida) 36, 37
Linguatulidae (Pentastomida) 36
Linguatuloidea (Pentastomida) 36
Linognathus (Phthiraptera) 184
Linshcosteus (Hemiptera) 492, 496, 500
   L. carnifex 492
   L. chota 492
   L. confumus 492
   L. costalis 492
   L. kali 492
Linyphiidae (Araneae) 664, 674
Liphistiomorphae (Araneae) 660, 664, 665, 666
Liponyssoides sanguineus (Acari) 633, 634, 635, 636
Lipoptena cervi (Diptera) 581
Listeria (Bacteria) 561
   L. monocytogenes 568

Listropsylla (Siphonaptera) 548, 560
 L. dorippae 560
 L. fouriei 541
Lithobiomorpha (Myriapoda) 36
Lithosiidae (Lepidoptera) 586
Lithyphantes (Araneae) 664, 667
Litomosoides carinii (Acari) 636
Loa (Nematoda: Filarioidea) 317, 326
 L. loa 310, 317, 325
Lobelia (Lobeliaceae) 193
Loboscraspis griseifusca (Lepidoptera) 587
Lonchaeidae (Diptera) 65
Lonchoptera (Diptera) 59
Lonchopteridae (Diptera) 59, 65
Lonomia achelous (Lepidoptera) 586, 587
Lophoceraomyia (Diptera) 123, 158, 193
Lophopodomyia (Diptera) 145
Loxaspis (Hemiptera) 488
Loxosceles (Araneae) 583, 667, 672, 673, 676, 678
 L. reclusa 660, 667
Loxoscelidae (Araneae) 667
Lucilia (Diptera) 41, 411, 419, 420, 421, 437, 449, 450–1
 L. cuprina 421, 425, 450
 L. illustris 459
 L. sericata 411, 423, 450, 451, 459
Lutzia (Diptera) 131, 188
Lutzsimulium (Diptera) 250
Lutzomyia (Diptera) 6, 78, 82, 86, 90, 91, 92, 93, 98–9, 102, 107, 109
 L. amazonensis 109
 L. anduzei 109
 L. aracuchensis 109
 L. aragaoi group 91
 L. ayrozai 109
 L. baityi group 91
 L. carrerai 91, 109
 L. christophei 109
 L. colombiana 104
 L. complexa 109
 L. delpozoi group 91
 L. diabolica 109
 L. dreisbachi group 91
 L. evansi 109
 L. flaviscutellata 98, 103, 109
 L. gomezi 109
 L. hartmanni 98, 109

 L. intermedia 98, 109
 L. lanei group 91
 L. llanosmartinsi 109
 L. longipalpis 82, 90, 101, 102, 106, 107, 109
 L. migonei 109
 L. migonei group 91
 L. olmeca 98, 103, 109
 L. olmeca nociva 109
 L. oswaldi group 91
 L. panamensis 83, 109
 L. paraensis 109
 L. peruensis 98, 109
 L. pessoai 109
 L. pilosa group 91
 L. rupicola group 91
 L. saulensis group 91
 L. spinicrassa 109
 L. trapidoi 104, 109
 L. umbratilis 98, 103, 109
 L. wellcomei 109
 L. whitmani 109
 L. verrucarum 104, 109
 L. verrucarum group 91
 L. ylephiletor 104, 109
 L. youngi 109
 L. yucumensis 91, 109
Lycosa (Araneae) 667, 672, 676
 L. tarentula 675
Lycosidae (Araneae) 195, 667, 670, 671, 672, 673, 675
Lygaeidae (Hemiptera) 507
Lymantriidae (Lepidoptera) 586
Lyperosia (Diptera) (= *Haematobia*) 391
Lytta vesicatoria (Coleoptera) 584, 585

Macaca (Mammalia: Primates) 205
Machadomyia (Diptera) 342
Macracanthorhynchus hirudinaceus (Acanthocephala) 585
Macrocephalini (Hemiptera) 490
Macronyssidae (Acari) 633, 634, 636–9
Macropsyllidae (Siphonaptera) 535
Macroscelidea (Mammalia) 553
Malacopsyllidae (Siphonaptera) 535, 536, 553
Malacopsylloidea (Siphonaptera) 536
Malaraeus (Siphonaptera) 551, 560

 M. sinomus 560
 M. telechinus 547, 549, 560
Malaya (Diptera) 123, 125, 140, 146, 147, 176, 178, 189, 193
Mallophaga (Phthiraptera) 518
Manangocorinae (Hemiptera) 490
Mandibulata (Arthropoda) 32, 33
Mansonella (Nematoda: Filarioidea) 241
 M. ozzardi 241, 267, 300
 M. perstans 300
 M. streptocerca 300
Mansonia (Diptera) 120, 127, 129, 135, 136, 137, 138, 139, 140, 141, 143, 145, 146, 151, 154, 155–6, 163, 169, 172, 173, 175, 178, 181, 188, 191, 196, 200, 205, 209, 211, 217, 218
 M. annulata 156, 203, 204, 205
 M. annulifera 156, 203, 205
 M. bonneae 156, 203, 204, 205
 M. dives 156, 203, 204, 205
 M. indiana 156, 203, 205
 M. titillans 155, 202
 M. uniformis 135, 156, 203, 204, 205
Mansoniini (Diptera) 140
Mansonioides (Diptera) 136, 155, 156, 181, 202
Mantodea (Dictyoptera) 43, 477
Maorigoeldia (Diptera) 140, 147
Margaropus (Acari) 601, 608, 619
Marmota (Mammalia: Rodentia) 558, 561, 563
Marsupialia (Mammalia) 105, 439, 553, 554, 555, 557, 558, 561, 562, 621
Mastigoproctus giganteus (Arachnida) 659
Mastophora (Araneae) 668
Mauritia (Palmae) 193
Mayacnephia (Diptera) 250
Mecoptera (Insecta) 40
Megabothris (Siphonaptera) 546, 560
 M. abantis 560
 M. calcarifer 560
 M. clantoni 560
 M. rectangulatus 545, 560
 M. turbidus 560
Megaconops (Diptera) 295
Megaloptera (Insecta) 40
Megalopyge (Lepidoptera) 586

Megalopygidae (Lepidoptera) 586
Megamerinidae (Diptera) 65
Megaselia (Diptera) 431, 435
   M. halterata 578
   M. rufipes 456
   M. scalaris 456, 579
   M. spiracularis 456
Meilloniellum (Diptera) 250
Melanoconion (Diptera) 158, 173
Melipona (Hymenoptera) 582
Meloidae (Coleoptera) 585
Melophagus ovinus (Diptera) 408, 579, 581
Mendanocorini (Hemiptera) 490
Meringis (Siphonaptera) 551, 560
   M. shannoni 542, 560
Merostomata (Arthropoda) 33, 38
Mesocyclops (Crustacea) 35
Mesobuthus (Scorpiones) 672, 673, 676, 677
   M. martensi 677
   M. tamulus 677
Mesomermis flumenalis (Nematoda: Mermithoidea) 278
Mesopsylla (Siphonaptera) 547, 560
   M. apscheronica 560
   M. tuschkan 560
   M. tuschkan andruschkoi 542
Mesostigmata (Acari) 600, 632, 634–9
Mesothelae (Araneae) 666
Metacnephia (Diptera) 250, 251
Metapterini (Hemiptera) 490
Metarhizium (Fungi) 508
Metastigmata (Acari) 600
Metastriata (Acari) 607
Metomphalus (Diptera) 250
Micropezidae (Diptera) 65, 430, 458
Micropsylla (Siphonaptera) 562
Micropsylloides (Siphonaptera) 562
Micropygomyia (Diptera) 91
Microspora (Protozoa) 195
Microtominae (Hemiptera) 490
Microtomus (Hemiptera) 485
Microtriatoma (Hemiptera) 491, 499
   M. borbai 491
   M. trinidadensis 491
Milichiidae (Diptera) 66
Mimomyia (Diptera) 129, 130, 133, 134, 135, 137, 140, 146, 154, 156, 157, 168, 172, 178, 188, 191, 193, 196, 218
   M. chamberlaini 130
   M. deguzmanae 130, 168
   M. hybrida 130, 135, 156
   M. jeansottei 193
   M. levicastilloi 156, 157
   M. luzonensis 168
   M. modesta 156
   M. pallida 129, 135, 156
   M. roubaudi 157
Miridae (Hemiptera) 507
Mochlonyx (Diptera) 121
Mochlostyrax (Diptera) 173
Monoculicoides (Diptera) 293
Monomorium (Hymenoptera) 583
   M. pharaoensis 583
Monopsyllus (Siphonaptera) 546, 551, 558
Monotremata (Mammalia) 553, 621
Montisimulium (Diptera) 250
Mormotomyiidae (Diptera) 66
Morops (Diptera) 250
Morphoidea (Lepidoptera) 586
Mucidus (Diptera) 151, 181, 188
Muridae (Mammalia: Rodentia) 79, 109, 557, 558, 560
Murinae (Mammalia: Rodentia) 558, 560, 561, 564, 568
Mus musculus (Mammalia: Muridae) (= *M. domesticus*) 636
Musca (Diptera) 20, 44, 391, 393, 398, 404, 409, 411, 417, 419, 422, 424, 426, 438, 455
   M. autumnalis 422, 423, 425
   M. biseta 418, 423
   M. confiscata 391
   M. crassirostris 389, 392, 394, 397–8, 418, 455
   M. domestica 13, 52, 403, 404, 419, 420, 421, 422, 424, 426, 434, 437, 455, 456, 459
   M. domestica calleva 422
   M. domestica complex 418
   M. domestica curviforceps 422
   M. fasciata 422
   M. pattoni 422
   M. sorbens 20, 418, 422, 423
   M. vetustissima 418, 422, 423, 634
   M. vitripennis 422
Muscidae (Diptera) 25, 41, 55, 57, 66, 68, 71, 73, 75, 340, 389–402, 403, 404, 409, 411, 417, 419, 422–3, 430, 434, 436, 437, 454–6, 459, 589
Muscina (Diptera) 409, 417, 455, 456
   M. stabulans 455, 456
Muscinae (Diptera) 70, 391
Muscoidea (Diptera) 66, 69, 70, 411, 419
Muscomorpha (Diptera) 53, 54, 55, 56, 58, 59, 61, 62, 63, 64, 65, 66, 67, 68, 69, 71, 72, 73, 75, 403, 404
Mustelidae (Mammalia: Carnivora) 562
Mycetophilidae (Diptera) 4, 61, 64, 70, 73
Mycobacterium leprae (Bacteria) 481
Mycteromyia (Diptera) 312, 314, 324
Mycteromyiini (Diptera) 316, 321, 323
Mydidae (Diptera) 65
Mygalomorphae (Araneae) 38, 659, 662, 663, 664, 665, 666, 673
Mylabris (Coleoptera)
   M. alterna 585
   M. cichorii 585
Myopsitta monacha (Aves) 505
Myriapoda (Arthropoda) 32, 33, 35–6
Myrmecia (Hymenoptera) 583
   M. gulosa 582
Myrmecinae (Hymenoptera) 583
Mystacinobiidae (Diptera) 67, 408
'*Myzomyia*' (Diptera) 145
'*Myzorhynchus*' (Diptera) 145

Nabidae (Hemiptera) 484
Nairovirus (virus) 626
Nannomonas (Protozoa: Sarcomastigophora) 373
Nebo hierichonticus (Scorpiones) 676
Necator americanus (Nematoda: Strongyloidea) 481
Nematocera (Diptera) 4, 51, 52, 55, 58, 59, 60, 61, 62, 63, 64, 67, 68, 72, 73, 74, 100, 120, 241, 289, 403, 578
Nemestrinidae (Diptera) 65
Nemopalpus (Diptera) 79

Nemorhina (Diptera) 340
Neoaplectana (Nematoda: Rhabditoidea) 508
'*Neocellia*' (Diptera) 145
Neomacleaya (Diptera) 136
Neomyia (Diptera) 417
'*Neomyzomyia*' (Diptera) 145
Neophlebotomus (Diptera) 91, 96
Neopsylla (Siphonaptera) 547, 560
   N. bidentatiformis 560
   N. inopina 560
   N. mana 560
   N. pleskei 560
   N, pleskei orientalis 560
   N. setosa 560
   N. setosa spinea 542
Neotheobaldia (Diptera) 185
Neotoma (Mammalia: Rodentia) 557
Neotrombicula autumnalis (Acari) 644, 646
Neotyphloceras (Siphonaptera) 552, 560
   N. crassispina 543
   N. rosenbergi 560
Nepenthes (Nepenthaceae, pitcher plants) 147, 154, 193
Neriidae (Diptera) 65
Nerioidea (Diptera) 65
Neurochaetidae (Diptera) 66
Neuroptera (Insecta) 40
Nevermannia (Diptera) 251
Niptus (Coleoptera) 583
Noctuidae (Lepidoptera) 586, 587, 588
Nosema algerae (Protozoa: Microspora) 195
Nosinius (Siphonaptera) 561
Nosopsyllus (Siphonaptera) 546, 561
   N. aralis 561
   N. consimilis 561
   N. fasciatus 534, 537, 538, 542, 561
   N. laeviceps 561
   N. mokrzeckyi 561
   N. nilgeriensis 561
   N. tersus 561
   N. turkmenicus 561
Nothobranchius (Pisces) 212
Nothybidae (Diptera) 65
Notodontidae (Lepidoptera) 586
Notoedres cati (Acari) 639
Notolepria (Diptera) 251
Nuttalliella namaqua (Acari) 601, 608

Nuttalliellidae (Acari) 601, 608
Nycteribiidae (Diptera) 66, 70, 74, 340, 408, 489, 581
Nymphalidae (Lepidoptera) 586
Nymphomyiidae (Diptera) 64
Nymphomyioidea (Diptera) 64
Nypa (Palmae) 193
Nyssa (Nyssaceae) 193
Nyssomyia (Diptera) 91, 98, 107, 109
Nyssorhynchus (Diptera) 145, 173, 202

Ochlerotatus (Diptera) 151, 202
Ochotona (Mammalia: Lagomorpha) 557
Ochotonidae (Mammalia: Lagomorpha) 557
Ochrogaster (Lepidoptera) 586
Octomyomermis muspratti (Nematoda: Mermithoidea) 195
Odiniidae (Diptera) 66
Odocoileus virginianus (Mammalia: Artiodactyla) 612
Odonata (Insecta) 39, 40, 43, 324
Odontomyia (Diptera) 59
Oecacta (Diptera) 293
Oeciacus (Hemiptera) 488
   O. hirundinis 487
Oedemagena (Diptera) 441, 462
   O. tarandi 441
Oedemeridae (Coleoptera) 585
Oestridae (Diptera) 58, 67, 70, 74, 75, 340, 409, 410, 419, 430, 431, 432, 436, 438–45, 460
Oestrinae (Diptera) 436, 438, 439, 440, 534
Oestroidea (Diptera) 66, 68, 69, 70, 73, 340
Oestrus (Diptera) 430, 431
   O. ovis 410, 439, 440
Onchocerca (Nematoda: Filarioidea) 21, 241, 271, 274, 279, 300
   O. cebei 300
   O. cervicalis 300
   O. gibsoni 300
   O. gutturosa 300
   O. lienalis 271
   O. sweetae (= *cebei*) 300
   O. tarsicola 271
   O. volvulus 8, 241, 255, 266, 267, 270, 271, 272, 273, 277, 279

Onchocercidae (Nematoda: Filarioidea) 241
Onthofagus (Coleoptera)
   O. bifasciatus 583, 584
   O. unifasciatus 583
Onychophora (Arthropoda) 32, 33, 34
Ophionyssus (Acari) 636
   O. natricis 636
Ophthalmopsylla (Siphonaptera) 544, 561
   O. kukuschkini 561
   O. volgensis 561
   O. volgensis palestinica 541, 545, 548
Opifex (Diptera) 130, 140, 148, 152
   O. fuscus 136, 152, 153, 160, 174, 181, 185, 192, 194
Opiliocarida (Acari) 600
Opiliones (Arachnida) 597, 659
Opisocrostis (Siphonaptera) 550, 561
Opisodasys (Siphonaptera) 551, 561
   O. keeni 548
   O. nesiotus 561
Opisthacanthus (Scorpiones) 669
Opomyzidae (Diptera) 66
Opomyzoidea (Diptera) 66
Orbivirus (virus) 626
Orchopeas (Siphonaptera) 550, 561
   O. howardi 538
   O. leucopus 548, 561
   O. neotomae 561
   O. sexdentatus 561
   O. sexdentatus pennsylvanicus 546
Orgizomyia (Diptera) 322
Oribatida (Acari) 600
Ornithocoris (Hemiptera) 488
   O. toledoi 487
Ornithodorinae (Acari) 601
Ornithodoros (Acari) 601, 602, 607, 608, 610, 616, 619, 621, 624, 629, 630, 648, 649, 652
   O. coriaceus 611
   O. erraticus 611, 630
   O. hermsi 611, 630
   O. moubata 603, 610, 611, 624, 630
   O. moubata porcinus 611, 630
   O. papillipes (= *tholozani*) 611
   O. parkeri 611, 630
   O. rostratus 611

O. rudis 611, 630
O. savignyi 610, 611, 622
O. talaje 611
O. tholozani 611, 630
O. turicata 611, 630
O. venezuelensis (= rudis) 611
Ornithonyssus (Acari) 636
O. bacoti 633, 636, 637
O. bursa 633, 636, 638
O. sylviarum 633, 637, 638
Oropsylla (Siphonaptera) 545, 550, 561
O. brunneri 561
O. hirsuta 561
O. idahoensis 561
O. ilovaiskii 561
O. labis 561
O. montana 546, 549, 561
O. rupestris 561
O. silantiewi 547, 549, 561
O. tuberculata 561
Orthellia (Diptera) (= Neomyia) 417
Orthognatha (Araneae) 665, 666
Orthopodomyia (Diptera) 134, 135, 137, 140, 146, 157, 161, 172, 175, 176, 193
O. alba 161
O. pulcripalpis 161, 163, 187
O. signifera 161, 187, 193
Orthopodomyiini (Diptera) 140
Orthoptera (Insecta) 39, 40, 474
Oryctolagus cuniculus (Mammalia: Lagomorpha) 568
Oryzomys (Mammalia: Rodentia) 103
Otitidae (Diptera) 65
Otobinae (Acari) 601
Otobius (Acari) 601, 602, 608, 610, 621
O. lagophilus 610
O. megnini 610
Otomyinae (Siphonaptera) 558
Oxytelus (Coleoptera) 584

Pachyneuridae (Diptera) 64
Pachyneuroidea (Diptera) 64
Paederus (Coleoptera) 584, 585
P. sabaeus 584
Pallopteridae (Diptera) 66
Palpomyiini (Diptera) 295
Pan troglodytes (Mammalia: Primates) 521
Pandanus (Pandanaceae) 193

Pangoniinae (Diptera) 311, 314, 316, 321, 322
Pangoniini (Diptera) 314, 316, 321, 323
Panstrongylus (Hemiptera) 492, 497, 500, 501
P. chinai 492
P. diasi 492
P. geniculatus 492, 503, 505
P. guentheri 492
P. herrei 492
P. howardi 493
P. humeralis 493
P. lenti, 493
P. lignarius 493
P. lutzi 493
P. megistus 493, 495, 496, 497, 501, 505, 509, 510
P. rufotuberculatus 493
P. tupynambai 493
Pantophthalmidae (Diptera) 65
Parabelminus (Hemiptera) 491, 499
P. carioca 491
P. yurupucu 491
Parabuthus (Scorpiones) 672, 673, 676, 677
P. fulvipes 677
P. granulatus 677
P. liosoma 677
P. pallidus 677
Parabyssodon (Diptera) 251
Paracimex (Hemiptera) 488
Paradoxopsyllus (Siphonaptera) 544, 561
P. curvispinus 561
P. dashidorzhii 561
P. teretifrons 546, 549, 561
Paragonimus (Trematoda: Plagiorchidea) 34
P. westermani 34
'Paramyzomyia' (Diptera) 145
Paraphlebotomus (Diptera) 83, 87, 91, 94, 97, 107, 108
Parasarcophaga crassipalpis (Diptera) 12, 422
Parasimuliinae (Diptera) 249, 250
Parasimulium (Diptera) 245, 247, 249, 250
Parasitiformes (Acari) 600
Paratriatoma (Hemiptera) 493, 500
P. hirsuta 493
Paraustrosimulium (Diptera) 250
Parrotomyia (Diptera) 86, 91, 96

Pauropoda (Arthropoda) 33, 35
Parvidens (Diptera) 91, 93, 95
Passicimex (Hemiptera) 488
Pectinoctenus (Siphonaptera) 560
Pediculidae (Phthiraptera) 518
Pediculus (Phthiraptera) 517–26
P. capitis 517–26
P. humanus 517–26
Pelecorhynchidae (Diptera) 64
Penicus (Siphonaptera) 561
Pentastomida 32, 33, 36, 37, 38
Pericoma (Diptera) 80
Peripatus (Onychophora) 34
Periplaneta (Dictyoptera) 475
P. americana 473, 475, 477, 478, 479
P. australasiae 473, 475, 477, 478
P. brunnea 479
P. fuliginosa 479
Periscelididae (Diptera) 66
Perissodactyla (Mammalia) 439, 443
Perissommatidae (Diptera) 64
Peromyscopsylla (Siphonaptera) 551, 561
P. hesperomys adelpha 561
P. scotti 543
Peromyscus leucopus (Mammalia: Rodentia) 612
Phacochoerus aethiopicus (Mammalia: Artiodactyla) 369
Phasmida (Insecta) 40, 474
Philaematomyia insignis (Diptera) (= Musca crassirostris) 398
Philipomyia (Diptera) 322
Philipotabanus (Diptera) 322
Philoliche (Diptera) 314, 321
P. magrettii 310, 311
Philolichini (Diptera) 316, 321
Phimophorinae (Hemiptera) 490
Phimophorini (Hemiptera) 490
Phlebotomidae (Diptera) 89
Phlebotominae (Diptera) 8, 68, 72, 74, 75, 78–119, 288, 457, 578
Phlebotomus (Diptera) 9, 17, 57, 78, 82, 84, 85, 88, 89, 90, 91, 93, 94, 96–8, 99, 107, 108, 113, 252, 302
P. alexandri 83, 97, 108
P. andrejevi 106
P. ansarii 108

P. arabicus 83
P. argentipes 9, 90, 97, 100, 101, 102, 108
P. ariasi 100, 108
P. caucasicus 106, 108
P. celiae 108
P. chinensis 86, 102, 108
P. duboscqi 108
P. kandelakii 86, 87, 108
P. langeroni 108
P. longicuspis 108
P. longiductus 108
P. longipes 100, 108
P. martini 108
P. mongolensis 106
P. neglectus 108
P. newsteadi 95
P. nuri 87
P. orientalis 97, 108
P. papatasi 80, 81, 86, 97, 100, 102, 103, 106, 108, 110
P. pedifer 108
P. perfiliewi 108
P. perniciosus 104, 108
P. rossi 98
P. salehi 108
P. sergenti 97, 102, 103, 107, 108
P. smirnovi 108
P. tobbi 88, 108
P. transcaucasicus 108
P. vansomerenae 108
Phoneutria (Araneae) 667, 672, 676
P. fera 660, 667
P. nigriventer 667
P. rufibarbis 667
Phoniomyia (Diptera) 125, 135, 140, 146, 148, 173
P. fuscipes 135
Phonolibini (Hemiptera) 490
Phoridae (Diptera) 65, 69, 430, 435, 456, 578, 579, 589
Phormia (Diptera) 412, 438, 449
P. regina 412, 449, 451, 459
Phrynichida (Arachnida) 662
Phthiraptera (Insecta) 39, 40, 518
Phthirus (Phthiraptera) (= *Pthirus*) 517
Phymatinae (Hemiptera) 490
Phymatini (Hemiptera) 490
Physoderinae (Hemiptera) 490
Piezosimulium (Diptera) 250
Pintomyia (Diptera) 91, 109
Piophila casei (Diptera) 458
Piophilidae (Diptera) 66, 430, 434, 436, 458
Pipiza (Diptera) 59

Pipunculidae (Diptera) 65
Piratinae (Hemiptera) 490
Pisauridae (Araneae) 664
Pistia (Araceae) 188, 211
Plasmodiidae (Protozoa: Apicomplexa) 241
Plasmodium (Protozoa: Plasmodiidae) 21, 197
  P. falciparum 197, 200
  P. malariae 197, 200
  P. ovale 197
  P. vivax 197, 200
Platypezidae (Diptera) 65
Platypezoidea (Diptera) 65
Platystomatidae (Diptera) 65
Plecoptera (Insecta) 40
Pleochaetis (Siphonaptera) 561
  P. exilis 561
Plocopsylla (Siphonaptera) 552, 561
  P. hector 561
  P. ulysses 548
  P. wolffsohni 547
Ploiariolini (Hemiptera) 490
Plusaetis (Siphonaptera) 551, 562
  P. dolens quitanus 562
  P. equatoris 562
Poecilia reticulata (Pisces) 212
Polistes (Hymenoptera) 582
Pollenia (Diptera) 589
Polyceroconas (Myriapoda) 35
Polyctenidae (Diptera) 483, 484, 485, 486, 489
Polygenis (Siphonaptera) 550, 551, 562
  P. brachinus 562
  P. gwyni 562
  P. litargus 538, 562
  P. platensis cisandinus 562
Pomeroyellum (Diptera) 251
Pongidae (Mammalia: Primates) 518
Pontoculicoides (Diptera) 293
Porocephalida (Pentastomida) 36
Porocephaloidea (Pentastomida) 36
Potamochoerus porcus (Mammalia: Artiodactyla) 369
Potamon (Crustacea: Decapoda) 34
Potamonautes (Crustacea: Decapoda) 35, 255, 262
Premolis semirufa (Lepidoptera) 586
Presbytis (Mammalia: Primates) 205

Pressatia (Diptera) 91
Primicimex (Hemiptera) 488
Primicimicinae (Hemiptera) 488
Proboscidea (Mammalia) 439
Proechimys (Mammalia: Rodentia) 103
Proleptoconops (Diptera) 295
Propicimex (Hemiptera) 488
Prosimuliinae (Diptera) 249, 251
Prosimuliini (Diptera) 249, 264
Prosimulium (Diptera) 242, 243, 248, 250, 254, 259, 261, 266, 268
  P. mixtum 269, 277
  P. ursinum 262
Prostigmata (Acari) 600, 642
Prostriata (Acari) 607
Protacarus crani (Acari) 597
Proteus (Bacteria)
  P. mirabilis 459
  P. vulgaris 481
Protomacleaya (Diptera) 151
Protophormia (Diptera) 412, 449
  P. terraenovae 412, 438, 449, 451, 459
Protozoa 104, 195, 398, 567, 625
Protura (Insecta) 33, 40
Psammolestes (Hemiptera) 492, 499, 503, 505
  P. arthuri 492
  P. coreodes 492
  P. tertius 492
Psammomys (Mammalia: Rodentia) 109
Psaroniocompsa (Diptera) 251
Psathyromyia (Diptera) 91
Psechridae (Araneae) 664
Pseudocetherini (Hemiptera) 490
Pseudoficalbia (Diptera) 157, 185, 193
Pseudolynchia (Diptera) 409
  P. canariensis 581
Pseudomonas (Bacteria) 583
  P. aeruginosa 481
Pseudoscione (Diptera) 323
Pseudoscorpiones (Arachnida) 38, 659
Pseudotabanus (Diptera) 316, 322
Psilidae (Diptera) 65
Psilopelmia (Diptera) 5, 251
  P. ochracea 5
  see also *Simulium ochraceum*
Psilozia (Diptera) 251

Psitticimex (Hemiptera) 488
Psocoptera (Insecta) 40, 589
Psorophora (Diptera) 131, 134, 136, 137, 140, 145, 148, 152, 171, 180, 181, 184, 185, 186, 188, 192, 211, 217, 220, 438
   P. ciliata 152, 177
   P. columbiae 152
   P. confinnis 152, 177, 191
   P. cyanescens 177
   P. discolor 152
   P. ferox 152, 178
Psychoda (Diptera) 434, 578
   P. albipennis 457
   P. alternata 457, 578
Psychodidae (Diptera) 57, 68, 78, 79, 80, 85, 92, 430, 433, 434, 457, 577, 578
Psychodinae (Diptera) 64, 79, 80, 92, 457
Psychodoidea (Diptera) 64, 68
Psychodomorpha (Diptera) 64, 67, 68
Psychodopygus (Diptera) 86, 90, 91, 98, 107, 109
Psyllidae (Hemiptera) 65, 483, 484
Pternaspatha (Diptera) 251, 259
Pteromalidae (Hymenoptera) 324
Pterygota (Insecta) 33
Pthiridae (Phthiraptera) 518
Pthirus pubis (Phthiraptera) 517–26
Ptinidae (Coleoptera) 583
Ptychoptera (Diptera) 59
Ptychopteridae (Diptera) 59, 64
Ptychopteromorpha (Diptera) 64
Pulex (Siphonaptera) 537, 550, 562
   P. irritans 534, 535, 536, 537, 540, 562, 563, 567
   P. simulans 537, 562
Pulicidae (Siphonaptera) 535, 536, 553, 554, 558, 559, 562, 563, 564
Pulicoidea (Siphonaptera) 533, 536
Pupipara (Diptera) 70, 340
Pycnogonida (Arthropoda) 33, 38
Pyemotes (Acari) 642
   P. tritici 634, 643, 647
   P. ventricosus (= *tritici*) 647
Pyemotidae (Acari) 634, 643, 647

Pygiopsyllidae (Siphonaptera) 535, 553, 562
Pyralidae (Lepidoptera) 586, 588
Pyralis farinalis (Lepidoptera) 586
'*Pyretophorus*' (Diptera) 145
Pyrgotidae (Diptera) 65
Pyroglyphidae (Acari) 634, 641, 642

Rachisoura (Diptera) 131, 188
Ralipsylla (Siphonaptera) 562
Rattus norvegicus (Mammalia: Rodentia) 636
Ravenala (Strelitziaceae) 193
Reduviidae (Hemiptera) 483, 484, 485, 486, 489–502
Reduviinae (Hemiptera) 490, 491
Reoviridae (virus) 626
Rhadinopsylla (Siphonaptera) 531, 547, 562
   R. cedestis 562
   R. li ventricosa 544, 562
   R. ucrainica 562
Rhagionidae (Diptera) 55, 64, 69, 579–80
Rhaphidosomini (Hemiptera) 490
Rhigioglossa (Diptera) 321, 322
Rhinocricus (Myriapoda) 35
Rhinoestrus purpureus (Diptera) 439
Rhinomyzini (Diptera) 316, 321
Rhinophoridae (Diptera) 67, 70, 73, 408
Rhipicentor (Acari) 609, 621
Rhipicephalinae (Acari) 601
Rhipicephalus (Acari) 601, 602, 606, 609, 613, 618, 621, 627
   R. appendiculatus 613, 621, 628, 648, 651, 652
   R. evertsi 613
   R. sanguineus 613, 615, 628, 652
Rhodniini (Hemiptera) 490, 492
Rhodnius (Hemiptera) 492, 497, 498, 499, 501, 503, 505, 506, 507
   R. brethesi 492
   R. dalessandroi 492, 498
   R. domesticus 492
   R. ecuadoriensis 492, 505
   R. nasutus 492
   R. neglectus 492, 498, 505
   R. neivai 492

   R. pallescens 492, 505
   R. paraensis 492
   R. pictipes 492, 498
   R. prolixus 492, 495, 498, 501, 502, 503, 505, 508, 509, 510
   R. robustus 492, 498
Rhombomys (Mammalia: Rodentia) 109, 110
   R. opimus 110
Rhopalopsyllidae (Siphonaptera) 535, 536, 553, 559, 562
Rhopalopsyllus (Siphonaptera) 551, 562
   R. lugubris 538
Rhynchophthirina (Phthiraptera) 518
Rhynchotaenia (Diptera) 155
Rhyparobia maderae (Dictyoptera) 479
Richardiidae (Diptera) 65
Ricinulei (Arachnida) 662
Rickettsia (Rickettsiales) 524, 525
   R. akari 636
   R. australis 612, 628
   R. conori 568, 612, 613, 615, 628
   R. mooseri 525
   R. orientalis (= *tsutsugamushi*) 646
   R. prowazeki 21, 524, 525
   R. quintana 525
   R. rickettsi 613
   R. sibirica 613, 628
   R. tsutsugamushi 646
   R. typhi 568
Risidae (Diptera) 66
Rodentia (Mammalia) 439
Romanomermis culicivorax (Nematoda: Mermithoidea) 195, 212
Rondanomyia (Diptera) 96
Ropalomeridae (Diptera) 66
Runchomyia (Diptera) 140, 147

Sabethes (Diptera) 125, 135, 136, 140, 146, 147, 163, 170, 193
   S. belisarioi 147
   S. chloropterus 148, 181, 192, 207
Sabethini (Diptera) 139, 140, 146–8, 172, 192
Saicinae (Hemiptera) 490, 496
Saicini (Hemiptera) 490
Salmonella (Bacteria) 424, 425, 481, 560, 561, 581, 583
   S. bovis morbificans 481

S. enteritidis 568
S. typhi 481
S. typhimurium 480, 481, 568
Salticidae (Araneae) 667, 671
Salvinia (Salviniaceae) 211
Salyavatinae (Hemiptera) 490
Sansevieria (Agavaceae) 193
Sarcophaga (Diptera) 415, 422, 437, 438, 450, 452, 464
   S. crassipalpis 12, 415, 422, 453
   S. cruentata 415, 422, 452, 453
   S. haemorrhoidalis (= *cruentata*) 452
Sarcophagidae (Diptera) 12, 61, 67, 70, 73, 409, 410, 414, 415, 419, 421–2, 429, 430, 431, 436, 437, 450, 452–3, 459
Sarcophaginae (Diptera) 421, 422
Sarcoptes scabiei (Acari) 631, 633, 639–41
S. scabiei form *hominis* 639
Sarcoptidae (Acari) 633, 634, 639–41
Sarracenia (Sarraceniaceae) 187, 190
   S. purpurea 193
Saturniidae (Lepidoptera) 586, 587
Sauroleishmania (Protozoa: Sarcomastigophora), 98
Scaptia (Diptera) 321, 323
Scarabaeidae (Coleoptera) 583, 584, 585
Scathophagidae (Diptera) 66, 73, 195, 411
Scatopsidae (Diptera) 64
Scelionidae (Hymenoptera) 324
Scenopinidae (Diptera) 60, 65, 430, 433, 434, 458
Scenopinus (Diptera) 60, 434, 458
Scepsidinae (Diptera) 321
Scepsis (Diptera) 321
Schizophora (Diptera) 65, 67, 69
Schoenbaueria (Diptera) 251
Sciadoceridae (Diptera) 65
Sciaridae (Diptera) 64
Sciaroidea (Diptera) 64
Sciomyzidae (Diptera) 66
Scione (Diptera) 321
Scionini (Diptera) 316, 321, 323
Sciopemyia (Diptera) 91

Sciuridae (Mammalia: Rodentia) 558, 560, 561, 563, 569
Scolopendromorpha (Myriapoda) 36
Scorpio maurus (Scorpiones) 663
Scorpiones (Arachnida) 38, 659–82
Scorpionidae (Scorpiones) 663, 669, 676
Scotophaeus blackwalli (Araneae) 674
Scutigeromorpha (Myriapoda) 36
Selasoma (Diptera) 322
Semiadalia undecimnotata (Coleoptera) 585
Sepsidae (Diptera) 60, 66, 430, 458
Sepsis (Diptera) 60
Sergentomyia (Diptera) 78, 82, 83, 84, 85, 86, 89, 90, 91, 93, 96, 98, 99
   S. babu 87
   S. garnhami 98
   S. palestinensis 86
   S. punjabensis 86
   S. schwetzi 84
   S. squamipleuris 86
Sericopelma (Araneae) 667
Serratia marcescens (Bacteria) 481
Sessinia collaris (Coleoptera) 585
Shannoniana (Diptera) 140, 147
Shigella (Bacteria) 424, 425, 481
   S. dysenteriae 481
Sibine (Lepidoptera) 586
Sicariidae (Araneae) 660, 667, 672
Silvius (Diptera) 321
Simuliidae (Diptera) 4, 5, 12, 14, 15, 41, 55, 57, 58, 61, 63, 64, 68, 71, 72, 73, 74, 75, 79, 90, 100, 241–87, 300
Simuliinae (Diptera) 4, 5, 249, 250, 251
Simuliini (Diptera) 4, 5, 249, 250, 251
Simulium (Diptera) 4, 5, 12, 21, 55, 57, 241, 244, 245, 246, 248, 250, 251, 252, 254, 255, 257–8, 259, 264, 266, 267, 268, 270, 272, 274, 276, 278
   S. adersi 257
   S. albivirgulatum 251, 258, 268, 275

S. amazonicum 269
S. amazonicum group 259, 267
S. aokii 270
S. arakawae 269
S. arcticum 266, 277
S. arcticum complex 266
S. argentiscutum 269
S. bovis 258
S. buissoni 268
S. callidum 269
S. cholodkovskii 269
S. chutteri 279
S. colombaschense 269, 279
S. damnosum 4, 57, 267, 268, 273, 275, 276, 278, 279
S. damnosum complex 4, 14, 246, 251, 252, 254, 255, 256, 257, 260, 261, 263, 264, 265, 266, 273, 274, 275, 276, 277, 279, 582
   'Jimma' sibling 268
   'Kapere' sibling 268
   'Ketaketa' sibling 268
   'Nkusi' sibling 268
S. decimatum 269
S. decorum complex 263
S. dentulosum 257
S. dukei 258
S. equinum 264
S. erythrocephalum 12, 269, 270
S. ethiopiense 258, 268, 273, 275
S. exiguum 256, 274, 277
S. exiguum complex 251, 252, 256, 269, 277
S. griseicolle 257
S. guianense 269, 274, 277
S. horacioi 269
S. indicum 269
S. jenningsi complex 269
S. jolyi 268
S. kilibanum 268, 275
S. laciniatum 268
S. lineatum 243
S. luggeri 266, 277
S. maculatum 269
S. mengense 268
S. metallicum 273, 274
S. metallicum complex 251, 252, 256, 269, 277
S. neavei 255, 258, 261, 262, 268, 273, 275, 276, 277, 278
S. neavei group 35, 245, 251, 252, 255, 258, 266, 273, 275, 276, 279
S. ochraceum 5, 242, 274, 278
S. ochraceum complex 251,

S. ochraceum complex (cont'd) 252, 256, 264, 266, 269, 273, 277, 278
S. ornatum 264
S. ornatum complex 265, 269
S. ovazzae 258
S. oyapockense 263, 269, 274, 277
S. parnassum 269
S. pertinax 269
S. pictipes 266
S. posticatum 261, 266, 269, 270
S. quadrivittatum 269, 274, 277
S. rasyani 268, 273
S. reptans 269
S. sanctipauli 268, 275
S. sanctipauli subcomplex 4, 255, 268, 275
S. sanguineum 269
S. sirbanum 253, 259, 268, 273, 275, 276, 278, 279
S. soubrense 4, 268
S. squamosum 268, 275
S. transiens 269
S. truncatum 269
S. tuberosum complex 269
S. venustum complex 252, 269, 270, 277
S. vittatum complex, 254, 259, 266, 269, 270
S. wellmanni 257, 258
S. woodi 255, 258, 263, 268, 273, 275
S. yahense 268
Sinorhadinopsylla (Siphonaptera) 562
Sintonius (Diptera) 82, 91, 93, 96, 98
Siphonaptera (Insecta) 14, 39, 40, 529–75
Siphunculata (Phthiraptera) 518
Siphunculina funicola (Diptera) 579, 580
Solenopsis richteri (Hymenoptera) 583
Solifugae (Arachnida) 38, 659, 662
Somatiidae (Diptera) 65
Spaniopsis (Diptera) 580
S. longicornis 579
Spelaeomyia (Diptera) 84, 91, 95
Spelaeophlebotomus (Diptera) 85, 90, 91, 93
Spermophilus (Mammalia: Rodentia) 558, 559, 561, 563, 626

Sphaeridopinae (Hemiptera) 490
Sphaeroceridae (Diptera) 66, 69, 73, 458, 579, 580, 581, 589
Sphaeroceroidea (Diptera) 66
Sphaeromyiini (Diptera) 295
Sphinctopsylla (Siphonaptera) 552, 562
S. ares 546, 548
S. inca 543
S. mars 562
Sphingidae (Lepidoptera) 586
Spilopsylla cuniculi (Siphonaptera) 555, 556, 568
Spirobolus (Myriapoda) 35
Spirostreptus (Myriapoda) 35
Staphylinidae (Coleoptera) 584, 585
Staphylococcus (Bacteria) 583
S. aureus 481, 560
Steatoda (Araneae) 667
S. albomaculata 664
S. ancoratus 667
S. andinus 667
S. payculliana 675
Stegomyia (Diptera) 149, 150, 151, 173, 193, 202
Stegopterna (Diptera) 250
Stenistomera (Siphonaptera) 550, 562
S. alpina 562
S. macrodactyla 541, 562
Stenopodainae (Hemiptera) 490
Stenoponia (Siphonaptera) 547, 562
S. conspecta 562
S. sidimi 544
S. tripectinata 562
S. vlasovi 562
Stenotabanus (Diptera) 322
Stenoxenini (Diptera) 295
Stephanocircidae (Siphonaptera) 535, 554, 558, 561, 562
Stephanocircinae (Siphonaptera) 554
Stephanofilaria stilesi (Nematoda: Filarioidea) 399
Sternorrhyncha (Hemiptera) 485, 486
Stethomyia (Diptera) 145
Stibasoma (Diptera) 322
Stilobezziini (Diptera) 295
Stivalius (Siphonaptera) 538, 544, 552, 562

S. athalae 562
S. cognatus 562
Stomoxyinae (Diptera) 74, 75, 340, 389, 390, 391, 398, 399, 403, 417, 418, 419, 422
Stomoxys (Diptera) 57, 340, 390, 391, 393, 395, 396, 397, 409, 422, 438, 456
S. calcitrans 389, 390, 392, 393, 395, 397, 398, 399, 455
S. niger (*nigra*) 393, 395, 397, 398
S. ochrosoma 397
S. sitiens 393, 395, 397
Stratiomyidae (Diptera) 59, 65, 430, 433, 458
Stratiomyoidea (Diptera) 65
Streblidae (Diptera) 66, 70, 74, 340, 408, 581, 534
Strelitzia (Strelitziaceae) 193
Strepsiptera (Insecta) 40, 51
Streptococcus (Bacteria) 583
S. faecalis 481
Stricticimex (Hemiptera) 488
Strongylophthalmyiidae (Diptera) 65
Styloconops (Diptera) 295, 302
Stypommisa (Diptera) 322
Suidae (Mammalia: Artiodactyla) 369, 370, 371
Sulcicnephia (Diptera) 250
Supella (Dictyoptera)
S. longipalpa 473, 476, 477, 479
S. superlectilium (= *longipalpa*) 479
Suragina (Diptera) 580
Sycoracinae (Diptera) 79, 85, 92
Sylvicola (Diptera) 434
S. fenestralis 457
Sylvilagus (Mammalia: Lagomorpha) 558
Symphoromyia (Diptera) 55, 69, 579
Symphyla (Myriapoda) 33, 35
Synceros caffer (Mammalia: Artiodactyla) 369
Synneuridae (Diptera) 364
Synopsyllus (Siphonaptera) 548, 562
S. fonquerniei 562
Synosternus (Siphonaptera) 548, 563
S. pallidus 535, 563
Synphlebotomus (Diptera) 91, 94, 97, 107, 108
Synthesiomyia (Diptera) 455
S. nudiseta 455, 456

Synxenoderus (Hemiptera) 488
Syringogastridae (Diptera) 65
Syrphidae (Diptera) 59, 65, 69, 73, 430, 433, 435, 457
Syrphinae (Diptera) 457
Syrphoidea (Diptera) 65
Syrphus (Diptera) 457
 S. ribesii 435

Tabanidae (Diptera) 55, 56, 57, 61, 64, 68, 69, 72, 73, 74, 75, 310–32, 430, 431, 433, 434, 568
Tabaninae (Diptera) 312, 314, 316, 322, 323
Tabanini (Diptera) 312, 316, 321, 322, 324
Tabanocella (Diptera) 322
Tabanoidea (Diptera) 64, 69
Tabanomorpha (Diptera) 64, 65, 67, 69
Tabanus (Diptera) 25, 55, 57, 252, 310, 314, 315, 316, 321, 322, 323, 434
 T. atratus 325
 T. fraternus 312
 T. iyoensis 322
 T. lineola 323
 T. taeniola 325
 T. trivittatus 323
Tachinidae (Diptera) 67, 70, 72, 73, 324, 409, 410, 422
Tachiniscidae (Diptera) 65
Taeniothrips inconsequens (Thysanoptera) 588
Tamias (Mammalia: Rodentia) 626
Tanyderidae (Diptera) 64
Tanypezidae (Diptera) 65
Tardigrada (Arthropoda) 32, 33, 38
Tarentula (Araneae) 667
Tegeini (Hemiptera) 490
Teichomyza (Diptera) 433
Teichomyza fusca (Diptera) 435, 458
Telmatoscopus (Diptera) 457
Tenebrio molitor (Coleoptera) 585
Tenebrionidae (Coleoptera) 583
Tephritidae (Diptera) 65, 458
Tephritoidea (Diptera) 65
Teratomyzidae (Diptera) 66
Tetragnathidae (Araneae) 664
Tetranychidae (Acari) 643
Thaumaleidae (Diptera) 64, 68

Thaumetopoea (Lepidoptera) 586
Thaumetopoeidae (Lepidoptera) 586
Theileria parva (Protozoa: Apicomplexa) 613
Thelazia (Nematoda: Spiruroidea) 424
Thelohania (Protozoa: Microspora) 195
Thelyphonida (Arachnida) 38, 659
Themonocorini (Hemiptera) 490
Theobaldia (Diptera) 159
Theraphosa (Araneae) 667
Theraphosidae (Araneae) 667, 670, 671
Thereva (Diptera) 434
Therevidae (Diptera) 65, 430, 433, 434, 458
Theridiidae (Araneae) 659, 667, 672, 675
Thermobia domestica (Insecta: Zygentoma) 589
Thrassis (Siphonaptera) 550, 563
 T. acamantis 549, 563
 T. arizonensis 564
 T. bacchi 564
 T. fotus 564
 T. francisis 564
 T. pandorae 564
 T. petiolatus 564
 T. stanfordi 564
Thrips (Thysanoptera) 588
 T. imaginis 588
 T. tabaci 588
Thysanoptera (Insecta) 40, 588
Thysanura (Insecta) 33, 39, 589
Tiamastus (Siphonaptera) 551, 564
 T. cavicola 564
Tipula (Diptera) 434, 457
 T. oneili 6
Tipulidae (Diptera) 58, 62, 63, 64, 68, 73, 430, 433, 434, 457, 576
Tipulomorpha (Diptera) 64
Tityus (Scorpiones) 672, 673, 676, 677
 T. bahiensis 677
 T. cambridgei 677
 T. serrulatus 677, 678
 T. trinitatis 675, 677, 678
 T. trivittatus 677
Tlalocomyia (Diptera) 250
Topomyia (Diptera) 131, 140, 147, 193

T. gracilis 135
Toxorhynchites (Diptera) 120, 121, 131, 134, 135, 136, 140, 141, 142, 152, 163, 175, 176, 181, 187, 188, 192, 193, 194, 206, 212, 213
 T. noctezuma 181
 T. rutilus 193
 T. splendens 142
Toxorhynchitinae (Diptera) 121, 139, 140, 141–2, 152, 167, 172
Tragelaphus (Mammalia: Artiodactyla)
 T. scriptus 369
 T. strepsiceros 369
Transphlebotomus (Diptera) 91, 94, 95
Trechona (Araneae) 667, 670
Triatoma (Hemiptera) 493, 497, 500, 501, 502
 T. amicitiae 493, 501
 T. arthurneivai 493
 T. barberi 493, 505
 T. bolivari 493, 498
 T. bouvieri 493, 501
 T. brailovskyi 493, 498
 T. brasiliensis 493, 495, 505
 T. breyeri 493
 T. bruneri 493, 498
 T. carrioni 493
 T. cavernicola 493
 T. circummaculata 493
 T. costalimai 493, 498
 T. deanei 493
 T. delpontei 493
 T. dimidiata 493, 495, 501, 505
 T. dispar 493
 T. eratyrusiformis 493
 T. flavida 493
 T. gallardoi (= *patagonica*) 498
 T. gerstaeckeri 493
 T. guasayana 493, 505
 T. guazu 493
 T. hegneri 493
 T. incrassata 493
 T. indictiva 493
 T. infestans 493, 495, 501, 502, 503, 505, 509, 510
 T. lecticularia 494
 T. lenti 494
 T. leopoldi 494, 501
 T. limai 494
 T. longipennis 494
 T. maculata 494, 509
 T. matogrossensis 494
 T. matsunoi 494, 498

T. mazzottii 494, 503
T. melanocephala 494
T. mexicana 494
T. migrans 494, 501
T. neotomae 494
T. nigromaculata 494
T. nitida 494
T. obscura 494
T. oliveirai 494
T. pallidipennis 494
T. patagonica 494, 498
T. peninsularis 494
T. petrochii 494
T. phyllosoma 494
T. picturata 494
T. platensis 494, 503
T. protracta 494, 503, 505
T. pseudomaculata 494, 505
T. pugasi 494, 501
T. recurva 494
T. rubida 494
T. rubrofasciata 494, 500, 501
T. rubrovaria 495, 505
T. ryckmani 495
T. sanguisuga 495, 502
T. sinaloensis 495
T. sinica 495, 501
T. sordida 495, 505, 509
T. spinolai 495, 502, 503, 505
T. tibiamaculata 495
T. venosa 495
T. vitticeps 495
T. williami 495
T. wygodzinskyi 495
Triatomidae (Hemiptera) 491
Triatominae (Hemiptera) 484, 486, 489–507, 508, 509
Triatomini (Hemiptera) 490, 491, 492
Tribelocephalinae (Hemiptera) 490
Trichoceridae (Diptera) 64
Trichoceroidea (Diptera) 64
Trichodagmia (Diptera) 251
Trichogaster trichopterus (Pisces) 212
Trichogrammatidae (Hymenoptera) 324
Trichomyiinae (Diptera) 79, 85, 92
Trichoprosopon (Diptera) 125, 135, 136, 137, 140, 141, 147, 148, 167, 169, 170, 173, 182, 188, 193, 194
T. digitatum 181, 189, 190, 191
T. pallidiventer 168
Trichoptera (Insecta) 40, 589
Trichopygomyia (Diptera) 91

Trichuris (Nematoda: Trichinelloidea) 424
T. trichiura 481
Trigona (Hymenoptera) 582
Trilobita (Arthropoda) 32, 33
Trilobitomorpha (Arthropoda) 32, 33
Tripteroides (Diptera) 131, 134, 135, 140, 147, 169, 170, 173, 176, 188, 193
T. bambusa 181, 193
T. filipes 133
T. powelli 133, 135
T. stonei 135
Trithecoides (Diptera) 293
Tritopsylla (Siphonaptera) 557
Trombicula (Acari) 642
Trombiculidae (Acari) 599, 634, 643, 644–7, 653
Tropocyclops (Crustacea) 35
Trypanosoma (Protozoa: Sarcomastigophora) 333, 372, 373
T. brucei 21
T. brucei brucei 373, 378
T. brucei gambiense 333, 373, 375, 376, 377, 378
T. brucei rhodesiense 333, 373, 376, 377, 378
T. congolense 333, 373
T. cruzi 21, 484, 489, 491, 495, 501, 505, 506, 507, 508, 509
T. dionisii 487
T. incertum 487
T. rangeli 506
T. simiae 333, 373
T. vespertilionis 487
T. vivax 333, 373
Trypanosomatidae (Protozoa: Sarcomastigophora) 104
Trypanozoon (Protozoa: Sarcomastigophora) 373
Tunga (Siphonaptera) 548, 551, 554, 564
T. penetrans 541, 564, 565, 566
Tungidae (Siphonaptera) 535, 536, 554, 564
Twinnia (Diptera) 250, 262
Typha (Typhaceae) 188
Tyrophagus (Acari) 634
T. putrescentiae 642, 643

Udaya (Diptera) 130, 140, 152, 154
Uloboridae (Araneae) 659, 664
Uniramia (Arthropoda) 32, 33
Uranotaenia (Diptera) 125, 134,
136, 137, 139, 140, 146, 154, 157, 172, 173, 176, 178, 185, 191, 193, 194
U. anhydor 157
U. caeruleocephala 128
U. garnhami 193
U. lowii 157
U. sapphirina 157
U. shillitonis 193
U. unguiculata 157
Uranotaeniini (Diptera) 140
Uropsylla tasmanica (Siphonaptera) 555
Utricularia (Lentibulariaceae) 195

Vaejovidae (Scorpiones) 663, 669, 676
Vaejovis (Scorpiones) 676
V. spinigerus 663
Vavraia culicis (Protozoa: Microspora) 195
Vermileonidae (Diptera) 65
Vermipsyllidae (Siphonaptera) 535, 536, 554
Vermipsylloidea (Siphonaptera) 530
Vesciinae (Hemiptera) 490
Vespa (Hymenoptera) 582
Vespidae (Hymenoptera) 581
Vespula (Hymenoptera) 582
Viannia (Protozoa: Sarcomastigophora) 106
Viannamyia (Diptera) 91
Visayanocorini (Hemiptera) 490
Vorticella (Protozoa: Ciliophora) 195

Warileya (Diptera) 85, 89, 90, 91, 92, 93
Wilhelmia (Diptera) 251
Wilhelmiini 5
Wohlfahrtia (Diptera) 73, 410, 414, 419, 422, 430, 437, 452
W. magnifica 415, 416, 452, 460
W. meigeni 452
W. nuba 416, 452, 459
W. opaca 452, 460
W. seguyi 6
W. vigil 415, 452, 460
Wolbachia (Rickettsia-like organism) 165
W. pipientis 165
Wuchereria (Nematoda: Filarioidea) 21, 205
W. bancrofti 154, 200, 201, 202–3, 204, 205

Wyeomyia (Diptera) 121, 129, 131, 135, 136, 140, 146, 148, 150, 170, 176, 178, 193
  W. circumcincta 130, 147
  W. confusa 131, 133, 147
  W. luna 131
  W. medioalbipes 187
  W. moerbista 168
  W. smithii 136, 146, 147, 163, 187, 189, 190, 193
  W. vanduzeei 187

Xanthosoma (Araceae) 193
Xenasteiidae (Diptera) 66
Xenopsylla (Siphonaptera) 530, 537, 543, 548, 552, 564, 565, 567
  X. astia 564, 565
  X. brasiliensis 564, 565
  X. buxtoni 564
  X. cheopis 532, 533, 534, 539, 555, 564, 565
  X. conformis 564
  X. eridos 564
  X. gerbilli 564
  X. hirsuta 564
  X. hirtipes 564
  X. nubica 564
  X. nuttalli 564
  X. philoxera 564
  X. phyllomae 564
  X. pirlei 564
  X. skrjabini 564
  X. versuta 564
  X. vexabilis 564
Xiphiopsylla (Siphonaptera) 549, 554, 564
  X. daemonicola 542
  X. lippa 564
Xiphiopsyllidae (Siphonaptera) 535, 554, 564
Xylocoris (Hemiptera) 507
Xylomyidae (Diptera) 65
Xylophagidae (Diptera) 65

Yersinia (Bacteria) 107
  Y. pestis 21, 481, 567
  Y. pseudotuberculosis 568

Zeugnomyia (Diptera) 140, 152, 154, 188
Zoraptera (Insecta) 40
Zygaenidae (Lepidoptera) 586
Zygentoma (Insecta) 39, 40, 589

Subject index

Acarines 597–658
  classification 597, 600
  overview 597–600
  see also Ticks; Mites
Adenotrophic viviparity 363
African honeybee (*Apis mellifera adansonii*) 582
African horn-fly (*Haematobia minuta*) 398
African horse-sickness 299, 301, 303
African swine fever 610, 611
Afrotropical region, defined 17
Age-grading
  in biting midges 299
  in mosquitoes 180
  in sandflies 102
Akabane virus 299, 301
Allergy
  allergens
    bee venom 581
    contact 300, 576, 642
    inhalant 483, 578, 586, 589, 641
  allergic reactions 300, 483, 489, 579, 586, 589
  allergic rhinitis 576, 641, 642
  bronchial asthma 576, 579, 589, 641
Allopatry 16
Altamira syndrome 270
American chigger mite (*Eutrombicula alfreduggesi*) 646
American cockroach (*Periplaneta americana*) 473, 475, 477, 478, 479
American dog tick (*Dermacentor variabilis*) 613, 625, 628
American human babesiosis 630
Amphibians 158, 178, 324
  see also Frogs
Anaemia 489
Anaphylactic shock 58l

Anaplasmosis 580
Anautogeny, an autogenous reproduction 176, 264
Animal reservoirs
  in Brugian filariasis 205
  in leishmaniasis 105, 106, 109–10, 112
  in loiasis 325
  in myiasis 446
  in onchocerciasis 267, 271
  in tick-borne infection 624
  in trypanosomiasis 333, 369–70, 373–8, 380
Animal trypanosomiasis 338
  clinical 373
  compared with sleeping sickness 376
  control 379–80
  transmission of 373, 376
Antbears 611
Antelopes 396
Anthrax 310, 325, 399, 424, 585
Anticoagulants
  in mosquitoes 138
  in moth larvae 586
Ants 581, 582–3
  as mechanical vectors 583
  and mosquitoes 147, 176, 178
  stings and venoms 581, 582–3
  vernacular names in text
    bulldog-ants (*Myrmecia*) 583
    driver-ants (Dorylinae) 397
    fire-ant (*Solenopsis richteri*) 583
    harvester-ants 583
    Pharaoh ant (*Monomorium pharaoensis*) 583
Apes 518
Aphids 483
  aphid honeydew 101, 175, 324
Aphrodisiacs, insects as 586
Aptery 485
Arachnids 595–680

affinities 32–3
classification of 36–8, 597, 659
see also Spiders; Scorpions; Ticks; Mites; Acarines
Arboviruses
  criteria for vectors 22
  detection in vector 206, 303
  infections mentioned
    African horse sickness (AHS) 299, 301, 303
    African swine fever 610, 611
    Akabane virus (AKA) 299, 301
    Bluetongue virus (BT) 299, 300, 301, 302, 303
    Bovine ephemeral fever (BEF) 301
    Buttonwillow virus 302
    Chagres virus 104
    Chandipur virus 104
    Chikungunya virus 151, 208
    Colorado tick fever (CTF) 613, 626
    Congo virus 301
    Crimean–Congo haemorrhagic fever (CHF) 23, 616, 626, 627
    D'Aguilar virus 302
    Dengue haemorrhagic fever (DHF) 207, 208
    Eastern equine encephalomyelitis (EEE) 155, 158, 160, 208, 301
    epizootic haemorrhagic disease of deer (EHD) 299, 301
    equine infectious anaemia virus (EIAV) 325
    Gamboa virus 206
    ilheus virus 152
    Isfahan virus 104
    Japanese encephalitis (JE) 208, 209, 214, 215

Subject index 707

Kotonkan virus 302
Kyasanur Forest disease (KFD) 612, 624, 626–7
La Crosse virus 161, 208
Lokern virus 302
Louping ill virus 612
Main Drain virus 302
Mitchell River virus 302
Murray Valley encephalitis (MVE) 159, 208, 215
Nairobi sheep disease 616
Omsk haemorrhagic fever virus 558, 568
O'Nyong-Nyong virus 208
Oropouche virus 301
Palyam virus 302
papataci fever 103–4
Punta Toro virus 104
Quaranfil virus 610
Rift Valley fever 154, 159, 208, 301
Ross River virus 151, 208, 215
Russian Spring Summer encephalitis (RSSE) 612, 626
Sabo virus 302
St Louis encephalitis (SLE) 159, 161, 176, 208, 214, 636
sandfly fever 23, 78, 103
Sango virus 302
Sathuperi virus 302
Shgamonda virus 302
Shuni virus 302
Simbu virus group 301
Sindbis virus 208
three-day fever 103–4, 301
tick-borne encephalitides (TBE) 558, 559, 560, 612, 624, 626
Toscana virus 104
Venezuelan equine encephalomyelitis (VEE) 152, 156, 161, 208, 215
Warrego virus 302
Western equine encephalomyelitis (WEE) 159, 160, 208, 215, 301
West Nile virus 159, 208
yellow fever 148, 206–7
Zirqua virus 611
life cycle in vector 206
in mosquitoes 206–8
Armadillos 178, 553

Arrow poison, beetles as 585
Arthropods
  classification 32–3
  introduction to 30–47
  principal groups of 33–9, 40
Ascariasis 481
Asian cockroach (*Blattella asahinai*) 479
Aspergillosis 481
Assassin bugs (Reduviidae) 483, 484, 485, 486, 489–502
Asthma 576, 579, 589, 641
Australasian region, defined 17
Autogeny 176, 264, 298

Babesiosis (*Babesia* piroplasms) 612, 629–31
Bacilli, see Bacteria
Bacteria, bacterial infections
  genera mentioned
    Bacillus 211, 212, 213, 278, 279, 302, 325, 399, 426, 585
    Bartonella 78, 104
    Brucella 568
    Clostridium 583
    Erysipelothrix 568
    Escherichia 424, 481
    Francisella 325, 568
    Klebsiella 481
    Listeria 561, 568
    Mycobacterium 481
    Proteus 459, 481
    Pseudomonas 481, 583
    Salmonella 424, 425, 480, 560, 561, 568, 581, 583
    Serratia 481
    Shigella 424, 425, 481
    Staphylococcus 481, 560, 583
    Streptococcus 481, 583
    Yersinia 21, 107, 481, 567, 568
Bacteria
  infections with
    anthrax 310, 325, 399, 424, 585
    bartonellosis 78, 104
    brucellosis 399, 458
    Carrion's disease 104
    cholera 424
    diphtheria 424
    erysipeloid 557, 558, 560, 561, 562, 564, 568
    impetigo 525

leprosy 424, 481
listeriosis 558, 559, 561, 568
Oroya fever 104
plague 21, 537, 557, 558, 559, 560, 561, 562, 564, 567–8, 569
  distribution 481, 563
pseudotuberculosis 558, 559, 560, 561, 568
rickettsial infections see Rickettsiae
salmonellosis (*Salmonella* infections) 424, 425, 481, 559, 560, 561, 568, 581, 583
streptococcal infections 580
tuberculosis 424
veruga peruana 104
yaws 424, 580
see also Scientific names index
Bacterial insecticides (BTi) 211, 212, 278, 302
Bag moths 586
Baltic amber, Diptera in 52
Bancroftian filariasis 121, 144, 145, 148
  clinical 204–5
  distribution of 201
  periodicity in 204
  vectors of 202–3
Bartonellosis 78, 104
Bat bugs (Polyctenidae) 483, 484, 485, 486, 489
Bat-flies (Nycteribiidae) 66, 70, 74, 340, 408, 489, 581
Bat trypanosomes 487
Bats, as hosts 178, 408, 484, 487, 489, 505, 553, 621
Bears 178, 241
Bedbugs 484, 486–9
  biology 488–9
  biting pests 488
  medical importance 489
  vernacular names
    common bedbug (*Cimex lectularius*) 484, 487, 488, 489
    tropical bedbug (*Cimex hemipterus*) 484, 487, 488
Bed nets 179
Bees
  stings and venoms 576, 581–2
  vernacular names in text

Bees (cont'd)
  African honeybee (*Apis mellifera adansonii*) 582
  European honeybee (*Apis m. mellifera*) 582
  Italian honeybee (*Apis m. ligustica*) 582
  stingless bees (*Melipona*) 582
  sweat-bees (*Trigona*) 582
Beetles
  biting by 585
  causing conjunctivitis 584
  causing dermatitis 584–5
  invasion human body 583
  forensic importance 583
  transmission of pathogens 585
  urinogenital infestation by 583
  vernacular names in text
    carpet beetles (*Anthrenus*) 583, 584, 585
    coconut beetles (*Ananta*) 585
    dung-beetles (Scarabaeidae) 583
    larder beetle (*Dermestes lardarius*) 583
    meal-worms (tenebrionid larvae) 583
    'Spanish fly' (*Lytta vesicatoria*) 584, 585
Bereavement fever 524–5
Biological control
  of blackflies 278–9
  of bugs 508
  of mosquitoes 212–13
Biological transmission 20
Birds, as hosts
  of biting midges 300
  of blackflies 241, 264, 273
  of bugs 486, 487, 491, 495, 505
  of fleas 552, 553, 559
  of horseflies 324
  of louse-flies 581
  of mites 631, 634, 635, 636, 639, 641, 645
  of mosquitoes 157, 158, 159, 160, 165, 178, 179
  of ticks 601, 609, 610, 612, 616, 621, 627
  of tsetse-flies 371, 438
Biosystematics 2
Birnavirus 302
Biting midges (Ceratopogonidae) 288–309
  biology 298–9

  biting behaviour 298–9
  breeding sites 298, 302, 303
  dispersal 299, 302
  mating 298
  oviposition 299
classification 294–5
  important genera 295
  phylogeny 295
control 302
faunal and taxonomic literature 297
identification of principal genera 296–7
immature stages 294, 295, 298, 299, 302, 303, 304
medical importance
  as biting pests 299–300, 302
  as vectors 300–1
preservation 304
rearing 304
recognition 289–90
sampling 303
structure 290–4
  mouthparts 290
  wing pattern 291, 292
  wing venation 291–2
veterinary importance
  as biting pests 300
  as vectors 300, 301–2
Black beetle (cockroach) 477, 478
Blackflies (Simuliidae) 241–87
  adults
    mouthparts 247–8, 264
    structure 242–3, 247–8
    wing venation 248
  biology 259–67
    biting behaviour 263–4, 272–3, 276
    dispersal 266, 275
    host preferences 263–4
    life-history 259–61
    longevity 264
    mating 262–3
    oviposition 264–6
  classification 249–56
  collecting 279–82
  control 213, 275, 277–9
    biological 278–9
  diagnosis 242–3
  eggs 261, 264–6
  faunal and taxonomic literature 258–9
  fossils 242
  identification 254–8
  larvae
    biology 259–62

    breeding sites 275–6, 277
    control 277–9
    structure 242, 244–6
  medical importance 267–77
    biting pests 267–70
  pupae
    pharate pupa 260–1
    structure 242, 247
  rearing 280–1
  species complexes 252–4, 255–6, 266, 268, 273–7, 277–9
  vector–parasite interactions 272–4, 277
  see also Scientific names index
Blackfly fever 270
Black-widow spider (*Latrodectus mactans*) 659, 667, 672, 674, 676, 678
Blandford fly (*Simulium posticatum*) 261, 266, 269, 270
Blepharitis 524
Blindness
  'Nairobi eye' 584
  river-blindness, *see* Onchocerciasis
Blindness in onchocerciasis 272, 275
Blood-feeding
  by biting midges 295, 299
  by biting muscids 389, 396, 397–8, 399
  by bloodsucking bugs 486
  by fleas 564
  by mosquitoes 176–80
  by sandflies 101–2
  by ticks 602, 616, 618, 621, 622–3, 624
Blood-meal volume
  in biting midges 299, 300
  in mosquitoes 177
  in ticks 602, 622, 623
  in triatomines 504
Bloodsucking bugs (Hemiptera) 483–516
  bat bugs 489
  bedbugs 486–9
  blood-feeding 484–5, 486
  classification 485–6
  eggs 485
  medical importance, minor families 507
  mouthparts 484–5
  recognition 484–5
  see also Bedbugs; Triatomine bugs
Bloodsucking moths 587–8

Bloodsucking muscid flies 389–402
  biology 396–8
    of *Haematobia* 396–7
    of *Musca* 397
    of *Stomoxys* 397
  collecting 400
  control 399–400
  disease and pathogen relations 398–9
  faunal and taxonomic literature 395–6
  identification man-biting species 391–5
  immature stages 390–1
  rearing 400
  recognition 389–90
  vernacular names in text
    African horn-fly (*Haematobia minuta*) 398
    biting house-fly (*Stomoxys calcitrans*) 389
    dog-fly (*Stomoxys calcitrans*) 540, 566, 567
    horn-flies (*Haematobia*) 73
    Indian cattle-fly (*Musca crassirostris*) 389, 394, 397–8
    see also Stable-fly (*Stomoxys calcitrans*)
Blow-flies
  biology 420–1
  identification 411–12, 413–14
  maggot therapy 459
  see also Synanthropic flies
Bluebottles (*Calliphora*) 420–1
Bluetongue disease (BT) 299, 300, 301, 302, 303
Body louse 517
Bont tick 614
Booklouse 589
Borelliosis, see Tick-borne relapsing fever; Lyme disease; Louse-borne relapsing fever
Bot-flies (Oestridae) 70
  deer bot-fly (*Hypoderma diana*) 441, 442, 443
  human bot-fly (*Dermatobia hominis*) 429, 436, 438, 440, 459, 460, 461
  see also Scientific names index
Boutonneuse fever 568, 612, 613, 628
Bovine ephemeral fever (BEF) 301
Brachyptery 474, 485
Brazilian pemphigus 267

Brill Zinsser disease 524–5
Brown-banded cockroach (*Supella longipalpa*) 473, 476, 477, 479
Brown dog tick (*Rhipicephalus sanguineus*) 613, 615, 628, 652
Brown ear tick (*Rhipicephalus appendiculatus*) 613, 621, 628, 648, 651, 652
Brown recluse spiders (*Loxosceles*) 667, 672, 676
Brown-striped wood mosquito (*Aedes excrucians*) 121, 167
Brown-tail moth (*Euproctis*) 586
Brucellosis 399, 458
Brugian filariasis 205
  distribution 201
  vectors 203–4
Buffaloes 300, 369, 370, 371, 396, 621
Buffalo-gnat (*Cnephia pecuarum*) 268
Bugs (Hemiptera) 483–516
  biology
    bedbugs (Cimicidae) 486–9
    triatomine bugs 489–510
  classification 485–6
  vernacular names in text
    assassin bugs (Reduviidae) 484
    bat bugs (Polyctenidae) 489
    bedbugs (Cimicidae) 484–9
    cone-nosed bugs (Triatominae) 489–510
    flower-bugs (Anthocoridae) 507
    kissing bugs (Triatominae) 489–510
    'toe-biters' (belostomatid water-bugs) 483
  see also Bloodsucking bugs
Bushbuck (*Tragelaphus scriptus*) 369
Bush-fly (*Musca vetustissima*) 418, 422, 423, 634
Bush pig (*Potamochoerus porcus*) 369
Butterflies 40, 586
Buttonwillow virus, 302

Caddisflies (Trichoptera) 40, 589

Calabar swellings 325
California viruses 152
Calyptrate flies 54, 56, 62, 66, 67, 69, 70, 390, 403–28, 432, 460
  classification 66–7
  identification to families 407–11
  general biology 419–20
  medical importance 419
  morphology 406–7, 409
Camels 439, 452, 610, 639
Camel-spiders (Solifugae) 38, 659, 662
Canthariasis 583
Cantharidin 585
Capitulum 598, 599, 602, 604–6, 623
Carbamates 211, 215
Caribou 178
Carnoy's fixative 280
Carrion's disease 104
Castor bean tick (*Ixodes ricinus*) 604, 611, 612, 619, 620, 621, 626, 629, 631, 652
Caterpillars (lepidopterous larvae) 585–6
  cause of urticaria 586
Cats
  as hosts 643
  as reservoirs 205
Cattle, as hosts
  of biting midges 298, 300, 301
  of bloodsucking muscids 397
  and myiasis 438, 447, 451, 452, 455, 581
  of ticks 610, 612, 619, 621, 627
  of tsetse-flies 371, 373
Cayenne tick (*Amblyomma cajennense*) 615, 621, 628
Centipedes 35–6
Central European tick-borne encephalitis 626
Cephalic fans 245, 262
Cephalopharyngeal skeleton (of Diptera) 61, 390, 431, 463
Cestodes, see Helminths
Chaetotaxy 53, 131, 294, 317, 406, 599
Chagas disease 21, 489, 495, 506–7
  clinical 506
  control 508–9
  distribution 501
Chagoma 506

Chagres virus 104
Chandipur virus 104
Cheese-skipper (*Piophila casei*) 458
Chelicerae 599, 604, 620, 632, 640, 661–2, 664, 665
Chemosterilants 508
Chemotaxonomy 2, 15
  see also Isoenzymes
Chewing lice 518
Chicken mite (*Dermanyssus gallinae*) 633, 635, 636
Chigger-borne rickettsiosis 646
Chiggers (Trombiculidae) 599, 634, 643, 644–7, 649, 653
Chikungunya virus 151, 208
Chimpanzee (*Pan troglodytes*) 521
Chipmunks (*Tamias*) 626
Cholera 424
Christophers' stages, see Gonotrophic cycles
Chromosomes 15
  biting midges 294
  blackflies 15, 249, 251, 252–4
  horse-flies 317
  mites 632
  mosquitoes 15, 139, 162–3
  triatomines 498
  tsetse-flies 632
Cibarium 79, 82, 137–8, 145, 290
Circadian rhythms 181, 671
Cladism, cladistic analysis 3, 518
Classification
  alphataxonomy 2
  betataxonomy 2
  categories 3–5
  defined 2
  gammataxonomy 2
  principles 1–5
Clegs (*Haematopota*) 56, 310, 312, 314, 317, 321, 322, 323
Clothing louse 517, 518–20, 521, 524, 526
Cluster fly (*Pollenia*) 589
Cockchafers, larvae of 585
Cockroaches 473–82
  biology 473, 479–80
  classification 474–77
  control 481
  identification 477
  medical importance 20, 480–1
  recognition 473–4, 475, 477
  vernacular names in text
    American cockroach
    (*Periplaneta americana*) 475, 478
  Asian cockroach (*Blattella asahinai*) 479
  Australian cockroach (*Periplaneta australasiae*) 475, 478
  brown-banded cockroach (*Supella longipalpa*) 476, 479
  common or Oriental cockroach (*Blatta orientalis*) 474, 477–8, 480, 481
  German cockroach (*Blattella germanica*) 475, 478, 480
  Madeira cockroach (*Rhyparobia maderae*) 479
Co-evolution
  fleas and hosts 534, 552
  ticks and hosts 601
Collecting
  general 23–7
  killing agents 25
Colorado tick fever 613, 626
Compound eyes
  in Diptera 53
  general structure 43
Cone-nosed bugs, see Triatomine bugs
Congo floor-maggot (*Auchmeromyia senegalensis*) 11, 412, 420, 446, 447, 452
Congo virus 301
Conjunctivitis 580, 584, 586
Control
  biting midges 302
  blackflies 213, 275, 277–9
  cockroaches 481
  fleas 568–9
  lice 523, 525–7
  mites 648–9
  mosquitoes 208–16
  myiasis 459–62
  sandflies 107–10
  stomoxyine flies 399–400
  synanthropic flies 425–6
  tabanids 327
  ticks 621, 648
  triatomine bugs 508–9
  tsetse-flies 379–81
Copepods 34–5
Copra itch 642
Coxal glands 599, 622
Crab louse (*Pthirus pubis*) 517, 519, 520–1, 522, 524, 526–7

Crabs (Decapoda) 34, 35, 181, 255, 261, 262, 275, 280
  African river-crabs (*Potamonautes*) 35, 255, 262
Crane-flies (Tipulidae) 58, 62, 63, 64, 68, 73, 430, 433, 434, 457, 576
Crimean–Congo haemorrhagic fever (CHF) 23, 616, 626, 627
Crimean tick typhus 628
Cross-mating experiments 12
  in bugs 503
  in lice 520
  in mosquitoes 161–7
  in sandflies 90
  in tsetse-flies 346
Cross-winged bugs 483
Crustaceans 32–5
  medical importance 34, 35
  see also Crabs; Prawns
Crypsis, in spiders 671
Cuticle 30, 31–2, 39–41
Cuticular hydrocarbons 91, 92, 141, 254, 317, 347, 367, 607
Cyclorrhapha 63, 65
  see also Muscomorpha; Diptera
Cytogenetics, see Cytotaxonomy
Cytoplasmic incompatibility 165, 213
Cytospecies 252, 255, 256
Cytotaxonomy 2, 15
  in blackflies 252–4
  in mosquitoes 162–3
  in ticks 607

D'Aguilar virus 302
Daylight mosquito (*Anopheles crucians*) 121, 142, 173
DDT 110, 180, 211, 214, 215, 278, 425, 508, 526
Deer, as hosts 299, 301, 552, 562, 581, 612, 619
Deer-flies (*Chrysops*) 310
  see also Scientific names index
Deer-ked (*Lipoptena cervi*) 581
Delusory parasitosis 589
Dengue 206, 207–8
  vectors 208, 210, 215
Dengue haemorrhagic fever (DHF) 207, 208
Dengue shock syndrome (DSS) 207
Dermatitis 270, 581, 584, 586, 639

dhobie itch 300
grocer's itch 642
scrub itch 646
trombidiosis 646
see also Urticaria
Describing new species 8–11
Deutonymph 631, 645
Dhobie itch 300
Diapause
  in biting midges 298
  in blackflies 251, 266
  in Diptera 72
  in horse-flies 323
  in mosquitoes 298
  in sandflies 100
  in ticks 618, 623
Dichoptic head 53, 54, 407
Diplopods 32, 33, 35, 37
Diphtheria 424
Diptera (flies) 51–77
  biology 72–5
    development 72–4
    larvae causing myiasis 429–69
    mating 74
  classification 63–70, 71–2
    Brachycera 63, 64–7, 68–70
    Cyclorrhapha 63, 65
    Muscomorpha 63, 65–7, 69–70
    Nematocera 63, 64, 67, 68
  distinguishing features 51
  faunal and taxonomic literature 71–2
  fossils 52
  general structure 52–8
    venation 53–4, 56–7
  immature stages
    eggs 58, 72
    larvae 58, 60–1, 72–4
    pupae 59, 61–2
Dirofilariasis 205
Disney trap 112
Dispersal
  biting midges 299
  blackflies 266–7
  horse-flies 325
  mosquitoes 183–4
  sandflies 102
  tsetse-flies 320
DNA probes, sequencing 92, 107, 254
Dog flea 540, 566, 567
Dog fly (*Stomoxys calcitrans*) 540, 566, 567
  see also Scientific names index
Dog heartworm (*Dirofilaria immitis*) 151, 205

Dogs
  heartworm infection 151, 205
  as hosts 397, 398, 613
  as reservoirs 109–10
  myiasis in 438, 446, 447, 452
  scabies 639
Dracunculiasis 34, 35
Driver-ants (Dorylinae) 397
Drone-flies 457
Dugbe virus 301
Dysentery 481

Earwigs (Dermaptera) 40, 589
East coast fever 613
Eastern equine encephalomyelitis (EEE) 155, 158, 160, 208, 301
Ectoparasitic adaptations 44
  fleas 533–4
  flies 74
  lice 518, 519–20
  ticks 622
Egg laying, see Oviposition
Electrophoresis 316
Elephant-flies 310
Encephalitides 208
  see also Eastern, Venezuelan and Western equine encephalomyelitis; Encephalitis
Encephalitis 152, 158, 159, 161, 176, 208, 214, 215, 558, 559, 560, 612, 624, 626, 636
Endophagic, defined 179
Endophilic, defined 179
Endopterygotes 4, 39, 40
'Enfermedad de Robles' 267
Entomophobias 576, 589
Environmental change
  effect on blackfly breeding 279
  effect on chagas disease 507
  effect on tsetse-flies 362, 380
Epizootic haemorrhagic disease (EHD) 299, 301
Equine infectious anaemia virus (EIAV) 325
Erysipeloid 557, 558, 560, 561, 562, 564, 568
Erythema chronicum migrans (ECM) 625
Espundia 105
European sheep tick (*Ixodes ricinus*) 611–12
  see also Scientific names index
Eurygamous mating 175
Evolutionary classification 3

Exophagic, defined 179
Exophilic, defined 179
Exopterygotes 39, 40, 479, 483, 485
Exoskeleton 30–2, 39–41
Extrinsic incubation period 20
Eye-flies 69, 580
Eyed tampan (*Ornithodoros savignyi*) 610, 611, 622
Eyeless tampan (*Ornithodoros moubata*) 603, 610–11, 624, 630
Eye-moths 586
Eyeworms (*Thelazia*) 424

False-scorpions (Pseudoscorpiones) 38, 659
Fascicle (= syntrophium) 125
  see also Mouthparts
Filariasis 200–5
  clinical 201
  distribution 201
  transmission 201
  vectors summary 202–4
  see also Brugian, Bancroftian and Timor filariasis; Dirofilariasis; Onchocerciasis; Loiasis
Fire-ants (*Solenopsis*) 583
Firebrat (*Thermobia domestica*) 589
Fish
  as hosts 178
  use in mosquito control 212
Five-day fever 525
Flannel-moths (*Megalopyge*) 586
Fleas 529–75
  biology 555–7
  host preferences 552–4, 556, 557–64
  blood-feeding 564
  classification 534–6
  phylogeny 552
  principal genera 557–64
  collecting 569–70
  control 568–9
  faunal and taxonomic literature 554–5
  geographical distribution 552–4
  identification 536–52
  immature stages
    larvae 530, 555, 565, 566
    pupae 555
  as intermediate hosts 20, 566–7
  medical importance

712 Subject index

Fleas (cont'd)
  flea bites 564
  principal genera 529, 557, 577
  vector incrimination 529, 568
  vector status 567–8
  morphology 529–34
    mouthparts 530
  preparation 570
  rearing 570
  vernacular names in text
    cat flea (Ctenocephalides felis) 540, 555, 558, 566, 567, 570
    dog flea (Ctenocephalides canis) 540, 566, 567
    human flea (Pulex irritans) 535, 537, 562, 563, 566, 567
    jigger flea (Tunga penetrans) 555–6
    northern rat flea (Nosopsyllus fasciatus) 537
    Oriental rat flea (Xenopsylla cheopis) 532, 533, 564, 565
    rabbit flea (Spilopsylla cuniculi) 555, 556, 568
    sand-fleas (Tunga) 564, 565–6
Flies (Diptera)
  vernacular names in text
    African horn-fly (Haematobia minuta) 398
    blackflies (Simuliidae) 241
    Blandford fly (Simulium posticatum) 269
    bluebottles (Calliphora) 451
    buffalo-flies (Tabanidae) 310
    bush-fly (Musca vetustissima) 423, 634
    cluster fly (Pollenia) 589
    Congo floor-maggot (Auchmeromyia senegalensis) 11, 412, 420, 452
    crane-flies (Tipulidae) 457
    deer-flies (Chrysops) 310
    drone-flies (Syrphidae) 457
    elephant-flies (Tabanidae) 310
    eye-flies (Chloropidae) 580
    face-fly (Musca autumnalis) 423
    filter-fly (Psychoda) 457, 578
    flat-flies (Hippoboscidae) 581
    flesh-flies (Sarcophagidae) 414–15, 452–3
    frit-flies 580
    fruit-flies (Tephritidae) 458
    gadflies (Tabanidae) 310
    Golubatz fly (Simulium colombaschense) 269
    greenbottles (Lucilia) 420, 450–1
    horn-flies (Haematobia) 73, 389–402
    horse-flies (Tabanidae) 310–32
    house-fly (Musca domestica) 404, 422
    hover-flies (Syrphidae) 457
    human bot-fly (Dermatobia hominis) 438, 440, 459
    Indian cattle-fly (Musca crassirostris) 389, 394, 397–8
    latrine-fly (Fannia scalaris) 421, 453–4
    lesser dung-flies (Sphaeroceridae) 485, 580–1
    lesser house-fly (Fannia canicularis) 404, 416–17, 421, 453–4
    louse-flies (Hippoboscidae) 581
    Lund's fly (Cordylobia rodhaini) 446–7
    mooseflies (Tabanidae) 310
    moth-flies (Psychodidae) 457, 577–8
    New World screw-worm flies (Cochliomyia) 447–8
    Old World screw-worm flies (Chrysomya) 448–9
    Oriental latrine-fly (Chrysomya megacephala) 449
    reindeer warble-fly (Oedemagena tarandi) 441
    sandflies (Phlebotominae) 78–119
    scuttle-flies (Phoridae) 456–7, 578–9
    shore-flies (Ephydridae) 458, 580
    snipe-flies (Rhagionidae) 579–80
    soldier-flies (Stratiomyidae) 458
    stable-flies (Stomoxys calcitrans) 389, 402, 455
    stiletto-flies (Therevidae) 458
    sweat-flies (Hydrotaea) 422, 580
    tsetse-flies (Glossina) 16, 52, 333
    Tumbu fly (Cordylobia anthropophaga) 405, 413, 446, 447, 461
    vinegar-flies (Drosophilidae) 458
    window-flies (Scenopinidae) 458
  see also Diptera
Flight speed
  in tsetse-flies 370
Flinders Island spotted fever 628
Forensic entomology 583
Fossils
  biting midges 288
  blackflies 52, 242
  Diptera 52
  Hemiptera 485
  mites 597
  mosquitoes 120
  tsetse-flies 52, 333
Frit-flies 580
Frogs 93, 157, 300
  see also Amphibians
Fungi
  Entomophthorales 419
  genera mentioned
    Aspergillus 481
    Coelomomyces 195
    Lagenidium 195
    Metarhizium 508

Gadflies 310
Gamboa virus 206
Gender, of specific names 12
Genetic control 213
Gerbils 79, 109, 110, 553, 558
German cockroach (Blattella germanica) 473, 475, 477, 478, 480
Gibbons 518
Giraffe 396
Gnats
  biting gnats, see Biting midges

window-gnats
    (Anisopodidae) 457, 578
  see also Mosquitoes; Biting
    midges
Gnu (*Connochaetes gnou*) 370
Golubatz fly (*Simulium
    colombaschense*) 269, 279
Gomez Núñez box 509
Gonotrophic cycles
  in biting midges 299
  in blackflies 264
  in mosquitoes 180–1
  in ticks 623
Gorilla (*Gorilla gorilla*) 518, 521
Grain mite (*Pyemotes tritici*)
    634, 643, 647
Greenheads (*Tabanus*) 310
Grocer's itch 642
Ground-squirrels
    (*Spermophilus*) 558, 559,
    561, 563, 626
Grubs 61
  distinction from maggots 58
Guinea-worm (*Dracunculus
    medinensis*) 34

Haemorrhagic exanthem of
    Bolivia 270
Haemorrhagic nephroso-
    nephritis 559, 560, 577
Haemorrhagic syndrome of
    Altamira (HSA) 270
Hair follicle mites (*Demodex*)
    38, 631, 632, 634, 642,
    643–4
Harara, urticarial reaction 103
Hard ticks 601, 604
  see also Ixodidae in
    Scientific names index
Hardy Weinberg equilibrium
    16
Hartebeest, red (*Alcelaphus
    bucelaphus*) 370
Harvest mites (Trombiculidae)
    599, 634, 643, 644–7, 653
Harvester ants 583
Harvestmen (Opiliones) 659
Hay itch mite (*Pyemotes tritici*)
    634, 643, 647
Head louse (*Pediculus*) 517,
    518–20, 521, 522, 523–4,
    525–6
Heartwater fever 615
Heartworm, dog (*Dirofilaria
    immitis*) 151, 205
Hedgehogs 627
Helminths, helminthic
    infections

  genera mentioned
    Ancylostoma 424, 481
    Ascaris 481
    Brugia 200–5
    Diphyllobothrium 35
    Dipylidium 20, 558, 559,
      562, 566
    Dirofilaria 151, 205, 241
    Dracunculus 34
    Enterobius 481
    Gnathostoma 35
    Gongylonema 585
    Habronema 399
    Helidomermis 298
    Hymenolepis 561, 567, 585,
      586, 589
    Icosiella 300
    Mansonella 241, 267, 300
    Mesomermis 278
    Neoaplectana 508
    Octomyomermis 195
    Onchocerca 8, 21, 241, 255,
      266, 267, 270, 271, 272,
      273, 277, 279, 300
    Paragonimus 34
    Romanomermis 195, 212
    Thelazia 424
    Trichuris 424, 481
    Wuchereria 21, 154, 200–5
Hemimetabolous
    metamorphosis 38, 39, 485
Hepatitis 424, 481
Hepatitis-B 484, 489
Hibernation, see Diapause
Histoplasmosis 111
HIV 20, 489
Holarctic realm, defined 19
Holometabolous
    metamorphosis 38, 39, 529
Holoptic head 53, 54, 407
Homonymy, of scientific
    names 11
Honeybees 582
Hookworms 481
  New World hookworm
    (*Necator americanus*) 481
  Old World hookworm
    (*Ancylostoma duodenale*)
    481
Hornets (*Vespa*) 582
Horn-flies (*Haematobia*) 73,
    389–402
  biology 396–7
  control 399–400
  identification 391–5
  medical importance
    biting pest 398
    mechanical vector 398–9
  preservation 400

  taxonomic literature 395–6
Horse bot-flies 410, 459
Horse-flies (Tabanidae) 310–32
  biology 323–5
    biting behaviour 324–5
    breeding sites 323, 327
    immature stages 314–16,
      317, 323–4, 327
    life history 323–4
    mating 324
    oviposition 325
    parasites and predators
      324
  classification 316, 321–2
  control 327
  faunal and taxonomic
    literature 322–3
  identification of principal
    genera 317–21
  medical importance
    biting pests 325, 326
    loiasis 325–6
    vectors 317–20, 323, 324,
      325–6
  recognition 310
  sampling 327
  structure 310–14
    mouthparts 313, 324
Horses 208, 300, 301, 397, 439,
    440, 447, 451, 452, 455,
    518, 619, 639
Host relations
  bedbugs 488
  biting midges 298–9
  blackflies 263–4
  fleas 556–7
  horse-flies 324
  mites 636, 641, 645
  mosquitoes 176–9
  myiasis-producing flies
    438–58
  sandflies 101
  stable-flies, horn-flies 396,
    397
  ticks 621–2
  triatomine bugs 503
  tsetse-flies 367–70
House dust allergy 641
House dust mite
    (*Dermatophagoides*) 599,
    634, 641, 649, 653, 654
House-flies (*Musca*) 404, 422
  biology 422–3
  classification 422
  control 425–6
  identification 417–18
  larvae 434, 455–6
  medical importance 20,
    423–5, 455

House mouse mite
    (*Liponyssoides sanguineus*)
    633, 634, 635, 636
Human flea (*Pulex irritans*)
    534, 535, 536, 537, 540,
    562, 563, 566, 567
Human immunodeficiency
    virus (HIV) 20, 489
Hyenas 601
Hygiene, insects 589
Hypopus mite nymph 631
Hyraxes 553, 558

Identification 10–11
    compared to speciation 2
    defined 2
Ilheus virus 152
Impala (*Aepyceros melampus*)
    370
Impetigo 525
Indian tick typhus 628
Infection rates of parasites
    in biting midges 300
    in mosquitoes 199
    in sandflies 103, 107
    in ticks 626
    in tsetse-flies 378
Insect growth regulators
    in bugs 508
    in fleas 569
    in mosquitoes 211
Insecticide resistance 212, 216
Insectivores, general 553, 558,
    559, 562, 621, 646
Insects
    classification 39, 40
        phylogeny 32–3
    external structure 42–6
        cuticle 39–41
        wing venation 43–4
    flight 43–4
    as human food 589
    and hygiene 589
    internal structure 45–6
    metamorphosis 38–9
    phobias induced by 576, 589
Integument, *see* Exoskeleton;
    Cuticle
International Code of
    Zoological Nomenclature
    7–8
Iridoviruses 302, 610
Irritating mosquito
    (*Coquillettidia perturbans*)
    121, 155
Isfahan virus 104
Isoenzymes 16, 498
    in biting midges 294
    in blackflies 253–4

in bugs 498
in fleas 537
in horse-flies 316
in mites 632
in mosquitoes 161, 164
in sandflies 91
in ticks 607
in tsetse-flies 347
Isomorphic species 253
Itch mite (*Sarcoptes scabiei*)
    631, 633, 639–41

Japanese encephalitis (JE) 208,
    209, 214, 215
Jiggers (*Tunga* fleas) 541, 554,
    564, 555–6

Kala-azar, *see* Leishmaniasis
Kaposi's sarcoma 267
Karoo tick (*Ixodes rubicundus*)
    612, 625
Kasen, biting midge allergy
    300
Keds (Hippoboscidae) 581
    see also Scientific names
    index
Kennel tick (*Rhipicephalus
    sanguineus*) 613, 615, 628,
    652
Kenya tick typhus 613
Killer bees 582
Kissing-bugs 483–516, 589–610
Kotonkan virus 302
Kudu, greater (*Tragelaphus
    strepsiceros*) 369
'Kungu cake' 577
Kyasanur Forest disease (KFD)
    612, 624, 626–7

La Crosse virus 161, 208
Ladybirds (Coccinellidae) 585
Larder beetles (*Dermestes*) 583
Larvipary 363, 421, 452
Larvivorous fish 211, 212
Latent period, of infection 20
Latin names, *see* Scientific
    names
Latrine-fly (*Fannia scalaris*)
    403, 416, 417, 421, 453–4
    medical importance 421, 453
Lectins 378–9
Leishmaniasis 104–6
    control of disease 107–8,
    109–10
    of vectors 110

cutaneous leishmaniasis 78,
    79, 104, 105–6
    vectors 97–9
mucocutaneous
    leishmaniasis 105
vectors of leishmaniasis
    97–9, 104–7, 108–9
visceral leishmaniasis 78,
    104, 105–6
    vectors 97–9
Lemurs 621
Lentic habitat 73, 241
Leprosy 424, 481
Lesser dung-flies
    (Sphaeroceridae) 69, 485,
    580–1
Lesser house-fly (*Fannia
    canicularis*) 403, 404,
    416–17, 421, 453–4
    medical importance 421, 453
Lice (Phthiraptera) 517–28
    biology 521–3
        oviposition 521–3
    chewing lice (Mallophaga)
        518
    classification 518
    clothing louse (*Pediculus
        humanus*) 518–20, 521, 524,
        526
    control 523, 525–7
    crab (pubic) louse (*Pthirus
        pubis*) 517, 519, 520–1, 522,
        524, 526–7
    head louse (*Pediculus capitis*)
        518–20, 522, 523–4, 525–6
    medical importance 523–5
        biting pests 523–4
        role as disease vectors
            524–5
    parasite relationships 524,
        525, 625
    recognition 518
    structure 518, 521
    sucking lice (Anoplura) 518
Life cycle, life history
    bedbugs 488
    biting midges 298
    blackflies 259–61
    cockroaches 476
    Diptera, general 72
    fleas 555
    horse-flies 323–4
    lice 522
    mites 636, 645, 647
    mosquitoes 187, 195
    scorpions 670
    spiders 670
    stable-flies, horn-flies 396,
        397

synanthropic flies 396
ticks 616–17
triatomine bugs 502–3
tsetse-flies 365–6
Listeriosis 558, 559, 561, 568
Lizards, see Reptiles
Loiasis 325–6
  periodicity 325
  vectors of 317–20, 323, 324, 325–6
Lokern virus 302
Lone star tick (*Amblyomma americanum*) 615, 621, 625
Longevity
  blackflies 264
  blackfly onchocerciasis vectors 272
  cockroaches 480
  horse-flies 324
  mosquitoes 182
Lotic habitat 73, 241, 261–2
  importance in control 278
Louping ill virus 612
Louse, see Lice
Louse-borne relapsing fever 20, 399, 517, 524–5
Louse-flies (Hippoboscidae) 581
  see also Scientific names index
Lund's fly (*Cordylobia rodhaini*) 412, 446, 447
Lung-books 32, 662
Lyme disease 325, 612, 624, 625, 629
Lymphocytic choriomeningitis 557, 559, 560

Macroptery 485
Macrotrichia 41, 56
Madeira cockroach (*Rhyparobia maderae*) 479
Maggots 61, 390–1, 430–2
  distinction from grubs 58
  infestation with (myiasis) 429–69
  morphology of 431, 432
  therapeutic use 449, 452, 459
  vernacular names in text
    Congo floor-maggot (*Auchmeromyia senegalensis*) 11, 412, 420, 446, 447, 452
    hairy maggots (*Chrysomya*) 449
    rat-tailed maggot (*Eristalis*) 457
Main Drain virus 302

Malaria 197–200
  control 164
  detection in mosquito 200
  life cycle 199–200
Malpighian tubules 45, 138
  polytene chromosomes in 253
Mange 599, 639
  see also Scabies
Mange mites 599
Mansonelliasis 267, 268, 300–1
Maria sensor 509
Marseilles fever 628
Marsupials, as hosts 105, 553, 554, 555, 558, 560, 562, 621
Mating
  biting midges 298
  blackflies 262–3
  Diptera, general 74
  horse-flies 324
  mosquitoes 174
  sandflies 101
  scorpions 670–1
  spiders 671
  stable-flies, horn-flies 396, 397
  synanthropic flies 419, 420
  ticks 616, 620, 640
  triatomine bugs 503
Mating plug 139
Meal-worms (Tenebrionidae) 583
Mechanical transmission 20
  by ants 583
  by biting muscids 398–9
  by bloodsucking larvae 452
  by cockroaches 480–1
  by eye-flies 580
  by fleas 567, 568
  by house-flies 424
  by lice 525
  mechanisms of 424–5
  by sandflies 104
  by snipe-flies 580
  by tabanids 310, 325
  by tsetse-flies 373
Medal of the mosquito 121
Meningitis 301
Microbial insecticides, see Bacterial insecticides
Micropetery 485
Microtrichia 41, 56
Midges
  biting (Ceratopogonidae) 288–309
  non-biting (Chironomidae) 288
  phantom midges (Chaoboridae) 577

Millipedes (Diplopoda) 33, 35, 37
Mitchell River virus 302
Mites (Acari, part) 631–49
  biology of principal groups
    Acaridida 639–42
    Actinedida 642–7
    Mesostigmata 634–9
  classification 597, 600, 632, 634
  faunal literature 634
  collecting and preserving 652–3
  compared to ticks 597–600
  control 648–9
  fossil 597
  identification 632–4
  medical importance
    allergy 641
    causing dermatitis 631, 639, 642, 644, 646
    disease transmission 646
    scabies 639–41
  rearing 653–4
  recognition 631
  taxonomy 600
  techniques with 649–52
  vernacular names in text
    American chigger mite (*Eutrombicula alfreduggesi*) 646
    chicken mite (*Dermanyssus gallinae*) 636
    chiggers 644–7, 649, 653
    European harvest mite (*Neotrombicula autumnalis*) 644, 646
    European house dust mite (*Dermatophagoides pteronyssinus*) 641, 642
    follicle mites (*Demodex*) 38, 631, 634, 642, 643–4
    grain mite (*Pyemotes tritici*) 647
    harvest bugs (Trombiculidae) 644–7, 653
    hay itch mite (*Pyemotes tritici*) 647
    house dust mite (*Dermatophagoides*) 599, 641, 649, 653
    house mouse mite (*Liponyssoides sanguineus*) 636
    itch mite (*Sarcoptes scabiei*) 639–41
    northern fowl mite

Mites (cont'd)
  (*Ornithonyssus sylviarum*) 637
  red poultry mite (*Dermanyssus gallinae*) 636, 649
  scabies mite (*Sarcoptes scabiei*) 599, 631, 632, 639–41
  scrub itch mites (Trombiculidae) 644–7
  tropical fowl mite (*Ornithonyssus bursa*) 636
  tropical rat mite (*Ornithonyssus bacoti*) 636
  velvet mites (Trombiculidae) 644
Monkeys as hosts
  of biting midges 300
  of lice 518
  of mosquitoes 152, 205, 206, 207
  in myiasis 438, 439
  of tabanids 324, 325, 326
  of ticks 612, 621, 627
Moose tick (*Dermacentor albipictus*) 613
Monolayers in mosquito control 210
Mooseflies (Tabanidae) 310
Morphospecies 14, 15, 252
Mosquito fish 211, 212
Mosquitoes 120–240
  biology 173–96
    autogeny 176
    biting cycles 178–9
    bloodfeeding 125, 176–7
    diapause 183, 186, 187, 196
    dispersal 183–4
    gonotrophic cycles 180–1
    host location 176, 177
    host preferences 177–8
    longevity 182–3
    mating 174–5
    oviposition 176, 180, 181–2, 184
    of pupae 195–6
    resistance to desiccation 152, 153, 185
    sugar-feeding 175
    tolerance to temperature 184
  breeding sites
    artificial containers 141, 146, 148, 149, 152, 153, 154, 165, 166, 181, 187
    bamboo 141, 147, 148, 149, 152, 153, 154, 157, 158, 161, 181
    classification of breeding sites 189–95
    crab holes 160, 175, 181
    ground water 144, 145, 148, 151, 157, 158, 160, 165, 166
    latrines 154, 158, 165
    pitcher plants 130, 147, 154, 187
    plant axils 130, 131, 136, 141, 145, 146, 147, 148, 149, 151, 152, 154, 156, 157, 166, 181, 187
    rice fields 152, 158, 160
    rock pools 141, 149, 152, 153, 157, 158, 160, 161, 181, 186, 187
    salt water 149, 151, 164
    tree holes 136, 141, 147, 148, 149, 150, 151, 152, 153, 157, 158, 160, 161, 181, 186, 187
    tyres 141, 148, 149
  classification 139–67
  control 208–16
    bed nets 179, 213, 214
    biological control 212–13
    genetic control 213
    insecticidal control 215–16
    larval control 209–11
  faunal and taxonomic literature 171–3
  fossils 120
  identification (adults) 167–71
  larvae 186–95
    breeding 141, 188–95
    classification of larval habitats 189–95
    feeding 131, 142, 187–8
    larval development 187
    morphology 131–6, 142, 143, 145–6
    plant attachment 154–5, 156
    predatory larvae 141, 147, 151, 187, 195
  medical importance
    arboviruses 206–8
    filariasis 200–5
    malaria 197–200
  preservation
    adults, 219–20
    eggs 217
    larvae and pupae 218
  rearing and colonization 200–21
  sampling 216–19
    adults 218–19
    eggs 217
    larvae and pupae 217–18
  species complexes 141, 145, 161–7
  structure
    adults 121–8
    eggs 136–7, 138, 142, 145, 149, 158, 161, 181, 184–6
    pupae 129–31, 142, 146
  vernacular names in text
    brown-striped wood mosquito (*Aedes excrucians*) 121
    daylight mosquito (*Anopheles crucians*) 121
    irritating mosquito (*Coquillettidia perturbans*) 121
    northern common house mosquito (*Culex pipiens*) 121
    pitcher-plant mosquito (*Wyeomyia smithii*) 121
    southern common house mosquito (*Culex quinquefasciatus*) 121
    western tree-hole mosquito (*Aedes sierrensis*) 121
    yellow fever mosquito (*Aedes aegypti*) 121, 149
Moth-flies (Psychodidae) 457, 577–8
Moths (Lepidoptera, part)
  bloodsucking 587–8
  eye-moths 586–7
  as intermediate hosts 586
  urtication
    by adults 586
    by caterpillars 586
  vernacular names in text
    bag-shelter moth 586
    brown-tail moth (*Euproctis*) 586
    eye-moths 586–7
    flannel-moths (*Megalopyge*) 586
Mouthparts
  biting midges 290
  blackflies, adult 247–8
  bloodsucking bugs 484–5
  bloodsucking muscid flies 392, 397–8
  fleas 530
  horse-flies (tabanids) 313, 324
  insects, general 42–3

lice 518
mosquitoes
  adults 123, 125
  larvae 131–2
  sandflies 79–82, 101
  ticks 602, 604–5
  tsetse-flies 334
Murine typhus 525, 559, 560, 564, 567, 568
Murray Valley encephalitis (MVE) 159, 208, 215
Muscalure 420
Muscomorpha 53, 54, 55, 56, 58, 59, 61, 62, 63, 64, 65, 66, 67, 68, 69–70, 71, 72, 73, 75, 403, 404
Myiasis 403, 429–69
  adults of myiasis-producing larvae
    biology 419–23
    identification 407–18
    recognition 403–6
  biology of larvae
    calliphorids (Tumbu-fly screw-worms, etc.) 446–52
    cuterebrines (bot-flies) 438–9
    families of minor importance 456–8
    fanniids (lesser house-flies, latrine-flies) 453–6
    gasterophilines (bot-flies) 443
    hypodermatines (warble-flies) 439–43
    muscids (house-flies) 454–6
    oestrines (nasal flies) 439
  clinical
    accidental 430, 453, 455, 457, 458
    classification 429–30
    cutaneous myiasis 430, 438, 443, 446, 448, 449, 450, 452, 455, 457, 460
    immunity 446, 460
    nasopharyngeal myiasis 430, 439, 455
    ophthalmomyiasis 430, 438, 439, 441, 457, 458
  control 459–62
  larval
    identification 407–18, 432–8
    morphology 430–1
    preservation 462–4
  maggot therapy 449, 452, 459

Myriapods 35–6
Myxomatosis 32, 33, 35, 36, 267, 628

'Nairobi eye' 584
Nairobi sheep disease 613
Names, scientific 5
Naming species, see description of species
Nearctic region, defined 17
Nematodes, see Helminths
Nematodes, in vector control 212, 213
Neotropical region, defined 17
Nidicoly 631
'Nimitti' 576
Nits 521
  see also Lice
Nomenclature 5–12
  changes of names 11–12
  defined 2
  International Code of Zoological Nomenclature 7–8
Non-biting midges (Chironomidae) 288, 576–7
'No-no' or 'nau-nau' (*Simulium buissoni*) 268
Northern common house mosquito (*Culex pipiens*) 121
  see also Scientific names index
Northern fowl mite (*Ornithonyssus sylviarum*) 633, 637, 638
Northern rat-flea (*Nosopsyllus fasciatus*) 534, 537, 538, 542, 561
Norwegian scabies 641
No-see-ums (biting midges) 288
Nulliparous, defined 180

Oiling, against mosquito larvae 210
Omsk haemorrhagic fever virus 558, 568
Onchocerciasis 241, 266, 270–7, 279
  bovine 241, 300
  clinical 267, 271–2
  control 22, 266, 273, 275, 278–9
  vectors 250–1, 252, 255, 263, 268–9
Onchocerciasis Control Programme 22, 273, 275

Onchotaxy 242, 247
O'Nyong Nyong virus 208
Opossums 505, 506, 507, 557
Oriental cockroach (*Blatta orientalis*) 473, 474, 477–8, 479–80, 481
Oriental latrine fly (*Chrysomya megacephala*) 405, 414, 421, 449
Oriental rat flea (*Xenopsylla cheopis*) 532, 533, 539, 555, 564, 565
Oriental region, defined 17
Oriental sore 78
Oropouche virus 301
Oroya fever 104
Osteomyelitis, maggot treatment 459
Oviposition
  bedbugs 488
  biting midges 299
  blackflies 264–6
  cockroaches 479
  Diptera, general 74
  fleas 555
  horse-flies 325
  lice 521
  mites 636, 640
  mosquitoes 181–2
  sandflies 102
  spiders 670
  stable-flies, horn-flies 396, 397
  synanthropic flies 419
  ticks 623
Ovitraps 217, 218

Paedogenesis 72
Palaearctic region, defined 17
Palyam virus 302
Pangolins 554
Papataci fever 103–4
Paper wasps (*Polistes*) 582
Parasitoids 73, 503, 508, 671
Paris green 210
Parous, defined 180
Post kala-azar dermal leishmaniasis (PKDL) 105
Pediculosis 517
Pellagra, *Simulium* theory of 267
Pemphigus 267
Peritrophic membrane 21, 45, 138, 264, 273, 274
Phantom midges (Chaoboridae) 60, 64, 120, 139, 195, 577
Pharaoh ant (*Monomorium pharaoensis*) 583

Pharate (hidden) phases 260
  adult blackfly 247, 260–1
  adult tsetse-fly 366
  pupal blackfly 260–1
Pheneticism 3
Pheromones
  alarm (bugs) 490
  assembly
    in bugs 489, 504
    in ticks 619
  invitation
    in bugs 504
    in mosquitoes 177
    in sandflies 102
  oviposition
    in blackflies 266
    in mosquitoes 181, 185
    in sandflies 100
  sex
    in bugs 503
    in house-flies 420
    in midges 298
    in mosquitoes 174
    in sandflies 82, 84, 90, 101
    in ticks 619, 620
    in tsetse-flies 347, 367
Phoresy 255, 261–2, 275, 438
Phytotelmata 189
Pigeon tick (*Argas reflexus*) 610
Pigs, as hosts 338, 369, 447, 452, 562, 611, 639
Pinworm (*Enterobius vermicularis*) 481
Piroplasms 630
Plague 21, 537, 557, 558, 559, 560, 561, 562, 564, 567–8, 569
  distribution 481, 563
Pneumonia infections 481
Poliomyelitis 399, 424, 481
Polypneustic lobes 61, 339, 360, 363
Polytene chromosomes 14
  in blackflies 249, 252–4, 255
  in mosquitoes 139, 162–3
  in tsetse-flies 346
  see also Chromosomes
Porcupines 611
Prawns 261
Preservation
  ethyl alcohol 25–6
  general 23–7
  pest control 26
  pinning 25
  slide-mounting 26
  specimen data 26
Proboscis, see Mouthparts
Processionary moths 586
Proline 371

Protonymph 631, 645
Protozoans
  genera and subgenera mentioned
    Ambylospora 195
    Babesia 612, 629, 630, 631
    Duttonella 373
    Entamoeba 424, 481
    Epistylis 195
    Gonyaulax 34
    Haemoproteus 300
    Hepatocystis 300
    Leishmania 6, 21, 78, 91, 97–110, 398, 399
    Leucocytozoon 241, 300
    Nannomonas 373
    Nosema 195
    Plasmodium 21, 197, 200
    Sauroleishmania 98
    Theileria 613
    Thelohania 195
    Trypanosoma 21, 333, 373, 375–8, 484, 487, 489, 491, 501, 505–9
    Trypanozoon 373
    Vavraia 195
    Viannia 106
    Vorticella 195
Pruritis 524, 564, 611, 641
Pseudotuberculosis 558, 559, 560, 561, 568
Ptilinum, ptilinal fissure 63, 69, 366, 404, 419
Pulvilli 57, 147, 639
Punkies (biting midges) 288
Punta Toro virus 104
Pupariation 61, 70
Puparium 61, 63, 339, 363, 366, 381, 390–1, 419
  identification in tsetse-flies 359–61
Pyrethroids 211, 214, 215, 508, 509, 526, 648

Q fever 560, 568, 628–9
Quaranfil virus 610
Queensland tick typhus 612, 628
Queensland itch 300

Rabbit flea (*Spilopsylla cuniculi*) 555, 556, 568
Rabbits as hosts
  of fleas 558, 568
  of myiasis larvae 438, 439
  of lice 521, 526
  of mites 639, 643
  of ticks 610, 628
Rabies 581

Raccoons 300
Rats, see Rodents
Rat-tailed maggot (*Eristalis*) 435, 457
Recluse spider (*Loxosceles reclusa*) 660, 667
Red-legged tick (*Rhipicephalus evertsi*) 613
Red poultry mite (*Dermanyssus gallinae*) 633, 635, 636, 649
Reindeer 178, 441, 449
Reindeer warble fly (*Oedemagena tarandi*) 441
Relapsing fever
  louse-borne 20, 399, 517, 524–5
  tick-borne 525, 610, 611, 624, 629, 630
Repellents 214, 327, 508
Reptiles 98, 102, 158, 160, 178, 369, 371, 601, 621, 631, 636
Reservoirs, see Animal reservoirs
Rhinitis 576, 641, 642
Rhinoceros 371, 396
Rickettsiae
  affecting susceptibility to pathogens 378
  causing cytoplasmic incompatibility 165, 167
  transmission of pathogens, see Scientific names index
Rickettsiae, rickettsial infections
  genera mentioned
    Aegyptianella 610
    Cowdria 615
    Coxiella 568, 628
    Rickettsia 524, 525
  boutonneuse fever 568, 612, 613, 628
  chigger-borne rickettsiosis 646
  Crimean tick typhus 628
  Flinders Island spotted fever 628
  Indian tick typhus 628
  Kenya tick typhus 613
  Marseilles fever 628
  murine typhus 525, 559, 560, 564, 567, 568
  Q fever 560, 568, 628–9
  Queensland tick typhus 612, 628
  rickettsial pox 636
  São Paulo fever 627
  scrub typhus 646, 653
  Siberian tick typhus 628

Tobia fever 627
Tsutsugamushi disease 646
  see also Typhus
Rickettsial pox 636
Rift Valley fever 154, 159, 208, 301
River blindness, see Onchocerciasis
Rocky mountain spotted fever (RMSF) 613, 615, 627–8
Rocky mountain wood tick (*Dermacentor andersoni*) 613, 614, 625, 626, 628
Rodents as hosts
  of bugs 446, 507
  of fleas, 552, 553, 554, 557, 558, 559, 560, 561, 562, 563, 564, 567, 568, 569
  of horse-flies 324
  as intermediate hosts 585
  of lice 525
  of mites 634, 636, 639, 645, 646
  of myiasis larvae 438, 439
  of sandflies 79, 97, 100, 102, 103, 105, 106, 109, 110
  of ticks, 611, 612, 613, 621, 627
Romanas sign 506
Ross River virus 151, 208, 215
Russian Spring Summer encephalitis (RSSE) 612, 626

Sabo virus 302
St Louis encephalitis (SLE) 159, 161, 176, 208, 214, 636
Saliva
  of blackflies 247, 264
  of bugs 485
  of mites 645
  of mosquitoes 125, 138–9, 199
  of sandflies 101
  of ticks 625
  of tsetse-flies 334
Salivary glands
  in mosquitoes 200
  in ticks 622–3
  in tsetse-flies 373
Salmonellosis (*Salmonella* infections) 424, 425, 481, 559, 560, 561, 568, 581, 583
Sand tampan (*Ornithodoros savignyi*) 610, 611, 622
Sandflies (ceratopogonid sense) 78
Sandflies (Phlebotominae) 78–119

biology 100–3
  breeding sites 100, 103, 106, 110, 112
  dispersal 102–3
  feeding behaviour 101–2
  hosts 102
  life history 100–1
  mating 101
  resting sites 100–1, 103, 111
  vector–parasite interactions 103, 106–7
classification 89–90
  principal genera 90, 91, 96–9
collecting methods 111–12
control 107–110
faunal and taxonomic literature 99–100
identification 92–6
medical importance
  as biting pests 103
  incrimination of vectors 107
  transmission of *Bartonella* 104
  transmission of *Leishmania* 97–9, 104–7
  transmission of viruses 103–4
preservation and preparation 112–13
rearing and colonization 113
recognition 78, 79
structure
  adult 79–87
  egg 87
  larva 87–9
  pupa 89
Sandfly fever virus 23, 78, 103
Sandrats (*Psammomys*) 109
Sango virus 302
São Paulo fever 627
Sathuperi virus 302
Scabies 639–41, 648–9
  'Norwegian scabies' 641
Scabies mite (*Sarcoptes scabiei*) 599, 631, 632, 639–41
  immunity to 640, 641
'Scaly leg' of birds 639
Scent glands 490
Scientific names 6–7
Sclerotin 41
Scorpions 38, 659–82
  avoidance of stings 679
  biology, 670–1
  classification and identification 668–9

literature 669–70
collecting and preserving 679
general structure 660–6
medical importance 672, 673, 676–8
stings 678
Screw-worms
  biology 420–1, 447–9
  identification
    adults 413–14
    larvae 437–8
  New World screw-worms (*Cochliomyia*) 447–8
  Old World screw-worms (*Chrysomya*) 448–9
Scrub itch 646
Scrub typhus 646, 653
Scuttle-flies (Phoridae) 65, 69, 430, 435, 456–7, 578–9, 589
Sensilla 82, 124, 125, 126, 127, 247, 248, 290–1, 334, 484, 504, 531, 601, 621, 662, 672
Sepik virus 154
Setae 41
Sewage flies (*Psychoda*) 578
Sheep, as hosts 439, 447, 450, 452, 581, 612, 613
Sheep-ked (*Melophagus ovinus*) 408, 579, 581
Sheep-nostril fly (*Oestrus ovis*) 410, 439, 440
Sheep strike 421, 449, 450
Shgamonda virus 302
Shiner (cockroach) 478
Shore-flies (Ephydridae) 458, 580
Shuni virus 302
Siberian tick typhus 628
Sibling species, definition 14
  see also Species complexes
Silk glands
  in blackflies 262
  in spiders 664
Silverfish (*Lepisma saccharina*) 589
Simbu virus group 301
Simuliids, see Blackflies
Sindbis virus 208
Sleeping sickness 338–9, 376–9
  clinical 376, 378
  control 380–1
  distribution 377
  transmission 373
Snipe-flies (Rhagionidae) 55, 64, 69, 579–80
Soft ticks 601
  see also Argasidae in Scientific names index

Soldier-flies (Stratiomyidae) 59, 65, 430, 433, 458
Sound production
  in mosquitoes 174
  in triatomines 496
South African bont tick (*Amblyomma hebraeum*) 614, 615, 619, 628, 651
South African tick typhus, 612, 615, 628
South American trypanosomiasis, *see* Chagas disease
Southern house mosquito (*Culex quinquefasciatus*) 121
  see also Scientific names index
'Spanish fly' (*Lytta vesicatoria*) 584, 585
Speciation 2
Species
  authors of 7, 9
  concepts 8, 11, 12–14
  descriptions 8–11
  isomorphic 14–16
  names 6
Species complexes
  in blackflies 252–4
  in mites 639, 643–4
  mosquitoes 141, 161–7
  overview 14–16
Spermatheca 45
  in biting midges 293
  in blackflies 248
  in bugs 503
  in fleas 532
  in flies 58
  in mosquitoes 139
  in sandflies 84, 86
  in tsetse-flies 367
Spermatophore
  biting midges 298
  blackflies 263
  mites 646
  scorpions 670–1
  spiders 671
  ticks 620
  tsetse-flies 367
Spiders 579, 659–82
  avoidance of bites 678–9
  biology 670–2
  classification and identification 666–8
  collecting and preserving 679
  faunal literature 669–70
  general structure 660–6
  medical importance 672–6

vernacular names in text
  American brown recluse spider (*Loxosceles reclusa*) 667, 672, 676
  Australian (or Sydney) funnel-web spider (*Atrax robustus*) 660, 672
  bird-eating spiders 659
  black-widow spider (*Latrodectus mactans*) 667, 672, 676, 678
  house spider (*Scotophaeus blackwalli*) 674
  recluse spiders (*Loxosceles*) 667, 672
  tarantulas 659, 667
  wandering spiders (*Phoneutria*) 667
Spinose ear tick (*Otobius megnini*) 610
Stable-flies (*Stomoxys*) 389–402
  biology 397
  control 399
  identification 391–5
  medical importance 398–9, 455
  preservation 400
  taxonomic literature 395–6
Steamer (cockroach) 478
Stenogamous mating 175
Stephanofilariasis 399
Stigmata 599, 600, 607, 631, 632, 662
Stiletto-flies (Therevidae) 65, 430, 433, 434, 458
Stingless bees 582
Stings
  hymenopteran 581–3
  death from 581
  scorpion 662, 678, 679
Stouts (tabanids) 310
Subspecies 13
  of fleas 557
  of house-flies 422
  nomenclature of 7, 14
  of sandflies 90
  of tsetse-flies 343–7, 363
Sucking lice 517–28
Sugar-feeding
  in sandflies 101
  in mosquitoes 175
  in tsetse-flies 371
Swarms, swarming
  in biting midges 298
  in blackflies 263, 266
  in horse-flies 324
  in house-flies 421
  in mosquitoes 174–5
  as nuisance 589

  in tsetse-flies 366, 367
Sweat-flies 69, 422, 455, 580
Sweet itch 300
Sympatric distribution 16
Synanthropic flies
  biology
    of blow-flies 420–1
    of fanniids 421
    of flesh-flies (Sarcophagidae) 421–2
  life cycle, general 419–20
  of muscids 422–3
  classification 63, 65–7, 69–70
  control 425–6
  faunal and taxonomic literature 418–19
  identification
    of families 408–11
    of major groups 403–6
    of medically important genera 411–18
  medical importance
    as mechanical vectors 423–5
    as myiasis agents, *see* Myiasis
  morphology of adults 406–8, 409
  sexing adults 406–7
  see also Calyptrate flies; Diptera
Synonymy, of scientific names 11
Syntrophium 43, 125
Systematics 2

Tabanids, *see* Horse-flies
Taiga tick (*Ixodes persulcatus*) 612, 619, 621, 626, 629
Tailless whip-scorpions (Amblypygi) 30, 659, 661, 662
Tamné 439
Tapeworms
  beetle intermediate hosts 585
  dog tapeworm (*Dipylidium caninum*) 20, 558, 559, 562, 566
  flea intermediate hosts 20, 558, 559, 561, 562, 564, 566–7
  moth intermediate hosts 586
  rat tapeworm (*Hymenolepis diminuta*) 561, 567, 585, 586, 589
Tapirs 639
Tarantulas 659, 667, 675

Taxon (pl. taxa) explained 3
Taxonomy 2
  defined 2
Termites (Isoptera) 39, 40, 97, 100
Theileriosis 613
Thimni 439
Three-day fever
  in cattle 301
  in man 103–4
Thrips (Thysanoptera) 40, 588
Thunderflies (Thrips) 588
Tick-borne encephalitis 558, 559, 560, 612, 624, 626
Tick-borne relapsing fever 525, 610, 611, 624, 629, 630
Tick paralysis 612, 625
Ticks (Acari, part) 568, 600–31
  biology 616–23
    blood-feeding, 618, 621, 622–3, 624
    host location, 617, 618, 622, 623
    host specificity 621, 624
    larvae 607, 609, 612, 619
    longevity 619, 624
    mating behaviour 619, 620–1
    nymphs 607, 609, 612, 622
    oviposition 616, 618, 623
    pheromones 619–20
  classification 600–1
    phylogeny 601
  collecting 649–51, 681
  compared to mites 597–600, 601–2
  control 621, 648
  identification 607–9
  faunal and taxonomic literature 609
  medical importance 623–31
    principal genera 609–16
    ticks as cause of disease 625–6
    transmission of pathogens 206, 625–31
  morphology 601–7
  number of hosts
    one 616, 618, 619
    two 616–19
    three 616–19
  preserving 651
  rearing 607, 651–2
  vernacular names in text
    American dog tick (*Dermacentor variabilis*) 613, 625, 628
    brown dog tick (*Rhipicephalus sanguineus*) 613, 615, 628, 652
    brown ear tick (*Rhipicephalus appendiculatus*) 601, 602, 606, 609, 613, 618, 621, 627
    castor bean tick (*Ixodes ricinus*) 604, 611–12, 619, 620, 621, 626, 629, 631, 652
    Cayenne tick (*Amblyomma cajennense*) 615, 621, 628
    European sheep tick (*Ixodes ricinus*) 604, 611, 612, 619, 620, 621, 626, 629, 631, 652
    eyed (or sand) tampan (*Ornithodoros savignyi*) 610
    eyeless tampan (*Ornithodoros moubata*) 603, 610, 611, 622
    hard ticks (Ixodidae) 601, 604
    see also Ixodidae in Scientific names index
    Karoo tick (*Ixodes rubicundus*) 612
    kennel tick (*Rhipicephalus sanguineus*) 613, 615, 628, 652
    Lone Star tick (*Amblyomma americanum*) 615, 621, 625
    moose tick (*Dermacentor albipictus*) 613
    pigeon tick (*Argas reflexus*) 610
    red-legged tick (*Rhipicephalus evertsi*) 613
    Rocky Mountain wood tick (*Dermacentor andersoni*) 613, 614, 625, 626, 628
    soft ticks (Argasidae) 601
    see also Argasidae in Scientific names index
    South African bont tick (*Amblyomma hebraeum*) 614, 615, 619, 628, 651
    spinose ear tick (*Otobius megnini*) 610
    taiga tick (*Ixodes persulcatus*) 612, 619, 621, 626, 629
    tropical bont tick (*Amblyomma variegatum*) 206, 615, 619, 627, 651
    tropical horse tick (*Anocentor nitens*) 613, 619, 651
    winter tick (*Dermacentor albipictus*) 613
    yellow dog tick (*Haemaphysalis leachi*) 604, 612
Timor filariasis 205
Tobia fever 627
Toe-biters (Belostomatidae) 483
Tongue-worms (Pentastomida) 32, 33, 36, 37, 38
Tortoises 621
Toscana virus 104
Tourism, impact of biting flies 288, 299, 326, 398
Toxins 576
Trachea 45
Trachoma 20, 424
Transmission mechanisms 20–1
  biological 20
  by contagion 20
  criteria for vectors 22–3
  horizontal 23
  mechanical 20
  vertical 23
  see also Trans-ovarial transmission; Trans-stadial transmission
Trans-ovarial transmission 23
  in biting midges 302
  in fleas 567
  in mites 631, 646
  in mosquitoes 206
  in sandflies 104
  in ticks 624, 626, 627, 628, 629
Trans-stadial transmission 23
  in fleas 567
  in mites 631, 646
  in ticks 624, 628, 629
Transverse suture 53
Trench fever 517, 525
Triatomine bugs 489–510
  blood-feeding 502, 503–5
  eggs 503
  habitats 502, 503, 505
  host preference 504
  infection rates 507
  life cycle 502–3
  mating 503
  nymphs 496, 498, 503–4

Triatomine bugs (*cont'd*)
  classification 498
    phylogeny 500–1
  collecting 509
  colonization/laboratory
    509–10
  common names 498
  control 508–9
  geographical distribution
    491–5, 500–2
  identification 499–500
  medical importance 506–7
  recognition 491
  structure 495–6, 497
Trichuriasis 481
Trickle filter flies (*Psychoda*)
  457, 578
Trilobites 30, 32–3
Tritonymph 631, 645
Trombidiosis 646
Tropical fowl mite
  (*Ornithonyssus bursa*) 633,
  636, 638
Tropical rat mite
  (*Ornithonyssus bacoti*) 633,
  636, 637
True bugs (Hemiptera) 483
Trypanosomiasis, *see* Sleeping
  sickness; Chagas disease;
  Animal trypanosomiasis
Tsetse-flies (Glossinidae) 16,
  52, 333–8
  adult morphology 334–9
    genitalia 335–8, 340–1,
      352–9
    mouthparts 334, 337
    wing venation 335, 349
  biology
    biting behaviour 367–70
    breeding sites 366, 381
    dispersal 370
    habitats 342, 343, 362, 370,
      374–5, 377
    hosts 367–70, 371
    life history 365, 366
    longevity 379
    mating, 366–7
    resting sites 372, 382
    vector–parasite
      relationships 373, 378–9
  chromosomes 346
  classification 340–7
    phylogeny 340, 342, 343,
      347
    species and subspecies
      343–7
    subgenera 340–2
  collecting 381–2
  colonization 382
  control 379–81
    environmental control 380
    by insecticides 381
    by trapping 381
  distribution 338, 344–5,
    361–4, 372
  fossils 338
  identification 347–61
  larvae 338–9, 363, 378
  medical importance
    biting pests 372
    trypanosomiasis 372–3,
      376–9
  preservation 382
  puparium 339, 359–61, 363,
    366, 381
  recognition 334, 389, 408
  see also Scientific names
    index
Tsutsugamushi disease 646
Tuberculosis 424
Tularaemia 310, 325, 560, 568,
  580
Tumbu fly
  adult 404, 413
  biology 446, 460
  control 460, 461–2
  larva 447, 461–2
Tungiasis 564, 565–6
Turkey malaria 241
Turtles 324
Type specimens, types 10–11
Typhoid 481
Typhus
  African tick typhus 613
  bereavement fever 524–5
  Brill Zinsser disease 524–5
  Crimean tick typhus 628
  epidemic (classical) 517, 521,
    524–5, 526
  Indian tick typhus 628
  Kenyan tick typhus 628
  murine typhus 525, 559, 560,
    564, 567, 568
  Queensland tick typhus 612,
    628
  scrub typhus 646, 653
  Siberian tick typhus 613
  South African tick typhus
    612, 615, 628

Ultra low volume 215
Urticaria 585
  by beetles 585
  by caterpillars and moths
    586

'Vagabond's disease' 524
Vector–parasite interactions
  in blackflies 272–4, 277
  in house-flies 424–5
  in mechanical vectors 424–5
  in mosquitoes 200
  in sandflies 106–7
  in ticks 624, 625
  in tsetse-flies 373
Vectorial capacity 22–3, 163,
  197
Vectorial efficiency 23, 197
Vectors 19–23
  incrimination 22–3, 107, 300,
    424–5
  primary 22, 197, 277
  secondary 22, 197, 277
Velvet mites (Trombiculidae)
  599, 634, 643, 644–7, 653
Venereal transmission of
  arboviruses 206
Venation, *see* Wing venation
Venezuelan equine
  encephalomyelitis (VEE)
  152, 156, 161, 208, 215,
  267, 301
Venoms 576, 581
  ant venoms 583
  bee and wasp venoms 581–2
  scorpion venoms 676, 678
  spider venoms 659, 673,
    675–6
Verruga 104
Vesiculo viruses 104
  vesicular stomatitis 104
Vinchuca 507
Vinegar-flies (Drosophilidae)
  66, 69, 430, 435, 436, 458
Vinegaroon (*Mastigoproctus
  giganteus*) 659
Viruses
  families/genera mentioned
    Alphavirus 208
    Bunyaviridae 626
    Fibromavirus 568
    Flaviviridae 626
    Flavivirus 626
    Iridovirus 610
    Nairovirus 626
    Orbivirus 626
    Reoviridae 626
  virus infections
    hepatitis 424, 481
    hepatitis-B 484, 489
    human immunodeficiency
      virus (HIV) 20, 489
    myxomatosis 32, 33, 35,
      36, 267, 628
    poliomyelitis 399, 424, 481
    trachoma 20, 424
    vesicular stomatitis
      viruses 104

see also Arboviruses
Viviparity 484

Wandering spiders
  (*Phoneutria*) 667
Warble-flies
  (Hypodermatinae) 439–43
  reindeer warble-fly
  (*Oedemagena tarandi*) 441
Warrego virus 302
Warthog (*Phacochoerus
  aethiopicus*) 369, 371, 452,
  554, 601, 611
Wasps 581–2
Waterbuck 370
West Nile virus 159, 208
Western equine
  encephalomyelitis (WEE)
  159, 160, 208, 215, 301
Western tick-borne
  encephalitis 626

Whip-scorpions (Amblypygi)
  38, 659, 662
Whipworm (*Trichuris
  trichiura*) 424, 481
Wildebeest (*Connochaetes
  gnou*) 370
Wind-scorpions (Solifugae)
  38, 659, 662
Window-flies (Scenopinidae)
  60, 65, 430, 433, 434, 458
Window-gnats
  (Anisopodidae) 64, 430,
  433, 434, 457, 578, 589
Wing venation
  in Diptera 53–4, 56–7
  in insects, general, 43–4
  see also Individual groups –
  Blackflies; mosquitoes,
  etc.
Winter tick (*Dermacentor
  albipictus*) 613

Woodlice (isopod crustaceans)
  70, 408
Wood rats 557

Yak (*Bos mutus*) 440
Yaws 424, 580
Yellow fever 148, 206–7
  control 207
  vectors 207
Yellow fever mosquito (*Aedes
  aegypti*) 121
  see also Scientific names
  index

Zebra, common or Burchell's
  (*Equus burchellii*) 396
Zirqua virus 611
Zoogeographical regions
  16–19
Zoonoses 105, 106, 159, 206,
  377, 525, 567, 625, 626, 646